# Lecture Notes in Artificial Intelligence     7906

Subseries of Lecture Notes in Computer Science

LNAI Series Editors

Randy Goebel
*University of Alberta, Edmonton, Canada*
Yuzuru Tanaka
*Hokkaido University, Sapporo, Japan*
Wolfgang Wahlster
*DFKI and Saarland University, Saarbrücken, Germany*

LNAI Founding Series Editor

Joerg Siekmann
*DFKI and Saarland University, Saarbrücken, Germany*

Moonis Ali   Tibor Bosse
Koen V. Hindriks   Mark Hoogendoorn
Catholijn M. Jonker   Jan Treur (Eds.)

# Recent Trends in Applied Artificial Intelligence

26th International Conference
on Industrial, Engineering and Other Applications
of Applied Intelligent Systems, IEA/AIE 2013
Amsterdam, The Netherlands, June 17-21, 2013
Proceedings

 Springer

Series Editors

Randy Goebel, University of Alberta, Edmonton, Canada
Jörg Siekmann, University of Saarland, Saarbrücken, Germany
Wolfgang Wahlster, DFKI and University of Saarland, Saarbrücken, Germany


Volume Editors

Moonis Ali
Texas State University
San Marcos, TX 78666, USA
E-mail: ma04@txstate.edu

Koen V. Hindriks
Catholijn M. Jonker
Delft University of Technology
2628 CD Delft, The Netherlands
E-mail: {k.v.hindriks; c.m.jonker}@tudelft.nl

Tibor Bosse
Mark Hoogendoorn
Jan Treur
VU University Amsterdam
1081 HV Amsterdam, The Netherlands
E-mail: {t.bosse; m.hoogendoorn; j.treur}@vu.nl

ISSN 0302-9743     e-ISSN 1611-3349
ISBN 978-3-642-38576-6   e-ISBN 978-3-642-38577-3
DOI 10.1007/978-3-642-38577-3
Springer Heidelberg Dordrecht London New York

Library of Congress Control Number: 2013938609

CR Subject Classification (1998): I.2, H.4, F.2, H.2.8, I.5, I.4, H.3, F.1, G.2, C.2, K.4.4, I.6

LNCS Sublibrary: SL 7 – Artificial Intelligence

*Typesetting:* Camera-ready by author, data conversion by Scientific Publishing Services, Chennai, India

Printed on acid-free paper

Springer is part of Springer Science+Business Media (www.springer.com)

# Preface

Research in intelligent systems has taken a huge flight and has resulted in a great variety of techniques, methodologies, and tools for developing systems with intelligent behavior. In the application of such techniques in more practical settings, however, one is often faced with new challenges that need to be tackled to obtain a system that works well in such a setting. The International Society of Applied Intelligence (ISAI), through its annual IEA/AIE conferences, provides a forum for the international scientific and industrial community in the field of applied artificial intelligence to interactively participate in developing intelligent systems that are needed to solve the twenty-first century's ever-growing problems in almost every field.

The 26th International Conference on Industrial, Engineering and Other Applications of Applied Intelligence Systems (IEA/AIE-2013), held in Amsterdam, The Netherlands, followed the IEA/AIE tradition of providing an international scientific forum for researchers in the diverse field of applied intelligence. Invited speakers and authors identified challenges in the applications of intelligent systems in real-life settings and proposed innovative solutions. We received 185 papers, out of which 71 were selected for inclusion in these proceedings, resulting in a selection rate of 38.4%. Each paper was reviewed by at least two members of the Program Committee, thereby facilitating the selection of high-quality papers. The papers in the proceedings cover a wide range of topics, including knowledge representation and reasoning, robotics, cognitive modeling, planning, machine learning, pattern recognition, optimization, text mining, crowd modeling, auctions and negotiations, and evolutionary algorithms. In addition, several special sessions were organized as part of the IEA/AIE-2013 conference, namely, sessions on innovations in intelligent computation and applications; intelligent image and signal processing; business process intelligence; machine learning methods applied to manufacturing processes and production systems; advances in recommender systems; and finally a special session on decision support for safety-related systems. These proceedings, consisting of 71 chapters authored by participants of IEA/AIE-2013, cover both the theory and applications of applied intelligent systems. Together, these papers highlight new trends and frontiers of applied intelligence and show how new research could lead to innovative applications of considerable practical significance. We expect that these proceedings will provide useful reference for future research.

The conference also invited three outstanding scholars to give plenary keynote speeches. They were Matthias Klusch from DFKI Saarbrucken in Germany, Emile Aarts from Eindhoven University of Technology, The Netherlands, as well as Jan-Peter Larsen from SENSE Observation Systems, based in Rotterdam, The Netherlands.

Of course, this conference would not have been possible without the help of many people and organizations. First and foremost we want to thank the authors who contributed to these high-quality proceedings. The selection of the papers was made based upon the recommendations of the Program Committee, and we thank you for devoting your time to provide high-quality reviews and enable the selection of papers to compose this volume. In addition, we would like to thank the Special Session Chairs for their efforts that truly contributed to the quality and diversity of the program. We thank VU University Amsterdam for its support in the organization of the conference. Of course we want to thank our main sponsor, ISAI, and the associate sponsors Almende B.V., the Benelux Association for Artificial Intelligence, and the Municipality of Amsterdam for their support. The following cooperating organizations also deserve our gratitude: the Association for the Advancement of Artificial Intelligence (AAAI), the Association for Computing Machinery (ACM/SIGART), the Austrian Association for Artificial Intelligence (GAI), the Catalan Association for Artificial Intelligence (ACIA), Delft University of Technology, the European Neural Network Society (ENNS), the International Neural Network Society (INNS), the Italian Artificial Intelligence Association (AI*IA), the Japanese Society for Artificial Intelligence (JSAI), the Lithuanian Computer Society - Artificial Intelligence Section (LIKS-AIS), the Slovenian Artificial Intelligence Society (SLAIS), the Spanish Society for Artificial Intelligence (AEPIA), the Society for the Study of Artificial Intelligence and the Simulation of Behaviour (AISB), the Taiwanese Association for Artificial Intelligence (TAAI), the Taiwanese Association for Consumer Electronics (TACE), and Texas State University-San Marcos. We would also like to thank the keynote speakers who shared their vision on applications of intelligent systems. Last, but certainly not least, we want to thank the Organizing Committee of the conference.

April 2013

Moonis Ali
Tibor Bosse
Koen V. Hindriks
Mark Hoogendoorn
Catholijn M. Jonker
Jan Treur

# Conference Organization

## General Chair

Moonis Ali            Texas State University-San Marcos, USA

## Organizing Chair

Tibor Bosse           VU University Amsterdam, The Netherlands

## Program Chairs

| | |
|---|---|
| Koen V. Hindriks | Delft University of Technology, The Netherlands |
| Mark Hoogendoorn | VU University Amsterdam, The Netherlands |
| Catholijn M. Jonker | Delft University of Technology, The Netherlands |
| Jan Treur | VU University Amsterdam, The Netherlands |

## Doctoral Consortium Chair

Michel C. A. Klein      VU University Amsterdam, The Netherlands

## Local Arrangements Chair

Elly Lammers          VU University Amsterdam, The Netherlands

## Publicity Chairs

| | |
|---|---|
| Reyhan Aydogan | Delft University of Technology, The Netherlands |
| M. Birna van Riemsdijk | Delft University of Technology, The Netherlands |

## Reception Chair

Charlotte Gerritsen      VU University Amsterdam, The Netherlands

## Web Chairs

| | |
|---|---|
| Jeroen de Man | VU University Amsterdam, The Netherlands |
| Gabriele Modena | VU University Amsterdam, The Netherlands |

## Special Sessions

1. Advances in Recommender Systems
   (Alexander Felfernig and Walid Maalej)
2. Business Process Intelligence
   (Giovanni Acampora, Uzay Kaymak, and Vincenzo Loia)
3. Decision Support for Safety-Related Systems
   (Paul Chung and Kazuhiko Suzuki)
4. Innovations in Intelligent Computation and Applications
   (Shyi-Ming Chen, Jeng-Shyang Pan, and Mong-Hong Horng)
5. Intelligent Image and Signal Processing
   (Wen-Yuan Chen)
6. Machine Learning Methods Applied to Manufacturing Processes
   and Production Systems
   (Andrés Bustillo-Iglesias, Luis Norberto Lopez-De-La-Calle-Marcaide, and
   César García-Osorio)

## Invited Speakers

Emile Aarts            Eindhoven University of Technology, The Netherlands
Matthias Klusch        DFKI Saarbrücken, Germany
Jan Peter Larsen       SENSE Observation Systems, The Netherlands

## International Organizing Committee

Althari, A. (Malaysia)              He, J. (China)
Altincay, H. (Turkey)               Herrera, F. (Spain)
Aziz, A.A. (Malaysia)               Jain, L. (Australia)
Belli, F. (Germany)                 Last, M. (Israel)
Benferhat, S. (France)              Lee, C.-H. (Korea)
Chen, S.-M. (Taiwan)                Levin, M. (Russia)
Chung, P. (UK)                      Lim, A. (Hong Kong)
Correa da Silva, F.S. (Brazil)      Lukowicz, P. (Germany)
Dounias, G. (Greece)                Monteserin, A. (Argentina)
Esposito, F. (Italy)                Neerincx, M. (The Netherlands)
Felfering, A. (Austria)             Pratihar, D. (India)
Ferscha, A. (Austria)               Sensoy, M. (Turkey)
Figueroa, G. A. (Mexico)            Skowron, A. (Poland)
Fujita, H. (Japan)                  Wang, J. (Australia)
García-Pedrajas, N. (Spain)         Wong, M.-L. (Hong Kong)
Gitchang (Japan)                    Yang, C. (Canada)
Guesgen, H. (New Zealand)           Yordanova, S. (Bulgaria)
Halteren, van, A. (The Netherlands)

# Program Committee

Al-Mubaid, H. (USA)
Althari, A. (Malaysia)
Altincay, H. (Turkey)
Ashok, G. (USA)
Aydogan, R. (The Netherlands)
Aziz, A.A. (Malaysia)
Bae, Y. (Korea)
Barber, S. (USA)
Belli, F. (Germany)
Benferhat, S. (France)
Bentahar, J. (Canada)
Beun, R.-J. (The Netherlands)
Borzemski, L. (Poland)
Braubach, L. (Germany)
Brézillon, P. (France)
Bustillo, A. (Spain)
Bustince, H. (Spain)
Campa Gómez, F. (Spain)
Carvalho, J.P.B. (Portugal)
Chan, Ch.-Ch. (USA)
Chan, C.W. (Hong Kong)
Charles, D. (UK)
Chen, L.-J. (Taiwan)
Chen, P. (USA)
Chen, S.-M. (Taiwan)
Cho, H. (USA)
Cho, S.-B. (Korea)
Chou, J.-H. (Taiwan)
Chung, P.W.H. (UK)
Cordón, Ó. (Spain)
Correa da Silva, F. S. (Brazil)
Da Costa, J.M. (Portugal)
Dapoigny, R. (France)
De Souto, M. (Brazil)
Dignum, V. (The Netherlands)
Dolan, J. (USA)
Dounias, G. (Greece)
Dreyfus, G. (France)
Erdem, E. (Turkey)
Esposito, F. (Italy)
Fatima, S. (UK)
Felfernig, A. (Austria)
Ferro, D. (The Netherlands)

Fernández de Vega, F. (Spain)
Ferscha, A. (Austria)
Folgheraiter, M. (Germany)
Frias-Martinez, E. (Spain)
Fujita, H. (Japan)
Fyfe, C. (UK)
Gambäck, B. (Norway)
García-Osorio, C. (Spain)
García-Pedrajas, N. (Spain)
Gitchang (Japan)
Gomide, F. (Brazil)
Grzenda, M. (Poland)
Guesgen, H.W. (New Zealand)
Haasdijk, E. (The Netherlands)
Hakura, J. (Japan)
Halteren, van, A. (The Netherlands)
Havasi, C. (USA)
He, J. (China)
Hendtlass, T. (Australia)
Herrera, F. (Spain)
Herrera-Viedma, E. (Spain)
Heylen, D. (The Netherlands)
Hirota, K. (Japan)
Hou, W.-J. (Taiwan)
Huang, Y.-P. (Taiwan)
Hung, Ch.-Ch. (USA)
Indurkhya, B. (India)
Ito, T. (Japan)
Jacquenet, F. (France)
Julián, V. (Spain)
Kaikhah, K. (USA)
Kacprzyk, J. (Poland)
Khosrow, K. (USA)
Kinoshita, T. (Japan)
Klawonn, F. (Germany)
Krol, D. (Poland)
Kumar, A.N. (USA)
Kumova, B. (Turkey)
Lamirel, J.-Ch. (France)
Last, M. (Israel)
Lee, C.-H. (Korea)
Lee, G. (Taiwan)
Lee, H.-M. (Taiwan)

Levin, M. (Russia)
Liu, H. (USA)
Liu, Q. (USA)
Loia, V. (Italy)
Lozano, M. (Spain)
Lu, Y. (USA)
Lukowicz, P. (Germany)
Madani, K. (France)
Mahanti, P. (Canada)
Mansour, N. (Lebanon)
Marcelloni, F. (Italy)
Marinai, S. (Italy)
Matthews, M.M. (USA)
Mehrotra, K.G. (USA)
Meléndez, J. (Spain)
Mertsching, B. (Germany)
Mizoguchi, R. (Japan)
Mohammadian, M. (Australia)
Monostori, L. (Hungary)
Monteserin, A. (Argentina)
Müller, J.P. (Germany)
Nedjah, N. (Brazil)
Neerincx, M. (The Netherlands)
Nguyen, N.T. (Poland)
Okada, S. (Japan)
Okuno, H.G. (Japan)
Olivas, J.Á. (Spain)
Özgür, A. (Turkey)
Pan, J.-Sh. (Taiwan)
Pena, J.-M. (Spain)
Peng, J. (USA)
Peregrin, A. (Spain)
Pokahr, A. (Germany)
Potter, W. (USA)
Prade, H. (France)
Pratihar, D. (India)
Rayward-Smith, V.J. (UK)
Ren. F. (Japan)

Ramaswamy, S. (USA)
Romero Zaliz, R. (Spain)
Rosso, P. (Spain)
Sadok, D.F.H. (Brazil)
Sadri, F. (UK)
Sainz-Palmero, G. (Spain)
Salden, A. (The Netherlands)
Sànchez-Marrè, M. (Spain)
Selim, H. (Turkey)
Sensoy, M. (UK)
Soomro, S. (Austria)
Sousa, J. (Portugal)
Splunter, van, S. (The Netherlands)
Sun, J. (UK)
Suzuki, K. (Japan)
Tamir, D. (USA)
Tan, A.-H. (Singapore)
Tao, J. (USA)
Tereshko, V. (UK)
Tseng, L.-Y. (Taiwan)
Tseng, V.Sh.-M. (Taiwan)
Valente de Oliveira, J. (Portugal)
Valtorta, M. (USA)
Verwaart, T. (The Netherlands)
Viharos, Z.J. (Hungary)
Visser, E. (The Netherlands)
Wal, van der, C.N. (The Netherlands)
Wang, D. (Australia)
Wang, H. (USA)
Wang, J. (Australia)
Warnier, M. (The Netherlands)
Weert, de, M. (The Netherlands)
Wijngaards, N. (The Netherlands)
Yang, D.-L. (Taiwan)
Yang, Y.-B. (China)
Yin, Z. (USA)
Yolum, P. (Turkey)
Yordanova, S. (Bulgaria)

## Additional Reviewers

d'Amato, C.
Budalakoti, S.
Chakraborty, S.

Chiang, T.-C.
Couto, P.
Franco, M.

Heule, M.
Kanamori, R.
Kosucu, B.

Lan, G.-C.

Meng, Q.

Nieves Acedo, J.

Ninaus, G.

Sánchez Anguix, V.

Schüller, P.

Teng, T.-H.

Vieira, S.

Yang, S.

# Table of Contents

## Distributed Systems and Networks

## Evolutionary Algorithms

## Knowledge Representation and Reasoning

## Pattern Recognition

# Planning

# Problem Solving

# Robotics

# Text Mining

# Special Session on Advances in Recommender Systems

## Special Session on Business Process Intelligence

## Special Session on Decision Support for Safetyrelated Systems

# Special Session on Innovations in Intelligent Computation and Applications

# Special Session on Intelligent Image and Signal Processing

## Special Session on Machine Learning Methods Applied to Manufacturing Processes and Production Systems

# Developing Online Double Auction Mechanism for Fishery Markets

Kazuo Miyashita

Center for Service Research, AIST
1-1-1 Umezono, Tsukuba, Ibaraki 305-8568, Japan
k.miyashita@aist.go.jp

**Abstract.** In spot markets for trading fishes, single-sided auctions are used for clearing the market by virtue of its promptness and simplicity, which are important in dealing with perishable goods. However, in those auctions, sellers cannot participate in price-making process. A standard double auction market collects bids from traders and matches buyers' higher bids and sellers' lower bids to find the most efficient allocation, assuming that values of unsold items remain unchanged. Nevertheless, in the spot fish market, sellers suffer loss when they fail to sell the fish, whose salvage value is lost due to perishability. To solve the problem, we investigate the suitable design of an online double auction for fishery markets, where bids arrive dynamically with their time limits. Our market mechanism aims at improving traders' profitability by reducing trade failures in the face of uncertainty of incoming/leaving bids. We developed a heuristic matching rule for the market to prioritize traders' bids based on their time-criticality and evaluated its performance empirically.

**Keywords:** Mechanism Design, Double Auction, Perishable Goods.

## 1 Introduction

Fishes are traded mainly in the *spot market* because their yield and quality are unsteady and unpredictable. In the spot markets, which by definition occur after production, the production costs of *perishable goods* are typically *sunk costs* since those costs are *irretrievable* [1] when the goods are unsold and perished. Therefore, sellers risk losing those costs if the trades fail in the markets [1, 2].

In traditional markets for perishable goods, a *one-sided auction* such as a *Dutch auction* has been put into practice widely because its simplicity and promptness are vital for consummating transactions of a large volume of perishable goods smoothly with many participating buyers. In a one-sided auction, a seller can influence price-making by declaring a *reserve price*, below which the goods are not to be sold, in advance of transactions [3]. However, the seller cannot always sell the goods at the best price because (1) finding optimal reserve prices is difficult in realistic auctions, and (2) high reserve prices might result in allocation failures. Consequently, in a traditional market of perishable goods,

---

[1] In other words, their *salvage value* reduces to zero.

M. Ali et al. (Eds.): IEA/AIE 2013, LNAI 7906, pp. 1–11, 2013.

sellers are set at a distinct disadvantage. This problem has been investigated in the fields of agricultural economics [4] and experimental economics [5].

To realize fair price-making among traders while reducing allocation failures, we develop a prototypical fishery market, which adopts an online *double auction* (DA) as a market mechanism. In the online DA, multiple buyers and sellers arrive dynamically over time with their time limits. Both buyers and sellers tender their bids for trading commodities. The bid expresses a trader's offer for valuation and quantity of the commodity to be traded. The arrival time, time limit, and bid for a trade are all private information to a trader. Therefore, the online DA is uncertain about future trades. It collects bids over a specified interval of time, then clears the market at the expiration of the bidding interval by application of pre-determined rules.

A standard DA market, either static or online, usually assumes that sellers never lose their utility when they fail to sell the commodity. In such a market, traders' utility can be maximized using a price-based mechanism that matches the highest buyer's bid and the lowest seller's bid iteratively until there remains no matchable bid. However, in the online DA market of perishable commodities, the market mechanism should decide which bids with different prices and time-limits should be matched to increase traders' utility and reduce trade failures in the face of uncertainty about the future trades.

## 1.1 Related Works

Until recently, few research activities have addressed online double auction mechanisms [6–8]. They examine several important aspects of the problem: design of matching algorithms with good worst-case performance within the framework of competitive analysis [6], construction of a general framework that facilitates a truthful dynamic double auction from truthful, static double auction rules [7], and development of computationally efficient matching algorithms using weighted bipartite matching in graph theory [8]. Although their research results are theoretically fundamental, we cannot readily apply their mechanisms to our online DA problem because their models incorporate the assumption that trade failures never cause a loss to traders, which is not true in our market - a spot market for perishable goods.

To increase seller's revenue in perishable goods industries such as airlines and accommodation, several methodologies have been studied in the field of revenue management [9]. Nevertheless, those techniques are hard to apply to fishery markets, where not only demands but also supplies are highly uncertain.

We propose a heuristic online DA mechanism for the spot markets of perishable goods, which improves traders' revenue by reducing allocation failures. The remainder of the paper is organized as follows. Section 2 introduces our market model and presents desiderata and objectives for our online DA market. Section 3 defines a criticality-based allocation policy and explains a pricing policy based on the well-known k double-auction rule. Section 4 shows empirical results of comparing our criticality-based DA mechanism with a price-based one in a competitive demand-supply market.

# 2  Market Model

In our model of a fishery market, we consider discrete *time rounds* $T = \{1, 2, \cdots\}$, indexed by $t$. For simplicity we assume the market is for a single commodity. Agents are either sellers $(S)$ or buyers $(B)$, who arrive dynamically over time and depart at the time limit. In each round, the agents trade multiple units of fish. The market is cleared at the end of every round to find new allocations.

We model our market as a wholesale market for B2B transactions. In the market, seller $i$ brings the fish by the arrival time $a_i$. Therefore, seller $i$ incurs production cost before the trade starts. Furthermore, at the seller $i$'s departure time $d_i$, the salvage value of the fish evaporates because of its perishability unless it is traded successfully. Because of advance production and perishability, sellers face the distinct risk of failing to recoup the production cost in the trade. Buyers in our market procure goods to resell them in retail markets. For buyer $j$, the arrival time $a_j$ is the first time when buyer $j$ values the item. And, buyer $j$ is assumed to gain some profit by retailing the goods if he succeeds to procure them before the departure time $d_j$. Each agent $i$ has a type, $\theta_i = (v_i, q_i, a_i, d_i)$, where $v_i, q_i, a_i, d_i$ are non-negative real numbers, $v_i$ is agent $i$'s valuation of a single unit of the fish, $q_i$ is the quantity of the fish that agent $i$ wants to trade, $a_i$ is the arrival time and $d_i$ denotes the departure time. The duration between the arrival time and the departure time defines a *trading period* for the agent, and agents repeatedly participate in the auction over several periods.

**Table 1.** Inadequacy of agents' misreporting their type information

| | Quantity | | Arrival Time | | Departure Time | |
|---|---|---|---|---|---|---|
| | Less | More | Earlier | Later | Earlier | Later |
| Seller | Loss of sales | Infeasible in spot markets | Infeasible | Less chance for matching | Less chance for matching | Perish |
| Buyer | Loss of resales | Risk of excess stock | Infeasible | Less chance for matching | Less chance for matching | Delay of resale |

Agents are self-interested and their types are private information. At the beginning of a trading period, agent $i$ bids by making a claim about its type $\hat{\theta}_i = (\hat{v}_i, \hat{q}_i, \hat{a}_i, \hat{d}_i) \neq \theta_i$ to the auction. Furthermore, in succeeding rounds in the period, the agent can modify the value of its unmatched bid[2]. But if agents once depart from the market, they are not allowed to make a reentry of the same bid. An agent's self-interest is exhibited in its willingness to misrepresent its type when this will improve the outcome of the auction in its favor. Agent $i$ has incentives to manipulate the value of $v_i$, but the other components of its type $(q_i, a_i, d_i)$ are determined exogenously. As explained in Table 1, misrepresenting those values is not beneficial or feasible to the agent, whether it is a seller or a buyer. Consequently, in this paper, we consider that agent $i$ can misrepresent

---

[2] When we must distinguish between claims made by buyers and claims made by sellers, we refer to the *bid* from a buyer and the *ask* from a seller.

only its valuation $v_i$ for improving its utility among all the components of its type information $\theta_i$.

## 2.1   Desiderata and Objectives

Let $\hat{\theta}^t$ denote the set of agent types reported in round $t$, $\hat{\theta} = (\hat{\theta}^1, \hat{\theta}^2, \ldots, \hat{\theta}^t, \ldots)$ denote a complete reported type profile, and $\hat{\theta}^{\leq t}$ denote the reported type profile restricted to agent with reported arrival no later than round $t$. A report $\hat{\theta}_i^t = (\hat{v}_i^t, \hat{q}_i^t, \hat{a}_i^t, \hat{d}_i^t)$ is a bid made by agent $i$ at round $t$, which represents a commitment to trade at most $\hat{q}_i^t$ units of fish at a limit price of $\hat{v}_i^t$ at round $t \in [\hat{a}_i^t, \hat{d}_i^t]$.

In our market model, an agent can make several bids of which the trading period $[\hat{a}_i, \hat{d}_i]$ overlaps each other and the agent might have multiple bids at round $t$. However, to simplify the notation in the rest of the paper, we assume that each bid in any round $t$ belongs to a different agent.

In the market, a seller's ask and a buyer's bid can be matched when they satisfy the following condition of *matchability*.

**Definition 1 (Matchability).** *A seller $i$'s ask $\hat{\theta}_i^t = (\hat{v}_i^t, \hat{q}_i^t, \hat{a}_i^t, \hat{d}_i^t)$ and buyer $j$'s bid $\hat{\theta}_j^t = (\hat{v}_j^t, \hat{q}_j^t, \hat{a}_j^t, \hat{d}_j^t)$ are matchable , when*

$$(\hat{v}_i^t \leq \hat{v}_j^t) \wedge ([\hat{a}_i^t, \hat{d}_i^t] \cap [\hat{a}_j^t, \hat{d}_j^t] \neq \emptyset) \wedge (\hat{q}_i^t > 0) \wedge (\hat{q}_j^t > 0). \tag{1}$$

Mechanism of an online DA, $M = (\pi, x)$, defines an *allocation policy* $\pi = \{\pi^t\}^{t \in T}$, where $\pi_{i,j}^t(\hat{\theta}^{\leq t}) \in \mathbb{I}_{\geq 0}$ represents a quantity that agents $i$ and $j$ trade in round $t$ given reports $\hat{\theta}^{\leq t}$, and a *payment policy* $x = \{x^t\}^{t \in T}$, $x^t = (s^t, b^t)$, where $s_{i,j}^t(\hat{\theta}^{\leq t}) \in \mathbb{R}_{\geq 0}$ represents a payment that seller $i$ receives from buyer $j$ as a result of the trade in round $t$ and $b_{i,j}^t(\hat{\theta}^{\leq t}) \in \mathbb{R}_{>0}$ represents a payment made by buyer $j$ as a result of the trade with seller $i$ in round $t$.

We assume that agent $i$ has a *quasi-linear* utility function $u_i = f(\theta_i, \pi_i) \pm x_i$, so that the utility for a seller and a buyer is defined as follows.

- For seller $i$, when $\pi_{i,j}^t$ units of fish are sold to buyer $j$ at the price $s_{i,j}^t$ in round $t$ within the time interval $[a_i, d_i]$, then seller $i$ obtains utility $s_{i,j}^t$, since seller's production cost $v_i$ is sunk. Therefore, subtracting the total sunk cost, seller $i$'s utility $u_i$ is

$$u_i = \sum_{t \in [a_i, d_i]} \sum_{j \in B} s_{i,j}^t - q_i v_i. \tag{2}$$

- When buyer $j$ receives $\pi_{i,j}^t$ units of fish at the price $b_{i,j}^t$ in round $t$ within the time interval $[a_j, d_j]$, buyer $j$ obtains utility $\pi_{i,j}^t v_j - b_{i,j}^t$. And, buyer $j$ is supposed to make the profit $\pi_{i,j} v_j r_j$ by retailing the procured fish. Therefore, buyer $j$'s utility $u_j$ is

$$u_j = \sum_{t \in [a_j, d_j]} \sum_{i \in S} (\pi_{i,j}^t v_j (1 + r_j) - b_{i,j}^t). \tag{3}$$

Based on the utility for sellers and buyers in our online DA market, utility maximization as the design objective of the market is defined as follows.

**Definition 2 (Utility maximization).** *An online DA, $M = (\pi, x)$ is utility maximizing, when among set of functions $\pi$ and $x$ that satisfy the other constraints, the mechanism selects $\pi$ and $x$ that maximize*

$$\sum_{i \in S} \left( \sum_{j \in B} \sum_{t \in [a_j, d_j]} \pi_{i,j}^t(\hat{\theta}^{\leq t}) v_j (1 + r_j) - q_i v_i \right). \tag{4}$$

Agents are modeled as risk-neutral and utility-maximizing. Because both sellers and buyers gain profits from successful trades, they have an incentive to offer reasonable bids while trying to increase profits by submitting strategic bids.

We require that the online DA satisfies *budget-balance*, *feasibility* and *individual rationality*. Budget-balance ensures that in every round the mechanism collects and distributes the same amount of money from and to the agents, meaning that an auctioneer earns neither profit nor loss. Feasibility demands that the auctioneer take no short position in the commodity traded at any round. Individual rationality guarantees that no agent loses by participating in the market[3].

## 3   Mechanism Design

In spot markets for perishable goods, sellers raise their asking price and buyers lower their bidding price as a rational strategy to improve their utilities as long as they can avoid trade failures. Agents have to manipulate their valuation carefully for obtaining higher utilities. Our goal is to design a market mechanism that enables agents to secure reasonable utility without strategic bidding.

The well-known result of Myerson & Satterthwaite [10] demonstrates that no exchange can be efficient, budget-balanced, and individually rational. Therefore, we design an online DA mechanism that imposes budget-balance, feasibility and individual rationality, and which promotes reasonable efficiency.

### 3.1   Allocation Policy

Many studies of the DA mechanism use *allocative efficiency* from successful trades as the objective function, with the assumption that agents never suffer a loss from trade failures, and a common goal of its allocation policy is to compute trades that maximize *surplus*, the difference between bid prices and ask prices.

Figure 1 shows demand and supply curves in a multi-unit DA market. On the sellers' side, seller $i$ reports a price $\hat{v}_i^S$ and quantity $\hat{q}_i^S$ for the commodity to trade. On the buyers' side, buyer $j$ reports a price $\hat{v}_j^B$ and quantity $\hat{q}_j^B$. Without loss of generality, we assume $\hat{v}_1^S < \hat{v}_2^S < \cdots < \hat{v}_M^S$ and $\hat{v}_1^B > \hat{v}_2^B > \cdots > \hat{v}_N^B$.

The allocation rule that maximizes allocative efficiency arranges the supply volumes according to the ascending order of the seller's price and the demand volumes according to the descendant order of the buyer's price. As shown in Figure 1, $n$ buyers and $m$ sellers are matched to trade the commodity until the demand and the supply meet at the critical quantity $q^*$, thereby producing

---

[3] To be noted is that sunk costs for sellers are lost *before* they participate in the trades.

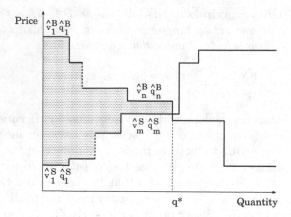

**Fig. 1.** A multi-unit double auction market, where $m$ buyers and $n$ sellers are matched

maximal surplus equal to the size of the shaded area in the figure. We designate this allocation rule as a *price-based* allocation rule.

The most important point to note is that many DA markets assume trades in future markets in which agents do not lose their utility even if they fail to trade. However, sellers of perishable goods in our spot market lose the salvage value of perished goods when they cannot be sold during the trading period. Consequently, as well as increasing surplus from trades, increasing the number of successful trades is important in the spot market for perishable goods. Although, for static DA markets, maximal matching mechanism is proposed in [11], there have been no such mechanisms for online DA markets.

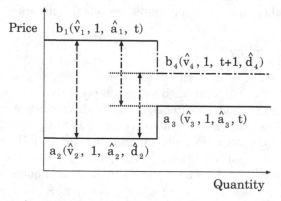

**Fig. 2.** An exemplary scenario of matching in an online DA

As shown in Figure 2, presuming that at round $t$ we have bid $b_1$ with $(\hat{v}_1, 1, \hat{a}_1, t)$ and two asks, $a_2$ with $(\hat{v}_2, 1, \hat{a}_2, \hat{d}_2)$ and $a_3$ with $(\hat{v}_3, 1, \hat{a}_3, t)$, and further presuming that $\hat{v}_1 > \hat{v}_3 > \hat{v}_2$ and $\hat{d}_2 > t$, then at round $t$, $b_1$ can match with either $a_2$ or $a_3$ and in price-based allocation, $a_2$ is matched with $b_1$ to gain a larger surplus, although

$a_3$ will then have to leave the market without being matched. If another bid $b_4$ with $(\hat{v}_4, 1, t+1, \hat{d}_4)$ comes at round $t+1$, then matching $b_1$ and $a_3$ at round $t$ and matching $b_4$ and $a_2$ at round $t+1$ increase a number of successful trades and if valuation in $b_4$ is higher than that in $a_3$ (i.e. $\hat{v}_4 > \hat{v}_3$), more surplus is gained in the matching.

To increase the number of bids that can trade successfully within their fixed time period, we developed a criticality based heuristic allocation policy. In the allocation policy, criticality of an agent's bid is calculated as described below: for agent $i$, the quantity of a commodity that remains unmatched until now is denoted as $r_i(\hat{\theta}^{\leq t})$. The quantity of remaining bids that are matchable with agent $i$'s bid is denoted as $m_i(\hat{\theta}^{\leq t})$. The time left for agent $i$ until it leaves the market is denoted as $l_i(\hat{\theta}^{\leq t})$. Consequently, agent $i$'s criticality is defined as $c_i(\hat{\theta}^{\leq t}) \equiv \frac{r_i(\hat{\theta}^{\leq t})}{m_i(\hat{\theta}^{\leq t})l_i(\hat{\theta}^{\leq t})}$.

At the beginning of allocation, bids from buyers are sorted in descending order with regard to their criticality. Furthermore, then, starting with the most critical bid, asks from the sellers that are matchable with the bid are sorted in descending order with regard to their criticality and matched in the sequence until the quantity requested by the bid is satisfied or there is no matchable ask in the sorted queue (i.e., no preemption is allowed in the matching process). The allocation process continues until there remains no bid that has matchable asks. We designate this allocation rule as a *criticality-based* allocation rule.

Because the criticality-based allocation policy does not incorporate consideration of the surplus produced in matching, the resultant allocations are not guaranteed to be allocative efficient. However, considering the possible loss that agents might suffer when they fail to trade using the price-based allocation rule, the criticality-based allocation policy is expected to earn more profit for the agents by increasing successful trades in certain market situations.

## 3.2 Pricing Policy

We impose budget-balance and promote reasonable efficiency in our online DA market, so that we adopt k-double auction [12] as our pricing policy.

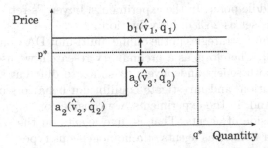

**Fig. 3.** An exemplary scenario of pricing in an online DA

As described in Section 3.1, the allocation policy repeatedly picks up the most critical bid and finds the matchable asks in the order of criticality. Figure 3 presents a situation in which a bid $b_1$ with valuation $\hat{v}_1$ can match with asks $a_2$ and $a_3$, whose respective valuations are $\hat{v}_2$ and $\hat{v}_3$. In this setting, $q^* = \hat{q}_2 + \hat{q}_3$ units of the commodity are traded and a price is determined as $p^* = k\hat{v}_1 + (1 - k)\hat{v}_3$. Results show that the buyer with a bid $b_1$ pays $p^*q^*$ amounts of payment, and the seller with an ask $a_2$ receives $p^*\hat{q}_2$ and the seller with an ask $a_3$ receives $p^*\hat{q}_3$. in this paper, we set 0.5 as the value of $k$.

## 4   Experiments

To evaluate performance of the online DA mechanism designed in Section 3, we conducted a set of multi-agent simulation experiments.

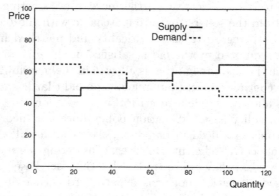

**Fig. 4.** Demand and supply curves used in the experiments

In each experiment, five sellers exist, each of which has 24 units of fish to trade, and five buyers, each of which has 24 units of demand. For the experiments, we use the demand-supply curves depicted in Figures 4, which suggests a competitive trading situation in which a demand curve and a supply curve intersect at the middle point. In the experiments, buyer $j$'s retailing profit rate, $r_j$ in Equation 3, is set as 25% of his valuation $v_j$.

Finding equilibrium strategies in the online multi-unit DA is an extremely difficult endeavor [13]. Therefore, as a preliminary research, we use simple agents of two types for both sellers and buyers to simulate different strategies in reporting their valuation, and investigate equilibrium behaviors of the agents in those situations. And, in the experiments, we consider only symmetric bidding strategies in each side of agents. That is, five agents on the seller side or the buyer side together behave as agents of a homogeneous type.

The strategies used in simulations are as follows: (1) *Truthfully reporting agent (TR)* always reports its valuation truthfully, and (2) *Aggressively reporting agent (AR)* reports its valuation randomly within a certain range. AR agents try to obtain a larger surplus when there is little risk of trade failure, and give up making a profit when there is little time left until their time limit. For seller agent $i$, the upper limit of its ask is 150% of its true valuation $v_i$ and the lower limit at time round $t$ is calculated as $v_i \frac{d_i - t}{d_i - a_i}$. For buyer agent $j$, the lower limit of its bid is 50% of its true valuation $v_j$ and the upper limit at time round $t$ is calculated as $v_j(1 - p_j \frac{d_j - t}{d_j - a_j})$, where $v_j(1 - p_j)$ equals the lower limit of the bid.

To evaluate the performance of a criticality-based DA mechanism developed in Section 3, we compare the results using the criticality-based DA mechanism with those produced by the price-based DA mechanism, which prioritizes bids and asks using reported valuation instead of criticality.

In each experiment, a simulation runs along six days and a market is cleared every two hours. Agent $i$ submits one bid or ask every day. Its arrival time $a_i$ is picked up randomly within a first half of the day, and departure time $d_i$ is determined as 48 hr plus its arrival time $a_i$. For each result, 100 randomized trials are executed to simulate diversified patterns of agents' arrival and departure.

## 4.1 Results

Tables 2 and 3 respectively show the results using the price-based DA and the criticality-based DA. The first line of each cell in the tables shows average matching rates and average clearing prices. And the second line reveals average of seller agents' utilities (refer to Equation 2) at the left and average of buyer agents' utilities (refer to Equation 3) at the right. Numbers in the third line (inside parentheses) are standard deviations of agents' utilities.

**Table 2.** Results of the price-based DA mechanism

| S\B | TR | | AR | |
|---|---|---|---|---|
| | 74%, 55.2 | | 30%, 48.2 | |
| TR | -5,873 \ | 9,236 | -18,781 \ | 6,259 |
| | (2,622) | (494) | (1,680) | (1,088) |
| | 94%, 52.1 | | 84%, 33.6 | |
| AR | -1,887 \ | 11,867 | -13,261 \ | 22,128 |
| | (586) | (432) | (861) | (1039) |

**Table 3.** Results of the criticality-based DA mechanism

| S\B | TR | | AR | |
|---|---|---|---|---|
| | 94%, 55.0 | | 31%, 48.1 | |
| TR | -845 \ | 9,720 | -18,581 \ | 6,336 |
| | (1,632) | (232) | (1,648) | (1,104) |
| | 96%, 53.3 | | 86%, 34.3 | |
| AR | -1,100 \ | 11,051 | -12,994 \ | 21,799 |
| | (580) | (393) | (764) | (1,062) |

Table 2 shows $(AR, AR)$ as a unique Nash equilibrium in the price-based DA mechanism. Agents behave aggressively to maximize their utilities in the price-based DA mechanism. However, Table 3 shows both $(AR, AR)$ and $(TR, TR)$ are Nash equilibria in the criticality-based DA mechanism. Sum of sellers' and buyers' utilities in those three equilibria are comparable. Therefore, they are similarly efficient, since we assume a quasi-linear utility function.

In the criticality-based DA, $(TR, TR)$ is Nash equilibrium because the sellers need not report strategically to achieve high matching rate when buyers report truthfully [4]. Furthermore, at the equilibrium of $(TR, TR)$ in the criticality-based DA, the difference of achieved utilities between buyers and sellers is smallest among all the achievable results. Thus, this equilibrium is preferable to other equilibria in Table 2 and 3 in terms of fairness between sellers and buyers.

## 5    Conclusions

We developed a criticality-based online DA mechanism for fishery markets to achieve efficient and fair allocations considering sellers' loss from trade failures. We explained that sellers have a high risk of losing value by trade failures because fish is perishable and must be produced in advance because of its unsteady yield and quality. To reduce trade failures in those spot markets for perishable goods, our criticality-based DA mechanism prioritizes the bids which have a smaller chance of being matched in their time period. Empirical results using multi-agent simulation showed that the criticality-based DA mechanism was effective in realizing efficient and fair allocations between sellers and buyers in a competitive market when compared with a price-based DA mechanism.

The results reported in the paper is very limited for any comprehensive conclusion on the design of online DA for perishable goods. For the purpose, we need to investigate other types of bidders' strategies in a wide variety of experimental settings. Additionally, behaviors of human subjects in the market must be examined carefully to evaluate effectiveness of the designed mechanism. And, the current criticality-based mechanism can be improved by incorporating some knowledge about future bids statistically predicted from past bidding records.

Fishery markets are usually attended only by local traders, and highly perishable nature of the fishery products prevents their trades being open to wider participants [14]. It caused fishermen's low incomes and fishery collapse in Japan. We hope our mechanism contributes to promote successful deployment of electronic markets for fisheries and improve fishermen's welfare.

**Acknowledgements.** This work was supported by JSPS KAKENHI Grant Number 24300101. I appreciate the precious comments and advices given to the paper by Prof. Atsushi Iwasaki at Kyushu University.

## References

1. Bastian, C., Menkhaus, D., O'Neill, P., Phillips, O.: Supply and Demand Risks in Laboratory Forward and Spot Markets: Implications for Agriculture. Jounal of Agricultural and Applied Economics, 159–173 (April 2000)
2. Wang, C.X., Webster, S.: Markdown money contracts for perishable goods with clearance pricing. European Journal of Operational Research 196(3), 1113–1122 (2009)

---

[4] The matching rate is 94% in this experiment, although $(TR, TR)$ in the price-based DA achieved a modest matching rate of 74%

3. Myerson, R.B.: Optimal Auction Design. Mathematics of Operations Research 6(1), 58–73 (1981)
4. Krogmeier, J., Menkhaus, D., Phillips, O., Schmitz, J.: An experimental economics approach to analyzing price discovery in forward and spot markets. Jounal of Agricultural and Applied Economics 2, 327–336 (1997)
5. Mestelman, S.: The Performance of Double-Auction and Posted-Offer Markets with Advance Production. In: Plott, C.R., Smith, V.L. (eds.) Handbook of Experimental Economics Results, vol. 1, pp. 77–82. Elsevier (2008)
6. Blum, A., Sandholm, T., Zinkevich, M.: Online algorithms for market clearing. Journal of the ACM 53(5), 845–879 (2006)
7. Bredin, J., Parkes, D.C., Duong, Q.: Chain: A dynamic double auction framework for matching patient agents. Journal of Artificial Intelligence Research 30(1), 133–179 (2007)
8. Zhao, D., Zhang, D., Perrussel, L.: Mechanism Design for Double Auctions with Temporal Constraints. In: Proceedings of IJCAI, pp. 1–6 (2011)
9. Talluri, K.T., van Ryzin, G.J.: The Theory and Practice of Revenue Management. Kluwer Academic Publishers, Norwell (2004)
10. Myerson, R., Satterthwaite, M.: Efficient Mechanisms for Bilateral Trading. Journal of Economic Theory 29(2), 265–281 (1983)
11. Zhao, D., Zhang, D., Khan, M., Perrussel, L.: Maximal Matching for Double Auction. In: Li, J. (ed.) AI 2010. LNCS, vol. 6464, pp. 516–525. Springer, Heidelberg (2010)
12. Satterthwaite, M.A., Williams, S.R.: Bilateral Trade with the Sealed Bid k-Double Auction: Existence and Efficiency. Journal of Economic Theory 48(1), 107–133 (1989)
13. Walsh, W., Das, R., Tesauro, G., Kephart, J.: Analyzing complex strategic interactions in multi-agent systems. In: AAAI 2002 Workshop on Game-Theoretic and Decision-Theoretic Agents, pp. 109–118 (2002)
14. Guillotreau, P., Jiménez-Toribio, R.: The price effect of expanding fish auction markets. Journal of Economic Behavior & Organization 79(3), 211–225 (2011)

# A Mediator-Based Agent Negotiation Protocol for Utilities That Change with Time

Keisuke Hara, Mikoto Okumura, and Takayuki Ito

Nagoya Institute of Technology,
Gokiso-cho, Showa-ku, Nagoya City, Aichi Pref., Japan
{hara.keisuke,okumura.mikoto}@itolab.nitech.ac.jp,
ito.takayuki@nitech.ac.jp
http://www.itolab.nitech.ac.jp

**Abstract.** Multiple-issue negotiation has been studied extensively because most real-world negotiations involve multiple issues that are interdependent. Our work focuses on negotiations with multiple interdependent issues in which agent utility functions are complex and nonlinear. Issues being interdependent means that it is not appropriate to negotiate over issues one-by-one. The decision on one issue is dependent on decisions about previous and subsequent issues. In the literature, several negotiation protocols are proposed: bidding-based protocol, constraints-based protocol, secure SA-based protocol, etc. However, all have assumed that utility does not change with time, whereas in reality, this may not be the case. In this paper, we focus on finding and following the "Pareto front" of the changing utility space over time. To find and follow the Pareto front effectively, we employ an evolutionary negotiation mechanism, in which a mediator takes the lead in negotiation based on GA. The experimental results show that our approach is suitable for the case where utility is dynamically changing over time.

**Keywords:** Multi-Agent, Negotiation, GA.

## 1 Introduction

Multi-issue negotiation protocols represent an important field of study because negotiation problems in the real world are often complex ones involving multiple issues. While there has been a lot of previous work in this area [1–4] these efforts have, to date, dealt almost exclusively with simple negotiations involving independent issues, and therefore linear (single optimum) utility functions. Many real-world negotiation problems, however, involve interdependent issues. When designers work together to design a car, for example, the value of a given carburetor is highly dependent on which engine is chosen. The addition of such interdependencies greatly complicates the agents utility functions, making them nonlinear, with multiple optima.

In the same way, studies related to negotiation problems involving interdependent issues [5, 6] have been carried out. Most studies, however, have not focused

M. Ali et al. (Eds.): IEA/AIE 2013, LNAI 7906, pp. 12–21, 2013.
© Springer-Verlag Berlin Heidelberg 2013

on the changes of the utility space over time. We have found there is a possibility that the temporal changes of utility are few during negotiations. This is because a lot of the studies on group decision-making research show the difficulty of identifying the temporal consistency of utility functions[7]. In addition, in economic theory, it is often assumed that utility function changes dynamically over time as well.

Therefore, in this paper, we propose a complex utility space that changes over time and negotiation protocols that can respond to these changes. In this paper, while decreasing or increasing only some issues, we discuss not only overall utility space change, but the shape of this change.

Human utility is likely to decrease over time [7]. In this paper, we represent this change by setting the decreasing or increasing rate for each issue. It is difficult for agents to predict in advance the entire change of the utility space because only some issues relating to utility are decreasing or increasing in an agent's utility space. Our experiments show that it is possible to track a consensus point that changes over time using the proposed method.

In addition, in the negotiation problem, it is important to seek the Pareto front, which refers to the set of Pareto optimal points. To find the Pareto front, the genetic algorithm that handles multiple solution candidates at the same time is better than the other methods, such as simulated annealing to obtain the single solution. In the field of multi-objective optimization, the genetic algorithm is used in obtaining the Pareto front for each objective function.

The remainder of the paper is organized as follows. First, we describe a model of non-linear multi-issue negotiation. Second, we describe the utility space that changes over time proposed in this paper. Third, we describe a method in which the mediator takes the lead in negotiation based on GA. Fourth, we describe the result of experiments and their evaluation. Finally, we conclude with a discussion of possible avenues for future work.

## 2   Negotiation with Nonlinear Utilities

We consider the situation where $N$ agents want to reach an agreement. There are $m$ issues, $s_j \in S$, to be negotiated. The number of dimensions of the utility space is the number of issues $+1$. For example, if there are 2 issues, the utility space has 3 dimensions. An issue $s_j$ has a value drawn from the domain of integers $[0, X]$, i.e., $s_j \in [0, X]$. A contract is represented by a vector of issue values $s = (s_1, \cdots, s_m)$.

An agent's utility function is described in terms of constraints. There are $l$ constraints, $c_k \in C$ . Each constraint represents a region with one or more dimensions, and has an associated utility value. A constraint $c_k$ has value $w_i(c_k, s)$ if and only if it is satisfied by contract $s$.

Figure 1 shows a model of a utility space in which issues are interdependent. A node indicates an issue and an edge indicates a constraint. This model can represent unary constraints, binomial constraints, ternary constraints. In this example, this constraint has a value of 50, and holds if the value for issue 2 is in the range [3,7] and the value for issue 3 is in the range [4,6]. Similarly, the constraint has a value of 100, and holds if the value for issue 1 is in the range [1,4] and the value for issue 2 is in the range [2,6] and the value for issue 3 is in the range [3,5].

An agent's utility for a contract $s$ is defined as $u_i(s) = \sum_{c_k \in C, s \in x(c_k)} w_i(c_k, s)$, where $x(c_k)$ is a set of possible contracts (solutions) of $c_k$. Every agent to participate in the negotiations has its own, typically unique, set of constraints.

In this paper, we assume the fundamental form of a decision-making problem, such as that of designing a car. As a specific example, we can cite the problem of deciding the hall style for an academic meeting or conference. In this example, there are specific issues like cost and capacity, with options of 500-700 thousand yen or 700-900 thousand yen, and 50-100 people or 100-150 people. We set the evaluation value considering the relationship of each issue and another issue against each choice; for example, we could use more cost if we could reserve the larger hall, and decide the answer for each issue. Now, the preference information that the agent should have is the combination between the alternative solution for each issue and that for another issue and the evaluation value. The constraint representation in this paper is sufficient to express this information, and can deal with the assumptive problem.

The objective function for our protocol can be described as equation (1).

$$\arg \max_{s} \sum_{i \in N} u_i(s) \qquad (1)$$

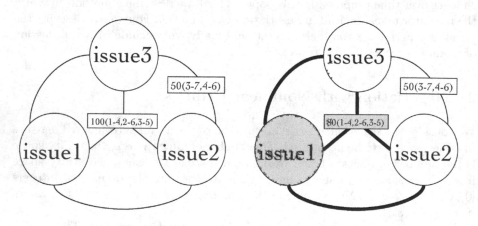

Fig. 1. Utility graph          Fig. 2. Utility decreasing on issue 1

## 3   Dynamic Utility Space That Changes over Time

Consider now the utility that changes over time by introducing a changing rate (decreasing or increasing rate). In this example, we discuss only decreasing. In general, a change in utility depends on certain issues in real-world negotiation. For example, consider the design of a car with which there are three issues: price, function, and design. It is reasonable to suppose that a person who has been focused on the price at the start of a negotiation might reduce the importance of price after a few iterations in a negotiation. However, it is less reasonable to think that the importance of all of the issues like price, function and design would decrease uniformly because many real-world negotiation problems involve interdependent issues.

Therefore, in this paper we consider a utility graph that changes over time depending on some of the issues. Figure 2 shows the influence of decreasing of issue 1 in Figure 1. By decreasing issue 1, the utility obtained from constraints that relate to issue 1 (bold edges) is reduced. In this example, the decreasing rate is 0.8 and a decreasing happens once. By comparing Figure 1 and Figure 2, the utility obtained from constraints that relate to issue 1 is reduced from 100 to 80. On the other hand, the utility obtained from the constraints that do not relate to issue 1 is unchanged and stays at 50.

Figures 3 and 4 show how the Pareto front has changed over time for each agent. While the vertical axis represents the agent utility value (UB), the horizontal axis represents the agent utility value (UA). Figure 3 shows the change of the Pareto front in case of decreasing all the issues. The Pareto front decreases overall because the utility of each agent is reduced equally. Figure 4 shows the change of the Pareto front in the case of decreasing only issue 1.

Figure 4 shows that the difference of the number of constraints related to issue 1 has an impact on change in the Pareto front. Consider this reason from the difference in the number of constraints. In this paper, each agent has a unique utility graph because utility graphs are created randomly. Also, the number of all constraints of each agent is equal. On the other hand, the number of constraints that relate to each issue is different among agents. Because the number of constraints represents the importance of the issue, the more constraints that relate to issue 1, the greater the importance of issue 1. This means issue 1 has more influence on the decreasing and there is greater decrease of the utility. In Figure 4, the Pareto front is reduced disproportionately to agent B because agent B has more than ten constraints related to issue 1.

Below, we show the general definition of the Pareto dominance, the Pareto optimality, and explain the Pareto front.

**Pareto dominance.** There are two utility vectors $x = (x_1, \cdots, x_n)$, $y = (y_1, \cdots, y_n)$, For all $i$, if $x_i > y_i$, $x$ (Pareto) dominates $y$. if for all $i$, $x_i \geqq y_i$, and $x_i > y_i$ at least one $i$, $x$ weakly (Pareto) dominates $y$.

**Pareto optimality.** If $x$ is not weakly dominated by any other utility vectors, $x$ is pareto efficient. If $x$ is not dominated by any other utility vectors, $x$ is weakly pareto efficiency.

**Pareto front.** The Pareto front is the set of Pareto optimal solutions.

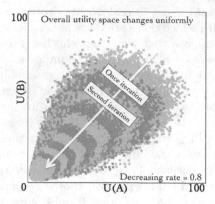

**Fig. 3.** Pareto front changes over time

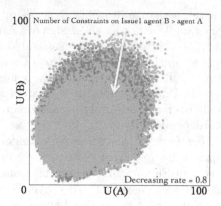

**Fig. 4.** Agent B with bias

## 4    Negotiation Protocol Based on GA

Ito[6] has illustrated a consensus optimization mechanism among agents based on a distributed genetic algorithm. The study, however, has not taken account of the change over time of an agent's utility. The purpose of this study is to show that a negotiation protocol based on GA is useful for consensus building even if the agent's utility changes over time.

As shown in Table 1, we can map the consensus point onto the negotiation as the chromosome, an issue as a genetic locus, and the value of the issue as the gene when we apply GA to a negotiation among the agents.

**Table 1.** Mapping a negotiation problem into a GA

| Negotiation | consensus point | issue | value of issue |
|:---:|:---:|:---:|:---:|
| GA | chromosome | genetic locus | gene |

In the proposed algorithm, a mediator facilitates negotiations while accepting the preference of each agent and attempts to obtain as high a consensus point as possible. Figure 5 shows the outline of the proposed algorithm. Figure 5 describes the case of two agents, but can be easily extended to n agents. First, the mediator sends a set of chromosomes to each agent. Each agent sorts the chromosomes based on its own utility space. That is, consensus points are sorted by each agent's values. Then, each agent submits the ranking information of the top half of chromosomes to the mediator. Then, the mediator calculates the Pareto dominance relations and creates a copy of chromosomes that are not Pareto-dominated (better chromosomes), saves and leaves them to the next generation. Then, the mediator does a crossover and a mutation. The above procedure is repeated a defined number of times.

The significant point of this algorithm is that it is possible to carry over Pareto dominant points to later generations (called "dominant inheritance"). Also, because each agent sends the additional ranking information, the mediator can decide Pareto dominance relations among chromosomes without knowing the specific utility value of each chromosome.

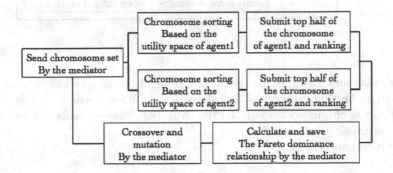

**Fig. 5.** Flow of proposed method

# 5    Experimental Result

## 5.1    Experimental Setting

We conducted several experiments to evaluate the effectiveness of our approach. In each experiment, we ran 100 negotiations between agents with randomly generated utility functions. 1 negotiation has 20 iterations that means change over time. For each run, we applied an optimizer to the sum of the utility functions of all agents to find the contract with the highest possible social welfare. This value was used to assess the efficiency (i.e. how closely optimal social welfare was approached) of the negotiation protocols. We used exhaustive search (EX) to find the optimum contract. We compared the algorithm called hill-climbing (HC) in Figure 6. HC is a method in which the mediator takes the lead in a negotiation without GA in this paper.

The parameters for our experiments are defined as follows:

- Number of agents is 2, 20, 100. Number of issues is 5. Domain for issue values is [0,9].
- Constraints for nonlinear utility spaces: 30 unary constraints, 30 binary constraints, 30 trinary constraints, etc. (A unary constraint relates to one issue, a binary constraint relates to two issues, and so on).
- The maximum value for a constraint: 100. Constraints that satisfy many issues thus have, on average, larger weights. This seems reasonable for many domains. In meeting scheduling, for example, higher order constraints concern more people than lower order constraints, so they are more important for that reason.

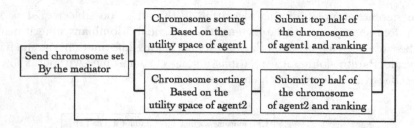

**Fig. 6.** Flow without GA (HC)

- The maximum width for a constraint: 7. The following constraints, therefore, would all be valid: issue 1 = [2,6], issue 3 = [2,9] and issue 7 = [1,3].
- Number of chromosomes is 10 to 100. Number of generations is 5 to 100.
- The decreasing rate is 0.8.
- The increasing rate is 1.1.

## 5.2   Results

### i. Evaluate Effectiveness of GA

Let us first consider the effectiveness of GA without influence of change in utility. In Figure 7, the vertical axis represents the agent utility value (UB), and the horizontal axis represents the agent utility value (UA). The area colored gray refers to the negotiable region, while Figure 7 shows the top of some consensus points of GA (black) and HC (white). The number of generations is 20. The number of chromosomes is 20. GA is able to search for the Pareto front, but HC cannot. That is, it is difficult to search for the Pareto front.

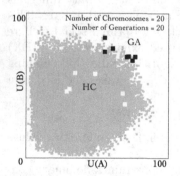

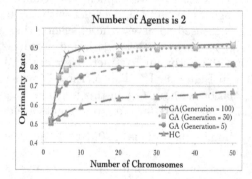

**Fig. 7.** Search for Pareto front     **Fig. 8.** Optimality with number of chromosomes

Figure 8 shows the optimality with a number of chromosomes. The greater the number of chromosomes and generations, the more optimality our protocol gets because the search domain becomes broad. Because our protocol searches for the Pareto front, it is not a problem that optimality is not 1.0.

## ii. The Case when Overall Utility Space Changes Uniformly

In Figure 9(a), the vertical axis represents the agent utility value (UB), and the horizontal axis represents the agent utility value (UA). The areas colored gray and dark gray refer to the negotiable region. Black dots represent points of agreement.

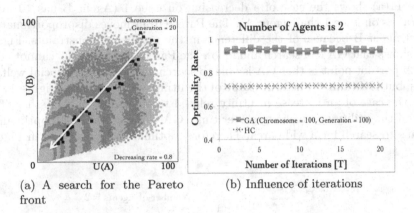

(a) A search for the Pareto front

(b) Influence of iterations

**Fig. 9.** The case when overall utility space changes uniformly

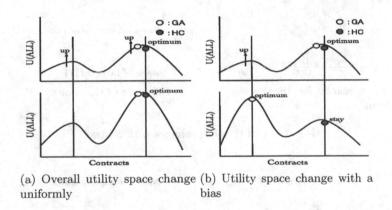

(a) Overall utility space change uniformly

(b) Utility space change with a bias

**Fig. 10.** The shape of the utility space

Figures 9 and 10(a) show the case of a decreasing or increasing on all issues. In this case, the utility space of each agent changes in the same way. That is, the shape of the utility space stays almost the same as in Figure 10(a). Therefore, the Pareto front is reduced but the shape stays almost the same. Thus, the optimality is not changed in Figure 9(b). We say that GA is able to search for the Pareto front and HC cannot in Figure 9(a).

### iii. The Case when Utility Space Changes with a Bias

Figure 10(b) and Figure 11 show the case when utility space changes with a bias. As Figure 10(b) indicates, the shape of the overall utility space changes in the case where only some of the issues are related to the changes of utility. Each agent has its own bias on which constraints are related to which issue. This makes it difficult to effectively follow the optimal points during the utility space changes. Figure 11(a) shows the case of a decreasing on issue 1 (Agent B has 10 more constraints on issue 1 than agent A). The Pareto front moves disproportionately against agent B because of the difference in the number of constraints. Figure 11(a) shows that GA can search and follow the Pareto front and HC cannot. The most important point is that GA is able to search and follow high social welfare points, but HC cannot when the shape of the utility space changes. Figure 11(b) shows the case of an increase in utility related to only some of the issues. The shape of the utility space changes over time. GA can maintain high optimality. HC fails to search for a wide area. Therefore, optimality decreases as in Figure 11(b).

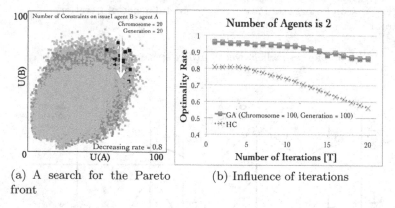

(a) A search for the Pareto front

(b) Influence of iterations

**Fig. 11.** Utility space changes with a bias

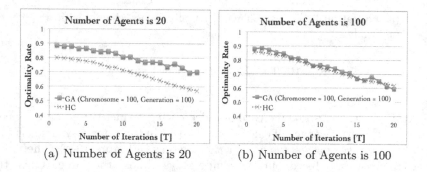

(a) Number of Agents is 20

(b) Number of Agents is 100

**Fig. 12.** Relation between optimum rate and number of agents

**IV. 20 Agents and 100 Agents**
Figure 12 shows the case of 20 agents and 100 agents and utility space changes with bias. Although we can see bumpy shapes, GA is a better result than HC when the number of agents is 20. However, when the number of agents is 100, it is more bumpy, and GA is similar to HC. The main reason is that the greater the number of agents, the more complex the shape of the utility space is. Thus, it is difficult to optimize with GA. We say that GA is able to search for higher social welfare points on average.

# 6    Conclusions and Future Work

In this paper, we proposed a mediator that takes the lead in negotiation based on GA and thus a nonlinear utility space that changes over time. Our experimental results show that our method is able to follow the change of the shape of utility space over time. Possible future work includes improving scalability by developing strategies of the mediator.

**Acknowledgement.** This work is partially supported by the Funding Program for Next Generation World-Leading Researchers (NEXT Program) of the Japan Cabinet Office.

# References

1. Faratin, P., Sierra, C., Jenning, N.R.: Using similarity criteria to make issue trade-offs in automated negotiations. In: Artificial Intelligence, pp. 205–237 (2002)
2. Soh, L.-K., Li, X.: Adaptive, confidence-based multi-agent negotiation strategy. In: Proceedings of the Third International Joint Conference on Autonomous Agents and Multi-agent Systems (AAMAS 2004) (2004)
3. Fatima, S., Wooldridge, M., Jennings, N.R.: Optimal negotiation of multiple issues in incomplete information settings. In: Proc. of Autonomous Agents and Multi-Agent Systems (AAMAS 2004) (2004)
4. Lau, R.Y.K.: Towards Genetically Optimised Multi-Agent Multi-Issue Negotiations. In: Proceedings of the 38th Annual Hawaii International Conference on System Sciences (HICSS 2005) (2005)
5. Ito, T., Klein, M., Hattori, H.: A multi-issue negotiation protocol among agents with nonlinear utility functions. In: Multiagent Grid Syst., vol. 4 (2008)
6. Ito, T., Klein, M.: A Consensus Optimization Mechanism among Agents based on Genetic Algorithm for Multi-issue Negotiation Problems. In: JAWS 2009, pp. 286–293 (2009)
7. Strotz, R.H.: Myopia and Inconsistency in Dynamic Utility Maximization. In: Proceedings of the 10th International Conference on Cooperative Information Agents, vol. 23 (1955)

# Learning Parameters for a Cognitive Model
# on Situation Awareness

Richard Koopmanschap[1,2], Mark Hoogendoorn[2], and Jan Joris Roessingh[1]

[1] National Aerospace Laboratory,
Department of Training, Simulation, and Operator Performance
Anthony Fokkerweg 2, 1059 CM Amsterdam, The Netherlands
{smartbandits,jan.joris.roessingh}@nlr.nl
[2] VU University Amsterdam,
Department of Computer Science
De Boelelaan 1081, 1081 HV Amsterdam, The Netherlands
m.hoogendoorn@vu.nl

**Abstract.** Cognitive models are very useful to establish human-like behavior in an agent. Such human-like behavior can be essential in for instance serious games in which humans have to learn a certain task, and are either faced with automated teammates or opponents. To tailor these cognitive models towards a certain scenario can however be a time-consuming task requiring a lot of domain expertise. In this paper, a cognitive model is taken as a basis, and the addition of scenario specific information is for a large part automated. The performance of the approach of automatically adding scenario specific information is rigorously evaluated using a case study in the domain of fighter pilots.

## 1    Introduction

In many settings, it is highly desirable that agents exhibit human-like behavior. In the serious gaming domain (see e.g. [1]) for example, having human-like computerized team-mates or opponents can contribute to a more realistic and therefore potentially more effective training experience. For instance, for the domain of fighter pilot training, in order for trainees to learn appropriate tactical avoidance maneuvers in a simulator it is essential to have an opponent fighter that attacks in a way a human pilot would.

A variety of cognitive models have been defined that facilitate the generation of this desired behavior, for instance models for Situation Awareness (SA, see e.g. [2]), theory of mind (see e.g. [3]), naturalistic decision making (e.g. [4]), and many more. Such models however require domain knowledge. For instance in an SA model, there needs to be knowledge on how to make interpretations of certain observations for the domain at hand, and for a decision making model knowledge should be present about the suitability of certain decisions in one way or the other. An approach to specify this knowledge is to involve domain experts, extract their knowledge, and incorporate this knowledge in the cognitive models. However, this requires a substantial effort from

M. Ali et al. (Eds.): IEA/AIE 2013, LNAI 7906, pp. 22–32, 2013.

experts which are most often scarce and expensive. An alternative approach is to avoid the utilization of cognitive models, and apply a straightforward machine learning approach. The disadvantage thereof is that this is not guaranteed to result in human-like behavior as the underlying mechanisms do not necessarily resemble (and are not constrained by) human or humanlike cognition. Hence, a combination of the two approaches might be the most appropriate, thereby reducing the expert effort required and still obtaining human-like behavior.

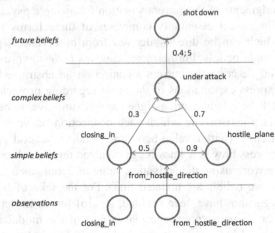

**Fig. 1.** Example mental model 'hostile plane' (cf. [5])

One example of an approach which combines a cognitive model with learning is presented in Gini, Hoogendoorn and van Lambalgen [5]. In particular, learning is applied to an SA model. In the described approach, it is assumed that knowledge about links between observations and judgments of the situation are provided by domain experts. Furthermore, it is assumed that the expert provides a correct judgment for each and every situation (i.e. all possible sets of observations) which the agent might encounter. The learning takes care of finding appropriate values for the strengths of connections between observations and the judgment of the situations. Although the results are promising and it takes away a substantial part of the burden for the domain expert, it still requires the domain expert to go through a large number of situations and provide a desired outcome.

In this paper, the research is taken a significant step further. Instead of assuming the desired output to be provided by an expert and learn the connection strengths based on this output, learning is based on the performance of the agent in scenarios, specifically designed to train SA. To enable this, state-of-the-art machine learning algorithms are used and refined. The approach has been applied to air-to-air combat by fighter pilots and has been systematically evaluated for a large number of scenarios.

This paper is organized as follows. In Section 2 the underlying cognitive model is explained. Section 3 presents the learning approach which is used to learn the domain knowledge encapsulated in the SA model. A case study is presented in Section 4, and the accompanying results of the application of the machine learning approach are included in Section 5. Finally, Section 6 provides a discussion.

## 2    Situation Awareness Model

The cognitive model which is subject to learning is the model presented in [6], which is a slightly modified version of [2]. The model is meant to create high-level judgments of the current situation following a psychological model from Endsley [7]. The model essentially comprises of three forms of beliefs, namely *simple beliefs*, which can be directly derived from observations performed by the agent, or other simple beliefs. Furthermore, *complex beliefs* express combinations of simple beliefs, and describe the current situation on an abstracted level, and finally, *future beliefs* express expectations of the agent regarding projected (future) events in the scenario. All these beliefs are assigned an activation value between 0 and 1 and are connected via a network, in which each connection between a pair of beliefs has a particular strength (with a value between -1 and 1). Several update properties are specified that express how new knowledge obtained through observations is propagated through the network using the activation value of beliefs as a rank ordering of priority (the most active beliefs are updated first). For the sake of brevity, the details of the updating algorithm have been omitted, see [6] for an in-depth treatment of the update rules. The domain knowledge that is part of the model is a so-called belief network. An example of such a network is shown in Figure 1.

The example shows that each connection between a pair of beliefs receives a connection strength. For instance, the connection between the simple belief *from hostile direction* and the simple belief *hostile plane* has a connection strength of 0.9. This relative high value will help to activate the belief that an approaching plane is indeed hostile when it is observed to come from a hostile direction. In larger, more meaningful, networks, the task of determining and assigning the appropriate values for connection strengths has to be done by domain experts, which consider this task as cumbersome and prone to error. Therefore, this task will be allocated to a learning algorithm. Note however that the conceptual belief structure of the network must still be provided, a task for which expert input is still needed, but which is considered less cumbersome and error prone.

In order to enable learning based upon the performance of the agent in scenarios, the agent needs to do more than just judging the situation. The agent additionally needs to decide on an action. To this end, the SA model has been extended with a simple decision making model (DM), namely a Finite State Machine model.

## 3    Learning Approach

In this section, the learning approach is presented. Hereby, it is assumed that the following elements are present within the agent integrated SA decision making model:

1. The structure of the SA belief network (i.e. the precise beliefs that should be present).
2. The full decision making model that maps activation values of beliefs to concrete actions for the specific domain at hand.

The outcome of the learning process will be connection strengths within the belief network. In order to learn these connection strengths in an appropriate manner, the following approach is proposed:

1. Domain experts provide a scenario which facilitates the learning of specific behavior. Hereby, the scenario allows for appropriate behavior and less appropriate or even unwanted behavior of the agent, and this can be measured by means of a performance metric (in this case called a *fitness*).
2. Using the scenarios, machine learning techniques are deployed that strive towards achieving the best (i.e. most appropriate) behavior of the agents within the scenarios by finding the best settings for the connection strengths.

The learning algorithm is expressed in a semi-formal manner below.

In order to perform learning, two different learning algorithms are tested and extended with several methods to enhance their performance: a simple hill climbing algorithm and an evolutionary algorithm. All algorithms represent the search space as a vector of all weights in the network.

| | Algorithm 1. Learning loop | Extension |
|---|---|---|
| 1 | **Algorithm 1. Learning loop** | **Extension** |
| 2 | initialize the connection strengths vector | (3),(4) |
| 3 | while (runs < max_runs) | |
| 4 | initialize the scenario | (5) |
| 5 | while (time < scenario_time) | |
| 6 | determine the current observations | |
| 7 | feed the observations to the agent | |
| 8 | determine activation values of the beliefs with connection | |
| 9 | strengths vector (SA model) | |
| 10 | derive the actions of the agent with the activation values of | (1) |
| 11 | beliefs (DM model) | |
| 12 | perform the actions in the world | |
| 13 | time++ | |
| 14 | end | |
| 15 | determine the performance in the scenario (fitness value) | (2) |
| 16 | apply a learning strategy based upon the current performance | (3),(4) |
| 17 | and the connection strengths vector | |
| 18 | runs++ | |
| 19 | end | |

The hill climbing and evolution algorithm use Gaussian mutation to generate offspring. Each scalar in the vector has a predetermined mutation probability and a sigma with which the Gaussian mutation is performed. The evolution algorithm applies tournament selection for both the survival selection and the parent selection. For crossover a uniform crossover operator is used.

Since the learning process takes place in a simulated environment, performance issues will quickly emerge. Therefore, five extensions to the above algorithms have been developed to speed up the learning process. These extensions can be divided into two categories: extensions that alter the fitness function and extensions that limit the search space. The following extensions are applied on the algorithms. Note that the location where the extensions are placed in the algorithm itself are shown in the description of algorithm 1.

**(1) Adding Output Noise.** In order to let the decision making model make a decision, the activation value of the required beliefs needs to exceed a certain threshold. This results in the effect that small changes in connection strengths between beliefs in the SA model are unlikely to cause beliefs to exceed the threshold and hence, do not result in any differences in fitness. Machine learning algorithms have difficulty finding an optimum in such cases. In order to deal with this problem an approach has been developed: a random Gaussian bias is assigned to the activation value of each belief at the start of each trial of a scenario. This random bias is added to the activation value of a belief whenever the decision making model uses the value of the belief. This may cause the activation value to pass the threshold (in either direction) when it otherwise would not have. The frequency of passing the threshold depends on the distance between the activation value of the belief in the scenario and the threshold in the decision making model. If this process is repeated a sufficient number of times for each connection strength vector this will smooth the decision surface for the learning algorithm as it will provide an average fitness of the values around the activation value currently achieved, and therefore show the average performance of the current region of activation values. The extension is expressed in algorithm 2 below, whereby the changes are shown in italics.

```
1    Algorithm 2. Learning loop extended with output noise
2    initialize the connection strengths vector
3    while (runs < max_runs)
4        initialize the scenario
         for(each belief x)
             generate a random number bias (b(x) = N(σ,μ))
5        while (time < scenario_time)
6            determine the current observations
7            feed the observations to the agent
8            determine activation values of the beliefs with connection strengths vector
9                (SA model)
10           derive the actions of the agent with the activation values of belief(x) + b(x)
11               (DM model)
12           perform the actions in the world
13               time++
14       end
15       determine the performance in the scenario (fitness value)
16       apply a learning strategy based upon the current performance
17           and the connection strengths vector
18       runs++
19   end
```

**(2) Expanding the Fitness Function with Scenario Specific Information.** It is preferable to use a fitness function that does not require a substantial amount of information from the domain, for instance by just providing a payoff for a successful run in a scenario or a non-successful one. However, sometimes a more fine-grained view is necessary to learn the finer details of the domain (e.g. providing fitness for milestones reached, etcetera), therefore an option for an extension is also to provide such a more fine-grained fitness function. In algorithm 1 this only influences the way in which the fitness is calculated (line 15).

(3) **Using Symmetry Information.** In conventional neural networks the precise meaning of the nodes is often difficult to determine. In cognitive models, on the other hand, the nodes are constructed for specific concepts that are often well understood. Although it may not be clear what the optimal connection strength between two beliefs is, it is often clear that many parts in a belief network are symmetrical. For example, the connection between beliefs that deal with situations to the North may be equivalent to similar connections between beliefs that deal with situations to the South. By explicitly defining such symmetries beforehand the learning algorithm can limit the search space by always assigning these connections the same weights. An additional advantage of defining symmetries is that it allows the agent to generalize its learned behavior to situations that have not been previously encountered. If the agent would never have seen a training example that activates the beliefs dealing with situations to the South, the agent may still be able to handle the situation correctly if it has seen similar examples to the North. In the algorithm it limits the initialization and updating possibilities of the algorithm (line 2 and 16).

(4) **Limiting Values in the Search Space.** Values in the search space of the fully learned belief networks of the situation awareness model do not emerge equally often. In particular, negative values for connection strengths are rare. The learning algorithm can be enhanced by reducing the set of values with which it can update the weight vector. For example, only the negative value -1 could be allowed in addition to the value range 0-1. By removing uncommon values, unpromising areas of the search space can be removed before the learning starts. Again, this influences the possibilities in line 2 and 16 of the algorithm.

(5) **Incremental Learning.** It has been shown in previous research that training an agent on a simple task before training it on a more difficult task can improve results (see e.g. [8]). As the meaning of the nodes in the network are well understood, it is fairly straightforward to use this technique here to train parts of the network in scenarios that are specifically designed to learn those simple tasks. The learned connection strengths can then be stored and used for more complex tasks that require simpler parts of the network to work correctly. For the purpose of this research a part of the complete belief network is first trained in a scenario. When the scenario is completed all connection strengths of that part of the belief network are stored. The algorithm then starts the next scenario with a larger part of the belief network and loads the stored weights from the first scenario to re-use them in the current one. Any number of scenarios can be linked, provided that the belief networks contain more nodes each subsequent time. In the approach, the algorithm does not change connection strengths that it has already learned and so it is left to the designer of the scenarios to ensure that the connection strengths accurately represent the connections between beliefs after being trained in the first scenario. Algorithm 3 describes the algorithm with the use of the incremental learning extension.

```
1    Algorithm 3. Learning loop with incremental learning
     for (each scenario)
       load all stored connection strengths
2      initialize the connection strengths vector with the still unknown connection strengths
3      while (runs < max_runs)
4        initialize the scenario
5        while (time < scenario_time)
6          determine the current observations
7          feed the observations to the agent
8          determine activation values of the beliefs with connection strengths vector
9                (SA model)
10         derive the actions of the agent with the activation values of beliefs
11               (DM model)
12         perform the actions in the world
13         time++
14       end
15       determine the performance in the scenario (fitness value)
16       apply a learning strategy based upon the current performance
17             and the connection strengths vector
18       runs++
19     end
       save all connection strengths
20   end
```

# 4    Case Study and Experimental Setup

In order to test the suitability of the approach, it has been applied in the domain of air combat. There is an interest from this domain in agents exhibiting realistic tactical behaviour so as to increase the value of simulation training for fighter pilots. Although the focus lies on demonstrating adversarial behaviour in air-to-air missions, the results are more widely applicable in the simulation domain. In this case, a 'one versus one' combat engagement scenario has been selected, whereby two opposing agents (pilots controlling fighter planes) try to eliminate each other. Good behavioral performance for such agent is defined as correctly detecting planes on the radar, correctly identifying such planes as being hostile or not, and subsequently engaging a hostile plane.

A scenario has been created in which this desired behavior can be displayed. The scenario thus consists of two agents, an intelligent (cognitive and adaptive) agent and an agent that behaves merely on the basis of a non-adaptive script. The intelligent agent contains the SA and DM model described above in addition to one of the described learning techniques. The scripted agent uses a small and static set of pre-defined rules. In the one versus one scenario the intelligent agent patrols until it encounters either a friendly or a hostile target, which it will have to identify and, in case it is hostile, destroy.

In order to be able to use the incremental learning described in Section 3, scenarios have been created for each of the behaviors the intelligent agent is required to be able to perform to be successful in the one versus one scenario. Each of the scenarios trains a different part of the belief network and only has access to that part of the network. The scenario that deals with identifying and remembering the location of nearby planes, the awareness scenario, is particularly suitable for experiments. Therefore this scenario will be described in greater detail here and used in several experiments of which the results are shown in the next section.

In the awareness scenario the belief network consists of nodes that store the location of nearby planes and whether this plane is friendly, hostile or unidentified. The DM model simply involves the action to fire a missile in any direction for which a sufficiently active belief is present that a hostile plane resides in that direction. A single run in the awareness scenario consists of multiple trials in which an intelligent agent is presented with a number of planes. Each time the agent destroys an enemy it will gain a fitness value of 100, each time it destroys a friendly target it will be given a fitness value of -100. At the end of a scenario the total fitness value is divided by the number of trials in the scenario. In order to make a comparison between the different learning algorithms the following steps are performed:

1. On the *awareness scenario*, find appropriate parameter settings for the two different learning algorithms.
2. Make a comparison between the two learning approaches on the aforementioned scenario, and also compare them with a benchmark, namely random search through the space of connection strengths.
3. Evaluate how much different extensions improve the performance of the learning algorithms using the *awareness scenario*, except for the incremental learning which is performed on a number of scenarios.
4. Evaluate the performance of the best algorithm on the most complete scenario (called one v. one).

The results are described in Section 5.


## 5    Results

Based on preliminary experiments, reasonably well performing parameter values were determined to use in the experiments, these values are given in Table 1. Figure 2 shows the performance on the awareness scenario of the three different algorithms without using any extensions. Each line in figure 2 represents the average fitness value over 80 runs. A t-test has shown that the performances of the approaches are statistically significant with a 5% confidence interval. The results indicate that the evolutionary algorithm performed best, better than random search through the space. Hill climbing was unable to surpass the benchmark, mainly due to the fact that it has difficulty escaping local optima. Random search is able to find a suitable area in the search space quickly, but converges to a lower fitness than the evolutionary algorithm as it lacks a local search operator.

**Table 1.** Algorithm parameters

| Parameter | Evolutionary Algorithm Setting | Hill Climbing Setting |
|---|---|---|
| Number of experiment repeats | 80 | 80 |
| Number of generations | 1000 | 1000 |
| Mutation Rate | 0.5 | 0.5 |
| Mutation sigma | 0.3 | 0.3 |
| Population size | 30 | - |
| Number replaced per generation | 1 | - |
| Tournament size | 5 | - |

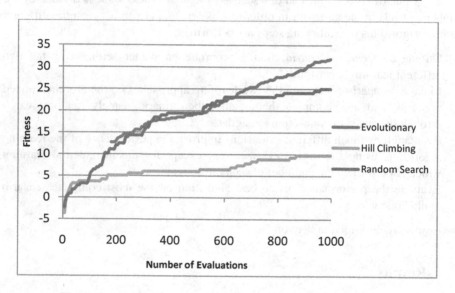

**Fig. 2.** Comparison between approaches using the awareness scenario

Table 2 shows the performance of the evolutionary algorithm with the extensions proposed in Section 3. Note that with exception of the last experiment, the extensions have not been tested together, but separate to judge their individual contributions. In the last experiment it is demonstrated that various extensions can work together to produce good results in a complex scenario. From the table it can be seen that a scenario specific fitness function, the limiting of the search space, and incrementally learning improved the results ($p < 0.05$), whereas adding output noise and the defining symmetry did not. The poor results for adding output noise likely stem from the fact that the amount of repeats on each scenario is kept very low as the computational effort for calculating through a scenario is quite high and therefore a larger number of repeats are not feasible. The lack of results produced by defining symmetry is most likely caused by the specific scenario used, in which it was not able to fully come to fruition.

**Table 2.** Experimental results using different extensions for the evolutionary algorithm

| Exp. # | Scenario | Extensions | Fitness |
|--------|----------|------------|---------|
| 1 | Awareness | No extensions | 31.6 |
| 2 | Awareness | Adding output noise | 21.7 |
| 3 | Awareness | Scenario specific fitness function | 42.6 |
| 4 | Awareness | Pre-defined symmetry | 29.3 |
| 5 | Awareness | Limited search space | 42.6 |
| 6 | Awareness | Incremental learning | 50 |
| 7 | Awareness | Incremental learning (equal time) | 37.5 |
| 8 | One v One | No extensions | 28 |
| 9 | One v One | Pre-defined symmetry / Limited search space / Incremental learning | 50 |

The difference between experiments 6 and 7 is that in the former the agent is allowed to learn until it achieves perfect results at the first task, before starting with the second task. In experiment 7 the agent splits the 1000 evaluations between both scenarios equally, learning each scenario for 500 evaluations. The table shows that providing a limited time per scenario results in worsened performance. Experiments 8 and 9 in the table show a comparison between the standard evolution algorithm and an extended one. As the table shows, the extended evolutionary algorithm was able to learn complete fighter behavior in every trail. The algorithm without the extensions performs much worse.

# 6    Discussion

In this paper, an evolutionary learning technique has been presented with which it is possible to learn aspects of a cognitive model. The determination of these aspects, i.e. connection strengths between beliefs in the model would previously require a substantial knowledge elicitation effort with domain experts, but these connection strengths can now be learned effectively from a simulated environment. The evolutionary technique is based upon scenarios which are configured to learn appropriate (domain dependent) connection strengths in the cognitive model. In order to test the approach, a case study has been performed in the domain of fighter pilot air combat. The results show that it is possible to apply the presented learning technique to optimize belief networks for cognitive models of intelligent agents (opponent fighters) in the aforementioned domain. The learning results in the required 'human-like' behavior. In fact, the technique combines the best of two worlds: it results in human-like behavior with a decreased need to consult domain experts. In order to further improve the learning behavior, several extensions have been proposed, of which some have shown to be successful in the current application. The work presented here, builds forth on the work reported in [5], but provides a significant augmentation as the approach presented here only requires appropriate scenarios whereas the technique presented in [5] requires domain experts to manually dictate the precise outcomes of the model being learned.

Learning approaches to obtain a sufficient level of situation awareness have been proposed before. In [9] for example, a genetic algorithm is used to obtain situation awareness. Alternative machine learning methods have been used for this purpose as well, such as particle filters that have been used for state estimation for Mars Rovers [10] and Adaptive Resonance Theory (ART) that has been used for information fusion [11]. None of these approaches however strives towards the development of human-like behavior.

Future work could attempt to further decrease the amount of expert knowledge necessary through additionally extending or simplifying the topology of such a cognitive model, rather than just adapting the connection strength between nodes. It would obviously be of interest to compare the relative performance of the different techniques and the initial level of domain-specific expertise that is needed to eventually create expert agents.

# References

1. Abt, C.: Serious Games. The Viking Press, New York (1970)
2. Hoogendoorn, M., van Lambalgen, R.M., Treur, J.: Modeling Situation Awareness in Human-Like Agents using Mental Models. In: Walsh, T. (ed.) Proceedings of the Twenty-Second International Joint Conference on Artificial Intelligence, IJCAI 2011, pp. 1697–1704 (2011)
3. Harbers, M., van den Bosch, K., Meyer, J.J.: Modeling Agent with a Theory of Mind. In: Baeza-Yates, R., Lang, J., Mitra, S., Parsons, S., Pasi, G. (eds.) Proceedings of the 2009 IEEE/WIC/ACM International Joint Conference on Web Intelligence and Agent Technology, pp. 217–224. IEEE Computer Society Press (2009)
4. Hoogendoorn, M., Merk, R.J., Treur, J.: An Agent Model for Decision Making Based upon Experiences Applied in the Domain of Fighter Pilots. In: Huang, X.J., Ghorbani, A.A., Hacid, M.-S., Yamaguchi, T. (eds.) Proceedings of the 10th IEEE/WIC/ACM International Conference on Intelligent Agent Technology, IAT 2010, pp. 101–108. IEEE Computer Society Press (2010)
5. Gini, M.L., Hoogendoorn, M., van Lambalgen, R.: Learning Belief Connections in a Model for Situation Awareness. In: Kinny, D., Hsu, J.Y.-J., Governatori, G., Ghose, A.K. (eds.) PRIMA 2011. LNCS, vol. 7047, pp. 373–384. Springer, Heidelberg (2011)
6. Bosse, T., Merk, R.J., Treur, J.: Modelling Temporal Aspects of Situation Awareness. In: Huang, T., Zeng, Z., Li, C., Leung, C.S. (eds.) ICONIP 2012, Part I. LNCS, vol. 7663, pp. 473–483. Springer, Heidelberg (2012)
7. Endsley, M.R.: Toward a theory of Situation Awareness in dynamic systems. Human Factors 37(1), 32–64 (1995)
8. Baluja, S.: Population-based incremental learning: A method for integrating genetic search based function optimization and competitive learning. CMU-CS-94-163. Carnegie Mellon University (1994)
9. Salerno, J., Hinman, M., Boulware, D., Bello, P.: Information fusion for Situational Awareness. In: Proc. Int'l Conf. on International Fusion Cairns, Australia (2003)
10. Dearden, R., Willeke, T., Simmons, R., Verma, V., Hutter, F., Thrun, S.: Real-time fault detection and situational awareness for rovers: report on the Mars technology program task. In: Proc. Aerospace Conference, vol. 2, pp. 826–840 (March 2004)
11. Brannon, N., Conrad, G., Draelos, T., Seiffertt, J., Wunsch, D., Zhang, P.: Coordinated machine learning and decision dupport for Situation Awareness. Neural Networks 22(3), 316–325 (2009)

# Fostering Social Interaction of Home-Bound Elderly People: The *EasyReach* System[*]

Roberto Bisiani, Davide Merico, Stefano Pinardi, Matteo Dominoni[1],
Amedeo Cesta, Andrea Orlandini, Riccardo Rasconi, Marco Suriano,
Alessandro Umbrico[2], Orkunt Sabuncu, Torsten Schaub[3], Daniela D'Aloisi,
Raffaele Nicolussi, Filomena Papa[4], Vassilis Bouglas, Giannis Giakas[5],
Thanassis Kavatzikidis[6], and Silvio Bonfiglio[7],[**]

[1] Università Milano Bicocca, Italy
[2] CNR – National Research Council of Italy, ISTC
[3] University of Potsdam, Germany
[4] Fondazione Ugo Bordoni, Italy
[5] CERETETH, Greece
[6] IKnowHow, Greece
[7] FIMI, BARCO, Italy
Silvio.Bonfiglio@barco.com

**Abstract.** This paper presents the *EasyReach* system, a tool that aims at getting the elderly and pre-digital divide population closer to new technologies by creating a simplified social enviroment that facilitates interaction, trying to allow them to (i) easily keep in contact with friends and relatives, (ii) share their lifetime expertise, and (iii) avoid isolation. The *EasyReach* tool creates for the elderly a special social TV channel accessed by means of their own TV set and a specialized remote control endowed with gesture recognition, video and audio capture capabilities. A hidden personal assistant reasons on user preferences in the background allowing better focalization on his/her social interests.

## 1 Introduction

In contrast with a widespread belief, older people (although primarily benefit-driven) do use computers and Internet [1] and there is also evidence that the use of computers indeed improves the performance in daily living activities, increases cognitive functioning, and decreases the level of depression [2]. Nevertheless, several barriers like access, performance and psychological issues still exist, at least for some individuals.

The *EasyReach* Project[1] aims at providing an innovative and sustainable ICT solution to allow elderly and less educated people (i.e., pre digital-divide population) to participate in the benefits of ICT-based social interactions when confined at home for several reasons. The *EasyReach* system is specifically targeted

---

[*] EasyReach is partially supported by the EU Ambient Assisted Living Joint Program (AAL-2009-2-117).
[**] Corresponding author.
[1] http://www.easyreach-project.eu

M. Ali et al. (Eds.): IEA/AIE 2013, LNAI 7906, pp. 33–42, 2013.
© Springer-Verlag Berlin Heidelberg 2013

towards those individuals who, because of poor scholarization, low income and linguistic barriers, still find it difficult to use computers to improve their socialization. The key motivation for the project stems from the known evidence that elderly people tend to isolate themselves, especially as physical deterioration constrains them to stay at home for medium/long periods. The solution pursued in **EasyReach** (see the sketchy impression in Fig. 1) consists in facilitating social connections among the elderly, and entails the cooperation of some state-of-the-art technologies to provide a non-intrusive solution with user-friendly and personalized access to services similar to those of social networks but designed to be close to the needs of an old person that is at home due to a physical impediment.

A key choice has been the use of the home TV to minimize the change of habits. The **EasyReach** systems comes with a set-top-box that introduces a "social channel" that makes available services to share the user experience with people outside (either single individuals or groups), while maintaining the access to all the usual TV channels. Great effort in the project is dedicated to the synthesis of a remote controller able to gather inertial and multimedia data to capture the behavior of the user. Such captured data

**Fig. 1.** Using EasyReach at home

are used for gesture-based interaction as well as for the creation of multi-media content from the user environment to be shared with the external world. The **EasyReach** Social Interaction Environment is the front-end of the social channel, and represents an interaction facilitator from the user to the external network (and vice versa). The social channel thus created provides some of the services usually attributed to state-of-the-art social media; yet these services have been here redesigned, extremely simplified and tailored to meet the needs of pre digital-divide population. A further aspect in the project is the use of a Personal Assistant (PA) to monitor both user activities and interests in the background, providing the user with proactive aid in the use of the **EasyReach** environment.

This paper describes the project goals and the mid-term prototype that has been demonstrated live at the Ambient Assisted Living (AAL) Forum in Eindhoven in September 2012, receiving the attention of diversified stakeholder representatives. The paper presents first the architecture of the system (Section 2), then proceeds to describing its three main building blocks: the Hardware (remote control and set-top-box) will be presented in Section 3, the Social Interaction Environment will be presented in Section 4, while the Personal Assistant will be presented in Section 5. Finally, Section 6 describes the current evaluation steps according to a user driven methodology, while some conclusions end the paper.

## 2    The EasyReach Architecture

The *EasyReach* architecture is presented in Figure 2. We identify the main hardware elements: a TV set constituting the main user interface; the *EasyReach* Remote Controller integrating (i) a pointing service endowed with an inertial controller, (ii) a Hi-Res camera, and (iii) an audio recorder; a set-top-box (STB) providing the "gluing factor" among all hardware components thus allowing the user to access the system functionalities and join the *EasyReach* Network; an *EasyReach* Cloud representing the core of the *EasyReach* Network, responsible to convey the system services and to manage all user information stored into a centralized database (DB). The set-top-box runs the remote controller server (see next section) and the so-called "EasyReach Client" that contains the software for (i) the social interaction environment, (ii) the personal assistant and (iii) the services required for a flexible connection to the system centralized database.

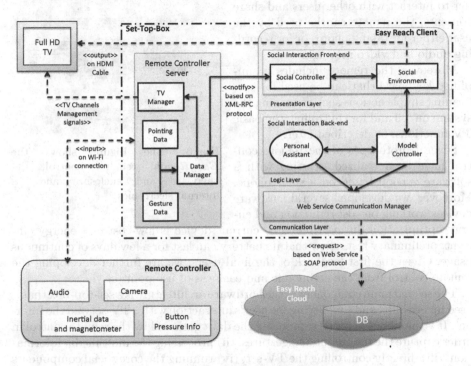

**Fig. 2.** The *EasyReach* architecture

## 3    Hardware Components and the Remote Control

A key design choice has been to provide a platform that features the TV-set as the media to access contents as well as to communicate over the Internet, making the inclusion of a remote control and a set-top-box the most natural choice to complete the hardware architecture of the system. The Figure 3 shows the *EasyReach* remote control, its enclosure and its internal electronics.

The remote control includes a complete three-dimensional inertial unit (with accelerometers, gyroscopes and magnetometers), a camera, a microphone, a keyboard and a rechargeable battery. The main characteristic of the remote control is that it gathers inertial and multimedia data to capture the behavior of the user.

In particular, the inertial data is used to track user movements as well as recognize particular gestures performed by the user for interacting with the system. A software component, running in the set-top-box, converts the low-level inertial data to pointer and gesture data in order to recognize higher level features such as particular movements (e.g., moving the remote left, right, up and down) and rotations (e.g., clockwise and counterclockwise). Figure 2 shows how both the HW and SW components of the remote control are integrated in the overall *EasyReach* architecture.

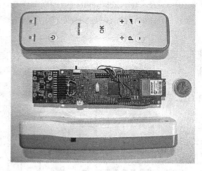

The camera and the microphone are used for gathering multimedia data. In order to interact with other users and share contents, the remote can be used by the user either for taking photos or for recording audio and video messages.

Moreover, the remote includes a simplified keyboard that can be used for performing simple actions such as turning the system on/off and for controlling the basic TV functions, as described later.

Several prototypes of the remote control have been realized, starting with a software version based on a smartphone. Moreover, we developed several hardware devices working on cost reduction and energy optimization during time.

**Fig. 3.** The Figure shows the *EasyReach* remote control, its keyboard and enclosure and its internal electronics

The current device is low-cost and energy efficient; preliminary tests show that its battery can last for a few days of continuous usage. Given the first feedback of the final users, we are further developing the remote control from the ergonomic and easiness of use point of view.

The second main component of the hardware architecture is a set-top-box that is used for enabling and simplifying the user's interaction with a TV-set and the Internet. It is mainly used for (i) computing the data gathered by the remote control in order capture the user input and gestures, (ii) processing the multimedia information, (iii) directly controlling the TV-set, (iv) running the core social components and (v) managing Internet connections and contents.

Therefore, it has storage, processing power and main memory capabilities in the ballpark of an average personal computer. In order to provide the main Social Interaction front-end, the set-top-box is directly connected with the user's TV-set (normally through a HDMI cable). Moreover, it provides Wi-Fi connectivity and it has a programmable infrared device used for controlling the TV-set basic functions (e.g., channel switching and volume management).

## 4  The Social Interaction Environment

The *EasyReach* Social Interaction Environment aims at providing the elderly and pre-digital divide people with a simple and clean environment to interact with, where the information is easy-to-find and presented in a structured way. The interfaces of today's digital devices are often too complicated and dispersive for non accustomed people (e.g., the social applications on the smartphones, designed for small displays, and with a lot of "crowded" icons, or even the ordinary web pages, often very distracting due to their unstructured nature and the huge quantity of information displayed). A sketchy view of the *EasyReach* front-end is shown in Figure 4, while its SW integration within the overall system architecture is shown in Figure 2, as part of the *EasyReach Client* component.

**Fig. 4.** The Social Environment

Two main sections can be identified on the screen:

1. *The "Frame"*: the static part of the environment, composed of the gesture-driven scrollable *contact bar* (at the bottom), containing the user's friends and relatives, as well as the people suggested by the Personal Assistant (see Section 5), the gesture-driven scrollable *group bar* (on the left), where the user can find his personal groups and the suggested ones, and a static *command bar* (on the right), displaying the list of the actions available to the user (e.g., take a picture/video, create a group and so on).

2. *The Information Area*: a dynamic part that changes on the base of the user's actions or selections: for instance, if the user selects a friend (as shown in the figure), or a group, the *Information Area* will show the messages exchanged between the user and the selected element; similarly, if the user visits his gallery or wants to create a group, the *Information Area* will switch according to the the user's selected action.

*Social Functionalities.* The main goal of the project is to build a system able to offer functionalities that prevent users from isolating themselves from the society. More specifically, the system intends to offer the possibilities to easily:

1. Keep in touch with friends and relatives in a more "immersive" way than can be achieved with a phone call or a letter.
2. Foster socialization with other people that have common interests with the user, in order to keep them socially active (e.g., sharing their experience and knowledge of a lifetime and giving/receiving help in any area of interest).
3. Create discussion groups or join existing ones.
4. "Reach" organizations even from home (both "communitary" ones, like user's church or senior center, and "official institutions").

The key idea of the Project is to be as simple as possible for the user, in both system installation and utilization. In fact, after connecting the STB to the TV, the user is immediately able to join the *Social Channel* and to enjoy the benefits of today's technology, even as a pre-digital divide user. With the **EasyReach** special remote control the user can take pictures and videos of a special event (e.g., a family dinner, a birthday etc.), share his/her knowledge and experience by creating a video-tutorial and posting it on thematic groups or, simply, have conversations with friends and relatives.

The remote control's gesture recognition mechanism helps the user to navigate the system, by giving him the possibility to scroll his list of users and groups through some simple gestures, like moving the controller right/left or up/down. After selecting one of the groups (or a friend) of interest, the user can immediately start exchanging messages and/or updates within the chosen context. Also, the system allows organizations and official institutions to join the *Social Channel*, in order to be easily reachable by the users, send them useful information/advises, and help them more directly. The opportunity to make the user's own expertise available to the whole **EasyReach** community can be invaluable to maintain the self-perception of being socially active. Also, the user himself can easily create new discussion groups or join one of the groups suggested by the Personal Assistant on the basis of the user's interests.

## 5    The Personal Assistant

The Personal Assistant (PA) is the software component that suggests new interactions through the social network, and is integrated as part of the Logic Layer in the Social Interaction Back-end component (see Figure 2). The PA not only analyses the profile of a user, but also monitors his activities and interactions within the social network, reasoning on these data to suggest new interactions. Basically, the PA suggests which items should be shown in the user's lists, where an item can be a person or a group; the suggestion may lead to a new interaction depending on the user's will to engage in an interaction regarding the selected person/group. A person who is not a friend of the user might be shown in the contact list because they have common interests. Similar reasoning can also apply to a group the user is not a member of. The user can check the suggested items and can send the person a message or check the group message board.

This implicit suggestion method is less confusing and annoying for the elderly than asking explicitly each time and making him feel forced to interact.

Next we explain how the PA works by intelligently selecting items to be shown in the user's lists. The PA module communicates with 2 modules through their respective interfaces: the presentation layer and the social engine database (see the architectural diagram in Figure 2). Note that the interface for the social engine database features a one way communication (i.e., the PA fetches data from the social engine to monitor interactions of the user), while the interface between the PA and the presentation layer manages two way communication. For instance, the presentation layer can invoke the PA when it needs to show the user's list items, and the selected items are returned to the representation layer. Additionally, the interface is designed so that the PA can get updates related to activities and interactions of the user within the social network; in this way some of inefficient polling of the social engine database are avoided.

We implemented a framework for the PA to suggest new interactions. Answer Set Programming (ASP; [3]), a popular declarative problem solving approach in the field of knowledge representation and reasoning, is utilized to implement the reasoning capabilities needed by the PA to intelligently select items to be shown in the user's lists (a similar approach to using ASP to implement suggesting interactions related to events in the social network is [4]). In this work we use the answer set solver from the Potassco answer set collection [5]. We refer the reader to [6] for the syntax and semantics of the ASP language.

In order to reason on user interests or group topics we need a model of interests; we utilize a taxonomy of users' interests for this purpose. The selection of interests is based on preliminary investigations about elderly needs, hobbies and expectations. Formally, we modeled our taxonomy as a forest of keywords of interests where the edges represent the subsumption relation among keywords. In the ASP program of the PA, the taxonomy is encoded by logic program facts for keyword nodes and subsumption relation. Figure 5 depicts a subset of the taxonomy used in *EasyReach*. Note that the keyword *Documentary* is subsumed by *Programs*, and *Programs* is subsumed by *TV*. Thus, a keyword at a deeper level of a tree in the taxonomy represents more specialized interest than one at a shallower level. A user creates a profile by specifying keywords corresponding to his interests. The taxonomy allows the PA to exploit the semantic information inherent in a user profile. For example, when a user specifies Formula 1 as his interest, PA can use not only *Formula 1* but also *Auto racing* or *Motorcycle racing* for reasoning to suggest new items in his list. The following explains how PA performs this reasoning.

When PA is asked for populating new items for the user's lists by the presentation layer, it first considers the interests mentioned in the user's profile. Later, starting from interest nodes in the taxonomy, it traverses to connecting nodes using the subsumption relation. A connecting node must be reachable with a path whose length is expressed as a parametric value denoting the maximum allowed path length. For instance, let {*Motorcycle racing, Formula* 1} be the set of interests mentioned in a user's profile and the taxonomy used be the one shown in Figure 5. Assuming that the maximum allowed path length is 2, the interest nodes

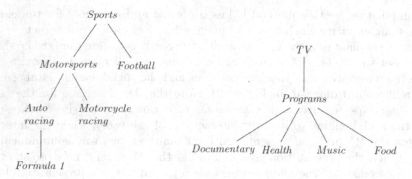

**Fig. 5.** A subset of the taxonomy of user's interests

reachable from *Motorcycle racing* are {*Motorsports, Auto racing, Sports*}. Considering all the interests in the user's profile, the PA takes the set {*Motorcycle racing, Formula 1, Motorsports, Auto racing, Sports*} into account when checking other users with common interests. Additionally, the PA assigns weight to the interests according to the depth of its node in the taxonomy tree. For our example the weight of *Formula* 1 is greater then *Motorsports*. The intuition is that the more specialized a common interest is, the better the suggestion. We encoded the taxonomy traversal and weighting in ASP and formally modeled the item suggestion problem as a quantitative optimization problem where the reasoner tries to maximize the total weight of a selected set of items. The selected items are fed back to the presentation layer to be shown in the user's lists. Besides user interests, we plan to integrate other sources of information for suggestion new items; the quantitative optimization-based suggestion framework of the PA can handle such additional information. The location of the user or his recent activities within the social network, such as sending a message to a person from a common group, are also important information to be used in suggesting items.

## 6  Evaluation Steps

The **EasyReach** project adopts the User Centred Design (UCD) approach since its beginning. User involvement has been very intense from the early stages of the development process of the ICT services [7]. Primary users (elderly people) and secondary users (e.g., caregivers and relatives) have been involved from the early stages of the development process of the technical solutions. The main objectives of user involvement are: (a) to gather user needs and preferences for user requirements identification, (b) to test and evaluate the intermediate and final **EasyReach** prototypes. The UCD approach in the project is applied through three main steps:

1. A preliminary investigation about elderly needs, expectations and preferences with respect to **EasyReach** solutions was realized using the group interview technique. Firstly, contacts were established with user organizations: Federazione Nazionale Pensionati (FNP), a retired and elderly union,

in Rome (Italy) and the Ancescao association, a no-profit association promoting the social inclusion of elderly people, in Milan. A first group interview
was realized (October 2011) in Rome using an appropriate methodology [8].
Further, two group interviews were carried out (February 2012) in two different Senior Centres in Milan.

2. A "second wave" of user involvement was realized to obtain feedback from end-
users about the early design versions (using low fidelity prototypes) of some
system components. A qualitative investigation using the group interview technique and involving a small number of potential users was realized in Rome
(June 2012). User reactions were collected and the idea behind the project was
judged good by participants to meet the need of the elderly "to remain in contact with the real world". Some requirements/critical points emerged, including: ease of use; adequate communication protocols; assistance to join groups
with similar interests; protection from illegal behaviors; moderation in groups
capable of settling a controversy when something unpleasant happens.

3. Three pilot studies are planned in order to assess the developed technological
solutions in real life environments. The relevant characteristic of these pilots
is to conduct field trials involving real users. The planned pilots are realized
in different national, geographic and cultural contexts.

The main objectives of the pilots are: to evaluate system usability by the elderly
people in different environments (e.g., home, senior center); to evaluate user experience [9] and user acceptance [10] of the ICT solution developed in the project; to
evaluate the effectiveness of the *EasyReach* solution in terms of social inclusion
and improvement of quality of life. Both subjective and objective techniques for
data collection are employed. They include: semi structured questionnaires for elderly people, observation of interaction sessions, log file analysis, interviews with
secondary users. Adequate procedures are adopted during pilots for training elderly. These procedures have the objective of getting the elderly person to think
to be able to master the system reducing the sense of fear and inadequacy [11]. During pilots a reference person (tutor) has the task to provide any kind of help about
the system use, to solve practical problems in the field trial and to motivate the
elderly in correctly participating to the pilot. The three pilots are realized in two
European countries: Italy (Pilot 1 in Rome and Pilot 2 in Milan) and Germany. The
pilot in Rome involves two Senior Centres; they are selected to be representative of
two different kinds of cultural areas. This pilot is realized placing the *EasyReach*
prototype in two different environments: the home of the elderly people and the
Senior Centre. The second Italian pilot is organized in the Castanese, an area in
the northern part of Milan. Even in this pilot the *EasyReach* system is placed at
the home of elderly people and at the senior centre. The German pilot is realised
in the Florencehort Seniorenzentrum at Stansdorf, a senior residence; it involves
people living in this senior residence.

# 7  Conclusions

This paper presents a comprehensive description of a system that aims at coun-
teracting the elderly's tendency to isolate themselves when they are constrained

at home for any reason. The *EasyReach* system proposes a new TV social channel aimed at creating a simplified interaction space between an old home-bound user and his/her network of connections spread over the internet. Special attention has been given to the particular communication bandwidth offered by the *EasyReach* remote control: the use of gestures, the photo, audio, video media, the very restrictive use of text. Goal of the system is to close the gap with people not familiar with state-of-the-art ICT technology and offer them an opportunity of using it. The following slogan might summarize what the project is trying to achieve: "using a pre digital-divide appliance to give access to post digital-divide opportunities". Our intensive tests with users in the final part of the project will assess the extent to which this goal has been achieved. Continuous intermediate tests have constantly encouraged us in pursuing the project objectives.

# References

1. Morrell, R.W., Dailey, S.R., Feldman, C., Mayhorn, C.B., Echt, K.V., Podany, K.I.: Older adults and information technology: A compendium of scientific research and Web site accessibility guidelines. National Institute on Aging, Bethesda (2003)
2. Bond, G.E., Wolf-Wilets, V., Fiedler, E., Burr, R.L.: Computer-aided cognitive training of the aged: A pilot study. Clinical Gerontologist 22, 19–42 (2002)
3. Baral, C.: Knowledge Representation, Reasoning and Declarative Problem Solving. Cambridge University Press (2003)
4. Jost, H., Sabuncu, O., Schaub, T.: Suggesting new interactions related to events in a social network for elderly. In: Proceedings of the Second International Workshop on Design and Implementation of Independent and Assisted Living Technology (2012)
5. Gebser, M., Kaminski, R., Kaufmann, B., Ostrowski, M., Schaub, T., Schneider, M.: Potassco: The Potsdam answer set solving collection. AI Communications 24(2), 105–124 (2011)
6. Simons, P., Niemelä, I., Soininen, T.: Extending and implementing the stable model semantics. Artificial Intelligence 138(1-2), 181–234 (2002)
7. Maguire, M., Bevan, N.: User requirements analysis. A review of supporting methods. In: Proceedings of IFIP 17th World Computer Congress, pp. 133–148. Kluwer Academic Publishers, Montreal (2002)
8. Papa, F., Sapio, B., Pelagalli, M.F.: User experience of elderly people with digital television: a qualitative investigation. In: Proceedings of the 9th European Conference on Interactive TV and Video (Euro ITV 2011), Lisbon, Portugal, pp. 223–226 (2011)
9. Hassenzahl, M., Law, E.L., Hvannberg, E.T.: User Experience, Towards a unified view. In: Proceedings of the 2nd Cost 294 - Mause International Open Workshop, Oslo, Norway (2006)
10. Venkatesh, V., Morris, M.G., Davis, G.B., Davis, F.D.: User Acceptance of Information Technology: Toward a Unified View. MIS Quarterly 27(3), 425–478 (2003)
11. Papa, F., Spedaletti, S.: Broadband Cellular Radio Telecommunication Technologies in Distance Learning: A Human Factors Field Study. Personal and Ubiquitous Computing 5, 231–242 (2001)

# Cooperative Games with Incomplete Information among Secondary Base Stations in Cognitive Radio Networks

Jerzy Martyna

Jagiellonian University, Institute of Computer Science
Faculty of Mathematics and Computer Science
ul. Prof. S. Lojasiewicza 6, 30-348 Cracow, Poland

**Abstract.** In this paper, we propose a model for coalition formation among Secondary Base Stations (SBSs) with incomplete information in cognitive radio (CR) networks. This model allows us to analyze any situation in which players are imperfectly informed about the aspect of their environment that is relevant to their decision-making. On the other hand, by using the proposed method based on the game theory with incomplete information, SBSs can collaborate and self-organize into disjoint independent coalitions. The simulation results show that the proposed method yields a performance advantage in terms of the average payoff per SBS reaching up to 145% relative to the non-cooperative case.

**Keywords:** game theory, cognitive radio networks, Bayesian equilibrium.

## 1 Introduction

Recent technological advances have led to the development of distributed and self-configuring wireless network architectures. This is seen particularly in the case of radio spectrum usage, which refers to frequency segment that have been licensed to a particular primary service, but is particularly utilised at a given location or time. The Federal Communication Commission (FCC) [4] has reported vast temporal and geographic variations in allocated spectrum usage, ranging from 15-85% in bands below 3 GHz that are favoured in non-line-of-sight radio propagation. On the other hand, a large portion of the assigned spectrum is used sporadically, leading to an under-utilisation of a significant amount of the spectrum.

Cognitive radio (CR) [1, 7, 11] has been extensively researched in recent years as a promising technology to improve spectrum utilization. Cognitive radio networks and spectrum-sensing techniques are natural ways to allow these new technologies to be deployed. These spectrum-sensing techniques and the ability to switch between radio access technologies are the fundamental requirements for transmitters to adapt to varying radio channel qualities, network congestion, interference, and quality of service (QoS) requirements.

The CR users can operate in both licensed and unlicensed bands. Therefore, we may categorise the CR application of spectrum ino three possible scenarios: (i) CR network on a licensed band, (ii) CR network on unlicensed band, (iii) CR network on both a licensed band and unlicensed band. We use Fig. 1 to illustrate the third scenario. Thus, from

M. Ali et al. (Eds.): IEA/AIE 2013, LNAI 7906, pp. 43–52, 2013.

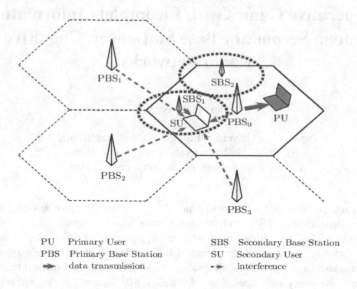

| PU | Primary User | SBS | Secondary Base Station |
|---|---|---|---|
| PBS | Primary Base Station | SU | Secondary User |
| ➤ | data transmission | ⇢ | interference |

**Fig. 1.** Downlink/uplink CR network

the users' perspective, the CR network coexists with the primary system in the same geographic location. A primary system operated in the licensed band has the highest priority to use that frequency band (e.g. 2G/3G cellular, digital TV broadcasts, etc.). Other unlicensed users and/or systems can neither interfere with the primary system in an obtrusive way nor occupy the licensed band. By using the pricing scheme, each of the primary service providers (operators) maximises its profit under the QoS constraint for primary users (PUs). All of the unlicensed secondary users (SUs) are equipped with cognitive radio technologies, usually static or mobile. The primary users (PUs) are responsible for throwing unused frequencies to the secondary users for a fee. While the existing literature has focused on the communications needed for CR system control, this paper assumes a network of secondary base stations (SBSs). Every SBS can only have information on a small number of PUs or channels. It causes interference to PUs and SBSs.

The key to the build the CR networks is the application of artificial intelligence (i.e. cognition) to communications devices. This will enable intelligent local decisions to made on network routing, spectrum and resources usage, etc. Such decisions can take into account mixed systems and applications, and even devices that break the rules. The study of artificial intelligence, learning, and reasoning has been around for a number of years, but it is only now that concepts such as reinforcement-based learning, game theory and neural networks are being actively applied to CR networks. Among the game-theoretic approaches to addressing resource management in these systems, the paper [10] proposed the use of Shapley value to power allocation games in CR networks.

The motivation of this paper is exactly to study how game theory [5] can be implemented in the situations where CR devices have limited information. Recently, games of complete information have been used in various types of communication networks, multiple input and multiple output channels [9], interference channels [2], and combination of them. Games with complete information have been studied in the distributed collaborative spectrum sensing [19], to solve of a dynamic spectrum sharing [16], interference minimalization in the CR networks [12], etc.

However, there are no existing methods of calculating the equilibrium policy in a general game with incomplete information. The imperfect information or partial channel state information (CSI) means that the CSI is not perfectly estimated/observed at the transmitter/receiver side. This is a common situation which usually happens in a real wireless communication, since it may be too "expensive" for every radio receiver/transmitter to keep the information from the channels of all other devices. Firstly, Harsanyi and Selten [8] proposed an extension of the Nash solution to Bayesian bargaining problems. A new generalization of the Nash bargaining solution for two player games with incomplete information was presented by Myerson [13].

In this paper, we use the game-theory approach to find an explicit expression for utility functions or the values of coalition among the set of players (the SBSs) with incomplete information in the CR networks. Thus, we propose an algorithm for coalition formation among SBSs with incomplete information. We provide simulation results proving the analytical results and also provide performance improvement percentages in terms of the average payoff per system player without cooperation among SBSs.

The rest of this paper is organized as follows. In section 2 we present the problem formulation. The coalition formation among SBS with incomplete information is provided in section 3. An algorithm for the coalition formation of SBSs with incomplete information in CR network is given in section 4. Simulation results are described in section 5. A summary and concluding remarks are given in section 6.

## 2  Problem Formulation

In this section, we present the model of the CR system consisting of the PUs and the SUs. We assume that the $N$ secondary base stations (SBSs) also belong to the CR network. Each $i$-th SBS can service number $L_i$ of SUs in a specific geographical area. It means that each SBS provides coverage area for a given cell or mesh. Let $\mathcal{N}$ be the set of all SBSs and $\mathcal{K}$ be the set of all PUs. Each PU can use a number of admissible wireless channels. We assume that each SU can employ the $k$-th channel of PU, if the $k$-th channel is not transmitting and this channel is available for the SU. According to the approach given by D. Niyato [15] each $i$-th SBS can be characterized by accurate statistics regarding a subset $\mathcal{K}_i \in \mathcal{K}$ of PUs during the period of time the channels remain stationary.

Let each $i$-th SBS use energy detectors which belong to the main practical signal detectors in the CR network. Assuming the Raleigh fading, the probability that the $i$-th SBS accurately received the signal from PU $k \in \mathcal{K}$ is given by [6]

$$P_{det,k}^i = e^{-\frac{\lambda_{i,k}}{2}} \sum_{n=0}^{m-2} \frac{1}{n!} \left( \frac{\lambda_{i,k}}{2} \right)^n + \left( \frac{1 + \overline{\gamma}_{k,i}}{\overline{\gamma}_{k,i}} \right)^{m-1}$$

$$\times \left[ e^{-\frac{\lambda_{i,k}}{2(1+\overline{\gamma}_{ki})}} - e^{-\frac{\lambda_{i,k}}{2}} \sum_{n=0}^{m-2} \frac{1}{n!} \left( \frac{\lambda_{i,k}\overline{\gamma}_{ki}}{2(1+\overline{\gamma}_{ki})} \right)^n \right] \tag{1}$$

where $\lambda_{i,k}$ is the energy detection threshold selected by the $i$-th SBS for sensing the $k$-th channel, $m$ is the time bandwidth product. $\overline{\gamma}_{ki}$ is the average SNR of the received signal from the $k$-th PU, where $P_k$ is the transmit power of the $k$-th PU, $g_{ki} = \frac{1}{d_{ki}^\mu}$ is the path loss between the $k$-th PU and the $i$-th SBS, $d_{ki}$ is the distance between the $k$-th PU and the $i$-th SBS, $\sigma^2$ is the Gaussian noise variance.

Thus, as was shown in [6] the false alarm probability perceived by the $i$-th SBS $i \in \mathcal{N}$ over the $k$-th channel, $k \in \mathcal{K}$, belonging to PU, is given by

$$P_{fal,k}^i = P_{fal} = \frac{\Gamma(m, \frac{\lambda_{i,k}}{2})}{\Gamma(m)} \tag{2}$$

where $\Gamma(.,.)$ is the incomplete gamma function and $\Gamma(.)$ is the gamma function.

We note that the non-cooperative false alarm probability depends on the position of SU. Thus, we can drop the index $k$ in Eq. (2) and we get the missing probability perceived by the $i$-th SBS, namely [3,6]

$$P_{mis,i} = 1 - P_{det,i} \tag{3}$$

Assuming a non-cooperative collaboration for every $i$-th SBS, $i \in \mathcal{N}$ we can obtain the amount of information which is transmitted to the SUs served by it over its control channel, namely

$$v(\{i\}) = \sum_{k \in \mathcal{K}_i} \sum_{j=1}^{L_i} [(1 - P_{fal,k}^i)\theta_k \rho_{ji} - \alpha_k(1 - P_{det,k}^i)$$

$$(1 - \theta_k)(\rho_{kr_k} - \rho_{kr_k}^j)] \tag{4}$$

where $L_i$ is the number of SUs served by the $i$-th SBS, $\alpha_u$ is the penalty factor imposed by the $k$-th PU for the SU that causes the interference. $(1 - P_{det,k}^i)$ defines the probability that the $i$-th SBS treated channel $k$ as available while the PU is actually transmitting. It means the probability that the SNR received by the $i$-th SBS is given by [17]

$$\rho_{ji} = e^{-\frac{v_0}{\overline{\gamma}_{ji}}} \tag{5}$$

where $v_0$ is the target SNR for all PUs, SUs, SBSs, $\overline{\gamma}_{ji}$ is the average SNR received by the $i$-th SBS from all SUs with the transmit power $P_j$ of the $j$-th SU. It is defined as

$$\overline{\gamma}_{ji} = \frac{P_k g_{kr_k}}{\sigma^2 + g_{jk}P_j} \tag{6}$$

where $g_{ji}$ is the channel gain between the $j$-th SU and the $i$-th SBS. The probability $(\rho_{kr_k} - \rho_{kr_k}^j)$ indicates the reduction of a successful transmission at its receiver $r_k$ of the $k$-th PU at its receiver $r_k$ caused by the transmission from the $i$-th SU over $k$-th channel.

Assuming Rayleigh fading and BPSK modulation within each coalition, the probability of reporting error between the $i$-th SBS and the $j$-th SU is given by [20]

$$P_{e,i,j} = \frac{1}{2}\left(1 - \sqrt{\frac{\overline{\gamma}_{ji}}{2 + \overline{\gamma}_{ji}}}\right) \tag{7}$$

Inside a coalition $C$ by a collaborative sensing, the missing and the false alarm probabilities of a coalition is given by

$$Q_{mis,C} = \prod_{i \in C} [P_{mis,i}(1 - P_{e,i,j}) + (1 - P_{mis,j})P_{e,i,j}] \qquad (8)$$

$$Q_{fal,C} = 1 - \prod_{i \in C} [(1 - P_{fal})(1 - P_{e,i,j}) + P_{fal}P_{e,i,j}] \qquad (9)$$

It is obvious that the reduction of the missing probability will decrease the interference on the PU and increase the probability of its detection. Within a coalition the SUs minimize their missing probabilities. We assume that in each coalition $C$ an SU is selected as coalition head. It collects the sensing bits from the coalitionŠs members and acts as the head to form a coalition.

Thus, assuming Rayleigh fading and BPSK modulation within each coalition, the probability of reporting error between an SU and the coalition head is given by [20]:

# 3 The Coalition Formation among Secondary Base Stations with Incomplete Information in the CR Networks

In this section presented is the SBS game with incomplete information in CR network.

The problem can be formulated with the help of using a cooperative game theory [14]. More formally, we have a $(\Omega, u)$ coalition game, where $\Omega$ is the set of players (the SBSs) and $u$ is the utility function or the value of the coalition.

Following the coalition game of Harsanyi [8], a possible definition for a Bayesian game [5] is as follows.

**Definition 1 (Bayesian Game).** *A Bayesian game G is a strategic-form game with incomplete information, which can be described as follows:*

$$\mathcal{G} = \langle \Omega, \{\mathcal{T}_n, \mathcal{A}_n, \rho_n, u_n\}_{n \in \mathcal{N}} \rangle \qquad (10)$$

*which consists of*
- *a player set:* $\Omega = \{1, \ldots, N\}$
- *a type set:* $\mathcal{T}_n(\mathcal{T} = \mathcal{T}_1 \times \mathcal{T}_2 \times \cdots \times \mathcal{T}_N)$
- *an action set:* $\mathcal{A}_n(\mathcal{A} = \mathcal{A}_1 \times \mathcal{A}_2 \times \cdots \mathcal{A}_N)$
- *a probability function set:* $\rho_n : \mathcal{T}_n \rightarrow \mathcal{F}(\mathcal{T}_{-n})$
- *a payoff function set:* $u_n : \mathcal{A} \times \mathcal{T} \rightarrow \mathcal{R}$, *where* $u_n(a, \tau)$ *is the the payoff of player* $n$ *when action profile is* $a \in \mathcal{A}$ *and type profile is* $\tau \in \mathcal{T}$.

The set of strategies depends on the type of the player. Additionally, we assume the type of the player is relevant to his decision. The decision is dependent on information which it possesses. A strategy for the player is a function mapping its type set into it action set. The probability function $\rho_n$ represents the conditional probability $\rho_n(-\tau_n|\tau_n)$ that is assigned to the type of profile $\tau_n \in \mathcal{T}_{-n}$ by the given $\tau_n$.

The payoff function of player $n$ is a function of strategy profile $s(.) = \{s_1(.), \ldots, s_N(.)\}$ and the type profile $\tau = \{\tau_1, \ldots, \tau_N\}$ of all players in the game and is given by

$$u_k(s(\tau), \tau) = u_n(s_1(\tau_1), \ldots, s_N(\tau_N), \tau_1, \ldots, \tau_N) \qquad (11)$$

We recall that in a strategic-form game with complete information, each player chooses one action. In a Bayesian game each player chooses a set or collection of actions (strategy $s_n(.)$).

A definition for a payoff of player in the Bayesian game as follows:

**Definition 2.** *The player's payoff in a Bayesian game is given by*

$$u_n(\tilde{s}_n(\tau_n), s_{-n}(\tau_{-n}), \tau) = u_n(s_1(\tau_1), \ldots, \tilde{s}_n(\tau_n),$$
$$s_{n+1}(\tau_{n+1}), \ldots, s_N(\tau_N), \tau) \tag{12}$$

*where $\tilde{s}_n(.), s_{-n}(.)$ denotes the strategy profile where all players play $s(.)$ except player $n$.*

Next, we define the Bayesian equilibrium (BE) as follows:

**Definition 3 (Bayesian Equilibrium).** *The strategy profile $s^*(.)$ is a Bayesian equilibrium (BE), if for all $n \in \mathcal{N}$, and for all $s_n(.) \in S_n$ and $s_{-n}(.) \in S_{-n}$*

$$E_\tau \left[u_n(s_n^*(\tau_{-n}, \tau)\right] \geq E_\tau \left[u_n(s_n(\tau_n, s_{-n}^*(\tau_{-n}), \tau)\right] \tag{13}$$

*where*

$$E_\tau \left[u_n(x_n(\tau_n, x_{-n}(\tau_{-n}), \tau)\right] \overset{\triangle}{=} \sum_{\tau_{-n} \in \mathcal{T}_{-n}} \rho_n(\tau_{-n} \mid \tau_n)$$
$$u_n(x_n(\tau_n), x_{-n}(\tau_{-n}), \tau) \tag{14}$$

*is the expected payoff of player $n$, which is averaged over the joint distribution of all players' types.*

For the proposed game the false alarm probabilities for the $i$-th and $j$ SBSs over channel $k$ are given by $P_{fal,k}^i$ and $P_{fal,k}^j$. Thus, the utility function or the value of the coalition is given by $u(C)$, namely

$$u(C) = (1 - Q_{mis,C}) - Cost(Q_{fal,C}) \tag{15}$$

where $Q_{mis,C}$ is the missing probability of coalition $C$.

For the cooperation problem we can provide the following definition [14].

**Definition 4 (Transferable Utility of Coalitional Game).** *A coalitional game $(\Omega, u)$ is said to have a transferable utility if value $u(C)$ can be arbitrarily apportioned between the coalition players. Otherwise, the coalitional game has a non-transferable utility and each player will have their own utility within coalition $C$.*

Based on these concerns, it is important to say, that the utility of coalition $C$ is equal to the utility of each SBS in the coalition. Thus, the used $(\Omega, u)$ coalitional game model has a non-transferable utility. We assume that in the coalitional game the stability of the grand coalition of all the players is generally assumed and the grand coalition maximizes the utilities of the players. Then, player $i$ may choose the randomized strategy $s$ which maximizes his expected utility. Informally, we could provide a Nash equilibrium here.

Assuming the perfect coalition of SBS $C_{per}$, the false alarm probability is given by

$$Q_{fal,C_{per}} = 1 - \prod_{i \in C_{per}} (1 - P_{fal}) = 1 - (1 - P_{fal})^{|C_{per}|} \tag{16}$$

## 4   The Coalition Formation Algorithm

In this section, we propose an algorithm for the coalition formation of SBS with incomplete information in CR networks.

The algorithm works on two levels: the possible coalition formation and the grand coalition formation. The first level is the basis for all the coalitions formation. Each member of group $C$ cooperates so as to maximize their collective payoff. At this level a maximum number of SBSs per coalition is defined. At the second level the grand coalition is formed. Firstly, the utility function of the formed coalition is calculated. If the utility function of formed coalition reaches the maximum value, the Bayesian equilibrium (BE) is tested for the coaltion. Finding the Bayesian equilibrium (BE) finishes the operation of the algorithm. If two or more coalitions possess the Bayesian equilibrium with the same value of the payoff, the normalized equilibrium introduced by Rosen [18] is proposed, in which it is shown that an unique equilibrium exists if the payoff functions satisfy the condition of the diagonal strictly concave.

Fig. 2. shows the pseudo-code of the proposed algorithm.

**procedure** *BSSs coalition formation*;
  **begin**
    $i := 0$; $| C_{per} | := 1$;
    *compute* $Q_{fal,C}$;
    **while** $Q_{fal,C} > Q_{fal,C_{per}}$ **do**
      **begin**
        $i := i + 1$;
        **if** *BE exists for given* $| C_{per} |$ **then**
          **begin**
            $| C_{per} | := i$; *compute* $Q_{fal,C}$;
          **end**;
      **end**;
  **end**;

**Fig. 2.** Coalition formation algorithm

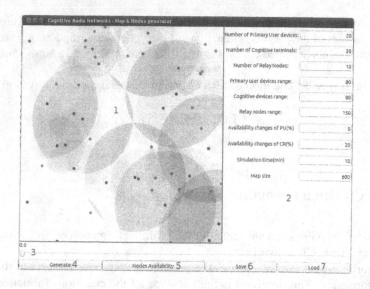

**Fig. 3.** A toolkit of our simulation program

# 5    Simulation Results

A simulation was used to confirm the above given algorithm for the coalition formation among the SBSs with incomplete information. In Fig. 3, we present a toolkit of our simulation study. The simulation of the CR network has four squares with the PU at the center. Each square is equal to 1 km × 1 km. In each square 4 SBSs and 8 SUs were randomly deployed. Initially, it was assumed that each SBS is non-cooperative and detects information from its neighbours by means of the common channels. The energy detection threshold $\lambda_{i,k}$ for an $i$-th SBS over channel $k$ was chosen following the false probability $P_{f,k}^i = 0.05$, $\forall i \in \mathcal{N}$, $k \in K$. The transmit power of all the SU was assumed as equal to to 10 mW, the transmit power of all the PUs was equal to 100 mW, the noise variance $\sigma^2 = -90$ dBm.

In Fig. 4 we plot the average payoff per SBS versus the number of SBSs for both the organization of the CR network with a coalition of SBSs and the non-cooperation of SBSs. Both results are averaged over random positions of all the nodes (SUs and PUs). In the case of non-cooperation game of SBS the average payoff per SBS has a smaller value than for the cooperation game of SBS. The proposed algorithm significantly increases the average payoff up to 140% relative to the non-cooperative case at the number of 15 SBSs.

Fig. 5 shows the number of SBSs versus the average maximum coalition number. The graph shows that the number of SBSs increases with the maximum average coalition number. It is due to the fact that as $\mathcal{N}$ increases, the number of potential members of the coalition increases. The graph indicates that the typical size of the SBS coalition is proportional to the number of SBSs for the certain value. A large number of SBSs does not allow for the formulation of a relatively large coalition.

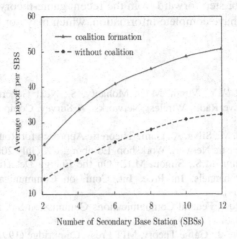

**Fig. 4.** The average payoff per SBS versus the number of SBSs

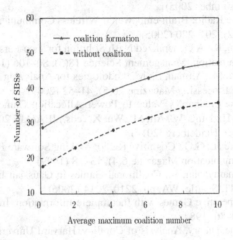

**Fig. 5.** The number of SBSs versus the average maximum coalition number of SBSs

## 6    Conclusion

This paper presents the game-theory model regarding games among the SBSs in the CR network. The Bayesian game and the solution concept of the Bayesian equilibrium are provided. An algorithm for coalition formation of SBSs with incomplete information has been proposed. Additionally, we show that the Bayesian game is a powerful tool to solve the problem of coalition formation among SBSs with incomplete information. The payoff from every coalition of SBSs with incomplete information allows us to determine whether to join or leave the coalition. Finally, the SBSs can reach a Bayesian equilibrium.

This study is a major step forward from the recent game-theory approach that only considers players that have complete information, which may not be realistic in many practical situations.

# References

[1] Akyildiz, I.F., Lee, W.Y., Vuran, M.C., Mohanty, S.: Next Generation Dynamic Spectrum Access/cognitive Radio Wireless Networks: A Survey. Computer Networks 50(13), 2127–2159 (2006)

[2] Altman, E., Debbah, M., Silva, A.: Game Theoretic Approach for Routing in Dense Ad Hoc Networks. In: Stochastic Network Workshop, Edinburg, UK (July 2007)

[3] Digham, F.F., Alouini, M.S., Simon, M.K.: On the Energy Detection of Unknown Signals over Fading Channels. In: Proc. Int. Conf. on Communications, Alaska, USA, pp. 3575–3579 (2003)

[4] First Report and Order, Federal Communications Commission Std. FCC 02-48 (February 2002)

[5] Fudenberg, D., Tirole, J.: Game Theory. MIT Press, Cambridge (1991)

[6] Ghasemi, A., Sousa, E.S.: Collaborative Spectrum Sensing for Opportunistic Access in Fading Environments. In: IEEE Symp. New Frontiers in Dynamic Spectrum Access Networks, Baltimore, USA (November 2005)

[7] Haykin, S.: Cognitive Radio: Brain-empowered Wireless Communications. IEEE J. Select. Areas Commun. 23(2), 201–220 (2005)

[8] Harsanyi, J.C., Selten, R.: A Generalized Nash Solution for Two-Person Bargaining Games with Incomplete Information. Management Science 18(5), 80–106 (1972)

[9] Lasaulce, S., Debbah, M., Altman, E.: Methodologies for Analyzing Equilibria in Wireless Games. IEEE Signal Processing Magazine 26(5), 41–52 (2009)

[10] Martyna, J.: The Use of Shapley Value to Power Allocation Games in Cognitive Radio Networks. In: Jiang, H., Ding, W., Ali, M., Wu, X. (eds.) IEA/AIE 2012. LNCS, vol. 7345, pp. 627–636. Springer, Heidelberg (2012)

[11] Mitola III, J., Maguire Jr., G.Q.: Cognitive Radio: Making Software Radios More Personal. IEEE Personal Communication Magazine 6(4), 13–18 (1999)

[12] Mathur, S., Sankaranarayanan, L.: Coalitional Games in Gaussian Interference Channels. In: Proc. of IEEE ISIT, Seattle, WA, pp. 2210–2214 (2006)

[13] Myerson, R.B.: Cooperative Games with Incomplete Information. International Journal of Game Theory 13, 69–86 (1984)

[14] Myerson, R.B.: Game Theory, Analysis of Conflicy. Harvard University Press, Cambridge (1991)

[15] Niyato, D., Hossein, E., Han, Z.: Dynamic Spectrum Access and Management in Cognitive Radio Networks. Cambridge University Press, Cambridge (2009)

[16] Perlaza, S.M., Lasaulce, S., Debbah, M., Chaufray, J.M.: Game Theory for Dynamic Spectrum Sharing. In: Zhang, Y., Zheng, J., Chen, H. (eds.) Cognitive Radio Networks: Architecture, Protocols and Standards. Taylor and Francis Group, Auerbach Publications, Boca Raton, FL (2010)

[17] Proakis, J.: Digital Communications, 4th edn. McGraw-Hill (2000)

[18] Rosen, J.B.: Existence and Uniqueness of Equilibrium Points for Concave $N$-person Games. Econometrica 33, 520–534 (1965)

[19] Saad, W., Han, Z., Debbah, M., Hjorungnes, A.: Coalitional Games for Distributed Collaborative Spectrum Sensing in Cognitive Radio Networks. In: Proc. IEEE INFOCOM 2009, pp. 2114–2122 (2009)

[20] Zhang, W., Letaief, K.B.: Cooperative Spectrum Sensing with Transmit and Relay Diversity in Cognitive Radio Networks. IEEE Trans. Wireless Commun. 7(12), 4761–4766 (2008)

# Ant Colony Optimisation for Planning Safe Escape Routes

Morten Goodwin, Ole-Christoffer Granmo,
Jaziar Radianti, Parvaneh Sarshar, and Sondre Glimsdal

Department of ICT, University of Agder,
Grimstad, Norway

**Abstract.** An emergency requiring evacuation is a chaotic event filled with uncertainties both for the people affected and rescuers. The evacuees are often left to themselves for navigation to the escape area. The chaotic situation increases when a predefined escape route is blocked by a hazard, and there is a need to re-think which escape route is safest.

This paper addresses automatically finding the safest escape route in emergency situations in large buildings or ships with imperfect knowledge of the hazards. The proposed solution, based on Ant Colony Optimisation, suggests a near optimal escape plan for every affected person — considering both dynamic spread of hazards and congestion avoidance.

The solution can be used both on an individual bases, such as from a personal smart phone of one of the evacuees, or from a remote location by emergency personnel trying to assist large groups.

## 1  Introduction

Evacuation planning is challenging due to the chaotic and un-organised situation occurring in a crisis situation. Unfortunately, decision makers often have an incomplete picture of hazards and potential escape routes. The situation is further complicated by the fact that people affected are often left alone without any contact with rescue personnel. Further, the chaotic nature of crises situations causes the best escape route to quickly change as hazards such as fires quickly develop.

There is no doubt that decision making in crises situations needs to be made in a timely manner to minimize the potential danger. However, it is difficult people affected to determine what are the best decisions in an evacuation situation. In fact, in most situations the evacuees are not aware of which path to follow for an escape because they either received insufficient information from the rescuers or are unfamiliar with both the architecture and of where the hazards are located. Similarly, the emergency personnel often do not have an overview of where people are located nor which rooms are affected by hazards. This makes evacuation planning particularly difficult [1] [2].

This paper is part of a larger project working on using smart phone technologies in emergency situations. In this project, the smart phones are used to communicate to and from people affected by crises situations. This enables

M. Ali et al. (Eds.): IEA/AIE 2013, LNAI 7906, pp. 53–62, 2013.

rescue personnel to get an overview of people escaping and how people move and communication on how to best escape can be given to the people escaping. In addition, the sensor information available in smart phones will communicate presence of hazards using the camera, position using GPS, gyroscope etc. This provides a threat map which is support both for rescue personnel and to automatically determine the best escape routes [3].

The project has three main steps:

1. Collect information.
2. Calculate escape plan.
3. Communicate the plan to the people affected

### 1.1   Collect Information

Initially in a crisis situation, it is essential to get an overview of the people affected and the hazards present.

The aim is that prior to a crisis situation, for example when people are embark on a cruise ship, they will have the option to download a mobile application for their phone. This application will in a crisis situation utilize the available sensors and communicate it to a central location. This way, the system, including emergency personnel, will be aware of locations of people, whether people are moving, the brightness in each room (indicating hazards such as fire or smoke).

### 1.2   Determine Plan

Computer and mathematical models have shown to be valuable for escape planning with large complex building with many people [4] [5] [6] [7] [8], but is mainly assuming a static representation of hazards. In contrast, our system will, based on available information, calculate the best escape plan and guide each affected person away from any potential hazard, as well as distributing the people to the most suited escape areas even when the hazards change.

### 1.3   Communicate Plan

When an adaptive plan is available, it should be communicated to the affected people, which can be done in two main ways. The primary method is for emergency personnel to actively communicate the plan to the affected people through any available means, such as loud speakers and communication directly to each affected person via the smart phone applications. Failing this, the smart phone applications can automatically present the plan using simple visual and verbal steps such as "turn around", "go left" and safely guide people to an escape area.

Unfortunately, even in situations where an optimal escape plan exists and every person affected are aware of the plan, the human mind is so perplexed that not all follow the plan [9]. Most significantly, in a crisis situation factors such as panic spread, people pushing, jamming up and overlooking alternative exits prevent a crowd from following an optimal plan [10] [11]. Therefore, it is important that information about both hazards and people are continuously updated to always provide the best plan.

## 2   Problem Formulation

Escape planning from a complex building or a large ship can be regarded as a combinatorial optimization problem. In line with common practice [6], we treat the escape as a bidirectional planar graph $G(V, E)$. Each possible location $i$ is connected with a vertex $v_i \in V$, and each potential flow from vertex $v_i$ to $v_j$ is represented by an edge $e_{i,j} \in E$.

In addition we define a function $h(v_i)$ representing the hazard for $v_i$, so that the function $h(v_i)$ returns probability values representing the likelihood of hazards.

The escape area is a vertex $v_t \in V$ (sink), and the people are located in any vertex $v_s \in V$ (any vertex is a source). Further, all search spaces from $v_s$ to $v_t$ is defined as $\mathbf{S}$.

The aim of the application is to find a search space $s* \in \mathbf{S}$ so that $f(s^*) \leq f(s)\forall_s \in \mathbf{S}$, where $f(s) = 1 - \Pi_{V_i \in s}(1 - h(v_i))$. I.e. minimizing the probability that a person encounters a hazard in at least one of the vertexes in the chosen search space.

### 2.1   Hazards

The hazard functions are populated by both known observations and based on indications and estimations. If a hazard $h(v_i) = 1$, it means that there is a known hazard and vertex $v_i$ is unsafe, and all evacuees should be routed away from the corresponding room. Similarly, $h(v_i) = 0$ means that $v_i$ is a known safe vertex. All other hazards are estimated based on known observations.

Definition of the hazard function is not part of this paper. This paper treats the hazard function as an unknown stochastic function returning a probability of a hazard in the room.

## 3   Ant Colony Optimisation (ACO)

Problem solving approaches inspired by nature and animals, so called swarm intelligence, have received a lot of attention due to their simplicity and adaptability. Ant Colony Optmisation (ACO) is one of the most popular swarm intelligence algorithms due to its general purpose optimization technique. ACO consists of artificial ants operating in a constructed graph. The ants release pheromones in favorable paths which subsequent ant members follow. This way, the colony of ants will walk more towards favorable paths and in consequence iteratively build the most favorable solution.[12].

ACO was first used to find shortest path from a source to a sink in a bidirectional graph. It has later increased in popularity due to its low complexity and its ability to work in dynamic environments. The flexibility of ACO is apparent as it has successfully been applied in a wide variety of problems such as finding solutions for NP hard problems [13], rule based classification [14], and is shown to be particularly useful for routing in real time industrial and telecommunication applications.

Finding the shortest path in a graph $G(V, E)$ using ACO in its simplest form works as follows. Artificial ants move from vertex to vertex. When an ant finds a route $s$ from the source $v_s$ to the sink $v_t$, the ant releases pheromones $\tau_{i,j}$ corresponding all edges $e_{i,j} \in s$. The pheromones for all ants $m$ is defined as

$$\tau_{i,j} \leftarrow (1-p)\tau_{i,j} + \sum_{k=1}^{m} \Delta\tau_{i,j}^{k} \tag{1}$$

The function is for ant $k$ is defined as

$$\Delta\tau_{i,j}^{k} = \begin{cases} Q/|s| & \text{if } e_{i,j} \in s \\ 0 & \text{otherwise} \end{cases} \tag{2}$$

where $Q$ is a constant.

The aim of each ant is to walk from $v_s$ to $v_t$ forming the path $s$. This is achieved by the following rule. When ant $k$ is in vertex $i$ it chooses to go to vertex $j$ with the probability $p_{i,j}^{k}$ defined as

$$p_{i,j}^{k} = \begin{cases} \dfrac{\tau_{i,j}^{\alpha}\eta_{i,j}^{\beta}}{\sum_{e_{i,l}} \tau_{i,j}^{\alpha}\eta_{i,j}^{\beta}} & \text{if } e_{i,j} \in N(s^p) \\ 0 & \text{otherwise} \end{cases} \tag{3}$$

where $s^p$ is the partial solution of $s$, and $N(s^p)$ are the possible vertexes to visit given $s^p$. $\eta_{i,j}$ is the inverse heuristic estimate of the distance between node $i$ and $j$, and $\alpha$ and $\beta$ are numbers between 0 and 1 to give the relevant importance between pheromones and the heuristics function.

In its simplest form, the $\beta = 1$ and $\alpha = 1$ so that the ants only consider the pheromones — and the heuristic function is ignored, giving:

$$p_{i,j}^{k} = \begin{cases} \dfrac{\tau_{i,j}}{\sum_{e_{i,l}} \tau_{i,j}} & \text{if } e_{i,j} \in N(s^p) \\ 0 & \text{otherwise} \end{cases} \tag{4}$$

In layman terms, the amount of pheromone released represent quality of the solution. This is achieved by each ant releasing a constant amount of pheromones. Consequently, the shorter the path found, the more pheromone per edge is released. Further, each ant is guided by a stochastic mechanism biased by the released pheromones. Thus, the ants walk randomly with a preference towards pheromones. In this way, the ants incrementally build up promising search space with means that a route $s$ converges towards the shortest route from $v_s$ to $v_t$, $s*$.

ACO has also successfully been applied for many network applications [15] [16] [17]. It has been empirically shown to have favourable results compared to other routing protocols with respect to short path routing and reduced load balancing. Therefore it seems particularly promising for the finding escape routes.

# 4   Solution

This section presents the ACO algorithms for finding escape routes in three distinct realistic environments. First, the algorithms interact with a **static environment** where the hazard functions remain unchanged, yet unknown. This resembles classical optimisation problems where the aim is to find a search space $s^*$ so that $f(s^*) \leq f(s)\forall_s \in \mathbf{S}$.

Subsequently, the problem is extended to interact with **dynamic environments** so that the probability of a hazard in $v_i$, $h(v_i)$, is no longer fixed but changes regularly according to some unknown stochastic function. This shows how well ACO works when environments change, such as fire spreading.

Last, ACO deals with the situation of **control flow**; each edge has a capacity $c(e_{i,j})$. This is a realistic environment where the doors between rooms have a limited capacity, which changes the problem dramatically as it is no longer sufficient to find an optimal solution for one person, but there is a need to consider the system as a whole.

This section presents empirical evidence for ACO working in all the above mentioned environments. In these experiments, all graphs are bidirectional, planar and connected — in line with common practice [6]. Without loss of generality, all experiments are carried out on randomly generated graphs of 1000 vertexes, and 5000 randomly distributed edges, of which 1000 edges are used to make sure the graph is connected.

All experiments are an average of 1000 runs.

## 4.1   Static Environments

ACO has been used for static routing in many situations before. In these experiments ACO is used in its simplest form, as described in section 3, with a slight adjustment. The constant $Q$ is replaced with a function of $s$:

$$Q(s) = 1 - \Pi_{v_i \in s}(1 - h(v_i)) \qquad (5)$$

I.e. $Q(s)$ represent the inverse hazard probability. The consequence of this is that safe paths are given large amounts of pheromones, and unsafe paths low amount. The pheromone updates are therefore as

$$\Delta\tau_{i,j}^k = \begin{cases} Q(s)/|s| & \text{if } e_{i,j} \in s \\ 0 & \text{otherwise} \end{cases} \qquad (6)$$

In the static environment $h(v_i)$ is defined as a random number between 1 or 0, and remain unchanged for each experiment.

Figure 1a and 1b show the behaviour of ACO static environments. Figure 1b shows the behaviour were there in addition to the normal setup the graph is manipulated so that there exists an $s$ so that $f(s) = 0$ — meaning that there always exists a safe path. The optimal solution is calculated using Djikstra's algorithm [18] by considering $h(v_i)$ as basis the cost function for edged $e_{*,i}$.

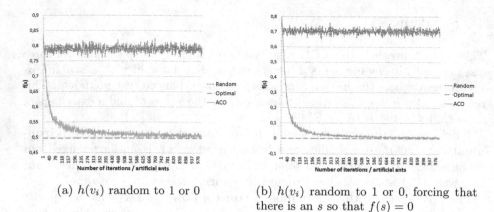

<table>
(a) $h(v_i)$ random to 1 or 0 | (b) $h(v_i)$ random to 1 or 0, forcing that there is an $s$ so that $f(s) = 0$
</table>

**Fig. 1.** Experiment results in static environments of randomly generated large graph comparing ACO to random and optimal. Hazards probabilities are randomly distributed as 0 or 1.

Figure 1b show the same experiment but with adjustments of hazard probabilities so that there is an $s$ so that $f(s) = 0$ — meaning that there always exists a safe path.

Both experiments show that ACO is able to find the near optimal solution with very few iterations.

## 4.2   Dynamic Environments

ACO has also been used for dynamic environments in many situations [19] [20] [21]. This is achieved by letting, for each time step, the pheromones evaporate with a defined probability, typically between 0.01 and 0.20 [12]. The evaporation probability is balance between convergence accuracy and adaptability. I.e. you choose to what extent the ants should work towards a more optimal solution or should be able to adapt to potential other solutions. [21] showed that ACO based routing works well in situations with significant dynamics and continuously broken and newly established connections, which resembles finding an escape route when hazards change.

More specifically, ACO has been used for dynamic environments to detect weakness in networks [22]. This is done by utilizing a mechanism so that problems are reported on a black board. Ants which found a solution to a problem report this on a black board, and update pheromone value on the black board with an inverse probability of how it is solved. This way, collectively, more ants will walk towards problems which have not been solved and thus putting more effort on the unsolved vulnerabilities. Hence, dynamically weaknesses in networks are iteratively solved.

Figure 2a and 2b show ACO in dynamic environments where the $h(v_i)$ hazards are updated every 200th iteration with the following rule:

$$h(v_i) = 1 - h(v_i) \forall_{v_i} \in V \qquad (7)$$

Thus, for every 200th iteration the environment is exactly opposite which in turn means that, if an algorithm has learnt an optimal route, the route changes to as far away from optimal as possible. The results show in figure 2a that when the evaporation rate is set to 0, the ACO learns an optimal solution which becomes outdated when $h(v_i)$ changes, and it is not able to adapt to the new optimal solution. Further, every time the hazard probabilities changes the algorithm is further away from the solution. On the other hand, figure 2b shows that when the evaporation rate is set to 0.2, the algorithm is able to quickly adapt to new environments — and is thus able to interact well with dynamic environments.

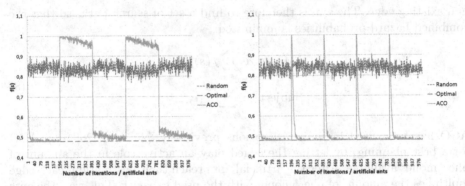

(a) Evaporation rate set to 0, ACO cannot adapt

(b) Evaporation rate set to 0.2

**Fig. 2.** Experiment results in dynamic environments. Hazard probabilities, $h(v_i)$, change for every 200th iteration.

## 4.3   Control Flow

The above situations only consider escapes where there are unlimited capacities — which is not a realistic situation.

While most existing work focus on computer model based escapes focus in finding the shortest safe path for each person [23] [24], this section innovatively extends the ACO to also consider control flow, and in this way avoid congestion of people. Thus an edge can at a certain time step either be full or have room for more people. We achieve this by letting the available capacity of an edge $e_{i,j}$ be defined by $c(e_{i,j})$. Any edge with $c(e_{i,j}) = 1$ cannot be used and is equal to a

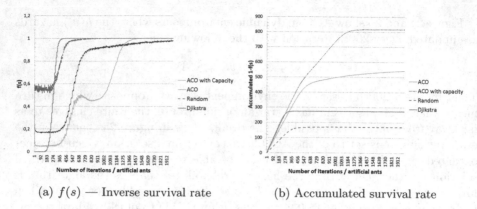

(a) $f(s)$ — Inverse survival rate          (b) Accumulated survival rate

**Fig. 3.** Experiment results comparing ACO capacity randomly distributed

non existing edge. The aim is therefore to find a set of search paths so that the combined hazard probabilities is minimized:

$$\text{Minimize} \sum_{s \in \mathbf{s}} f(s)$$
$$\text{Subject to} \sum_{e_{i,j} \in s} c(e_{i,j}) \tag{8}$$

ACO has been used in similar situations previously. [25] and [26] used ACO for a best planning, by letting the edged play an active role in the amount of pheromones available. After the artificial ants reach their destination, each edge multiplies the amount of pheromones with the used capacity. This way, the ants have two mechanisms to guide them: (1) The quality level of the path as noted by the amount of pheromones released by previous ants, and (2) the available capacity. [25] showed empirically that in most situation the ACO reached the same path as a brute force approach.

In line with state-of-the-art, we update the function $\Delta \tau_{i,j}^k$ from equation 2 as following.

$$\Delta \tau_{i,j}^k = \begin{cases} \dfrac{Q}{|s|} \dfrac{c(e_{i,j})}{\max(c(e_{i,j}))} & if\ e_{i,i} \in s \\ 0\ otherwise \end{cases} \tag{9}$$

where $c(e_{i,j})$ is the used capacity of edge $e_{i,j}$ and $\max(c(e_{i,j}))$ is the total capacity available at $e_{i,j}$.

Figure 3a and 3b show the behavior of ACO in the environment which includes control flow. The experiments show that ACO considering capacity has a better over all performance than ACO without capacity consideration. It is noteworthy that Djikstra, which is identical to the algorithm used in previous experiments, does not perform well because it is not able to plan ahead, and is therefore not able to handle control flow well. In fact, both ACO approaches outperform the much more costly brute force. Lastly, figure 3b shows the accumulated survival rate for each algorithm. This can be read as: How many people will survive if all followed the suggested plan perfectly.

## 5   Conclusion

This paper presents an application using ACO for finding safe escape routes in emergency situations. ACO is used in three main environments. Firstly, ACO operates in a stationary environment where it quickly reaches a near optimal solution. Secondly, this paper empirically shows that ACO is able to cope with dynamic situations were hazards rapidly change. Lastly, ACO is extended to handle vertex capacity. This enables ACO to find escape routes for groups of people. The application is an essential part of a tool mapping threats and planning escape routes.

For our further work, we plan to investigate ACO in more realistic scenarios, including which setup of ACO works best — such as pheromone distribution and evaporation probabilities. In this paper we assume that all participants can report their location. In practice, only some escapees will be able to provide this — rendering the problem more difficult. We have also assumed that all evacuees will follow a suggested plan. In a real evacuation situation, most people will escape in groups together with their family and friends. Further, when panic spreads people will be less likely to follow plans, which will affect overall results.

We also plan to examine how well ACO works compared to traditional control flow optimisation algorithms, and develop a smart phone application which can be used for safe escape planning.

## References

1. Li, Q., Rus, D.: Navigation protocols in sensor networks. ACM Transactions on Sensor Networks (TOSN) 1(1), 3–35 (2005)
2. Li, Q., De Rosa, M., Rus, D.: Distributed algorithms for guiding navigation across a sensor network. In: Proceedings of the 9th Annual International Conference on Mobile Computing and Networking, pp. 313–325. ACM (2003)
3. Radiante, J., Granmo, O.-C., Bouhmala, N., Sarshar, P., Yazidi, A., Gonzalez, J.: Crowd models for emergency evacuation: A review targeting human-centered sensing. In: Proceeding of 46th Hawaii International Conference for System Sciences (2013)
4. Thompson, P.A., Marchant, E.W.: A computer model for the evacuation of large building populations. Fire Safety Journal 24(2), 131–148 (1995)
5. Thompson, P.A., Marchant, E.W.: Testing and application of the computer model simulex. Fire Safety Journal 24(2), 149–166 (1995)
6. Hamacher, H.W., Tjandra, S.A.: Mathematical modelling of evacuation problems– a state of the art. Pedestrian and Evacuation Dynamics 2002, 227–266 (2002)
7. Kim, D., Shin, S.: Local path planning using a new artificial potential function composition and its analytical design guidelines. Advanced Robotics 20(1), 115–135 (2006)

8. Braun, A., Bodmann, B., Musse, S.: Simulating virtual crowds in emergency situations. In: Proceedings of the ACM Symposium on Virtual Reality Software and Technology, pp. 244–252. ACM (2005)

9. Liu, S., Yang, L., Fang, T., Li, J.: Evacuation from a classroom considering the occupant density around exits. Physica A: Statistical Mechanics and its Applications 388(9), 1921–1928 (2009)

10. Wang, J., Lo, S., Sun, J., Wang, Q., Mu, H.: Qualitative simulation of the panic spread in large-scale evacuation. Simulation (2012)

11. Helbing, D., Farkas, I., Vicsek, T.: Simulating dynamical features of escape panic. Nature 407(6803), 487–490 (2000)

12. Dorigo, M., Birattari, M., Stutzle, T.: Ant colony optimization. IEEE Computational Intelligence Magazine 1(4), 28–39 (2006)

13. Gutjahr, W.: A graph-based ant system and its convergence. Future Generation Computer Systems 16(8), 873–888 (2000)

14. Liu, B., Abbas, H., McKay, B.: Classification rule discovery with ant colony optimization. In: IEEE/WIC International Conference on Intelligent Agent Technology, IAT 2003, pp. 83–88. IEEE (2003)

15. Dorigo, M., Stützle, T.: Ant colony optimization: overview and recent advances. In: Handbook of Metaheuristics, pp. 227–263 (2010)

16. Ducatelle, F., Di Caro, G.A., Gambardella, L.M.: An analysis of the different components of the antHocNet routing algorithm. In: Dorigo, M., Gambardella, L.M., Birattari, M., Martinoli, A., Poli, R., Stützle, T. (eds.) ANTS 2006. LNCS, vol. 4150, pp. 37–48. Springer, Heidelberg (2006)

17. Murtala Zungeru, A., Ang, L., Phooi Seng, K.: Performance evaluation of ant-based routing protocols for wireless sensor networks (2012)

18. Dijkstra, E.W.: A note on two problems in connexion with graphs. Numerische Mathematik 1(1), 269–271 (1959)

19. Di Caro, G., Dorigo, M.: Antnet: Distributed stigmergetic control for communications networks. arXiv preprint arXiv:1105.5449 (2011)

20. Di Caro, G., Ducatelle, F., Gambardella, L.M.: Anthocnet: an adaptive nature-inspired algorithm for routing in mobile ad hoc networks. European Transactions on Telecommunications 16(5), 443–455 (2005)

21. Dressler, F., Koch, R., Gerla, M.: Path heuristics using ACO for inter-domain routing in mobile ad hoc and sensor networks. In: Suzuki, J., Nakano, T. (eds.) BIONETICS 2010. LNICST, vol. 87, pp. 128–142. Springer, Heidelberg (2012)

22. Selvan, G., Pothumani, S., Manivannan, R., Senthilnayaki, R., Balasubramanian, K.: Weakness recognition in network using Aco and mobile agents. In: 2012 International Conference on Advances in Engineering, Science and Management (ICAESM), pp. 459–462. IEEE (2012)

23. Zhan, A., Wu, F., Chen, G.: Sos: A safe, ordered, and speedy emergency navigation algorithm in wireless sensor networks. In: 2011 Proceedings of 20th International Conference on Computer Communications and Networks (ICCCN), pp. 1–6. IEEE (2011)

24. Li, S., Zhan, A., Wu, X., Yang, P., Chen, G.: Efficient emergency rescue navigation with wireless sensor networks. Journal of Information Science and Engineering 27, 51–64 (2011)

25. Hsiao, Y., Chuang, C., Chien, C.: Ant colony optimization for best path planning. In: IEEE International Symposium on Communications and Information Technology, ISCIT 2004, vol. 1, pp. 109–113. IEEE (2004)

26. Sim, K., Sun, W.: Ant colony optimization for routing and load-balancing: survey and new directions. Systems, Man and Cybernetics, Part A: Systems and Humans 33(5), 560–572 (2003)

# A Spatio-temporal Probabilistic Model
# of Hazard and Crowd Dynamics
# in Disasters for Evacuation Planning

Ole-Christoffer Granmo[1], Jaziar Radianti[1], Morten Goodwin[1], Julie Dugdale[1,2],
Parvaneh Sarshar[1], Sondre Glimsdal[1], and Jose J. Gonzalez[1]

[1] Centre for Integrated Emergency Management,
University of Agder Grimstad, Norway
[2] Grenoble 2 University/Grenoble Informatics Laboratory (LIG), France

**Abstract.** Managing the uncertainties that arise in disasters – such as ship fire –
can be extremely challenging. Previous work has typically focused either on
modeling crowd behavior or hazard dynamics, targeting fully known
environments. However, when a disaster strikes, uncertainty about the nature,
extent and further development of the hazard is the rule rather than the
exception. Additionally, crowd and hazard dynamics are both intertwined and
uncertain, making evacuation planning extremely difficult. To address this
challenge, we propose a novel spatio-temporal *probabilistic* model that
integrates crowd with hazard dynamics, using a ship fire as a proof-of-concept
scenario. The model is realized as a *dynamic* Bayesian network (DBN),
supporting distinct kinds of crowd evacuation behavior – both descriptive and
normative (optimal). Descriptive modeling is based on studies of physical fire
models, crowd psychology models, and corresponding flow models, while we
identify optimal behavior using Ant-Based Colony Optimization (ACO).
Simulation results demonstrate that the DNB model allows us to track and
forecast the movement of people until they escape, as the hazard develops from
time step to time step. Furthermore, the ACO provides safe paths, dynamically
responding to current threats.

**Keywords:** Dynamic Bayesian Networks, Ant Based Colony Optimization,
Evacuation Planning, Crowd Modeling, Hazard Modeling.

# 1    Introduction

Evacuating large crowds of people during a fire is a huge challenge to emergency
planners and ill-conceived evacuation plans have resulted in many potentially
avoidable deaths over the years [1]. However, accurately evaluating evacuation plans
through real world evacuation exercises is disruptive, hard to organize and does not
always give a true picture of what will happen in the real situation. These challenges
are further aggravated by the uncertainty of how a hazard will evolve; requiring
evacuation plans to be dynamically adapted to the situation at hand.

Formal crowd models have previously been found useful for off-line escape
planning in large and complex buildings, frequented by a significant number of

M. Ali et al. (Eds.): IEA/AIE 2013, LNAI 7906, pp. 63–72, 2013.

people [9-15]. The focus is typically either on crowd behavior or hazard dynamics, for fully known environments. However, when a disaster strikes, uncertainty about the nature, extent and evolution of the hazard is the rule rather than the exception. Additionally, crowd- and hazard dynamics will be both intertwined and uncertain, making evacuation planning extremely difficult. To address this challenge, we propose a novel spatio-temporal probabilistic model that integrates crowd dynamics with hazard dynamics. The overall goal is to build an integrated emergency evacuation model comprising hazard and threat maps, crowd evacuation, and path planning.

The research reported here is conducted as part of the *SmartRescue* project, where we are also currently investigating how to use smartphones for real-time and immediate threat assessment and evacuation support, addressing the needs of both emergency managers and the general public. Smartphones are equipped with ever more advanced sensor technologies, including accelerometer, digital compass, gyroscope, GPS, microphone, and camera. This has enabled entirely new types of smartphone applications that connect low-level sensor input with high-level events to be developed. The integrated crowd evacuation- and hazard model reported here is fundamental for the smartphone based reasoning engine that we envision for threat assessment and evacuation support.

We take as our case study the emergency evacuation of passengers from a ship, triggered by a major fire. Managing uncertainty in such a scenario is of great importance for decision makers and rescuers. They need to be able to evaluate the impact of the different strategies available so that they can select an evacuation plan that ensures as ideal evacuation as possible.

The organization of this paper is as follows: Sect. 2 presents the ship scenario, forming the environment for our approach. We then introduce the novel integrated hazard- and crowd evacuation model in Sect. 3, covering a detailed DBN design for intertwined modeling of hazard- and crowd dynamics, including congestion. A pertinent aspect of this approach is that we apply Ant-Based Colony Optimization (ACO) to configure the DBN with optimal escape paths. In Sect. 4, we present and discuss simulation results that demonstrate that the DNB model allows us to track and forecast the movement of people until they escape, as the hazard develops from time step to time step. Furthermore, the ACO approach provides safe paths, dynamically responding to current threats. We conclude the paper in Sect. 5 by providing pointers for further research.

# 2    Ship Fire Scenario and Modeling Approach

## 2.1    Ship Scenario

We use an onboard ship fire as an application scenario for our model. Fig. 1(a) depicts the ship layout and 1(b) shows the layout as a directed graph. The ship consists of compartments, stairways, corridors and an embarkation area. $A$, $B$ and $C$ are the compartments, each with doors connecting them to $D$ - a corridor. $D_1$, $D_2$ and $D_3$ represent the corridor area that directly links to the different compartments. $E$ is an embarkation area (muster or assembly area) where in an emergency, all passengers gather before being evacuated and abandoning the ship. $S_1$ and $S_2$ are the stairways connecting the corridor to $E$. In (b) the directed graph edges specify the possible direction of movement for the passengers, including the option of remaining in one's current location, for instance due to congestion in adjacent rooms/stairways, panic or confusion, debilitating health conditions, or simply obliviousness to the hazard.

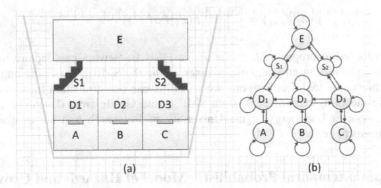

(a)               (b)

**Fig. 1.** Hypothetical ship layout (a), represented as a directed graph (b)

## 2.2 Modeling Approach

A novel aspect of our approach is that we integrate crowd dynamics with hazard dynamics and model the entire emergency evacuation process using a Dynamic Bayesian Network (DBN) [2]. A DBN integrates concepts from graph theory and probability theory, capturing conditional independencies between a set of random variables, $X = (x_1, \dots, x_n)$ [3] by means of a directed acyclic graph (DAG) [4]. Each directed edge in the DAG typically represents a cause-effect relationship. This allows the joint probability distribution of the variables to be decomposed based on the DAG as follows, with $pa_i$ being the parents of $x_i$ in the DAG:

$$P(x_1, \dots, x_n) = \prod_{i=1}^{n} p(x_i \mid pa_i) \tag{1}$$

Thus, the full complexity of the joint probability distribution can be represented by a limited number of simpler conditional probability distributions.

To capture temporal dynamics, variables are organized as a sequence of time slices $(1, \dots, t)$, $Z_{i:t} = (Z_1, \dots, Z_t)$. This organization allows us to forecast both hazard and crowd behavior, $P(Z_{t+n} | Z_{1:t})$, where $n > 0$ indicates how far into the future the forecasting is projected [2]. A compact definition of a DBN consists of the pair $(B_1, B \rightarrow)$, where $B_1$ is a traditional Bayesian network that defines the prior distribution of state variables $p(Z_1)$. $B \rightarrow$ is a two slice temporal BN (2TBN) that defines the transition model $p(Z_t | Z_{t-1})$ as follows:

$$p(Z_t | Z_{t-1}) = \prod_{i=1}^{n} p(Z_t^i | Pa(Z_t^i)) \tag{2}$$

Here, $Z_t^i$ is the $i$-th node at time step $t$. $Pa(Z_t^i)$ are the parents of $Z_t^i$, which can be from a previous time slice. The structure repeats itself, and the process is stationary, so parameters for the slices $t = 2,3 \ldots$ remain the same. Accordingly, the joint probability distribution for a sequence of length $T$ can be obtained by unrolling the 2TBN network:

$$p(Z_t|Z_{t-1}) = \prod_{t=1}^{T} \prod_{i=1}^{N} p(Z_t^i|Pa(Z_t^i)) \tag{3}$$

We now proceed to proposing how the above DBN framework can capture hazard- and crowd dynamics in an integrated manner, using the SMILE reasoning engine as the implementation tool. Furthermore, we will illustrate our simulation findings using the GeNIe modeling environment. Both SMILE and GeNIe were developed by the Decision Systems Laboratory at the University of Pittsburgh and are available at http://genie.sis.pitt.edu/.

# 3      Spatio-temporal Probabilistic Model of Hazard- and Crowd Dynamics

In this section, we describe the rescue planning model and the integrated evacuation model. The rescue planning model is a brute force search of all possible escape paths to find paths that minimize hazard exposure for the full range of hazard scenarios. We apply the ACO algorithm to efficiently find optimal paths for large disasters.

The application of BNs for decision support in maritime accidents has been discussed by Datubo et al. [5]. Their focus is to model and simulate accident scenarios such as fire, flooding and collision. The purpose is to identify optimal decision policies, including alarm- and evacuation strategies, based on maximizing expected utility under different hazard scenarios. Our approach is quite different from Datubo et al.'s work, since we are focusing on tracking both passenger and hazard, dealing with the uncertainty that arises when one has to rely more or less on limited sensor information. In addition we forecast crowd behavior and hazard development for the purpose of producing real-time risk maps, evacuation paths, and dynamically identifying the optimum movements of passengers as the hazard evolves.

## 3.1      Rescue Planning

Rescue planning can be regarded as a combinatorial optimization problem using graph traversal as a starting point. We let each location $i$ be connected with a vertex $v_i$, and let each potential flow from location $i$ to $j$ be represented by an edge $e_{i,j}$. In addition to a common graph representation, we let each vertex $v_i$ have a hazard $h_i$, so that the hazard values $h_i \in H$ are probability values representing the likelihood of hazards. Thus, the vertices, edges and hazards are a constructed graph with hazards $G(V, E, H)$.

Further, we let one of the vertices, $v_n \in V$, represent the escape area, while $v_0 \in V$ represents the starting locations for passengers. We then define a search space $S$ as a finite set of discrete decision variables, so that any $s \in S$ is a possible route from $v_0$

to $v_N$. We further define a function $f$ representing the hazard probability of $s$. The objective is therefore to find an $s^* \in S$ so that $f(s^*)$ is minimized. This way $f(s^*)$ can be read as an inverse survival rate by choosing the escape route $s^*$.

By calculating the escape route for every possible combination of $h_i \in H$ in the scenario in Fig.1, we can generate at least one escape route. Our approach is thus similar to the problem of finding the shortest path between two nodes in a graph, for which many optimization algorithms exist [7]. However, we are focusing on finding multiple paths that both address congestion and minimizes hazard exposure.

ACO has some attractive properties, namely that ACO algorithms can efficiently handle dynamic situations, stochastic optimization problems, multiple objective optimization, parallel distributed implementations and incremental optimization [6]. For route-based escape planning this is particularly useful since we expect:

- the parts of the layout graph for escape to change dynamically as the hazard evolves,
- decisions to be based on stochastic information as the hazards are represented by stochastic functions,
- to face multiple objectives such as multiple escape nodes, and
- parallel distributed implementations for online and offline computation, supporting for instance smartphone based sensing and computing.

Finding the shortest path in a graph $G(V, E)$ using ACO works as follows: Artificial ants move from vertex to vertex. When an ant finds a route $s$ from $v_o$ to $v_N$, the ant deposits pheromones in all $v_i \in s$. The amount of pheromone deposited depends on the quality of the solution. Consequently, the shorter the found path, the more pheromone per vertex is deposited. Furthermore, each ant is guided by a stochastic mechanism biased by the amount of pheromone on the trail; the higher the amount of pheromone, the more attractive the route will be to the ant. Thus, the ants walk randomly, but with a preference towards higher pheromone routes. In this way, the ants incrementally build up a promising search space.

For rescue planning, we propose a slight adjustment to the ACO. Instead of deposited a constant amount of pheromone, we let the ants deposit pheromones equivalent to the inverse of the "size" of the perceived hazard in $s$, i.e. $1 - f(s)$. Thus, when an artificial ant arrives at $v_N$ will have a combined probability value from the hazard probabilities from the visited vertexes as $(s) = 1 - \prod_{v_i \in s}(1 - h_i)$. Hence, a low $f(s)$ signifies a low probability of a perceived hazard in route $s$, meaning more pheromone will be deposited. In Sect. 4 we provide empirical support for ACO finding the path with the smallest probability of hazards; a minimized $f(s)$.

## 3.2   Integrated Evacuation Model

The integrated evacuation model (IEM) is designed to keep track of the location of people, their flow between locations, as well as the corresponding hazard status of the locations, from time step to time step. The overall structure of IEM is shown in Fig. 2. The IEM consists of a *Hazard Model*, a *Risk Model*, a *Behavior Model*, a *Flow Model* and a *Crowd Model*. Each model encapsulates a DBN, with the *Hazard Model* being detailed in Fig. 3 for example purposes (the other models have a similar structure).

For each location $X$, the *Hazard Model* contains a variable $Ha_X(t)$ that represents the status of the hazard, for that location, at time step $t$. These variables capture the dynamics of the hazard, which for fire involves its development and spreading from location to location. The *Hazard Model* for our particular scenario thus consists of nine hazard nodes, each referring to a particular part of the ship layout from Fig. 1. We model the fire hazard based on physical fire properties, abstracting the progressive stages of fire using the states: *Dormant, Growing, Developed, Decaying,* and *Burnout.* Note that depending on the nature of the barriers separating locations, a *Developed* fire may potentially spread to neighboring locations, triggering a transition from *Dormant* to *Growing* in the neighboring location.

**Fig. 2.** Macro view of DBN model for Integrated Evacuation Model (IEM)

As illustrated in Fig. 2, the *Risk Model* at time step $t$ depends on the *Hazard Model*. For each location $X$, the Risk Model contains a variable $R_X(t)$ with three states: *Low Risk, Medium Risk,* and *High Risk*. Briefly stated, the *Risk Model* maps the state of each node in the *Hazard Model* to an appropriate risk level, to allow a generic representation of risk, independent of the type of hazard being addressed. For simulation purposes, the dormant fire stage is considered to be *Low Risk*, while a fire in the *Growing* or *Burnout* stage introduces *Medium Risk*. Finally, if the fire is *Developed,* then we have *High Risk*.

Risk levels, in turn, are translated into a *Behavior Model*, with optimal response to each possible risk scenario being determined by ACO for normative crowd behavior in larger scenarios. We develop descriptive models based on studies of crowd psychology models. The *Behavior Model* is simply a single DBN variable, $B_X(t)$, with each state mapping to a particular flow graph (global evacuation plan), as defined in Sect. 3.1. For instance, one particular flow graph could suggest the escape routes: "$A$-$D_1$-$S_1$-$E$", "$B$-$D_2$-$D_1$-$S_1$-$E$" and "$C$-$D_3$-$S_2$-$E$". Such paths are selected based on pre-computed risk assessments, based on overall survival rate. Note that we study both the ideal evacuation and typical evacuation of the crowd for each hazard scenario.

The *Flow Model* in Fig. 2 manages the flow of incoming and outgoing people along each edge in the layout graph in Fig. 1, represented by three mutually supporting variables for each location $X$. The variables $I_X(t)$ interleaves incoming flows by alternating between neighbor locations, while $O_X(t)$ routes the incoming flow into an outgoing flow by selecting an appropriate destination location. Finally, the variable $S_X(t)$ is used to calculate the resulting number of people in location X, having $I_X(t)$ and $O_X(t)$ as parents. *In other words, our DBN can keep track of the actual number of people at each location by just counting!* This organization allows us to model the flow of incoming and outgoing people from different locations, as well as congestion, flow efficiency, and crowd confusion, all within the framework of DBNs. Here, flow efficiency reflects how quickly the evacuation process is implemented, and how quickly people are reacting. Confusion models suboptimal movement decisions, and can potentially be extended to be governed by local factors such as smoke, panic, and so on.

Finally, the *Crowd Model* keeps track of the amount of people at each location $X$ at time step $t$. It also serves as storage for the crowd movement calculations performed by the *Flow Model*. Currently, we apply a rough counting scheme that for each location $X$ keeps track of whether the location is *Empty*, contains *Some* people or contains *Many* people, using the variable $C_X(t)$. Basically, an *Empty* location or one that only contains *Some* people, can receive *Some* people from a neighboring location, while a location with *Many* people must be unloaded before more people can enter. This can easily be extended to more fine granular counting and flow management as necessary for different application scenarios.

# 4    Simulations, Results and Discussion

## 4.1    Simulations of Integrated Evacuation Model

We have run several simulations based on the IEM. In one particularly challenging and representative scenario, we assumed that a fire had started concurrently at locations $S_2$ and $A$ at time step $t_1$. This means that the shortest escape route from location $B$ and $C$ will be hazardous, and people should be rerouted through staircase $S_1$. We used 50 time slices to forecast and analyze hazard dynamics, as well as the behavior of people reacting to this complex fire scenario. To obtain estimates of the posterior probabilities given the evidence of fire, we applied Adaptive Importance Sampling (AIS). This algorithm is designed to tackle large and complex BNs [7], and is used here for forecasting purposes. For tracking, the use of more advanced variants of this algorithm, which support resampling, such as particle filtering, is appropriate. The obtained results are presented in Fig. 3 and Fig. 4.

Fig. 3 illustrates the development of the hazard probability distribution $H_X(t)=\{Dormant, Growing, Developed, Decaying, Burnout\}$ over time, for each location $X$. In brief, the bar charts inside each node in the Hazard Model summarize the probability (y-axis) associated with each phase of the corresponding hazard, for each time step (x-axis). Note how the development of the fire reflects our initial evidence, where we, for instance, can observe when the fire is likely to spread to neighbors, such as location $D_1$ in time step *7*, and at time step *14* for location $D_2$. To conclude, our DBN provides us with a global probabilistic threat picture, allowing us to assess the current hazard situation, forecast further development, and relating cause and effect, despite potentially arbitrary and limited evidence. In addition to reasoning about hazards, we can also track and forecast crowd behavior using our DBN. The goal of an evacuation is to transfer people from unsafe to safe areas. Figs. 4 (a), (b) and (c) show the simulation results of people moving along the path $A$-$D_1$-$S_1$. At $t_0$ we have *Many* people in each of the compartments $(A,B,C)$, while corridors and stairways are *Empty*.

As can be seen from Fig. 4, it is not until after time step *36* that the probability that room $A$ is completely vacated starts approaching *1.0*. Furthermore, Fig. 4 (b) and (c) show that the probability that the number of people in $D_1$ and $S_1$ are increasing initially, as people start arriving from locations $A,B,$ and $C$ . In general, under the simulated conditions, where people follow the optimal plan without panicking, most

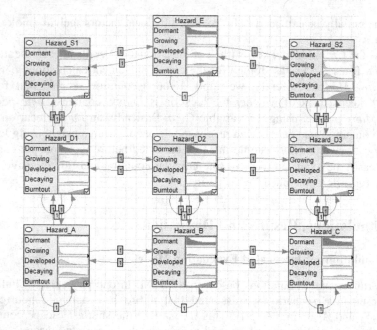

**Fig. 3.** Inferred probabilistic hazard dynamics for a particular hazard scenario

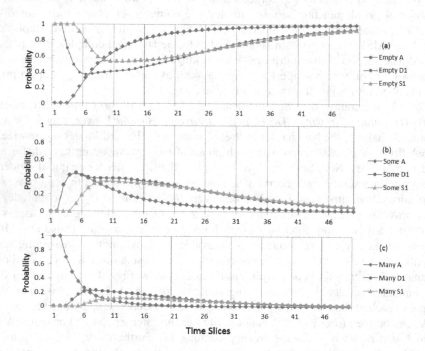

**Fig. 4.** Crowd dynamics for fire in location $A$ and $S_2$

people are able to proceed to the exit area. Therefore, from time step 36 it is likely that most people have evacuated, or succumbed to the hazard. To conclude, we notice that the IEM can track and forecast hazard development and people flow, taking into account uncertainty and the intertwined relationship between crowd- and hazard dynamics.

## 4.2    Route Planning with ACO Approach

Fig. 5 shows the simulation results from the ACO route planning, used for identifying optimal flow graphs for the ship fire scenario shown in Fig. 1. To evaluate evacuation route redundancy, we assigned hazard probabilities randomly and uniformly to either 0 or 1 for each location. The result suggests that low risk routes existed in about 40% of the fire scenarios. The ACO finds these safe routes with probability close to 1.0 using merely 10 ants. The optimal safe routes contain nodes where $f(s)=0$, i.e. nodes were no hazard is present. The results are obtained from the average of 10000 runs. This shows that the adjusted ACO is able to quickly and adaptively find a safe path when it exists.

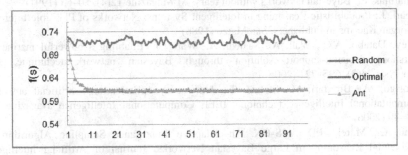

**Fig. 5.** ACO for Route Planning with 9 Nodes

In addition, a scheme for finding the optimal escape paths for all distributions of hazards has been implemented. The scheme is based on minimizing overall risk, assigning a risk minimizing plan to each hazard scenario. When integrated with the *Behavior Model* of the DBN, the DBN supports dynamically adjusting escape plans that take into account present uncertainty.

## 5    Conclusions and Future Directions

In this paper we have proposed a spatio-temporal probabilistic model of hazard- and crowd dynamics in disasters, with the intent of supporting real-time evacuation planning by means of situation tracking and forecasting. We have applied brute force and ACO based techniques to identify safest paths, and a DBN for forecasting flow of people and hazard development. Empirical results reveal that ACO is able to quickly find the safest path, using only 10 ants, adaptively for the various hazard scenarios. The IEM allows us to keep track of, and predict, the location of people and hazard status, through each time step, taking into account uncertainty as well as the intertwined relationship between crowd and hazard dynamics. Future directions of the SmartRescue project include calibrating and refining our model based on feedback from a group of practitioners. We also intend to investigate how a collection of smartphones can form a distributed sensing and computation platform. This involves distributed schemes for tracking, forecasting and planning; allowing the IEM and ACO schemes to run seamlessly in the smartphone network.

**Acknowledgement.** We thank Aust-Agder Utviklings- og Kompetansefond for funding the SmartRescue research. We also wish to express our thanks to the external SmartRescue reference group for valuable feedback and advice.

# References

1. World Fire Statistics Center.: World fire statistics bulletin. World Fire Statistics Bulletin, Valéria Pacella, Geneva (2012)
2. Murphy, K.P.: Dynamic Bayesian Networks: Representation, inference and learning. PhD Dissertation, University of California, Berkeley (2002)
3. Charniak, E.: Bayesian networks without tears. AI Magazine 12(4), 50–63 (1991)
4. Pearl, J.: Probabilistic Reasoning in Intelligent Systems: Networks of Plausible Inference. Morgan Kaufmann Publishers, San Mateo (1988)
5. Eleye-Datubo, A.G., Wall, A., Saajedi, A., Wang, J.: Enabling a powerful marine and offshore decision-support solution through Bayesian network technique. Risk Analysis 26(3), 695–721 (2006)
6. Dorigo, M., Birattari, M., Stutzle, T.: Ant colony optimization - artificial ants as a computational intelligence technique. IEEE Computational Intelligence Magazine 1(4), 28–39 (2006)
7. Jian, C., Marek, J.D.: AIS-BN: An Adaptive Importance Sampling Algorithm for Evidential Reasoning in Large Bayesian Networks. Journal of Artificial Intelligence Research 13, 155–188 (2000)
8. Thompson, P.A., Marchant, E.W.: A computer model for the evacuation of large building populations. Fire Safety Journal 24(2), 131–148 (1995)
9. Thompson, P.A., Marchant, E.W.: Testing and application of the computer model simulex. Fire Safety Journal 24(2), 149–166 (1995)
10. Hamacher, H., Tjandra, S.: Mathematical modelling of evacuation problems – a state of the art. Pedestrian and Evacuation Dynamics 2002, 227–266 (2002)
11. Kim, D., Shin, S.: Local path planning using a new artificial potential function composition and its analytical design guidelines. Advanced Robotics 20(1), 115–135 (2006)
12. Braun, A., Bodmann, B., Musse, S.: Simulating virtual crowds in emergency situations. In: Proceedings of the ACM Symposium on Virtual Reality Software and Technology, pp. 244–252. ACM (2005)
13. Liu, S., Yang, L., Fang, T., Li, J.: Evacuation from a classroom considering the occupant density around exits. Physica A: Statistical Mechanics and its Applications 388(9), 1921–1928 (2009)
14. Wang, J.H., Lo, S.M., Sun, J.H., Wang, Q.S., Mu, H.L.: Qualitative simulation of the panic spread in large-scale evacuation. Simulations 88, 1465–1474 (2012)
15. Helbing, D., Farkas, I., Vicsek, T.: Simulating dynamical features of escape panic. Nature 407(6803), 487–490 (2000)

# Predicting Human Behavior in Crowds:
# Cognitive Modeling versus Neural Networks

Mark Hoogendoorn

VU University Amsterdam, Department of Computer Science
De Boelelaan 1081, 1081 HV Amsterdam, The Netherlands
m.hoogendoorn@vu.nl

**Abstract.** Being able to make predictions on the behavior of crowds allows for
the exploration of the effectiveness of certain measures to control crowds. Taking effective measures might be crucial to avoid severe consequences in case
the crowd goes out of control. Recently, a number of simulation models have
been developed for crowd behavior and the descriptive capabilities of these
models have been shown. In this paper the aim is to judge the predictive capabilities of these complex models based upon real data. Hereby, techniques from
the domain of computational intelligence are used to find appropriate parameter
settings for the model. Furthermore, a comparison is made with an alternative
approach, namely to utilize neural networks for the same purpose.

## 1    Introduction

As numerous incidents such as the disaster during the Love Parade in Duisberg,
Germany in 2010 and the incident on the Dam Square in the Netherlands in that same
year have shown: crowds can easily go out of control. Mostly only a small trigger is
needed to cause a panic that can end in catastrophe. Therefore, paying special attention to the control of these crows is of utmost importance. This control could involve
the strategic positioning of fences, of people, etcetera. However, to decide upon these
strategic positions is not a trivial matter as it is not easy to make predictions on the
behavior of such crowds.

Recently, within multi-agent systems, models have been developed that are able to
simulate crowd behavior, making it possible to explore how well particular measures
to control a crowd work. These models are sometimes based on more physics oriented
approaches such as the well-known model of Helbing [7], but also cognitive oriented
models have been developed, for example the ASCRIBE model [8]. The latter
for instance takes the emotions of the individual agents in the crowd into consideration when determining their exact movement. Although validation of these models is
essential, only few have been rigorously validated using real data. In [1; 9], actual
human movement data obtained from video footage of a panicked crowd has been
used to show that the model is able to reproduce such movements. These experiments
have however been more targeted towards showing that the models are able to *describe* the data well, and not on their *predictive* capabilities, which is of course crucial
to obtain a model that can be used for the purposes listed above. As a consequence, a
lot of parameters of the individuals are highly tailored (in fact, each individual has a

M. Ali et al. (Eds.): IEA/AIE 2013, LNAI 7906, pp. 73–82, 2013.

set of unique parameters), making it hard to utilize for predictions of behavior of individuals with different characteristics. Secondly, the cognitive models specified are quite complex and the question is whether techniques from computational intelligence such as neural networks are not able to do the job just as well, or perhaps even better.

In this paper, the main purpose is: (1) to see whether a generic set of parameters can be found for an existing cognitive model to make robust predictions on crowd movement, and (2) to see how well a neural network would describe and predict crowd behavior compared to that tailored cognitive model. In order to do so, learning techniques from the domain of evolutionary algorithms have been utilized to tune the parameters of the existing cognitive model based upon training data, and a dedicated neural network is designed that is trained using a standard back propagation algorithm. Tests are performed using an existing dataset (cf. [1]).

This paper is organized as follows. The existing cognitive model used is described in Section 2, followed by the neural network design in Section 3. Section 4 presents the learning algorithm for the cognitive model, and Section 5 discusses the experimental setup and results. Finally, Section 6 is a discussion.

## 2      Existing Model

As mentioned in the introduction, one of the purposes of this paper is to judge the predictive capabilities of an existing model, namely the ASCRIBE model. This section discusses the model on a high level. Note that the precise mathematical details of the model have been omitted for the sake of brevity, see [8] for more details. In Figure 1, the main concepts underlying the ascribe model are shown.

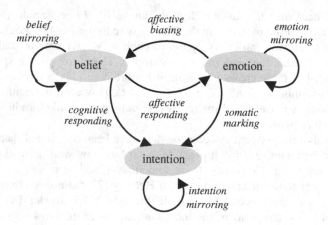

**Fig. 1.** Concepts and their relationship in ASCRIBE (cf. [8])

The states within the agent comprise of beliefs, emotions, and intentions. In the mathematical model, each state has a value between 0 and 1. Internally, the beliefs influence the emotion (i.e. an emotional response to a belief) and the intentions of the agent (i.e. responding to a certain belief by intending a certain action). The emotions in turn influence the way in which beliefs are perceived by means of an affective bias, and influence the value of intentions by associating a feeling with a certain option (referred to as Somatic Marking, following Damasio (cf. [4])). Besides these internal influences, the agent is also influenced by external factors, namely the agents in their neighborhood, resulting in (to a certain degree) mirroring the beliefs, emotions, and intentions of the others.

The model is highly generic in the sense that it describes social contagion in general, but an instance of the model has been specified that involves movements in crowds and panic situations (cf. [1]). Hereby, specific beliefs, intentions, and emotions have been inserted. More precisely, these states now concern beliefs, intentions and emotions associated with the options that the agents has. The options are the wind directions the agent can move in (N, NE, etcetera). In addition, a belief is present on the seriousness of the current situation, and an extra emotion, namely a general emotion of fear. The agent moves into the wind direction with the highest intention value (in case this direction is not blocked, otherwise the second highest intention is taken) and the speed of movement depends on the strength of the intention, ranging from a certain minimum speed to a set maximum speed.

This given instantiation of the model has a number of parameters that can be set on an individual basis. In previous research, the following parameters were found to work better in case they were to be set individually: (1) the distance within which agents influence each other (i.e. when mirroring takes place); (2) the way in which the initial belief about the positivity of the situation is obtained (by means of a parameter expressing from how far an agent is able to obtain information based on sight); (3) the minimum travel speed of an agent, and (4) the maximum travel speed. Therefore, only these parameters will be considered in the remainder of this paper, and the other values will remain at the value they have been given during previous work.

## 3    Neural Network Design

Next to the approach to express a complete and highly complex cognitive model, an alternative approach would be to utilize a neural network (representing the class of pure learning-based approach) for the learning of the behavior of agents within crowds. In this section, a dedicated neural network design is introduced to control agents in a crowd. The architecture is shown in Figure 2. Essentially, the input of the neural network comprises of different groups of nodes: (1) information about the agent's own prior movements; (2) information about possible obstacles in the direction the agent is currently heading at, (3) information about the movements and locations of other agents, and (4) information about the current time point. The information about the movement of other agents contains information on the $i$ closest agents as well as the source of a certain panic. For the former, input is received on the distance of the other agent compared to the agent itself, and the direction of movement of the other agent (x and y movement). For the source of the panic, the same input is provided except now the direction of movement does not concern the movement of the panic source, but the direction compared to the current position of the agent itself. All the values provided to these nodes are scaled to the values [-1,1] for directions and [0,1] for obstacles and time. As output, the neural network provides the x-and y-movement. The hidden neurons have a hyperbolic tangent sigmoid function, whereas the output nodes contain a linear function. The number of hidden neurons and layers is left variable; different values will be tried when running experiments.

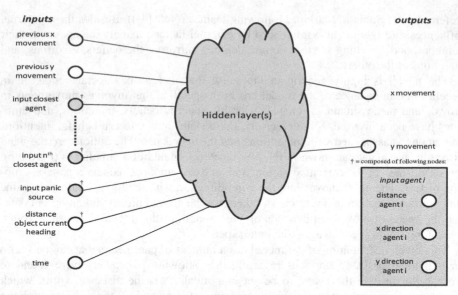

**Fig. 2.** Neural network architecture

## 4    Learning Algorithms

It is assumed that a dataset is available which expresses the coordinates of agents, sources of panic, and obstacles at each time point of a crowd panic. Using this information, the parameters of the two models can be learned. For the neural network this learning is relatively straightforward: the dataset is translated into the corresponding input and desired output of the neural network and a time point in combination with a single agent is considered a training sample for the Levenberg-Marquardt back propagation algorithm [3; 11]. For the learning part of the ASCRIBE model things are a bit more complicated. In order to tune the parameters of ASCRIBE, an evolutionary approach is used. Hereby, parameters are not tuned on an individual agent level, but parameters for the population of agents as a whole are determined. This facilitates prediction of movement of the agents. A "standard" genetic algorithm is used:

- *Individual representation.* The population is composed of individuals that are represented by a binary string which represents real values (i.e. the parameters of the ASCRIBE model) with a certain precision.
- *Population initialization.* The population is initialized randomly.
- *Selection.* The selection of individuals is performed by first ranking the individuals using linear ranking and then selecting individuals based upon stochastic universal sampling.
- *Mutation.* The mutation operator used is straightforward: each bit is simply mutated with a certain probability.
- *Crossover.* A single point crossover function is used to combine the individuals.

Crucial for the evolutionary algorithm is the specification of the fitness function. In order to determine the fitness, an entire simulation is performed using the model with the set of parameters represented by the individual subject to evaluation. Hereby, the agents are placed at their initial positions and follow the path dictated by the model with the selected parameter settings. As the main goal is to reproduce the movements of the people involved in the scenario, it was decided to take the average (Euclidean) distance (over all agents and time points) between the actual and simulated location:

$$\varepsilon = \Sigma_{agents\ a}\ \Sigma_{timepoints\ t}\ \frac{\sqrt{(x(a,t,sim)-x(a,t,data))^2 + (y(a,t,sim)-y(a,t,data))^2}}{\#agents\ \#timepoints}$$

Here, $x(a, t, sim)$ is the x-coordinate of agent $a$ at time point $t$ in the simulation, and $x(a, t, data)$ the same in the real data (similarly the y-coordinates). Both are in meters. This fitness measure complies with the measure used in [1].

# 5    Case Study: 4 May 2010

This section presents the results for a case study using a dataset from an incident on the 4th of May in Amsterdam, the Netherlands. First, the dataset is described, followed by the experimental setup and results.

## 5.1    Dataset Description

As a case study, data from an incident which took place on the Dam Square in Amsterdam, the Netherlands on May 4th 2010 has been used. The incident involves a person that starts shouting during a two minute period of silence during the national remembrance of the dead. As a result of the shouting person, people start to panic and run away[1]. A dataset has been composed that describes the coordinates of movement of 35 people during the incident (see [1]). Next to these coordinates, the coordinates of the source of the panic (i.e. the shouting man) as well as the objects on the square have been logged as well. This data can be used for training the various models and judging the outcome of the models. In order to train the neural network a slight translation of the data needs to take place, but this translation is quite trivial (translating coordinated over time into movement and direction, etcetera) and therefore not further described here.

## 5.2    Experimental Setup and Results

This section describes the experimental setup used, and presents and discusses the results obtained in detail.

### Experimental Setup
During the experiments, comparisons will be made on the descriptive performance (i.e. the performance on the training set, which contains the full set of individuals) of the various approaches as well as the prescriptive performance. For judging the

---

[1] See http://www.youtube.com/watch?v=1AEXcxwHJVw for a movie clip

predictive performance a 5-fold cross validation approach is used, whereby each fold consists of $1/5^{th}$ of all agents. The performance is averaged over the performance on the five configurations of the folds. Based upon this approach, the following models will be compared:

1.  *ANN.* An artificial neural network as detailed in Section 3 and 4.
2.  *ASCRIBE generic.* The ASCRIBE model with a single set of parameters for all agents, whereby the learning takes place using evolution (see Section 2 and 4).
3.  *ASCRIBE individual.* The ASCRIBE model with individual parameter per agent. This approach is used as a benchmark, it can be seen as the optimal way of describing the individual agents with ASCRIBE. Note that using such individual parameters is not realistic for judgment of predictive capabilities, but it is used nonetheless.
4.  *No movement.* Assumes that all agents remain at their position.

For the neural network based approach various configurations of the neural network have been used:

1.1  *Isolated NN.* Using no information of movement of surrounding agents;
1.2  *Isolated NN with time.* Using time and no additional information of movement of surrounding agent;
1.3  *Neighbors NN.* Using information of movement of the closest two surrounding agents;
1.4  *Neighbors NN with time.* Using information of movement of the closest two surrounding agents and time.

For each of the learning algorithms 10 runs have been performed for each setting (the choice for 10 runs was made due to the costly fitness function of the evolutionary algorithm, making it very time consuming to run for the predictive case as 50 runs are required). For the neural network a run consisted of 1000 epochs of the Levenberg-Marquardt back propagation algorithm with an initial $\mu$ set at 0.001, the $\mu$ increase factor is set to 10 and the decrease factor to 0.1. Furthermore, in experiments not reported here for the sake of brevity different settings for the hidden layers in the neural network have been used, thereby showing that a single hidden layers with three hidden neurons worked best for the first two networks, and two hidden layers with five and three neurons in the respective layers for the third and fourth neural network configurations. For the evolutionary approach to learn the parameters of the ASCRIBE model (setting 2) a population size of 50 was used, and 100 generations. A mutation rate of 0.01 has been used and a crossover rate of 0.5. These parameters are based on the "standard" settings as proposed by DeJong [5], but some modifications were made (the process converges earlier resulting in a need for fewer generations, and the crossover and mutation rate have been altered based on test runs).

**Results**

The results are detailed in this section. First, the descriptive capabilities are shown, followed by the predictive capabilities.

*Performance Training Set.* Figure 3a shows an overview of the average performance of the different neural network configurations on the training set. Here, the x-axis shows the time of the incident, whereas the y-axis shows the average difference between the simulated and actual position of the agents in the crowd (cf. Section 4, except the value is not averaged over all time points). It can be seen that the Neighbors NN performs worst, whereas the other network configurations perform very similar. The averages over time points and standard deviations are shown in Table 1. The differences are not statistically significant (using a t-test with 5% significance level).

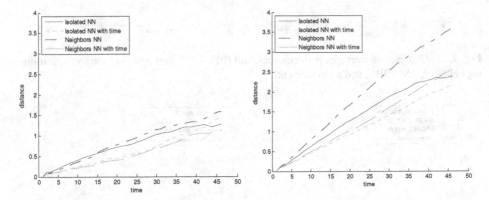

**Fig. 3.** (a) Comparison average training set performance different NN configurations; (b) Comparison best runs on training set of NN's

Figure 3b shows the runs during which the performance was best. Here, the performances are again very close to each other. Of course, this comparison is interesting, but the most interesting is to see how well the neural networks perform compared to the benchmark, and also how well the ASCRIBE model with a single set of parameters for all agents is able to describe the behavior. Figure 4a compares the average performances of the different approaches (the precise results are again shown in Table 1). It can be seen that the ASCRIBE model with individual parameter performs best, which is obvious, it is highly tailored to the case at hand. The ASCRIBE model with evolution performs somewhat worse, whereas the average performance of the best neural network configuration is even worse than predicting no movement (whereby the latter gives a good impression of the average movement of the crowd as the error found at the end is equal to the average final distance traveled per agent).

When comparing the best runs, the outcome is suddenly completely different (see Figure 4b). The neural network performs even better than the ASCRIBE model with individual parameters (note that in this case the neural network which considers the movement of neighboring agents and time is best). Hence, the neural network based approach has huge fluctuations in outcome, from a very bad performance, to excellent performance.

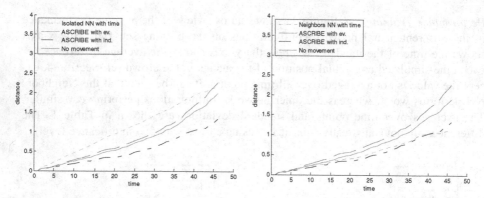

**Fig. 4.** (a) Comparison averages performances, and (b) comparison best performances on training set NNs, ASCRIBE, and no movement

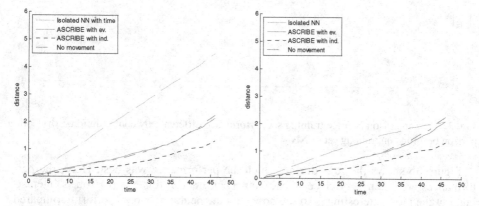

**Fig. 5.** (a) Comparison averages performance, and (b) comparison best runs performances on test set NNs, ASCRIBE, and no movement

*Performance Test Set.* As said, in this paper the predictive capabilities of the various approaches are also subject of investigation as this would facilitate the utilization of models for simulating hypothetical scenarios. Figure 5a shows the predictive capabilities of the average performance of the best neural network configuration with respect to the predictive capabilities (in this case the isolated neural network with time). It can be seen that the performance of ASCRIBE with generic parameters is the best of the more complex predictive models (although the performance is not too good, it is close to no movement), but for the neural network approach the performance is very bad. In this case, there is a statistically significant difference between the best neural network approach and the ASCRIBE model with generic parameters (p-value of 0.0001).

When comparing the best runs again, the predictive capabilities of some neural networks trained (in this case with neighbors and time considered as that provides the best predictive run) are a bit better (see Figure 5b), but still performing worst in comparison with the other approaches.

**Table 1.** Detailed results various algorithms

| Approach | Training set | | | Test set | | |
|---|---|---|---|---|---|---|
| | Avg. perf. | Std. dev. | Best perf. | Avg. perf. | Std. dev. | Best perf. |
| Isolated NN | 1,4002 | 0.9678 | 0.7777 | 2.4158 | 0.5574 | 1.1893 |
| Isolated NN with time | 1.1051 | 1.0956 | 0.619 | 2.2519 | 0.5422 | 1.6082 |
| Neighbors NN | 1.9134 | 1.7985 | 0.8531 | 2.6535 | 0.5124 | 1.9511 |
| Neighbors NN with time | 1.2477 | 1.2195 | 0.551 | 2.6269 | 0.8239 | 1.3894 |
| ASCRIBE generic | 0.7584 | 0.0507 | 0.7228 | 0.8482 | 0.0234 | 0.8082 |
| ASCRIBE individual | 0.5231 | 0 | 0.5231 | 0.5231 | 0 | 0.5231 |
| No movement | 0.8394 | 0 | 0.8394 | 0.8394 | 0 | 0.8394 |

# 6    Discussion

In this paper, two approaches have been used to both describe as well as predict behavior of humans within crowds: a neural network based approach and an approach based on theories from the domain of social psychology called ASCRIBE [8]. To make it possible for ASCRIBE to predict behavior of individuals an evolutionary approach has been developed that tunes the parameters of the model based on a training set, and uses those parameters settings for making predictions. Such parameters are generic, and not specifically tailored towards each individual agent. In order to evaluate the approaches on their suitability they have been tested against a real-world dataset (cf. [1]). The results showed that both for the descriptive as well as the prescriptive behavior the performance of the ASCRIBE model with learning of generic parameters showed reasonable performance, although for the case of prediction it does not move a lot beyond predicting no movements. The neural network based approach performed significantly worse especially for the predictive case, and showed a large variation in performance. One can thus conclude that it is quite difficult to find a generic set of parameters that is able to predict human movement in crowds well due to the large individual differences.

Of course, more work has been done in this area. As said, Helbing [7] has proposed a model based upon particle physics that is able to describe panic in crowds, the model is able to generate realistic crowd behavior, but has not been evaluated for its predictive capabilities. In [1] a comparison has been made regarding the descriptive capabilities of the models, showing that ASCRIBE is able to describe this behavior best. In [9] a comparison between three models is made (ASCRIBE, an epidemiological-based Durupinar model [6], and the ESCAPE model) by evaluation of the descriptive capabilities, showing that ASCRIBE perform superior. Again, no evaluation based on the prescriptive capabilities is made, which is the main focus of this paper, nor have the parameters been rigorously tunes as done here.

# References

1. Bosse, T., Hoogendoorn, M., Klein, M.C.A., Treur, J., van der Wal, C.N.: Agent-Based Analysis of Patterns in Crowd Behaviour Involving Contagion of Mental States. In: Mehrotra, K.G., Mohan, C.K., Oh, J.C., Varshney, P.K., Ali, M. (eds.) IEA/AIE 2011, Part II. LNCS (LNAI), vol. 6704, pp. 566–577. Springer, Heidelberg (2011)
2. Braun, A., Musse, S.R., de Oliveira, L.P.L., Bodmann, B.E.J.: Modeling Individual Behaviors in Crowd Simulation. In: The 16th International Conference on Computer Animation and Social Agents, CASA 2003, pp. 143–147. IEEE Press, New Jersey (2003)
3. Marquardt, D.: An algorithm for least squares estimation of non-linear parameters. J. Ind. Appl. Math., 431–441 (1963)
4. Damasio, A.: The Somatic Marker Hypothesis and the Possible Functions of the Prefrontal Cortex. Phil. Transactions of the Royal Society: Biological Sciences 351, 1413–1420 (1996)
5. DeJong, K.A., Spears, W.M.: An Analysis of the Interacting Roles of Population Size and Crossover in Genetic Algorithms. In: Schwefel, H.-P., Männer, R. (eds.) PPSN 1990. LNCS, vol. 496, pp. 38–47. Springer, Heidelberg (1991)
6. Durupinar, F.: From Audiences to Mobs: Crowd Simulation with Psychological Factors. PhD dissertation, Bilkent University, Dept. Comp. Eng. (July 2010)
7. Helbing, D., Farkas, I., Vicsek, T.: Simulating Dynamical Features of Escape Panic. Nature 407(6803), 487–490 (2000)
8. Hoogendoorn, M., Treur, J., van der Wal, C.N., van Wissen, A.: Modelling the Interplay of Emotions, Beliefs and Intentions within Collective Decision Making Based on Insights from Social Neuroscience. In: Wong, K.W., Mendis, B.S.U., Bouzerdoum, A. (eds.) ICONIP 2010, Part I. LNCS (LNAI), vol. 6443, pp. 196–206. Springer, Heidelberg (2010)
9. Tsai, J., Bowring, E., Marsella, S., Tambe, M.: Empirical evaluation of computational emotional contagion models. In: Vilhjálmsson, H.H., Kopp, S., Marsella, S., Thórisson, K.R. (eds.) IVA 2011. LNCS (LNAI), vol. 6895, pp. 384–397. Springer, Heidelberg (2011)
10. Tsai, J., Fridman, N., Bowring, E., Brown, M., Epstein, S., Kaminka, G., Marsella, S., Ogden, A., Rika, I., Sheel, A., Taylor, M.E., Wang, X., Zilka, A., Tambe, M.: ESCAPES - Evacuation Simulation with Children, Authorities, Parents, Emotions, and Social comparison. In: Tumer, K., Yolum, P., Sonenberg, L., Stone, P. (eds.) Proceedings of the 10th International Conference on Autonomous Agents and Multiagent Systems, AAMAS 2011 (2011), Innovative Applications Track
11. Levenberg, K.: A method for the solution of certain problems in least squares. Quart. Appl. Math. 2, 164–168 (1944)

# Analyzing Grid Log Data with Affinity Propagation

Gabriele Modena and Maarten van Someren

Universiteit van Amsterdam
Amsterdam, The Netherlands

**Abstract.** In this paper we present an unsupervised learning approach
to detect meaningful job traffic patterns in Grid log data. Manual anomaly
detection on modern Grid environments is troublesome given their increasing complexity, the distributed, dynamic topology of the network
and heterogeneity of the jobs being executed. The ability to automatically
detect meaningful events with little or no human intervention is therefore
desirable. We evaluate our method on a set of log data collected on the
Grid. Since we lack a priori knowledge of patterns that can be detected
and no labelled data is available, an unsupervised learning method is followed. We cluster jobs executed on the Grid using Affinity Propagation.
We try to explain discovered clusters using representative features and we
label them with the help of domain experts. Finally, as a further validation step, we construct a classifier for five of the detected clusters and we
use it to predict the termination status of unseen jobs.

## 1  Introduction

Grid computing provides a paradigm for combining on-demand computational
and storage power of geographically distributed groups of computers. Resources
are shared within a homogeneous group of users and used to solve a common
goal. This computing methodology is used by research and industrial institutions
to approach computational problems that are untreatable on a single machine
or even an HPC cluster [3]. There is a need to improve the stability of grid
systems. System administrators need to manually analyze and correlate large
amounts of log data, coming from multiple sources, to find possible causes of
a problem. This is a challenge because of the complex and dynamic settings
of those environments: there are many components and interactions; resources
may join and leave at any time; resources are shared by multiple organizations;
resources are heterogeneous and distributed; the components undergo continuous
improvements and changing standards; the kind computational problems being
addressed on a Grid characterizes its job traffic.

A number of patterns can be seen as symptoms of anomalies [11]. These
include: job abortion rate, resubmitted jobs rate, cross-domain traffic patterns,
resource starvation, unreachable or black hole nodes. Observing such patterns
in isolation, by means of monitoring and logging tools, is a necessary but not
sufficient condition for detecting and addressing anomalies.

M. Ali et al. (Eds.): IEA/AIE 2013, LNAI 7906, pp. 83–91, 2013.

We define clusters of job traffic patterns that share common properties as *events*. Our goal is to detect events that could indicate anomalies or failures in the environment. Events detection needs to be carried out without explicit human analysis of the logs and manual correlation of data. Therefore, we lack a priori knowledge of the kind of patterns that can be detected in log data at a given time, nor we know how they relate to each other in forming anomalous situations. Since no categories, and thus labelled data, are available an unsupervised learning method is followed. In this paper we represent the lifecycle of the job in the environment as a feature vector and use *Affinity Propagation* [5] to cluster job traffic. We consider the emerging clusters as candidates meaningful events in the environment and we try to explain them in terms of representative features of the data selected via Recursive Feature Elimination (RFE) [6]. The detected clusters have been validated and labeled with the help of domain experts. As a further validation step, we report the accuracy of a prediction experiment in which a classifier is trained to map unseen jobs to a selection of five previously detected clusters.

This paper is organized as follows. In Section 2 we introduce key properties of the grid infrastructure and of the jobs being executed. In Section 3 we present an unsupervised learning method based on Affinity Propagation to cluster patterns in job traffic. In Section 4 we report the results of experiments performed using data from the EGEE grid and we position our research among related work. Section 5 concludes the paper.

## 2   Grid Infrastructure

For the purpose of discovering meaningful events in log data, we observe the grid as a federation of distributed computer clusters, deployed in decentralized administrative domains. We describe this environment using the terminology and conventions adopted by the gLite middleware[1]. At the highest level of abstraction we want to analyze the behaviour of three types of entities: computing elements (CEs), resource brokers (RBs) and users (Us). At a lower level of abstraction we are interested in describing a job lifecycle in the environment. A job can be in any of the following states: submitted, waiting, ready, scheduled, running, done, cleared, aborted, cancelled [4]. Each state is associated with the job presence on or transition to a grid component. A job is first submitted by a user, then waiting for a Resource Broker (RB) to find a matching Computing Element (CE). Once a CE is found, the job is first marked ready for transfer, then transferred to the resource and scheduled (inserted in the local batch queue). When selected from the queue, the job is running, until successful termination (done: OK), or failure (done: failed). Upon failure, the job is resubmitted and goes through the whole process. Early termination (aborts or cancels) triggered by either the user or the network components can occur at any step in the job lifecycle.

---

[1] http://glite.cern.ch/

# 3    Job Traffic Analysis

## Data Collection

The dataset used in this work has been provided by the Grid Observatory. The Grid Observatory is part of the EGEE-III EU project INFSO-RI-222667, an open project that collects, publishes and analyzes data on the behaviour of the European Grid for E-science (EGEE). We analyze a job centric database of log data recorded at the GRIF/LAL site of the EGEE Grid. These logs contain information about the lifecycle of jobs submitted to GRIF/LAL in the period from 1/1/2008 to 31/12/2008. Data is provided as a MySQL dump of the Logging&Bookkeeping log service. From this database we obtained a year worth of log information. An extensive analysis of the EGEE-II Grid, of which GRIF/LAL is a subset, can be found in [8].

## Data Preparation

We view the environment from the perspective of a job as it is traced by the logging service. We extract the transitions from and to components of the grid, the job status at each transition and job requirements specified by a user. We discretize these attributes and represent a job lifecycle as a feature vector. We extend the job centric view of database with some attributes of the environment resources (brokers, computing elements and users) that a job interacts with during its lifecycle. We do so by counting the number of jobs recorded by these components at the different stages of the lifecycle described in Section 2. We consider jobs in the submitted, scheduled, transferred, aborted, terminated and done status. An example of these features is shown in Table 1. Assuming that brokers and computing elements that process an higher number of jobs than the others play a more important role in the system, we weight the performance of each component by an estimate of their importance. The resulting vector for a job consists of over 40 features. A full list and description is provided in an external appendix[2].

## Method

We use *Affinity Propagation Clustering* (APC) [5] to group patterns of job lifecycles. APC is an application of the belief propagation method [12]. The algorithm operates by viewing each data point as a node in a network and then recursively transmits real-valued messages along the edges that convey how well a data point would fit into a cluster with a different data point as center. The algorithm identifies exemplars among data points and forms clusters around them. Such information is then sent across the network and, at each iteration, exemplars and clusters are updated locally. The message passing procedure may terminate after a given number of iterations, after changes in the messages fall below a certain threshold, or when local decisions about exemplars and clusters remain constant for some number of iterations. A key property of this method is

---

[2] https://www.dropbox.com/s/pcbeomuqeu9wnh4/ieaaie13_appendix.pdf

**Table 1.** Examples of jobs and resource brokers attributes

| Label | Element | Description | Type |
|---|---|---|---|
| job_length | job | total time spent by the job in the system | int |
| job_path_length | job | number of transitions to components | int |
| job_last_recorded_exit | job | latest status code for the job | int |
| job_last_seen | job | last component visited by a job | int |
| resubmission_count | job | number of times a job has been resubmitted | int |
| rb_num_jobs | broker | number of incoming (submitted) jobs | int |
| rb_resubmitted_jobs | broker | number of incoming (submitted) jobs | int |
| rb_scheduled_jobs | broker | number of outgoing (scheduled) jobs | int |
| rb_done_jobs | broker | number of successfully terminated jobs | int |
| rb_aborted | broker | number of aborted jobs (by the system) | int |
| rb_avg_job_p_time | broker | average job permanence time in the broker queue | float |
| rb_rc | broker | number of jobs resubmitted by the broker | int |
| ce_num_jobs | CE | number of incoming (submitted) jobs | int |
| ce_done_ok_jobs | CE | number successfully terminated (done: ok) jobs | int |
| ce_average_job_p_time | CE | average job permanence time in the CE queue | float |

that the number of clusters to detect does not need to be specified a priori. This enables automatic model selection and serves our purpose of detecting patterns when we don't know the number of categories, and thus labels, present in the data.

APC takes as input a set of real-valued similarities between data points. The similarity $s(i,k)$ indicates how well the data point with index $k$ is suited to be the exemplar for data point $i$. For each data point $k$ the method takes as input a real number $s(k,k)$ such that data points with larger values of $s(k,k)$ are more likely to be chosen as exemplars. These values are called preferences. The number of identified exemplars (number of clusters) is influenced by the input preferences and the message passing procedure. Messages can be combined at any stage to decide which points are exemplars (responsibility) and, for every other point, which exemplar it belongs to (availability).

A single iteration involves computing all responsibility messages based on the current availability messages, the input similarities and the input preferences. Finally all availability messages are computed based on the responsibility messages. Computing responsibilities and availabilities according to simple update rules will often lead to oscillations. To address this problem the messages are damped so that $msg_{new} = \lambda msg_{old} + (1 - \lambda)(msg_{new})$. Where the damping factor $\lambda$ is between 0 and 1. The algorithm monitors the exemplar decisions that would be made after each iteration and if these don't change over a certain threshold (*convits*) the procedure terminates. If the maximum number of allowed iterations (*maxits*) is reached the procedure terminates in any case, eventually without converging to a solution.

We used Manhattan distance to determine the similarity between jobs. We initialize the value for preferences to the median similarity. Empirically we observed that the best cluster separation is obtained by setting the damp factor to 0.95, the maximum number of iterations to 4000 and convergence iterations threshold to 400.

On each of the discovered clusters we apply a feature selection algorithm, RFE, to identify predictive features. The purpose of this step is to explain and label clusters in terms of relevant features of the jobs belonging to it. The method we used operates by first training an SVM on a given training set (a detected cluster). Features are then ordered using the weights of the resulting classifier and the ones with smallest weight are eliminated. The process is repeated with the training set restricted to the remaining features. In this work we used the RFE implementation provided by the mlpy package [2].

## 4    Evaluation

We evaluate our method on a random sample of the log database with entries recorded over 7 days. In this sample we observe 2356 unique jobs, 16 users, 18 brokers and 56 computing elements. In this dataset we observe an average job path length of 4 hops, average job length of 7.5 hours and an average job abortion rate close to 50%. These metrics are in line with the measures reported on a large scale analysis of the EGEE-III Grid given in [8], of which GRIF/LAL is a subset. We further consider a single day of activity and we collect data at time intervals of size $\Delta_t = 1, 4, 8, 12, 16, 24$ hours each. For each time window we update the feature vectors by re-computing the attributes of jobs, users, brokers and computing elements with the information available.

With this setup we perform the following experiments. First we compare running Affinity Propagation Clustering on the dataset as a whole and on the daily samples. We want to observe how the number and size of cluster differs when data is collected over time spans of different length. At the same time we want to identify what is a good amount of time for detecting meaningful events. We then perform RFE on each cluster. We submit the identified clusters and the selected features to two domain experts for evaluation and to label the clusters. Once clusters are labelled we use them in a prediction task. We are interested in determining if a job that has not yet terminated is more likely to complete successfully or fail due to an anomaly. The goal of this experiment is to further validate the predictive quality of the detected clusters.

**Clustering Experiment**
When using 1 to 4 hours worth of data the algorithm detected 2 clusters. With and 12 to 24 hours, 12 clusters are detected. With the 8 hours sample, 5 clusters

have been detected. The jobs in the dataset are 723 for the one hour sample and their number remains constant to 1112 within 24 hours. 2356 jobs are present in the whole week sample and 16 clusters have been detected. We observe a change in the number of clusters before and after 8 hours. This may be interpreted as a symptom of a change in the environment or job status. Moreover we observe a change in the total number of elements in each clusters at 8, 16 and 24 hours. This may also be interpreted as a change in job status (previously running jobs may have terminated or been cancelled). We asked two domain experts, a Grid scientist and a network administrator, to evaluate the results of our method. We supplied them with a description of the clusters in terms of features selected by RFE as well as relevant log information about the transitions from component to component (network hops) and job status. Two of the clusters were found to represent normal (good) situations on the network. Four are anomalies of which two seem to be user related problems. Three clusters were difficult to explain but possibly interesting (worth investigating) the remaining cases were not understandable or not meaningful. Domain experts validated our interpretations of clusters detected using different time windows and reported that the most meaningful and easy to interpret clusters are the ones discovered using 8 to 24 hours worth of data.

**Prediction Experiment**

As a final step we want to predict if an unseen, not yet terminated, job is likely to fail or terminate successfully. In this experiment we use the clusters detected in the unsupervised learning step to train a multi-class classifier. Our goal is to map unseen jobs to events discovered by Affinity Propagation Clustering and further validate the predictive quality of the clusters. We selected 5 out of the 12 detected by APC. Such clusters are the ones that domain experts have been able to identify and label most clearly. They have been recognized to carry valuable information in regard to the network status. For this experiment we consider a dataset of 991 jobs. A summary of the five classes is given in Table 2. Table 3 depicts the five clusters in terms of the characteristic features selected by feature elimination.

**Table 2.** Termination classes. The thrid column indicates the number of jobs belonging to each class.

| C_1:    ok job | Successfully terminated jobs | 557 |
|---|---|---|
| C_2: failed #1 | Describes a failure involving computing elements and brokers | 189 |
| C_3: failed #2 | High job resubmission count and low resubmission time intervals. Possible black hole node. | 133 |
| C_4: failed #3 | This cluster seems to represent a user related problem; maybe a user submitting jobs to a broker unable to match her requirements. | 11 |
| C_5: failed #4 | Anomaly possibly caused by resource exhaustion on computing elements | 101 |

**Table 3.** Job clusters. Representative features are listed in the second column.

| cluster | characteristic features |
|---------|------------------------|
| C_1 | job_length, user, rb_num_jobs, ce_nu m_jobs |
| C_2 | job_length, rb_num_jobs, ce_num_jobs |
| C_3 | resubmission_count, resubmission_time_interval, rb_average_job_permanence |
| C_4 | user, rb_aborted, rb_terminated, rb_average_job_permanence, rb_num_jobs |
| C_5 | ce_average_job_permanence, average_terminated, ce_num_jobs |

We model prediction as a classification task using a Support Vector Machine (SVM). We compare the accuracy of the classifier in boolean (successful vs. failed jobs) and multi class (successful vs. four failure classes) scenarios. Multi class classification is carried out as a pairwise comparison of the SVM output [1]. We use 75% of the previously clustered data points for training the classifier and 25% for testing. We reported a classification accuracy of 0.84 for the boolean case and 0.59 for the multi class case.

## 5  Related Work

We have been able to discover and characterize clusters that match failure classes reported in [11] without requiring any explicit expert knowledge nor time consuming human analysis of the logs. Our work has similarities with [9], where the authors use rule extraction to find associations in log data to detect faulty components. In this paper we address the problem of detecting anomalies as a clustering task and look at general events rather than isolated patterns.

There are more papers researching the problem of detecting anomalies in Grid log data. In [10] the authors propose an agent based approach to analyze cross domain traffic patterns. In [7] the authors suggest a symptoms based model to diagnose malfunctioning resources in the Grid. The method identifies indicators of possible failure and extracts a set of rules to match observations from log files to diagnose situations of potential malfunctioning (symptoms). The work is focused on low level, domain specific job information that may not be feasible to observe in a cross-domain setup. Our approach operates at a higher level of abstraction and considers the Grid as a whole, without concentrating on a specific site and without requiring a priori knowledge of the types of anomalies that are to be found in the dataset. In [13] a method is proposed that mines the Logging&bookkeeping database with k-means and three classes of failures labelled a priori. The authors report that additional clusters unknown to the algorithm have been discovered. These new clusters are characterized as mixtures of data points that show features of the known failure classes and are labelled accordingly. In our approach we observe the emergence of clusters withouth the need of specifying the number of clusters to detect a priori, nor to include domain knowledge about the type of failures in the clustering step. We are able to provide a tailored and refined overview that includes information about components of the environment. A further contribution of our work is the validation of detected clusters by domain experts.

## 6    Conclusion

In this paper we used an unsupervised learning method, Affinity Propagation Clustering, to discover meaningful events in Grid log data. Feedback and validation received from domain experts on the experiments we reported is encouraging. We have been able to automatically recognize anomalies without manual intervention. In a further experimet aimed at validating the predictive quality of the detected clusters, we show how clusters detected by APC can be used to train a classifier to identify jobs likely to terminate correctly or fail. We reported accuracy better than random choice for both boolean and multi-class classification tasks.

Clusters detected by APC can be used to assist decision making in troubleshooting the environment by providing domain experts with an overview of the events that happened on the network within different time windows. Automated event discovery results in less human time to be spent in analyzing and correlating log data. We plan to extend our work as follows. First we want to collect more data withn the considered time windows to scale up our experiments and possibly analyze different type of Grid environments, with different job traffic patterns. In future work we will also investigate methods to model time and components interaction more explicitly.

## References

1. Abe, S.: Support Vector Machines for Pattern Classification (Advances in Pattern Recognition). Springer-Verlag New York, Inc., Secaucus (2005)
2. Albanese, D., Visintainer, R., Merler, S., Riccadonna, S., Jurman, G., Furlanello, C.: mlpy: Machine learning python (2012)
3. Baker, M., Buyya, R., Laforenza, D.: Grids and grid technologies for wide-area distributed computing. Software: Practice and Experience 32, 1437–1466 (2002)
4. Laure, E., et al.: Programming the Grid with gLite. Technical Report EGEE-TR-2006-001, CERN, Geneva (March 2006)
5. Frey, B.J., Dueck, D.: Clustering by passing messages between data points. Science 315, 972–976 (2007)
6. Guyon, I., Weston, J., Barnhill, S., Vapnik, V.: Gene selection for cancer classification using support vector machines. Mach. Learn. 46(1-3), 389–422 (2002)
7. Hofer, J., Fahringer, T.: Grid application fault diagnosis using wrapper services and machine learning. In: Krämer, B.J., Lin, K.-J., Narasimhan, P. (eds.) ICSOC 2007. LNCS, vol. 4749, pp. 233–244. Springer, Heidelberg (2007)
8. Ilijašić, L., Saitta, L.: Characterization of a computational grid as a complex system. In: Proceedings of the 6th International Conference Industry Session on Grids Meets Autonomic Computing, GMAC 2009, pp. 9–18. ACM, New York (2009)
9. Maier, G., Schiffers, M., Kranzlmueller, D., Gaidioz, B.: Association rule mining on grid monitoring data to detect error sources. Journal of Physics: Conference Series 219(7), 072041 (2010)

10. Mulder, W., Jacobs, C.: Grid management support by means of collaborative learning agents. In: Proceedings of the 6th International Conference Industry Session on Grids Meets Autonomic Computing, GMAC 2009, pp. 43–50. ACM, New York (2009)
11. Neocleous, K., Dikaiakos, M.D., Fragopoulou, P., Markatos, E.P.: Failure management in grids: the case of the egee infrastructure. Parallel Processing Letters 17(4), 391–410 (2007)
12. Yedidia, J.S., Freeman, W.T., Weiss, Y.: Understanding belief propagation and its generalizations. In: Exploring Artificial Intelligence in the New Millennium, pp. 239–269. Morgan Kaufmann Publishers Inc., San Francisco (2003)
13. Zhang, X., Sebag, M., Germain, C.: Toward behavioral modeling of a grid system: Mining the logging and bookkeeping files. In: Proceedings of the Seventh IEEE International Conference on Data Mining Workshops, ICDMW 2007, pp. 581–588. IEEE Computer Society, Washington, DC (2007)

# Using Ising Model to Study Distributed Systems Processing Capability

Facundo Caram, Araceli Proto, Hernán Merlino, and Ramón García-Martínez

Facultad de Ingeniería, Universidad de Buenos Aires, Argentina
Grupo de Investigación en Dinámica y Complejidad de la Sociedad de la Información y Grupo
de Investigación en Sistemas de Información, Departamento Desarrollo Productivo y
Tecnológico, Universidad Nacional de Lanús, Argentina
{aproto,rgarcia}@unla.edu.ar

**Abstract.** Quality of service based on distributed systems must be preserved
throughout all stages of life cycle. The stage in which this feature is critical is in
stage planning of system capacity. Because this is an estimate, the traditional
approach is based on the use of queues for capacity calculation. This paper
proposes the use of Ising traditional model to capacity study.

**Keywords:** Distributed Systems, Processing Capability, Ising Model.

## 1 Introduction

Capacity planning of distributed systems presents the challenge of having to give
an estimate of the resources used by the system based on the requirements defined
at an early stage of life cycle system. To address this, the community has used
queuing networks models (Network Queuing Models - QN) QN models are a set
of interconnected queues, each of which includes the waiting time to meet every
user [1].

The Users move between these queues to complete your request. The input
parameters for a QN model are divided into two groups: [a] load current: this
parameter provides the system load at any given time, [b] service demand: is the
average time of service provided by a resource.

One important aspect of QN models is performance system. Technologies for
building software systems such Web Services, XML RPC, Grid Computing allow
building more robust and complex systems, but use of these technologies make
capacity planning more complex, and the community needed tools to make a precise
estimation, for this in this paper authors present an alternative to estimate the
performance of a distributed system based on the traditional Ising model.

In distributed architecture each component has a finite processing capacity, these
components may be redundant, in this particular, traditional Ising model introduces a
mechanism for resource cooperation. Through the simulation are determined "virtual"
temperature, magnetization and critical system parameters.

The Performance Engineering (PE) encompasses a set of tasks, activities and
deliverables that are applied to each phase of the life cycle in order to achieve quality

M. Ali et al. (Eds.): IEA/AIE 2013, LNAI 7906, pp. 92–101, 2013.

of service as detailed in the non-functional requirements. PE is a collection of methods that supports the development of performance oriented systems over the entire life cycle. In performance-oriented methodologies, the different stages of the life cycle are linked with models of engineering performance: (a) load model: simulates the real burden to be borne by the system, (b) performance model: is the response time to be answered a job application with respect to a given workload and (c) availability model: is the standard by which to evaluate how long the system is available to work.

## 2    Distributed Systems Modeling Using Ising

Our interest is to find an analogy of distributed systems based on grid computing with a well-known physics model [3]. We believe that the Ising model [4] is particularly suitable for this purpose as described below.

In our model each site or traditional spin Ising model represents a cell or computer processing grid-computing. The computers with available resources are considered if = -1 (black site), and those without computers or resources are saturated with if = +1 (target site), (see Figure 1).

**Fig. 1.** Cell's black have available resources, cell's white not, these are congested

Energy of the Ising system in specified configuration {si} is defined as:

$$E_I\{s_i\} = -\sum_{\langle ij \rangle} e_{ij}s_is_j - H\sum_{i=1}^{N} s_i$$

Where the subscript I refer to Ising and symbol <ij> denotes a pair of nearest neighbor spins; don't have distinction between <ij> and <ji>. Thus the sum of <ij> has $\gamma.N/2$ terms, where $\gamma$ is the number of nearest neighbors of any given site and the geometry of the grid (lattice) is described by $\gamma$ y $e_{ij}$. Situation very close to reality is use a grid of Ising LxL, with all $e_{ij}$= e, $\gamma$ = 4 and H= 0. Time between service requirements (TMER), corresponds to a temperature value "virtual" grid, calculated using the parameter estimation technique [7, 8].

## 3    Parameter Estimation in Ising Model

If defined $\sigma_{i,j} = (s_{i-1,j} + s_{i+1,j} + s_{i,j-1} + s_{i,j+1})$ and calculate the conditional probability for a fixed site at position i, j, we obtain the following:

$$P\left(s_{i,j}/\{s_{i,j}\} - s_{i,j}\right) =$$
$$P\left(s_{i,j}/s_{i-1,j}, s_{i+1,j}, s_{i,j-1}, s_{i,j+1}\right) =$$
$$= \frac{\exp\left(\beta s_{ij}\sigma_{i,j}\right)}{\exp\left(\beta\sigma_{i,j}\right) + \exp\left(-\beta\sigma_{i,j}\right)} \tag{1}$$

The $\beta$ parameter estimate is made by using a common technique of parameter estimation in Markov random fields [5], which is the maximum like hood estimator pseudo [7, 9]. This criterion states that $\hat{\beta}$, (the estimator $\beta$) is one that maximizes the following product of probabilities:

$$\prod_{i,j=1}^{N} P\left(s_{i,j}/s_{i-1,j}, s_{i+1,j}, s_{i,j-1}, s_{i,j+1}\right) \tag{2}$$

As it is usual to apply the natural logarithm of the above expression, using the expression 1, we achieve the following objective function parameter $\beta$:

$$J\left(\beta\right) = \sum_{i,j=1}^{M} \left[\beta s_{ij}\sigma_{i,j} - \ln\left(2\cosh\left(\beta\sigma_{i,j}\right)\right)\right] \tag{3}$$

Now we are defined respect to $\beta$, and the above expression is found that the value of $\hat{\beta}$ makes $J(\hat{\beta}) = 0$.

$$\frac{dJ\left(\beta\right)}{d\beta} = \sum_{i,j=1}^{M} \left[s_{ij}\sigma_{i,j} - \sigma_{i,j}\tanh\left(\beta\sigma_{i,j}\right)\right] = 0 \tag{4}$$

To analyze the expression achieved and develop an algorithm for the estimation of $\hat{\beta}$, we make the following definitions novice: is $N_\alpha$ the number of sites for which the sum of its neighbors is equal to $\alpha$. Note that the possible values for $\alpha$ are restricted to $0$; $+2$; $-2$; $+4$ y $-4$ and of course we have to $L^2 = N_0 + N_{-2} + N_{+2} + N_{-4} + N_{+4}$. Now we decompose sum in 4 as follows:

$$\underbrace{-2\sum^{N_{-2}} s_{ij} + 2\sum^{N_{+2}} s_{ij} - 4\sum^{N_{-4}} s_{ij} + 4\sum^{N_{+4}} s_{ij} -}_{A}$$

$$\underbrace{-2\left(N_{-2} + N_{+2}\right)\tanh\left(2\beta\right)}_{B}$$

$$\underbrace{-4\left(N_{-4} + N_{+4}\right)\tanh\left(4\beta\right)}_{C} = 0 \tag{5}$$

Where the notation $\sum^{N\alpha} s_{ij}$ means taking the sum of all sites whose neighbors join $\alpha$. Numbers A, B, and C are quantities that can easily calculate the sample $\{s_{i,j}\}$ and the estimator is obtained as the $\hat{\beta}$ satisfies:

$$A - 2B\tanh(2\beta) - 4C\tanh(4\beta) = 0 \tag{6}$$

Once we have calculated A, B, and C, find the zero function using a numerical technique such as the optimal method Myller [6].

# 4     Simulations and Results

This section presents the modes of operation of a computer network as a grid (section 3.1), an analysis of the cooperative and non-cooperative scenarios (section 3.2), and discusses the dynamic scenario (section 3.3).

## 4.1     Modes of Operation of a Computer Network as a Grid

To model operation of a computer network (grid computing - GC) in a 2D Ising grid, author simulates the following cases: when the cells (computers) interact with their nearest neighbors "cooperative case" when cells don`t interact "non-cooperative case" (normal operation), and finally when no predetermined relationships between cells and these demand dynamically among its neighbors to find a free resource, temperature grid evolution $(1/\beta)$ as a function of time between requirements transactions. It is also observed that when modeling the cooperative case, the number of lost requirements is lower. In latter case establishing a rule demands that gives rise to a type adaptive dynamic interaction; when a computer is congested, start looking among its first neighbors with available resources. If found a neighbor with availability, it's establishing a link between two that remain fixed for future operations. Failure to find any resources available neighbor, the computer will lose that requirement and remain unattached until the next time the need for a new search for resources to process requirements. Thus as a random graph in [3] we see that the network evolves or grid, for example at the beginning of the simulation computer is not previously connected with each other. Then during the processing requirements, according to the level of congestion, computers begin to look for resources in their neighbors. This produces a transition in the operation state of the grid, which generates links between computers. The number of possible links for each computer in this model goes from zero to four. Rate of generation of links depends on TMER. Once this is overcome transient and steady state is reached, most computers come to have four links created and the grid behaves as if in a cooperative scenario. The difference with pure cooperative case is that this scenario almost cooperative was created dynamically based on demand. Dynamically network involve static scenario (without cooperation) to start another static scenario (with cooperation) in ready state.

## 4.2     Analysis of Cooperative and Non-cooperative Scenarios

To evaluate temperature $(1/\beta)$ of grid explained before, we define the network parameters: number of computers (cells), amount of computer resources (processing power), number of requirements, mean time between requirements and mean duration of processing of each request, generating the same amount of requests for each computer, using an exponential distribution for the arrival of requirements. The duration of each requirement is defined by a normal distribution. Then enter TMER different values for both scenarios. Finally these parameters with the parameter β are determined temperature corresponding to "virtual" grid in each case. In Figure 2 have different values of the parameter β according to each value of TMER. Besides the cases with and without cooperation, in the same figure also shows the theoretical value of the critical temperature of the Ising model. In Figure 3 you can see the percentage of unprocessed requests for both cases.

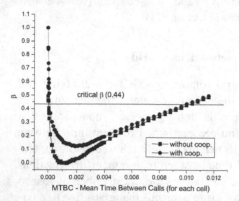

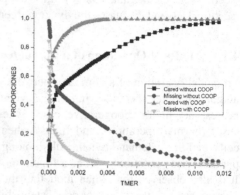

**Fig. 2.** β is TMER function, for cases with and without cooperation, versus critical temperature of Ising model, $\beta = 0; 44$. Tested with Lattice of 25x25 computers with 10 resource each one and requests media of 0,1 and dispersion of 0,05.

**Fig. 3.** Proportions of missing requirements and cared for both scenarios

Figure 3 shows that there are three zones or regions; one when the TMER ≤ 0.0001, and is very short in relation to the amount of resources and computers, and the average duration of each requirement. This situation is reflected in the case of a high rate of arrival of requests, so high that the grid processing cannot reach in time, due to lack of capacity and resources therefore quickly becomes congested, we call this situation, by high congestion magnetization (magnetization 2). There is another area where TMER ≤ 0.0001 ≤ 0.01, where the grid is within the range of operation, no magnetization area. Finally this zone where 0.01 ≤ TMER, here the grid is never high congestion, however this is on average rather low or zero. This situation is called for vacancy magnetization (magnetization 1), because the network has almost all its resources to process requests that are presented. All states are shown for each value stationary TMER. It is easy to see that in the non-cooperative case, the percentage of lost or unattended requirements is high and is always above the number of un-served requirements for the cooperative scenario, which is a fairly logical. Analyzing the cooperative case, because of its relevance, you can clearly see that for low values of TMER, where the grid is very congested, the number of missing requirements is very high in both cases, although it is higher in case non-cooperative. After this analysis, we can conclude that the cooperative scenario is more convenient, because the number of missing requirements is lower than for non-cooperative as the simulation shows.

## 4.3    Dynamic Scenario Analysis

In this section we simulate the behavior of the grid when there is no predefined neighboring links in the initial state. Here the links are generated according to the need of each computer and by TMER. We can see the evolution of the state grid non-cooperative to cooperative state, according to the values of TMER. Therefore the bonds are generated in a dynamic manner according to requirements and demand as a function of the parameter TMER (see Figure 4). During simulation it's taking pictures of the evolution of the grid during the search process explained above, and observing a sample TMER for different values, we can determine how long the transition for each value of TMER (see Figures 5 and 6).   Graphically this process is a gradual Represented By change of color of black, where the computer does not have ties, through shades of gray to reach the target, where the computer has generated four possible links (see Figure 4).   To know how many links has every computer in every step, we add counters that indicate the number of computers 12 without links L0, the number of computers that have a single link L1 and so on up to four computers with links L4.   Therefore we need only observe the evolution of these counters during simulation to determine when each computer has reached steady state for each TMER. For a simulation of 1500 requirements, have been sampled at times proportional counters, one every 50 requirements for each computer. This operation results in 30 samples. From Figures 5 and 6 shows that the steady state shown between samples 20 and 30 (when the counter reaches L4 almost about 100%). Then for small values of TMER (magnetization 2), at first glance one might say that in a very high level of congestion, the speed of relationships should be very high, but the simulation indicates otherwise. This means that within this range the generation of links is very slow, but the explanation is reflected in the amount of lost requirements, which is about 45% as shown in Figure 7. This interpret rate of arrival of requests is so high that the grid does not have processing power, so the computer does not have enough time to meet all the requirements that are presented. Thus only a few links and computers can generate most of the requirements are discarded.

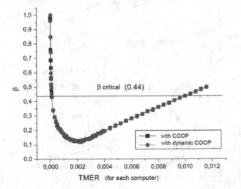

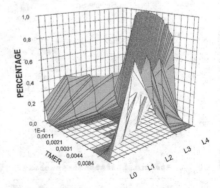

**Fig. 4.** In function TMER, scenarios for static and dynamic cooperation, versus the actual critical temperature in Ising model

**Fig. 5.** Links distribution for different TMER in a 30 size sample

Another way of looking at this phenomenon is to observe the evolution of each counter during the simulation, as shown in Figures 8 and 9, there are some more representative values TMER where the behavior is different grid.

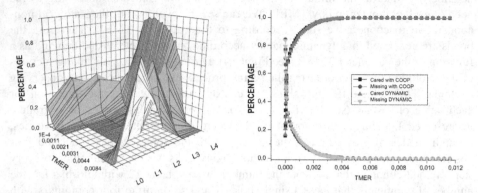

**Fig. 6.** Distribution of Links in 20 samples of different TMER

**Fig. 7.** Proportion of requests serviced and lost in the dynamic scenario

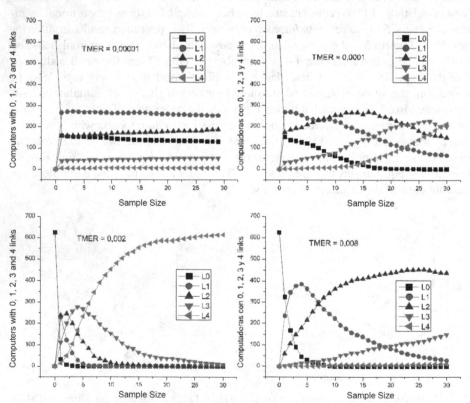

**Fig. 8a.** Links evolution in computers for TMER

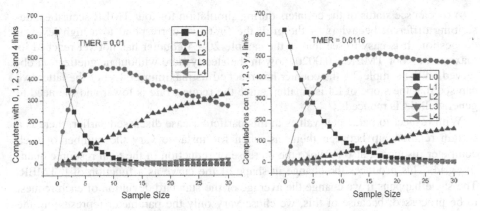

**Fig. 8b.** Links evolution in computers for TMER

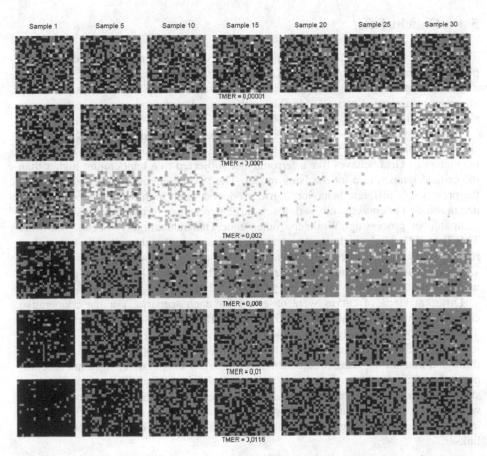

**Fig. 9.** Evolution of grid during processing requirements 1500 on each computer, for 6 representatives TMER

You can see status of the counters during simulation for four TMER securities, describing different behaviors of the grid: the first two correspond to a high level of congestion. It is easy to see that in the sample 20 L4 counter has not yet reached its maximum.   For TMER = 0.0020, now in the steady state without magnetization observed on the sample 20, the counter has reached its maximum L4.  For the latter, we can see that the slope of L4 is smaller, since the request rate is lower and the need to generate links is reduced.

With respect to parameter values in this particular case discussed earlier, there is a certain relationship between them, as when for instance vary the number of each computer resources, also provoke the increase or reduction of the total resource quantity grid. This produces the change in shape of the curve as a function of β TMER. The same happens if we change the average or the standard deviation of each request to be processed; because of this, we chose vary only the parameter representing the rate of arrival of requests, leaving the other parameters fixed.

## 5    Conclusions

The result of this work was to find an analogy between the Ising model and dynamics processing requirements within a grid of computers. It was observed that grid reaches a specific virtual operating temperature as a function of TMER (in this case, leaving all other parameters fixed). To do this, authors used a method of pseudo maximum similarity estimation.

Authors found that, there are three areas of operation listed in this analogy analysis: two of them where the grid is "magnetized" and the other where it is "non-magnetized". Grid behaves like a ferromagnet subjected to a temperature higher than the critical temperature for the Ising model. The two magnetisation zones indicating the presence of different behavior, a grid produced when high traffic supports (Magnetization 2) and consequently will lose a large percentage of requirements, since it is incapable of processing. While the other zone (magnetization 1) occurs when the TMER is sufficiently high, so the congestion level is low and practically only a few requirements are discarded.  States mentioned above was analyzed for two separate scenarios: cooperative and non-cooperative.  In both cases the range is not similar magnetization varies between TMER ≤ 0.0001 ≤ 0.01.

Different kinds of curves were observed for each scenario. For the cooperative case, the minimum was at TMER ≤ 0.0019. For the other case, the minimum was at TMER ≤ 0.0011.  Finally, authors performed an analysis of a dynamic process of relationships, where the grid at the start was completely uncooperative and after going through a transitional state, became almost completely cooperative, since not all computers have generated the four links.  Here, we also identified three zones: the transient state, where the number of un-served requests is very high, and another state where it became clear that the grid is within the area of "no magnetization" and the last state where depending on time elapsed, the grid has generated almost 100% of the links.

# References

[1] Fortier, P., Howard, M.: Computer Systems Performance Evaluation and Prediction. Digital Press (2003)

[2] Menascé, D., Almeida, V., Dowdy, L.: Performance by Design: Computer Capacity Planning by Example. Prentice Hall PTR (2004)

[3] Albert, R., Barabási, A.-L.: Statistical mechanics of complex networks. Reviews of Modern Physics 74, 47–97 (2002)

[4] Huang, K.: Statistical Mechanics, 2nd edn. Wiley (April 1987)

[5] Greaffeath, D.: Introduction to Random Fields. In: Denumerable Markov Chains, pp. 425–458. Springer, New York (1976)

[6] Press, W.H., Flannery, B.P., Teukolsky, S.A., Vetterling, W.T.: Numerical Recipes in FORTRAN: The Art of Scientific Computing, 2nd edn., p. 364. Cambridge University Press, Cambridge (1992)

[7] Bouman, C.A.: ICIP 1995 Tutorial: Markov Random Fields and Stochastic Image Models (1995)

[8] Peierls, R.E.: On Ising´s model of ferromagnetism. Proc. Camb. Phil. Soc. 32, 477 (1936)

[9] Besag, J.: Eficiency of pseudo likely-hood estimation for simple Gaussian Fields. Biometrica 64, 616 (1977)

# Computing the Consensus Permutation in Mallows Distribution by Using Genetic Algorithms

Juan A. Aledo, Jose A. Gámez, and David Molina

Departamento de Matemáticas, Departamento de Sistemas Informáticos
Universidad de Castilla-La Mancha
{juanangel.aledo,jose.gamez}@uclm.es, d.molina@estudiante.uam.es

**Abstract.** We propose the use of a genetic algorithm in order to solve the rank aggregation problem, which consists in, given a dataset of rankings (or permutations) of $n$ objects, finding the ranking which best *represents* such dataset. Though different probabilistic models have been proposed to tackle this problem (see e.g. [12]), the so called *Mallows model* is the one that has more attentions [1]. Exact computation of the parameters of this model is an NP-hard problem [19], justifies the use of metaheuristic algorithms for its resolution. In particular, we propose a genetic algorithm for solving this problem and show that, in most cases (specially in the most complex ones) we get statistically significant better results than the ones obtained by the state of the art algorithms.

**Keywords:** Mallows model, rank aggregation, genetic algorithms.

## 1 Introduction

In this paper we focus on a machine learning problem in which instances in the dataset are *permutations* or *rankings* of $n$ objects. This is a problem which has gained popularity in the last years because it finds applications on several fields: preference lists [14], voting in elections [8], information retrieval [4], combinatorial optimization [3], etc.

Basically, the problem we face is the following one: given a dataset containing $N$ rankings representing the preferences of $N$ judges (or the output of several search engines, etc.), can we tackle these data in a compact way for making some future decisions? That is, can we identify the consensus ranking?

The answer to this question is the well known *rank aggregation problem*, where the goal is to obtain the ranking which best represents all the input ones.

Although different probabilistic models have been proposed to deal with this problem (see e.g. [12]), there is one whose use has gained popularity in the specific literature: the *Mallows model* [18]. This model has certain resemblance to the Gaussian distribution, as it is specified by two parameters: a permutation $\pi_0$ which can be seen as the consensus ranking, and a dispersion parameter $\theta$ (see Section 2.2 for details). Exact computation of these parameters can be done from

M. Ali et al. (Eds.): IEA/AIE 2013, LNAI 7906, pp. 102–111, 2013.

a dataset of rankings, though worst-case exponential time is required because of the NP-hardness of the problem [19].

If $\pi_0$ is known, then $\theta$ can be easily estimated by using a binary search over a suitable real interval. For this reason, many approaches to estimate the Mallows model parameters focus on the problem of computing the consensus ranking. This is the case of the study carried out by Ali and Meila [1], where they analyzed a big deal of heuristics and exact search algorithms to solve this problem. The main conclusion is that for small values of $n$, integer linear programming (ILP) should be used because it returns the exact solution. However, for large values of $n$, in general, an approximate version of the $A^*$ (branch and bound) algorithm presented in [19] is the best choice.

Nevertheless, in that study we missed the use of some metaheuristic approaches. Concretely, the segment of problems with low values of $\theta$ and large $n$ are specially suitable for this kind of algorithms, as our work will show for the particular case of genetic algorithms. In fact, for this values of the parameters, the consensus among the input rankings is small, making the search particularly difficult. In these cases, ILP for attaining the exact solution is infeasible, and also the number of nodes for the branch and bound algorithm $A^*$ grows so much that an approximate version of it must be used. More concretely, this approximate version *limits* the size of the queue of nodes to explore, so increasing the risk of loosing promising paths which eventually could drive to the best solution.

Thus, our aim is to study the applicability of metaheuristics algorithms to the problem of obtaining the consensus ranking. Concretely, as a first approximation we propose to use *genetic algorithms* [13] to guide the search process (see Section 4). In this sense, the authors have developed a comparison study (not included in this paper) among several genetic algorithms in order to determine the most suitable one for solving the rank aggregation problem. As a result of such study, we conclude that the algorithm which we will call GA presents the best behavior (in comparison to a wide family of generic algorithms included in [17]) to solve our particular problem. Then, we compare it with the best algorithms tested in [1], showing that in most of the cases, and specially in the most complex ones, the GA algorithm obtains statistically significant better results.

The paper is organized as follows. In Section 2 we provide background knowledge on the problem under study and the Mallows model. In Section 3 we briefly review related proposals and concretely the outstanding algorithms used in the study carried out in [1]. Section 4 is devoted to describe the details of the POS-ISM genetic algorithm that we propose to approach the problem. In Section 5 we present an experimental study to test the proposed approach and discuss the obtained results. Finally in Section 6 we provide some conclusions.

## 2    Preliminaries

Next we provide the notation to be used and some background.

## 2.1   Ranks/Permutations

Suppose we have $n$ items labeled $1, 2, \ldots, n$ that are to be ranked. Then, any permutation $\pi$ of these items represents a ranking. The space of all the possible rankings agrees with the *Symmetric group* $\mathbb{S}_n$:

**Definition 1.** *(Symmetric Group)[9] The symmetric group $\mathbb{S}_n$ is the group whose elements are all the permutations of the $n$ symbols $\{1, 2, \ldots, n\}$, and whose group operation is the composition of such permutations, which are treated as bijective functions from the set of symbols to itself. Since there exists $n!$ different permutations of size $n$, the order of the symmetric group is $n!$.*

We will use $n$−tuples to represent rankings. Thus, by $\sigma = (x_1, x_2, \cdots, x_n)$ we denote the ranking whose first position is $x_1$, the second is $x_2$, etc, and we use $\sigma(j)$ to denote the $j$-th element of $\sigma$.

To solve the rank aggregation problem and also to estimate the consensus ranking, we need to establish a way for measuring the difference between rankings. To do this we use a *distance* which allows us to know how similar they are. Although different distances are available in the literature, we describe here the *Kendall tau* distance, because it is usually considered for the definition of Mallows distribution.

**Definition 2.** *(Kendall Distance[16]) The Kendall distance $d(\pi, \sigma)$ between two rankings $\pi$ and $\sigma$ is defined as the total number of item pairs over which they disagree. There is disagreement over an item pair $(i, j)$ if the relative ordering of $i$ and $j$ is different in $\pi$ and $\sigma$.*

## 2.2   The Mallows Model

The Mallows model [18] is a distance-based probability distribution over permutation spaces which belongs to the exponential family. Given a distance over permutations, it can be defined by two parameters: the central permutation $\pi_0$, and the spread parameter $\theta$.

**Definition 3.** *(Mallows model) [18] The Mallows Model is the probability distribution that satisfies, for all rankings $\pi \in \mathbb{S}_n$,*

$$P(\pi) = \frac{e^{-\theta \cdot d(\pi, \pi_0)}}{\psi(\theta)}, \tag{1}$$

*where rankings $\pi_0$ and $\theta \geq 0$ are the model parameters and $\psi(\theta)$ is a normalization constant (see [11]).*

The parameter $\theta$ of the Mallows model quantifies the concentration of the distribution around its peak $\pi_0$. For $\theta > 0$, the probability of $\pi_0$ is the one with the highest probability value and the probability of the other $n! - 1$ permutations decreases with the distance from the central permutation (and the spread parameter $\theta$). For $\theta = 0$, we just get the uniform distribution. Because of these properties, the Mallows distribution on the space of permutations is considered analogous to the Gaussian distribution on the space of permutations.

## 2.3   The Kemeny Ranking Problem

As aforementioned, computing the consensus ranking is equivalent to the rank aggregation problem, also known as the Kemeny ranking problem [15].

**Definition 4.** *(Kemeny Ranking Problem) [1]*
*Given a set of $N$ rankings (permutations) $\pi_1, \pi_2, \cdots, \pi_N$ of size $n$, the Kemeny ranking problem consists in finding the ranking $\pi_0$ that satisfies*

$$\pi_0 = argmin_\pi \frac{1}{N} \sum_{i=1}^{N} d(\pi_i, \pi), \qquad (2)$$

*where $d(\pi, \pi')$ stands for the Kendall distance between $\pi$ and $\pi'$. $\pi_0$ in Equation (2) is the permutation that minimizes the total number of disagreements with the rankings contained in the set, and is called the Kemeny ranking of the set.*

Finding the Kemeny ranking is an NP-hard problem for $N \geq 4$ (see e.g. [10]). The problem of computing this raking is nowadays an active field of research and several proposals for its exact computation have been presented, of course, without polynomial running time guarantees. Besides exact algorithms, many of heuristic nature have been also proposed (see e.g. [1]).

In this work we propose to test the use of genetic algorithms to guide the search. As we will see in Section 5, by means of this kind of algorithms we are able to get good solutions to complex problems; more specially to those with large dimension $n$ and small degree of consensus $\theta$.

## 3   Related Works

In this paper we rely on the excellent experimental comparison carried out by Ali and Meila [1]. In their study they compare a wide family of algorithms including exact Integer Linear Programming, Branch and Bound, specific heuristic and voting algorithms, and some others approximate algorithms. From their results/conclusions and taking into account our motivation, we have selected the best ones which we describe below. In order to do that, let us to introduce a data structure which make easier their description.

As in [1], given permutations $\pi_1, \pi_2, \ldots, \pi_N$, the *precedence matrix* $Q = [Q_{ab}]_{a,b=1:n}$ is defined as

$$Q_{ab} = \frac{1}{N} \sum_{i=1}^{N} 1(a \prec_{\pi_i} b), \qquad (3)$$

where $1(\cdot)$ is the indicator function and $\prec_\pi$ means "precedes in the ranking $\pi$". In other words, $Q_{ab}$ represents the fraction of times that item $a$ is ranked before than item $b$ across all the input rankings.

The following algorithms show the best behaviour according to [1]:

- **Integer Linear Programming (ILP).** It is an exact approach based on a mathematical formulation of the problem [6]. We do not include it in our experimentation because it runs out of memory for large values of $n$.
- **Branch and Bound (B&B) Search.** This algorithm relies on the use of the well known $A^*$ algorithm jointly with the admissible heuristics proposed in [19], which guarantees to obtain the exact solution. However, in the worst case the search tree has $n!$ paths and the search becomes intractable. As was showed in [1], it works fine when there is a strong agreement between the rankings $\pi_1, \pi_2, \ldots, \pi_N$, that is, when $\theta$ is large, because in these cases the algorithm only expands a limited number of nodes.
- **B&B with Beam Search.** The same algorithm described above but using a limited size for the queue of nodes to be expanded. In this way, the algorithm does not run out of memory, but now there is a hight risk, specially in complex problems, of pruning good paths. Thus, the algorithm becomes an approximate one, looking for a good tradeoff between memory requirements and accuracy. Several beam sizes are experimented in [1] and according to it we have set the beam size as 1000.
- **CSS.** A graph-based approximate algorithm that implements a greedy version of the *Branch and Bound* method introduced in [5].
- **Borda.** This is an approximate algorithm which computes the sums of the columns of $Q$ (3), i.e. $q_a = \sum_b Q_{ab}$, and then returns the permutation that sorts $q_a$ in descending order [2].
- **DK** This algorithm is a *Branch and Bound* solver for the Integer Program described in [7], enhanced with the improved heuristics presented in [6].

## 4    Proposed Genetic Algorithm

In this work we study the competence of *genetic algorithms* (GAs) [13] to face the problem under study. Undoubtedly, there is room for their application, because due to the global search they carry out, we can expect to find better solutions in complex problems. Of course, we know that more CPU time will be required, but we think this is not a problem, as the estimation/learning of $\pi_0$ can be done *off-line*, being time important only for posterior *inference* over the learnt model.

GAs are the best representative of *evolutionary computation* and they work by maintaining a population of solutions which evolves according to natural selection principles. In practice they are stochastic optimization algorithms where natural selection is guided by three main operators: *selection, crossover* and *mutation*. Figure 1 shows the scheme of a canonical GA. Now, we present the main design decisions and parameter setting for the proposed GA:

- **Individual/Chromosome Representation.** As any ranking can be the consensus ranking $\pi_0$, our chromosomes or potential solutions are permutations of the $n$ items. Therefore we search in $\mathbb{S}_n$, whose cardinality is $n!$.

```
BEGIN GA
  Make initial population at random
  WHILE NOT stop DO
BEGIN
    Select parents from the population.
    Produce children from the selected parents.
    Mutate the individuals.
    Extend the population adding the children to it.
    Reduce the extended population.
END
  Output the best individual found.
END GA.
```

**Fig. 1.** Pseudocode of a canonical GA

- **Fitness Function.** The objective of our GA is to solve the Kemeny ranking problem (2), therefore the fitness of a given individual $\pi$ is:

$$f(\pi) = \frac{1}{N} \sum_{i=1}^{N} d(\pi_i, \pi),$$

- **Population.** Looking for a tradeoff between efficiency and population diversity, we have set the population proportional to the problem dimension (complexity). Thus, our GA will have a population of $k \cdot n$ individuals, $n$ being the number of objects to rank, and $k > 1$ an appropriate integer. After preliminary experiments with different values, we decided to set $k = 20$. The initial population is randomly generated.
- **Selection.** In order to maintain diversity, we use a selection mechanism with low selective pressure, concretely a tournament selection [20] of size 2. Thus, at each iteration we randomly select $k \cdot n$ pairs of chromosomes and the individual of each pair with better (small) fitness is selected.
- **Crossover.** To select the crossover (and mutation) operators we have considered a great deal of choices successfully tested for other problems in the search space of permutations. Concretely, we have experimented with several combinations of crossover-mutation operators taken from the study carried out in [17] for the TSP problem. From our study[1] we finally have chosen the pair of operators POS and ISM for crossover and mutation respectively.
  Roughly speaking, the *Position based crossover operator (POS)* works as follows: it starts by selecting a random set of positions; the values for these positions are kept in both parents; the remaining positions are filled by using the relative ordering in the other parent. For example, consider the parents (1 2 3 4 5 6 7) and (4 3 7 2 6 5 1), and the positions {2, 3, 5} are selected.

---

[1] We have tested different combinations in the design of the GA: several schemes (standard and steady-step), selection mechanisms and crossover-mutation pairs of operators were studied. Because of the lack of space here we only show the results for our *winner* model, letting the full comparison for a long version of this paper.

Then, in the first step children are created as (* 2 3 * 5 * *) and
(* 3 7 * 6 * *), while the unused items have the following relative ordering in
the other parent: (4761) and (1245). In the second step the empty positions
are filled: (4 2 3 7 5 6 1) and (1 3 7 2 6 4 5). The pairs for crossover application
are randomly formed from the selected individuals.

- **Mutation.** ISM (*Insertion mutation*) operator is the one that best combines
  with POS crossover operator according to our study for this problem. It
  randomly chooses an element in the permutation, which is removed from its
  current position and inserted in a new one randomly selected.
- **Next Population Construction.** We use a *truncation* operator, that is,
  the population obtained after crossover and mutation and the previous population are put together in a common pool, and the best $k \cdot n$ adapted
  individuals are selected.
- **Stopping Criterion.** We stop the algorithm when we detect it has converged or is stagnated. Concretely, we stop if after $p$ generations the best
  individual has not changed. After several experiments, we have set $p = 60$.

# 5    Experiments

In this section we describe the experiments carried out to test the goodness of
our proposal with respect to those showing an outstanding behaviour in [1].

## 5.1    Datasets

All the datasets in the experiments have been generated using the Mallows
model. For each case we choose a permutation $\pi_0$ of size $n$ and a value for
$\theta$. Then, $N$ permutations are generated from the resulting Mallows model by
sampling according to the procedure described in [19]. In our case, we have set
$\pi_0$ as the identity permutation $\pi_0 = (1\ 2 \ldots n)$. Regarding $\theta$, we have tested
four different values: 0.2, 0.1, 0.01 and 0.001. Remember that the greater the
value, the stronger the consensus in the data, and therefore the easier to solve
the resulting Kemeny ranking problem. As our goal is to test the algorithm in
complex problems, we have set $n$ to the following four values: 50, 100, 150 and
200. Regarding the number of instances, in all the cases we have set $N = 100$
in order to focus on the Mallows model parameters. In a future work we plan to
study the impact of the number of instances (ranks) in the dataset.

Finally, for each of the 16 combinations of the previous parameters (4 $\theta$'s and
4 $n$'s) we generate 20 different datasets of $N$ instances, in order to average the
results and avoid sampling effects. That is, we experiment with 320 datasets.

## 5.2    Methodology

The GA and the last four algorithms described in Section 3 are run for each one
of 320 generated dataset. In the case of B&B only the approximate version (beam
search) is used, because the exact one runs out of space in most of the cases.

**Table 1.** Mean Kendall distance for each algorithm over different parameter values

| n=50 | | | | |
|---|---|---|---|---|
| | $\theta$=0.2 | $\theta$=0.1 | $\theta$=0.01 | $\theta$=0.001 |
| GA | 18781.5* | 32010.4* | 55876.9* | 56846.9* |
| B&B | 18781.5* | 32010.4* | 55892.8⁻ | 56866.2⁻ |
| CSS | 18834.2⁻ | 32088.3⁻ | 56072.0⁻ | 57049.9⁻ |
| Borda | 18783.7⁻ | 32019.4⁻ | 55991.5⁻ | 56970.1⁻ |
| DK | 18781.6 | 32012.8 | 55958.2⁻ | 56954.6⁻ |

| n=100 | | | | |
|---|---|---|---|---|
| | $\theta$=0.2 | $\theta$=0.1 | $\theta$=0.01 | $\theta$=0.001 |
| GA | 41215.4* | 78802.6* | 215224.7* | 230323.1* |
| B&B | 41255.4* | 78810.2⁻ | 215298.6⁻ | 230498.3⁻ |
| CSS | 41320.1⁻ | 79012.6⁻ | 215745.0⁻ | 231026.6⁻ |
| Borda | 41257.1⁻ | 78827.9⁻ | 215530.1⁻ | 230827.7⁻ |
| DK | 41255.4* | 78805.8⁻ | 215429.4⁻ | 230803.8⁻ |

| n=150 | | | | |
|---|---|---|---|---|
| | $\theta$=0.2 | $\theta$=0.1 | $\theta$=0.01 | $\theta$=0.001 |
| GA | 63717.6* | 126058.3* | 458967.2* | 519673.1* |
| B&B | 63717.7 | 126061.0⁻ | 459091.7⁻ | 520123.3⁻ |
| CSS | 63890.3⁻ | 126417.1⁻ | 459943.1⁻ | 521001.5⁻ |
| Borda | 63724.5⁻ | 126096.4⁻ | 459513.7⁻ | 520699.8⁻ |
| DK | 63717.7 | 126064.5⁻ | 459349.8⁻ | 520834.0⁻ |

| n=200 | | | | |
|---|---|---|---|---|
| | $\theta$=0.2 | $\theta$=0.1 | $\theta$=0.01 | $\theta$=0.001 |
| GA | 86264.8* | 173430.3* | 769227.1* | 923284.0* |
| B&B | 86268.5⁻ | 173439.5⁻ | 769463.9⁻ | 924155.7⁻ |
| CSS | 86515.4⁻ | 173942.9⁻ | 770632.3⁻ | 925365.5⁻ |
| Borda | 86270.7⁻ | 173481.0⁻ | 769999.5⁻ | 925021.0⁻ |
| DK | 86265.0 | 173433.6⁻ | 769713.6⁻ | 925602.1⁻ |

Regarding the GA, because of its stochastic nature, we carry out 5 independent runs for each dataset and the average of the five runs is used for comparison.

The code for B&B, Borda, CSS and DK is the one provided by Ali and Meila [1] and is written in Java. Starting from that package we have also coded our GA in Java. Experiments have been carried out in the clusters of the supercomputing service of the University of Castilla-La Mancha (Spain). Concretely, they run under Linux operating system and we have been allowed to book a maximum of 20 GB of RAM memory.

## 5.3   Results

Table 1 shows the obtained results. We have organized them from the easiest to the most difficult case. Thus, results for the smallest $n$ are in the first rows, and results for largest $\theta$ are on the left. The content of each cell in the table accounts for the average of the 20 different datasets sampled for the corresponding $(n, \theta)$ pair. These *mean values* are expressed in terms of the Kendall distance between the output (permutation) provided by the algorithm and all the permutations in the generated dataset. So, the smaller the value, the better the permutation.

To make easier the interpretation of the results, the algorithm (some times more than one) obtaining the best result for each $(n, \theta)$ combination is marked with a star symbol. Furthermore, for obtaining sound conclusions a statistical analysis has been carried out. Thus, Wilcoxon test has been used to ascertain whether the behavior of the tested algorithms is statistically significant. The algorithm with the best average is used as reference and compared with the remaining ones using a significance level $\alpha = 0.05$. Algorithms showing a statistically significant performance worse than the best one are marked with a minus symbol.

## 5.4   Results Discussion

The first conclusion is clear: the GA always obtains the best result, beating in all the cases to CS and Borda algorithms. Regarding B&B and DK, as we can expect they are competitive with respect to the GA only in the *less complex* cases, that is those having large $\theta$ and/or small $n$. However, they perform significantly worse in the remaining ones.

## 6   Conclusions

A study about the applicability of GAs to the problem of consensus permutation estimation in Mallows parameter estimation has been carried out. The proposal obtains very good results in all the cases, being statistically significantly better than competing approaches in most cases, specially in the harder ones, that is, large number of items to be ranked and/or few consensus among the instances (permutations) in the dataset.

In the near future we plan to go on with this research by following several lines: (1) increasing the experimental study by testing larger values for $n$ and considering the impact of the data set number of instances; (2) studying the behaviour of the GA and competing approaches when the data does not come from a pure Mallows distribution, but from a *mixture* of them; and (3) extending the approach to the *generalized* Mallows model, while consensus permutation and $\theta$'s values should be estimated simultaneously.

**Acknowledgements.** The authors want to thank Alnur Ali and Marina Meila for facilitating us the code of their algorithms used in the experimental comparison. This work has been partially funded by FEDER funds and the Spanish Government (MICINN) through project TIN2010-20900-C04-03.

# References

1. Ali, A., Meila, M.: Experiments with kemeny ranking: What works when? Mathematical Social Sciences 64(1), 28–40 (2012)
2. Borda, J.: Memoire sur les elections au scrutin. Histoire de l'Academie Royal des Sciences
3. Ceberio, J., Mendiburu, A., Lozano, J.A.: Introducing the mallows model on estimation of distribution algorithms. In: Lu, B.-L., Zhang, L., Kwok, J. (eds.) ICONIP 2011, Part II. LNCS, vol. 7063, pp. 461–470. Springer, Heidelberg (2011)
4. Chen, H., Branavan, S.R.K., Barzilay, R., Karger, D.R.: Global models of document structure using latent permutations. In: Proceedings of Human Language Technologies, NAACL 2009, pp. 371–379 (2009)
5. Cohen, W.W., Schapire, R.E., Singer, Y.: Learning to order things. J. Artif. Int. Res. 10(1), 243–270 (1999)
6. Conitzer, V., Davenport, A.J., Kalagnanam, J.: Improved bounds for computing kemeny rankings. In: AAAI, pp. 620–626 (2006)
7. Davenport, A.J., Kalagnanam, J.: A computational study of the kemeny rule for preference aggregation. In: AAAI, pp. 697–702 (2004)
8. Diaconis, P.: A generalization of spectral analysis with application to ranked data. The Annals of Statistics, 949–979 (1989)
9. Diaconis, P.: Group representations in probability and statistics. Lecture Notes - Monograph Series, vol. 11. Institute of Mathematical Statistics, Harvard (1988)
10. Dwork, C., Kumar, R., Naor, M., Sivakumar, D.: Rank aggregation methods for the web. In: WWW, pp. 613–622 (2001)
11. Fligner, M.A., Verducci, J.S.: Distance based ranking models. Journal of the Royal Statistical Society 48(3), 359–369 (1986)
12. Fligner, M.A., Verducci, J.: Probability models and statistical analyses for ranking data. Springer (1993)
13. Goldberg, D.: Genetic algorithms in search, optimization and machine learning. Addison-Wesley (1989)
14. Kamishima, T.: Nantonac collaborative filtering: Recommendation based on order responses. In: The 9th International Conference on Knowledge Discovery and Data Mining (KDD), pp. 583–588 (2003)
15. Kemeny, J.L., Snell, J.G.: Mathematical models in the social sciences. Blaisdell, New York
16. Kendall, M.G.: A new measure of rank correlation. Biometrika 30, 81–93 (1938)
17. Larrañaga, P., Kuijpers, C.M.H., Murga, R.H., Inza, I., Dizdarevic, S.: Genetic algorithms for the travelling salesman problem: A review of representations and operators. Artif. Intell. Rev. 13(2), 129–170 (1999)
18. Mallows, C.L.: Non-null ranking models. I. Biometrika 44(1-2), 114–130 (1957)
19. Meila, M., Phadnis, K., Patterson, A., Bilmes, J.: Consensus ranking under the exponential model. In: 22nd Conf. on Uncertainty in Artificial Intelligence (2007)
20. Miller, B.L., Goldberg, D.E.: Genetic algorithms, tournament selection, and the effects of noise. Complex Systems 9, 193–212 (1995)

# A Hybrid Genetic Algorithm
# for the Maximum Satisfiability Problem

Lin-Yu Tseng and Yo-An Lin

Department of Computer Science and Communication Engineering,
Providence University, 200, Sec. 7, Taiwan Boulevard, Taichung 43301 Taiwan
lytseng@cs.nchu.edu.tw, alimen@gmail.com

**Abstract.** The satisfiability problem is the first problem proved to be NP-complete and has been one of the core NP-complete problems since then. It has many applications in many fields such as artificial intelligence, circuit design and VLSI testing. The maximum satisfiability problem is an optimization version of the satisfiablity problem, which is a decision problem. In this study, a hybrid genetic algorithm is proposed for the maximum satisfiability problem. The proposed algorithm has three characteristics: 1. A new fitness function is designed to guide the search more effectively; 2. A local search scheme is designed, in which a restart mechanism is devised to help the local search scheme escape from the solutions near an already searched local optimum; 3. The local search scheme is hybridized with a two-layered genetic algorithm. We compared the proposed algorithm with other algorithms published in the literature, on the benchmarks offered by Gotllieb, Marchion and Rossi [12] and Hao, Lardeux and Saubion[18].

**Keywords:** Maximum Satisfiability Problem, Local Search Method, Hybrid Genetic Algorithm.

## 1    Introduction

The satisfiability problem is a decision problem whose definition is as follows. Given a Boolean formula $\Phi: B^n \to B$, where $B = \{0, 1\}$, is there an assignment $X = (x_1, x_2, \ldots, x_n) \in B^n$ which satisfies this Boolean formula?

Stephen Cook proved this problem NP-complete in 1971. Since then, this problem has been a very important one in the computation theory research. Besides that, many applications such as artificial intelligence, hardware design and VLSI testing are also related to this problem.

Algorithm solving the satisfiability problem can be divided into two categories: complete algorithms and incomplete algorithms. A representative algorithm of the former category is the Davis-Putnam algorithm, which is based on the branch-and-bound strategy. It sequentially determines the value of each variable and tries to examine every solution except those can be bounded. Although a complete algorithm can find the assignment that satisfies the Boolean formula if there exists one, the worst case time complexity is exponential and in general not affordable. Therefore

M. Ali et al. (Eds.): IEA/AIE 2013, LNAI 7906, pp. 112–120, 2013.

many incomplete algorithms, which include heuristic algorithms and meta-heuristic algorithms were proposed. Heuristic algorithms start with a solution (an assignment) then based on some heuristic strategy, they iteratively flip one on some variables to do the local search. Noting the success of applying the local search method to the constraint satisfaction problem [1], in 1992 Gu [12] and Selman et al. [19] were the first ones to propose these kinds of algorithms to solve the satisfiability problem. Their algorithms were called SAT1.1 and Greedy SAT (GSAT) respectively. In the following year, Gent and Walsh proposed a framework of the local search algorithm call GenSAT [8]. After that, several local search based algorithms including Walk SAT (WSAT) [20][21] and Historic SAT (HSAT) [8] were presented. Furthermore, some meta-heuristic methods including variable neighborhood search [13], tabu search [17] and simulated annealing [3] were also proposed to solve this problem. The above mentioned methods in general had two drawbacks: long search time and tending to get trapped in local optima.

As early as in 1989, de Jong and Spears had proposed the genetic algorithm for the satisfiability problem [5]. Because the performance of that genetic algorithm was worse than that of the complete algorithms, no other genetic algorithms were proposed for years. In 1997, Eiben and van der Hauw proposed a new advance through a genetic approach called SAWEA [6] in which an automatic adapting fitness function was invented. This study inspired several studies of applying the genetic algorithm to solve the satisfiability problem. Among them are RFEA [9], EFEA2 [10], FlipGA [16] and ASAP [18], and FlipGA was especially noteworthy. FlipGA and ASAP hybridized the genetic algorithm and the local search scheme. These kinds of algorithms are called the genetic local search algorithms or the memetic algorithms. In addition to FlipGA and ASAP, GASAT proposed by Hao et al. in 2003 [14] and then improved in 2007 [15] is another genetic local search algorithm. Surveying these incomplete algorithms, we observed that there were three points that might be improved. First, the search guiding ability of the fitness function is not good enough. Second, the local search method may spend too much time searching the same area. Third, the balance between the global search and the local search is still not very good. Hence, in this study we proposed a hybrid genetic algorithm for solving the maximum satisfiability problem and tried to improve these three points. The experimental results showed some improvements.

The remaining part of the paper is organized as follows. In Section 2, we introduce the formal definition of the maximum satisfiability problem. In Section 3, we describe the proposed hybrid genetic algorithm. The experimental results and the conclusions are given in Section 4 and Section 5, respectively.

## 2    Definition of the Maximum Satisfiability Problem

The definition of the maximum satisfiability problem (MAX-SAT) is given in the following.

Give a Boolean formula $\Phi: B^n \rightarrow B$, where $B = \{0, 1\}$ and $\Phi$ is in conjunctive normal form (CNF), we want to find an assignment $X = (x_1, x_2, \ldots, x_n) \in B^n$ such that the number of satisfied clauses is maximized.

By a Boolean formula $\Phi$ is in conjunctive normal form, we mean $\Phi = \bigwedge_{i=1}^{m} C_i$, that is, $\Phi$ is the conjunction of $m$ clauses $C_1, C_2, \ldots, C_m$. Furthermore, each clause is the disjunction of literals, that is, $C_i = \bigvee_{j=1}^{|C_i|} l_{ij}$. A literal $l$ may be either a variable x or its negation $\neg$x.

# 3     The Hybrid Genetic Algorithm

In this section, we describe the proposed hybrid genetic algorithm. We first introduce the new fitness function. Then, we elaborate the proposed local search method reSAT. Finally, we describe the hybrid genetic algorithm.

## 3.1     The Fitness Function

We define a new fitness function $f_{Total}$ for the genetic algorithm. $f_{Total}$ is given the following.

$$f_{Total}(\mathrm{X}) = \left(m - f_{MAXSAT}(X)\right) + f_{lit}(X), \tag{1}$$
$$\text{where } f_{lit}(X) = \sum_{i=1}^{m} t_i / (\sum_{i=1}^{m} |C_i| + 1).$$

$m$ is the number of clauses. $f_{MAXSAT}(X)$ is the number of non-satisfied clauses and this number is the smaller the better. $t_i$ is the number of literals that are true in clause $C_i$. We don't wish to have many true literals in the same clause because such an assignment may falsify other clauses. So $\sum t_i$ is also the smaller the better. The denominator of $f_{lit}$ is to scale the value such that $f_{lit}$ is between 0 and 1. Since the fitness function in the genetic algorithm is the larger the better, we define the fitness function as $\bar{f}_{Total} = (m + 1) - f_{Total}$.

"Minimizing breaks" was proposed by WSAT [20][21]. "Minimizing breaks" tries to find the variable $x$ in an assignment such that flipping $x$ will minimize the number of clauses that are changed from satisfied to unsatisfied. In the following, we used this concept and $f_{lit}$ to define a score function to be used in our local search method. In the following equation, number_of_breaks($X$) is the number of clauses that are changed from satisfied to unsatisfied after flipping variable $x$ in the assignment.

$$\text{score}(x_j) = number\_of\_breaks(x_j) + f_{lit}\left(flip(X, x_j)\right) - f_{lit}(X) \tag{2}$$

Of course, score($x_j$) is the smaller the better.

## 3.2     The Local Search Method

In the local search method, a search step is to flip a variable in the assignment (the solution). Because a search step is "small" (change a little), the drawback of such a local search method is tending to be trapped in local optima. In the proposed local search method reSAT, we set three conditions to restart the search from a new randomly generated solution. The first condition: If after *stuck-limit* times of flips, the current best solution *pBest* has not been improved, we restart the search. The second condition: If we visit the current best solution *pBest loop-limit* times, we restart the search. The third condition: If we visit any solution that is in the tabu list, we restart the search. The pseudo code for reSAT is given in Figure 1.

```
Procedure reSAT(Φ, T, Max-flips, p)
    pBest ← T,   stuck ← 0,   loop ← 0,   f ← 0
    for i ← 1 to Max-flips do
        if T satisfies Φ then return (T, f)
        Unsat-clause ← select-an-unsat-clause(Φ, T)   /*Randomly select
                                                        /*an unsatisfied clause.
        r ← random([0, 1])
        if r < p then V ← random-pick(Unsat-clause)   /*Randomly select
                                                        /* a variable in this clause.
        else for each variable xⱼ in Unsat-clause do
                 scoreⱼ ← score(xⱼ)
             end for
             V ← minimum(score)       /* Select the variable that has the
                                       /*minimum score.
        end if
        T ← flip(T, V)
        if T is in tabu list then
                f ← 1, return (T, f)
        else if T is better than pBest then
                pBest ← T,   loop ← 0,   stuck ← 0
        else if T = pBest then
                loop ← loop+1
                if loop > loop_limit then f ← 1,   exit for
        else if T is worse than pBest then
                stuck ← stuck+1
                if stuck > stuck_limit then f ← 1,   exit for
        end if
    end for
    if f = 1 and pBest is not in tabu list then
        add pBest into tabu list
    end if
    return (T, f)
```

Fig. 1. The pseudo code of reSAT

## 3.3  The Hybrid Genetic Algorithm

The proposed hybrid genetic algorithm hybridizes the genetic algorithm and the local search method reSAT. This algorithm consists of the outer layer and the inner layer. The outer layer is a genetic algorithm whose job is to generate solutions to replace those solutions whose restart-flags are set in the inner layer. The inner layer is called the region search, which is a hybridization of reSAT and the crossover operator. We first describe the outer layer in the following.

Step 1.  (Initialization) Randomly generate *p-size* solutions $X_1, X_2, ..., X_{p-size}$ as the initial population.

Step 2.  For each $X_i, i = 1, 2, ..., p - size$, execute the region search (the inner layer).

Step 3.  Execute the genetic algorithm that includes selection, crossover and mutation to generate solution to replace those solutions whose restart-flags are set in the inner layer.

Step 4.  If the optimal solution is found or the termination condition is met, stop and output the best solution found. Otherwise go to step 2.

We use the roulette wheel selection with the fitness function $\bar{f}_{Total} = (m + 1) - f_{Total}$. The crossover operator used is the uniform crossover. Each variable in a solution has 5% of probability to change its value in the mutation operation.

Next let us describe the region search (the inner layer) in the following. Suppose the solution T is given to do the region search.

Step 1.  Restart-count $\leftarrow 0$;  i $\leftarrow 0$

Step 2.  Do the shaking operation to T to obtain another solution T'. (Shaking operation will be described later.)

Step 3.  Apply the local search method reSAT to T' for *flips* step. (i.e. Call reSAT($\Phi$, T', flips, p))

Step 4.  If the restart-flag of T' is set then go to step 5 else go to step 6.

Step 5.  restart-count $\leftarrow$ restart-count + 1

If restart-count > restart-count-limit then return the best solution found and set the restart-flag.

else go to step 2.

Step 6.  i $\leftarrow$ i+1, save T' in $T_i$

if i = 2 then go to step 7.

else go to step 2.

Step 7.  Apply the uniform crossover operator to $T_1$ and $T_2$ to obtain $T_{new}$.

Step 8.  Apply the local search method reSAT to $T_{new}$ for *flips* steps. (i.e. Call reSAT($\Phi$, $T_{new}$, flips, p)).

Step 9.  Return the best solution found and reset the restart-flag.

The shaking operation is explained in the following. Suppose we are doing the shaking operation to a solution T, we first collect the clauses that are not satisfied by T, then we collect all the variables that appear in these unsatisfied clauses. After that, let us random reassign values to these variables. This action changes T to T' and is called the shaking operation.

## 4    Experimental Results

We tested the proposed method on two benchmarks. Each part of the first benchmark was originally proposed by Back et al. [2], Gottlieb an Voss [10], Marchiori and Rossi [16] and de Jong and Kosters [4]. These parts are then merged into a benchmark by Gottlieb et al. [12]. The details of the first benchmark are shown in Table 1.

**Table 1.** The details of the first benchmark

| Prob. name | # of prob. | # of prob. for each # of var. | # of var. |
|:---:|:---:|:---:|:---:|
| A | 12 | 3 | 30, 40, 50, 100 |
| B | 150 | 50 | 50 |
| | | | 75, 100 |
| C | 500 | 100 | 20, 40, 60, 80, 100 |

The second benchmark was collected by Hao et al.[18]. There were fifteen problems. Some of them were randomly generated and others were dervied from practical problems. The number of variables and the number of clauses in the second benchmark are much larger than those in the first benchmark. The details of the second benchmark are shown in Table 2.

**Table 2.** The details of the second benchmark

| Prob. name | # of var. | # of claus. | Prob. name | # of var. | # of claus. |
|:---:|:---:|:---:|:---:|:---:|:---:|
| par-16-4-c | 324 | 1292 | color-18-4 | 1296 | 95904 |
| mat25.shuffled | 588 | 1968 | glassy-a | 399 | 1862 |
| mat26.shuffled | 744 | 2464 | glassy-b | 450 | 2100 |
| par32-5-c | 1339 | 5350 | hgen2-a | 500 | 1750 |
| difp_19_0_arr_rcr | 1201 | 6563 | hgen2-b | 500 | 1750 |
| difp_19_99_arr_rcr | 1201 | 6563 | f1000 | 1000 | 4250 |
| par32-5 | 3176 | 10325 | f2000 | 2000 | 8500 |
| color-10-3 | 300 | 6475 | | | |

The values of parameters were determined by empirical experience. They are listed as follows. p-size was set to 10; Max-flips (in Procedure reSAT) was set to 4 *(number of variables); p (in Procedure reSAT) was set to 0.3; stuck-limit and loop-limit (both in Procedure reSAT) were set to 10*(number of variables) and 0.03 * (number of variables). Table 3 shows the comparison of the proposed method and seven other methods on the first benchmark. The termination condition is 300,000 flips and every problem is executed 50 times. In Table 3, SR denotes success rate. The comparision results reveal that the proposed method outperforms all other seven methods. Table 4 shows the comparison of the proposed method with FlipGA and GASAT. In Table 4, var and cls denote number of variables and number of clauses, respectively. f.c. (false clause) denotes number of clauses that are not satisfied. s.d.

denotes standard deviation. flips denotes number of flips taken to find the best solution and this value had been divided by 1000. sec. denotes the CPU time spent. The comparison results show that the proposed method obtains better solutions on some problems, especially difp_19_0_arr_rcr, difp_19_99_arr_rcr, color-10-3 and color-18-4, but it obtains inferior solutions on some other probelms, especially par32-5-c, par32-5 and f2000. In general, the proposed algorithm takes more computation time and it works well for small to medium size problems, whereas solution quality degenerates when problem size is large.

**Table 3.** Comparison of the proposed algorithm and seven other algorithms on the first benchmark

| Instances | Suite | A | A | A | A | B | B | B | C | C | C | C | C |
|---|---|---|---|---|---|---|---|---|---|---|---|---|---|
| | n. p. | 3 | 3 | 3 | 3 | 50 | 50 | 50 | 100 | 100 | 100 | 100 | 100 |
| | n. var. | 30 | 40 | 50 | 100 | 50 | 75 | 100 | 20 | 40 | 60 | 80 | 100 |
| Hybrid GA | SR(%) | 100 | 100 | 100 | 100 | 100 | 99 | 83 | 100 | 100 | 100 | 98 | 84 |
| | flips | 829 | 590 | 9201 | 19426 | 5358 | 26779 | 53356 | 229 | 1946 | 11352 | 38838 | 53274 |
| GASAT | SR(%) | 99 | 100 | 91 | 95 | 96 | 83 | 69 | 100 | 100 | 97 | 66 | 74 |
| | flips | 1123 | 1135 | 1850 | 7550 | 2732 | 6703 | 28433 | 109 | 903 | 9597 | 7153 | 1533 |
| SAWEA | SR(%) | 100 | 93 | 85 | 72 | - | - | - | 100 | 89 | 73 | 52 | 51 |
| | flips | 34015 | 53289 | 60743 | 86631 | - | - | - | 12634 | 35988 | 47131 | 62859 | 69657 |
| REFA2 | SR(%) | 100 | 100 | 100 | 99 | 100 | 95 | 77 | 100 | 100 | 99 | 92 | 72 |
| | flips | 3535 | 3231 | 8506 | 26501 | 12053 | 41478 | 71907 | 365 | 3015 | 18857 | 50199 | 68053 |
| REFA2+ | SR(%) | 100 | 100 | 100 | 97 | 100 | 96 | 81 | 100 | 100 | 99 | 95 | 79 |
| | flips | 2481 | 3081 | 7822 | 34780 | 11350 | 39396 | 80282 | 365 | 2951 | 19957 | 49312 | 74459 |
| FlipGA | SR(%) | 100 | 100 | 100 | 89 | 100 | 82 | 57 | 100 | 100 | 100 | 73 | 62 |
| | flips | 25490 | 17693 | 127900 | 116653 | 103800 | 29818 | 20675 | 1073 | 14320 | 127520 | 29957 | 20319 |
| ASAP | SR(%) | 100 | 100 | 100 | 100 | 100 | 87 | 59 | 100 | 100 | 100 | 72 | 61 |
| | flips | 9550 | 8960 | 68483 | 52276 | 61186 | 39659 | 43601 | 648 | 16644 | 184419 | 45942 | 34548 |
| WSAT | SR(%) | 100 | 100 | 100 | 100 | 95 | 84 | 60 | 100 | 100 | 94 | 72 | 63 |
| | flips | 1631 | 3742 | 15384 | 19680 | 16603 | 33722 | 23853 | 334 | 5472 | 20999 | 30168 | 21331 |

**Table 4.** Comparison of the proposed algorithm and two other algorithms on the second benchmark

| Second Benchmark | | | Hybrid GA | | | | FlipGA | | | | GASAT | | | |
|---|---|---|---|---|---|---|---|---|---|---|---|---|---|---|
| instances | var | cls | f.c. avg. | f.c. s.d. | flips | sec. avg. | f.c. avg. | f.c. s.d. | flips | sec. | f.c. avg. | f.c. s.d. | flips | sec. |
| par-16-4-c | 324 | 1292 | 15.10 | 1.57 | 2470 | 898 | 10.15 | 0.85 | 1026 | 9 | **5.85** | 1.01 | 137 | 35 |
| mat25.shuffled | 588 | 1968 | 8.00 | 0.00 | 2744 | 1675 | 10.10 | 2.41 | 674 | 28 | **7.60** | 0.80 | 113 | 161 |
| mat26.shuffled | 744 | 2464 | **8.00** | 0.00 | 53 | 41 | 12.89 | 3.54 | 9143 | 44 | **8.00** | 0.00 | 41 | 749 |
| par32-5-c | 1339 | 5350 | 112.40 | 24.99 | 3581 | 5245 | 22.40 | 1.68 | 5746 | 77 | **19.60** | 1.69 | 311 | 129 |
| difp_19_0_arr_rcr | 1201 | 6563 | **18.30** | 0.45 | 3464 | 7464 | 87.60 | 7.23 | 20340 | 109 | 84.25 | 6.13 | 657 | 661 |
| difp_19_99_arr_rcr | 1201 | 6563 | **14.00** | 0.00 | 5439 | 11714 | 87.95 | 9.75 | 20172 | 107 | 81.40 | 7.14 | 639 | 658 |
| par32-5 | 3176 | 10325 | 181.12 | 8.28 | 6007 | 15531 | 50.65 | 3.35 | 62149 | 290 | **41.25** | 5.02 | 755 | 813 |
| color-10-3 | 300 | 6475 | **1.20** | 2.56 | 2765 | 5093 | 41.65 | 1.80 | 24664 | 86 | 46.53 | 3.08 | 18 | 1343 |
| color-18-4 | 1296 | 95904 | **55.90** | 2.19 | 569 | 13751 | 2064.35 | 363.65 | 2818 | 1150 | 248.50 | 0.50 | 27 | 33128 |
| glassy-a | 399 | 1862 | 8.40 | 1.11 | 1778 | 1095 | 7.60 | 1.06 | 411 | 16 | **5.00** | 0.00 | 153 | 18 |
| glassy-b | 450 | 2100 | 11.40 | 1.06 | 2743 | 1890 | 11.45 | 1.28 | 479 | 21 | **8.95** | 0.22 | 166 | 86 |
| hgen2-a | 500 | 1750 | **0.00** | 0.00 | 992 | 577 | 6.24 | 1.19 | 579 | 16 | 1.40 | 0.49 | 294 | 22 |
| hgen2-b | 500 | 1750 | **0.00** | 0.00 | 357 | 220 | 7.00 | 1.34 | 575 | 18 | 1.80 | 0.68 | 268 | 22 |
| f1000 | 1000 | 4250 | 7.50 | 0.80 | 5749 | 7900 | 8.90 | 1.67 | 1480 | 47 | **2.30** | 0.90 | 408 | 45 |
| f2000 | 2000 | 8500 | 19.60 | 2.05 | 7588 | 20779 | 16.90 | 2.07 | 3641 | 122 | **7.35** | 1.80 | 709 | 97 |

## 5     Conclusions

In this study, a hybrid two-layer genetic algorithm was proposed for the maximum satisfiability problem. First, a new fitness function was proposed. Also, a new local search scheme with restart mechanism was proposed. This local search scheme was hybridized with the crossover operator to form the inner layer, namely the region search. And the outer layer was a genetic algorithm aiming to generate new solutions to replace those solutions whose restart flags were set when running the inner layer.

The experimental results reveal that the proposed algorithm works well for small to medium size problems, but does not work very well for large size problems. In the future study, we plan to enhance the capability of the global search to improve the performance of the algorithm when it is applied to solve large size problems.

## References

1. Adorf, H.M., Johnston, M.D.: A Discrete Stochastic Neural Network Algorithm for Constraint Satisfaction Problems. In: Proceedings of the International Joint Conference on Neural Networks, vol. 3, pp. 917–924 (June 1990)
2. Back, T., Eiben, A., Vink, M.: A Superior Evolutionary Algorithm for 3-SAT. In: Porto, V.W., Saravanan, N., Waagen, D., Eiben, A.E. (eds.) EP 1998. LNCS, vol. 1447, pp. 123–136. Springer, Heidelberg (1998)
3. Beringer, A., Aschemann, G., Hoos, H.H., Metzger, M., Weiss, A.: GSAT versus Simulated Annealing. In: Proceedings of the European Conference on Artificial Intelligence, pp. 130–134 (1994)
4. de Jong, M., Kosters, W.: Solving 3-SAT using adaptive sampling. In: Proceedings of 10th Dutch/Belgian Artificial Intelligence Conference, pp. 221–228 (1998)

5. de Jong, K.A., Spears, W.M.: Using Genetic Algorithms to Solve NP-Complete Problems. In: Proceedings of the Third International Conference on Genetic Algorithms, pp. 124–132 (1989)
6. Eiben, A., van der Hauw, J.: Solving 3-SAT with Adaptive Genetic Algorithms. In: Proceedings of the 4th IEEE Conference on Evolutionary Computation, pp. 81–86 (1997)
7. Gent, I.P., Walsh, T.: Towards an Understanding of Hill-climbing Procedures for SAT. In: Proceedings of AAAI 1993, pp. 28–33 (1993)
8. Gent, I.P., Walsh, T.: Unsatisfied Variables in Local Search. In: Hybrid Problems, Hybrid Solutions (AISB 1995), pp. 73–85 (1995)
9. Gottlieb, J., Voss, N.: Representations, Fitness Functions and Genetic Operators for the Satisfiability Problem. In: Hao, J.-K., Lutton, E., Ronald, E., Schoenauer, M., Snyers, D. (eds.) AE 1997. LNCS, vol. 1363, pp. 55–68. Springer, Heidelberg (1998)
10. Gottlieb, J., Voss, N.: Adaptive Fitness Functions for the Satisfiability Problem. In: Deb, K., Rudolph, G., Lutton, E., Merelo, J.J., Schoenauer, M., Schwefel, H.-P., Yao, X. (eds.) PPSN 2000. LNCS, vol. 1917, pp. 621–630. Springer, Heidelberg (2000)
11. Gottlieb, J., Marchiori, E., Rossi, C.: Evolutionary Algorithms for the Satisfiability Problem. Evolutionary Computation 10(1), 35–50 (2002)
12. Gu, J.: Efficient Local Search for Very Large-Scale Satisfiability Problems. ACM SIGART Bulletin 3(1), 8–12 (1992)
13. Hansen, P., Mladenovic, N.: An Introduction to Variable Neighborhood Search. In: Advances and Trends in Local Search Paradigms for Optimization, pp. 433–458 (1999)
14. Hao, J.-K., Lardeux, F., Saubion, F.: Evolutionary Computing for the Satisfiability Problem. In: Raidl, G.R., et al. (eds.) EvoWorkshops 2003. LNCS, vol. 2611, pp. 258–267. Springer, Heidelberg (2003)
15. Hao, J.K., Lardeux, F., Saubion, F.: GASAT: A Genetic Local Search Algorithm for the Satisfiability Problem. Evolutionary Computation 14(2), 223–253 (2006)
16. Marchiori, E., Rossi, C.: A Flipping Genetic Algorithm for Hard 3-SAT Problems. In: Proceedings of Genetic and Evolutionary Computation Conference, pp. 393–400 (1999)
17. Mazure, B., Sais, L., Gregoire, E.: Tabu Search for SAT. In: Proceedings of the 14th National Conference on Artificial Intelligence, pp. 281–285 (1997)
18. Rossi, C., Marchiori, E., Kok, J.: An Adaptive Evolutionary Algorithm for the Satisfiability Problem. In: Proceedings of ACM Symposium on Applied Computing, pp. 463–469 (2000)
19. Selman, B., Levesque, H.J., Mitchell, D.: A New Method for Solving Hard Satisfiability Problems. In: Proceedings of the Tenth National Conference on Artificial Intelligence, pp. 440–446 (1992)
20. Selman, B., Kautz, H.A., Cohen, B.: Noise Strategies for Improving Local Search. In: Proceedings of the Twelfth National Conference on Artificial Intelligence, vol. 1, pp. 337–343 (1994)
21. McAllester, D., Selman, B., Kautz, H.: Evidence for Invariants in Local Search. In: Proceedings of the National Conference on Artificial Intelligence, pp. 321–326 (1997)

# An Object Representation Model
# Based on the Mechanism of Visual Perception

Hui Wei* and Ziyan Wang

Laboratory of Cognitive Model and Algorithm, Department of Computer Science,
Fudan University, Shanghai, China
weihui@fudan.edu.cn

**Abstract.** In areas of artificial intelligence and computer vision, object representation and recognition is an important topic, and lots of methods have been developed for it. However, analysis and obtain the knowledge of object's structure at higher levels are still very difficult now. We draw on the experience of pattern recognition theories of cognitive psychology to construct a compact, abstract and symbolic representing model of object's contour based on components fitting which is more consistent with human's cognition process. In addition, we design an algorithm to match components between different images of the same object in order to make the feature extraction at higher levels possible.

**Keywords:** Object representation, visual perception, recognition-by-components, components matching.

## 1 Introduction

Object representation and recognition is an important topic on visual information processing of artificial intelligence. To implement human's ability of perception, pattern recognition is one of the main tasks. And to get further understanding for complicated visual information and achieve the goal of recognition finally, finding a appropriate representation is a key point.

Currently, popular methods of visual pattern recognition contain the method of feature extraction by descriptors such the scale-invariant feature transform (SIFT) [1], the ones based on the statistic characteristics [2], some discriminant analysis methods based on the kernel functions [3][4] and so on. But features extracted by most of these methods pay much attention to the description of physical characteristics of object's image. Commonly, they map an object to a point in a high-dimensional feature space to analyze or build the model. Although these methods obtain good results to a large extent on recognition, the representation information is not extracted from the respect of human's cognition. So it's not easy to use them to cognize and understand the image at a higher level and accomplish more difficult tasks, especially on forming the knowledge and concepts of object's shape such as relevance between parts and the hierarchy

---

* Corresponding author.

M. Ali et al. (Eds.): IEA/AIE 2013, LNAI 7906, pp. 121–130, 2013.
© Springer-Verlag Berlin Heidelberg 2013

between the whole and the part. To mine more knowledge about the object from its image using the approach of machine learning, we have to think about potential models for visual information of object. To get inspired, we need to learn the thoughts derived from the discussion on the ability of human's cognition from cognitive psychology.

## 2    Representation and Recognition Based on Cognitive Psychology

Psychologists began to discuss about how humans process the image received by visual system and finally form our perception long before. It is what we are interested in that humans can recognize the specified object easily although the stimulates they received are quite complicated. As the result, cognitive psychologists build several models of pattern recognition based on some experimental phenomenons and the primary explanations for them.

Template matching, recognition by prototypes and feature analysis are three psychological theories of pattern recognition at early stage. Their ideas are widely drown by current computer applications of recognition. However, they are just simple models which refer to a specified local step or character during the process of cognition superficially. Accordingly, their related representations only have simple structures, and compared with human's powerful memory system, they are incompetent at many complex tasks in reality. So we need more frameworks with further mechanism.

Some visual perception frameworks with more complete process are put forward one after another to satisfy the demand of more complex visual analysis tasks since computational theory of vision created by D.Marr [6] came out. I.Biederman presents the recognition-by-components (RBC) theory [7][8]. Biederman use geometrical ions (geons) as the key of representation and recognition. Geons are view-independent simple geometries. Biederman abstracts 24 types of geons and 180 types of relationships between two geons. The rough shape of each object can be treated as a structure assembled by these components, geons. While recognizing, these geons are key properties. Main steps of RBC contains edge extraction, detection of view-independent properties, determination of components, matching of components to object representation and object identification. As the main geons are detected, the corresponded pattern can be recognized then. In fact, diverse objects are simplified as highly abstract geometric modeling by this method, and knowledge of the whole object is acquired by the composition of object's components.

The theories introduced above, including Marr's computational theory and Biederman's RBC theory, are bottom-up frameworks based on feature analysis. However, these frameworks encounter a lot of difficulties while explaining human's recognition phenomenons, such as object-superiority effect [5]. Who holds the opposite opinion with the though of bottom-up recognition is top-down recognition theory group. Chen Lin's topological theory of vision [9][10] is one of top-down theories. Compared with local features, his theory pays more attention to topological properties in a large scale. He emphasizes that human's vision system prefers the global properties of image to the local ones, and provides a series of recognitive experimental results which is difficult to explained by the traditional feature detection model to support his opinions. But there is a great difficult to impede the top-down theories to become practical applications, which is that they are not easy to be implemented by computer programs.

Most of the recent pattern recognition theories introduced above from Marr's computation theory are just psychological models and there's a long distance for them to be clear enough and modeled by computers. It's rare to find examples of practical computer systems built on these theories currently. However, these theories will bring a lot of inspirations to the areas of artificial intelligence, pattern recognition and so on. So, trying to build the system who have better understanding of visual image and is easier to be implemented based on them will bring important significance.

## 3    An Object Representation Model and a Method for Matching Components

### 3.1    Basic Design for Representation Model

Drawing on the though of RBC theory, we use some components to fit the image of object to compact its shape with the aim of analysis at a higher level. But 24 types of geons defined by RBC, a psychological theory, cannot be built by computers easily. Firstly, using cylinder, sphere, or other geons to fit the object is not easy to implemented by computers. Secondly, even if fitting is finished successfully, computer still have difficulty on analyzing characters of these geons and relationships between them. In order to make it possible, we should consider to choose another scheme of geons.

Here we use a single-type geon to fit the image. It is the 2-dimensional Gaussian distribution whose transverse section is an ellipse. We take several these ellipses which is always long and narrow to fit the contour of an object. Weighted sum of multiple 2-dimensional Gaussian distributions forms the 2-dimensional Gaussian mixture model, a probability model. Making use of probability model to fit the image is because of small movement or deformation of contour between different images of the same object. In fact, Gaussian model who has a high probability at the center of components and a low probability at the point far away reflect a possible probability distribution of appearances of the contour's features.

Formula (1) is the expression of 2-dimensional Gaussian function with translation vector $(x_0, y_0)$ and angle of rotation $\theta$, as single Gaussian components. The 2-dimensional Gaussian function which is the weighted sum of multiple components is expressed by Formula (2).

$$g(x,y) = \frac{1}{2\pi\sigma_x\sigma_y} e^{-\left(\frac{((x-x_0)cos\theta-(y-y_0)sin\theta)^2}{2\sigma_x^2} + \frac{((x-x_0)sin\theta+(y-y_0)cos\theta)^2}{2\sigma_y^2}\right)} \tag{1}$$

$$G(x, y) = \sum_k w_k g_k(x, y) \tag{2}$$

We can see from the expression formula of 2-dimensional Gaussian mixture model that although we use single-type components, the components' shape can be controlled by a group of parameters $(x_0, y_0, \theta, \sigma_x, \sigma_y)$. So not only is the model's expression ability not too limited but also a compact and isomorphic fitting expression is gained using the single-type component. As the result, complicated geometrical shapes are converted into one expression, or we say a group of parameter. Each component with a group of parameter can be expressed as a vector who has fixed length, so all components from one image can be expressed as a parameter matrix. There's no doubt that this matrixing representation is convenient to store and analyze for computers. After learn from multiple sample images, computers can compare different parameter matrix and extract the constant components from different images. And these constant components are the key to achieve the goal of recognition.

Figure. 4 shows the representation of this fitting model.

Current method we use is bottom-up until this fitting model is built. But the purpose of this bottom-up phase is to form a compact, abstract and symbolic data structure which is easy to analyze by computers. On the basis of this compact representation, we can use the parameters' values to analyze features of each component and those of relationships between components to get a further representation with graph structure. The graph can reflect larger-scale structural features, that is, on the basis of the representation, understanding and analysis at higher levels can be carried out and top-down analysis becomes possible. We can construct cognitive framework combining two directions, both bottom-up and top-down, using this representation. The framework will be accord with characters of human's cognition more, so it will be significant.

## 3.2 Machine Learning for Components Representation

Now we introduce a learning method from sample image to Gaussian mixture model. First of all edge detection is processed on objects' images and fitting by components is based on it. Since the count number of components fit for the image is not constant all the time, to qualify for this job, using expectation-maximization (EM) algorithm to estimate parameters is a common choice. We can deduce the iterative formula from the definition formula of Gaussian mixture model using EM algorithm.

While fitting the object, object's contour represented by binary image can be regard as a set of sampling points which is lighted pixel in the image. Suppose $(x_i, y_i)$ represents the $i$-th sampling point. At the $t$-th iteration, $\omega_j^{(t)}$ is the learnt group of parameters of the $j$-th component, and $p_{ji}^{(t)}$ is the posterior probability calculated by the group $\omega_j^{(t)}$ at the position of the $i$-th sampling point.

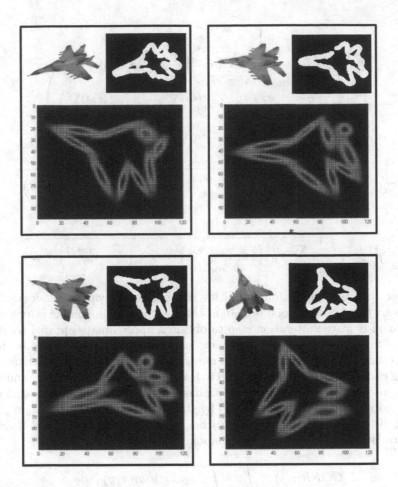

**Fig. 1.** The representation of this fitting model: in each subgraph, the top left corner is the original image of the object, the top right corner is the bold edge image, the bottom is the fitting image of Gaussian components

Let $p_j^{(t)} = \sum_i p_{ji}^{(t)}$. Then the iteration formula for calculating from the $i$-th to $(i+1)$-th is as follow:

$$x_{0j}^{(t+1)} = \frac{1}{p_j^{(t)}} \sum_{i=1}^{n} p_{ji}^t x_i \tag{3}$$

$$y_{0j}^{(t+1)} = \frac{1}{p_j^{(t)}} \sum_{i=1}^{n} p_{ji}^t y_i \tag{4}$$

$$\theta_j^{(t+1)} = \begin{cases} \frac{\pi}{8} & (J = 0) \\ \frac{1}{4}arctan\frac{-4BC}{J} & (J > 0) \\ \frac{1}{4}arctan\frac{-4BC}{J} + \frac{\pi}{4} & (J < 0) \end{cases}$$

$$\left( \begin{array}{c} J = B^2 - 4C^2 \\ B = -\sum_{i=1}^{n} p_{ji}^{(t)} \left( \left( x_i - x_{0j}^{(t+1)} \right)^2 - \left( y_i - y_{0j}^{(t+1)} \right)^2 \right) \\ C = \sum_{i=1}^{n} \left( p_{ji}^{(t)} \left( x_i - x_{0j}^{(t+1)} \right) \left( y_i - y_{0j}^{(t+1)} \right) \right) \end{array} \right) \tag{5}$$

$$\sigma_{xj}^{(t+1)} = \sqrt{\frac{1}{p_j^{(t)}} \sum p_{ji}^{(t)} \left( \left( x_i - x_{0j}^{(t+1)} \right) cos\theta - \left( y_i - y_{0j}^{(t+1)} \right) sin\theta \right)^2} \tag{6}$$

$$\sigma_{yj}^{(t+1)} = \sqrt{\frac{1}{p_j^{(t)}} \sum p_{ji}^{(t)} \left( \left( x_i - x_{0j}^{(t+1)} \right) sin\theta + \left( y_i - y_{0j}^{(t+1)} \right) cos\theta \right)^2} \tag{7}$$

Consider that EM algorithm initializes the parameter values randomly, and makes sure that the likelihood probability of the whole fitting model is increasing monotonously at each iteration step to obtain a approximate optimal solution. In this situation, components are learnt with randomness and may be terrible sometimes. We use split-reduction mechanism as supplement to enhance the learning effect. First use EM algorithm to learn components whose count number is much more than the set value at split phase, then reduce two components with highest correlation value into one step by step. The result after reduction phase is much better than the original algorithm on representativeness and stability. The correlation value between two components is defined as follow.

$$CORR(i,j) = \int_{-\infty}^{+\infty} \int_{-\infty}^{+\infty} g_i(x,y)g_j(x,y)dxdy \tag{8}$$

In Figure. 3.2,(d) shows the result using split-reduction mechanism, compared with (b) who does'nt use it.

### 3.3  A Matching Method for Components between Different Samples

Fitting can be finished for each sample images by the method introduced above. But the each fitting model we get currently is just built on single sample image, that is, for each image we extract one parametric representation of components. However, it's not enough. To extract constant components and relationships, we need to establish components' matching relationship between represents of different samples. Since one component of object seen from one viewpoint may not appear in the image from another view, if we success to match components between samples, we can mine the components which appears frequently as important features or prototypes of the object. So we design a components'

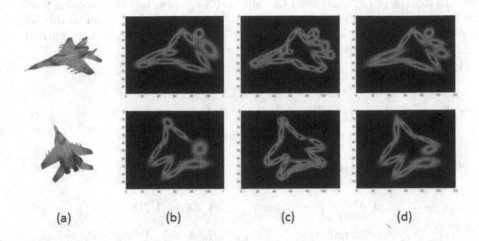

(a)                    (b)                    (c)                    (d)

**Fig. 2.** (a) original images of the object(b) 9-component fitting result using EM algorithm directly only after split phase; (c) 20-component fitting result; (d) 9-components fitting result reduced from (c) using split-reduction mechanism

matching algorithm between samples' fitting results who has the same number of components with others.

Movements and deformations always exist between different sample images of the same object. After size normalization, movements contains translations and rotations. In this situation, the difference vectors of specified component-pairs between samples are still similar. Although there's small deformation, we can also use all difference vectors related to the component to establish correct matching relationship jointly. Big deformation may need to be processed at higher level of recognition, so we don't discuss about it here. In a word, the matching algorithm is based on comparisons of difference vectors. The difference vector defined here contains 5 dimensions: suppose two groups of parameters from the same sample image are respectively $\omega_m$ and $\omega_n$, so their difference vector is $(log(\sigma_{xm}/\sigma_{xn})$, $log(\sigma_{ym}/\sigma_{yn})$, $\sqrt{(x_{0m} - x_{0n})^2 + (y_{0m} - y_{0n})^2}$, $\theta_m - \theta_n$, $\beta)$, and $\beta$ in it means difference angle between the direction of the component which is corresponded to $\omega_m$ and the connecting line between two components. Suppose each sample to be matched has $s$ components, then calculate pairwise difference vectors between all components from current sample to form a matrix called difference matrix whose size is $s \times s \times 5$.

Following is the formalized description of the alignment algorithm. Suppose there're two samples $S_A$ and $S_B$, whose corresponded difference matrices are respectively $A$ and $B$. The algorithm will swap the order of components, that is, bring elementary row/colum transformations on the matrices, to make the top-left corners of two matrices similar as soon as possible. The processed difference matrices are respectively $A'$ and $B'$. At the same time, we can get two matching sequence which is the sequence storing the serial numbers of components.

The $i$-th item in $v^A$, $v_i^A$, and the $i$-th item in $v^B$, $v_i^B$, are the $i$-th matching pair, and the order of components in processed difference matrice is equal to this sequence. The algorithm will execute by $(s-1)$ steps and $k$ represents current step. Here is the description of the algorithm.

(1) When $k = 1$:

$$(v_1^A, v_2^A, v_1^B, v_2^B) = \underset{1 \le i^A, j^A, i^B, j^B \le s}{\operatorname{argmin}} \left\{ \mid A_{i^A, j^A} - B_{i^B, j^B} \mid + \mid A_{j^A, i^A} - B_{j^B, i^B} \mid \right\}$$
$$X'_{i,j} = X_{v_i^X, v_j^X}. \ (1 \le i, j \le 2; X = A, B)$$

(2) When $k = 2, 3, \ldots, (s-1)$:

$$U^X = \{ i = v_j^X \mid j = 1, \ldots, k \}. \ (X = A, B)$$
$$(v_{k+1}^A, v_{k+1}^B) = \underset{m \in \overline{V^A}, n \in \overline{V^B}}{\operatorname{argmin}} \left\{ \sum_{t=1}^{k} \mid A_{m,t} - B_{n,t} \mid + \sum_{t=1}^{k} \mid A_{t,m} - B_{t,n} \mid \right\}$$
$$X'_{k+1,j} = X_{v_{k+1}^X, v_j^X}. \ (j = 1, \ldots, k+1; X = A, B)$$
$$X'_{i,k+1} = X_{v_i^X, v_{k+1}^X}. \ (j = 1, \ldots, k; X = A, B)$$

The more front the component-pair in the matching sequence in the result, the better they match. We can set a threshold to filter the result based on this character since not all the components can match appropriately sometimes.

## 4  Results

We use the simulation toy model of MIG-29, the Soviet battleplane, as the test object. Above Figure. 4 has shown the fitting results using 2-dimensional Gaussian mixture model. Figure. 4 shows the execution results of above matching algorithm.

## 5  Discussion

As we can see from the fitting results of our model, we provide a compact, abstract and symbolic representation of object's contour. Each components can be seen as an assembled unit of object's contour. And it's very possible that components with similar parameters which can be treated as the same component w often appear in fitting results of different contour images of the same object. These components can be extracted as features with constancy of the object. And the constant relationships between constant components also can be described by components' difference vectors. We mine these constant components and relationships and can use them as the prototype of recognition to make it easier to understand the image further by the machine.

On one side, the features themselves are expressed as vectors constituted by groups of model parameters and all features of one object form a parameter

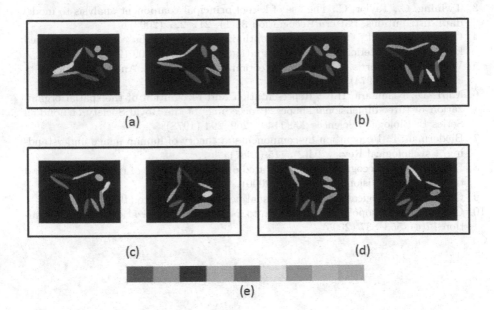

**Fig. 3.** (a) (d) shows 4 teams of matching result. Same color represents the components matched together. (e) shows the order of color from left to right which is same as the order of matching sequence. It means that the more left the color locates here, the better its corresponded components match.

matrix. It's very easy to store and calculate for computers. On the other hand, when we regard a component as a vertex and the component difference as an edge, we obtain a graph structure to describe the prototype of the object. The topological structure of this graph can reflect the large-scale structural features. In fact, we use bottom-up method to get this representation, then based on this graph structure, we can use recognition method which is top-down or combination with two directions. Thus it becomes possible to recognize images using the large-scale structural descriptions of object which is usually ignored by current popular algorithms. As a conclusion, this components-based representation and the graph descriptor at its further step have important potential value.

**Acknowledgments.** This work was supported by the 973 Program (Project No.2010CB327900), the NSFC major project (Project No. 30990263), and the NSFC project (No. 6115042).

# References

1. Lowe, D.G.: Distinctive image features from scale-invariant keypoints. International Journal of Computer Vision 60(2), 91–110 (2004)
2. Jain, A.K., Duin, R.P.W., et al.: Statistical pattern recognition: A review. IEEE Transactions on Pattern Analysis and Machine Intelligence 22(1), 4–37 (2000)

3. Twining, C., Taylor, C.: The use of kernel principal component analysis to model data distributions. Pattern Recognition 36(1), 217–227 (2003)
4. Chen, B., Liu, H., et al.: A kernel optimization method based on the localized kernel Fisher criterion. Pattern Recognition 41(3), 1098–1109 (2008)
5. Weisstein, N., Harris, C.S.: Visual detection of line segments: An object-superiority effect. Science (1974)
6. Marr, D., Nishihara, H.K.: Representation and recognition of the spatial organization of three-dimensional shapes. Proceedings of the Royal Society of London. Series B. Biological Sciences 200(1140), 269–294 (1978)
7. Biederman, I.: Recognition-by-components: a theory of human image understanding. Psychological Review 94(2), 115 (1987)
8. Biederman, I.: Recognizing depth-rotated objects: A review of recent research and theory. Spatial Vision 13(2-3), 2–3 (2000)
9. Chen, L.: Topological structure in visual perception. Science (1982)
10. Chen, L.: The topological approach to perceptual organization. Visual Cognition 12(4), 553–637 (2005)

# Web Metadata Extraction and Semantic Indexing for Learning Objects Extraction*

John Atkinson[1], Andrea Gonzalez[1], Mauricio Munoz[1], and Hernan Astudillo[2]

[1] Department of Computer Sciences, Universidad de Concepcion, Chile
atkinson@inf.udec.cl
[2] Department of Informatics, Universidad Tecnica Federico Santa Maria, Chile
hernan@inf.utfsm.cl

**Abstract.** In this work, a new approach to automatic metadata extraction and semantic indexing for educational purposes is proposed to identify learning objects that may assist educators to prepare pedagogical materials from the Web. The model combines natural language processing techniques and machine learning methods to deal with semi-structured information on the web from which metadata are extracted. Experiments show the promise of the approach to effectively extract metadata web resources containing educational materials.

## 1 Introduction

Academic curricula for primary and secondary schools provide a set of goals, mandatory contents, and expected outcomes in order to assess the different themes covered in each course. Teachers must then plan the way they will achieve these results and their evaluation by designing a teaching strategy that enables students to fullfil the expected learning and the minimum contents. To this end, educators must search for and select suitable educational resources, specific contents to be discussed, and necessary skills and knowledge for student to acquire during a course.

Most of the search for educational resources is carried out by using typical search engines. However, while these systems are quite effective to look for general-purpose materials, they are not designed to effective index, get and prepare suitable educational materials and learning objects timely for non-textual information. As a consequence, educators must spend a long time to collect material for course preparation. While there are fair techniques to index relevant documents, searching for and selecting educational resources is a time-consuming task which is more demanding in terms of effectiveness. This may partially be due to that actual information searching technologies do not consider the educator's specific methodological and educational needs so as to prepare effective lectures.

---

* This research was partially supported by the National Council for Scientific and Technological Research (FONDECYT, Chile) under grant number 1130035: *"An Evolutionary Computation Approach to Natural-Language Chunking for Biological Text Mining Applications"*.

M. Ali et al. (Eds.): IEA/AIE 2013, LNAI 7906, pp. 131–140, 2013.

Accordingly, this paper describes a simple multi-strategy approach for educational metadata extraction and semantic indexing. It aims to identify *Learning Objects* (LO) and so assist further tasks of educational material searching. Our model combines machine learning methods and *Natural Language Processing* (NLP) techniques to deal with semi-structured information usually existing on candidate web resources containing LO. The main claim of this work is that a combination of simple methods for educational metadata extraction and indexing may be effective so as to fullfil teachers' educational goals.

## 2    Related Work

In the context of educational technology assisting teachers, identifying *Learning Objects* (LO) and resources based on information extracted from the web must be an efficient and effective task. However, searching for billions of webpages on blind basis so as to select and extract LO makes it a very demanding task. In order to make it efficient, systems use focused crawlers [3] to collect and guide the process of information retrieval for webpages relevant to an educational target domain (i.e., History, Geography, etc).

In order to extract metadata, techniques must be capable of recognizing key implicit/explicit information contained on the extracted documents so as to identify and index LOs. They usually include webpages, text files, audio files, digital presentations, simulations, graphics, etc. Thus, obtaining LOs from the web can be seen as the problem of analyzing and extracting information from web documents [10, 8].

While these tasks have been approached sequentially, for several domains, this strategy spreads the classification error towards the information extraction task which reduces the overall effectiveness [10]. Hence retrieved web documents should contain common features so as to extract coherent metadata and to differentiate a target document from non-relevant information. However, this situation does not hold completely whenever different kinds of webpages are obtained such as are blogs, semi-structured documents with educational contents, forums, etc. In order to address this issue, text classification techniques using statistical and machine learning methods have been used. Modern approaches use more sophisticated methods based on terms weighting, statistical n-gram modeling, linguistic features, features combination or feature extraction and selection techniques [1].

Once relevant documents are retrieved, metadata extraction from LOs is performed. This is a complex and demanding task as learning materials do not contain only textual data but also multimedia, images, audio, video, etc. On the other hand, metadata do not only describe educational material but also they include knowledge on teaching methods and strategies, domains, relationship with other LOs, and so on [4, 7, 2].

Information extraction methods for educational purposes can be classified into two categories: resource-based methods and context-based methods [2]. Resource-based methods collect, extract, classify and spread metadata. The collection task involves harvesting previously-tagged metadata existing in the resource [5]. In order to extract metadata from LOs, several approaches have been applied from text mining and NLP [2]. These range from simple techniques such as regular expression matching to parsers and stochastic named-entity recognizers. For multimedia material and images containing text, OCR methods have been used to extract the underlying text which may contain key information to generate metadata from the LO [9, 7]. However, the methods depend also on the nature of the source [2, 11, 7].

## 3   A Multi-strategy Approach to Educational Metadata Extraction

In this section, a model to extract educational metadata and semantic indexing is proposed in the context of an existing framework for intelligent access to educational materials and LOs from the Web. The overall approach that combines several techniques starts off with a teacher's topic (query) which is used by a specific-purpose crawler to search for LOs on the Web. A metadata extractor then produces the contents for the LOs of the retrieved documents, and a semantic indexer assesses and feeds the crawler by using data provided by the metadata extractor. Finally, an *Ingester* builds educational strategies from the obtained domain metadata and the indexed resources so as to generate faceted educational material.

Initially, the crawler collects educational documents from the web by using a single seed and detects those webpages which might be relevant to the educational topic. For this, obtained webpages are indexed so as to get a relevance measure of each source (webpage or site) via semantic indexing facilities. This allows the crawler to determine whether is worth continuing to search for documents within a website. A key task of the relevance metrics is computed from lexicosemantical information provided by the educational metadata extractor.

The metadata extraction task combines state-of-the-art classifiers, simple NLP techniques and pattern matching methods to fill in data slots in the form of XML documents in which LOs are identified. As a side-effect, key underlying semantic information is extracted from the documents in order to guide the crawler. Unlike other models to educational metadata extraction, this is a first approach to deal with LOs in Spanish in educational domains in which additional resources are used (i.e., ontologies, educational strategies, etc). Thus, in the context of the general framework, two main tasks are proposed: *Semantic Indexing*, and *Metadata Extraction*.

### 3.1   Semantic Indexing

A focused crawler searches for LOs on the web by using a *Best-First* search method, in which a webpage hyperlink's structure is used to first collect those most promising webpages based on previously-generated seeds, and then it carries on with the inner webpages. To this end, an evaluation is performed for both a webpage's relevance according to the topics of interest, and the relationship of its hyperlinks with those topics.

This indexing task provides content-based information which is used to assess the relevance of a retrieved web document and its hyperlinks. In order to feed the crawler, a list of a document's topics is created containing weights associated to each webpage or hyperlink for such topics. For this, key information regarding the web documents is automatically identified such as main topic, key terms, and relevant passages of the text. On the other hand, hyperlinks are evaluated according to terms existing in the hyperlink's sourounding text and close keywords.

In order to generate a weighted list of topics, a hierarchy of themes was extracted from an ontology in the domain of History and Geography. This is based on the educational contents required by the Academic Curricula into different levels of education, and is organized by starting from domain general themes (i.e, World History, Geography, etc) downto specific topics. The ontology also contains concepts about educational resources, learning goals and teaching strategies, which are linked to domain's topics by using specific-purpose semantic relationships.

Next, in order to recognize a domain's topics, web documents are used to train a corpus-based semantic analysis method so as to further infer semantic closeness between unseen documents and categorized topics. Specifically, *Latent Semantic Analysis* (LSA) [6] was used to represent and relate semantic text data. Here, each topic has a vector representation (*semantic vector*) in a multi-dimensional space. LSA is used to generate semantic vectors representing implicit co-occurrence relationships between concepts within a corpus of textual documents. This knowledge is then used by our methods to determine the semantic closeness between topics and texts.

For our approach, each topic is seen as set of a text's features, and so LSA builds semantic vectors for each relevant word and topic of a text. Thus, a topic's semantic vector is represented as the vectorial sum of its defining terms. Next, semantic spaces for a web document itself and its hyperlinks are generated. It is then used to compute the semantic closeness to the vectors representing each topic. This similarity measure is proportional to the weight assigned to each topic so the crawler will be receiving a list as: $[topic_1(weight_1), topic_2(weight_2), ...]$

Weights are computed from information provided in the hyperlink's text and the neighbor terms by using two distinct methods. The final weights of the list $(L_t)$ are obtained by linearly combining the list of topics of the hyperlinks' text $(L_p)$ and the list of neighbor terms $(L_n)$ as: $L_t = \alpha * L_p + \beta * L_n$, where $\alpha$ and $\beta$ represent experimentally set weighting factors. Generally, the hyperlink's text should have a direct relation with its contents by expressing a phrase about it, whereas neighbor terms can refer to other elements of the webpage so they are less reliable to represent the hyperlink's context. Accordingly, weighting factors were experimentally assigned differently subject to $\alpha \geq \beta$:

- *Computing $L_p$:* uses the hyperlink's text obtained from the document retrieved by the crawler. This text is stemmed and then used to obtain its LSA semantic vector for each of its topics. This semantic vector provides information regarding the closeness between the hyperlink's text and the document's list of topics. Since topics are hierarchically related with each other according to an available ontology, it is easy to see that those closer to the leaves are more specific and therefore contribute more knowledge than those of the higher levels. Hence, this value is computed as $\frac{Similarity\_LSA(topic,hyperlink\_text)}{Reverse\_Position(topic)}$, where $Similarity\_LSA(text_1, text_2)$ represents the *cosine* distance between the semantic vectors of $text_1$ and $text_2$, respectively, and $Reverse\_Position(topic)$ is the reverse order of the position (level) of *topic* in the hierarchy.

- *Computing* $L_n$: LSA semantic vectors for all the neighbour terms is generated in order to compute the similarity between each term and the extracted topics. The weight for each topic is then calculated as the average LSA similarity between the semantic vectors of the neighbor $n$ terms and the topic, dividing it by the reverse position of the topic in the hierarchy as $\sum_{i=1}^{n} \frac{\frac{similarity\_LSA(topic,term_i)}{\#neighbor\_Terms}}{Reverse\_Position(topic)}$.

## 3.2 Metadata Extraction

Each retrieved document contains information which may be of interest to teachers in the target domain. Once the candidate webpage has been obtained according to the ranking and relevance, metadata are generated so as to identify properties of the underlying LOs. Some properties include the learning object's title, the terms characterizing the documents or more complex (or implicit) ones such as the document's language, the summary, scope of the LO, persons/organizations contributing the educational object, type of educational object, etc. In order to extract these kinds of implicit properties, our extraction approach combines techniques and methods for information collection, extraction and classification in order to fill the metadata's properties. Accordingly, the following tasks have been performed so as to extract metadata into an XML document:

- **Metadata Harvesting:** this task searches for key information contained in the metadata related to the current educational resource. For this, documents collected by the crawler are first provided to the semantic indexer by using an XML input document containing key basic data of the resource such as URL, title, etc. Next, the webpage referenced by the resource is looked for (i.e., RDF, OWL, or other resource format). Obtained documents are analyzed using an RDF parser which extracts all the basic and explicit required properties.
- **Metadata Extraction:** this task extract key basic and compound terms from available texts in the web documents or resources by carrying out terms selection and extraction. This has been applied to extract and rank terms based on their frequency in the corresponding texts, and so selected candidate terms become those exceeding some threshold. In addition, a *Named-Entity Recognition* (NER) task was carried out to identify more complex terms of phrases containing key information of the resource.

- **Metadata Classification:** Some properties of the document can not be directly extracted. This kind of information includes properties such as the document's language, scope, etc. An effective strategy to deal with this involves using text classifiers so as to infer the target properties of the documents. Thus, an n-gram language model was used to recognize language using the *TextCat*[1] tool. For other properties such as the document's scope, a Bayesian classifier was trained by using a training corpus built from web documents contained defined 'labels' for target scopes (i.e., History and Geography, World History, etc).

## 4    Experiments and Results

In order to assess the effectiveness of our model for educational metadata extraction and semantic indexing, two kinds of experiments were conducted:

1. *Classsification Accuracy:* different proposed extraction and classification methods were evaluated so as to classify and generate accurate metadata from educational webpages and resources.
2. *Quality of the extracted metadata:* a group of teachers was provided with metadata generated by our model and they were then required to assess the quality of the outcome.

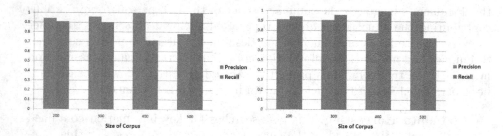

**Fig. 1.** Performance of a Naive Bayes Classifier for two different categories: History (Left) versus Others (Right)

For experimental purposes, web documents were extracted from the educational repository: *www.memoriachilena.cl* which keeps educational materials on Chilean history and geography. In addition, some cleaning and preprocessing tasks were carried out in order to generate metadata and enable semantic indexing:

- *Preparation of corpus for training classifiers and extractors:* a set of web documents were extracted and annotated in order to extract training and testing datasets for the classifier.

---

[1] http://odur.let.rug.nl/~vannoord/TextCat/

– *Corpus pre-processing:* features extraction and NER tasks were performed so as to extract and represent input vectors for the classifiers.

Some adjusting experiments for the classifiers using different features recognized by the NER task into different domains (History/Geography -**HG**- and Others -**O**-) are shown in figure 2. This shows classification accuracy (*Acc*) for the two classifiers using nouns and adjectives. Overall, a Bayes classifier show the best results compared to the SVM method.

| No. of Documents | Acc | Bayes | | Acc | SVM | |
|---|---|---|---|---|---|---|
| | | *HG* | *O* | | *HG* | *O* |
| 300 | 93% | 270 | 30 | 80% | 300 | 0 |
| | | 122 | 88 | | 199 | 101 |
| 500 | 92.6% | 475 | 25 | 0% | 500 | 0 |
| | | 43 | 372 | | 415 | 0 |

**Fig. 2.** Accuracy of Classifiers for Different NER Methods

In addition, different configurations for a Naive Bayes classifier were evaluated in order to predict document classes. The results of the experiments conducted for different corpus sizes and features can be seen at figure 1.

Figure shows good performance for both configurations (History and other domains). However, some slight decrease is observed for other domains, which may be due to the natural ambiguity: history of Chilean football can be labeled as a history document or as a 'sport' document. In order to investigate the extent to which our approach is effective to provide useful metadata to teachers in order for them to prepare a simple lecture, a series of experiments was also conducted. For this, metadata were extracted from web documents and they were then displayed in a friendly webpage. Each teacher (out of 16) checked and assessed the extracted metadata (100 XML-style metadata documents) according to certain defined quality criteria. Experiments analysed 9000 source webpages and extracted 3300 LOs, and it took about 30 minutes for each teacher to assess the extracted metadata.

Data were extracted from the original web documents where the metadata were extracted from. In order to evaluate the obtained data, teachers were asked to rank the quality with labels ranging from 'Very Good' to 'Very Bad' (figure 3 (left)). In addition, they included comments regarding the reasons behind their evaluations. For each source document, the types of data provided to teachers included: title of the educational resources, key terms, scope of the document, summary of the educational resource, etc. The ranking was based on a 5-level scale with the following meaning: Very Bad (expected data were not extracted but the information was indeed contained in the source resource), Bad Fair, Good, Very Good (extracted data matched the information of the original document, and they are complete and correct).

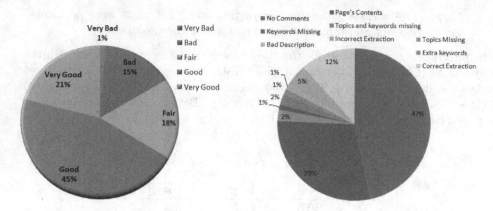

**Fig. 3.** Assessing Metadata Quality for the Metadata Extraction task (left) and Teachers' Reasons for Assessment (right)

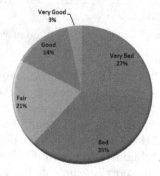

**Fig. 4.** Assessing Metadata Quality from using a traditional Text Retrieval method

For evaluations below 'Very Good', teachers provided reasons behind their assessment as summarized in data of figure 3 (right). Overall, the results can be grouped as follows:

- **Metadata Quality:** results indicate an average fair satisfaction ('Good' or 'Very Good') for 66% of the evaluations (see distribution in graphic 3 (left)). For below-average evaluations, they were mainly due to the fact that either source documents did not provide information enough for the teacher or they were not relevant. Reasons for this were the unaccurate extraction of data or the data incompleteness.
- **Source Documents Relevance:** results indicate a teacher's low satisfaction on delivered documents showing that the topic assignment may not be of high quality. Note that automatic evaluation of the relevance was provided by the semantic indexing component of our approach. In addition, teacher comments suggested that the educational material extracted was very specific so they could not be used to prepare a full lecture. This evaluation was

carried out by displaying topics sorted by their computed ranking of relevance according to the indexing task, thus the used model may need to be
adapted to use metrics beyond semantic similarity on individual topics.

On the other hand, Teachers also assessed the 'quality' of data retrieved in
terms of their ability to 'fill in' further metadata required to prepare a simple
lecture by using resources provided by a state-of-the-art text retrieval method.
Overall results of the assessment can be seen in figure 4. This experiment shows
low results in terms of the quality of the prepared data (almost 35% of 'Bad'
evaluations) and a fair quality was only seen for nearly 17% of the cases ('Good'
and 'Very Good') compared to almost 66% when using our model. In addition,
despite the low results for the traditional method, individual evaluations showed
that a few number of metadata extracted for the 'Very Good' evaluations by
our approach (10%), were also in the results retrieved by the search engine. This
may suggest that some documents and resources retrieved by the search engine
are structured enough to allow accurate metadata extraction.

## 5   Conclusions

In this paper, a new approach to metadata extraction and semantic indexing
aiming at identifying learning objects was presented in an educational context.
This allows to assist secondary teachers to search for teaching materials. The approach combines automatic classifiers, simple natural language processing techniques and corpus-based semantic in order to extract semi-structured data from
educational sources on the Web. This included both extracting metadata from
documents which has RDF-like structure and identifying key pieces of data in
plain webpages.

Different experiments show the promise of the approach to effectively provide
with relevant knowledge to teachers looking for educational materials. However,
a lot of information can be missed due to the different formats available on
the web documents which make it difficult to interpret and provide understandable information to users. Furthermore, the overall model shows promise when
comparing with traditional search facilities to teachers by using current search
engine.

## References

[1] Alpaydin, E.: Introduction to Machine Learning. The MIT Press (2004)
[2] Bauer, M., Maier, R., Thalmann, P.: Metadata generation for learning objects:
    An experimental comparison of automatic and collaborative solutions. e-Learning,
    181–195 (2010)
[3] Gauch, S., Wang, Q.: Ontology-based focused crawling. In: International Confer-
    ence on Information, Process, and Knowledge Management, pp. 123–128 (2009)
[4] Jain, S., Pareek, J.: Keyphrase extraction tool (ket) for semantic metadata an-
    notation of learning materials. In: International Conference on Signal Processing
    Systems, Singapore (2009)

[5] Jain, S., Pareek, J.: Automatic topic(s) identification from learning material: An ontological approach. In: Second International Conference on Computer Engineering and Applications, Bali Island, Indonesia (2010)

[6] Landauer, T., McNamara, D., Dennis, S., Kintsch, W.: Handbook of Latent Semantic Analysis. University of Colorado Institute of Cognitive Science Series. Lawrence Erlbaum Associates (2007)

[7] Meire, M., Ochoa, X., Duval, E.: Samgi: Automatic metadata generation v2. 0. In: Proceedings of World Conference on Educational Multimedia, Hypermedia and Telecommunications, vol. 2007, pp. 1195–1204 (2007)

[8] Nugent, G., Kupzyk, K., Riley, S., Miller, L.: Empirical usage metadata in learning objects. In: 39th ASEE/IEEE Frontiers in Education Conference, San Antonio, TX, USA (2009)

[9] Park, J., Lu, C.: Application of semi-automatic metadata generation in libraries: Types, tools, and techniques. Library and Information Science Research 31, 225–231 (2009)

[10] Ping, L.: Towards combining web classification and web information extraction: a case study. In: Proceedings of the 15th ACM SIGKDD International Conference on Knowledge Discovery and Data Mining, Paris, France, pp. 1235–1244 (2009)

[11] Xiong, Y., Luo, P., Zhao, Y., Lin, F.: Ofcourse: Web content discovery, classification and information extraction. In: The 18th ACM Conference on Information and Knowledge Management, Hong Kong (2009)

# Approximately Recurring Motif Discovery Using Shift Density Estimation

Yasser Mohammad[1,2] and Toyoaki Nishida[2]

[1] Assiut University, Egypt
[2] Kyoto University, Japan
yasserm@aun.edu.eg, nishida@i.kyoto-u.ac.jp

**Abstract.** Approximately Recurring Motif (ARM) discovery is the problem of finding unknown patterns that appear frequently in real valued timeseries. In this paper, we propose a novel algorithm for solving this problem that can achieve performance comparable with the most accurate algorithms to solve this problem with a speed comparable to the fastest ones. The main idea behind the proposed algorithm is to convert the problem of ARM discovery into a density estimation problem in the single dimensionality shift-space (rather than in the original time-series space). This makes the algorithm more robust to short noise bursts that can dramatically affect the performance of most available algorithms. The paper also reports the results of applying the proposed algorithm to synthetic and real-world datasets.

## 1 Introduction

Consider a robot watching a human communicating with another using free hand gestures to achieve some task [13]. The ability to automatically discover recurring motion patterns allows the robot to learn important gestures related to this domain. Consider an infant listening to the speech around it. The ability to discover recurring speech patterns (words) can be of great value in learning the vocabulary of language. In both of these cases, and in uncountable others, the patterns do not recur exactly in the perceptual space of the learner. These cases motivate our search for an unsupervised algorithm that can discover these kinds of approximately recurring motifs (ARMs) in general time-series. Several algorithms have been proposed for solving this problem [10] [7] [2],[15],[4] [6], [17].

In this paper we propose a novel algorithm for solving ARM discovery directly. The proposed algorithm achieves high specificity in discovered ARMs and high correct discovery rate and its time and space complexities can be adjusted as needed by the application. The main insight of the proposed algorithm is to convert the problem of subsequence density estimation which is multidimensional in nature into a more manageable single dimensional shift density estimation. This allows the algorithm to discover complete ARMs (with potential don't-care sections). The paper also reports a quantitative 6-dimensions evaluation criteria for comparing ARM discovery algorithms.

M. Ali et al. (Eds.): IEA/AIE 2013, LNAI 7906, pp. 141–150, 2013.

## 2    Problem Statement and Related Work

A time series $x(t)$ is an ordered set of $T$ real values. A subsequence $x_{i,j} = [x(i) : x(j)]$ is a contiguous part of a time series $x$. Given two subsequences $\alpha_{i,j}$ and $\beta_{k,l}$, a distance function $D(.,.)$ and a positive real value $R$, we say the two subsequences *match up to $R$* if and only if $D(x_{i,j}, x_{k,l}) < R$. We call $R$ the range following [16]. In this paper, we assume that the distance function $D(.,.)$ is normalized by length of its inputs. Moreover, our algorithm will only apply $D$ to pairs of subsequences of the same length. In most cases, the distance between overlapping subsequences is considered to be infinitely high to avoid assuming that two sequences are matching just because they are shifted versions of each other (these are called trivial motifs [5]).

An approximately recurrent motif (ARM) or motif for short is a set of subsequences that are similar in some sense. In most cases similarity between subsequences is measured as the inverse of their distance [14]. Either the Euclidean distance or dynamic time wrapping (DTW) could be used for this calculation. Relying on these distance functions in the definition of a motif implies that a predefined motif length must be given to the algorithm. Several algorithms were suggested to discover distance based motifs [10] [7] [2] [15] [4] [6] [8] [17]. Many of these algorithms are based on the PROJECTIONS algorithm proposed in [18] which uses hashing of random projections to approximate the problem of comparing all pairwise distances between $n$ subsequences to achieve linear rather than quadratic space and time complexities. Because this algorithm works only with discrete spaces, the time series must be discretized before applying any of PROJECTIONS variants to it. A common discretization algorithm employed for this purpose is SAX [6]. An unsupervised method for finding a sensible range parameter for these algorithms was proposed in [7]. The proposed algorithm differs from all of these approaches (even with automatic range estimation) in requiring no discretization step and being able to discover motifs in a range of lengths rather than a single length. The proposed algorithm also has adjustable space and time complexity and is linear in the worst case, while all PROJECTIONS algorithms require good selection of the discretization process parameters to lead to sparse collision matrices in order to avoid being quadratic.

Another approach for finding these motifs was proposed in [1] that uses random sampling from the time series (without any disretization). This algorithm requires an upper limit on the motif length and also is not guaranteed to discover any motifs or to discover them in order. An explicit assumption of this algorithm is that the motifs are frequent enough that random sampling will has a high probability of sampling two complete occurrences in candidate and comparison windows of lengths just above the maximum motif length. The sampling process was improved in MCFull [10] by utilizing a change point discovery algorithm to guide the sampling process with reported significant increase in discovery rate. Even though no clear definition of what is actually discovered by these algorithms (other than being frequent), they actually discover ARMs. The proposed algorithm has higher discovery rate than MCFull with comparable speed as will be shown in section 4.

The MK algorithm for discovering exact motifs was proposed in [14] and it is the most cited motif discovery algorithm since its appearance. In MK, the Euclidean distance between pairs of subsequences of length $l$ is used to rank motif candidates. This has the problem of requiring a predefined motif length (while our approach requires only a motif length range). It is also sensitive to short bursts of noise that can affect the distance. The main difference between the proposed approach and this algorithm is that we rely on multiple distance estimations between short subsequences rather than a single distance calculation of the predefined motif length. This has three major advantages: First we need not specify a specific length. Secondly, the distance function is not required to be a metric (i.e. it is not required to satisfy the triangular inequality). Finally, the proposed algorithm can *ignore* short bursts of noise inside the subsequences (because of its multi-distance calculations) which is not possible if MK is directly used. Nevertheless, the MK algorithm can be used to speedup finding best matches during shift-density estimation. This was not tried in this work but will be compared with the current implementation in future work. Hereafter, we will use the word *motif* and ARM interchangeably as long as the context is clear.

## 3   Proposed Algorithm

The algorithm uses three types of windows. The candidate window is a subsequence $s$ that is being considered for similar subsequences in the time series. The candidate window should be wide enough to contain a complete occurrence of any ARM to be discovered. The length of this window is called $w$. The random window is a time series of the same length as the candidate window and is constructed by randomly selecting $w$ values from the time-series. This window is used by the algorithm to discover an upper limit of distances that can be considered *small* during processing. The idea of using a random window for this purpose can be found in [1]. The third type of windows is the comparison window. Comparison windows are subsequences of $w$ points that are to be compared to the current candidate window in search for ARM occurrences within them.

The algorithm proceeds in three major steps: firstly, candidate locations of ARM occurrences are discovered using a change point discovery algorithm [11] and candidate windows are sampled around these points. This set is called $\mathsf{C}$ hereafter. Secondly, each one of these windows is compared with the rest of them (acting as comparison windows) and best matching windows are found as well as the best time-shifts in the best comparison windows to get it to best match the candidate window. This is the core step of the algorithm and is the point at which shift-density estimation is carried-out. Finally, the shifts required are analyzed in order to remove partial ARM occurrences, multiple ARM occurrences, and out-of-ARM parts of the candidate window and a new ARM is announced if a long enough occurrence could be found at that stage. The following subsections present the final two stages. For more details about the first stage please refer to [10].

## 3.1 Finding Best Matches

This step is the core of the proposed algorithm. Each member of $C$ is treated as a candidate window while the ones after it are treated as a comparison window set until all members of $C$ are considered. The current candidate window ($c$) is divided into $w - \bar{w}$ ordered overlapping subwindows ( $x_{c(i)}$ where $w$ is the length of the window, $\bar{w} = \eta l_{min}$, $1 \leq i \leq w - \bar{w}$ and $0 < \eta < 1$) The discovery accuracy is not sensitive to the choice of $\eta$ and we select $\eta = 0.5$ for the rest of this paper. The same process is applied to every window in the current comparison set. The following steps are then applied for each comparison window ($j$) for the same candidate window ($c$):

Firstly, the distances between all candidate subwindows and comparison sub-windows are calculated ($D\left(x_{c(i)}, x_{j(k)}\right)$ for $i, k = 1 : w - \bar{w}$). The distance found is then appended to the list of distances at the shift $i - k$ which corresponds to the shift to be applied to the comparison window in order to get its $k$-th subwinodw to align with the $i$-th subwindow of the candidate window. By the end of this process, we have a list of distances for each possible shift of the comparison window. Our goal is then to find the best shift required to minimize the summation of all subwindow distances between the comparison window and the candidate window. Our main assumption is that the candidate and comparison windows are *larger* than the longest ARM occurrence to be discovered. This means that some of the distances in every list are not between parts of the ARM occurrences (even if an occurrence happens to exist in both the candidate and comparison windows). For this reason we keep only the distances considered *small* from each list. This can be achieved by keeping the smallest $K$ distances from the list (where $K$ is a user-defined parameter). In this paper, we utilize a different approach that was first proposed in [1]. The idea is to generate a window of length $w$ from the time series by concatenating randomly selected samples from it. This window which is called the random window, is then compared to all the candidate windows and the mean distances between the $\bar{w}$ subwindows is then used as a measure of smallness. The algorithm also keeps track of the comparison subwindow indices corresponding to these *small* distances.

Finally, the comparison windows are sorted according to their average distance to the candidate window, with the best shift of each of them recorded. Comparison windows with an average distance greater than the *small* distance limit (found as described in the previous paragraph) are removed from the list to reduce the required processing time.

At the end of this process and after applying it to all candidate windows, we have for each member of $C$ a set of best matching members with the appropriate shifts required to align the ARM occurrences in them (if any).

An important advantage of this technique over the one proposed in [1] is that we need not have the complete ARM inside both the candidate and comparison windows because even if a part of the ARM occurrence is contained in one of them, the alignment process implicit in calculating the shifts will still discover the similarity between contained parts of the two occurrences. This is an important advantage of the proposed algorithm because it remedies any localization inaccuracies in the change point discovery step. All what we need is that the CPD algorithm discovers locations within $w - \bar{w}$ points from the true beginning or ending of the ARM occurrence. Even if parts of multiple occurrences are contained within the candidate or comparison windows (or both), the algorithm can automatically select the appropriate occurrence to consider from each.

In practice, it is not necessary to use the complete $w - \bar{w}$ set of subwindows, as long as the number of subwindows selected is large enough to cover the complete window. As a limiting case, we can select the number of subwindows to be $\frac{w}{\bar{w}}$.

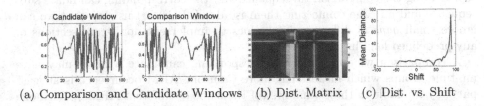

(a) Comparison and Candidate Windows     (b) Dist. Matrix     (c) Dist. vs. Shift

**Fig. 1.** Processing Steps During Best Matches Finding

Fig. 1 shows the processing steps of the proposed algorithm. Fig. 1-a shows a candidate window and a comparison window during the execution of the algorithm. The ARM occurrence in the candidate window is partial, yet the algorithm will be able to find the best fit between the two windows. Fig. 1-b shows the distances between pairs of $\bar{w}$ subwindows. Distances that correspond to subwindows of the occurrence in the two windows are much smaller than the distances elsewhere. During actual execution, this matrix need not be built but is shown here for illustration only. Fig. 1-c shows the mean distance as a function of the shift needed to align the subwindows. It is clear that the minimum of the distance happens when shifting the comparison window left by 47 positions and considering Fig. 1-a this is the correct shift required to match the two occurrences.

After finishing this step, we have for each one of the candidate subwindows a list of nearest comparison subwindows and the shift required to minimize the distance between them. This will be needed in the final step of the algorithm.

## 3.2   Stitching ARM Occurrences

The final step of the algorithm is to generate a set of ARMs each containing two or more ARM occurrences from the outputs of the best match finding stage. The output of this stage is an ARM graph where each clique corresponds to an ARM and each node to an occurrence. This graph is initialized to an empty graph and is filled incrementally as will be shown in this section.

The core data structure of this stage is the *matching matrix* which is constructed for each candidate window in order. Assuming we have $m$ candidate windows and $n$ candidate subwindows in every candidate window ($n = w - \bar{w}$), then this matrix is a square $m - 1 \times n$ matrix with each row corresponding to a comparison window and each column corresponding to a subwindow. The value at element $(i, j)$ is the shift required to align the subwindow represented by the column $i$ with its nearest subwindow in the comparison window $j$.

Using the matching matrix, we calculate the number of *contiguous* subwindows of each comparison window that match a contiguous set of subwinodws in the candidate window (e.g. having the same shift value in the matrix for more than one column). If the lengths of the resulting comparison and candidate subsequences (which is called a group) are larger than or equal to $l_{min}$ (the minimum acceptable ARM length), a node is added (if not existing) to the ARM graph representing the comparison subsequence and the corresponding candidate subsequence and an edge connecting them is added. It is at this step that we can ignore small *gaps* of different shift values to implement don't-care sections of any predefined length.

A comparison window and the corresponding candidate window may have multiple groups which means that more than one ARM occurrence is at least partially available in these windows (of the same or different ARMs). Each one of these groups is added the graph (a new node is added only if the subsequence it represents is not existing in the graph).

By the end of this process, the ARM graph is populated and each clique of this graph represents an ARM.

# 4   Evaluation

The proposed algorithm was evaluated in comparison with other ARM discovery algorithms using synthetic data for which the exact locations of ARM occurrences is known. The algorithm was then applied to detection of gestures from accelerometer data and motion pattern discovery for a mobile robot simulation. This section presents these experiments. The source code of the proposed algorithm and the other four algorithms it was compared with as well as the data of these experiments are available from the authors as a part of a complete change point discovery and ARM discovery MATLAB/Octave toolbox in the supporting page of this paper at [3].

Comparing ARM discovery algorithms is not a trivial task due to the large number of possible errors that these algorithms can fall into. Discovered ARMs may cover a complete real ARM, a part of it, multiple real ARMs or nothing at all. Another problem is that an algorithm may succeed in covering all real ARMs but on the expense of adding extra parts from the time series around their occurrences to its discovered ARMs. There is no single number that can capture all of these possible problems. Nevertheless, a quantitative comparison is necessary to assess the pros and cons of each algorithm for specific tasks or types of time series. In this paper we compare algorithms along six performance dimensions.

Assume that we have a time-series $x$ with $n_T$ embedded ARMs ($\{\Xi_i\}$) where $1 \leq i \leq n_T$ and each ARM ($\Xi_i$) contains $\mu_i$ occurrences ($\{\xi_k^i\}$) for $1 \leq k \leq \mu_i$. Assume also that applying an ARM discovery algorithm to $x$ generated $n_D$

ARMs ($\{M_j\}$) where $1 \leq j \leq n_D$ and each *discovered* ARM ($M_j$) contains $o_j$ occurrences ($\{m_p^j\}$) for $1 \leq p \leq o_i$. Given this notation we can define the following performance measures for this algorithm:

*Correct Discovery Rate (CDR):* The fraction of discovered ARMs ($M_j$s) for which each occurrence $m_p^j$ is overlapping one and only one true ARM occurrence $\xi_k^i$ and all these covered true ARM occurrences ($\xi_k^i$s) are members of the same true ARM ($\Xi_i$).

*Covering Partial-ARMs Rate (CPR):* The fraction of discovered ARMs ($M_j$s) not covered in CDR because at least one occurrence is not covering any real motif occurrence.

*Covering Multiple-ARMs Rate (CMR):* The fraction of discovered ARMs ($M_j$s) for which at least one $m_p^j$ is overlapping one real motif $\Xi_i$ and at least one other occurrence $m_q^j$ is covering a different motif $\Xi_k$ .

*Covering No-ARMs Rate (CAR):* The fraction of discovered ARMs ($M_j$s) for which all occurrences $m_p^j$ are overlapping no true ARM occurrences.

*Covered (C):* The fraction of the time-series sequences represented by true ARM occurrences ($\xi_k^i$s) that are covered by at least one discovered ARM occurrence $m_p^j$. C will always be between zero and one and represents the sensitivity of the algorithm.

*Extras (E):* The length of the time-series sequences represented by discovered ARM occurrences ($m_p^j$s) that cover no true ARM occurrence $\xi_k^i$. E will always be a positive number and represents the specificity of the algorithm.

The higher CDR and C and lower CPR, CMR, CAR, and E, the better the algorithm.

## 4.1   Synthetic Data

As a first experiment, we evaluated the proposed algorithm against four ARM discovery algorithms using synthetic data with embedded ARMs. The first comparison algorithm is MCFull which was proposed in [10] as an improvement of the basic sampling algorithm of [1]. The second and third algorithms are variations of the GSteX (Greedy Stem Extension) system proposed in [9]. This algorithm utilizes a different approach as it builds the distance graph directly from short subwindows without the shift estimation step. GSteX uses a large distance matrix that can easily become superlinear depending on the number of change points discovered. We also compare the proposed system with PROJECTIONS as explained in [16] and [2]. This algorithm utilizes SAX [5] to discritize the timeseries then applies random projections based on the work of [18]. This algorithm does not require the CPD step but it requires the specification of an exact ARM length as well as a range parameter of *near* distances.

The test data consisted of 50 timeseries of length between 2000 and 4000 points each (depending on the total number of embedded ARM occurrences) that were generated randomly from a uniform distribution ranging from -1 to 1. Depending on the experiment, a number of random ARM patterns were generated and embedded into the database and noise was then added to the complete time series. Random ARM patterns are very challenging for our CPD as there

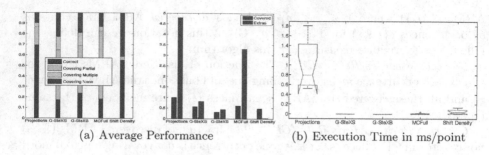

(a) Average Performance         (b) Execution Time in ms/point

**Fig. 2.** Average Performance of all Algorithms

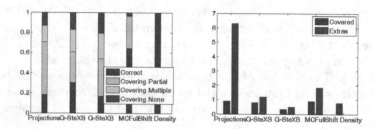

**Fig. 3.** Performance of all algorithms averaged over localization errors

is no underlying structure to be utilized in finding the change points. Because PROJECTIONS can only discover ARMs of a fixed predefined length, we selected $l_{min} = l_{max} = 60$ for all of our experiments.

Fig. 2(a) shows the overall performance of the five algorithms averaged over all noise levels in the previous data-set. The best performing algorithm in terms of correct discovery rate was PROJECTIONS with an average CDR of 0.69 followed by the proposed algorithm with 0.58 CDR. In terms of covered fractions/extras balance PROJECTIONS showed the highest covered fraction (highest sensitivity) but with highest extras fraction (lowest specificity). MCFull followed in terms of covered fraction but with still low specificity. The proposed algorithm showed comparable sensitivity to MCFull but with much higher specificity. Notice that these results are averaged over all noise levels. Fig. 2(b) shows the execution time in milliseconds per point for each algorithm. The proposed algorithm achieved an order of magnitude increase in speed compared with PROJECTIONS. For long time series that are encountered in real-world situations, this improvement in speed and the high specificity of the algorithm may compensate for the small reduction in correct discovery rate.

Fig. 3 shows the overall performance of all algorithms averaged over localization error. Even though the figure shows that the proposed algorithm outperforms the other algorithms in terms of correct discovery rate and specificity (E) and with comparable sensitivity to the other best performing algorithms (C), these results should not be taken at face value. Because in this experiment, CPD was performing better than what we would expect in real applications, PRO-JECTIONS was unfairly penalized due to its inability to utilize this information. These results represent more of an *asymptotic* behavior as the performance of the CPD algorithm improves. Nevertheless, the comparison of the proposed algorithm with MCFull, GSteXS, and GSteXB was fair and show that the proposed algorithm has higher potential of employing improvements in change point discovery accuracy.

## 4.2   Real World Evaluations

The first application of the proposed algorithm to real world data was in gesture discovery. Our task is to build a robot that can be operated with free hand-gestures without any predefined protocol. The way to achieve that is to have the robot *watch* as a human subject is guiding another robot/human using hand gestures. The learner then discovers the gestures related to the task by running our proposed ARM discovery algorithm to the data collected from an accelerometer attached to the tip of the middle finger of the operator's dominant hand. We collected only 13 minutes of data during which seven gestures were used. The data was sampled 100 times/second leading to a 78000 points 3D time-series. The time-series was converted into a single space time series using PCA as proposed in [12]. The proposed algorithm as well as GSteXS were applied to this projected time-series. The proposed algorithm discovered 9 gestures, the top seven of them corresponded to the true gestures (with a discovery rate of 100%) while GSteXS discovered 16 gestures and the longest six of them corresponded to six of the seven gestures embedded in the data (with a discovery rate of 85.7%) and five of them corresponded to partial and multiple coverings of these gestures.

As another proof of concept experiment, we employed a simulated differential drive robot moving in an empty arena of area $4m^2$. The robot had the same dimensions as an e-puck robot and executed one of three different motions at random times (a circle, a triangle and a square). At every step, the robot selected either one of these patterns or a random point in the arena and moved toward it. The robot had a reactive process to avoided the boundaries of the arena. Ten sessions with four occurrences of each pattern within each session were collected and the proposed algorithm was applied to each session after projecting the 2D time-series into a 1-D time-series as in the previous case. The algorithm discovered 3 motifs corresponding to the three motion patterns. In this case there were no partial motifs or false positives and discovery rate was 100%.

## 5   Conclusions

In this paper, we proposed a new algorithm for discovering approximately recurrent motifs in time series based on shift density estimation. The main insight behind the algorithm is to convert the problem from a density estimation in the high-dimensionality time-series subsequences space into a more manageable density estimation in the single dimensional shift space. The proposed algorithm require only the specification of a lower limit on ARM occurrence lengths and can discover multiple ARMs at the same time.

# References

1. Catalano, J., Armstrong, T., Oates, T.: Discovering patterns in real-valued time series. In: Fürnkranz, J., Scheffer, T., Spiliopoulou, M. (eds.) PKDD 2006. LNCS (LNAI), vol. 4213, pp. 462–469. Springer, Heidelberg (2006)
2. Chiu, B., Keogh, E., Lonardi, S.: Probabilistic discovery of time series motifs. In: KDD 2003, pp. 493–498. ACM, New York (2003)
3. CPMD Toolbox, http://www.ii.ist.i.kyoto-u.ac.jp/~yasser/cpmd/cpmd.html
4. Jensen, K.L., Styczynxki, M.P., Rigoutsos, I., Stephanopoulos, G.N.: A generic motif discovery algorithm for sequenctial data. BioInformatics 22(1), 21–28 (2006)
5. Keogh, E., Lin, J., Fu, A.: Hot sax: efficiently finding the most unusual time series subsequence. In: IEEE ICDM, p. 8 (November 2005)
6. Lin, J., Keogh, E., Lonardi, S., Patel, P.: Finding motifs in time series. In: The 2nd Workshop on Temporal Data Mining, pp. 53–68 (2002)
7. Minnen, D., Starner, T., Essa, I., Isbell, C.: Improving activity discovery with automatic neighborhood estimation. Int. Joint Conf. on Artificial Intelligence (2007)
8. Minnen, D., Essa, I., Isbell, C.L., Starner, T.: Detecting Subdimensional Motifs: An Efficient Algorithm for Generalized Multivariate Pattern Discovery. In: IEEE ICDM (2007)
9. Mohammad, Y., Ohmoto, Y., Nishida, T.: G-steX: Greedy stem extension for free-length constrained motif discovery. In: Jiang, H., Ding, W., Ali, M., Wu, X. (eds.) IEA/AIE 2012. LNCS, vol. 7345, pp. 417–426. Springer, Heidelberg (2012)
10. Mohammad, Y., Nishida, T.: Constrained motif discovery in time series. New Generation Computing 27(4), 319–346 (2009)
11. Mohammad, Y., Nishida, T.: Robust singular spectrum transform. In: Chien, B.-C., Hong, T.-P., Chen, S.-M., Ali, M. (eds.) IEA/AIE 2009. LNCS, vol. 5579, pp. 123–132. Springer, Heidelberg (2009)
12. Mohammad, Y., Nishida, T.: On comparing SSA-based change point discovery algorithms. In: 2011 IEEE/SICE IIS, pp. 938–945 (2011)
13. Mohammad, Y., Nishida, T., Okada, S.: Unsupervised simultaneous learning of gestures, actions and their associations for human-robot interaction. In: IEEE/RSJ IROS, pp. 2537–2544. IEEE Press, Piscataway (2009)
14. Mueen, A., Keogh, E., Zhu, Q., Cash, S.: Exact discovery of time series motifs. In: Proc. of 2009 SIAM (2009)
15. Oates, T.: Peruse: An unsupervised algorithm for finding recurring patterns in time series. In: IEEE ICDM, pp. 330–337 (2002)
16. Patel, P., Keogh, E., Lin, J., Lonardi, S.: Mining motifs in massive time series databases. In: IEEE ICDM, pp. 370–377 (2002)
17. Tang, H., Liao, S.S.: Discovering original motifs with different lengths from time series. Know.-Based Syst. 21(7), 666–671 (2008)
18. Tompa, M., Buhler, J.: Finding motifs using random projections. In: 5th Intl. Conference on Computational Molecular Biology, pp. 67–74 (April 2001)

# An Online Anomalous Time Series Detection Algorithm for Univariate Data Streams

Huaming Huang, Kishan Mehrotra, and Chilukuri K. Mohan

Department of EECS, Syracuse University
{hhuang13,mehrotra,ckmohan}@syr.edu

**Abstract.** We address the online anomalous time series detection problem among a set of series, combining three simple distance measures. This approach, akin to control charts, makes it easy to determine when a series begins to differ from other series. Empirical evidence shows that this novel online anomalous time series detection algorithm performs very well, while being efficient in terms of time complexity, when compared to approaches previously discussed in the literature.

**Keywords:** Outlier detection, anomalous time series detection, real time detection, online algorithms.

## 1 Introduction

Anomaly detection is important in multiple domains, and a special case is the identification of time series whose characteristics are substantially different from a large set of other time series. For example, the behavior over time of the stock price of a company may exhibit significant deviations from those of other companies in a comparison group, suggesting an underlying problem that should trigger actions by stock traders. In order to carry out such actions, it is important to detect such deviations as they occur in time, else the trader may be too late and may suffer losses due to delays in analysis and decision-making. Hence the focus of our research is on *online anomalous time series detection*, also known as real-time detection. In the research reported in this paper, a time series is identified as anomalous if it satisfies one of the following conditions:

- It contains subsequences that vary significantly from other time series in the database; or
- The time series is generated by a mechanism or process that is substantially different from the others.

Typical anomaly detection problem requires detecting the occurrence of an unusual subsequence within a time series. This problem has been addressed by various researchers [1, 6, 7, 11–14].

- Yamanishi, et al. [14], address this problem using a probability-based model in non-stationary time series data.
- Wei, et al. [13], use symbolic aggregation representation (SAX) of time series.

M. Ali et al. (Eds.): IEA/AIE 2013, LNAI 7906, pp. 151–160, 2013.

- Keogh, et al., [6,7] employ suffix tree comparison, HOT SAX algorithm, and nearest non-self match distance.
- Toshniwal, et al., [12] extended HOT SAX algorithm to detect outliers in streaming data time series.

Sometimes, the same external cause may result in abrupt changes in most of the time series being considered, e.g., increases in stock prices of most defense contractors as a result of an announcement of impending military conflict. This should not be considered as a case of one time series being anomalous with respect to other time series.

The online anomaly detection problem can be contrasted with a simpler problem that has been addressed by many researchers, in which a training set consisting of non-anomalous time series is available for the application of a learning algorithm, resulting in a trained model that represents the behavior of non-anomalous time series. A new time series can then be compared with this model, using a distance metric and thresholding to detect whether the new series is anomalous. Chandola [2] has summarized such anomalous time series detection techniques into four types: (1) Kernel based techniques; (2) Window based; (3) Predictive; and (4) Segmentation based.

- Leng, et al. [9], use a new time series representation method, based on key points and dynamic time warping to find the anomalous series.
- Chan, et al. [1], propose to use the Greedy-Split algorithm to build up a box model from multiple training series, and then use the model so obtained to assign the anomaly score for each point in a series.
- Sadik, et al. [11], developed an algorithm based on adaptive probability density functions and distance-based techniques to detect outliers in the data streams.

Unfortunately, these methods require training data which is not usually available, and are also not useful in many scenarios in which detection is required while data is being collected, since the primary purpose of anomalous time series detection is to take appropriate action in a timely manner.

In contrast with the above methods, our focus is on the important variation where multiple time series are being observed simultaneously, and detection should occur as soon as one (or more) series begins to differ from the rest. No training set is required, and no model is constructed *a priori*. Some methods mentioned earlier, such as [11, 13], can be used directly for online detection; other methods, such as [2], can be adapted to online detection. But the main problems found in these methods are:

- Finding the optimal values of parameters of the algorithms requires the user to have expert domain knowledge which may not be available.
- High amount of computational effort is required, so that these methods are not suitable for online detection in large data streams.
- Concept drift and data uncertainty are common problems in the data streams, which prevent successful application of fixed model based methods.

To overcome these problems, we propose an algorithm that utilizes information from multiple distance metrics. This unsupervised method requires only one parameter, viz., the number of nearest neighbors ($k$). Experimental results show that this approach is very successful in anomalous time series detection, and is also computationally efficient enough to permit online execution.

In Section 2, we begin with the multiple distance approach for time series anomaly detection, and then show how we can extend it to online detection. Experiments and results are discussed in Section 3, followed by conclusions in Section 4.

## 2   The Online MUDIM Algorithm

In earlier work [4], we proposed MUDIM, an anomaly detection method based on the following three distance measures:

- $dist_f$: Standard deviation of differences between two time series (DiffStD);
- $dist_t$: Same trend (STrend), which identifies the degree of synchronization of a series with another series; and
- $dist_s$: Proposed by SAXBAG [10], which captures patterns of a time series using a histogram of possible patterns.

Normalizing (by a linear transformation to the [0,1] interval) and averaging these three metrics results in a comprehensive and balanced distance measure that is more efficient and sensitive than the constituent (or other) individual measures.

### 2.1   Naive Online MUDIM Algorithm(NMUDIM)

All three distance measures (mentioned above) can be updated for incremental changes, as follows.

- **DiffStD:** Since the variance (square of the standard deviation) of differences between series $i$ and $j$ at time $n$ can be calculated as:

$$\text{dist}_f(i,j) = \frac{n \times ssq(i,j) - (sqs(i,j))^2}{n \times (n-1)}, \qquad (1)$$

$$ssq(i,j) = \sum_{t=1}^{n}(x_i(t) - x_j(t))^2; sqs(i,j) = \sum_{t=1}^{n}(x_i(t) - x_j(t)). \qquad (2)$$

and the numerator can be updated for the $(n+1)$th observations by adding $(x_i(n+1) - x_j(n+1))^2$ and $(x_i(n+1) - x_j(n+1))$, respectively.
- **STrend:** Let $x_i'(n) = x_i(n) - x_i(n-1)$. Then, by definition,

$$S_{i,j}(n) = \begin{cases} 1 & \text{if } x_i'(n) \cdot x_j'(n) > 0 \text{ or } x_i'(n) = x_j'(n) = 0 \\ 0 & \text{otherwise} \end{cases}$$

Consequently,

$$\text{dist}_t(i,j) = \frac{\sum_{t=2}^{n} S_{i,j}}{n-1}. \tag{3}$$

Therefore, to update it, we fix the numerator by adding the last trend $S_{i,j}(n)$ and accordingly modify the denominator as well.

- **SAXBAG** is based on SAX. It converts the data segment in the sliding window of size $w$ to a single SAX word and then counts the frequencies $f_i$ of each word. When data at time $n+1$ is observed, a new SAX word will be generated based on $\{x_i(n+2-w), x_i(n+3-w), x_i(n+4-w), ..., x_i(n+1)\}$. The stored data set can be updated to account for the new SAX word.

Based on the above equations, we first propose the "naive" online detection algorithm NMUDIM presented below. Anomaly scores, $O_i$'s, for each time series can be plotted, for example see Figure 1.

---

## Algorithm NMUDIM

**Require:** a positive integer $k$ (number of nearest neighbor) , $l$ (initial length for preprocessing) and data stream sets $\mathcal{D}$.
**Ensure:** anomaly score $O$ for each series in $\mathcal{D}$.

1: $O = \emptyset$;
2: Initially length of all series in $\mathcal{D}$ is $l$.
3: **for** $i \in \mathcal{D}$ **do**
4:    Calculate and store $sax_i, f_i$;
5:    **for** $j \in \mathcal{D}$ **do**
6:       Calculate $\text{dist}_{msm}(i,j)$;
7:       Calculate and store $S_{i,j}, ssq(i,j), sqs(i,j)$;
8:    **end for**
9: **end for**
10: **while** new observations at time $\tau$ arrive **do**
11:    **for** $i \in \mathcal{D}$ **do**
12:       Update $f_i, sax_i$;
13:       **for** $j \in \mathcal{D}$ **do**
14:          Update $S_{i,j}, ssq(i,j), sqs(i,j)$ and $\text{dist}_s(i,j), \text{dist}_t(i,j), \text{dist}_f(i,j)$ based on eqs 1 to 3;
15:          Update $\text{dist}_{msm}(i,j)$;
16:       **end for**
17:    **end for**
18:    **for** $i \in \mathcal{D}$ **do**
19:       $A_i(k) = \frac{\sum_{j \in \mathcal{N}_k(i)} \text{dist}_{msm}(i,j)}{|\mathcal{N}_k(i)|}$;
20:       $O_i(\tau) = A_i(k)$;
21:    **end for**
22: **end while**

---

Ignoring the length of a time series, the time complexity of NMUDIM is $O(n^2)$, because in MUDIM and NMUDIM we calculate distances $dist_l(i,j)$, $l = s, t, f$ for all $i \neq j$. In addition, $kNN$ of series $i$ are evaluated for each $i$. In the next subsection we propose a method to reduce this complexity (of $kNN$).

## 2.2    Faster Online Detection of MUDIM (OMUDIM)

Evaluation of $k$-nearest neighbor is computationally more complex than using any $k$-neighbors of a time series. Moreover, if $A'_j(k)$ denotes the average distance of any $k$-neighbors, then, clearly,

$$A_j(k) \leq A'_j(k). \tag{4}$$

In addition, if we can find a threshold $\lambda$ such that $A'_j(k) < \lambda$ implies $j$ is not an anomalous series; then most of the non-anomalous series can be excluded from anomaly score calculations. This is the essence of OMUDIM. To find an estimate of the threshold, $\lambda$, we apply the following sampling procedure: we calculate $A_j(k)$'s for $j \in \mathcal{S}$, where $\mathcal{S}$ contains $\alpha \times 100$ percent of $\mathcal{D}$, randomly selected. Then $\lambda$ is chosen as the top $(\beta \times 100)$th percentile of $A_j(k)$'s in descending order. Based on the above observations, we propose OMUDIM, an improved online version of MUDIM, whose key steps are as follows:

1. Find $\lambda$ as described above.
2. For $x_i \in \mathcal{D} - \mathcal{S}$, maintain a binary max heap consisting of $dist_{msm}(i,j)$ where $j$'s are selected $k$ neighbors of $i$. If the average of these $k$ neighbors is less than $\lambda$, series $i$ is declared as non-anomalous. Else $dist_{msm}(i,j)$ is calculated for next selected value of $j$, heap is updated by keeping only the smallest $k$ values of $dist_{msm}$'s; and anomalousness of series $i$ is tested using the above criterion. This process stops if at any stage, the series is found to be non-anomalous or no $j$ is left.
3. Calculate the anomaly scores of all potential anomalous series (found in Step2) and find the anomalous series, if any.
4. The above step are repeated once new observations arrive.

The details of OMUDIM are given in Algorithm OMUDIM. By applying these techniques, the time complexity of MUDIM is considerably reduced.

## 3    Experiments

In the experiments, we use eleven datasets, consisting of three synthetic datasets and eight modified real datasets from [8] and [5], to assess the effectiveness of the proposed algorithm. Key characteristics of the datasets are shown in the Table 1. Performance of our algorithms, measured in terms of capturing anomalous series, with existing online algorithms is presented in Table 2. We also calculate the running time of the algorithm for synthetic data streams. The comparison of computational effort between the NMUDIM and OMUDIM are also shown in Table 3.

---

**Algorithm OMUDIM**

---

**Require:** a positive integer $k$ (number of nearest neighbor) , $l$ (initial length for preprocessing), a real number $\alpha \in [0..1]$( proportion of random sampling),$\beta \in [0..1]$(proportion of top used for threshold selection) and data stream sets $\mathcal{D}$.

**Ensure:** anomaly score $O$ for each series in $\mathcal{D}$.

1: $O = \emptyset$.
2: Initially length of all series in $\mathcal{D}$ is $l$.
3: **for** $i \in \mathcal{D}$ **do**
4:     Calculate and store $sax_i, f_i$.
5:     **for** $j \in \mathcal{D}$ **do**
6:         Calculate $\text{dist}_{msm}(i,j)$
7:         Calculate and store $S_{i,j}, ssq(i,j), sqs(i,j)$.
8:     **end for**
9: **end for**
10: **while** new observations at time $\tau$ arrive **do**
11:     Find $\mathcal{S}$ which is randomly selected $\alpha$ proportion of the series from $\mathcal{D}$;
12:     **for** $i \in \mathcal{S}$ **do**
13:         **for** $j \in \mathcal{D}$ **do**
14:             Calculate $A_i(k)$ based on Algorithm **NMUDIM**.
15:         **end for**
16:     **end for**
17:     Set threshold $h = A_\beta(k)$, where $A_\beta(k)$ is the top $(\beta \times 100)$th percentile of $A_i(k)$'s in descending order.
18:     Mark $i$ as non-anomalous if its $A_i(k)$ is less than $h$.
19:     **for** $i \in \mathcal{D} - \mathcal{S}$ **do**
20:         **for** $j \in \mathcal{D}$ **do**
21:             Update $S_{i,j}, ssq(i,j), sqs(i,j)$ and $\text{dist}_{msm}(i,j)$ based on eqs 1 to 3
22:             Update $heap_i$ and $heap_j$ with $\text{dist}_{msm}(i,j)$ where $heap_i$ and $heap_j$ contain the $k$ minimum $\text{dist}_{msm}$ for series $i,j$
23:             **if** $average(heap_j) < h$ and $heap_j.size = k$ **then**
24:                 mark $j$ as non-anomalous;
25:             **end if**
26:             **if** $average(heap_i) < h$ and $heap_i.size = k$ **then**
27:                 mark $i$ as non-anomalous; **break;**
28:             **end if**
29:         **end for**
30:     **end for**
31:     **for** $i \in \mathcal{D}$ **do**
32:         **if** $i$ is marked **then**
33:             $A_i(k) = 0$.
34:         **else**
35:             $A_i(k) = \frac{\sum_{j \in \mathcal{N}_k(i)} \text{dist}_{msm}(i,j)}{|\mathcal{N}_k(i)|}$.
36:         **end if**
37:         $O_i(\tau) = A_i(k)$
38:     **end for**
39: **end while**

---

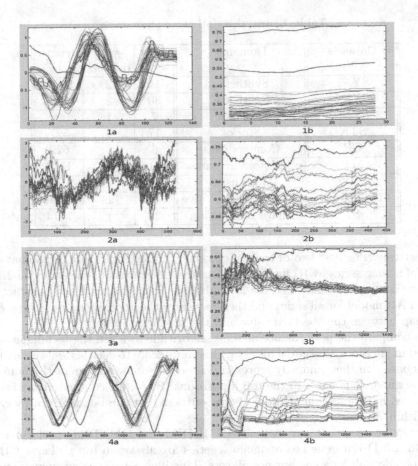

**Fig. 1.** Four datasets (from top to bottom) : SYN2, stocks, motor, and shape1. (a) Original time series; (b) We plot the online anomaly scores for each series at time 100+x. Anomalous series are marked as red.

## 3.1   Results

In all experiments, the initialization is performed at $l = 100$ observations and the rest of the observations of the series are used to test the effectiveness of the proposed algorithm. Number of nearest neighbors is set at $k = 5$ for all datasets. All figures are obtained based on results of NMUDIM.

We show details for four datasets in Figure 1; the data series are plotted in (a) and the performance of the algorithm in (b). As more data arrives and if more unusual patterns occur, the anomaly score increases and the gap between anomalous and normal series becomes larger. Normal series' anomaly scores converge if they are similar to each other.

We compare the performance of our algorithms with three other online detection algorithms based on (a) Euclidean distance, (b) Dynamic Time Warping (DTW), and (c) Autoregressive (AR) approach, proposed by [2], [7] and [3]

**Table 1.** Brief descriptions of all datasets

| Datasets | Domains | # of series | # of outliers | Length |
|----------|---------|-------------|---------------|--------|
| SYN1 | Synthetic | 14 | 2 | 500 |
| SYN2 | Synthetic | 30 | 2 | 128 |
| SYN3 | Synthetic | 7 | 1 | 500 |
| STK | Stocks | 17 | 1 | 527 |
| CP | Commodities | 6 | 1 | 629 |
| MT | Motor current | 21 | 1 | 1500 |
| PW | Power usage | 51 | 7 | 672 |
| T14/T16/T17 | NASA Shuttle Valve | 5 | 1 | 1000 |
| S1 | Image recognition | 21 | 1 | 1614 |

respectively. The first two of these methods calculate a measure of anomalousness of a time series by (i) finding the $k$-nearest neighbors of the series, and (ii) use the average distance of these $k$-neighbors. The third method constructs a global AR model for all series and then measures the anomaly score at time $t$ as the gap between the observed value and the predicted value.

To compare the performance of these algorithms, we use the technique suggested in [2]. Since all algorithms assign anomaly score to each series, we sort the series based on their anomaly scores in descending order and count the number of true anomalous series in the top $p$ positions. Performance of an algorithm is measured in terms of the number of true outliers that can be identified by each algorithm.

It can be seen that our methods (NMUDIM and OMUDIM) perform very well for all 11 datasets; i.e., anomalous series are always in top $p$ places. Other methods do well for some but not all sets. This illustrates that no single "pure" metric is sufficient for capturing multiple types of anomalies.

**Table 2.** Performance of all algorithms. Numbers show the true outliers that are identified by algorithms within top $p$ positions.

| Datasets | $p =$# of true outliers | Euclid | AR | DTW | NMUDIM | OMUDIM |
|----------|------------------------|--------|-----|-----|--------|--------|
| SYN1 | 2 | 2 | 1 | 2 | 2 | 2 |
| SYN2 | 2 | 2 | 2 | 2 | 2 | 2 |
| SYN3 | 1 | 0 | 1 | 0 | 1 | 1 |
| STK | 1 | 0 | 0 | 0 | 1 | 1 |
| CP | 1 | 0 | 1 | 0 | 1 | 1 |
| MT | 1 | 1 | 0 | 1 | 1 | 1 |
| PW | 7 | 7 | 1 | 7 | 7 | 7 |
| T14 | 1 | 1 | 0 | 1 | 1 | 1 |
| T16 | 1 | 1 | 0 | 1 | 1 | 1 |
| T17 | 1 | 1 | 0 | 1 | 1 | 1 |
| S1 | 1 | 1 | 0 | 1 | 1 | 1 |

## 3.2   Time Complexity

To find the 'real' time complexity of the algorithm six synthetic datasets were created. The algorithm was implemented by Matlab R2010b, and was run on a machine with Core i7, 6G memory, Windows 7 system.

The time complexity of the algorithm is $O(m \times n^2)$ because it updates stored data structures when new data arrive and then inter-time series distances are obtained for each pair of series. In addition, the $k$-nearest neighbors are obtained in order to calculate the anomaly scores.

In all four figures the anomalous series begins to differ from the rest of the group within as few as 100 additional observations. The hardest detection problem is in 2a of Figure 1 consisting of stock prices; the anomalous series is difficult to visualize. Even in this case, our algorithm succeeds in identifying the anomalous series. Experimental results describing computational effort are shown in Figure 3. The parameters for OMUDIM are set as follows: $k$ is 1, $\alpha$ is 0.05 if number of series is less than 200, otherwise 0.1. $\beta$ is 0.1. OMUDIM is about 60% faster than NMUDIM.

**Table 3.** Running time of NMUDIM and average computation workload comparison between NMUDIM and OMUDIM. "Workload" represents the average number of comparisons performed in each iteration.

| Running Time | Length | # of Series | NMUDIM (Seconds) | NUMDIM Workload | OMUDIM Workload | Ratio |
|---|---|---|---|---|---|---|
| Synthetic1 | 1500 | 20 | 3.90 | 190 | 63 | 33.2% |
| Synthetic2 | 1500 | 40 | 8.20 | 780 | 405 | 51.9% |
| Synthetic3 | 1500 | 80 | 18.28 | 3160 | 1138 | 36.0% |
| Synthetic4 | 1500 | 200 | 53.60 | 19900 | 6535 | 32.8% |
| Synthetic5 | 1500 | 400 | 152.69 | 79800 | 28387 | 35.6% |
| Synthetic6 | 1500 | 1000 | 706.64 | 499500 | 113304 | 22.7% |

## 4   Concluding Remarks

In this paper, we have proposed an algorithm for online anomaly detection of time series. This approach is efficient and detects anomalous series as soon as it begins to drift away from the other (non-anomalous) series, a substantial advantage over other anomaly detection algorithms for time series. Our approach can handle data uncertainty very well, and its online version only requires an initial length of data to start and doesn't require any 'training datasets'. Compared with other methods, it requires less domain knowledge. Future research will focus on reducing the time complexity of the algorithm to handle large datasets.

# References

1. Chan, P.K., Mahoney, M.V.: Modeling multiple time series for anomaly detection. In: 15th IEEE International Conf. on Data Mining, pp. 90–97 (2005)
2. Chandola, D.C.V., Kumar, V.: Detecting anomalies in a time series database. Technical Report 09-004 (2009)
3. Fujimaki, R., Yairi, T., Machida, K.: An anomaly detection method for spacecraft using relevance vector learning. In: Ho, T.-B., Cheung, D., Liu, H. (eds.) PAKDD 2005. LNCS (LNAI), vol. 3518, pp. 785–790. Springer, Heidelberg (2005)
4. Huang, H., Mehrotra, K., Mohan, C.K.: Detection of anomalous time series based on multiple distance measures. Technical Report, EECS, Syracuse University (2012)
5. Yahoo Inc. Stocks price dataset (2012),
   http://finance.yahoo.com/q/hp?s=WMT+Historical+Prices
6. Keogh, E., Lonardi, S., Chiu, B.Y.: Finding surprising patterns in a time series database in linear time and space. In: Proc. of the 8th ACM SIGKDD International Conf. on Knowledge Discovery and Data Mining, pp. 550–556 (2002)
7. Keogh, E., Ratanamahatana, C.A.: Exact indexing of dynamic time warping. Knowledge Inf. Syst., 358–386 (2005)
8. Keogh, E., Zhu, Q., Hu, B., Hao, Y., Xi, X., Wei, L., Ratanamahatana, C.A.: The ucr time series classification/clustering (2011),
   http://www.cs.ucr.edu/~eamonn/time_series_data/
9. Leng, M., Lai, X., Tan, G., Xu, X.: Time series representation for anomaly detection. In: IEEE International Conference on Computer Science and Information Technology, ICCSIT 2009, August 8-11, pp. 628–632 (2009)
10. Lin, J., Khade, R., Li, Y.: Rotation-invariant similarity in time series using bag-of-patterns representation. J. of Intelligent Information Systems, 1–29 (2012)
11. Sadik, M. S., Gruenwald, L.: DBOD-DS: Distance based outlier detection for data streams. In: Bringas, P.G., Hameurlain, A., Quirchmayr, G. (eds.) DEXA 2010, Part I. LNCS, vol. 6261, pp. 122–136. Springer, Heidelberg (2010)
12. Toshniwal, D., Yadav, S.: Adaptive outlier detection in streaming time series. In: Proceedings of International Conference on Asia Agriculture and Animal, ICAAA, Hong Kong, vol. 13, pp. 186–192 (2011)
13. Wei, L., Kumar, N., Nishanth Lolla, V., Keogh, E., Lonardi, S., Ratanamahatana, C.A.: Assumption-free anomaly detection in time series. In: Proc. 17th International Scientific and Statistical Database Management Conf., SSDBM 2005 (2005)
14. Yamanishi, K., Takeuchi, J.: A unifying framework for detecting outliers and change points from non-stationary time series data. In: Proc. of the Eighth ACM SIGKDD, pp. 676–681. ACM (2002)

# Assisting Web Site Navigation through Web Usage Patterns

Oznur Kirmemis Alkan and Pinar Karagoz

METU Computer Eng. Dept. 06810 Ankara Turkey
{oznur.kirmemis,karagoz}@ceng.metu.edu.tr

**Abstract.** Extracting patterns from Web usage data helps to facilitate better Web personalization and Web structure readjustment. There are a number of different approaches proposed for Web Usage Mining, such as Markov models and their variations, or models based on pattern recognition techniques such as sequence mining. This paper describes a new framework, which combines clustering of users' sessions together with a novel algorithm, called PathSearch-BF, in order to construct smart access paths that will be presented to the Web users for assisting them during their navigation in the Web sites. Through experimental evaluation on well-known datasets, we show that the proposed methodology can achieve valuable results.

**Keywords:** Web Usage Mining, clustering, PathSearch-BF, access path.

## 1   Introduction

In order to facilitate users' navigation in the Web sites, analysis of the usage information that is recorded in web logs reveal valuable information. There are various web usage mining techniques including clustering, association rule mining, classification and sequential pattern mining [18, 19]. In this paper, we present a new model that constructs smart access paths that will assist the web users for getting to the target page easily. The solution combines clustering with a novel algorithm, called Path-Search-BF, for access path discovery from server logs.

Our overall approach can be summarized as follows. The user sessions are clustered based on the similarity of the web pages accessed during the sessions. For each cluster, smart access paths are discovered by PathSearch-BF algorithm. This new algorithm constructs access paths in a stepwise fashion as in constructing a search tree. However, the tree constructed has BF (branching factor) nodes at each level, which are the best child nodes according to the evaluation function. The novelty of our solution lies in this novel approach for constructing smart access sequences. These access sequences can be viewed as sequential patterns; however, different from sequential pattern mining solutions, the paths are constructed by using an evaluation function, which considers the co-occurrence of accesses, time spent on each access, and the distinct number of users that are traversing the access sequence. These patterns are sequences of accesses such that they may not be directly linked in the navigational organization of the web site; however, they are conceptually related according to the users' interests. Here, the motivation behind is that, users generally

M. Ali et al. (Eds.): IEA/AIE 2013, LNAI 7906, pp. 161–170, 2013.

follow similar sequences of accesses in order to get what they need from the web site. However, sometimes, since they are lost in the Web site, or due to the complexity of the navigational organization of the site, they have to traverse through unnecessary accesses. In order to understand the idea better, consider a simple shopping e-commerce Web site, where the users examine similar items through (A, B, C, F) accesses and buy these items through (G, H) accesses. However, the site may have been organized such that, there may not be direct links between these accesses. Therefore, users with the aim of examining and then buying these items will have to traverse through many other accesses between these six requests. Here, for this scenario, our system tries to identify the smart access path (A, B, C, F, G, H), which is actually a "conceptually related" but "not directly linked" sequence of accesses. After access paths are discovered for each cluster, the system is ready to present related paths to the users. The discovery of access paths is done offline and the presentation of the paths to the users is done online.

The experimental results show that this novel method produces satisfactory results in terms of several well-known evaluation metrics from the literature, and using distinct number of users in the evaluation process improves the accuracy in addition to the frequency of co-occurrence of Web pages in users' sessions. Equally important, these results are robust across sites with different structures.

The rest of the paper is organized as follows: In Section 2, summary of related work in this area is presented. In Section 3, the proposed framework is described in detail. In Section 4, the evaluation of the system and the discussion of the results are given. Finally, concluding remarks are presented in Section 5.

## 2     Related Work

In recent years, there has been an increase in the number of studies on web usage mining [12-14]. The main motivation behind these studies is to get a better understanding of the reactions and of the users' navigation and then to help them in their interaction with the Web site. Considering the techniques applied to solve the Web mining problem, several approaches exist including mining association rules from click-stream data [10], and extracting frequent sequences and frequent generalized sequences [9].

Learning users' navigation behaviors for Web page recommendation has attracted attention in data mining community. Markov model and its variants have been proposed to model users' behaviors in several studies [7], [23]. Web surfing has become an important activity for many consumers and a web recommender that models user behavior by constructing a knowledge base using temporal web access patterns is proposed in [3]. As in the case for Web page recommendation, various attempts have been exploited to achieve Web page access prediction by preprocessing Web server log files and analyzing Web users' navigational patterns [15].

Sequential pattern discovery is also a widely studied technique for web usage mining. In [16], Prefix-tree structure is presented for sequential patterns, following the basic principles of the Generalized Sequential Patterns (GSP) algorithm [8]. Frequent tree patterns can be applied to web log analysis. In [2], an algorithm to find frequent patterns from rooted unordered trees is introduced.

As mentioned, we have used clustering so as to group sessions into related units. Hay et al. in [17], uses sequence alignment to measure similarity, while in [15] belief functions are exploited. In [21], a technique for fine-tuning the clusters of similar web access patterns by approximating through least square approach is presented. In [20], the authors propose a recommendation model where they take the visiting order of the current user into account and active user session is assigned to the most similar cluster of user sessions. To compute frequent click paths, an algorithm is presented in [24], in order to isolate sub sessions that will give a granular view of user activity.

## 3    Proposed Framework

The proposed framework involves three phases: data preparation and transformation, pattern discovery, and pattern presentation. The pattern discovery phase is the main part of the solution, which uses a hybrid approach in order to construct the smart access paths.

Data preparation step includes data cleaning, user identification, session identification, and page visit duration calculation. In the data cleaning part, the irrelevant log entries like erroneous accesses, requests with the filename suffixes such as, gif, jpeg, and jpg are eliminated. We identify a user by using the IP address and browser information. For the session identification task, a new session is created when a new user is encountered or if the visiting page time exceeds 30 minutes for the same user [11].

Page visit duration, which is defined as the time difference between consecutive page requests, is used in the pattern discovery phase. The calculation of the page visit duration is straightforward except for the last page in a session. For the last page, the visit duration is set to the average of the durations for that page across all the sessions in which it is not the last page request. One other step that is performed before pattern discovery phase is the session elimination, in which sessions that include less than 5 pages are removed in order to eliminate the effect of random accesses in the web site.

In pattern discovery phase, a hybrid approach is employed. This approach combines clustering with a novel pattern extraction algorithm. Clustering is used for partitioning the set of sessions based on the user interests in terms of the web page requests. Smart access paths are then extracted within each cluster. Partitioning the data before pattern discovery phase has the advantage especially for the web sites with many different usage characteristics.

**Clustering of Users' Sessions.** In order to cluster the sessions, K-Means algorithm is used under city-block distance metric [22]. K-Means algorithm is chosen due to its linear time complexity and its computational efficiency over other clustering methods such as hierarchical methods. For the clustering process, a session-pageview matrix is constructed where each column is a pageview and each row is a session of a user represented as a vector. The matrix entries are integer values that represent the frequency of web page visits in that session.

**Discovering Smart Access Path: PathSearch-BF Algorithm.** This algorithm constructs paths in a recursive manner. It can be viewed as the construction of a tree where the root node of the tree is null, and at each recursive call, one level of the tree is built, which corresponds to growing a smart path through adding a new page link to it. At each level of the tree, BF number of nodes exist. The algorithm is presented in Figure 1.

Given a cluster, the output of the algorithm is the smart paths constructed for that cluster. For the initial call, parent array includes the top-BF accessed pages in terms of the number of distinct sessions in which they exist. This array keeps the smart access paths that have been constructed up to the current call of the function. The child array (line 7) however, keeps either the smart paths whose construction have been completed or who will be carried to the next call of the function as a parent node. The algorithm always keeps the best BF nodes in the child array according to the value returned by the *Evaluate* function (line 15). Therefore, if a parent node cannot produce a child that will be carried to the next call, that node's smart path is outputted and it does not grow any more (lines 23-31).

Evaluation function examines a parent node's possible child. The value of the child node's path depends on the number of sessions in which the child node path exists, the number of distinct users that traverse child node path in their sessions, and the total duration of the child node path in user sessions. Evaluation value is calculated by using the Formula (1).

$$
\begin{aligned}
\textit{Evaluation Value} = \\
& (\textit{NumberofSessionsConfirm}/\textit{NumberofSessions})* w\_session\_count + \\
& (\textit{Noof Distinct Users} /\textit{Total Noof Users})*w\_user\_count + \\
& (\textit{Duration}/\textit{NumberofSessionsInClno}) * w\_duration
\end{aligned}
\tag{1}
$$

The value of a smart access path increases as it exists in many user sessions. However, it is also possible that these sessions belong to a small number of users who access the Web page very frequently. A smart access path that claims to shorten long access sequences of Web users should be useful for as many users as possible in the same cluster. Therefore, not only it should exist in many user sessions, but also it should exist in many distinct users' access sequences. In addition, the aim of the smart path is to shorten long access sequences in a goal-oriented fashion, therefore, the value of a smart path increases as the length of the path it shortens increases.

To realize these ideas, all the sessions that has the child node path in their access sequence in the same order, are found. These sessions are examined in order to find out how many distinct users own these sessions and the total duration of the child node path in them. Then the value of the path is evaluated by normalizing these three values and weighting them with three parameters; w_session_count, w_user_count and w_duration.

**Presentation of the Patterns.** The usage scenario of the system is designed as follows. Whenever a new user enters the Web site and visits a sufficient number of Web pages, which is considered as three pages for the current version of the solution, we look for the most similar cluster to the user's access history by using these three accesses. Top-N smart access sequences generated for that cluster are recommended to the user. The smart paths can be presented as a list of recommended links or products to the users. As in adaptive web studies, they can be used to adapt web pages in order to better suit users' preferences. This can be done, for instance, by inserting links that exist in the constructed paths, but does not exist in the Web sites' current navigational structure.

# 4     Experimental Evaluation

In this study, we use two different data sets. NASA dataset [6] originally contains 3,461,612 requests. After the data preparation phase, we have 737148 accesses, 90707 sessions, 2459 distinct requests, and 6406 distinct users. The second dataset that is from the University of Saskatchewan's WWW server [5] originally contains 2,408,625 requests. After preprocessing, 288727 accesses, 40784 sessions, 2285 distinct requests, and 16664 distinct users are left.

For the evaluation of the system, 30% of the cleaned sessions are randomly selected as the test set, and the remaining part as the training set. For each user session in the test set, we have used 70% of the accesses to find the correct cluster that the test session belongs to. Then we check whether any smart path match with that session. We aim to find whether the last access of the smart path also exists in the rest of that test session as well. We recommend best six smart paths that have the highest evaluation values.

For measuring the performance of the system, we have adapted and used coverage and accuracy metrics [4]. In order to describe the metrics' adaptation to our environment, following definitions are used: SP_EXC_L: smart access path except its last access. SC: total number of sessions. SC_SP: number of sessions in which SP_EXC_L exists. SC_SP_SUCC: number of sessions that any of the smart access paths succeeds.

Coverage measures the system's ability in terms of the number of the test sessions that any of the produced smart access paths exist without considering their last access. It is calculated as given in (2).

$$Coverage = SC\_SP/SC \tag{2}$$

Accuracy measures the system's ability to provide correct predictions. Considering our system, we look at each test session and find out whether any of the smart paths except their last access (SP_EXC_L) exist in that session. If exists, we look at the rest of the test session in order to find out whether the last access of our smart path exists in the rest of the test session as well. It is calculated as given in (3).

$$Accuracy = SC\_SP\_SUCC/SC\_SP \tag{3}$$

We have conducted two sets of experiments. The first set of experiments aims to find out the effect of BF and number of clusters on the accuracy and coverage of the solution. The second set of experiments is performed in order to analyze how session count, user count, and duration parameters affect accuracy.

In the first set of experiments, we perform tests with different number of clusters and BF values. In order to find out the effect of the clustering, we also conduct experiments when no clustering of user sessions is performed. The results for these

```
1:input:
2:Parent_Array[BF] ←array of Parent States  that will
   create children at next iteration
3:cl_no ← the cluster number; the algorithm creates
   smart access sequences using the sessions in
   cluster numbered cl_no
4:output:
5:Smart_Access_Path_List
6:Request_Array[NumberOfRequests] ← array of all
  possible accesses
7:Child_Array[BF]← array of child nodes created at
  each level, in descending order of value
8:for i = 0 to BF do
9:   parent_state = Parent_Array[i]
10:    source_access = parent_state.destination_access
11:    pre_access_pattern = {parent_state.
        pre_access_pattern} +{source_access}
12:    for j = 0 to NumberOfRequestsdo
13:       destination_access = Request_Array[i]
14:       new_link←Link(source_access,
                         destination_access)
15:       value ← Evaluate(new_link, pre_access_pattern,
                          cl_no)
16:       ind = PlaceToInsert(ChildArray,value)
17:       if ind> 0then
18:          state ← NewState(pre_access_pattern,
                              new_link, value)
19:          Child_Array[ind] = state
20:       end if
21:    end for
22:end for
23:for i = 0 to NumberOfParents do
24:    parent_state = ParentArray[i]
25:    if parent_state != null and
          parent_state.child_count = 0 then
26:        smart_path.path = {parent_state.
                             pre_access_pattern} +
27:                         {parent_state.
                             destination_access}
28:        smart_path.value = parent_state.value
29:     Smart_Access_Path_List .Add(smart_path)
30:    end if
31:end for
32:if Child_Array[0] != null then
33:    PathSearch-BF(Child_Array,cl_no)
34:end if
```

**Fig. 1.** PathSearch-BF Algoritm

experiments are obtained under session count, user count, and duration parameters as 0.33, 0.33 and 0.34, respectively. Figures 2 and 3 plot the accuracy results, whereas Figures 4 and 5 plot the coverage of the solution for the NASA and University of Saskatchewan's dataset, respectively. For both of the datasets, when no clustering is performed, the performance of the system is significantly lower than all other confi-gurations independent of the number of clusters used.

Considering both of the datasets, the accuracy of the solution is lowest for all BF values when 5 clusters are used. In addition, the system gets the best performance (95%) for the NASA dataset when the number of clusters is 10 and (98%) when 30 clusters is utilized for BF value set to 100 for the University of Saskatchewan's dataset.

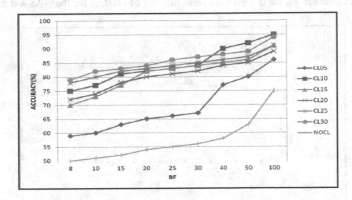

**Fig. 2.** Accuracy versus BF with different number of clusters for the NASA dataset

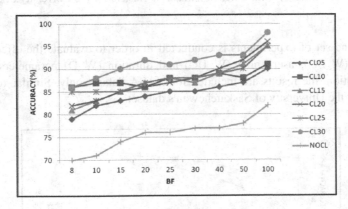

**Fig. 3.** Accuracy versus BF with different number of clusters for the University of Saskatche-wan's dataset

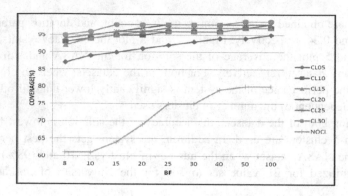

**Fig. 4.** Coverage versus BF with different number of clusters for the NASA dataset

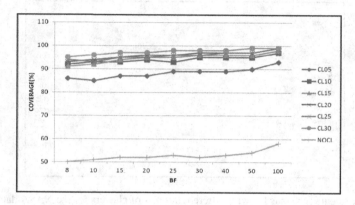

**Fig. 5.** Coverage versus BF with different number of clusters for the University of Saskatchewan's dataset

The second set of experiments is conducted in order to evaluate the effect of session count (W_SC), user count (W_UC) and duration (W_D) parameters. Due to space limitation, the results are included for NASA dataset only. Similar results are obtained for the University of Saskatchewan's dataset.

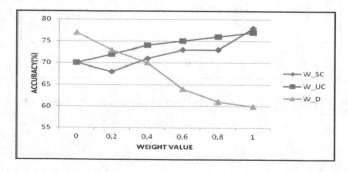

**Fig. 6.** Accuracy versus different values of weight parameters for the NASA dataset

The sum of these parameters are always 1 in the system, therefore, while we are experimenting on one of these three weights, we set the other two weights to the same value. The results for these experiments are obtained by fixing the number of clusters and BF to 10. Figure 6 displays the accuracy of the solution as we change the weight parameters. As session count and user count increase, accuracy increases. However, as the weight of the duration increases, the accuracy decreases. Therefore, we can conclude that, as we cover more distinct user sessions with the smart paths, the performance of the system is better.

## 5    Conclusion

The presented framework is based on a hybrid approach that combines clustering with a search based pattern extraction algorithm. The role of clustering in the framework is to group user sessions according to similar navigational behaviors. The system is evaluated by using two well-known datasets and the results are very promising. Concerning the notion of the Web sites, in addition to statistical evaluation, human evaluation of the results is also needed. Therefore, as the future work, it is planned to adapt the system to a real-life Web site and present the smart paths to the Web site users and let them evaluate the system. Additively, we plan to develop variations of the algorithm to extract smart access paths where instead of BF nodes at each level of the tree, each parent produces BF nodes, which will increase the time and space complexities of the offline phase, however can result in better performance.

**Acknowledgements.** This work is supported by METU BAP project BAP-03-12-2013-001.

## References

1. Mobasher, B., Dai, H., Luo, T., Nakagawa, M.: Effective personalization based on Association Rule Discovery from Web usage data. In: Proceedings of the 3rd ACM Workshop on Web Information and Data Management, New York, pp. 9–15 (2001)
2. Chehreghani, M.H.: Efficiently mining unordered trees. In: Proceeding of the 11th IEEE International Conference on Data Mining, pp. 33–52 (2011)
3. Baoyao, Z.: Effective Web content recommendation based on consumer behavior modeling. In: Proceedings of the 11th IEEE International Conference on Consumer Electronics, pp. 471–472 (2011)
4. Gery, M., Haddad, H.: Evaluation of Web usage mining approaches for user's next request prediction. In: Proceedings of the 5th ACM International Workshop on Web Information and Data Management, pp. 74–81 (2003)
5. University of Saskatchewan's WWW server Log,
   http://ita.ee.lbl.gov/html/contrib/Sask-HTTP.html
6. NASA Kennedy Space Center Log,
   http://ita.ee.lbl.gov/html/contrib/NASA-HTTP.html

7. Bonnin, G., Brun, A., Boyer, A.: A low-order Markov Model integrating long-distance histories for collaborative recommender systems. In: IUI 2009, pp. 57–66. ACM, New York (2009)

8. Ren, J.D., Cheng, Y.B., Yang, L.L.: An algorithm for mining generalized sequential patterns. In: Proceedings of International Conference on Machine Learning and Cybernetics, vol. 2, pp. 1288–1292 (2004)

9. Gaul, W., Schmidh-Thieme, L.: Mining web navigation path fragments. In: Proceedings of the Workshop on Web Mining for E-Commerce – Challenges and Opportunities, Boston, MA, pp. 319–322 (2000)

10. Frias-Martinez, E., Karamcheti, V.: A prediction model for user access sequences. In: WEBKDD Workshop: Web Mining for Usage Patterns and User Profiles (2002)

11. Cooley, R., Mobasher, B., Srinivastava, J.: Web mining: Information and Pattern Discovery on the World Wide Web. In: International Conference on Tools with Artificial Intelligence, pp. 558–567. IEEE, Newport Beach (1997)

12. Zhou, B., Hui, S.C., Chang, K.: An intelligent recommender system using sequential Web access patterns. In: IEEE Conference on Cybernetics and Intelligent Systems, vol. 1, pp. 393–398 (2004)

13. Spiliopoulou, M., Faulstich, L.C., Winkler, K.: A data miner analyzing the navigational behavior of web users. In: Proceeding of the Workshop on Machine Learning in User Modeling of the ACAI 1999 International Conference, Creta, Greece (1999)

14. Bonchi, F., Giannotti, F., Gozzi, C., Manco, G., Nani, M., Pedreschi, D., Renso, C., Ruggieri, S.: Web log data warehousing and mining for intelligent web caching. Data Knowledge Engineering 39(2), 165–189 (2001)

15. Srivastava, J., Cooley, R., Deshpande, M., Tan, P.: Web usage mining: Discovery and applications of usage patterns from web data. SIGDD Explorations 1(2), 12–23 (2000)

16. Masseglia, F., Poncelet, P., Cicchetti, R.: An Efficient algorithm for Web usage mining. Networking and Information Systems Journal (NIS) 2(5-6), 571–603 (1999)

17. Hay, B., Wets, G., Vanhoof, K.: Clustering navigation patterns on a website using a sequence alignment method. In: Intelligent Techniques for Web Personalization: 17th Int. Joint Conf. on Artificial Intelligence, Seattle, WA, USA, pp. 1–6 (2001)

18. Nasraoui, O., Saka, E.: Web Usage Mining in noisy and ambiguous environments: Exploring the role of concept hierarchies, compression, and robust user profiles. In: From Web to Social Web: Discovering and Deploying User and Content Profiles: Workshop on Web Mining, WebMine, Berlin, Germany (2006)

19. Mobasher, B.: Data mining for web personalization. In: Brusilovsky, P., Kobsa, A., Nejdl, W. (eds.) The Adaptive Web. LNCS, vol. 4321, pp. 90–135. Springer, Heidelberg (2007)

20. Gunduz, S., Ozsu, M.T.: A web page prediction model based on clickstream tree representation of user behavior. In: SIGKDD 2003, USA, pp. 535–540 (2003)

21. Patil, S.S.: A Least square approach to analyze usage data for effective web personalization. In: Proceedings of International Conference on Advancs in Computer Science, pp. 110–114 (2011)

22. Han, J., Kamber, M.: Data Mining Concept and Techniques. Morgan Kaufman (2002)

23. Yang, Q., Fan, J., Wang, J., Zhou, L.: Personalizing Web page recommendation via collaborative filtering and topic-aware Markov Model. In: 10th International Conference on Data Mining (ICDM), Sydney, NSW, 1145–1150 (2010)

24. Ruiz, E.M., Millán, S., Peña, J.M., Hadjimichael, M., Marbán, O.: Subsessions: A granular approach to click path analysis. Int. J. Intell. Syst. 19(7), 619–637 (2004)

# GRASPing Examination Board Assignments for University-Entrance Exams

Jorge Puente-Peinador[1], Camino R. Vela[1], Inés González-Rodríguez[2], Juan José Palacios[1], and Luis J. Rodríguez[3]

[1] A.I. Centre and Department of Computer Science,
University of Oviedo, Spain
{puente,crvela,palaciosjuan}@uniovi.es

[2] Department of Mathematics, Statistics and Computing,
University of Cantabria, Spain
ines.gonzalez@unican.es

[3] Department of Statistics and O.R.,
University of Oviedo, Spain
luisj@uniovi.es

**Abstract.** In the sequel, we tackle a real-world problem: forming examination boards for the university-entrance exams at the autonomous region of Asturias (Spain). We formulate the problem, propose a heuristic GRASP-like method to solve it and show some experimental results to support our proposal.

## 1 Introduction

In Spain, students who finish Secondary School must take a University Entrance Exam (UEE), consisting of several modules or subjects (Maths, Chemistry, Art...), some of which are optional. The mark obtained in this exam is key for entering university. It is therefore the responsibility of the organisers to provide the best possible environment for the examination process. Its organisation is undertaken independently in each of the 17 Spanish autonomous regions by an Organising Committee, and it includes deciding on aspects such as the number of days allocated to the exam, the number and location of exam venues or the optional exam-modules offered in each venue.

A critical issue for the committee is to decide for each venue on the composition of an examination board in charge of invigilation. Exam venues with varying sizes require a different number of examiners in order to provide proper invigilation. It is also necessary that examiners cover different areas of expertise, so they can adequately reply to students' questions. While the board's composition must guarantee high standards in terms of expertise and invigilation capacity, examiners' travelling expenses incur an economical cost which should be kept to a minimum.

In the following, we shall tackle the Examination Board Assignment (EBA) problem. After modelling the problem in Section 2, we shall propose in Section 3 a GRASP algorithm to solve it. To evaluate our proposal, in Section 4 we will

M. Ali et al. (Eds.): IEA/AIE 2013, LNAI 7906, pp. 171–180, 2013.

present some detailed results analysing the different parts of the algorithm as well as compare it with the solution provided by the human experts on some real-world instances of the problem. Finally, Section 5 presents some conclusions.

## 2     Problem Definition and Formulation

The Organising Committee in the Principality of Asturias (the region where the University of Oviedo lies) is in charge of a varying number of UEE examination venues geographically distributed throughout the region's territory. Students are assigned to a specific venue depending on their school of origin. The comittee takes into account the total number of students registered for that particular UEE session as well as the optional modules they intend to take to select a set of teachers (examiners hereafter) for invigilation. These examiners can be university staff or secondary-school teachers (these include education supervisors, on special duty at the Education Department of the Administration). Given each examiner's academic profile, he/she is associated to a particular subject and is also meant to have different levels of affinity to every other subject, which make him/her more or less adequate to answer student questions during the examination process. The Committee appoints for each board a president and a secretary, who must be university academic staff and imposes some additional constraints on the composition of the boards, including their size. The goal in the EBA is to assign all the examiners from the initial set to different examination boards given the predefined assignment of president and secretary, in such a way that all constraints hold and the following objectives are optimised:

1. It is not always possible to assign an expert on each subject to every board; instead, the affinity concept is used and the objective is to *maximise the expertise affinity coverage* provided by the examiners on each board.
2. Examiners get paid for their travel expenses if they are assigned to a venue located elsewhere than their usual workplace. The total amount depends on the distance and may include accommodation expenses. Clearly, travel costs should be minimised.

Having outlined the problem statement, we give some more detail about its different components in the following.

The *input data* available when organising each UEE session are the following:

1. **Examiners** Let $E$ denote the set of examiners. For each examiner $e \in E$, we are given the following information: subject in which he/she is specialised, denoted Subj($e$); type of examiner, which can be university lecturer, education supervisor or secondary school teacher on active duty; school, if he/she is a secondary school teacher, denoted School($e$); and city or town where his/her workplace is located, denoted Town($e$).
2. **Venues** Let $V$ denote the set of venues. For each venue $v \in V$ we are given the following information: preassigned president and secretary from $E$; set of the secondary schools associated to this venue, denoted Lschools($v$); set

of subjects for which exams will take place at this venue, denoted Lsubj($v$); location (city or town), denoted Location($v$); and size of the examination board (i.e. number of examiners) for that venue, denoted Size($v$). It is assumed that $\sum_{v \in V}$ Size($v$) = $|E|$.
3. **Subjects** Let $S$ be the set of subjects where each subject $s \in S$ is characterised by a textual description.
4. **Affinity array**, where rows correspond to examiners and columns to subjects, so $A(e, s) \in [0, 1]$ is the affinity degree between examiner $e$ and subject $s$, elicited by the Organising Commitee based on the area of expertise of each examiner so it usually corresponds to the affinity between Subj($e$) and $s$.
5. Data related to **travelling costs**, namely: table with distances between different towns or cities where exam venues, schools or university campus may be located, so distance($a, b$) denotes the distance between locations $a$ and $b$; travelling cost per kilometre ($c_{km}$); distance threshold for staying overnight and claiming accommodation expenses ($\Delta$); accommodation cost per night ($h$); travelling allowance per day ($a$); and duration (in days) of the exam ($n_d$).

The problem *solution* is the **Examination Board Assignment** (EBA for short), specifying the set Board($v$) of examiners composing the board for each venue $v \in V$. This set is the disjoint union of the sets of university and secondary-school teachers, denoted UBoard($v$) and SBoard($v$) respectively. The latter includes both secondary-school teachers on active service and education supervisors, that is, there is a subset ESBoard($v$) $\subset$ SBoard($v$) formed by those examiners in Board($v$) on special duty as education supervisors. The solution must also provide the following information for each venue $v \in V$: the total cost associated to its board, denoted $TC(v)$, and the degree of coverage for each subject $s \in$ Lsubj($v$), denoted Coverage($v, s$), as well as the overall coverage $AC(v)$.

Additional requirements of the Organising Committee translate into five *constraints* for this problem; the first two are hard, while the remaining ones are soft and may be relaxed if needed.

1. The predefined and final size of each exam board must coincide:

$$C_1 : \forall v \in V \; |\text{Board}(v)| = \text{Size}(v) \tag{1}$$

where $|\text{Board}(v)|$ denotes the cardinality of set Board($v$).
2. Teachers from secondary schools associated to a venue may not be assigned to it (since their students are taking the exam in that same venue):

$$C_2 : \forall v \in V \; \forall e \in \text{SBoard}(v) \; \text{School}(e) \notin \text{Lschools}(v). \tag{2}$$

3. It is preferred that an equilibrium exists between the number of university lecturers and secondary-school teachers in each venue:

$$C_3 : \forall v \in V \; \frac{|\text{UBoard}(v)|}{|\text{Board}(v)|} \in [0.4, 0.6]. \tag{3}$$

4. The number of venues where a subject is taken and the number of examiners expert on that subject may differ substantially. To distribute experts as homogenously as possible, an upper bound is established for each each venue $v \in V$ and subject $s \in \mathrm{Lsubj}(v)$ as follows:

$$M(v, s) = \left\lceil 1 + |\{e \in E : s = \mathrm{Subj}(e)\}| * \frac{\mathrm{Size}(v)}{\sum_{v' \in V : s \in \mathrm{Lsubj}(v')} \mathrm{Size}(v')} \right\rceil \quad (4)$$

where $\lceil x \rceil$ denotes the nearest integer greater than or equal to $x$. The constraint is then expressed as:

$$C_4 : \forall v \in V, \forall s \in \mathrm{Lsubj}(v) \ |\{e \in \mathrm{Board}(v) : s = \mathrm{Subj}(e)\}| \leq M(v, s). \quad (5)$$

5. It is preferred that education supervisors are equally distributed across all venues:

$$C_5 : \max_{v \in V} |\{e \in \mathrm{ESBoard}(v)\}| - \min_{v \in V} |\{e \in \mathrm{ESBoard}(v)\}| \leq 1. \quad (6)$$

Regarding the *objectives*, the most important one is to maximise the degree of expertise coverage provided by the members of each board. For any venue $v \in S$ and any subject $s \in \mathrm{Lsubj}(v)$, the degree of coverage of $s$ at $v$ is given by $\mathrm{Coverage}(v, s) = \max_{e \in \mathrm{Board}(v)}\{A(e, s)\}$ and the overall coverage at a venue $v \in V$ is calculated as:

$$AC(v) = \frac{\sum_{s \in \mathrm{Lsubj}(v)} \mathrm{Coverage}(v, s)}{|\mathrm{Lsubj}(v)|} \quad (7)$$

Notice that $AC(v) \in [0, 1]$ for every venue $v \in V$, with $AC(v) = 1$ meaning a perfect coverage. We can then formalise the first objective as:

$$\min f_1 = |V| - \sum_{v \in V} AC(v). \quad (8)$$

Regarding the second objective, the cost of assigning an examiner $e \in E$ to a venue $v \in V$ is computed as:

$$\mathrm{Cost}(e, v) = \begin{cases} 0 & \text{if } d_{ve} = 0, \\ 2d_{ve}c_{km}n_d + an_d & \text{if } 0 < d_{ve} \leq \Delta, \\ 2d_{ve}c_{km}n_d + an_d + (n_d - 1)h & \text{if } d_{ve} > \Delta, \end{cases} \quad (9)$$

where $d_{ve} = \mathrm{distance}(\mathrm{Location}(v), \mathrm{Town}(e))$ is the distance between the location of venue $v$ and the workplace of examiner $e$. The overall cost for a venue $v \in V$ is then given by $TC(v) = \sum_{e \in \mathrm{Board}(v)} \mathrm{Cost}(e, v)$ so the second objective is:

$$\min f_2 = \sum_{v \in V} TC(v) \quad (10)$$

Since the most important objective for the Organising Committee is to provide a good expertise coverage, we adopt a lexicographic approach [1] and define the examination board assignment problem as follows:

$$\text{lexmin}(f_1, f_2)$$
$$\text{subject to: } C_i, 1 \leq i \leq 5. \tag{11}$$

where lexmin denotes lexicographically minimising the vector $(f_1, f_2)$: if $f(S)$ denotes the value of objective $f$ for a feasible solution $S$, $S$ is preferred to $S'$, denoted $s \preceq_{lexmin} s'$, iff $f_1(S) < f_1(S')$ or $f_1(S) = f_1(S')$ and $f_2(S) \leq f_2(S')$. Lexicographic minimisation is well-suited to seek a compromise between conflicting interests, as well as reconciling this requirement with the crucial notion of Pareto-optimality [2].

The problem above is related to personnel scheduling problems [3],[4]. It shares with this family of problems objective functions and constraints, for example, coverage constraints, which are among the most important soft constraints [4]. However, it is not a typical problem from this family: we do not need to schedule shifts or days off, there is no single workplace (as in nurse rostering), all "tasks" in our problem demand "specific skills" and despite having groups of employees, these have individual profiles. All in all, the specific characteristics of the problem under consideration allow for no comparisons with other proposals from the literature. Finally, it is possible to prove that the resulting problem is $NP$-hard (we omit the proof due to lack of space).

## 3  Solving the Problem with a GRASP-Like Algorithm

The complexity of the problem under consideration suggests using meta-heuristic methods to solve it. Such methods allow problem-specific information to be incorporated and exploited, as well as making it easy to deal with complex objectives, in particular with "messy real world objectives and constraints" [3]. Here, we propose a solving method inspired in GRASP (Greedy Randomized Adaptive Search Procedure) meta-heuristics [5]. GRASP methods have proved popular for solving personnel scheduling problems (see [6] and references therein). This is in part due to the fact that these problems can be modelled as allocation problems and neighbourhood structures can be naturally defined by moving a single allocation or swapping two or more allocations. It is also natural to fix one or more allocations at each step of the construction phase.

Our method, in Algorithm 1, consists of a construction phase followed by an improvement phase. The construction phase is in itself divided in two steps: first, a Sequential-Greedy-Randomized-Construction algorithm (SGRC) builds an initial solution and then, if this solution is not feasible, a Solution-Repair algorithm (SR) obtains a feasible one by relaxing some of the soft constraints. Once a complete solution is available, the second phase performs a Local Search (LS) where the solution's neighbourhood is systematically explored until a local optimum is found. As random decisions are present in every phase, the process is

---

**Input** An instance $P$ of an EBA problem, $NRestarts$
**Output** A solution $S = \{Board(v) : v \in V\}$ for instance $P$
  **for** $k = 1, \ldots, NRestarts$ **do**
    $S \leftarrow$ SGRC();
    **if** $S$ is not complete **then** $S \leftarrow$ SR($S$);
    $S \leftarrow$ LS($S$);
    Update best solution $S^*$ so far if needed;
  **return** $S^*$, the best solution.

---

**Algorithm 1:** GRASP-like algorithm for the EBA problem

repeated $NRestarts$ times. The output is the best solution from all runs chosen as follows: for a pair of feasible solutions, we use $\preceq_{lexmin}$; else, feasible solutions are preferred to unfeasible ones and, between two unfeasible solutions, preference is established in terms of objective values as well as constraint satisfaction.

### 3.1 Sequential Greedy Randomized Construction

The SGRC starts by assigning the president and the secretary to each board as specified in the input data. Then, examiners from $E$ not assigned yet are allocated to boards in tree consecutive steps: assignment of education supervisors, assignment of "covering" examiners and assignment of "non-covering" examiners. Non-determinism situations are in some cases solved using heuristic criteria and in others introducing randomness, thus increasing the diversity of the greedy algorithm. In all cases, all constraints must be complied with, so it is possible that, at the end of SGRC some examiners from $E$ still remained unassigned. In this case, SGRC yields a partial solution that needs to be repaired.

In the first step, provided that $C_5$ holds, each education supervisor is assigned to that venue $v$ where its contribution to the coverage of Board($v$) is maximum. Ties between venues are broken using costs. In the second step, venues are sorted in increasing-size order and, following that order, for each venue, *valid* examiners are assigned to it as long as the venue's board is not full and the examiner's contribution to the venue's coverage is positive and maximal. An *examiner* is *valid* if he/she satisfies constraints $C_2$ and $C_4$ as well as two partial constraints $C_1' : |\text{Board}(v)| \leq \text{Size}(v)$ and $C_3' : |UBoard(v)|, |SBoard(v)| \leq 0.6Size(v)$ (satisfying $C_1'$ and $C_3'$ along SGRC means that $C_1$ and $C_3$ will be satisfied in the end). Ties are broken at random. At the end of the second step, there may be examiners in $E$ not assigned because of their null contribution to the covering of compatible boards. In the third step those non-covering examiners are assigned to venues (ordered at random) keeping costs to their minimum.

### 3.2 Solution Repair

SGRC may end without assigning all the examiners in $E$ because doing so would violate some constraints. In this case, SR builds a full solution by relaxing some of the soft constraints, with a possible deterioration in the coverage objective.

First, $C_4$ is relaxed in as few venues as possible, this being the constraint that has empirically proved to be more conflicting. For each venue $v$ with free positions in its board we pick a non-assigned examiner $e$ at random. Then we consider all pairs $(e', v')$ such that either $v' = v, e' = e$ or $v' \neq v, e' \in \mathrm{Board}(v')$ and such that assigning $e$ to $\mathrm{Board}(v') - \{e'\}$ and $e'$ to $\mathrm{Board}(v)$ satisfies constraints $C_1', C_2, C_3'$ and $C_5$. If more than one of such pairs exist, we select the pair maximising $AC(v) + AC(v')$ after the "swap" and exchange the examiners. This procedure continues until $\mathrm{Board}(v)$ is complete or no swaps are possible. In the case that relaxing $C_4$ were not enough, we would start all over again asking for the swap to satisfy only $C_1'$ and $C_2$ (i.e. we relax the remaining soft constraints simultaneously). It is possible to prove that this process always terminates with a full solution where all venues are complete and all examiners have been assigned.

### 3.3 Local Search

The LS starts from the full solution provided by SRGC+SR and explores its neighbourhood trying to improve on objective values and constraint satisfaction. To this end, it considers the list, ordered by venue, of examiners who are not education supervisors nor a prefixed president or secretary. Then, the neighbourhood is generated using a Forward Pairwise Interchange (FPE) [7]: for each examiner $e$, we consider exchanging it with another examiner $e'$ in another board following $e$ in the list; if the reassignment improves the solution, the exchange is made, otherwise, it is discarded. After trying all exchanges, the process starts again from the first examiner in the list. This continues until no exchange has been made for a whole round of trials.

In this process, the degree to which a solution $S$ does not satisfy $C_4$ is:

$$D_{ns}(S, C_4) = \sum_{v \in V} \sum_{s \in \mathrm{LSubjects}(v)} |\{e \in \mathrm{Board}(v) : s = \mathrm{Subj}(e)\}| - M(v, s) \quad (12)$$

Analogously, the degree of dissatisfaction of $C_3$, $D_{ns}(S, C_3)$, is the number of venues $v$ where the equilibrium between university and secondary-school teachers is broken, and the degree of dissatisfaction of $C_5$, $D_{ns}(S, C_5)$, is the number of venues where the number of supervisors does not correspond to a uniform distribution. Then, a solution $S$ is *better than* or *preferred to* another solution $S'$ (denoted $S \preceq_{LS} S'$) in the sense of Pareto-dominance [8], that is, iff $f_1(S) \leq f_1(S')$, $f_2(S) \leq f_2(S')$, $D_{ns}(S, C_i) \leq D_{ns}(S', C_i)$ for $i = 3, 4, 5$ and at least one of these inequalities is strict.

## 4    Experimental Results

To evaluate the proposed algorithm, we shall use three real instances of the EBA problem, corresponding to UEE celebrated in June 2010, 2011 and 2012. Unfortunately, the historical data available before 2010 cannot be used, because the UEE format was different and so was the problem definition.

**Table 1.** Evaluation of the stages of the proposed GRASP

| Instance | | Best | | Average | | Std. Dev. | |
|---|---|---|---|---|---|---|---|
| | | $AC(\%)$ | $TC(\text{\euro})$ | $AC(\%)$ | $TC(\text{\euro})$ | $AC$ | $TC$ |
| June2010 | SGRC+SR | 74.80 | 38444 | 73.16 | 36209 | 0.0054 | 599.5 |
| | GRASP | 77.46 | 27824 | 77.20 | 28600 | 0.0011 | 451.9 |
| June2011 | SGRC+SR | 70.31 | 37067 | 68.97 | 34644 | 0.0045 | 669.4 |
| | GRASP | 71.56 | 27846 | 71.32 | 27837 | 0.0009 | 672.9 |
| June2012 | SGRC+SR | 73.25 | 34087 | 71.80 | 33629 | 0.0042 | 648.38 |
| | GRASP | 75.27 | 26281 | 74.96 | 26372 | 0.0011 | 585.23 |

First, we shall analyse the behaviour of the two main modules of the algorithm, the constructive phase (including repair) and the local search. Then, we will compare the solutions of our algorithm to the solutions provided by the human experts in years 2010 and 2011; in year 2012 the actual solution has already been obtained with the proposed GRASP algorithm and, therefore, no experts' solution is available.

The prototype is programmed in C++ on a Xeon E5520 processor running Linux (SL 6.0). In all cases, the number of restarts for the GRASP algorithm is $NRestarts = 50$. Also, given its strong stochastic nature, the algorithm is run 50 times to obtain statistically significant results. The average runtime for one run is 391.2, 292.8 and 386.46 seconds for 2010, 2011 and 2012 respectively.

### 4.1   Analysis of the Algorithm

The first set of experiments provides a better insight into the algorithm's behaviour, namely, into the constructive phase — denoted SGRC+SR and consisting of SGRC followed by SR — and the local search phase. Since SGRC+SR has $NRestarts = 50$ restarts in each run of the GRASP algorithm and we consider results of 50 runs, SGRC+SR is executed 2500 times for fair comparisons.

A summary of the results is as follows: at the end of the first step, in average 70% of the examiners from $E$ have already been allocated to a venue (between 66% and 77%). At the end of the second step, between 93 and 100% (97.8% in average) of the members of $E$ have already been assigned. When solution repair is needed, all pending examiners are assigned after relaxing constraint $C_4$, so $C_3$ and $C_5$ need not be relaxed. Once SR is finished, the percentage of solutions not satisfying $C_4$ is, respectively, 82%, 89% and 95% for years 2010, 2011 and 2012. Also, the value of $f_1$ (expertise covering) remains unchanged during SR.

We now evaluate the contribution of LS to the GRASP algorithm, by comparing the results obtained after the constructive phase (SGRC+SR) (the first set of complete solutions) to those obtained at the end of the GRASP algorithm, after applying local search. Table 1 contains the best, average and standard deviation of the two objective values Average Coverage ($AC$) and Total Cost of travel expenses ($TC$) obtained in both cases. It does not report data related to constraint satisfaction because LS always yields solutions where all 5 constraints

**Table 2.** Comparison between expert and GRASP results

| Instance | Method | Problem Objectives | | Constraint satisfaction degree(%) | | | | |
|---|---|---|---|---|---|---|---|---|
| | | $AC(\%)$ | $TC(e)$ | $C_1$ | $C_2$ | $C_3$ | $C_4$ | $C_5$ |
| June2010 | Expert | 73.04 | 32032 | 100 | 100 | 60.00 | 93.40 | 80 |
| | GRASP Best | 77.46 | 27824 | 100 | 100 | 100 | 100 | 100 |
| | GRASP Avg | 77.20 | 28600 | 100 | 100 | 100 | 100 | 100 |
| June2011 | Expert | 68.20 | 32748 | 100 | 100 | 69.23 | 100 | 61.53 |
| | GRASP Best | 71.56 | 27846 | 100 | 100 | 100 | 100 | 100 |
| | GRASP Avg | 71.32 | 27837 | 100 | 100 | 100 | 100 | 100 |

hold. We can see that LS noticeably reduces the total cost as well as improving the expertise coverage. Clearly, the greatest improvement is in costs, since this objective received considerably less attention in the constructive phase. Costs are reduced in average 9215€ for the best solution and 7224€ for the average solution. Regarding the expertise coverage, it is not surprising that local search makes little difference: coverage levels around 70% are already quite high, considering the proportion of small venues with few examiners. The standard deviation values obtained in all cases support the robustness of the results.

## 4.2 Comparison with Experts' Solutions

We will now assess the quality of the best and average GRASP solutions compared to the solution provided each year by human experts. Table 2 shows for all cases the objective values of expertise coverage ($AC$) and total cost ($TC$), as well as a measure of the degree of constraint satisfaction, defined as follows:

$C_3$: Percentage of venues $v \in V$ where $\frac{|\text{UBoard}(v)|}{|\text{Board}(v)|} \in [0.4, 0.6]$.

$C_4$: Percentage of venues $v \in V$ where for every subject $s \in \text{Lsubj}(v)$ $|\{e \in \text{Board}(v) : s = \text{Subj}(e)\}| \leq M(v, s)$.

$C_5$: Percentage of venues $v \in V$ where the number of education supervisors does not follow a uniform distribution.

Unlike the experts' solutions, both the best and average solutions provided by the GRASP algorithm fully satisfy constraints in all cases. At the same time, they improve in expertise coverage and costs, despite the fact that constraint satisfaction is usually obtained at the cost of worsening objective function values. Furthermore, comparing both Tables 1 and 2 we can see that even SGRC+SR improves the experts' coverage whilst simultaneously satisfying $C_3$ and $C_5$.

The solution provided by the proposed GRASP algorithm has another advantage: a considerable saving in working hours and effort for the team in charge of finding the EBA at hand. The differences between instances of the EBA problem made it impossible for the experts to somehow reuse previous solutions, so solutions had to be built from scratch, requiring full-time commitment from a team of experts for days. Time is specially critical because the input data are usually available only a few days before the UEE takes place. Our GRASP proposal allows to obtain a better solution with considerable less effort in a much shorter time. Additionally, the modular design of the algorithm allows it to be used as a replanning procedure in the case of unexpected incidences. It is also possible to perform some of the phases at hand, should it be the wish of the experts, or simply start from a human-made assignment and improve it. These character-istics add to the value of the proposed solution. Finally, it must be highlighted that, as from 2012, our method has substituted the human experts who can now employ their time and effort in more fruitful tasks.

## 5　Conclusions

We have tackled a real-world problem which consists in assigning a set of exam-iners to different boards in order to maximise expertise coverage in each board and also minimise the total costs incurred, subject to a series of hard and soft constraints. We have modelled this problem in the framework of multiobjective combinatorial optimisation with constraints and proposed a solving method, based on GRASP techniques, which significatively improves the results obtained by human experts: not only does it obtain better objective values while satisfy-ing more constraints, but it has also proved a considerable saving both in time and effort for the team in charge of solving the problem.

**Acknowledgments.** This research has been supported by the Spanish Govern-ment under research grants FEDER TIN2010-20976-C02-02, MTM2010-16051 and MTM2011-22993.

## References

1. Ehrgott, M., Gandibleux, X.: A survey and annotated bibliography of multiobjective combinatorial optimization. OR Spektrum 22, 425–460 (2000)
2. Bouveret, S., Lemaître, M.: Computing leximin-optimal solutions in constraint net-works. Artificial Intelligence 173, 343–364 (2009)
3. Ernst, A.T., Jiang, H., Krishnamoorthy, M., Sier, D.: Staff scheduling and rostering: A review of applications, methods and models. European Journal of Operational Research 153, 3–27 (2004)
4. Van den Bergh, J., Beliën, J., De Bruecker, P., Demeulemeester, E., De Boeck, L.: Personnel scheduling: A literature review. European Journal of Operational Re-search 226, 367–385 (2013)
5. Resende, M.G.C., Ribeiro, C.C.: Greedy randomized adaptive search procedures advances and applications. In: Gendreau, M., Potvin, J. (eds.) Handbook of Meta-heuristics, 2nd edn., pp. 283–320. Springer (2010)
6. Ernst, A.T., Jiang, H., Krishnamoorthy, M., Owens, B., Sier, D.: An annotated bib-liography of personnel scheduling and rostering. Annals of Operations Research 127, 21–144 (2004)
7. Della Croce, F.: Generalized pairwise interchanges and machine scheduling. Euro-pean Journal of Operational Research 83(2), 310–319 (1995)
8. Talbi, E.G.: Metaheuristics. From Design to Implementation. Wiley (2009)

# A Greedy Look-Ahead Heuristic
# for the Container Relocation Problem

Bo Jin*, Andrew Lim, and Wenbin Zhu

Department of Management Sciences, City University of Hong Kong,
Tat Chee Avenue, Kowloon Tong, Hong Kong
{msjinbo,limandrew}@cityu.edu.hk, i@zhuwb.com

**Abstract.** This paper addresses a classic problem in the container storage and transportation, the container relocation problem. For a given layout of a container bay, the containers should be retrieved in a predefined order, the best operation plan with fewest crane operations is going to be determined. We develop a greedy look-ahead heuristic for this particular purpose. Comparing with existing approaches presented from literature, our heuristic provides better solutions than best known solutions in shorter computational time.

**Keywords:** Transportation, Logistics, Heuristics, Container Relocation Problem.

## 1 Introduction

The container relocation problem we are investigating is first studied by Kim and Hong [1]. It models the retrieval of containers in container port. In a typical port, containers are stacked to form bays and blocks in container yards (Figure 1). The retrieval of containers is performed by a yard crane. The objective is to minimize the total number of yard crane operations to retrieve all containers in a bay according to given retrieval priority.

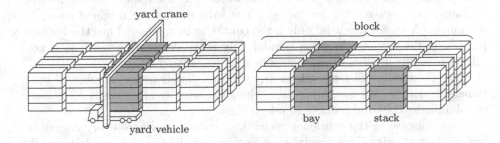

**Fig. 1.** Container yard overview

---

* Corresponding author.

M. Ali et al. (Eds.): IEA/AIE 2013, LNAI 7906, pp. 181–190, 2013.
© Springer-Verlag Berlin Heidelberg 2013

Each bay consists of $S$ stacks of containers labeled $1, \ldots, S$ from left to right and $T$ tiers of containers labeled $1, \ldots, T$ from bottom up. Figure 2 illustrate a bay with 5 stacks and 4 tiers, where every box represents a container and the number in the box represents the priority of the container with smaller number indicating higher priority. For example, heavy containers are usually assigned higher priority than light containers, so that they will be retrieved and loaded onto a ship before light containers. So that light containers ends up on the top of heavy containers in a ship, which increases stability and safety. There are other practical considerations such as the discharging port of the container that affect priority. Without loss of generality, we assume the priority of each container is given as input and there are a total of $P$ priorities $1, \ldots, P$. We introduce a dummy stack with label 0 to represent the location of the truck.

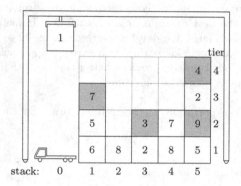

**Fig. 2.** Container bay overview

A yard crane operation that moves the container on top of the stack $s_1$ to the top of the stack $s_2$, is denoted by $\langle s_1 \to s_2 \rangle$ and called a *move*. Moving a container to the truck located at the dummy stack is called a *retrieval*, as the container will be transported away by the truck immediately; moving a container to any other stacks is called a *relocation*, which will not reduce the number of containers in the bay. A move $\langle s_1 \to s_2 \rangle$ is valid only if the origin stack $s_1$ is not empty. A retrieval is valid only if the container being moved has the highest priority in the bay. A relocation is valid only if the destination stack $s_2$ is not full, that is, the number of containers in $s_2$ is less than $T$.

Since the number of retrievals in any solution equals to the number of containers initially in the bay, minimizing the total number of moves is equivalent to minimizing the total number of relocations.

A special case of the container relocation problem is where every container has distinct priority. This special case was first studied by Kim and Hong [1], who limited the set of possible moves and derived a branch and bound algorithm. Caserta et. al. [2] improved the results by proposing a dynamic programming model and solved the model by considering only a subset of possible moves that lies in a "corridor". The "corridor" method was subsequently improved by Caserta and Voß [3,4]. Zhu et. al. [5] devised a strong lower bound and developed an iterative deepening A* algorithm (IDA*) that substantially improved the solution quality.

The general case of the container relocation problem was first studied by Kim and Hong [1] who restricted their attention to only a subset of all valid moves when constructing a solution. They proposed a branch and bound algorithm and a heuristic algorithm. Forster and Bortfeldt [6] proposed a tree search algorithm where only promising branches are explored.

Lee and Lee [7] extended the container relocation problem to a multiple bay scenario, which aims to retrieve all containers in multiple bays and minimize the total execution time for the yard crane including the bay-changing time cost and intra-bay job time cost.

## 2    The Greedy Look-Ahead Heuristic

Our approach to the container relocation problem can be break down into three levels as illustrated in Figure 3. At the top level, our approach is a greedy heuristic. We try to make the most promising move in each step until all container are retrieved. At the middle level, we use a lookahead tree search to decide the most promising move for each step. We limited the depth of the tree search to control the total computation effort. When the depth limit is reached, we invoke a probing heuristic to evaluate the potential of each node.

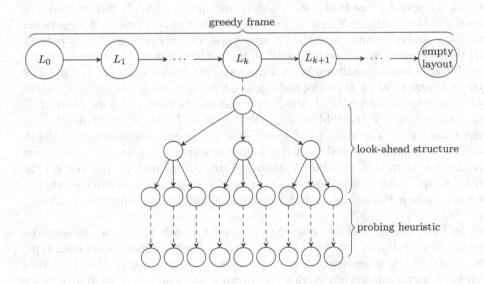

**Fig. 3.** The greedy look-ahead heuristic

## 2.1 The Top Level Greedy Heuristic

The top level greedy heuristic is given in Algorithm 1. Given a layout of the bay, if a container with the highest priority is at the top of some stack, we direct retrieve it. Otherwise, all containers with the highest priority are beneath some other containers. In this case we can only make a move that is a relocation. We try to perform a lookahead tree search with limited search depth to identify the most promising move. We then perform the move to obtain a new layout and proceed to next step. We repeat the process until all containers are retrieved.

---

**Algorithm 1.** Greedy-Lookahead Algorithm

---

Greedy($L$)

1    $relo\_no = 0$
2    **while** $L$ is not empty
3        **if** a container with the highest priority is at the top of stack $s$ in $L$
4            $m = \langle s \rightarrow 0 \rangle$
5        **else**
6            $m = \text{Next-Move}(L, 0)$
7            $relo\_no = relo\_no + 1$
8        $L = $ the layout resulted by applying move $m$ to $L$
9        **return** $relo\_no$

---

## 2.2 Look-Ahead Tree Search

Given a layout $L$ consists of $S$ stacks, there are up to $S \times (S-1)$ valid relocations, each valid relocation will lead to a new layout. In the new layout, if a container with the highest priority is at the top of some stack, we can immediately retrieve it. We will repeat such retrievals until all the containers with the highest priority are beneath other containers. We call the resultant layout a child of $L$. Therefore there is one child of $L$ corresponding to each valid relocation of $L$. Children of $L$ correspond to nodes at depth 1 in the search tree. For each child of $L$, we can generate grandchildren. Grandchildren of $L$ are nodes at depth 2 in the search tree. The size of the search tree grows exponentially as the depth increases, therefore we will limit the maximal search depth to the user defined parameter $D$ to control total computation time. Similarly, we can control the total computation time by limiting the branching factor of the search tree, that is, we can make better use of computation time by only considering most promising moves.

To limit the number of children generated for each layout, we classify the moves of each layout into 6 types, and select a few candidates from each type. So that the set of move considered are sufficiently diversified. Which in turn leads to better chance of covering the various parts of the entire search space and increase the likelihood of finding high quality solution.

A container is called a *badly-placed* container if and only if it is placed above some container with higher priority in the same stack, which means this container has to be relocated at least once before it is retrieved. Other containers are called *well-placed* containers. We will call a relocation *Bad-Good* move if the container before the container before the relocation is badly-placed and after the relocation is well-placed. Similarly, we define *Bad-Bad* (BB) move, *Good-Bad* (GB) move and *Good-Good* (GG) move.

We call the containers with the highest priority the *targets*. Moves that remove a non-target container above a target are called *freeing-target* (FT) moves, and other moves are called *non-freeing-target* (NT) moves. According to this criterion, we further classify BG and BB moves into 4 subcategories, FT_BG, NT_BG, FT_BB and NT_BB moves. We define the *priority of a stack* as the highest priority among all containers in the stack. We measure a move by computing the absolute difference between the priority of the moved container and the priority of the receiving stack. And we sort FT_BG, NT_BG, FT_BB, NT_BB moves in ascending order by this measure.

Let $f(s)$ be the highest priority among all containers except the top one in the stack $s$. We measure a GB move $\langle s_1 \rightarrow s_2 \rangle$ by computing the absolute difference between $f(s_1)$ and the priority of stack $s_2$. We sort GB moves in ascending order by this measure. The GG moves are sorted in ascending order by the priority of the receiving stack.

The look-ahead tree search is given in Algorithm 2, the procedure Next-Move returns the best move and the best score obtained if we follow this move.

---

**Algorithm 2.** Look-ahead Tree Search

---

Next-Move($L, depth$)

```
 1   if L is empty or depth equals to D
 2        return {NULL, depth + Probe(L)}
 3   {best_next, best_score} = {NULL, INF}
 4   Classify moves into six types: FT_BG, FT_BB, NT_BG, NT_BB, GG and GB
 5   move_list = select first few candidates from each of the six types
 6   for each move m in move_list
 7        L' = the layout resulted by applying move m to L
 8        while a container with the highest priority is at the top of stack s in L'
 9             L' = the layout resulted by applying move ⟨s → 0⟩ to L'
10        {next2, score2} = Next-Move(L', depth + 1)
11        if score2 < best_score
12             {best_next, best_score} = {m, score}
13   return {best_next, best_score}
```

---

The motivation of including a few moves from each type can be briefly explained as follows.

Firstly, FT_BG and FT_BB moves are certainly necessary since they remove the containers on the targets. And eventually we will be able to retrieve the targets.

Secondly, Figure 4 and 5 illustrate why NT_BG and GG moves are useful in certain scenarios. In the layout shown in Figure 4 with $(S, T) = (3, 4)$, it is better to move container 5 from stack 3 to stack 1, so that one additional move can be saved. In the layout shown in Figure 5 with $(S, T) = (3, 4)$, if GG moves are not employed, it needs 4 relocations for retrieving containers 1, 2, 3. However, if container 2 is relocated first onto the top of container 3, only 3 relocations are required.

**Fig. 4.** Example for NT_BG move      **Fig. 5.** Example for GG move

Lastly, Figure 6 and 7 illustrate why GB and NT_BB moves are useful in certain scenarios. We consider the layout shown in Figure 6 with $(S, T) = (3, 5)$, GB move which relocates container 3 to the third stack would lead to a solution of 8 relocations, however it needs 9 relocations when GB moves are not involved. Extend this situation to the layout illustrated in Figure 7 with $(S, T) = (3, 6)$, if we move container 4 and 3 from stack 1 to stack 3 first, only 12 relocations are required to retrieve all the containers. However, if NT_BB moves are not employed, 13 relocations are needed.

**Fig. 6.** Example for GB move      **Fig. 7.** Example for NT_BB move

## 2.3   Probing Heuristic

When the tree search process in Algorithm 2 reaches the depth limitation, we evaluate the quality of this leaf node by using a probing heuristic. The function Probe returns the number of relocations required to retrieve all the containers in the given layout, the number of relocations is then used as the score of this leaf node. Zhu et. al. [5] presented an efficient heuristic (named "PU2" in their paper) for the CRP variant where each container has a unique priority. We extend it to handle the situation where multiple containers share a priority.

The heuristic repeatedly chooses a target, moves the containers above it onto other stacks and then retrieves the target, until all the containers are retrieved.

Firstly, select the stack $s^*$ which contains a target $c^*$ with the least number of non-target containers above the target, tie breaking by smaller stack label. Then we remove the containers above $c^*$ one by one. For each non-target container $c$ at the top of stack $s^*$, we determine its destination stack. The destination stack is chosen from these stacks whose heights are less than $T$ under the following criteria:

1. If there are one or more stacks whose priority is not less than the priority of $c$, select the stack with the smallest priority, tie breaking by smaller stack label.

2. Otherwise, container $c$ will result in additional relocation for itself in the future. Before moving $c$, we can consider first relocating other container in other stacks. If stack $s'$ and its top container $c'$ satisfy the following three conditions,
   (i) the priority of container $c'$ equals to the priority of stack $s'$,
   (ii) it is possible to relocated $c'$ without incurring an additional relocation for itself in the future, and
   (iii) $c$ is well-placed by replacing $c'$ in stack $s'$,
   then we could first relocate $c'$, then relocate $c$ onto stack $s'$. If more than one such stack $s'$ exists, the one with largest priority is chosen.

3. Otherwise, let stack $s_2$ be the stack with the largest priority, here we perform an additional check, if
   (i) the priority of container $c$ is not the second smallest among all containers in stack $s$, and
   (ii) the height of stack $s_2$ equals to $T - 1$,
   then pick the stack with the second largest priority. Otherwise, choose stack $s_2$.

## 2.4   Termination Conditions

Before the greedy look-ahead heuristic executes, the probing heuristic is applied to obtain an initial solution. And each time the probing heuristic finishes, a solution is also yielded. We record new solutions obtained, and the best solution is updated if possible. The best found solution during the algorithm is finally reported. Our algorithm terminates until any of the three termination conditions is reached:

(i) The greedy look-ahead heuristic finishes,
(ii) Executing time limitation is reached,
(iii) Optimal solution is found, i.e., the number of relocations equals to the lower bound.

## 3   Computational Experiments

The greedy look-ahead heuristic is tested on two categories of test cases: CVS test cases provided by Caserta et. al. [2] and BF test cases provided by Bortfelft and Forster [8]. Our algorithm is implemented in Java and runs on the Intel

Core2 Quad CPU with 2.66GHz and 4GB RAM, and the computational results are compared with current best results.

We limit the search depth by parameter $D$, and the branching factor by 6 parameters, $n_{FT\_BG}$, $n_{FT\_BB}$, $n_{NT\_BG}$, $n_{NT\_BB}$, $n_{GG}$ and $n_{GB}$, corresponding to the numbers of children generated by 6 types of moves, respectively. For each test case, the mean lower bound of relocations (Lower Bound), the mean number of relocations per solution (*no.*), the mean total computational time (*time*) are presented.

## 3.1    Results on CVS Test Cases

Caserta et. al. [2] provided 21 test cases for the CRP. Each test case consists 40 instances. In these instances, containers have unique priorities and the stacks are not constrained in height. The test cases are labeled CVS *number of containers per stack–number of stacks*.

For the first 20 CVS test cases, Forster and Bortfelft [6] (Core2 Duo, 2 GHz, 2 GB RAM) provided the current best results, except the best result for CVS 10–10 which is provided by the "corridor" method of Caserta and Voß [3,4] (Pentium IV, 512 RAM).

For CVS test cases, we set $D = 3, 4$, $n_{FT\_BG}=5$, $n_{FT\_BB}=3$, $n_{NT\_BG}=5$, $n_{NT\_BB}=3$, $n_{GG}=1$ and $n_{GB}=1$. The running time is limited by 60 seconds. Table 1 compares the current best solution (Current Best) for the CVS test cases and the computational results of the greedy look-ahead heuristic (GLA).

**Table 1.** Comparing results on CVS test cases

| Test Case | Lower Bound | Current Best | | GLA ($D = 3$) | | GLA ($D = 4$) | |
|---|---|---|---|---|---|---|---|
| | | *no.* | *time* (s) | *no.* | *time* (s) | *no.* | *time* (s) |
| CVS 3–3 | 3.875 | 4.950 | <0.1 | 4.975 | <0.1 | 4.950 | <0.1 |
| CVS 3–4 | 4.825 | 6.050 | <0.1 | 6.025 | <0.1 | 6.025 | <0.1 |
| CVS 3–5 | 5.775 | 6.850 | <0.1 | 6.850 | <0.1 | 6.850 | 0.1 |
| CVS 3–6 | 7.250 | 8.275 | <0.1 | 8.275 | <0.1 | 8.275 | 0.3 |
| CVS 3–7 | 8.175 | 9.200 | <0.1 | 9.125 | 0.1 | 9.125 | 0.5 |
| CVS 3–8 | 9.250 | 10.450 | <0.1 | 10.325 | 0.1 | 10.300 | 0.9 |
| CVS 4–4 | 7.375 | 9.900 | <0.1 | 9.850 | <0.1 | 9.750 | 0.1 |
| CVS 4–5 | 10.250 | 12.625 | <0.1 | 12.375 | <0.1 | 12.250 | 0.4 |
| CVS 4–6 | 11.075 | 13.700 | <0.1 | 13.475 | 0.1 | 13.400 | 0.9 |
| CVS 4–7 | 13.400 | 15.775 | <0.1 | 15.450 | 0.1 | 15.450 | 1.7 |
| CVS 5–4 | 10.750 | 15.000 | <0.1 | 14.700 | <0.1 | 14.650 | 0.2 |
| CVS 5–5 | 13.400 | 18.625 | <0.1 | 17.850 | 0.1 | 17.800 | 0.8 |
| CVS 5–6 | 16.900 | 21.825 | <0.1 | 21.200 | 0.2 | 21.050 | 1.8 |
| CVS 5–7 | 19.200 | 23.575 | <0.1 | 23.050 | 0.3 | 22.800 | 3.1 |
| CVS 5–8 | 22.225 | 26.725 | <0.1 | 26.250 | 0.4 | 26.175 | 4.8 |
| CVS 5–9 | 25.075 | 29.450 | 0.1 | 28.950 | 0.6 | 28.750 | 7.5 |
| CVS 5–10 | 27.725 | 31.850 | 0.2 | 31.450 | 0.8 | 31.300 | 9.8 |
| CVS 6–6 | 21.525 | 29.675 | 0.4 | 28.800 | 0.3 | 28.575 | 2.8 |
| CVS 6–10 | 36.075 | 43.600 | 2.3 | 42.775 | 1.3 | 42.250 | 16.4 |
| CVS 10–6 | 42.500 | 75.650 | 45.5 | 69.950 | 1.2 | 68.775 | 12.1 |
| CVS 10–10 | 70.700 | 105.500 | 60.0 | 100.125 | 5.0 | 99.250 | 57.5 |
| Mean | 18.445 | 25.27 | 5.2 | 23.896 | 0.5 | 23.702 | 5.8 |

## 3.2  Results on BF Test Cases

Bortfelft and Forster [8] introduced 32 new test cases for the container pre-marshalling problem, each test case consists of 20 instances, which can be used as CRP test cases. In the BF test cases, the layouts consist of either 16 or 20 stacks, the stacks have a maximum height of either 5 or 8 containers. The utilization of the layout is either 60% or 80%, and the group number is either 20% or 40% of the container number.

Computational results are illustrated in the rest columns. We set the parameters by $n_{FT\_BG}=5$, $n_{FT\_BB}=3$, $n_{NT\_BG}=5$, $n_{NT\_BB}=3$, $n_{GG}=1$ and $n_{GB}=1$ and time limitation is 60 seconds.

**Table 2.** Comparing results on BF test cases

| Test Case | No. of stacks | No. of tiers | No. of containers | No. of priorities | Lower Bound | Current Best no. | time | Greedy Look-ahead no. | time |
|-----------|---------------|--------------|-------------------|-------------------|-------------|------------------|------|------------------------|------|
| BF 1  | 16 | 5 | 48  | 10 | 21.80 | 21.80 | <0.1 | 21.80 | <0.1 |
| BF 2  | 16 | 5 | 48  | 10 | 25.80 | 25.80 | <0.1 | 25.80 | <0.1 |
| BF 3  | 16 | 5 | 48  | 20 | 22.55 | 22.55 | <0.1 | 22.55 | <0.1 |
| BF 4  | 16 | 5 | 48  | 20 | 26.80 | 26.80 | <0.1 | 26.80 | <0.1 |
| BF 5  | 16 | 5 | 64  | 13 | 29.65 | 29.90 | <0.1 | 29.90 | 0.9 |
| BF 6  | 16 | 5 | 64  | 13 | 35.80 | 35.80 | <0.1 | 35.80 | 0.2 |
| BF 7  | 16 | 5 | 64  | 26 | 30.10 | 30.90 | <0.1 | 30.80 | 1.2 |
| BF 8  | 16 | 5 | 64  | 26 | 36.80 | 36.80 | <0.1 | 36.80 | 0.5 |
| BF 9  | 16 | 8 | 77  | 16 | 38.00 | 39.45 | 4.8  | 39.35 | 3.2 |
| BF 10 | 16 | 8 | 77  | 16 | 45.05 | 45.55 | 7.4  | 45.50 | 2.7 |
| BF 11 | 16 | 8 | 77  | 31 | 38.20 | 40.60 | 12.3 | 40.55 | 3.1 |
| BF 12 | 16 | 8 | 77  | 31 | 45.90 | 46.00 | 4.7  | 46.00 | 1.2 |
| BF 13 | 16 | 8 | 103 | 21 | 50.40 | 58.30 | 23.0 | 56.55 | 6.3 |
| BF 14 | 16 | 8 | 103 | 21 | 62.15 | 68.30 | 40.1 | 65.60 | 8.9 |
| BF 15 | 16 | 8 | 103 | 42 | 51.15 | 60.70 | 43.9 | 58.35 | 5.2 |
| BF 16 | 16 | 8 | 103 | 42 | 63.05 | 72.25 | 53.8 | 68.50 | 6.6 |
| BF 17 | 20 | 5 | 60  | 12 | 27.25 | 27.25 | <1.0 | 27.25 | <0.1 |
| BF 18 | 20 | 5 | 60  | 12 | 31.70 | 31.70 | <1.0 | 31.70 | <0.1 |
| BF 19 | 20 | 5 | 60  | 24 | 27.55 | 27.60 | <1.0 | 27.60 | 0.1 |
| BF 20 | 20 | 5 | 60  | 24 | 34.35 | 34.35 | <1.0 | 34.35 | <0.1 |
| BF 21 | 20 | 5 | 80  | 16 | 37.25 | 37.70 | 5.5  | 37.65 | 2.4 |
| BF 22 | 20 | 5 | 80  | 16 | 44.55 | 44.55 | <1.0 | 44.55 | <0.1 |
| BF 23 | 20 | 5 | 80  | 32 | 37.30 | 38.05 | 6.3  | 37.95 | 2.4 |
| BF 24 | 20 | 5 | 80  | 32 | 43.95 | 43.95 | <1.0 | 43.95 | <0.1 |
| BF 25 | 20 | 8 | 96  | 20 | 46.15 | 47.45 | 11.6 | 47.20 | 5.9 |
| BF 26 | 20 | 8 | 96  | 20 | 55.65 | 55.85 | 3.2  | 55.70 | 0.4 |
| BF 27 | 20 | 8 | 96  | 39 | 46.70 | 48.75 | 26.5 | 48.40 | 5.3 |
| BF 28 | 20 | 8 | 96  | 39 | 57.00 | 57.40 | 18.0 | 57.15 | 2.5 |
| BF 29 | 20 | 8 | 128 | 26 | 63.10 | 73.05 | 60.0 | 70.10 | 11.7 |
| BF 30 | 20 | 8 | 128 | 26 | 76.15 | 83.90 | 57.0 | 79.35 | 15.1 |
| BF 31 | 20 | 8 | 128 | 52 | 64.25 | 74.95 | 56.9 | 71.40 | 8.9 |
| BF 32 | 20 | 8 | 128 | 52 | 77.40 | 86.25 | 58.9 | 80.85 | 12.8 |
| | | Mean | | | 43.55 | 46.07 | 15.9 | 45.18 | 3.4 |

## 4  Conclusions

As shown in the two tables, the greedy look-ahead heuristic provides better solutions for existing test cases in shorter running time. The greedy look-ahead heuristic is a very efficient algorithm to find a solution of high quality, although it cannot guarantee the optimality even if given infinite running time.

The efficiency of the greedy look-ahead heuristic is mainly affected by the look-ahead depth and width. However, the increment of the time consuming by raising look-ahead depth is exponential, hence we have to limit a proper lookahead depth depending on the instance scale. Looking back to the test case CVS 3–3, the greedy look-ahead heuristic with $D = 3$ does not outperform the current best known result, since the look-ahead scope is not large enough. In the last two columns, the greedy look-ahead heuristic with $D = 4$ provides better results than that with $D = 3$ while takes more executing time.

At the beginning period of the whole process, the layout obviously has more containers that would yields more possible successors. So that, the heuristic should put more effect on the beginning period than the hind period. Further improvement will be studied in the future work.

# References

1. Kim, K.H., Hong, G.P.: A heuristic rule for relocating blocks. Computers & Operations Research 33(4), 940–954 (2006)
2. Caserta, M., Voß, S., Sniedovich, M.: Applying the corridor method to a blocks relocation problem. OR Spectrum 33(4), 915–929 (2011)
3. Caserta, M., Voß, S.: A cooperative strategy for guiding the corridor method. In: Krasnogor, N., Melián-Batista, M.B., Pérez, J.A.M., Moreno-Vega, J.M., Pelta, D.A. (eds.) NICSO 2008. SCI, vol. 236, pp. 273–286. Springer, Heidelberg (2009)
4. Caserta, M., Voß, S.: Corridor selection and fine tuning for the corridor method. In: Stützle, T. (ed.) LION 3. LNCS, vol. 5851, pp. 163–175. Springer, Heidelberg (2009)
5. Zhu, W., Qin, H., Lim, A., Zhang, H.: Iterative deepening a* algorithms for the container relocation problem. IEEE Transactions on Automation Science and Engineering 9(4), 710–722 (2012)
6. Forster, F., Bortfeldt, A.: A tree search procedure for the container relocation problem. Computers & Operations Research 39(2), 299–309 (2012)
7. Lee, Y., Lee, Y.J.: A heuristic for retrieving containers from a yard. Computers & Operations Research 37(6), 1139–1147 (2010)
8. Bortfeldt, A., Forster, F.: A tree search procedure for the container pre-marshalling problem. European Journal of Operational Research 217(3), 531–540 (2012)

# Integrating Planning and Scheduling in the ISS Fluid Science Laboratory Domain

Amedeo Cesta[1], Riccardo De Benedictis[1], Andrea Orlandini[1], Riccardo Rasconi[1], Luigi Carotenuto[2], and Antonio Ceriello[2],*

[1] CNR – National Research Council of Italy, ISTC
name.surname@istc.cnr.it
[2] Telespazio S.p.A., Italy
name.surname@telespazio.com

**Abstract.** This paper describes a Planning and Scheduling Service (Pss) to support the Increment Planning Process for the International Space Station (ISS) payload management. The Pss is described while targeting the planning of experiments in the Fluid Science Laboratory (FSL), an ISS facility managed by Telespazio User Support and Operation Centre (T-USOC) that has been identified as a representative case study due to its complexity. The timeline-based approach inside the Pss is evaluated against realistic planning benchmark problems.

## 1 Introduction

The collaboration among the authors started within the activities of ULISSE[1] a project funded by EU and indicated by REA as an example of successful FP7 project in the Space area [1]. The project's main goal has been (a) the data valorization around the International Space Station (ISS) experiments, but also (b) the synthesis of new tools to favor ISS management activities in a broader sense. One of these new tools has been studied to support the USOCs efforts during planning processes around ISS Payloads.

The USOCs (User Support and Operation Centres) are a network of scientific space facility operation centres that are established in various European countries with the support of national space agencies and are engaged by the European Space Agency to conduct the operations for European scientific experiments on board the Columbus as well as other modules of the ISS. Each USOC is responsible for a particular ISS on-board facility that is to be operated in order to perform scientific experiments and generate the related scientific data. In their ordinary operations, USOCs have to periodically interact with the Columbus European Planning Team (EPT), the center that coordinates nominal operations.

As a consequence of the ISS's long operational lifetime, the mission planning process is performed in several distinct steps – Strategic, Tactical and Execution/Increment

---

* Authors have been partially supported by EU under the ULISSE project in [2009-2011]. CNR authors are currently supported by group internal funds.
[1] "USOCs Knowledge Integration and dissemination for Space Science Experimentation" – GA.218815 – www.ulisse-space.eu

M. Ali et al. (Eds.): IEA/AIE 2013, LNAI 7906, pp. 191–201, 2013.
© Springer-Verlag Berlin Heidelberg 2013

Planning – with distinct planning products covering different time intervals and ranging from several years to just a few days [2]. In general, an *increment period* lasts three months and it is defined as the time between two launches with ISS crew exchanges. Our focus has been on *Increment Planning* whose goal is to develop increment-specific operations products and the associated information necessary to prepare and conduct real-time operations on payloads (also called facilities). The information generated through this process is exploited by users, ground controllers, flight crew, etc. to plan the preparation and execution of an increment plan and to help make management decisions. In producing the increment plan, an effective interaction between EPT and single USOCs is important. In common practice such a problem is addressed and solved without any decision support tool and dedicating a specialized human resource when needed. Indeed, after several interaction phases between the two centers, the need for a either complete or partial recomputation of the plan was identified. Manually generating a complete plan is not only time consuming but also an error-prone task, hence the idea of using automated planning and scheduling techniques. Our collaboration concerned the experiments in the Fluid Science Laboratory (FSL), an ISS facility managed by the Telespazio USOC on behalf of ESA. The FSL is considered a representative case study for Increment Planning due to the complexity of the involved constraints.

This paper describes our proposal for a Planning and Scheduling Service (Pss) for the Increment Planning of the ISS payload management by using a timeline-based environment for planning and scheduling. Section 2 gives more details on the addressed problem and its constraints, Section 3 describes the timeline based approach we have used, Section 4 describes the current experimental performance of the Pss against realistic benchmarks. A concluding section ends the paper.

## 2   The Increment Planning Process

In current ISS practice, the Increment Planning process is divided into two main phases: the Pre-Increment Planning and the Short-Term Planning. As usual in space practice they have different time granularities and levels of detail. The Pre-Increment Planning delivers a list of all activities to be carried out during the increment and, possibly, identifies bottlenecks on the most critical on-board resources. The Short-Term Planning details portions of the increment, usually one or two weeks, developing more detailed plans and schedules. The main product of the Pre-Increment is the On-Orbit Operations Plan (OOS) consisting of a list of activities to be performed on a weekly basis, and if necessary on a daily basis. The Short-Term Plan is defined two weeks before the beginning date of actual execution, detailing one week of the OOS and giving a quite detailed view of the activities to be carried out during this week (Short-Term Plans are also named Weekly Look-ahead Plans (WLP)). Our work is currently devoted to solve the Short-Term Planning by capturing first a realistic domain model and then synthesizing the related complete plan. Once a WLP is identified, the subsequent plan management phase is also very important. As a matter of fact, the daily activities of USOC engineers involve two aspects: (a) to *synthesize* in details feasible[2] WLPs originating from the

---

[2] i.e., compliant with a set of hard constraints provided by both the facility and the ISS.

PI's requests and that will have to be communicated to the Columbus European Planning Team (EPT) for future execution on board the scientific facility controlled by the USOC; (b) to efficiently *adapt* such plans in case of need, i.e., to be able to promptly produce alternative WLP to face of a number of modifications that are usually communicated to the USOC during the plan refinement process. The two capabilities of synthesizing new plans and managing changes are the main activities to be supported by a decision-making aid for the USOC engineers.

*The Fluid Science Laboratory.* As previously stated, we are applying the Pss to the planning of experiments for the Fluid Science Laboratory (FSL), an ISS facility managed by the Telespazio USOC. The FSL is a multi-user facility designed for the execution of experiments on fluid physics under microgravity conditions. It consists of different modules and equipment functionally integrated into one of the Columbus Orbital Facility (COF) *racks*. The FSL is equipped with a number of optical instruments that allow to separately implement a wide variety of diagnostic techniques (e.g., Electronic Speckle Pattern Interferometry, Wollaston interferometry, etc.); however, different diagnostic techniques can be combined together, and each combination is called *optical mode*. An optical mode is therefore composed of a specific set of diagnostic techniques. Currently, the FSL provides 86 different optical modes. The main FSL functionalities such as power conversion and distribution, communication and data processing (including data recording and playback), facility commanding and monitoring by both ground and flight operators, as well as telescience performed via ground centers, are guaranteed by exploiting the COF resources. Generally, an experiment execution consists of several runs, a run is a part of the experiment that uses a defined configuration and setting of the facility (as for example a specific optical mode). The data recorded by the experiments are either sent in a real-time manner to the ground or stored on a local mass memory (with a 30GB capacity). In both cases, data is routed to the Columbus for communicating to the ground through the two downlink channels: *Low Rate*, mainly intended for housekeeping and small quantity of scientific data and *High Rate*, mainly intended for video images and for high rate science data. In order to avoid the loss of experimental results due to memory oversubscription, data downlink operations guaranteeing that all data and images stored in the FSL mass memory are downloaded to earth, must be continuously planned and executed. The FSL is always in one defined status, among the following: Off, Stand-by, Configuration & Checkout and Nominal. A set of operations on FSL has been identified that require an initial status of FSL and may eventually lead to a transition to a new status of the facility. Each activity is characterized by several parameters, e.g., the team that operates the FSL, the duration of the activity, resource consumptions, etc. A complete specification of the FSL activities is given in [3].

*Constraints on Feasible Solutions.* In general, each plan executed on the FSL has to comply with two different kinds of constraints: Facility constraints and Operative constraints. Facility constraints (**F-constraints**) are related to the FSL functioning conditions: max memory storage; selectable frame rates; max number of simultaneous video channels; max data rate for downlink; single image size, max number of files per single download to ground. Moreover, constraints on resources usage must be considered (e.g., crew time, FSL power consumption, links, etc). Operative constraints (**O-constraints**) are related to general operational requirements: prior to performing scientific experiments and/or diagnostic tests correctly, the FSL must follow a precise sequence of steps in order to be fully operative: initially, it must be mechanically configured according to the experiment/test to be performed; subsequently, the operative rack has to be activated; finally, the set of diagnostics must be executed before the experiment can commence. At the end of each operative cycle, the previous operations must be planned to be executed in the reverse order. At the beginning of each operative period, an optical mode test has to be performed to check whether the mechanical configuration has been correctly completed (avoiding to perform experiments with optical targets wrongly set); the High Rate Data Link (HRDL) has to be allocated during each run that requires real time data transmission and during each download of recorded data; a run cannot be interrupted and has to be executed continuously. Other operative constraints are required for safety issues: during non operative periods, the status of both the FSL and the Rack must be off, FSL mechanical configuration and de-configuration activities have to be performed with both FSL Rack and FSL switched off. Finally, also some USOC local restrictions have to be considered. At present, we considered the following: the total duration of operations for each day must not exceed 12 hours (including switch on and off of any system); no operations have to be planned during weekends.

Generally, the execution of an experiment consists of several runs; a run is a segment of the experiment that uses a defined configuration and setting of the facility (as for example a specific optical mode). The objective function for the problem is a composition of the following: all the planned experiments must be performed; the use of the FSL mass memory must be maximized; crew activity must be minimized.

## 3   A Timeline-Based Approach for Increment Planning Problems

In approaching domain modeling for planning, we use a timeline based approach [4,5]. According to this approach, a planning problem is modeled in terms of temporal functions that describe the *dynamic evolution* of the relevant domain features over a finite temporal horizon. The variables may assume a finite set of values over intervals of time. The domain causality is described as patterns of temporal constraints among values, usually called *compatibilities*, that characterize the legal physical behavior of the domain features (values on the same or among different dynamic variables) [4]. Solving a timeline-based planning problem equals to synthesizing, among the legal ones, those temporal functions that satisfy additional patterns of temporal constraints (the *planning goals*) over their temporal behavior. In a previous work [6] we have build a first solution to the increment planning using a general-purpose development environment for timeline-based planners called the TRF (the Timeline Representation Framework [7]). That solution consisted of a combination of different solvers on top of the TRF that turned out to be quite limited in scalability. In the current paper we present an approach that uses a different planner, called J-TRE, that integrates to a finer grain the planning and scheduling features needed in this domain and turns out to offer better performance. The rest of this section introduces first the J-TRE then shows how the FSL domain is modeled through it.

## 3.1 The J-TRE Solving Environment

The most direct way to describe a *timeline* is to consider it as a collection of time-bounded predicates called *tokens*. The task of a planner is to find a legal sequence of tokens that bring the timelines into a final configuration that verifies both the *domain theory* (i.e., the set of compatibilities that model the domain's dynamic behavior) as well as a determined set of desired *goals* conditions. Following a refinement search strategy, the J-TRE planner starts from an initial state and traverses the search space by adding or removing tokens and/or relations (i.e., changing the current state) until all goals are satisfied. The predicates (i.e., the tokens) that can be accommodated on a timeline, as well as the behavior assumed by the planner when a new token is added, depend on the nature of the timeline itself and on the modeled domain. J-TRE allows the utilization of families of timelines which provide different modeling power, such as *multi-valued state variables a la* [4] as well as *renewable and consumable resources* like those commonly used in constraint-based scheduling [8]. A plan is consistent when all goals are satisfied and all the values of the domain variables satisfy the domain constraints over the time horizon. If we call *flaw* every possible inconsistency of the current plan, the planning process can be reduced to the identification and the resolution of each flaw in the plan until a consistent plan is found, i.e., the propagation of the solving constraints succeeds and all flaws are eliminated (see Algorithm 1 as a reference). At each iteration, the previous solving strategy entails the following steps: (i) the identification of a set of flaws, (ii) the choice of a flaw according to a determined *selection strategy*, and (iii) the resolution of the flaw through a *resolution strategy*.

**Algorithm 1.** General solving procedure

```
1:  function SOLVE(goals)
2:      for all n ∈ fringe do
3:          n.enqueue(goals);
4:      end for
5:      while fringe ≠ ∅ do
6:          node ← fringe.poll();
7:          if isConsistent(node) then
8:              if isSolution(node) then
9:                  return node;
10:             end if
11:             f ← selectFlaw(node);
12:             fringe ← fringe ∪ f.solve(node);
13:         end if
14:     end while
15:     return ⊥;
16: end function
```

In more details, the solver works as follows: initially, all pending goals are enqueued in the search space *fringe*[3] (lines 2 – 4). Then, at each solving iteration a node is extracted from the fringe (lines 5 – 6) and a consistency check routine is called (line 7); if the node is evaluated to be *consistent* (i.e., no constraint propagation inconsistency is acknowledged for that node), and if the same node represents a *solution* (line 8), then the node is returned and the procedure exits with success. In the opposite case, the node is analyzed for all the present flaws, and one of them is selected for resolution (line 11). Subsequently, the selected flaw is solved (line 12); this operation may inherently generate new nodes (i.e., the current node is *expanded*) that are added to the current fringe. The solving procedure ends when a consistent node (i.e., containing no flaws) is found. Finally, if the fringe gets empty during the solving process (i.e., the procedure reaches a dead end) the procedure exits with no solution (line 15).

While flaws can be of different types and can arise for different reasons, what they all have in common is that a *search choice* is necessary to solve each of them. There can

---

[3] The *fringe* is the collection of the search nodes that are generated but not yet expanded.

basically be four kinds of flaws: (i) *goal flaws* arise when a new token is added to a state variable to satisfy a *compatibility* requirement (see Section 3.2), (ii) *disjunction flaws* and (iii) *preference flaws* respectively arise when a disjunction or a preference statement is found while enforcing the domain rules (expressed in DDL code), and (iv) *timeline inconsistency flaws* arise when inconsistencies (a.k.a. contention peaks) are detected on some domain objects. Among the most remarkable advancements offered by the J-TRE software infrastructure w.r.t. to previous P&S representation frameworks such as the TRF, we underscore the following: (i) the "unification" of the concept of *flaw* (i.e., a plan inconsistency) into a single entity that is uniformly treated (and reasoned upon) throughout the whole J-TRE infrastructure. In J-TRE, flaw analysis and management is no longer spread across specialized reasoners depending on the flaw type, thus allowing to introduce more effective search heuristics that exploit the cross-comparisons among flaws of different types; (ii) the possibility to express constraints of increasing complexity among different domain parameters (e.g., modeling the dependency between resource quantity to be produced and the production activity duration, etc.); (iii) the introduction of the consumable resources among the active domain component types.

### 3.2   A Timeline Specification for the FSL Domain

To obtain a timeline-based specification of the FSL domain both types of timelines have been considered: *multi-valued state variables* and *renewable and consumable resources*.

Multi-valued state variables are those representing time varying features for the temporal occurrence of mechanical configurations, rack activations as well as both FSL status and activities. In this regard, we consider four different state variables, i.e., *Mechanical Configuration*, *Rack*, *FSL* and *FSL Activities*. The Figure 1 depicts a detailed view of the values that can be assumed by these state variables, their durations and the allowed value transitions in accordance with **O-constraints**. Resources are used in order to represent the actual resources whose use is necessary in order to accomplish the mentioned tasks, i.e., the HRDL, S and KU communication channels availability as well as the involvement of crew and some generic tools while performing tasks. Additionally, the on-board memory (30GB) is represented using a resource type. Then, five binary resources are considered, i.e., *HRDL*, *S-Chan*, *KU-Chan*, *Crew* and *Tools*. Additionally, also a consumable multi-capacity resource is defined, i.e., *Memory*[4]. The above resources are defined according to the **F-constraints**.

In general, before any operative period, the FSL requires a *Mechanical Configuration*. In particular, during not operative periods no optical target is mounted on the FSL (*Deconfigured*) while, before starting an optical checkout, a suitable optical target is to be mounted to result mechanically configured (*Configured*). The *Rack* can assume a *Active* status when switched on while, when switched off, it assumes a *Not Active* value. The FSL can assume different status as well as perform different activities. The *FSL* may be in one of the following status: *Off* while not operating, *StandBy* after initialization and in operative mode when ready for Control and Checkout (i.e., *CC*).

---

[4] The design of the FSL ensures that every activities is fulfilling the power consumption limit assigned to the facility hence the power consumption has not been modeled.

The *FSL_Activities* represents the full set of activities that can be performed on-board. Namely, the FSL activities are the following: installation or removal of an optical target (respectively, *OPT_TGT_INST(x)*[5] and *OPT_TGT_RMV*); activation and deactivation of the rack (*RACK_ACT* and *RACK_DEACT*); initialization and activation of the optical component (*FSL_STBY* and *FSL_CC*); finally, while running an optical checkout the FSL may assume either the *RT_OPT_CO(y)* or *REC_OPT_CO(z)* (the variables $y$ and $z$ are representing a particular run that should be performed) according to the possibility of downloading in real time the result rather than storing it on the on board memory; finally, executing a downlink activity results in assuming the *DATA_DNLK* value. In Figure 1, we detail the values that can be assumed by these state variables, their durations and the allowed value transitions in accordance with the facility and operative requirements. It is worth underscoring that some durations are not known in advance. In fact, the mechanical configuration depends on the kind of mounted optical target (i.e., the duration $d0$ is dependent by the parameter $x$). Similarly, the duration of both the recorded ($d1$) and real time experiments ($d2$) is influenced by the type of run to be executed. Finally, the duration of the downlink operation ($d3$) is clearly dependent by the amount of Gbytes to be transmitted stored on the on board memory.

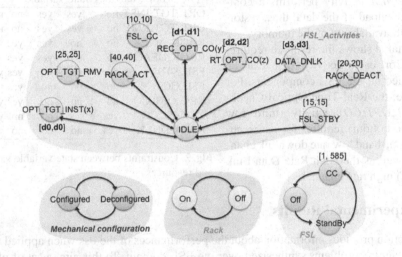

**Fig. 1.** Value transitions for state variables describing the FSL (temporal durations in minutes)

In the FSL domain, the following compatibilities are considered (not shown in Figure 1 not to overload the representation): (1) *OPT_TGT_INST* values must occur DURING a *Deconfigured* value on the Mechanical Configuration variable; (2) *OPT_TGT_RMV* as well as *Active* and *RACK_ACT* must occur DURING a *Configured* value on the Mechanical Configuration state variable; (3) *RACK_DEACT*, *RACK_STBY* and *RACK_CC* must occur DURING a *Active* value on the Rack state variable; (4) *REC_OPT_CO*, *RT_OPT_CO*, *DATA_DNLK* must occur during a *CC* value on the FSL state variable;

---

[5] The variable x represents one of the 86 available optical modes.

(5) each *REC_OPT_CO* must be BEFORE a *DATA_DNLK*. The first two compatibilities globally express the circumstance that all FSL activities must be performed within a Rack activation/deactivation cycle, and that such cycle must be performed once an optical target has been configured. The third compatibility enforces that the FSL is supposed to be initialized before being fully operative. Then, constraint (4) enforces that data stored on the FSL mass memory must be downloaded on some earth ground station. Finally, compatibility (5) states that every download task should be performed after a recorded experiment.

The table in Fig. 2 provides the considered constraint relations between the state variable's values (i.e., the activities of the final FSL schedule) and the resources that are necessary for the execution of each activity (binary), according to the FSL scheduling model. The *Memory* resource is used by the *REC_OPT_CO* in a size that is dependent by the duration of the same task. Each *DATA_DNLK* activity performs a complete download of the data, thus, restoring a full availability of the memory.

The table shows the resource requirements for each activity that can be performed on the FSL components. For example, the Real-Time Optical Checkout (*RT_OPT_CO*) activity (third row from the bottom) requires for its execution the *ku*-band low rate downlink channel, as well as the High Rate Data Link (HRDL) high rate channel.

| Value | HRDL | Tools | Crew | S | ku |
|---|---|---|---|---|---|
| Mechanical Configuration State Variable | | | | | |
| NotConfigured | no | no | no | no | no |
| Configured | no | no | no | no | no |
| Rack State Variable | | | | | |
| Not Active | no | no | no | no | no |
| Active | no | no | no | no | no |
| FSL State Variable | | | | | |
| OFF | no | no | no | no | no |
| StandBy | no | no | no | no | no |
| CC | no | no | no | no | no |
| FSL_Activities State Variable | | | | | |
| OPT_TGT_INST | no | yes | yes | no | no |
| OPT_TGT_RMV | no | yes | yes | no | no |
| RACK_ACT | no | no | yes | yes | yes |
| RACK_DEACT | no | no | yes | yes | yes |
| FSL_STBY | no | no | no | yes | yes |
| FSL_CC | no | no | no | yes | yes |
| REC_OPT_CO | no | no | no | yes | yes |
| RT_OPT_CO | yes | no | no | no | yes |
| DATA_DNLK | yes | no | no | no | yes |

**Fig. 2.** Constraints between state variable values and resources

## 4  Experimental Results

This section provides information about the performances of the Pss when applied to a set of planning problems synthesized over the FSL domain. To this aim, a set of planning experiments have been performed by running the Pss on 30 problem instances of increasing size relatively to a particular experiment type called *Optical Check-Out*. The Optical Check-Out is an activity aimed at verifying and optimizing the functionalities of one or more FSL optical instruments. Like for any other experiments, the optical checkouts require the FSL to be set in a specific mechanical configuration and status, i.e., the FSL has to be *mechanically configured* according to the experiment requirements, while the operative rack has to be *activated*. Lastly, an Optical Check-Out can be performed in a Real-Time setting (*RT_OPT_CO*), i.e., directly communicating the experimental results to earth as data are collected, or in a Recorded setting (*REC_OPT_CO*), i.e., produced data are stored in the FSL mass memory and downloaded to earth in a separate operative session. Therefore, in the *RT_OPT_CO* case the experiment operations can be planned for only if the High Rate Data Link (HRDL) channel for real-time data downloading is available, whereas in the *REC_OPT_CO* case the final plan will have to schedule for all the data-dumping operations so as to prevent loss of precious data from the limited capacity on-board memory, which makes the planning problem slightly harder to solve. In this test, all 30 problem instances have been resolved in two different experimental settings, one relatively to recorded optical checkout experiment goals (*REC_OPT_CO* wide-dashed line), and the other related to real-time optical checkout experiment goals (*RT_OPT_CO* narrow-dashed line).

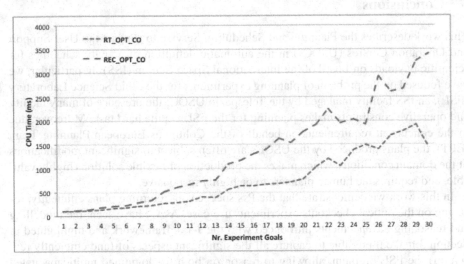

**Fig. 3.** Pss planning performance graph

The Pss performance results are presented in Figure 3. In the figure, the $X$ axis shows the problem size, expressed in terms of number of goals pertaining to each problem (varying from 1 to 30), that must all be satisfied by the planning process. The $Y$ axis provides the CPU time required by the solving procedure expressed in milliseconds. All experimental runs have been performed on a Dell Optiplex 980, 2.93 Ghz Intel Core i7 CPU with 4Gb RAM, under Windows 7 32 bit Operating System.

As Figure 3 shows, the Pss solving performances are rather good, as even the largest instances are solved in a reasonably short amount of time. More specifically, the Pss planner succeeds in solving the most complex instances in the set (which entails producing a plan for 30 optical checkout tests, corresponding to a continuous execution time of about 13 days on board of the FSL) in less than 2 seconds in the *RT_OPT_CO* case, and in less than 3.5 seconds in the *REC_OPT_CO* case. As this is the first attempt to introduce AI-based P&S techniques in the FSL domain (all FSL planning activities are actually performed manually) it is not possible to provide "structured" data about the performance improvements obtained w.r.t. previous approaches. However, given the complexity of the problem at hand, e.g., the presence of a high number of existing domain constraints and/or the problem variability inherent to the increment planning process, introducing a tool able to provide within seconds correct solutions on a two weeks' planning horizon, represents a remarkable advancement. Another important aspect is related to the fact that the proposed system can be used as a *Mixed Initiative* tool; the fast response time allows to provide the user with a set of consecutive solutions, leaving to the human operator the "last word" on the final decision. Last but not least, the proposed Pss is particularly effective to cope against uncertainty, as it allows a fast recomputation of alternative solutions in the face of unexpected changes of the environmental conditions.

## 5   Conclusions

This work describes the Planning and Scheduling Service to support the User Support and Operation Centres (USOCs) in the automatic definition of activity schedules for scientific payloads on-board of the International Space Station (ISS). In particular, we have focused on the problem of planning experiments for the Fluid Science Laboratory (FSL), an ISS facility managed by the Telespazio USOC; the presence of many facility and operative constraints makes planning for the FSL a quite hard task. Moreover, due to the concurrent requirements on behalf of the Columbus European Planning Team (EPT), the plans produced by the USOCs are often subject to significant modifications of the domain conditions, which makes the production of flexible solutions highly valuable, and requires the human planners to be highly responsive.

In this work we demonstrate that the Pss succeeds in returning plans efficiently, focusing on the optical checkouts experiments use case. As we have seen, the modelling and reasoning advancements introduced with the J-TRE framework and highlighted in Section 3 make it possible to capture all the significant aspects of (and efficiently reason on) the FSL problem, allowing to reason on both the logic and multicapacitated resource aspects of the domain, in a truly interleaved planning and scheduling perspective. Moreover, this reasoning power is currently being exploited in the FSL domain with an optimization perspective for the production of plans with minimized makespan: the results obtained so far are promising, and will be further pursued with the introduction of improved search heuristics.

## References

1. Kuijpers, E., Carotenuto, L., Cornier, C., Damen, D., Grimbach, A.: The ULISSE Environment for Collaboration on ISS Experiment Data and Knowledge Representation. In: 61st International Astronautics Congress, IAC-10-D5.2.2, Prague, CZ (2010)
2. Leuttgens, R., Volpp, J.: Operations planning for the international space station. ESA Bulletin (1998)
3. Carotenuto, L., Ceriello, A., Cesta, A., De Benedictis, R., Orlandini, A., Rasconi, R.: Planning and Scheduling Services to Support Facility Management in the ISS. In: 63rd International Astronautics Congress, IAC-12-B5.2.10, Naples, IT (2012)
4. Muscettola, N.: HSTS: Integrating Planning and Scheduling. In: Zweben, M., Fox, M.S. (eds.) Intelligent Scheduling. Morgan Kauffmann (1994)
5. Jonsson, A., Morris, P., Muscettola, N., Rajan, K., Smith, B.: Planning in Interplanetary Space: Theory and Practice. In: Proceedings of the Fifth Int. Conf. on AI Planning and Scheduling, AIPS 2000 (2000)

6. Cesta, A., Fratini, S., Rasconi, R., Orlandini, A.: A Planning and Scheduling Service for the ULISSE Platform. In: IEEE Fourth International Conference on Space Mission Challenges for Information Technology, SMC-IT 2011, pp. 35–42. IEEE Computer Society (2011)
7. Cesta, A., Cortellessa, G., Fratini, S., Oddi, A.: Developing an End-to-End Planning Application from a Timeline Representation Framework. In: Proceedings of the Twenty-First Conference on Innovative Applications of Artificial Intelligence, IAAI (2009)
8. Cesta, A., Oddi, A., Smith, S.F.: A Constraint-based Method for Project Scheduling with Time Windows. Journal of Heuristics 8(1), 109–136 (2002)

# Efficient Identification of Energy-Optimal Switching and Operating Sequences for Modular Factory Automation Systems

Sebastian Mechs[1], Steffen Lamparter[1], Jörn Peschke[1], and Jörg P. Müller[3]

[1] Siemens AG, Munich/Nuremberg, Germany
{sebastian.mechs.ext,steffen.lamparter,
joern.peschke}@siemens.com
[2] Clausthal University of Technology, Department of Informatics, Clausthal-Zellerfeld,
Julius-Albert-Str. 4, Germany
joerg.mueller@tu-clausthal.de

**Abstract.** In order to enable energy-efficient operation of factory automation systems during non-productive (idling) phases, the energy-optimal sequence of operating modes has to be calculated. Due to modular structures and runtime constraints, the combinatorial optimization problems that have to be solved to calculate energy-minimizing schedules for today's automation systems become extremely complex. In this paper, a novel domain-specific branch-and-bound algorithm is proposed that takes the structural knowledge about the automation system into account in order to attenuate the exponential complexity. Relying on a network of automation subsystems representing the switching and energetic operating behavior of the automation system, the method uses the minimal energy demand for unrelated subsystems, which can be efficiently calculated using off-the-shelf constraint optimization techniques, as lower bounds for the energy demand of the complete automation system. Evaluations indicate that the proposed procedure outperforms a complete enumeration in terms of computational time by more than 70 percent while assuring to identify the energy-optimal switching and operating sequences.

**Keywords:** energy efficiency, factory automation systems, constraint optimization, branch-and-bound algorithm.

## 1 Introduction

In industrial settings, automation infrastructure can support optimal energetic operation in non-productive (idling) phases of the production system. Energy may serve in this context as universal cost quantifier of the demand for resources like compressed air or laser output. Energy management systems lack of tool support for guiding systems to energy-efficient modes during non-productive phases [1]. This problem has not been addressed analytically for factory automation systems yet. Besides the determination of the correct sequence of modes, it is necessary to calculate the time spent in each mode in order to identify energy-optimal sequences for non-productive intervals. Calculating energy-optimal sequences based on optimization models suffers from an exponential

M. Ali et al. (Eds.): IEA/AIE 2013, LNAI 7906, pp. 202–211, 2013.

complexity leading to unsatisfying runtime efficiency. Therefore, the inherent problem complexity should be attenuated by exploiting structural information in this domain. Modularity aspects can be used to selectively calculate the energy demand for complete systems based on the energy demand for modules (subsystems). Beyond the formulation of the optimization problem [2], the search effort for energy-optimal switching and operating schedules for a complete system can be reduced by a structure-exploiting branch-and-bound procedure [3], [4].

The correspondence between unrelated (on subsystem level) and related (on system level) sequence combinations and their effects on computational time is evaluated by modeling and scheduling the switched operation of a real-world factory automation system for non-productive phases. The outline of the paper is as follows. First, we give an insight in related work (Section 2) relevant for the domain of factory automation systems. After having introduced the model of structure and behavior (Section 3), the transformation into a constraint optimization problem is shown (Section 4). The algorithm for efficient identification of energy-optimal strategies is presented in section 5. The performance and the guarantee for optimality are illustrated in the context of a real-world factory automation system in the evaluation part (Section 6). Finally, a summary and outlook to future work is given (Section 7).

## 2   Related Work

On production planing and control (PPC) level, energetic aspects of manufacturing are considered in the scope of machine scheduling and capacity planing. In [5], [6] the production process is separated into segments with specific power input. With the help of a mathematical description of the power input of production modes, the energy input of a production process can be evaluated in early stages. [7] links material flow simulation with energetic considerations. Energy states are defined linking machine operation to input power. The authors state that for actually controlling machines, more detailed models are required. The abstraction level of PPC approaches do not describe the inner machine/component behavior in a granularity necessary for the problem of this contribution. Discrete event simulation which is often used for handling complexity requires profound expertise in the problem domain and does not guarantee to find the energy optimum.

On a low level (machine/component level), automata theory provides well founded mathematical and graphical support. Addressing minimal cost reachability, [8] and [9] discuss a priced timed automata (PTA) approach. Identifying energy-optimal sequences by techniques relevant for PTA has rather poor runtime performance for our problem. A composite model based on discrete event systems that takes process-related and functional dependencies into account is proposed in [10]. The objective of the approach is to give decision support how to operate machines by a state-based (discrete event) model. Moreover, the proposed model is intended to predict the energy input during production as a mathematical sequence of operating modes. However, the comparison of alternative sequences is not presented.

## 3  Structural and Behavioral Model

A model based on automata theory is proposed for describing structural and behavioral aspects of the automation system that serves as basis for the optimization model introduced in Section 4. In order to model the temporal switching and the energetic operating behavior [11], the factory automation system is formally represented as a network of priced (weighted) timed automata [8], [9]. The extension of automata theory by time and cost annotations joins a well founded mathematical model with capability for graphical representation.

Subsystem interdependencies are formally described as guards on edges of the automaton (logical constraints) referring to shared variables. Shared variables can be set (assigned) by one automaton and can be read by other automata. This detail enables a comprehensive formulation of hardware-specific and process-related dependencies relevant for switched operation in form of networks of automation subsystems.

Describing the behavioral aspects of an automation subsystem requires the introduction of time and energy information. Staying for a time period in a specific mode demands a certain amount of energy input expressed by input power (pc). The subsystems may switch between operating modes. In general, this procedure is time-dependent.

**Definition 1 (Networked Timed Automation Subsystem).** *The temporal transition and the energetic operating behavior of a single automation subsystem is represented by a networked priced timed automaton $subs_i = (M^i, \Sigma^i, m_0^i, c, Inv^i, E^i, \Omega^i, SV)$:*

- *$M^i$ is a finite set of modes consisting of a set of operating modes $OM^i$ and a set of transitional modes $TM^i$ with $M^i \subseteq OM^i \cup TM^i$ and $OM^i \cap TM^i = \emptyset$*
- *An alphabet $\Sigma^i$*
- *The initial operating mode $m_0^i \in OM^i$*
- *$sv^i \subset SV$ as a shared variable with $sv_{initial}^i = m_0^i$*
- *A set of assignments to the shared variable $A^i(sv^i) : sv^i \to m^i$*
- *A set of guards on shared variables $G^i(sv^k)$ with $G^i(sv^k) := g^i(sv^l) \wedge ... \wedge g^i(sv^n)$ in the form of $g^i := sv^l \sim m_j$ with $\sim \in \{==, \neq\}$ with $g_{neq}^i := sv^l \neq m_j$ and $g_{eq}^i := sv^l == m_j$*
- *A time variable $c \in \mathbb{R}_0^+$ called clock*
- *A set of invariants $Inv: M \to F(c)$ with $f := c \sim t$ and $\sim \in \{<, \leq\}$, and $t \in \mathbb{R}^+$*
- *A set of clock guards $G^i(c)$: $g^i := c \sim t$ and $\sim \in \{<, \leq, =, \geq, >\}$*
- *A set of clock resets $R^i(c)$: $r^i := c = 0$*
- *A finite set of edges $E^i \subseteq M^i \times \Sigma^i \times G^i(c) \times R^i(c) \times A^i(sv^i) \times G^i(sv^k) \times M^i$*
- *A cost function $\Omega^i: M^i, E^i \to \mathbb{N}_0^+$ mapping to each mode and transition an input power value $pc_i$*

Based on this formalism, feasible sequences satisfying a specification and the minimum-switch property can be determined by symbolic reachability analysis.

**Definition 2 (Switching Sequence).** *A switching sequence $k$ $seq_k^i$ of a subsystem $subs^i$ is the finite succession of symbolic states in the form of $(m_0, Z_0) \xrightarrow{\sigma_1, \tau_1} (m_1, Z_1) \xrightarrow{\sigma_2, \tau_2} ... \xrightarrow{\sigma_n, \tau_n} (m_{tar}, Z_{tar})$ with $m_{tar}$ as target mode, $m_0$ as initial mode, and $Z_{tar}$ as target zone, $m_i \in M$ and $Z_i \neq \emptyset$. The minimum-switch property requires the sequence $seq_k^i$ not to contain one mode more than twice.*

For instance, a feasible sequence 1 of subsystem 2 can have the following form [12]: $\text{seq}_1{}^2 : (m_1, t_1) \xrightarrow[d_1]{pc_1} (m_1, t_2) \to (m_2, t_1) \xrightarrow[d_2]{pc_2} \dots$ with discrete transitions as $(m, t) \to (m', t')$, with an edge $(m, g, r, m')$ so that the guard $g$ evaluates to true in $(m, t)$, $t'$ is $t$ where all clocks in the set $r$ are reset as well as delay transitions represented by $(m, t)$ $\xrightarrow[d]{pc} (m, t')$ where all clocks are incremented by $d := t' = t + d$ and $pc$ as cost rate. The symbolic reachability identifies the set of switching and operating sequences that are feasible in terms of time for each subsystem for a given specification. The comparison of different sequences in terms of energy input is part of the next section.

## 4   Calculation of Energy-Optimal Strategies

The objective of the approach is to find combinations of sequences within different subsystems that are optimal in terms of the given objective function (Subsection 4.3). In this paper, it is distinguished between unrelated $\text{seq}_k^i$ (Subsection 4.1) and related $\text{seq}*_k^i$ (Subsection 4.2) sequences. In an unrelated (related) sequence the switching between modes is independent (dependent) from respectively on other subsystem's switching behavior. Independence means that the decision variables of the subsystems do not occur in common constraints of the constraint optimization problem (COP). An energy-optimal specificity of a sequence $\text{seq}_k^i$ is denoted as $\text{seq}_{\text{opt}}^i$ and can be determined for unrelated and related sequences. On system level, an exponential amount of feasible sequences exists which is addressed by the following definition.

**Definition 3 (Alternative Strategies L).** *A strategy $l$ is a tuple of (subsystem) sequences in the form of $l = (\text{seq}*_a^i, \text{seq}*_b^j, \dots, \text{seq}*_c^n) \in L \mid a, b, c, i, j, n \in \mathbb{N}_0^+ \wedge i \neq j \neq n$, and $\text{seq}_a^i \in \text{Alt}^i$, $\text{seq}_b^j \in \text{Alt}^j$, and $\text{seq}_c^n \in \text{Alt}^n$. In general, $\text{seq}*_a^i$, $\text{seq}*_b^j$, and $\text{seq}*_c^n$ exhibit common constraints to be respected. If common constraints are omitted, a strategy $l_{p,unrel} = (\text{seq}_a^i, \text{seq}_b^j, \dots, \text{seq}_c^n) \in L_{unrelated}$ is called unrelated, otherwise the strategy $l_p \in L$ is called related. A finite set of strategies L exists with $|L| = \text{Alt}^i \times \dots \times \text{Alt}^n$.*

The maximum number of possible strategies $|L|$ is based on the Cartesian product of the domains of alternative switching sequences $\text{Alt}^i$ and increases exponentially with the number of subsystems $\text{subs}^i$.

$$|L| = \prod_{i=1}^{n} \text{Alt}^i \approx [\text{Seq}_{av}]^n \qquad (1)$$

with $i = 1, 2, \dots, n$ as the number of subsystems, $\text{Alt}^i$ as the quantity of alternative switching sequences in subsystem $\text{subs}^i$ and $\text{Seq}_{av}$ as the average number of switching sequences feasible within subsystem $\text{subs}^i$. Each strategy $l$ can be evaluated according to its energy demand value.

**Definition 4 (Minimum Energy Demand of Unrelated/Related Strategies).** *The minimum energy demand of an unrelated strategy is given by $e(l_{p,unrel}) = e(\text{seq}_k^i, \text{seq}_l^j) = e(\text{seq}_l^j) + e(\text{seq}_l^j)$. The minimum energy demand of a related strategy is greater-equal the linear superposition of the minimum energy demands of its switching sequences: $e(l) = e(\text{seq}*_k^i, \text{seq}*_l^j) \geq e(\text{seq}_l^j) + e(\text{seq}_l^j)$.*

### 4.1    Constraints of Unrelated Strategies

Defining the COP for unrelated sequences, the succession set of modes $M^{seq^i_k}$ of a sequence $seq^i_k$ is mapped to the set of interval variables $V^{seq^i_k}$ by function $I$.

$$I : M^{seq^i_k} \rightarrow V^{seq^i_k} \tag{2}$$

Clock guards $G^i(c)$ on an outgoing edge $e$ of the mode $m$ with $g^i := c \sim t$ with $\sim \in \{<, \leq, =, \geq, >\}$ and $t \in \mathbb{N}^+$ are transformed to temporal constraints.

$$g^i \rightarrow length(v(seq^i_k)) \sim t \tag{3}$$

The mode succession within a sequence is expressed by the set of precedence constraints $Pre$. The precedence of modes calculated during symbolic reachability is transformed into a precedence of interval variables. The formulation of $m_p(seq^i_k)$ precedes $m_{p+1}(seq^i_k)$ with $p \in \mathbb{N}^+$ is transformed into a precedence constraint of interval variables.

$$pre := end(v_p(seq^i_k)) \leq start(v_{p+1}(seq^i_k)) \tag{4}$$

### 4.2    Constraints of Related Strategies

In addition to the constraints for unrelated strategies, dependencies between subsystems cause to take the constraints between switching sequences of different subsystems into account. Interdependencies of subsystems affect the order of switching operations within a sequence. Therefore, guards $g^a_{eq}$, $g^a_{neq}$ on shared variables $sv^b$ defined on transitions of the switching sequence of subsystems $a$ with $a \neq b$ are transformed into precedence constraints.

For guards $g^a_{eq} := sv^b == m^b$ holds:

$$start(v(seq*^a_k)) \leq end(v(seq*^b_l)) \land start(v(seq*^a_k)) \geq start(v(seq*^b_l)) \tag{5}$$

For $g^a_{neq} := sv^b \neq m^b$ holds:

$$end(v(seq*^a_k)) < start(v(seq*^b_l)) \lor start(v(seq*^a_k)) > end(v(seq*^b_l)) \tag{6}$$

In this way, logical guards referring on shared variables expressing a precedence of operating modes within different subsystems are mapped into the COP. Including dependencies between subsystems causes an essential modification. Sequences may not be considered as independent as in the previous subsection.

### 4.3    Objective Function

The *energy-minimizing* objective $obj_{energy}$ accounts for reaching the given target mode $m_{tar}$ within a specified deadline $D$. Addressing the energy-optimal problem, the given target mode has to be reached within a specified time interval with deadline $D$. The objective function is stated as follows with $length(v_p(seq_k^i))$ denoting the interval length of an interval variable $v_p(seq_k^i)$. The input power of a mode is denoted by $pc_p(v_p(seq_k^i))$.

$$\text{Minimize obj}_{energy} = \sum_p pc_p(v_p(seq_k^i)) \cdot length(v_p(seq_k^i)) \qquad (7)$$

with $i = 1, ..., n$ and $pc$ as input power of interval variable $v$.

For objective $obj_{energy}$, the domain of interval variables needs to be constrained according to a predefined global time deadline $D$.

$$\sum_p length(v_p(seq_k^i)) \leq D \qquad (8)$$

with $i = 0, 1, 2, ..., n$ and $k \in \mathbb{N}_0^+$.

## 5   Identification of Energy-Optimal Related Strategies

Since the investigation of $L_{unrelated}$ is much less complex than the investigation of $L$, an approach for benefiting from the information contained in $L_{unrelated}$ is discussed. The proposed algorithm exploits energy demand of unrelated strategies to decide which related strategy should be examined next. The decision is guided by a branch-and-bound procedure which examines strategies according to the energy demand of unrelated strategies in an increasing order applying a *stop criterion* for terminating the investigation of $L$ as soon as the energy-optimal strategy was found.

> **Data:** $L_{unrelated}$ ordered with regard to $e(l_{p, unrel})$
> $\quad e_{low} = \infty$
> **Result:** Optimal, related strategy $l_{opt}$
> **foreach** $l_{p, unrel}$ in $L_{unrelated}$ **do**
> $\quad$ generate $l_p$ and $e(l_p)$
>
> $\quad$ **if** $e(l_p) \leq e_{low}$ **then** $e_{low} = e(l_p)$ ; // save lowest $e(l_p)$
> $\quad$ **if** $e_{low} \leq e(l_{p, unrel})$ **then** return $l_{opt} = l_p$ ; // stop criterion
> **end**
>
> $\quad$ **Algorithm 1.** Branch-and-bound algorithm for guided investigation of $L$

Applying the stop criterion of Algo. 1 identifies the energy-optimal strategy, if the ordered set of $L_{unrelated}$ is monotonically increasing and represents the lower bound of energy demand for each $l_p \in L$.

**Theorem 1.** *The energy demand $e(l_{p, unrel})$ of an unrelated strategy $l_{p, unrel} \in L_{unrelated}$ is the lower bound (Lower bound property) for the energy demand $e(l_k)$ of related strategies $l_k \in L$ with $k \geq p$, if $e(l_{p, unrel}) \leq e(l_{p+1, unrel}) \; \forall \; p, k \in \mathbb{N}_0^+$ (Monotonicity property).*

Theorem 1 requires Lemma 1 (monotonicity property) and Lemma 2 (lower bound property).

**Lemma 1.** MONOTONICITY: *Unrelated strategies* $l_{p,unrel} \in L_{unrelated}$ $\forall\, p = 1, 2, ...,$ *n are ordered with regard to their energy demands with* $e(l_{1,unrel}) \leq e(l_{2,unrel}) \leq ... \leq e(l_{n,unrel})$.

*Proof (Monotonicity).* Switching sequences $seq_k^i$ are evaluated by their optimal energy demand $e(seq_k^i)$ so that switching sequences $seq_1^i$, $seq_2^i$, ..., $seq_n^i$ can be ordered according to $e(seq_1^i) \leq e(seq_2^i)$ $leq$ ... $leq$ $e(seq_n^i)$. Since the energy demand of an unrelated strategy is calculated by $e(l_{p,unrel}) = \sum_{i=1}^{N} e(seq_k^i)$ with k = 1, 2, ... n, the set of unrelated strategies can be ordered in a monotonically increasing way in the form of $e(l_{1,unrel}) \leq e(l_{2,unrel}) \leq ... \leq e(l_{n,unrel})$.

**Lemma 2.** LOWER BOUND: *For related strategies holds that* $e(l_p) \geq e(l_{p,unrel})$ $\forall\, l_p$ $\in L$ *and* $l_{p,unrel} \in L_{unrelated}$.

*Proof (Lower bound).* The energy demand $e(seq_k^i)$ of each sequence $seq_k$ represents the lower bound in terms of optimal energy demand. An unrelated strategy $l_{p,unrel} \in L_{unrelated}$ preserves the optimality property because of $e(l_{p,unrel}) = e(seq_k^i) + ... + e(seq_k^N)$. In this way, an unrelated strategy $l_{p,unrel}$ provides the achievable lower bound of energy demand $e(l_{p,unrel})$ for a specific combination of switching sequences. This complies with Bellman's principle of optimality. Related strategies $l_p \in L$ show additional common constraints between subsystems. By means of additional constraints for related strategies, the optimal energy demand $e(l_p)$ of a related strategy $l_p$ can not be lower than its correspondent in form of the optimal energy demand $e(l_{p,unrel})$ of the unrelated strategy $l_{p,unrel}$.

Lemma 1 and Lemma 2 establish the basis for using Theorem 1 to implement Algorithm 1 which terminates if $e_{low} \leq e(l_{p,\,unrel})$ holds. This is exemplified in Figure 1. There, the investigation of L ($|L| = 728$) can be terminated after having investigated unrelated Strategy 16, since $e_{low} = e(l_1) \leq e(l_{16,\,unrel})$. That Algorithm 1 finds the optimal strategy is one aspect to be ensured. Investigating practical problems, the algorithm based on Theorem 1 needs to find energy-optimal strategies more efficiently than a complete enumeration of the state space.

## 6    Evaluation

The following test cases are based on a factory automation system with interdependent subsystems similar to [2]. The four instances (Subsection 6.1) correspond to the energy-minimal objective defined in Subsection 4.3 with different system scope. Subsystems are identical and show pairwise dependencies.

In all test cases, the proposed algorithm is checked if it identifies the energy-optimal strategy. Performance is evaluated by indicating the percentage of L that has to be investigated to find the energy-optimal strategy.

The evaluation of Algo. 1 is executed on a Microsoft Windows 7 operating system with 3,5 GHz CPU and 4 GB RAM. The setup is tested in a C# .NET implementation using IBM ILOG CP Optimizer (version 12.4) as commercial, state-of-the-art implementation for constraint optimization problems with discrete variables.

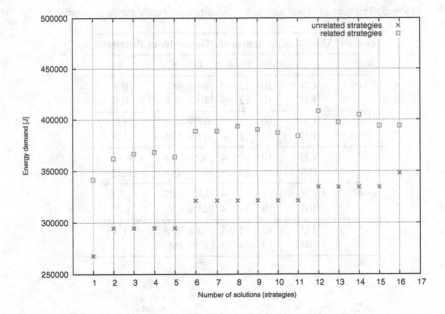

**Fig. 1.** Energy demand of related and unrelated strategies (scenario 4)

## 6.1 Test Scenarios

The parameter variations (scenarios) are shown in Table 1. For complete enumeration (*CE*) and the proposed branch-and-bound procedure (*BB*), the system scale (number of modes *Mod*, of transitions *Tra*, and of subsystems *Sub*) as well as the distribution of subsystem dependencies (*D* as dependency density, *C* as clustering coefficient, and *V* as average degree of a subsystem) are varied for evaluation of their impact on the number of strategies and the performance of applied algorithms.

## 6.2 Comparison

A crucial property of the proposed branch-and-bound procedure consists in the capability to investigate efficiently the search space *L* with a guarantee to find the energy-optimal strategy. The performance of the branch-and-bound algorithm compared to a complete enumeration is shown using the following characteristics: memory consumption, computational runtime and the percentage of investigated *L* in Figure 2.

The runtime for finding the energy-optimal strategy is significantly reduced by the proposed algorithm. In all scenarios, the branch-and-bound procedure finds the energy-minimal strategy and searches a maximum percentage of 47 percent of *L*. More complex subsystem behavior (Scenarios 3 to 5) results in an exhaustive search by CE. Even in a setting with complex subsystem interaction (Scenario 5), the BB procedure reduces computational time by more than 70 percent compared to a CE search.

**Table 1.** Test scenarios and parameter variations for performance comparison

| Scenario | Algorithm | Mod | Tra | Sub | D | C | V |
|----------|-----------|-----|-----|-----|-----|-----|------|
| | | System scale | | | Dependency distribution | | |
| 1 | CE | 10 | 12 | 5 | 40% | 0% | 0,8 |
| | BB | 10 | 12 | 5 | 40% | 0% | 0,8 |
| 2 | CE | 10 | 12 | 4 | 67% | 75% | 1,0 |
| | BB | 10 | 12 | 4 | 67% | 75% | 1,0 |
| 3 | CE | 19 | 24 | 4 | 50% | 0% | 0,75 |
| | BB | 19 | 24 | 4 | 50% | 0% | 0,75 |
| 4 | CE | 19 | 24 | 4 | 83% | 83% | 1,25 |
| | BB | 19 | 24 | 4 | 83% | 83% | 1,25 |
| 5 | CE | 19 | 24 | 5 | 70% | 76% | 1,4 |
| | BB | 19 | 24 | 5 | 70% | 76% | 1,4 |

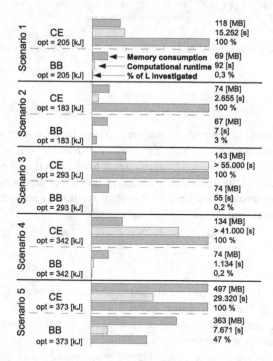

**Fig. 2.** Memory consumption, computational runtime for CE and BB investigation of $L$, Scenarios 1 to 5

# 7   Conclusion and Future Work

A branch-and-bound algorithm for efficient investigation of the set of strategies was introduced in order to benefit from energy savings potentials during non-productivity. In the presented scenarios with varied system scale and structure, the proposed structure-exploiting procedure is capable to identify efficiently the energy-minimal strategies for automation systems that is beyond previous work. Besides the presented problem, the simulation-based feasibility analysis of strategies is a major concern. For this reason, future work will focus on the verification of strategy feasibility.

# References

1. Niemann, K.H.: Energiemanagement in Automatisierungssystemen. In: VDI-Berichte 2171, VDI Wissensforum GmbH, pp. 99–102 (2012)
2. Mechs, S., Lamparter, S., Müller, J.P.: On Evaluation of Alternative Switching Strategies for Energy-Efficient Operation of Modular Factory Automation Systems. In: Proc. of the 17th IEEE Conference on Emerging Technologies and Factory Automation (ETFA 2012). IEEE Press (2012)
3. Land, A.H., Doig, A.G.: An automatic method of solving discrete programming problems. Econometrica 28(3), 497–520 (1960)
4. Lawler, E.L., Wood, D.E.: Branch-and-bound methods: A survey. Operations Research 14, 699–719 (1966)
5. Chiotellis, S., Weinert, N., Seliger, G.: Simulation-based, energy-aware production planning. In: CIRP International Conference on Manufacturing Systems, pp. 964–971. Neuer Wissenschaftlicher Verlag, Vienna (2010)
6. Weinert, N.: Vorgehensweise für Planung und Betrieb energieeffizienter Produktionssysteme. Dissertation, TU Berlin (2010)
7. Kulus, D., Wolff, D., Ungerland, S., Dreher, S.: Energieverbrauchssimulation als Werkzeug der Digitalen Fabrik. zwf-online, inpro-Innovationsakademie 106, 585–589 (2011)
8. Alur, R., La Torre, S., Pappas, G.J.: Optimal paths in weighted timed automata. In: Di Benedetto, M.D., Sangiovanni-Vincentelli, A.L. (eds.) HSCC 2001. LNCS, vol. 2034, pp. 49–62. Springer, Heidelberg (2001)
9. Behrmann, G., Fehnker, A., Hune, T., Larsen, K., Pettersson, P., Romijn, J., Vaandrager, F.: Minimum-cost Reachability for Priced Timed Automata. In: Di Benedetto, M.D., Sangiovanni-Vincentelli, A.L. (eds.) HSCC 2001. LNCS, vol. 2034, pp. 147–161. Springer, Heidelberg (2001)
10. Dietmair, A., Verl, A.: A generic energy consumption model for decision making and energy efficiency optimisation in manufacturing. International Journal of Sustainable Engineering 2(2), 123–133 (2009)
11. Mechs, S., Müller, J.P., Lamparter, S., Peschke, J.: Networked priced timed automata for energy-efficient factory automation. In: Proc. of the 2012 American Control Conference, IEEE Transactions on Automatic Control, American Automatic Control Council, pp. 5310–5317 (2012)
12. Behrmann, G., Larsen, K.G., Rasmussen, J.I.: Priced timed automata: Algorithms and applications. In: de Boer, F.S., Bonsangue, M.M., Graf, S., de Roever, W.-P. (eds.) FMCO 2004. LNCS, vol. 3657, pp. 162–182. Springer, Heidelberg (2005)

# The Two-Dimensional Vector Packing Problem with Courier Cost Structure

Qian Hu*, Andrew Lim, and Wenbin Zhu

Department of Management Sciences, City University of Hong Kong,
Tat Chee Ave, Kowloon Tong, Hong Kong
{huqian,limandrew}@cityu.edu.hk, i@zhuwb.com

**Abstract.** The two-dimensional vector packing problem with courier cost structure is a practical problem faced by many manufacturers that ship products using courier service. The manufacturer must ship a number of items using standard-sized cartons, where the cost of a carton quoted by the courier is determined by a piecewise linear function of its weight. The cost function is not necessarily convex or concave. The objective is to pack all items into cartons such that the total delivery cost is minimized while observing both the weight limit and volume capacity constraints. In this study, we investigate solution methods to this problem.

**Keywords:** Heuristic, Application, Bin packing, Two-dimensional vector packing, Courier cost structure.

## 1 Introduction

We consider a new extension to the bin packing problem, the two dimensional vector packing problem with piecewise linear cost function (2DVPP-PLC). In the 2DVPP-PLC, a set of items are to be packed into identical bins so that the total cost of utilized bins is minimized. Each item has two attributes: the weight and the volume. A set of items can be packed into a bin if their total weight and volume do not exceed the capacities of the bin. The cost of a bin is a piecewise linear function of the total weight of packed items.

This study is motivated by the real operations in a manufacturer of children's apparel with several production bases and hundreds of retail stores located across the globe. The manufacturer distributes its products from its production bases to retail stores through an express courier company such as Federal Express or DHL under a long-term contract. Articles of children's apparel (such as shorts, jackets and rompers) of the same style and size are bundled to form items; for example, one item may be a bundle of two dozen rompers. These items are subsequently packed into cartons that are picked up by the courier company for distribution. In each delivery, cartons are charged according to the cost structure negotiated by the manufacturer and the courier company.

---

* Corresponding author.

M. Ali et al. (Eds.): IEA/AIE 2013, LNAI 7906, pp. 212–221, 2013.

The courier company groups delivery destinations into zones, and provides a pricing scheme for each. For a given production base and the destination zone, the delivery cost for a carton is a function of the total weight of the items packed into that carton. We obtained from the manufacturer the cost structure offered by the courier company to deliver one carton from a certain production base to each of the seven zones, as shown in Figure 1(a), where the $x$-axis represents the total weight of the items in pounds (lb) in a carton and the $y$-axis corresponds to the delivery cost of that carton in US dollars.

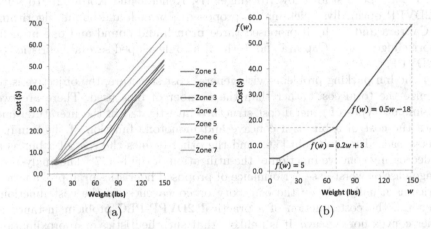

(a)                                    (b)

**Fig. 1.** (a) Pricing schemes for 7 zones. (b) Piecewise linear function for Zone 3.

In the past, the manufacturer packed items in cartons as fully as possible. The managers did not realize they could save money through packing in a wiser way. Actually, if the delivery destination is in Zone 3 (cost structure approximated in Figure 1(b)), a carton packed to 70 lb will achieve the minimum unit cost. Since the cost structure varies, pack to the fullest might not remain an optimal strategy! We have conferred with the courier company on the rationale behind this peculiar pricing scheme, which was devised based on practical considerations. Delivering each carton incurs a fixed transaction cost, which is US\$5 for a carton with weight no more than 10 lb. When the carton weight exceeds 10 lb, the delivery cost increases with weight at a rate of 20 cents per lb. However, this rate increases to about 50 cents per lb when the carton weight is greater than 70 lb. One of the reasons is that light cartons can be handed manually by single worker with ease in the logistics, while overweight cartons require special handling equipment or more than one worker which increases the operation costs considerably.

Moreover, such logistic activity is common to many other manufacturers. As the shipping cost is directly determined by a packing plan, an efficient and effective solution method to the 2DVPP-PLC would be of great practical value. Courier service as whole is a multi-billion industry. Even small percentage saving translates into huge dollar figure. Despite its economic value to practitioners, there is no prior literature dedicated to this problem to the best of our knowledge.

The 2DVPP-PLC generalizes the 2DVPP. While the 2DVPP aims to minimize the number of bins used, the 2DVPP-PLC extends and targets to minimize the total cost which depends on the given piecewise linear cost function. In the 2DVPP-PLC, minimizing cost does not imply that the number of bins used is minimized. In fact, splitting an overly filled bin into two bins may reduce the total cost. In existing literature, the 2DVPP has been solved mainly using two classes of algorithms: exact algorithms and heuristics. Approximation algorithms to the 2DVPP have been studied by Woeginger [1], Kellerer and Kotov [2]. To solve the 2DVPP optimally, Spieksma [3] proposed a branch-and-bound algorithm, and Caprara and Toth [4] proposed three branch-and-bound and one branch-and-price algorithms. Caprara and Toth [4] also developed several heuristics to the 2DVPP.

For the bin packing problems with general cost structure, the objective is to minimize the total cost rather than the number of bins used. There are two predominant types of general cost structure investigated in the literature: one defines the cost of a bin as a concave and monotone function of the number of items packed into the bin [5,6], and the other defines the cost of a bin as a non-decreasing concave function of the utilization of the bin [7]. The analysis of average-case or worst-case performance of proposed heuristics or approximation algorithms usually rely on the convexity or concaveness of the cost function. In general, the cost function of a practical 2DVPP-PLC problem instance is neither convex nor concave. It is unlikely that such heuristics or approximation algorithms can be directly extended to the 2DVPP-PLC. In our study, we derive its formulation and propose an iterated local search heuristic.

## 2   Problem Definition and an IP Formulation

The 2DVPP-PLC is defined as follows. There are $n$ types of items denoted by the set $I = \{1, 2, \ldots, n\}$, and there are $d_i$ items of type $i \in I$ with volume $v_i$ and weight $w_i$. We are given an unlimited number of identical bins, each with a volume capacity $V$ and a weight limit $W$. All problem input values $w_i$, $v_i$, $W$ and $V$ are assumed to be integers. The cost of each bin is a piecewise linear function $f(x)$ of the total weight $x$ of the items packed. It is not restricted to be concave or convex. The objective is to identify a feasible packing plan that places all items into bins such that the total cost of all used bins is minimized. Without loss of generality, we assume every item is small enough to be packed into a bin, so that a feasible solution always exists.

Let $B$ be the set of bins available. Suppose the cost function has $q$ pieces; the $k$-th piece is given by $f(x) = s_k x + b_k$ for $x \in [e_{k-1}, e_k]$, where $e_{k-1} <= e_k$ for all $k \in K = \{1, 2, \ldots, q\}$. Let $y_j^k$ be a binary variable that equals 1 if bin $j$ is used and the $k$-th piece of the cost function is activated, and 0 otherwise. Let $x_{i,j}^k$ be an integer variable that represents the number of type $i$ items packed into bin $j$ subject to the $k$-th piece of the cost function. We formulate the 2DVPP-PLC as follows.

**(2DVPP-PLC)**:

$$\text{Minimize} \quad \sum_{j \in B} \sum_{k \in K} \left( s_k \sum_{i \in I} w_i x_{i,j}^k + b_k y_j^k \right) \tag{1}$$

$$\text{Subject to} \quad \sum_{j \in B} \sum_{k \in K} x_{i,j}^k \geq d_i, \qquad\qquad \forall\, i \in I \tag{2}$$

$$e_{k-1} y_j^k \leq \sum_{i \in I} w_i x_{i,j}^k \leq e_k y_j^k, \qquad \forall\, j \in B, k \in K \tag{3}$$

$$\sum_{k \in K} y_j^k \leq 1, \qquad\qquad\qquad \forall\, j \in B \tag{4}$$

$$\sum_{i \in I} \sum_{k \in K} v_i x_{i,j}^k \leq V, \qquad\qquad \forall\, j \in B \tag{5}$$

$$x_{i,j}^k \geq 0 \text{ and integer}, \qquad \forall\, i \in I, j \in B, k \in K$$

$$y_j^k \in \{0,1\}, \qquad\qquad\qquad \forall\, j \in B, k \in K$$

For any bin $j$, at most one piece of the cost function is activated to evaluate its cost (Constraints 4). A bin is used and the $k$-th piece of the cost function is activated if and only if the total weight of the items packed lies in the interval $(e_{k-1}, e_k]$ (Constraints 3). In a feasible solution, all items must be packed into some bin (Constraints 2), and the capacities of each bin cannot be exceeded (Constraints 5 and Constraints 3 considering $e_k y_j^k \leq W$).

Standard solver such as ILOG CPLEX can be employed to solve the model. But, such approach is viable only for small instances according to our experiments (Section 4). It is therefore necessary to investigate efficient heuristic approaches for large instances.

## 3 An Iterative Local Search Algorithm

We design an iterative local search (ILS) algorithm to solve the 2DVPP-PLC. Our ILS algorithm is to explore the permutations of input items, where each permutation can be decoded into a feasible solution. Further, we introduce a short-path decoder and three neighborhood operators.

Our shortest-path decoder (SP) is based on the idea of [8] for the variable bin-size bin packing problem. Given a sequence of items, we try to partition the sequence into non-overlapping segments such that each segment is a set of items in a bin. The optimal partition of the sequence corresponds to a shortest path in a graph associated with the input sequence.

Given an input sequence $\pi$, we first construct a weighted acyclic graph $G = (V, E)$. The vertex set $V = \{1, \ldots, n, n+1\}$, where vertex $i$ represents the $i$-th item in the sequence and $n+1$ is a dummy vertex. The directed edge set $E$ is constructed as follows: (1) add an edge $(i, i+1)$ for $i = 1, \ldots, n$, the cost of which is $f(w_{\pi(i)})$; (2) add an edge $(i, j)$ for $1 \leq i < j \leq n+1$ if $\sum_{k=i}^{j-1} v_{\pi(k)} \leq V$

and $\sum_{k=i}^{j-1} w_{\pi(k)} \leq W$, the cost of which is $f\left(\sum_{k=i}^{j-1} w_{\pi(k)}\right)$ as calculated by the given piecewise linear cost function. We call $G$ the *cost graph* of $\pi$.

It is apparent that any path in $G$ from vertex 1 to vertex $n + 1$ corresponds to a feasible 2DVPP-PLC solution, where each edge $(i, j)$ in the path indicates that all items $\pi(i), \pi(i + 1), \ldots, \pi(j - 1)$ are in the same bin, and the cost of the solution is the total cost of all edges in the path. Furthermore, since every possible partition of the input sequence is captured by a path in graph $G$, the best feasible partition of the sequence can be found by solving a shortest path problem from vertex 1 to $n + 1$ on $G$ in $O(|E|)$ time.

Based on the shortest-path decoder, we develop three dedicated neighborhood operators: *bin-shuffle*, *bin-shake* and *shuffle-shake*. Given a sequence (permutation) of items, all three operators will transform it into a new sequence and the solution derived from the new sequence using the shortest-path decoder will never be worse than that from the original sequence.

The first operator *bin-shuffle* changes the relative position of some bins while preserving the relative position of the items inside each bin. Let $SP(\pi) = (s_1, s_2, \ldots, s_k)$ be the solution produced by the SP decoder on sequence $\pi$, where each element $s_i$ is a segment of $\pi$ corresponding to all items in a single bin. We can produce a solution $S'$ from $SP(\pi)$ by randomly permuting the segments of $SP(\pi)$, while the order of the items in each segment is unchanged. Clearly the cost of $S'$ is the same as $SP(\pi)$. We can concatenate all the segments in $S'$ to obtain a new sequence $\pi'$. We then apply shortest-path decoder on sequence $\pi$ to obtain a new solution $SP(\pi')$. Since $S'$ corresponds to a path in the cost graph, whereas the solution $SP(\pi')$ corresponds to a shortest path, the cost of $SP(\pi')$ will never be higher than $SP(\pi)$.

The second operator *bin-shake* preserves relative position of all the bins but changes the relative position of some items inside each bin. Let $SP(\pi) = (s_1, s_2, \ldots, s_k)$ be the solution produced by the SP decoder on sequence $\pi$. For each segment $s_i$, we randomly shuffle the elements in it to obtain a new segment $s_i'$. We then concatenate $s_1', s_2', \ldots, s_k'$ to obtain a new sequence $\pi'$. Let $SP(\pi')$ be the solution found by shortest-path decoder on sequence $\pi'$, following the similar argument as bin-shuffle, the cost of $SP(\pi')$ will never be higher than $SP(\pi)$.

The third operator *shuffle-shake* combines the previous two operators. It changes the relative position of some bins and the relative position of some items inside each bin. That is, given a sequence $\pi$ and its solution $SP(\pi)$, the new sequence $\pi'$ is generated by randomly permuting the segments in $\pi$, and the items in each segment are randomly permuted. Once again, the solution produced by the SP decoder for this new sequence will be no worse than the original.

We use an example to illustrate the bin-shuffle and the bin-shake operators. The small instance provides bins with $W = V = 150$ and 10 items with weights (30, 30, 30, 30, 30, 30, 20, 20, 20, 20) and volumes (10, 20, 30, 40, 50, 60, 20, 30, 40, 50). The cost function is given by Figure 1(b). Figures 2(a) and 2(b) depict the process of the bin-shuffle and bin-shake operators, respectively; the bins in the solutions for each sequence are demarcated by dotted lines.

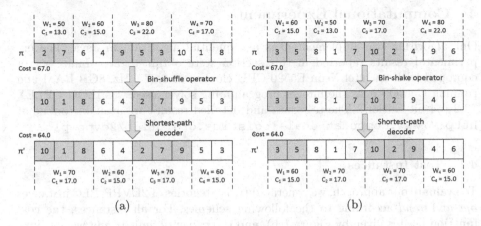

**Fig. 2.** (a) Bin-shuffle operator. (b) Bin-shake operator.

Our iterative local search algorithm iteratively employs the SP decoder and neighborhood operators to explore multiple packing sequences (Algorithm 1):

---

**Algorithm 1.** Iterative Local Search (ILS) for the 2DVPP-PLC

---

**Require:** maximum number of iterations $N_i$;

1: Generate a set of initial sequences $\Pi$;
2: $i \leftarrow 1$;
3: **while** $i \leq N_i$ **do**
4:    **for all** $\pi \in \Pi$ **do**
5:       Apply the *NeighborhoodSearch* on $\pi$ to produce a new sequence $\pi'$;
6:       **if** $(SP(\pi')$ cost less than $SP(\pi)$ **then**
7:          $\Pi \leftarrow \Pi \cup \{\pi'\} \setminus \{\pi\}$;
8:       **end if**
9:    **end for**
10:   $i \leftarrow i + 1$;
11: **end while**
12: **return** the best solution in $\Pi$;

---

The ILS begins by randomly generating a set of initial sequences. For each initial sequence, we evaluate it by applying the shortest-path decoder.

Next, the ILS starts the iterative exploration in the solution space. For each sequence $\pi \in \Pi$, we employ the *NeighborhoodSearch* procedure on $\pi$ to produce a new sequence $\pi'$. The *NeighborhoodSearch* randomly selects a neighborhood operator from the bin-shuffle, bin-shake and shuffle-shake. The chosen operator produces a neighbor of $\pi$, which is denoted as $\pi'$. $\pi'$ is then decoded using the SP decoder. $\pi$ will be replaced by $\pi'$ in $\Pi$ if the corresponding solution of $S_B$ has a lower cost. This process is repeated $N_i$ times and finally the best solution in $\Pi$ is returned.

# 4    Computational Experiments

Our algorithm was implemented as a sequential algorithm in Java. The experimental results reported in this section were obtained on a Linux server equipped with an Intel Xeon E5520 CPU clocked at 2.27 GHz, 8GB RAM and running CentOS 5.4. The integer programming solver used was ILOG CPLEX 12.2 (64-bit edition). The test data and detailed results are available online at http://www.computational-logistics.org/orlib/2dvpp-plc.

## 4.1    Test Instances

To evaluate our approach, we generated two categories of 2DVPP-PLC instances *opt* and *rand* according to the following schemes. For all instances, the cost function used is given by Figure 1(b), and the capacity limit $V$ and weight limit $W$ of each bin are both set to 150.

The first category *opt* contains instances with known perfect optimal solutions. Given the cost function $f(x)$ described in Figure 1(b), we can compute the average cost per unit weight $g(x) = f(x)/x$. The minimum of $g(x)$ is achieved when the bin weight is exactly 70 lb. Hence, if there exists a solution where the total weight of the items inside each bin is exactly 70 lb, then this packing plan must be optimal. Based on this observation, we generated the instances in this category as follows.

We take the number of bins $|B|$ from $\{25, 50, 100, 200\}$, resulting in four sets of instances. For each bin, we generate some items with total weight exactly equal to 70 and total volume less than 150 using the following process. The number of items to be constructed, denoted by *num*, is selected from the discrete uniform distribution $U_d[2, 6]$. Next, we invoke the procedure *Cut(num, 70)* (Algorithm 2) to generate the weight of each item; the *Cut(a, b)* procedure produces an array with $a$ integer elements whose sum is $b$. To generate the volume of each item, we first select a value *total* from the discrete uniform distribution $U_d[70, 150]$ and then invoke the *Cut(num, total)* procedure. If two randomly generated item types share the same weight and volume, we combine them into a single item type. For each instance set, we randomly generated ten instances, for a total of 40 instances. All of these instances have optimal solutions where the total weight of the items in each bin is 70. Each set is labeled "opt$x$", where $x$ is the number of bins in the optimal solution.

The second category *rand* consists of four classes of instances where the total number of items $n \in \{128, 256, 512, 1024\}$, respectively. For each class, there are three sets of instances, for a total of 12 sets. In the tree sets, there is an equal number of items of each type; the number of type $i$ items is $d_i \in \{1, 8, 32\}$, respectively. For all instances in this category, the weight and volume of each item are both selected from the discrete uniform distribution $U_d[1, 70]$. These instance sets are labeled "rand$n$-$s$", where $n$ is the number of items and $s$ is the set index. In this category, there are 120 instances in total.

---

**Algorithm 2.** *Cut(num, total)*

1: Create an array *array* with *num* elements;
2: For each element in *array*, set $array[i]$ = integer from discrete uniform distribution $U_d[1, total]$;
3: $a\_total = \sum_{j=1}^{num} array[j]$;
4: For each element in *array*, set $array[i] = \lceil (array[i] \times total)/a\_total \rceil$
5: Randomly select $(\sum_{i=1}^{num} array[i] - total)$ elements from *array*, and subtract 1 from each of the selected elements;
6: **return** *array*

---

## 4.2 Results

We proceed to present the overall results of our two different approaches: CPLEX and ILS. In our first approach, the model is solved by commercial ILOG CPLEX solver with default settings. Each instance is solved once with the time limit set to 3600 seconds. In our second approach, each instance is solved once using ILS with maximum iterations $N_i = 5000$ and random seed 3.

Table 1 reports the results for "opt" instances. The column *#inst* gives the number of instances in a test set. The columns under the heading *CPLEX* and *ILS* report the performance of CPLEX and ILS respectively. The column *Avg. Gap (%)* reports the percentage gap between the best solution found and the known optimal solution averaged over the solved instances in the test set. The column *Avg. Time (s)* reports the total CPU time in seconds averaged over the solved instances. For CPLEX, the number of instances successfully solved is reported in the column *#Feasible* and the number of instances with optimal solution found and proven is given in the column *#Optimal*. For ILS, all instances are successfully solved.

**Table 1.** Comparison on "opt" instances

| Set | #inst | CPLEX | | | | ILS | |
|---|---|---|---|---|---|---|---|
| | | #Feasible | #Optimal | Avg. Gap (%) | Avg. Time (s) | Avg. Gap (%) | Avg. Time (s) |
| opt25 | 10 | 10 | 10 | 0.00 | 524.62 | 0.56 | 1.42 |
| opt50 | 10 | 10 | 7 | 0.01 | 2535.18 | 0.37 | 4.56 |
| opt100 | 10 | 10 | 0 | 2.92 | 3600.44 | 0.35 | 16.62 |
| opt200 | 10 | 6 | 0 | 11.90 | 3600.83 | 0.40 | 80.65 |
| Grand Total | 40 | 36 | 17 | N/A | N/A | 0.42 | 25.81 |

We can see that for the small instances CPLEX clearly outperform ILS. However, the performance of CPLEX deteriorates dramatically for larger instances. In contrast, the performance of ILS increases as the size of instances increases – the average gap to optimal dropped from 0.56% for opt25 instances to 0.40% for opt200 instances. Hence, ILS is a better choice than CPLEX for solving large instances from practice.

We then investigate the convergence of ILS on "OPT" instances by controlling the number of maximum iterations $N_i$. Table 2 summarizes the average performance of ILS with $N_i$ set to 500, 1000, 2000 and 5000, respectively. Compare the average percentage gap between "$N_i = 500$" and "$N_i = 5000$", we can see that ILS already achieves good solutions within the first 500 iterations. Further, increasing the number of iterations only marginally improves the solution quality.

**Table 2.** Performance of ILS with different $N_i$ on "opt" instances

| Set | Avg. Gap (%) | | | | Avg. Time (s) | | | |
|---|---|---|---|---|---|---|---|---|
| | $N_i = 500$ | $N_i = 1000$ | $N_i = 2000$ | $N_i = 5000$ | $N_i = 500$ | $N_i = 1000$ | $N_i = 2000$ | $N_i = 5000$ |
| opt25 | 0.64 | 0.64 | 0.64 | 0.56 | 0.14 | 0.29 | 0.57 | 1.42 |
| opt50 | 0.46 | 0.46 | 0.39 | 0.37 | 0.46 | 0.91 | 1.82 | 4.56 |
| opt100 | 0.48 | 0.44 | 0.43 | 0.35 | 1.68 | 3.33 | 6.66 | 16.62 |
| opt200 | 0.47 | 0.45 | 0.42 | 0.40 | 8.30 | 16.44 | 32.58 | 80.65 |
| Grand Total | 0.51 | 0.50 | 0.47 | 0.42 | 2.65 | 5.24 | 10.41 | 25.81 |

We, therefore, set $N_i = 500$ and report the results of ILS on "rand" instances in Table 3. The columns $m$ give the number of test instances in an instance set. The columns *Total Cost* report the cost of the best solution summed over all instances in an instance set. The columns *Avg. Time (s)* report the total CPU time in seconds averaged over all instances in an instance set.

**Table 3.** Performance of ILS with $N_i = 500$ on "rand" instances

| Set | $m$ | Total Cost | Avg. Time (s) | Set | $m$ | Total Cost | Avg. Time (s) |
|---|---|---|---|---|---|---|---|
| inst128-1 | 10 | 11085.2 | 0.17 | inst512-1 | 10 | 43673.1 | 2.33 |
| inst128-2 | 10 | 10624.3 | 0.18 | inst512-2 | 10 | 44612.7 | 2.32 |
| inst128-3 | 10 | 11712 | 0.17 | inst512-3 | 10 | 44027.4 | 2.34 |
| inst256-1 | 10 | 21944.5 | 0.62 | inst1024-1 | 10 | 87389.6 | 12.92 |
| inst256-2 | 10 | 22059.3 | 0.61 | inst1024-2 | 10 | 87974.6 | 12.47 |
| inst256-3 | 10 | 22481.7 | 0.64 | inst1024-3 | 10 | 87654.4 | 12.92 |

From Table 2 and 3, we can conclude that our ILS ($N_i = 500$) takes only a few seconds to produce a high quality solution. As a result, our ILS algorithm can be used as a subroutine in a decision support system which needs to evaluate multiple scenarios quickly.

## 5    Conclusions and Future Work

In this paper, we introduced a new and practical two-dimensional vector packing problem with a piecewise linear cost function (2DVPP-PLC) based on the price structure from a practical courier company. It models the problem of shipping a number of items in standard-sized cartons where the cost of a carton is a piecewise linear function of the weight of the items in the carton. We presented an integer programming formulation for this problem. Solving the model directly using commercial ILOG CPLEX is viable only for small instances. We adapted a shortest-path decoder to convert a sequence of items into a solution, and then we built an iterative local search (ILS) algorithm upon the decoder to improve the solutions. Computational results suggest ILS can produce high quality solutions efficiently for instances of practical size. Our results serve as benchmark for future use.

In addition, the ILS approach has a potential to solve a set of bin packing problems with various cost structures. In future work, we will investigate its performance on some other problems. Also, the shortest-path decoder and the operators can be integrated into sequence-based metaheuristics, such as genetic algorithm. Another potential avenue of research is to design exact algorithms for the 2DVPP-PLC using more advanced techniques.

# References

1. Woeginger, G.J.: There is no asymptotic PTAS for two-dimensional vector packing. Information Processing Letters 64(6), 293–297 (1997)
2. Kellerer, H., Kotov, V.: An approximation algorithm with absolute worst-case performance ratio 2 for two-dimensional vector packing. Operations Research Letters 31(1), 35–41 (2003)
3. Spieksma, F.C.R.: A branch-and-bound algorithm for the two-dimensional vector packing problem. Computers & Operations Research 21(1), 19–25 (1994)
4. Caprara, A., Toth, P.: Lower bounds and algorithms for the 2-dimensional vector packing problem. Discrete Applied Mathematics 111(3), 231–262 (2001)
5. Anily, S., Bramel, J., Simchi-Levi, D.: Worst-case analysis of heuristics for the bin packing problem with general cost structures. Operations Research 42(2), 287–298 (1994)
6. Epstein, L., Levin, A.: Bin packing with general cost structures. Mathematical Programming 132(1-2), 355–391 (2012)
7. Leung, J.Y.T., Li, C.L.: An asymptotic approximation scheme for the concave cost bin packing problem. European Journal of Operational Research 191(2), 582–586 (2008)
8. Haouari, M., Serairi, M.: Heuristics for the variable sized bin-packing problem. Computers & Operations Research 36(10), 2877–2884 (2009)

# Penguins Search Optimization Algorithm (PeSOA)

Youcef Gheraibia[1] and Abdelouahab Moussaoui[2]

[1] Computing Department, Mohamed Cherif Messadia Univercity,
SOUK AHRAS 41000, Algeria
`youcef.gheraibia@gmail.com`
[2] Computing Department, Ferhat Abbas Univercity, SETIF 19000, Algeria

**Abstract.** In this paper we propose a new meta-heuristic algorithm called penguins Search Optimization Algorithm (PeSOA), based on collaborative hunting strategy of penguins. In recent years, various effective methods, inspired by nature and based on cooperative strategies, have been proposed to solve NP-hard problems in which, no solutions in polynomial time could be found. The global optimization process starts with individual search process of each penguin, who must communicate to his group its position and the number of fish found. This collaboration aims to synchronize dives in order to achieve a global solution (place with high amounts of food). The global solution is chosen by election of the best group of penguins who ate the maximum of fish. After describing the behavior of penguins, we present the formulation of the algorithm before presenting the various tests with popular benchmarks. Comparative studies with other meta-heuristics have proved that PeSOA performs better as far as new optimization strategy of collaborative and progressive research of the space solutions.

**Keywords:** Meta-heuristic, Optimization, Penguins, Bio-inspiration, NP-hard.

## 1 Introduction

The development of new artificial systems based on meta-heuristics, inspired by natural phenomena, has proved its effectiveness in solving NP-hard problems. These new strategies find their applications in many fields such as aeronautics, computer science, artificial intelligence, medical imagering, biology, etc. These techniques are as varied as the sources of inspiration from which they emerge (physical, biological, ecological behavior, etc..).

The first work in the field of optimization began around the year 1952 with the use of stochastic methods [16]. Rechenberg in 1965 designed the first algorithm using evolution strategies in the optimization [14]. The year 1970 was marked by the revolution in optimization algorithms inspired by natural and biological phenomena with cellular automata [6]. From that time many optimization algorithms have been proposed, John Holland [10] with genetic algorithms until this

M. Ali et al. (Eds.): IEA/AIE 2013, LNAI 7906, pp. 222–231, 2013.

recent years with the wide use of inspiration by various domains in the optimization algorithms [24,12,22], such as Bacterial Foraging Optimization Algorithm [13], Bat algorithm [25], Firefly algorithm [22], and cuckoo search (CS) [23].

Meta-heuristics can be classified in different ways [12,27,1,2]. Most of methods use the concept of population, in which a set of solutions are computed in parallel at each iteration, such as genetic algorithms, PSO (Particle Swarm Optimization Algorithm) [3] and the ACO (Ant Colony Optimization) [5]. Other meta-heuristics use the history of their search in order to guide the optimization in subsequent iterations by adding a learning phase of intermediate results that will lead to an optimal solution. However some algorithms such as guided local search, modify the representation of the problem. This allows us to classify the meta-heuristics into two categories: the first class gathers methods with dynamic objective function and the second one groups methods with static objective function.

In this paper we propose a new meta-heuristic, called PeSOA (Penguins Search Optimization Algorithm), based on hunting behavior of penguins. The hunting strategy of penguins is more than fascinating since they can collaborate their efforts and synchronize their dives to optimize the global energy in the process of collective hunting and nutrition. First we formulate the algorithm by the idealization of collaborative hunting behavior of penguins, then we present the results obtained by comparison of the PeSOA with other methods used in the optimization as PSO and GA.

## 2  Hunting Strategy of Penguins

The optimality theory of foraging behavior was modeled in the works of [15,16]. These two studies hypothesized that dietary behavior may be explained by economic reasoning: when the gain of energy is greater than the expenditure required to obtain this gain, so it comes to a profitable food search activity. Penguins, as biological beings, use this assumption to extract information about the time and cost of food searches and energy content of prey, on one hand, and the choice to hunt or not in the selected area, depending on its high resource and the distance between feeding areas, on the other.

The behavior of air-breathing aquatic predators including penguins was noticed by [11]. The surface is a place for penguins as they are forced to return after each foraging trip. A trip implies immersion in apnea. The duration of a trip is limited by the oxygen reserves of penguins, and the speed at which they use it, that is to say their metabolism [7,26]. The works in the field of animal behavioral ecology of penguins has given us clear and motivating ideas for the development of a new optimization method based on the behavior of penguins.

Penguins are sea birds, unable to fly because of their adaptation to aquatic life [7,28,21,9]. The wings are ideal for swimming and can be considered as fins: penguins fly through water and can dive more than 520m to search for food. Although this is more efficient and less tiring to swim underwater than at the surface, they must regularly return to the surface to breath. They are able to keep breathing while swimming rapidly (7 to 10 km / h) [8]. During the dives, the penguin's heart rate slows down. Under water, the hunting eyes of the penguin are wide open, his cornea is protected by a nictitating membrane. The retina allows him to distinguish shapes and colors.

Penguins feed on fish and squid. For this, they must hunt in group and synchronize their dives to optimize the foraging [21]. Penguins communicate with each other with vocalizations. These vocalizations are unique to each penguin (like fingerprints in humans). Therefore, they allow the unique identification of each penguin and the recognition of penguins to each other [7]. This factor of identification and recognition is important since there is a large size of the colonies and a great similarity of the penguins. The amount of the necessary food for a penguin is variable depending on species, age, variety and quantity of food available in each region. Studies had shown that a colony of 5 million of penguins may eat daily 8 million pounds of krill and small fish [7].

## 3   The PeSOA Algorithm

The optimization algorithm based on the hunting behavior of penguins can be described in numerous ways. While all methods agree to optimize their objective functions such as maximizing the amount of energy extracted from the energy invested, we propose to simplify the optimization function by using rules, described below, to guide the search strategy by the penguins:

**Rule 1:** A penguin population is made up of several groups.
**Rule 2:** Each group is composed of a variable number of penguins that can vary depending on food availability in a specific location.
**Rule 3:** They hunt in group and move randomly until they find food when oxygen reserves are not depleted.
**Rule 4:** They can perform simultaneous dives to a depth identical.
**Rule 5:** Each group of penguins starts searching in a specific position (hole "i") and random levels (levels "j1, j2, ..., jn").
**Rule 6:** Each penguin looks for foods in random way and individually in its group, and after rough number of dives, penguins back on the ice to share with its affiliate's, the location (represented by the level or depth of the dive) where he found food and plenty of it (represented by the amount of eaten fich). This rule ensures intra-group communication.
**Rule 7:** At one level, one can have from 0 to N penguins (penguin or any group) according to the abundance of food.
**Rule 8:** If the number of fish in a hole is not enough (or none) for the group, part of the group (or the whole group) migrates to another hole. (This rule ensures inter-group communication)
**Rule 9:** The group who ate the most fish delivers us the location of rich food represented by the hole and the level.

In the algorithm each penguin is represented by the hole "i" and level "j" and the number of fish eaten. The distribution of penguins is based on probabilities of existence of fish in both holes and levels. The penguins are divided into groups (not necessarily the same cardinality) and begin searching in random positions. After a fixed number of dives, the penguins back on the ice to share with its affiliate's depth (level) and quantity (number) of the food found (Intergroup Communication). The penguins of one or more groups with little food, follow at the next dive, the penguins who chased a lot of fish.

---

**Pseudocode of the algorithm PeSOA :**
Generate random population of P solutions (penguins) in groups ;
Initialize the probability of existence of fish in the holes and levels;
**For i=1 to number of generations;**
**For each individual i ∈ P do**
**While oxygen reserves are not depleted do**
- Take a random step.
- Improve the penguin position using Eqs. (1)
- Update quantities of fish eaten for this penguin.
**End**
**End**
- Update quantities of eaten fish in the holes, levels and the best group.
- Redistributes the probabilities of penguins in holes and levels (these probabilities are calculated based on the number of fish eaten).
- Update best-solution
**End**

---

All Penguins ( i ) represent a solution (Xi)are distributed in groups, and each group search food in defined holes (Hj) with differences levels (Lk). In this process penguins sorted in order to their groups and start search in a specific hole and level according to food disponibility probability (Pjk).In each cycle, Accordingly, the position of the penguin with each new solution is Adjusted as follows:

$$D_{new} = D_{LastLast} + rand() \; |X_{LocalBest} - X_{LocalLast}| \qquad (1)$$

Where Rand() is a random number for distribution; and we have three solution, the best local solution ,the last solution and the new solution. the calculations in update solution equation (equation 1) are repeated for each penguins in each group, and after several plunged , penguins communicate to each other the best solution witch represented by the number of eaten fish, and we calculate the new distribution probability of holes and levels.

## 4  Implementation and Validation

### 4.1  Implementation and Numerical Simulation

After the mathematical formulation of the problem and for a visual representation of the results, we implemented the algorithm of penguins (PeSOA) in Linux with the open source tool, Scilab [18]. The validation of the algorithm is done with different benchmarks test function which we know in advance the solution. One of the functions commonly used for validation of new algorithms is the Rastrigin function [4], described by:

$$f(x) = 10.n + \sum_{i=1}^{n} x_i^2 - 10. \cos(2.\pi.x_i) - 5.12 \le x_i \le 5.12 \qquad (2)$$

The function of Rastringin is a non-convex function; it is a typical example of non linear multimodal function widely used to test the performance of optimization algorithms. The global minimum of Rastingin function is the origin, where its value is zero:

$$f(x) = 0 \quad x(i) = 0, i = 1 : n \qquad (3)$$

Figure 1 represents a 3D graph of the Rastingin function to optimize. The optimization algorithm of penguins (PeSOA) reaches the global optimum ($f(x) = 0$) without difficulty as shown in Figure 2.

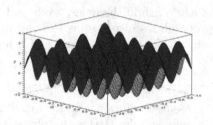

**Fig. 1.** Rastringin function, its global minimum is at the origin, where its value is zero

As shown in Figure 2, the penguin population reached a global optimum after a certain number of iterations without stuck in local optima. The two key points of any optimization algorithms are actually operated by PeSOA, on one hand, diversification and exploration of the search space and on the other hand; the exploitation of the best positions encountered by other penguins.

For the variability of tests, we randomly vary the number of penguins between 10 and 300, the number of holes between 3 and 30, and the number of levels between 2 and 15. The number of dives is fixed at 5, in all iterations.

According to the result sets of our simulation, we have found that there are no ideal parameters for our algorithm. However, we have also seen that the parameters haven't any influence on the convergence of the algorithm unless they are attached to very small values. Indeed, the number of penguins must be large enough to find all optima. The number of levels and holes are proportional to the number of penguins. For the rest of the simulation, we set the number of penguins, the number of holes and the number of levels at 40, 5 and 5, respectively.

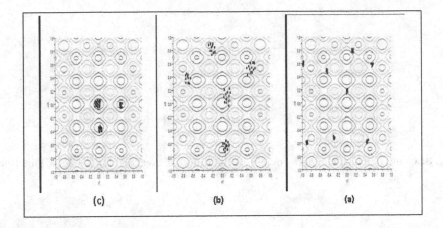

**Fig. 2.** The initial population of penguins (2.a) with their initial position of groups, Figure (2.b) is the distribution of penguins after 10 iterations and the third dive, (2.c) after 30 iterations with the Rastringin function.

## 4.2 Testing and Comparison of PeSOA with PSO and GA

The PeSOA algorithm is population based. The individuals representing the population are handling a set of solutions in parallel, in each iteration. In our simulation we compared our algorithm with two well-known optimization algorithms: PSO (Particle Swarm Optimization) and GA (Genetic Algorithm).

In recent years, many variations of these two algorithms have been proposed to enrich the research strategies and to optimize the number of iterations required to ensuring a better convergence to the optimal solution. In our case we opted for the two standard versions.

Each new algorithm must obviously be tested and validated through standardized benchmarks [5,19,20] with a set of test functions. Several test functions exist where each expresses a certain number of criteria to verify the characteristics of the optimization algorithm such as robustness, sensitivity, etc. In our simulation, we used the following test functions:

**The Function of De Jong.** De Jong function is a test function also known as the sphere function. De Jong function is continuous, convex and uni-modal. It is defined by:

$$f(x) = \sum_{i=1}^{n} x_i^2 \quad -5.12 \leq x_i \leq 5.12 \tag{4}$$

De Jong function has a global minimum; $f(x) = 0$, $x(i) = 0$, $i = 1: n$.

**The Function of Rosenbrock.** Classical Rosenbrock function is also known under the name of the function banane, is a non-convex function, widely used as typical test function for optimization problems. Rosenbrock function is defined by:

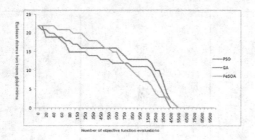

**Fig. 3.** Comparison of the performance of the different methods for De Jong Function

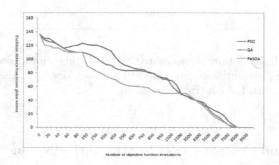

**Fig. 4.** Comparison of the performance of the different methods for Rosenberg function

$$f(x) = \sum_{i=1}^{n-1} 100.(x_{i+1} - x_i^2)^2 + (1 - x_i)^2 \qquad -2.048 \le x_i \le 2.048 \qquad (5)$$

and has a global minimum at: $f(x) = 0$, $x\ (i) = 1$, $i = 1{:}n$.

From the two figures (4 and 5), we can see that, with the 80 tested for the three algorithms and for both functions (De Jong and Rosenberg), the algorithms reached the global minimum in a similar path, as both functions are simple and contain only one global minimum.

**The Function of Schwefel.** This function is special, because the global minimum is rather remote (the geometric point of view) for the best local minimum. It is defined by:

$$f(x) = \sum_{i=1}^{n} -x_i.\sin\sqrt{\|x_i\|} \qquad -500 \le x_i \le 500 \qquad (6)$$

Schwefel Function has a global minimum: $f(x) = -n.418.9829 \quad x(i) = 420.9687$, $i = 1{:}\ n$.

The figure below shows that the 80 tests for the three algorithms with Schwefel function , the PeSOA shows a rapid convergence relative to the two other PSO algorithms and GA.

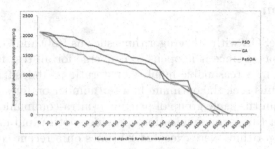

**Fig. 5.** Comparison of the performance of the different methods for Schwefel function

**The Function of Michalewicz.** Michalewicz function is a multimodal test function also having many local minima. It is defined through the following formula:

$$f(x) = \sum_{i=1}^{n} \sin x_i \cdot \left( \sin \left( \frac{i.x_i^2}{\Pi} \right) \right)^{2.m} \qquad m = 10, 0 \leq x_i \leq \Pi \qquad (7)$$

Michalewicz function has a global minimum: $f(x) = -1.8013 \ (m = 10), \ i = 1: n.$

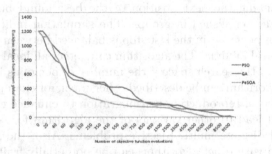

**Fig. 6.** Comparison of the performance of the different methods for Michalewicz function

Finally, the figure below shows the Michalewicz function witch is a function with n! local minima. While the performances of the PSO algorithm relapse significantly for this function, the PeSOA converges to the global optimum, but very slowly. This can be explained by the fact that penguins are divided between local minima and the global minimum. That said, the three algorithms converge to the optimum despite that there are many local minima, with a slight advantage to PeSOA.

## 5    Discussion

Comparison of PeSOA with other algorithms such as PSO and GA, has exhibited a strong performance since it happens to converge for all test functions, to the global optimum, in a reasonable number of iterations. Indeed, the principle of letting the penguins seek the optimum in a set number of dives, communicating information within the same group of penguins (intra-communication), then the entire population (inter-communication) provides both rapid and optimal convergence. The algorithms were executed 80 times on a recent computer clocked at 3.4 GHz. The stopping criterion of algorithms is defined by a tolerance of 0.00001. The penguins share their positions after a number of dives, which is, among many other advantages, one of the most important of PeSOA since the right way does not necessarily always starts with the right first stage.

## 6    Conclusion

In this paper we have formulated a new meta-heuristic for continuous optimization problems. The new approach based on the collaborative strategy of hunting in penguins. The proposed algorithm is validated on well-known benchmarks from the literature and a comparison is made with both GA and PSO algorithms. Simulation and comparison showed that PeSOA is more robust and efficient compared to other algorithms because its research strategy does not rely only on changing the next position of the best found but the course of the algorithm is accomplished in groups. The simulation indicated, also, that the distribution of penguins in the last step is balanced between the global minimum and other local minima. The algorithm can detect all local minima and the global minimum in the search space if the number of penguins is large enough.

The PeSOA algorithm can be described other ways, for example we can introduce the principles of reproduction and migration to enhance the search mechanism, which will lead to other variants of the algorithm for solving problems optimization and discrete multi-objective problems. Also, the hybridization of PeSOA algorithm with other algorithms can be potentially fruitful.

## References

1. Blum, C., Roli, A.: Metaheuristics in combinatorial optimization: Overview and conceptural comparision. ACM Comput. 35, 268–308 (2003)
2. Bonabeau, E., Dorigo, M., Theraulaz, G.: Swarm Intelligence: From Natural to Artificial Systems. Oxford University Press (1999)
3. Bratton, D., Kennedy: Defining a standard for particle swarm optimization. Elsevier Publishing (2007)
4. Chattopadhyay, R.: A study of test functions for optimization algorithms. J. Opt. Theory Appl. 3, 231–236 (1971)
5. Colorni, A., Dorigo, M., Maniezzo, M.: Distributed Optimization by Ant Colonies, pp. 134–142. Elsevier Publishing (1991)

6. Gardner, M.: Mathematical Games - The fantastic combinations of John Conway's new solitaire game "life", 120–123 (1970) (archived from the original on June 3, 2009)
7. Simpson, G.: Penguins: Past and Present, Here and There. Yale University Press (1976)
8. Green, K., Williams, R., Green, M.G.: Foraging ecology and diving behavior of Macaroni Penguins (Eudypteschrysolophus) at Heard Island. Arine Ornithology 26, 27–34 (1998)
9. Hanuise, N., Bost, C.A., Huin, W., Auber, A., Halsey, L.G., Handrich, Y.: Measuring foraging activity in a deep-diving bird: comparing wiggles, oesophageal temperatures and beak-opening angles as proxies of feeding. The Journal of Experimental Biology 213, 3874–3880 (2010)
10. Holland, J.H.: Adaptation in Natural and Artificial Systems. University of Michigan Press, Ann Arbor (1975)
11. Houston, A., McNamara, J.M.: A general theory of central place foraging for single-prey loaders. Theoretical Population Biology 28, 233–262 (1985)
12. Jason, B: Clever Algorithms Nature-Inspired Programming Recipes. Lulu Enterprises (January 2011)
13. Liu, Y., Passino, K.: Biomimicry of social foraging bacteria for distributed optimization: Models, principles, and emergent behaviors. Journal of Optimization Theory and Applications 115(3), 603–628 (2002)
14. Rechenberg, I.: Cybernetic Solution Path of an Experimental Problem, Royal Aircraft. Establishment Library Translation (1965)
15. MacArthur, R.H., Pianka, E.: On optimal use of a patchy environment. The American Naturalist 100, 603–609 (1966)
16. Mori, Y.: Optimal diving behavior for foraging in relation to body size. The American Naturalist 15, 269–276 (2002)
17. Robbins, H., Monro, S.: A Stochastic Approximation Method. Annals of Mathematical Statistics 22, 400–407 (1951)
18. Scilab Consortium (DIGITEO). SCILAB 5.3.2 (2010)
19. Schoen, F.: A wide class of test functions for global optimization. Global Optimization Journal 3, 133–137 (1993)
20. Shang, Y.W., Qiu, Y.H.: A note on the extended rosenrbock function. Evolutionary Computation 14, 119–126 (2006)
21. Takahashi, A., Sato, K., Nishikawa, J., Watanuki, Y., Naito, Y.: Synchronous diving behavior of Adelie penguins. Journal of Ethology 22, 5–11 (2004)
22. Yang, X.S.: Nature-Inspired Metaheuristic Algorithms. Luniver Press (2008)
23. Yang, X.S., Deb, S.: Cuckoo search via Levy flight, vol. 9, pp. 210–214. IEEE Publications (2009)
24. Yang, X.S.: Biology-derived algorithms in engineering optimization. In: Handbook of Bioinspired Algorithms and Applications, pp. 589–600 (2005)
25. Yang, X.S.: Bat algorithm for multi-objective optimization. IJBIC 5, 267–274 (2011)
26. Yann, T., Yves, C.: Synchronous underwater foraging behavior in penguins. Cooper Ornithological Soc. 101, 179–185 (2005)
27. Yang, X.S.: Engineering Optimization: An Introduction with Metaheuristic Applications. Wiley (2010)
28. Wayen, L.: Penguins of the World. Firefly Books (October 1, 1997)

# A Bidirectional Building Approach for the 2D Guillotine Knapsack Packing Problem

Lijun Wei*, Andrew Lim, and Wenbin Zhu

Department of Management Sciences, City University of Hong Kong,
Tat Chee Ave, Kowloon Tong, Hong Kong
lijunwei522@gmail.com, lim.andrew@cityu.edu.hk, i@zhuwb.com

**Abstract.** We investigate the 2D guillotine knapsack packing problem, where the objective is to select and pack a set of rectangles into a sheet with fix size and maximize the total profit of packed rectangles. We combine well known two methods namely top-down approach and bottom-up approach into a coherent algorithm to address this problem. Computational experiments on benchmark test sets show that our approach could find optimal solution for almost all the instances with moderate size and outperform all existing approaches for the larger instances.

**Keywords:** Cutting, Packing, guillotine-cut, 2D knapsack, block-building.

## 1 Introduction

The *2D knapsack packing problem* (2DKP) is a fundamental problem in cutting and packing literature. We are given a rectangular sheet of width $W$ and height $H$, and $m$ types of rectangles with dimensions $w_i \times h_i$, $i = 1, \ldots, m$. There are $b_i$ pieces of rectangles of type $i$ with a profit $p_i$ for each piece. The objective is to select and orthogonally pack a set of rectangles into the sheet and maximize total profit of packed rectangles. In addition, we assume both the size of sheet and input rectangles are integer, and the orientation of each rectangle is fixed. In many cutting applications, due to the physical constraint of the cutting blade, each cut must be parallel to the sides of the sheet and separate the sheet into two completely separated sheets. Such a cut is called *guillotine-cut*. A packing pattern observes the guillotine-cut constraint if all rectangles can be cut off the sheet by a series of guillotine-cuts. We call the 2D knapsack packing problem following the guillotine-cut constraint the *2D guillotine knapsack packing problem* (2DGKP).

Two general techniques have been widely used to solve the 2DGKP problem exactly: the top-down approach and the bottom-up approach. The top-down approach generates all the feasible packing patterns by repeatedly cutting a sheet into two new smaller sheets. The first top-down approach was introduced by Christofides & Whitlock [1], where all possible packing patterns are enumerated

---

* Corresponding author.

M. Ali et al. (Eds.): IEA/AIE 2013, LNAI 7906, pp. 232–241, 2013.

by tree search. In the search tree, branches correspond to all possible cuts on some sheet.

The bottom-up approach generates all the feasible cutting patterns by combining two patterns horizontally or vertically. Wang [2] first proposes the bottom-up concept. Later, Viswanathan & Bagchi [3] develop a best-first branch-and-bound algorithm based on the bottom-up concept; Hifi [4] improve this method by using more tighter lower and upper bounds; Cung et al. [5] improve its efficiency further by enhancing the initial lower bound and the upper bound at each internal node and suppress certain type of symmetrical patterns that are redundant. Recently, Yoon et al. [6] present an improved best-first branch-and-bound algorithm and obtained the best results for the 2DGKP; Dolatabadi et al. [7] present a recursive exact procedure which could also be seen as a bottom-up approach.

We combine top-down and bottom-up approach into a coherent algorithm for the 2DGKP. Similar to the bottom-up approach, we first arrange the rectangles into blocks where a block is a set of rectangles enclosed by its minimum bounding rectangle. We will only generate a limit number of blocks that are likely to be a part of high quality solutions. The packing process is carried out in a top-down fashion. Starting with the input sheet and the blocks list; we will first select a block and place it at a corner of the sheet. The sheet will be divided into two smaller sheets after the block is placed. In each subsequent step, we will select a smaller sheet and repeat the packing process until all remaining sheets are too small to accommodate any rectangles.

Although our algorithm is not an exact approach, experiment on test sets shows that our algorithm finds the optimal solutions for most of the instances of moderate size and obtains better solutions than existing algorithms for the larger instances.

## 2    Bidirectional Building Approach for the 2DGKP

### 2.1    Bidirectional Building Approach

In our algorithm, a partial solution is represented by a state with follow attributes:

  - *rectList*: a list of available rectangles
  - *blockList*: a list of candidate blocks we can use
  - *spaceList*: a list of free space
  - *packedRects*: a list of rectangles already packed
  - *packedProfit*: the total profit of packed rectangles

We say a packing is *complete* if either the list of available rectangles *rectList* is empty or the list of free space *spaceList* is empty. A space is a rectangular region $R$ described by the following four attributes:

  - $x,y$: the coordinates of bottom left corner of $R$
  - $w,h$: the width and height of $R$

---

**Algorithm 1.** Bidirectional building approach

---

BIDIRECTIONALBUILDING($W, H, Rects, blockList$)

    // Input: $W,H$: the width, height of the input sheet
    //        $Rects$: a list of input rectangle
    //        $blockList$: a list of candidate block we can use

    // create a free space representing the empty sheet
1   $R.x = 0; R.y = 0; R.w = W; R.h = H$
    // create an initial state
2   $S.rectList = Rects; S.blockList = blockList; S.spaceList = \{R\}$
3   $S.packedRects = \emptyset; S.packedProfit = 0$

4   Create a dummy solution $bestSol$ with $bestSol.packedProfit = 0$
5   **while** $S.spaceList \neq \emptyset$ and $S.Rects \neq \emptyset$
6       $R$ = space with the smallest area in $S.spaceList$
7       Delete $R$ from $S.spaceList$
        // find the block that best fits $R$
8       $bestNextState = \text{NULL}; maxProfit = 0$
9       **for** each block $b \in S.blockList$ fits into $R$
10          **for** each method of dividing space $div \in \{\text{VERTICAL}, \text{HORIZONTAL}\}$
11             $nextState = S$
12             $nextState = \text{PLACEBLOCK}(nextState, R, b)$
13             $newSpaces$ = space generated by using the method $div$
14             add each space in $newSpaces$ into $nextState.spaceList$
15             $finalState = \text{GREEDYPACK}(nextState)$
16             **if** $finalState.packedProfit > maxProfit$
17                 $bestNextState = nextState$
18                 $maxProfit = finalState.packedProfit$
19                 update $bestSol$ if $finalState$ is a better solution
20       **if** $bestNextState \neq \text{NULL}$
21          $S = bestNextState$
22  **return** $bestSol$

PLACEBLOCK($S, R, b$)

    // Input: $S$, a state representing a partial packing
    //        $R$: a rectangular region to be filled with
    //        $b$: a block to be packed
1   **for** each rectangle $i$ in $b.rectList$
2       Add rectangle $i$ to $S.packedRects$
3       $S.packedProfit = S.packedProfit + p_i$
4       delete $i$ from $S.rectList$
5   Delete all blocks from $S.blockList$ that contains more rectangles
    than available in $S.rectList$
6   **return** $S$

---

Our bidirectional building approach BIDIRECTIONALBUILDING is given in Algorithm 1 and works as follows. We first create a free space $R$ to represent the empty
sheet and an initial state $S$ whose *spaceList* only contain the space $R$ (line 1 - 3).
Then, we start the packing process. At each iteration of the packing process (line 6 -
21), we first get the space $R$ with the smallest area from the space list in the current
state $S$. We then try to select a block that fits $R$ the most and place the block at
the bottom left corner of $R$ and move to the next step.

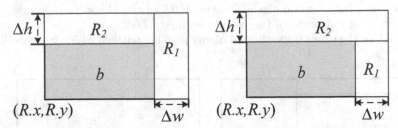

(a) Divide remaining space vertically   (b) Divide remaining space horizontally

**Fig. 1.** Two possible ways to divide the space after placing a block

To evaluate the fitness of a block $b$ with respect to a space $R$, we first place
$b$ at the bottom left corner of $R$. We then consider two possible ways to divide
the remaining space as illustrated in Figure 1. To obtain a complete packing for
a partial packing, we use a greedy heuristic GREEDYPACK(line 15). The partial
packing leading to the best complete packing will be recorded (line 17) and
selected as our next state (line 21).

Given a partial packing $S$, we generate a complete packing iteratively using
GREEDYPACK($S$) as follows. In each iteration, we first get the space $R$ with the
smallest area from the free space list in current state. Then, we try to find the first
block $b$ in the block list fitting into $R$. If a block $b$ is found, we place it at the bottom
left corner of the region $R$ using PLACEBLOCK, call the process GENNEWSPACE
($S, R, b$) to generate new spaces and add new space generated into the free space
list. The above process is repeated until we get a complete packing.

After we place a block $b$ at the bottom left corner of the region $R$, there
are two possible ways to divide the remaining space in the given region as illustrated in Figure 1. We use the procss GENNEWSPACE($S, R, b$) to select the
dividing method and generate new spaces, where $S$ is a state representing a partial packing. The process works as follow. If both the remaining width $\Delta w$ and
remaining height $\Delta h$ are too small to accommodate any unpacked rectangle, we
will not generate any new space. If either $\Delta w$ or $\Delta h$ is too small to accommodate
any unpacked rectangle, we will divide the remaining space horizontally (vertically) if the $\Delta h$ ($\Delta w$) is too small. Otherwise, for each method of generating
space, we use the area of generated space with larger area to evaluate it; we
prefer the method whose evaluated value is larger.

## 2.2  Block Generation

*Blocks* are recursively defined and generated. Every individual rectangle is a block, any two blocks can be joined either along $x$-axis as in Figure 2(a) or along $y$-axis as in Figure 2(b) to form a larger block. A block has the following attributes:

- *MBR*: the minimum bounding rectangle that encloses all rectangles inside the block. The dimensions of the block are defined by it.
- $w$, $h$: the width and height of the block, which is defined by its *MBR*.
- *rectList*: a list of rectangles inside the *MBR* of the block.
- *profit*: the total profit of the rectangles in *rectList*.
- *profitUB*: the profit upper bound of any packing pattern which use this block.

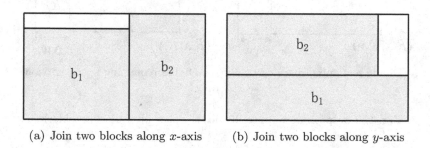

(a) Join two blocks along $x$-axis         (b) Join two blocks along $y$-axis

**Fig. 2.** Join two blocks to form a larger block

We use the process GENBLOCKS($W, H, Rects, maxCount$) to generate block, where *maxCount* is a number used to control the total number of blocks generated. The process works as follow. First, we generate blocks that consist of one rectangle. Then, we iteratively combine smaller blocks along $x$- and $y$-axis to generate larger blocks, until sufficient numbers of blocks are generated. The maximum number of blocks generated is controlled by the input parameter *maxCount*. Since, our objective is to maximize the profit packed rectangles in the sheet, we will only generate blocks whose *profitUB* is not below a user defined threshold *minProfit*. The method of calculating profit upper bound *profitUB* for a block will be described in section 2.4. Of course, blocks that are too wide or too tall to fit into the sheet are useless and therefore not generated. Furthermore, the blocks that consist of more rectangles than available in the input for certain type of rectangles will not be generated either. Finally, two blocks with same *MBR* and consisting of the same set of rectangles, even if the positions of rectangles differ, are equivalent for our purpose. Therefore, if a generated block is equivalent to an existing block, we will discard it.

Clearly, the number of blocks generated is a decreasing function of *minProfit*. However, it is hard to predict the exact number. To control the actual number of block generated, we use binary search on *minProfit* to find the maximum value of *minProfit*, on which the number of generated blocks is not less than *maxCount*.

## 2.3   Overall Algorithm

---

**Algorithm 2.** Iterative application of block building approach

---

IBBA($W, H, Rects$)

    // Input: $W$, $H$: width and height of sheet
    //          $Rects$: list of rectangles to be packed
    // Parameters: $timeLimit$: CPU time limit
1    $UB$ = the maximal profit of the relaxed one dimensional knapsack problem
2    Create a dummy solution $bestSol$ with $bestSol.packedProfit = 0$
3    $maxCount$ = number of types of rectangles in $Rects$
4    **while** $bestSol.packedProfit < UB$ and total execution time $< timeLimit$
5        $blockList$ = GENBLOCKS($W, H, Rects, maxCount$)
6        sort blocks in $blockList$ by decreasing profit of packed rectangle,
            then by increasing area of their $MBR$,
            then by increasing height of their $MBR$,
            then by increasing width of their $MBR$
7        $S$ = BLOCKBUILDINGAPPROACH($W, H, Rects, blockList$)
8        update $bestSol$ if $S$ is a better solution
9        $maxCount = 2 \times maxCount$
10  **return** $bestSol$

---

The solution produced by BIDIRECTIONALBUILDING is fully determined by the list of candidate blocks $blockList$. Our approach calls this procedure several times using different lists of candidate blocks and records the best solutions found (Algorithm 2). Since the number of all possible blocks grows exponentially as the number of input rectangles, it is simply too huge for any input of reasonable size. We, therefore, use the variable $maxCount$ to limit the total number of blocks generated. In the first iteration, we set $maxCount$ to the number of types of input rectangles in $Rects$. We sort the generated blocks in descending order of profit and call the process BIDIRECTIONALBUILDING to construct a solution based on the sorted block list. In subsequent iteration, we will double $maxCount$ to allow more blocks to be considered. The above process is repeated until we find a solution that equals to the upper bound **1DKP** as described in section 2.4 or the time limit is reached.

### 2.4   Calculation of Upper Bound

The 2D guillotine knapsack packing problem can be formulated as:

$$\mathbf{2DGKP}(W, H): \qquad \max \sum_{i=1}^{m} p_i \cdot x_i \qquad (1)$$

$$\sum_{i=1}^{m} w_i \cdot h_i \cdot x_i \leq W \cdot H \qquad (2)$$

$$x_i \in \{0, 1, \ldots, b_i\} \tag{3}$$

$$\text{x}_i \text{ copy of rectangle type } i \text{ could be}$$
$$\text{guillotine cut from the sheet with size } W \times H \tag{4}$$

If we delete the constraint 4, we get the relaxed one dimensional knapsack problem:

$$\mathbf{1DKP}(A): \quad \max \sum_{i=1}^{m} p_i \cdot x_i \tag{5}$$

$$\sum_{i=1}^{m} w_i \cdot h_i \cdot x_i \leq A \tag{6}$$

$$x_i \in \{0, 1, \ldots, b_i\} \tag{7}$$

where $A$ is capacity of the knapsack (i.e., $A = W \times H$ for an instance of 2DGKP). So for any instance of 2DGKP, the value of relaxed one dimensional knapsack problem is an upper bound of the original problem. We have used this upper bound in line 1 in Algorithm 2.

For a given block $b$, we can calculate profit upper bound of a packing that use the block $b$ as follows. Since $b$ occupy a space with area $b.w \times b.h$, so the area of remaining space in the input sheet is $leftArea = W \times H - b.w \times b.h$. The maximum profit we can get from the remaining space will not exceed $\mathbf{1DKP}(leftArea)$. So the profit of any packing using $b$ will not exceed $b.profit + \mathbf{1DKP}(leftArea)$.

We can improve this upper bound further by calculating the remaining space more accurately. As we assume the size of sheet and rectangle are integer, we can divide the input sheet into $W$ vertical strip of size $1 \times H$. For any packing solution using $b$, there are $b.w$ strips occupied by block $b$ in total. Consider any one of such strip, the total occupied height of this strip must be a combination of the height of $b$ and the height of input rectangles, which will not exceed

$$maxH = \max\{\sum_{i=1}^{m} h_i x_i \mid b.h + \sum_{i=1}^{m} h_i x_i \leq H, \ x_i \in \{0, 1, \ldots, b_i\} \text{ for } i = 1, 2, \ldots, m\} \tag{8}$$

So at least $H - maxH$ units of area will be wasted in such a strip, resulting in a total waste of $vWaste = b.w \times (H - maxH)$. We divide the sheet into horizontal strips follow a similar argument, we can establish the minimum waste $hWaste$. Therefore, the minimum waste due to block $b$ is $vWaste + hWaste$. The total useful area in the remaining sheet after $b$ is placed is given by $leftArea = W \times H - b.w \times b.h - vWaste - hWaste$, and the profit upper bound of any packing using $b$ is given by $b.profit + \mathbf{1DKP}(leftArea)$.

## 3    Computational Experiments

There are six sets in existing literature. In the first three sets, the profit of a rectangle is equal to its area:

- **Set1:** 46 instances used by Hifi [8] to test their tabu search.
- **Set2:** 13 instances used by Dolatabadi et al. [7] to test their exact algorithm.
- **Set3:** 21 instances generated by Hopper & Turton [9].

In the other three sets, the profit of a rectangle is not equal to its area:

- **Set4:** 36 instances used by Hifi [8].
- **Set5:** 21 small instances used by Bortfeldt & Winter [10] to test their genetic algorithm.
- **Set6:** 630 larger instances, which were introduced by by Beasley [11].

Our **IBBA** is implemented as a sequential algorithm in C++ and compiled by GCC 4.1.2, and no multi-threading was explicitly utilized. It was executed on an Intel Xeon E5430 clocked at 2.66GHz (Quad Core) with 8 GB RAM running the CentOS 5 linux operating system. The *timeLimit* is set to 60 seconds for each instance. We compare our approach with the following leading algorithms in the literature:

- **H2004:** a hybrid approach by Hifi [8].
- **C2008:** a recursive algorithm by Chen [12].
- **B2009:** a genetic algorithm by Bortfeldt & Winter [10].
- **D2012:** an exact algorithm by Dolatabadi et al. [7].

**Table 1.** comparision between IBBA and existing algorithms on six benchmark test sets. The algorithm obtain best result on a test set is highlighted in bold.

| Set | #Inst | #opt | | | | | average relative gap (%) | | | | |
|-----|-------|-------|-------|-------|-------|------|-------|-------|-------|-------|-------|
|     |       | H2004 | C2008 | B2009 | D2012 | IBBA | H2008 | C2008 | B2009 | D2012 | IBBA |
| 1 | 46 | 37 | 43 | - | - | **44** | 0.031 | 0.006 | - | - | **0.001** |
| 2 | 13 | - | - | - | 12 | 12 | - | - | - | 0.180 | **0.154** |
| 3 | 21 | - | - | 5 | - | **7** | - | - | 1.211 | - | **0.592** |
| 4 | 36 | 30 | 28 | - | - | **33** | 0.175 | **0.171** | - | - | 0.174 |
| 5 | 21 | - | - | 14 | - | **21** | - | - | 1.246 | - | **0** |
| 6 | 630 | - | - | 194 | - | **315** | - | - | 1.260 | - | **0.920** |

The comparisons between IBBA and other algorithms on the six sets are summarized in the table 1, where *#opt* is the number of optimal solution found by each algorithm. The *relative gap* is calculated as the percentage gap between the found solution and the optimal solution/upper bound. Let the profit of the found solution be $p$ and the optimal solution/upper bound be $ub$, then the relative gap is calculated as $(ub - p) / ub$. For the instance whose optimal solution is known, $ub$ is set to the optimal value, otherwise, $ub$ is set to the value of relaxed one-dimensional knapsack problem.

For the set1, the optimal solution of all the instances is known based on the results of Yoon et al. [6]. Our algorithm **IBBA** finds the optimal solution for all the instances except the instances $ATP34$ and $ATP35$. However, the gap between our solution and the optimal solution is very small.

For the set4, the optimal solutions of all the instances except $ATP42$ and $ATP43$ are found by Yoon et al. [6]. **IBBA** finds the optimal solution for all the instances except $ATP42, ATP43$ and $ATP47$, which is the best among all the compared heuristic method.

For the larger instances in the set6, **IBBA** finds the optimal solution for almost half of the instances, which is much better than that of **B2009**. The average relative gap of **IBBA** is smaller than that of **B2009**.

## 4    Conclusions

In this paper, we study the 2D guillotine knapsack packing problem. While existing algorithm use either top-down or bottom-up approach, our algorithm combine both ideas. We construct a solution in top-down fashion: the input sheet is successively divided into small sheets in each step. We leverage the benefit of bottom-up approach: we combine rectangles into blocks and progressive combine small blocks into larger ones. We pack in the unit of block instead of rectangle. The computational results on well-known test sets show that our approach could find the optimal solution for almost all the instances of moderate size and gain best results for the larger instances compared with existing know algorithms.

## References

1. Christofides, N., Whitlock, C.: An algorithm for two-dimensional cutting problems. Operations Research 25(1), 30–44 (1977)
2. Wang, P.Y.: Two algorithms for constrained two-dimensional cutting stock problems. Operations Research 31(3), 573–586 (1983)
3. Viswanathan, K.V., Bagchi, A.: Best-first search methods for constrained two-dimensional cutting stock problems. Operations Research 41(4), 768–776 (1993)
4. Hifi, M.: An improvement of Viswanathan and Bagchi's exact algorithm for constrained two-dimensional cutting stock. Computers & Operations Research 24(8), 727–736 (1997)
5. Cung, V.D., Hifi, M., Le Cun, B.: Constrained two-dimensional cutting stock problems a best-first branch-and-bound algorithm. International Transactions in Operational Research 7(3), 185–210 (2000)
6. Yoon, K., Ahn, S., Kang, M.K.: An improved best-first branch-and-bound algorithm for constrained two-dimensional guillotine cutting problems. International Journal of Production Research, 1–14 (June, October 2012)
7. Dolatabadi, M., Lodi, A., Monaci, M.: Exact algorithms for the two-dimensional guillotine knapsack. Computers & Operations Research 39(1), 48–53 (2012)
8. Hifi, M.: Dynamic programming and hill-climbing techniques for constrained two-dimensional cutting stock problems. Journal of Combinatorial Optimization 8, 65–84 (2004)

9. Hopper, E., Turton, B.C.H.: An empirical investigation of meta-heuristic and heuristic algorithms for a 2D packing problem. European Journal of Operational Research 128(1), 34–57 (2001)
10. Bortfeldt, A., Winter, T.: A genetic algorithm for the two-dimensional knapsack problem with rectangular pieces. International Transactions in Operational Research 16(6), 685–713 (2009)
11. Beasley, J.E.: A population heuristic for constrained two-dimensional non-guillotine cutting. European Journal of Operational Research 156(3), 601–627 (2004)
12. Chen, Y.: A recursive algorithm for constrained two-dimensional cutting problems. Computational Optimization and Applications 41(3), 337–348 (2008)

# Increasing the Antibandwidth of Sparse Matrices by a Genetic Algorithm

Petrica C. Pop[1] and Oliviu Matei[2]

[1] Dept. of Mathematics and Informatics, Technical Univ. of Cluj-Napoca, Romania
petrica.pop@ubm.ro
[2] Dept. of Electrical Engineering, Technical University of Cluj-Napoca, Romania
oliviu.matei@ubm.ro

**Abstract.** The antibandwidth problem consists in finding a labeling of the vertices of a given undirected graph such that among all adjacent node pairs, the minimum difference between the node labels is maximized. In this paper, we formulate the antibandwidth problem in terms of matrices and propose an efficient genetic algorithm based heuristic approach for increasing the corresponding antibandwidth. We report computational results for a set of 30 benchmark instances. The preliminary results point out that our approach is an attractive and appropriate method to explore the solution space of this complex problem and leads to good solutions in reasonable computational times.

**Keywords**: antibandwidth problem, genetic algorithms, combinatorial optimization.

## 1 Introduction

The antibandwidth problem (ABP) was introduced by Leung et al. [7] and is also known in the literature as separation number problem or dual bandwidth problem.

The problem was introduced originally as a dual variation of the well-known bandwidth minimization problem. Even though it was not so intensively studied as the bandwidth minimization problem, the antibandwidth problem finds many practical applications: multiprocessor scheduling problem, radio frequency assignment problem [5], obnoxious facility location problem [8], obnoxious center problem, tournament scheduling, etc.

Leung et al. [7] proved that the antibandwidth problem is NP-hard, but there exist efficient polynomial algorithms for finding the antibandwidth parameter in complements of interval graphs, treshold graphs and arborescent comparability graphs [3].

At the beginning the researches focused on theoretical aspect of the antibandwidth problem: Leung et al. [7] proved the NP-completeness of the problem and gave as well some polynomial time algorithms for some special classes of graphs, Yixun and Jinjiang [9] derived several upper bounds for antibandwidth, Raspaud et al. [8] solved the problem in the case of the following special classes

M. Ali et al. (Eds.): IEA/AIE 2013, LNAI 7906, pp. 242–251, 2013.

of graphs: two dimensional meshes, tori and hypercubes and Dobrev et al. [2] proposed tight upper bounds for antibandwidth in the case of general Hamming graphs and provided optimal solutions for a special class of these graphs.

Recently, some heuristic approaches have been proposed in order to obtain high-quality solutions for the antibandwidth problem on general graphs: Duarte et al. [4] developed some heuristics based on GRASP and evolutionary path relinking and Bansal and Srivastava [1] elaborated a memetic algorithm.

In this paper, we first define the antibandwidth problem in the context of matrices and then we propose a novel genetic algorithm based heuristic in order to provide high-quality solutions to the problem. We test our approach on a set of 30 benchmark instances and study its efficiency in comparison to the existing heuristics for solving the problem.

## 2    Definition of the Problem

Let $G = (V, E)$ be an undirected finite graph, where $V$ denotes the set of vertices, $E$ the set of edges and $|V| = n$.

A labeling $f$ of the vertices of $G$ is a one-to-one mapping from the vertices set $V$ onto the set of integers $\{1, 2, ..., n\}$ (i.e. each vertex $v \in V$ has a unique label $f(v) \in \{1, 2, ..., n\}$). Obviously, each labeling can be identified with a permutation of $\{1, 2, ..., n\}$.

The antibandwidth of the graph $G$ under the labeling $f$ is defined as:

$$AB_f(G) = \min\{AB_f(v) : v \in V\}$$

where

$$AB_f(v) = \min\{|f(u) - f(v)| : u \in N(v)\}$$

is the antibandwidth of the vertex $v$ and by $N(v)$ we denoted the set of adjacent vertices of the vertex $v$.

Then the *antibandwidth problem* consists in finding a labeling $f$ which maximizes $AB_f(G)$:

$$AB(G) = \max_{f \in \pi_n} AB_f(G).$$

where by $\pi_n$ we denoted the set of all permutations of $\{1, 2, ..., n\}$.

In the context of the matrices the antibandwidth problem can be stated as follows: given a sparse matrix $A = [a_{ij}]_{n \times n}$, we are interested in finding permutations of the rows and columns that maximizes the distance of any non-zero entry from the center diagonal.

The graph and the matrix versions of the antibandwidth problem are equivalent. The equivalence is clear if we replace the nonzero entries of the matrix by 1s and interpret the result as the adjacency matrix of a graph.

In what it follows we present an example that shows this equivalence.

**Example.** Suppose we have an undirected graph $G = (V, E)$ with $|V| = 5$ and the given labeling $f$: $f(v_1) = 3, f(v_2) = 2, f(v_3) = 1, f(v_4) = 5, f(v_5) = 4$.

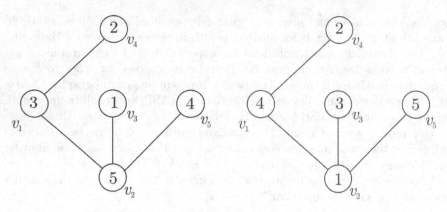

**Fig. 1.** Two different labelings $f$ and $f'$ of a graph $G$

Then the antibandwidths of each of the vertices of the graph $G$ under the labeling $f$ are:

$$AB_f(v_1) = \min\{|5-3|, |2-3|\} = 1$$
$$AB_f(v_2) = \min\{|3-5|, |1-5|, |4-5|\} = 1$$
$$AB_f(v_3) = \min\{|5-1|\} = 4$$
$$AB_f(v_4) = \min\{|3-2|\} = 1$$
$$AB_f(v_5) = \min\{|5-4|\} = 1$$

and the antibandwidth of the graph $G$ under labeling $f$ is:

$$AB(f) = \min_{v \in V} AB_f(v) = \min\{1, 1, 4, 1, 1\} = 1.$$

The adjacency matrix of the graph under labeling $f$ is:

$$A(f) = \begin{pmatrix} 1 & 0 & 0 & 0 & 1 \\ 0 & 1 & 1 & 0 & 0 \\ 0 & 1 & 1 & 0 & 1 \\ 0 & 0 & 0 & 1 & 1 \\ 1 & 0 & 1 & 1 & 1 \end{pmatrix}$$

If we consider the new labeling $f'$ described as well in figure 1, its antibandwidth is calculated as follows:

$$AB_{f'}(v_1) = \min\{|2-4|, |1-4|\} = 2$$
$$AB_{f'}(v_2) = \min\{|4-1|, |3-1|, |5-1|\} = 2$$
$$AB_{f'}(v_3) = \min\{|1-3|\} = 2$$
$$AB_{f'}(v_4) = \min\{|4-2|\} = 2$$
$$AB_{f'}(v_5) = \min\{|1-5|\} = 4$$

and the antibandwidth of the graph $G$ under $f'$ is:

$$AB(f') = max_{v \in V} AB_{f'}(v) = max\{2, 2, 2, 2, 4\} = 2.$$

We observe that the antibandwidth has been increased and the corresponding adjacency matrix $A(f')$ of the graph under labeling $f'$ is:

$$A(f') = \begin{pmatrix} 1 & 1 & 0 & 0 & 0 \\ 1 & 1 & 1 & 1 & 0 \\ 0 & 1 & 1 & 0 & 1 \\ 0 & 1 & 0 & 1 & 0 \\ 0 & 0 & 1 & 0 & 1 \end{pmatrix}.$$

# 3 The Genetic Algorithm for Solving the Antibandwidth Problem

The Genetic Algorithms (GAs) were introduced by Holland [6] in the early 1970s and provide search mechanisms in the solutions space following the Darwin's theory of evolution. For finding the solution, it is used a population.

In GAs, a potential solution to an optimization problem is represented as a set of parameters known as a gene. These parameters are joined together to form a string of values known as a chromosome. Each chromosome represents a possible solution to the optimization problem. Attached to each chromosome is defined a fitness value, which defines how good a solution that chromosome represents. A crossover operator plays the role of reproduction and a mutation operator is assigned to make random changes in the solutions. A selection procedure, simulating the natural selection, selects a certain number of parent solutions, which the crossover uses to generate new solutions, also called offspring. At the end of each iteration the offspring together with the solutions from the previous generation form a new generation, after undergoing a selection process to keep a constant population size. By using crossover, mutation and selection, the population will converge to one containing only chromosomes with a good fitness. The solutions are evaluated in terms of their fitness values identical to the fitness of individuals.

Genetic algorithms have proven to be a powerful and successful problem solving strategy for solving complex optimization problems, demonstrating the power of evolutionary principles.

Next we give the description of our genetic algorithm heuristic for solving the antibandwidth problem.

## 3.1 Genetic Representation

An individual is represented as a list of interchanges of lines or columns:

$$I = (W_1 < k_s^1, k_d^1 >, W_2 < k_s^2, k_d^2 >, ..., W_m < k_s^m, k_d^m >), \tag{1}$$

where $W_i \in 'L', 'C'$ ('L' means an interchange of lines, and 'C' is an interchange of columns) and $k_s^i$ respectively $k_d^i$ are the two lines/columns to be interchanged. The permutations are applied successively, in order. Given the following matrix:

$$A = \begin{pmatrix} a_{11} & a_{12} & a_{13} \\ a_{21} & a_{22} & a_{23} \\ a_{31} & a_{32} & a_{33} \end{pmatrix} \tag{2}$$

and the individual $I = (C < 1,3 >, L < 2,1 >)$, the matrix undergoes the following permutations:

1. the columns 1 and 3 are interchanged and results the following matrix:

$$A_1 = \begin{pmatrix} a_{13} & a_{12} & a_{11} \\ a_{23} & a_{22} & a_{21} \\ a_{23} & a_{32} & a_{31} \end{pmatrix}, \tag{3}$$

2. the lines 2 and 1 are interchanged and the result is:

$$A_2 = \begin{pmatrix} a_{23} & a_{22} & a_{21} \\ a_{13} & a_{12} & a_{11} \\ a_{23} & a_{32} & a_{31} \end{pmatrix}. \tag{4}$$

Therefore the resulted matrix after applying the individual $I$ is $A_2$.

In our algorithm the initial population is generated randomly. The length of each individual is chosen at random, up to twice the size of the matrix.

Every solution has a fitness value assigned to it, which measures its quality. In our case, the fitness value is given by the antibandwidth of the matrix resulted after a program is applied to the original matrix.

## 3.2   Genetic Operators

We considered three operations for modifying structures in our genetic algorithm: crossover, mutation and pruning. The most important one is the crossover operation. In the crossover operation, two solutions are combined to form two new solutions or offspring.

**Crossover.** Two parents are selected from the population by the binary tournament method. The two parents can undergo two types of crossover with the same probability: one-cut crossover, respectively a concatenation.

**One-Cut Crossover Operator.** Offspring are produced from two parent solutions using the following crossover procedure: it creates offspring which preserve the order and position of symbols in a subsequence of one parent while preserving the relative order of the remaining symbols from the other parent. It is implemented by selecting a random cut point. Given the two parents:

$$P_1 = (M_1^1 M_2^1 | M_3^1 M_4^1), \tag{5}$$

$$P_2 = (M_1^2 M_2^2 | M_3^2 M_4^2 M_5^2), \tag{6}$$

where the superior index represents the parent (first or second), the number of elements of the parent represent the number of permutations (interchanges of lines or columns) and "|" defines the cutting point, then the offspring are:

$$O_1 = (M_1^1 M_2^1 | M_3^2 M_4^2 M_5^2), \tag{7}$$

$$O_2 = (M_1^2 M_2^2 | M_3^1 M_4^1). \tag{8}$$

**Concatenation Operator.** The concatenation operator concatenates two parents. The first offspring is formed by adding the second parent at the end of the first one. The other offspring is made of the second parent followed by the first one.

Given the same two parents:

$$P_1 = (M_1^1 M_2^1 M_3^1 M_4^1), \tag{9}$$

$$P_2 = (M_1^2 M_2^2 M_3^2 M_4^2 M_5^2), \tag{10}$$

the offspring are:

$$P_1 = (M_1^1 M_2^1 M_3^1 M_4^1 | M_1^2 M_2^2 M_3^2 M_4^2 M_5^2), \tag{11}$$

$$P_2 = (M_1^2 M_2^2 M_3^2 M_4^2 M_5^2 | M_1^1 M_2^1 M_3^1 M_4^1). \tag{12}$$

The concatenation operator is of great importance because it is the most important way of evolving individuals to longer ones (with more moves).

**Mutation.** Mutation is another important feature of genetic algorithms. We use in our GA six random mutation operators chosen with the same probability:

1. addition of a move: an allele (position in an individual) is chosen at random and a new random move in inserted in that place. This way, the length of the individual increases.
2. removal of a move: a move is chosen randomly and removed from the individual. This way, the length of the individual decreases.
3. exchange of two moves: two alleles randomly selected are swapped.
4. replacement of the item which undergoes the move: a column is replaced by a line or the other way around: e.g. $L < 1, 4 >$ is replaced by $C < 1, 4 >$.
5. replacement of the position of the items which undergo the move: e.g. $L < 2, 3 >$ by $L < 1, 5 >$.
6. replacement of the entire move: a randomly selected allele is replaced by a new one, yet generated randomly.

**Pruning.** It is often the case that longer individuals contain shorter sequences of moves which lead to better results. This is the reason for introducing a new operator, called *pruning*.

Given an individual:

$$P = M_1M_2M_3...M_kM_{k+1}...M_n,$$ (13)

the pruning operator generates a new individual:

$$P = M_1M_2M_3...M_k$$ (14)

holding the following condition:

$$f([M_1M_2M_3...M_k]) = \min_{i=\overrightarrow{1,n}} f([M_1...M_i]),$$ (15)

where $f([M_1...M_i])$ is the fitness value of the sequence of moves $[M_1...M_i]$.

The pruning operator is deterministic and it is applied for all individuals.

### 3.3   Selection

The selection process is deterministic. In our algorithm we use the $(\mu + \lambda)$ selection, where $\mu$ parents produce $\lambda$ offspring. The new population of $(\mu + \lambda)$ is reduced again to $\mu$ individuals by a selection based of the "survival of the fittest" principle. In other words, parents survive until they are suppressed by better offspring. It might be possible for very well adapted individuals to survive forever.

### 3.4   Genetic Parameters

The genetic parameters are very important for the success of a GA, equally important as the other aspects, such as the representation of the individuals, the initial population and the genetic operators. Based on preliminary experiments, the most important genetic parameters chosen are: the population size $\mu$ has been set to 5 times the size of the matrix, the intermediate population size $\lambda$ was chosen twice the size of the population: $\lambda = 2 \cdot \mu$ and the mutation probability was set at 10%.

In our algorithm the termination strategy is based on a maximum number of generations to be run if no improvement of the maximum antibandwidth value is observed within 15 consecutive generations.

## 4   Computational Results

In this section we present some computational results in order to asses the effectiveness of our proposed genetic algorithm heuristic for solving the antibandwidth problem.

We tested our algorithm on a set of 30 random connected graphs (RCG) from RND/BUP (GDToolkit), denoted by ug$m\_n$, $m = 1, 2, 3, 4, 5$ and $n = 50, 60, 70, 80, 90, 100$, where $n$ denotes the number of vertices. In our experiments we performed 10 independent runs for each instance.

The testing machine was an an Intel i7, 2,6 GHz and 4 GB RAM. with Windows XP Professional as operating system. The algorithm was developed in Java, JDK 1.6.

Preliminary experiments were conducted to set the GA parameters namely the population size and the genetic operators. Due to page restrictions, we present here just the results of the experiments for assessing the effect of using the 6 mutation operators on the quality of the solutions returned by our GA, we conducted trials in the case of 6 instances. The comparison with the case of using a single mutation operator and respectively 3 mutation operators is shown in the next figure.

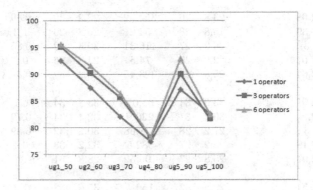

**Fig. 2.** The effect of different mutation operators on the quality of the solutions

In Table 1 we present the computational results obtained with our genetic algorithm heuristic in comparison with those obtained by Bansal and Srivastava [1] using their memetic algorithm (MA) for a set of 30 random connected graphs (RCG) from RND/BUP with the edge densities between 2.52% and 5.77%.

The first column in the table gives the name of the graph instance, the second one the corresponding density, the third column contains the upper bounds provided by Yixun and Jinjiang [9] and the last columns are presenting the results obtained by Bansal and Srivastava [1] using their MA and our results obtained with the proposed GA: the mean values, the highest value of the antibandwidth, the corresponding standard deviation and the mean elapsed time.

Analyzing the computational results, we observe that our genetic algorithm based heuristic provides higher antibandwidth values than the state-of-the-art approach considered by Bansal and Srivastava [1]. We have been able to improve the quality of the solution in 27 out of 30 instances and we obtained the same solutions in rest.

**Table 1.** Comparison of GA heuristic with MA for RCGs instances from RND/BUP collection

| Graphs | Edge density % | Upper Bound | Results of MA | | | Results of GA | | | |
|---|---|---|---|---|---|---|---|---|---|
| | | | Mean | Best | Time (sec.) | Mean | Best | Std. dev. | Time (sec.) |
| ug1_50 | 5,22 | | 17,4 | 18 | 65,28 | 18,2 | 20 | 0,89 | 37,685 |
| ug2_50 | 5,38 | | 16,8 | 18 | 33,51 | 17,8 | 20 | 1,1 | 21,615 |
| ug3_50 | 5,14 | 24 | 16,8 | 17 | 52,04 | 17,8 | 18 | 0,09 | 25,375 |
| ug4_50 | 5,71 | | 18,4 | 19 | 59,55 | 18,6 | 19 | 0,19 | 36,47 |
| ug5_50 | 5,06 | | 19,8 | 20 | 83,44 | 20,5 | 21 | 0,25 | 50,15 |
| ug1_60 | 4,46 | | 20,2 | 21 | 155,98 | 22,6 | 24 | 0,7 | 70,14 |
| ug2_60 | 4,57 | | 21,8 | 23 | 21,8 | 23,8 | 25 | 0,6 | 17,425 |
| ug3_60 | 4,46 | 29 | 20,2 | 22 | 116,12 | 22,6 | 23 | 0,19 | 70,14 |
| ug4_60 | 4,46 | | 21,2 | 22 | 235,2 | 21,4 | 22 | 0,29 | 131,145 |
| ug5_60 | 4,51 | | 20,8 | 21 | 70,2 | 22,6 | 23 | 0,19 | 39,315 |
| ug1_70 | 4,05 | | 23,8 | 26 | 354,52 | 28,2 | 30 | 0,89 | 160,37 |
| ug2_70 | 3,39 | | 15,8 | 28 | 54,2 | 26,4 | 29 | 1,3 | 30,415 |
| ug3_70 | 3 | 34 | 26,2 | 27 | 116,31 | 30,3 | 32 | 0,85 | 60,18 |
| ug4_70 | 4,01 | | 22,6 | 23 | 74,5 | 24,8 | 26 | 0,6 | 36,465 |
| ug5_70 | 3,64 | | 25,5 | 26 | 323,45 | 30,2 | 32 | 0,89 | 178,32 |
| ug1_80 | 2,91 | | 33,6 | 34 | 61,07 | 33,8 | 34 | 0,1 | 39,14 |
| ug2_80 | 2,94 | | 30,2 | 31 | 226,9 | 32,4 | 34 | 0,79 | 125,415 |
| ug3_80 | 3 | 39 | 27,2 | 28 | 138,68 | 30,2 | 32 | 0,89 | 81,415 |
| ug4_80 | 3,19 | | 25,2 | 27 | 582,5 | 32,2 | 33 | 0,4 | 337,4 |
| ug5_80 | 2,97 | | 25,8 | 27 | 190,52 | 28,8 | 31 | 1,1 | 100,62 |
| ug1_90 | 2,54 | | 29,8 | 32 | 150,4 | 35,6 | 38 | 1,2 | 86,18 |
| ug2_90 | 2,84 | | 34,6 | 35 | 469,62 | 35,2 | 36 | 0,39 | 232,915 |
| ug3_90 | 2,77 | 44 | 29,8 | 31 | 316,29 | 31,8 | 35 | 1,59 | 164,185 |
| ug4_90 | 2,47 | | 33,3 | 34 | 381,2 | 33,8 | 36 | 1,09 | 210,415 |
| ug5_90 | 2,59 | | 31,8 | 34 | 815,47 | 34,2 | 37 | 1,4 | 516,135 |
| ug1_100 | 2,3 | | 32,2 | 34 | 499,2 | 36,8 | 38 | 0,59 | 275,375 |
| ug2_100 | 2,3 | | 38,4 | 40 | 715,9 | 39,2 | 42 | 1,4 | 375,16 |
| ug3_100 | 2,34 | 49 | 31,6 | 33 | 256,4 | 38,3 | 40 | 0,85 | 137,15 |
| ug4_100 | 2,84 | | 33,8 | 35 | 419,8 | 39,5 | 41 | 0,75 | 215,19 |
| ug5_100 | 2,52 | | 30,6 | 31 | 693,65 | 37,1 | 39 | 0,95 | 365,14 |

Regarding the computational times, it is difficult to make a fair comparison between algorithms, because they have been evaluated on different computers and they have different stopping criteria. However, it should be noted that our heuristic is faster than the memetic algorithm.

# 5    Conclusions

In this paper, we defined the antibandwidth problem in terms of matrices and described an efficient genetic algorithm heuristic for increasing the corresponding antibandwidth. The proposed heuristic integrates a number of original features: we considered two crossover operators, six mutation operators and an additional operator called pruning. The diversity of these operators provide our algorithm with a good tradeoff between intensification and diversification.

Comparing our GA heuristic with the state-of-the-art memetic algorithm for solving the antibandwidth problem in the case of a set of 30 random connected graphs (RCG), we observed that our GA heuristic is superior in terms of solution quality and as well in terms of computational times.

In the future, we plan to explore the possibility of building a parallel implementation of the system in order to improve the execution time. In addition, we will need to asses the generality and scalability of the proposed heuristic by testing it on larger instances.

**Acknowledgments.** This work was supported by a grant of the Romanian National Authority for Scientific Research, CNCS - UEFISCDI, project number PN-II-RU-TE-2011-3-0113.

# References

1. Bansal, R., Srivastava, K.: Memetic algorithm for the antibandwidth maximization problem. Journal of Heuristics 17, 39–60 (2011)
2. Dobrev, S., Kralovic, R., Pardubska, D., Torok, L., Vrto, I.: Antibandwidth and cyclic antibandwidth of Hamming graphs. Electronic Notes in Discrete Mathematics 34, 295–300 (2009)
3. Donnelly, S., Isaak, G.: Hamiltonian powers in threshold and arborescent comparability graphs. Discrete Mathematics 202, 33–44 (1999)
4. Duarte, A., Marti, R., Resende, M.G.C., Silva, R.M.A.: GRASP with path relinking heuristics for the antibandwidth problem. Networks 58(3), 171–189 (2011)
5. Hale, W.K.: Frequency assignment: Theory and applications. Proceedings of the IEEE 68, 1497–1514 (1980)
6. Holland, J.: Adaptation in Natural and Artificial Systems. University of Michigan Press, Ann Arbor (1975)
7. Leung, J., Vornberger, O., Witthoff, J.: On some variants of the bandwidth minimization problem. SIAM Journal on Computing 13(3), 650–667 (1984)
8. Raspaud, A., Schroder, H., Sykora, O., Torok, L., Vrto, I.: Antibandwidth and cyclic antibandwidth of meshes and hypercubes. Discrete Mathematics 309, 3541–3552 (2009)
9. Yixun, L., Jinjiang, Y.: The dual bandwidth problem for graphs. J. of Zhengzhou University 35, 1–5 (2003)

# A New GA-Based Method for Temporal Constraint Problems

Reza Abbasian and Malek Mouhoub

Department of Computer Science
University of Regina
Regina, Canada
{abbasiar,mouhoubm}@cs.uregina.ca

**Abstract.** Managing numeric and symbolic temporal information is very relevant for a wide variety of applications including scheduling, planning, temporal databases, manufacturing and natural language processing. Often these applications are represented and managed with the well known constraint-based formalism called the Constraint Satisfaction Problem (CSP). We then talk about temporal CSPs where constraints represent qualitative or quantitative temporal information. Like CSPs, temporal CSPs are NP-hard problems and are traditionally solved with a backtrack search algorithm together with constraint propagation techniques. This method has however some limitations especially for large size problems. In order to overcome this difficulty in practice, we investigate the possibility of solving these problems using Genetic Algorithms (GAs). We propose a novel crossover specifically designed for solving TCSPs using GAs. In order to assess the performance of our proposed crossover over the well known heuristic based GAs, we conducted several experiments on randomly generated temporal CSP instances. In addition, we evaluated the performance of an integration of our crossover within a Parallel GA (PGA) approach. The test results clearly show that the proposed crossover outperforms the known GA methods for all the tests in terms of success rate and time needed to reach the solution. Moreover, when integrated within the PGA, our crossover is very efficient for solving very large size hard temporal CSPs.

**Keywords:** Parallel Genetic Algorithms, Constraint Satisfaction Problems, Temporal Constraints, Scheduling and Planning.

## 1 Introduction

A Constraint Satisfaction Problem (CSP) consists of a finite set of variables with finite domains, and a finite set of constraints restricting the possible combinations of variable values [3]. A solution to a CSP is a complete assignment of values to all the variables such that all the constraints are satisfied. In [12] a CSP-based model, TemPro, has been developed for representing and solving CSPs involving numeric and symbolic temporal constraints. This is the case of a wide variety of applications including scheduling, planning, temporal databases,

M. Ali et al. (Eds.): IEA/AIE 2013, LNAI 7906, pp. 252–261, 2013.

manufacturing and natural language processing. More formally, TemPro translates an application involving temporal information into a binary CSP[1] where variables are temporal events defined on domains of numeric intervals and binary constraints between variables correspond to disjunctions of Allen primitives [2]. This latter is called a Temporal CSP (TCSP)[2].

Like a CSP, a TCSP is known to be an NP-hard problem in general[3], a backtrack search algorithm of exponential time cost is needed to find a complete solution. In order to overcome this difficulty in practice, a systematic solving method based on constraint propagation techniques has been proposed in [12]. The goal here is to reduce the size of the search space before and during the backtrack search. While the proposed method has a lot of merits when tackling small and medium size problems, it suffers from its exponential time cost especially for large size hard problem instances. An alternative is to use approximation methods such as Genetic Algorithms (GAs). Despite some success of GAs when tackling CSPs [5, 8, 9], they generally suffer from poor crossover operators in solving combinatorial optimization problems. The main reason for such a phenomenon is that in CSPs (and TCSPs) changing the value of a variable can have direct effects on other variables that are in constraint relation with the changing variable and indirect effect on other variables. As a result, performing a random crossover can often reduce the quality of the solution.

In this paper we propose a novel crossover, that we call Parental Success Crossover (PSC), specially designed for solving TCSPs. In order to assess the performance of our proposed crossover over the basic one-point crossover, as well as known heuristic based GAs [5, 9], we conducted several experiments on TCSP instances randomly generated using the model RB proposed in [19]. This model is a revision of the standard Model B [6, 18], has exact phase transition and the ability to generate asymptotically hard instances. In addition, we evaluated the performance of an integration of our crossover within a Parallel GA (PGA) [10]. The test results clearly demonstrate that the proposed crossover outperforms the other known GAs on all problem instances in terms of success rate and time needed to reach the solution. In the case of over constrained (inconsistent) TCSPs, our method is better in terms of the quality of the returned solution. The quality of the solution is measured here by the number of solved constraints. Finally, when integrated within the PGA, our crossover is very efficient for solving large size and hard TCSPs.

The rest of the paper is structured as follows. The TemPro model is first introduced in the next section. Our proposed crossover is then covered in section3.

---

[1] A binary CSP is a CSP where all the constraints are either unary or binary relations between variables.

[2] Note that this name and the corresponding acronym was used in [4]. The TCSP, as defined by Dechter et al, is a quantitative temporal network used to represent only numeric temporal information. Nodes represent time points while arcs are labeled by a set of disjoint intervals denoting a disjunction of bounded differences between each pair of time points.

[3] There are special cases where CSPs and TCSPs are solved in polynomial time, for instance, the case where the related constraint network is a tree [7, 11].

Section 4 reports the results of comparative experiments we conducted on randomly generated TCSPs. Finally, concluding remarks and future directions are listed in section 5.

## 2    TemPro

TemPro[12] transforms a temporal problem under qualitative and quantitative constraints into a binary CSP where constraints are disjunctions of Allen primitives (see Figure 1 for the definition of the Allen primitives) and variables, representing temporal events, are defined on domains of time intervals. Each event domain (called also temporal window) contains the Set of Possible Occurrences (SOPO) of numeric intervals the corresponding event can take. The SOPO is the numeric constraint of the event. It is expressed by the fourfold: [earliest start, latest end, duration, step] where: earliest start is the earliest start time of the event, latest end is the latest end time of the event, duration is the duration of the event and step is the discretization step corresponding to the number of time units between the start time of two adjacent intervals. For some applications, the consistency of the problem depends on the discretization step. In this particular case, if the solution is not found, the user can decrease the value of the step and run again the solving algorithm. Decreasing the discretization step will however increase the complexity of the problem. Indeed, the total number of combinations (potential solutions) of a TCSP is $D^N$ where $N$ is the number of variables and $D$ their domain size. $D$ is computed as follows: $D = Max_{1 \leq i \leq N}(\frac{sup_i - inf_i - d_i}{s_i})$ where $sup_i$, $inf_i$, $d_i$ and $s_i$ are respectively the latest end time, earliest start time, duration and step of a given event $Evt_i$. As we can easily see, decreasing the value of $s_i$ will increase the domain size $D$ which increases the total number of possibilities of the search space. Note that begintime, endtime, duration and step can be constant values or variables taking values from a discrete and finite domain. Constraints can also be used, in the form of equations or inequalities, in order to restrict the values these variables can take [12]. In order to better understand TemPro and its related components, let us consider the following temporal constraint problem.

**Example 1.**

1. *John, Mary and Wendy separately rode to the soccer game.*
2. *It takes John 30 minutes, Mary 20 minutes and Wendy 50 minutes to get to the soccer game.*
3. *John either started or arrived just as Mary started.*
4. *John left home between 7:00 and 7:10.*
5. *Mary arrived at game between 7:55 and 8:00.*
6. *Wendy left home between 7:00 and 7:10.*
7. *John's trip overlapped the soccer game.*
8. *Mary's trip took place during the game or else the game took place during her trip.*
9. *The soccer game starts at 7:30 and lasts 105 minutes.*

10. *John either started or arrived just as Wendy started.*
11. *Mary and Wendy arrived together but started at different times.*

In order to be able to check whether the above story is consistent and answer queries such as: "what are the different arrival times of Mary ?", we can use the modeling framework TemPro [12], to translate the problem into the TCSP represented by the graph in figure 1. The nodes of the graph correspond to the four events of our story, namely: John, Mike and Wendy are going to the soccer game and the soccer game itself. To each node we associate the domain of possible time intervals corresponding to the related event. For instance, in the case of John's event we have the following SOPO: [0,40,30,1] respectively representing the earliest start time, latest end time, duration and step of the event (we assume that the discretization step for all the events is equal to 1). This SOPO corresponds to the following domain of possible interavls: $\{(0\ 30), (1\ 31), \ldots, (10\ 40)\}$. Finally, each arc is labeled by the disjunctive Allen relation representing the relative position between the corresponding pair of events. For instance, the relation "ESSiM" between Jonh's and Mary's events represents the information listed above in example 1 (item 3).

# 3 Proposed Crossover for Temporal Constraints

## 3.1 TCSP Individual Representation

Figure 1 shows a graph corresponding to the TCSP of example 1. An individual corresponds to a given temporal scenario and represents a complete assignment of intervals to all the events of the problem. For each individual, the fitness function is computed by counting the number of conflicts (constraint violations) due to the complete assignment. An individual with a fitness equal to zero is a solution to the TCSP (since all the temporal constraints are satisfied). For example, the fitness of the individual shown in Figure 1 is equal to 3.

## 3.2 Parental Success Crossover (PSC)

To perform a One Point Crossover (OPC), a crossing point is randomly choosen. All the genomes from the first parent that are before the crossing point will be included in the offspring. Also, all the genomes in the second parent that are after the crossing point, will be included the offspring. OPC (or any random crossover) performs poorly for TCSPs and CSPs in general. Indeed, in the case of TCSPs changing the value (interval) of a given event $E$ has direct effect on other events that are in constraint relation with $E$. It can also have indirect effects on other events. As a result, a random crossover (having a random crossing point) does not provide interesting results. To improve the crossover for TCSPs, we propose the following crossing method.

Each individual in the population maintains two records; the total number of times it has participated in reproduction ($N_p$), and the number of times the offspring it produced was fitter. We refer to the latter as Parental Success

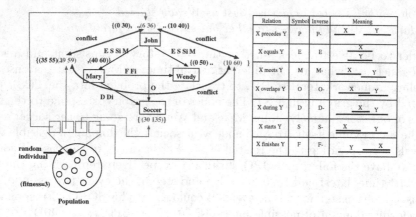

**Fig. 1.** GA representation of the TCSP in example 1

Number and denote it by $P_s$. The parental success ratio, denoted by $S$, can then be calculated as follows: $S = \frac{P_s}{N_p}$. Furthermore, we define the term Fitness Around Variable (FAV) as the number of conflicts between a certain variable (event) and its neighbors in the temporal constraint graph. Each individual keeps a record of the fitness around all of its variables. Using these new parameters, we create a Crossover Mask. The offspring is then produced according to this mask. Let $p_1$ and $p_2$ be two parents chosen for the crossover. The crossover mask specifies which allele (entry) in the new individual should inherit from which parent. Since the crossover is performed here using two parents, the crossover mask consists of binary digits. Let 1 specify choosing the allele from $p_1$ and 0 specify choosing it from $p_2$. To create the mask, we compare each allele in $p_1$ with its correspondence in $p_2$. If the FAV of the allele in $p_1$ is less than the one in $p_2$, we put a 1 in the mask. If the FAV of the allele in $p_2$ is less than the one in $p_1$, we put a 0 in the mask. In case of equality, we use the following probabilities for choosing the allele: $P(\text{choosing from } p_1) = 1/2 + (S_{p_1} - S_{p_2}) \times 1/2$ $P(\text{choosing from } p_2) = 1/2 + (S_{p_2} - S_{p_1}) \times 1/2$ Where, $S_{p_1}$ and $S_{p_2}$ respectively are the parental success ratios of $p_1$ and $p_2$.

## 3.3 Reproduction

Reproduction is performed amongst a number of fittest individuals in the population. To generate new offspring we randomly chose two individuals among the fittest ones in the population as the parents. We then pass them to the Parental Success Crossover. Also, every $I$ iterations, we choose the parents randomly as a mean to preserve the diversity in the population.

## 3.4 Mutation

In genetic algorithms, the mutation operator is used to help escape local optima. The mutation happens with a small probability. When an individual is passed to the mutation operator, some of its alleles are randomly changed to new values. This way a totally new individual is produced after the mutation. The mutation operator diversifies the population and helps the algorithm to work on different regions of the search space. We propose two different methods for the mutation.

**Mutation to Minimize the Number of Conflicts.** In this type of mutation, $N_{mutation}$ random vertices of the individual are selected and the number of conflicts around the chosen vertices and their adjacent vertices are minimized. Say vertex $A$ is randomly chosen for the mutation. Then, according to the adjacency matrix of the temporal constraint graph, for every vertex $B$ that shares a constraint with vertex $A$, if there is a conflict, $B$ will take a new value such that the constraint between $A$ and $B$ is satisfied. $N_{mutation}$ is computed as follows using another parameter that we introduce called Allele Mutation Percentage. Suppose we have an individual of size 100 and an allele mutation percentage of 20%. Then we have 20%*100=20. This number 20 is now the maximum possible value for $N_{mutation}$. Each time we perform a mutation, $N_{mutation}$ takes a random value between 1 and 20.

**Stochastic Value Change.** This mutation method randomly chooses $N_{mutation}$ vertices and assigns a random value (interval) to each.

## 4 Experimentation

The algorithms are implemented in Java programming language. In the experiments we used a machine with 2.5 GHz Core 2Duo CPU, 4 Gb of RAM running JDK 1.6. Our PGA is implemented using the Master-Slave architecture. The number of slaves is fixed to 10 and the population size per slave is equal to 300. The mutation is implemented as described in Section 3.4. The number of variables to change is also determined randomly from the following range $[2, individualLength/10]$ where $individualLength$ is the length of each individual. The mutation chance is set to 0.2. For all the tests reported in this section, each problem instance is solved 20 times by the given method and the average running time needed to return the solution is computed.

TCSPs are randomly generated using the model RB. This model is a revision of the standard Model B, and has exact phase transition and the ability to generate asymptotically hard instances. Following the model RB, we generate each TCSP instance in two steps as shown below and using the parameters $n$, $p$, $\alpha$ and $r$ where :

- $n$ is the number of events,
- $p$ ($0 < p < 1$) is the constraint tightness which can be measured, as shown in [16], as the fraction of all possible pairs of intervals from the domain of two events that are not allowed by the constraint,
- and $r$ and $\alpha$ ($0 < \alpha < 1$) are two positive constants (respectively set to 0.5 and 0.8).

1. Select with repetition $r\,n\,\log n$ random constraints. Each random constraint is formed by selecting without repetition 2 of $n$ events.
2. For each constraint we uniformly select without repetition $pd^2$ incompatible pairs of intervals from the domains of the pair of events involved by the constraint. $d = n^\alpha$ is the domain size of each event.

Tests are conducted on random consistent TCSP instances with tightness values varying from 0.05 to 0.65. First, we compared our proposed PSC against the One Point Crossover. Table 1 lists the results of running both methods on different random problems where the number of variables, $n$, is set to 80. As the table suggests, the success rate of PSC is higher than OPC. Furthermore, OPC is not able to find the solution to TCSP instances that have a tightness more than 0.4 in the given time (the timeout is set to 60 seconds).

**Table 1.** PSC and OPC for Consistent TCSPs

| Tightness | PSC | | | OPC | | |
|---|---|---|---|---|---|---|
| | Success Rate | Time (s) | Best Fitness | Success Rate | Time (s) | Best Fitness |
| 0.05 | 100% | 0.118 | 0 | 100% | 0.310 | 0 |
| 0.1 | 100% | 0.139 | 0 | 100% | 0.445 | 0 |
| 0.15 | 100% | 0.204 | 0 | 100% | 0.550 | 0 |
| 0.2 | 100% | 0.938 | 0 | 100% | 1.693 | 0 |
| 0.25 | 100% | 0.986 | 0 | 100% | 2.911 | 0 |
| 0.3 | 100% | 1.080 | 0 | 100% | 6.805 | 0 |
| 0.35 | 100% | 1.312 | 0 | 100% | 10.194 | 0 |
| 0.4 | 100% | 2.640 | 0 | 67.98% | 22.939 | 0 |
| 0.45 | 100% | 3.911 | 0 | 0% | - | 1 |
| 0.50 | 100% | 4.039 | 0 | 0% | - | 3 |
| 0.55 | 78.31% | 9.242 | 0 | 0% | - | 9 |
| 0.6 | 63.73% | 37.185 | 0 | 0% | - | 13 |

Over constrained temporal constraint problems are those where a complete solution satisfying all the constraints does not exist. In this case the goal is to find an assignment that satisfies most of the constraints. This is a known optimization problem called max TCSP. Approximation methods such as GAs can be used to efficiently solve the max TCSP but do not guarantee the optimality of the solution returned. In order to evaluate the performance of our proposed PSC (over the OPC) to solve the max TCSP we have conducted several experiments on unsolvable TCSPs randomly generated with the model RB and where the number of variables $n$ is ranging from 100 to 400. The timeout is set to 60 seconds. Figure 2 illustrates the results of these experiments where the x-axis corresponds to the number of variables and the y-axis corresponds to the number of conflicts reached by each method for 60 seconds of running time. As we can see from the figure, the Parental Success Crossover outperforms the conventional One Point crossover. Finally, we compared our proposed PSC to the best heuristic based GA, called Sawing GA, for solving TCSPs (as reported in [9]). The Sawing GA [5] is one of the adaptive fitness algorithms where the search is guided by changing the way the fitness is computed so that individuals are evaluated based on some special characteristics they may have. The Sawing GA is based on the SAW-ing mechanism where the idea is that constraints that are not satisfied or variables causing constraint violations after a certain number of steps must be hard, thus must be given a high weight (penalty). The results of this comparison are listed in Table 2. Our proposed algorithm obviously outperforms the other methods both in terms of running time and success rate.

**Table 2.** PSC, OSC and Sawing GA for consistent TCSPs

| Tightness | PSC | | | OPC | | | Sawing GA | | |
|---|---|---|---|---|---|---|---|---|---|
| | Success Rate | Time (s) | Best Fitness | Success Rate | Time (s) | Best Fitness | Success Rate | Time (s) | Best Fitness |
| 0.24 | 100% | 0.959 | 0 | 100% | 2.817 | 0 | 20% | 50 | 0 |
| 0.16 | 100% | 0.213 | 0 | 100% | 0.595 | 0 | 90% | 18 | 0 |
| 0.05 | 100% | 0.118 | 0 | 100% | 0.310 | 0 | 100% | 3 | 0 |

The Parallel Genetic Algorithm (PGA) is an extension of the GA. The well-known advantage of PGAs is their ability to facilitate different sub populations to evolve in diverse directions simultaneously [10]. It is shown that PGAs speed up the search process and can produce high quality solutions on complex problems [17]. We use the Master-Slave architecture for designing the PGA. In this architecture there is one single population divided into fractions. Each fraction is assigned to one slave process on which genetic operations are performed [10].

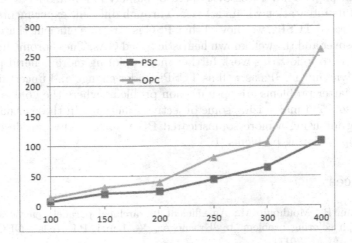

**Fig. 2.** Comparison of Best Fitness reached by PSC and OPC

In order to assess the performance of the integration of our PSC within the PGA we conducted experimental tests on consistent TCSPs with the tightness set to 0.35 as well as inconsistent TCSPs with tightness equal to 0.65. Figure 3 reports the time in seconds needed by the PGA to return the consistent solution for TCSPs instances with variables ranging from 100 to 1000 (values on the x-axis of the figure). Note that with the tightness fixed to 0.35 these are hard to solve problems especially for large size variables. Thanks to our PSC, the PGA was able to solve completely all the instances in a reasonable time. For 1000 variables for instance, a complete solution is returned in less than a minute which is very impressive. We are not aware of any solving method that can solve these large instances in the times shown by the figure.

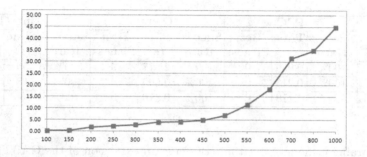

**Fig. 3.** PGA for Consistent TCSPs

## 5    Conclusion and Future Work

In order to overcome the difficulty when solving large size TCSPs, we have proposed in this paper a novel crossover operator namely Parental Success Crossover (PSC) and its integration within a PGA. Through different experiments on randomly generated TCSPs, we showed that PSC is far more efficient than the One Point Crossover and the well known heuristic based GAs. These promising results motivated us to follow this work further into exploring more general problems including dynamic TCSPs as well as TCSPs under change and uncertainty [13–15]. These latter problems are optimization problems where the goal is to come up with a solution maximizing some objective functions. In these situations we are looking for using a more sophisticated PGA such as the one designed for solving graph coloring problems [1].

## References

1. Abbasian, R., Mouhoub, M.: An efficient hierarchical parallel genetic algorithm for graph coloring problem. In: Krasnogor, N., Lanzi, P.L. (eds.) GECCO, pp. 521–528. ACM (2011)
2. Allen, J.: Maintaining knowledge about temporal intervals. CACM 26(11), 832–843 (1983)

3. Dechter, R.: Constraint Processing. Morgan Kaufmann (2003)
4. Dechter, R., Meiri, I., Pearl, J.: Temporal Constraint Networks. Artificial Intelligence 49, 61–95 (1991)
5. Eiben, A.E., van der Hauw, J.K., van Hemert, J.I.: Graph coloring with adaptive evolutionary algorithms. J. Heuristics 4(1), 25–46 (1998)
6. Gent, I., MacIntyre, E., Prosser, P., Smith, B., Walsh, T.: Random constraint satisfaction: Flaws and structure (1998)
7. Haralick, R., Elliott, G.: Increasing tree search efficiency for Constraint Satisfaction Problems. Artificial Intelligence 14, 263–313 (1980)
8. van der Hauw, J.: Evaluating and improving steady state evolutionary algorithms on constraint satisfaction problems (1996),
   citeseer.ist.psu.edu/vanderhauw96evaluating.html
9. Jashmi, B.J., Mouhoub, M.: Solving temporal constraint satisfaction problems with heuristic based evolutionary algorithms. In: Proceedings of the 2008 20th IEEE International Conference on Tools with Artificial Intelligence, vol. 2, pp. 525–529. IEEE Computer Society, Washington, DC (2008)
10. Lim, D., Ong, Y.S., Jin, Y., Sendhoff, B., Lee, B.S.: Efficient hierarchical parallel genetic algorithms using grid computing. Future Gener. Comput. Syst. 23(4), 658–670 (2007)
11. Mackworth, A.K., Freuder, E.: The complexity of some polynomial network-consistency algorithms for constraint satisfaction problems. Artificial Intelligence 25, 65–74 (1985)
12. Mouhoub, M.: Reasoning about Numeric and Symbolic Time Information. In: The Twelfth IEEE International Conference on Tools with Artificial Intelligence (ICTAI 2000), pp. 164–172. IEEE Computer Society, Vancouver (2000)
13. Mouhoub, M.: Dynamic path consistency for interval-based temporal reasoning. In: 21st International Conference on Artificial Intelligence and Applications (AIA 2003), Applied Informatics, pp. 393–398. ACTA Press (2003)
14. Mouhoub, M.: Systematic versus non systematic techniques for solving temporal constraints in a dynamic environment. AI Communications 17(4), 201–211 (2004)
15. Mouhoub, M., Sukpan, A.: Conditional and Composite Temporal CSPs. Applied Intelligence 36(1), 90–107 (2012)
16. Sabin, D., Freuder, E.C.: Contradicting conventional wisdom in constraint satisfaction. In: Proceedings of the Eleventh European Conference on Artificial Intelligence, pp. 125–129. John Wiley and Sons, Amsterdam (1994)
17. Sena, G.A., Megherbi, D., Isern, G.: Implementation of a parallel genetic algorithm on a cluster of workstations: Traveling salesman problem, a case study. Future Gener. Comput. Syst. 17(4), 477–488 (2001)
18. Smith, B., Dyer, M.: Locating the phase transition in binary constraint satisfaction problems. Artificial Intelligence 81, 155–181 (1996)
19. Xu, K., Li, W.: Exact Phase Transitions in Random Constraint Satisfaction Problems. Journal of Artificial Intelligence Research 12, 93–103 (2000)

# On Using the Theory of Regular Functions to Prove the ε-Optimality of the Continuous Pursuit Learning Automaton

Xuan Zhang[1], Ole-Christoffer Granmo[1], B. John Oommen[2,1,*], and Lei Jiao[1]

[1] Dept. of ICT, University of Agder, Grimstad, Norway
[2] School of Computer Science, Carleton University, Ottawa, Canada

**Abstract.** There are various families of Learning Automata (LA) such as Fixed Structure, Variable Structure, Discretized etc. Informally, if the environment is stationary, their ε-optimality is defined as their ability to converge to the optimal action with an arbitrarily large probability, if the learning parameter is sufficiently small/large. Of these LA families, Estimator Algorithms (EAs) are certainly the fastest, and within this family, the set of *Pursuit* algorithms have been considered to be the pioneering schemes. The existing proofs of the ε-optimality of all the reported EAs follow the same fundamental principles. Recently, it has been reported that the previous proofs for the ε-optimality of *all* the reported EAs have a *common flaw*. In other words, people have worked with this flawed reasoning for almost three decades. The flaw lies in the condition which apparently supports the so-called "monotonicity" property of the probability of selecting the optimal action, explained in the paper. In this paper, we provide a new method to prove the ε-optimality of the Continuous Pursuit Algorithm (CPA), which was the pioneering EA. The new proof follows the same outline of the previous proofs, but instead of examining the monotonicity property of the action probabilities, it rather examines their *submartingale* property, and then, unlike the traditional approach, invokes the theory of *Regular* functions to prove the ε-optimality. We believe that the proof is both unique and pioneering, and that it can form the basis for formally demonstrating the ε-optimality of other EAs.

**Keywords:** Pursuit Algorithms, Continuous Pursuit Algorithm, ε-optimality.

## 1 Introduction

Learning automata (LA) have been studied as a typical model of reinforcement learning for decades. They have found applications in a variety of fields, including game playing [1], parameter optimization [2], solving knapsack-like problems and utilizing the solution in web polling and sampling [3], vehicle path control [4], assigning capacities in prioritized networks [5], and resource allocation [6]. They have also been used in language processing, string taxonomy [7], graph partitioning [8], and map learning [9].

An LA is an adaptive decision-making unit that learns the optimal action from among a set of actions offered by the Environment it operates in. At each iteration, the LA

---

* *Chancellor's Professor*; *Fellow: IEEE* and *Fellow: IAPR*. The Author also holds an *Adjunct Professorship* with the Dept. of ICT, University of Agder, Norway.

M. Ali et al. (Eds.): IEA/AIE 2013, LNAI 7906, pp. 262–271, 2013.

selects one action, which triggers either a reward or a penalty as a response from the Environment. Based on the response and the knowledge acquired in the past iterations, the LA adjusts its action selection strategy in order to make a "wiser" decision in the next iteration. In such a way, the LA, even though it lacks a complete knowledge about the Environment, is able to learn through repeated interactions with the Environment, and adapts itself to the optimal decision.

Among the families of LA, Estimator Algorithms (EAs) (a more detailed survey of the families is found in [19]) work with a noticeably different paradigm, and are certainly the fastest and most accurate. Within this family, the set of *Pursuit* Algorithms (PAs) were the pioneering schemes, whose design and analysis were initiated by Thathachar and Sastry [10]. EAs augment an action probability updating scheme with the use of estimates of the reward probabilities of the respective actions. The first Pursuit Algorithm (PA) was designed to operate by updating the action probabilities based on the $L_{R-I}$ paradigm. By the same token, being an EA in its own right, the PA maintains running Maximum Likelihood (ML) reward probability estimates, which further determines the current "Best" action for the present iteration. The PA then pursues the current best action by linearly increasing *its* action probability. As the PA considers both the *short-term* responses of the Environment and the *long-term* reward probability estimates in formulating the action probability updating rules, it outperforms traditional VSSA schemes in terms of its accuracy and its rate of convergence.

The most difficult part in the design and analysis of LA consists of the formal proofs of their convergence accuracies. The mathematical techniques used for the various families (FSSA, VSSA, Discretized etc.) are quite distinct. The proof methodology for the family of FSSA is the simplest: it quite simply involves formulating the Markov chain for the LA, computing its equilibrium (or steady state) probabilities, and then computing the asymptotic action selection probabilities. The proofs of convergence for VSSA are more complex and involve the theory of small-step Markov processes, distance diminishing operators, and the theory of Regular functions. The proofs for Discretized LA involve the asymptotic analysis of the Markov chain that represents the LA in the discretized space, whence the *total* probability of convergence to the various actions is evaluated. However, understandably, the most difficult proofs involve the family of EAs. This is because the convergence involves two intertwined phenomena, namely the convergence of the reward estimates *and* the convergence of the action probabilities themselves. Ironically, the combination of these vectors in the updating rule is what renders the EA fast. However, if the accuracy of the estimates are poor because of inadequate estimation (i.e., if the sub-optimal actions are not sampled "enough number of times"), the convergence accuracy can be diminished. Hence the dilemma!

The ε-optimality of the EAs have been studied and presented in [11] [12] [13] [14] [15]. The basic result stated in these papers is that by utilizing a sufficiently small value for the learning parameter, the CPA will converge to the optimal action with an arbitrarily large probability. However, these proofs have a common flaw, which involves a very fine argument. In fact, the proofs reported in these papers "deduced" the ε-optimality based on the conclusion that after a sufficiently large time instant, $t_0$, the probability of selecting the optimal action is monotonically increasing, which, in turn, is based on the condition that the reward probability estimates are ordered properly *forever* after $t_0$.

This ordering is, indeed, true by the law of large numbers if all the actions are chosen *infinitely often*. But if such an "infinite" selection does not occur, the ordering cannot be guaranteed for *all* time instants after $t_0$.

As a consequence of this misinterpretation, the condition supporting the monotonicity property is false, which further leads to an incorrect proof for the CPA being $\varepsilon$-optimal. Even though this has been the accepted argument for almost three decades (even by the third author of this present paper who was the principal author of many of the above-mentioned papers), we credit the authors of [16] for discovering this flaw. While a detailed explanation of this is found in [16], a brief explanation on this issue is also included in this paper, in Section 3.

This paper aims at correcting the above-mentioned flaw used in the earlier proofs. As opposed to these proofs, we will show that while the so-called monotonicity property is sufficient for convergence, it is not really *necessary* for proving that the CPA is $\varepsilon$-optimal. Rather, we will present a completely new proof methodology which is based on the convergence theory of submartingales and the theory of Regular functions [17].

## 2  Overview of the CPA

Since this paper concentrates on the intricate nature of the CPA, it is mandatory that the reader has a fundamental understanding of it. It is, thus, briefly surveyed here. To do this, first of all, we present below the notations used:

$\alpha_i$: The $i^{th}$ action that can be selected by the LA, and is an element from the set $\{\alpha_1, \ldots \alpha_r\}$.
$p_i$: The $i^{th}$ element of the action probability vector $P$.
$\lambda$: The learning rate, where $0 < \lambda < 1$.
$u_i$: The number of times $\alpha_i$ has been rewarded when it has been selected.
$v_i$: The number of times $\alpha_i$ has been selected.
$\hat{d}_i$: The $i^{th}$ element of the reward probability estimates vector $\hat{D}$, $\hat{d}_i = \frac{u_i}{v_i}$.
$m$: The index of the optimal action.
$h$: The index of the greatest element of $\hat{D}$.
$R$: The Environment's response, where $R = 0$ corresponds to a Reward, and $R = 1$ to a Penalty.

The CPA follows a "pursuit" paradigm of learning, which consists of three steps. Firstly, it maintains an action probability vector $P = [p_1, p_2, ..., p_r]$ to determine the issue of which action is to be selected, where the sum of the $p_i$'s is unity, and where $r$ is the number of actions. Secondly, it maintains running ML reward probability estimates to determine which action can be reckoned to be the "best" in the current iteration. Thus, it updates $\hat{d}_i(t)$ based on the response from the Environment as below:

$$u_i(t) = u_i(t-1) + (1 - R(t));$$
$$v_i(t) = v_i(t-1) + 1$$
$$\hat{d}_i(t) = \frac{u_i(t)}{v_i(t)}.$$

Thirdly, based on the response of the Environment and the knowledge of the current best action, the CPA increases the probability of selecting the current best action as per the continuous $L_{R-I}$ rule. So, if $\hat{d}_h(t)$ is the largest element of $\hat{D}(t)$, we update $P(t)$ as:

**If** $R(t) = 0$ **Then**

$$p_j(t+1) = (1 - \lambda)p_j(t), j \neq h$$
$$p_h(t+1) = 1 - \sum_{j \neq h} p_j(t+1)$$

**Else**

$$P(t+1) = P(t)$$

We now visit the issue of the proof of the CPA's convergence.

## 3 Previous Proofs for CPA's ε-Optimality

The formal assertion of the ε-optimality of the CPA is stated in Theorem 1.

**Theorem 1.** *Given any* $\varepsilon > 0$ *and* $\delta > 0$, *there exist a* $\lambda^* > 0$ *and a* $t_0 < \infty$ *such that for all time* $t \geq t_0$ *and for any positive learning parameter* $\lambda < \lambda^*$,

$$Pr\{p_m(t) > 1 - \varepsilon\} > 1 - \delta.$$

The earlier reported proofs for the ε-optimality of the CPA follow the "four-step" strategy. Firstly, given a sufficiently small value for the learning parameter $\lambda$, all actions will be selected enough number of times before a finite time instant, $t_0$. Secondly, for all $t > t_0$, $\hat{d}_m$ will remain to be the maximum element of the reward probability estimates vector, $\hat{D}$. Thirdly, suppose $\hat{d}_m$ has been ranked as the largest element in $\hat{D}$ since $t_0$, the action probability sequence of $\{p_m(t)\}$, with $t > t_0$, will be monotonically increasing, whence one concludes that $p_m(t)$ converges to 1 with probability 1. Finally, given that the probability of $\hat{d}_m$ being the largest element in $\hat{D}$ is arbitrarily close to unity, and that $p_m(t) \rightarrow 1$ w.p. 1, ε-optimality is proven based on the axiom of total probability. All of these are listed below.

1. The first step of the CPA's proof of convergence is formalized by Theorem 2.

    **Theorem 2.** *For any given constants* $\hat{\delta} > 0$ *and* $M < \infty$, *there exist a* $\lambda^* > 0$ *and a* $t_0 < \infty$ *such that under the CPA algorithm, for all positive* $\lambda < \lambda^*$,

    $Pr\{All\ actions\ are\ selected\ at\ least\ M\ times\ each\ before\ t_0\} > 1 - \hat{\delta}$, *for all* $t > t_0$.
    The detailed proof for this result can be found in [14].
2. The sequence of probabilities, $\{p_m(t)_{(t>t_0)}\}$, is stated to be *monotonically* increasing. The previous proofs attempted to do this by showing that:

    $$|p_m(t)| \leq 1, \text{ and}$$

    $$\Delta p_m(t) = E[p_m(t+1) - p_m(t)|\bar{A}(t_0)] = d_m\lambda(1 - p_m(t)) \geq 0, t > t_0, \quad (1)$$

    where $\bar{A}(t_0)$ is the condition that after time $t_0$, for any $j \in (1, 2, ..., r)$, $\hat{d}_j$ remains within a small enough neighborhood of $d_j$ so that $\hat{d}_m$ remains the greatest element in $\hat{D}$. If this step of the "proof" was flawless, $p_m(t)$ can be shown to converge to 1 w.p. 1. But, as we shall see, the flaw lies here!

3. Since $p_m(t) \to 1$ w.p. 1, if it can, indeed, be proved that $Pr\{\bar{A}(t_0)\} > 1 - \delta$, by the axiom of total probability, one can then see that:

$$Pr\{p_m(t) > 1 - \varepsilon\} \geq Pr\{p_m(t) \to 1\}Pr\{\bar{A}(t_0)\} > 1 - \delta,$$

and $\varepsilon$-optimality is proved.

According to the sketch of the proof above, the key is to prove $Pr\{\bar{A}(t_0)\} > 1 - \delta$, i.e.,

$$Pr\{\bar{A}(t_0)\} = Pr\{\bigcap_{t>t_0}\{\hat{d}_j(t)_{\forall j} \text{ is within a } \frac{w}{2} \text{ neighborhood of } d_j \text{ at time } t\}\} > 1 - \delta.$$

(2)

In Eq. (2), $w$ is defined as the difference between the two *highest* reward probabilities.

In the proofs reported in the literature, Eq. (2) is considered to be true according to the law of large numbers, i.e., if each $\alpha_j$ has been selected enough number of times, then for $\forall j$,

$Pr\{\hat{d}_j(t) \text{ is within a } \frac{w}{2} \text{ neighborhood of } d_j \text{ at time } t\} > 1 - \delta'$, with $\delta' = 1 - \sqrt[r]{1 - \delta}$,

so that

$$\prod_{j=1,2,...,r} Pr\{\hat{d}_j(t) \text{ is within a } \frac{w}{2} \text{ neighborhood of } d_j \text{ at time } t\} > 1 - \delta.$$

However, there is a flaw in the above argument. In fact, if we define

$$A(t) = \{\hat{d}_j(t)_{\forall j} \text{ is within a } \frac{w}{2} \text{ neighborhood of } d_j \text{ at time } t\},$$

then the result that can be deduced from the law of large numbers when $t > t_0$ is that

$$Pr\{A(t)\} = \prod_{j=1,2,...,r} Pr\{\hat{d}_j(t) \text{ is within a } \frac{w}{2} \text{ neighborhood of } d_j \text{ at time } t\} > 1 - \delta.$$

But, indeed, the condition which Eq. (1) is based on is:

$$\bar{A}(t_0) = \bigcap_{t>t_0} A(t),$$

which means that for every single time instant in the future, i.e., $t > t_0$, $\hat{d}_j(t)_{(\forall j)}$ needs to be within the $\frac{w}{2}$ neighborhood of $d_j$. The flaw in the previous proofs reported in the literature is that they made a mistake by reckoning that $A(t)$ is equivalent to $\bar{A}(t_0)$. This renders the existing proofs for the CPA being $\varepsilon$-optimal, to be incorrect.

The flaw is documented in [16], which further provided a way of correcting the flaw, i.e., by proving $Pr\{\bar{A}(t_0)\} > 1 - \delta$ instead of proving $Pr\{A(t)\} > 1 - \delta$. However, their proof requires a sequence of *decreasing* values of the learning rate $\lambda$. We applaud the authors of [16] for discovering this flaw, and for submitting a more accurate proof for the CPA.

The proof methodology that we use here is quite distinct (and uses completely different techniques) than the proof reported in [16]. We seek an alternate proof because in their proof, the authors of [16] have required the constraint $\bar{A}(t_0)$, which is, indeed, a very strong condition. This, in turn, requires that for the CPA to achieve its $\varepsilon$-optimality, one must rely on an additional assumption that the parameter, $\lambda$, is gradually decreased during the learning process. We would like to remove this. Our new proof also follows a four-step sketch, but is rather based on the convergence theory of submartingales, and on the theory of Regular functions.

## 4    The New Proof for the CPA's ε-Optimality

### 4.1    The Moderation Property of CPA

The property of moderation can be described by Theorem 2, which has been proved in [14]. This implies that under the CPA, by utilizing a sufficiently small value for the learning parameter, $\lambda$, each action will be selected an arbitrarily large number of times.

### 4.2    The Key Condition $\bar{B}(t_0)$ for $\{p_m(t)_{t>t_0}\}$ Being a Submartingale

In our proof strategy, instead of examining the condition for $\{p_m(t)_{t>t_0}\}$ being *monotonically increasing*, we will investigate the condition for $\{p_m(t)_{t>t_0}\}$ being a *submartingale*. The latter is based on the condition, $\bar{B}(t_0)$, defined as follows:

$$q_j(t) = Pr\{|\hat{d}_j(t) - d_j| < \frac{w}{2}\},$$

$$q(t) = Pr\{|\hat{d}_j(t) - d_j| < \frac{w}{2}, \forall j \in (1,2,...,r)\} = \prod_{j=1,2,...,r} q_j(t), \tag{3}$$

$$B(t) = \{q(t) > 1 - \delta\}, \delta \in (0,1),$$

$$\bar{B}(t_0) = \{\bigcap_{t>t_0} \{q(t) > 1 - \delta\}\}. \tag{4}$$

Our goal in this step is to prove the following result, formulated in Theorem 3.

**Theorem 3.** *Given a* $\delta \in (0,1)$, *there exists a time instant* $t_0 < \infty$, *such that* $Pr\{\bar{B}(t_0)\} = 1$. *In other words, for this given* $\delta$, *there exists a* $t_0 < \infty$, *such that* $\forall t > t_0$: $q(t) > 1 - \delta$ *w. p. 1.*

**Sketch of Proof:** The proof of this is quite detailed. It includes the following steps:

1. By setting $\delta' = 1 - \sqrt[r]{1 - \delta}$, we observe that $\forall t > t_0$, if for $\forall j$, $q_j(t) > 1 - \delta'$, then $q(t) = \prod_{j=1,2,...,r} q_j(t) > \prod_{j=1,2,...,r} (1 - \delta') = 1 - \delta$. Therefore, if we define $B_j(t) = \{q_j(t) > 1 - \delta'\}$, and $\bar{B}_j(t_0) = \{\bigcap_{t>t_0} B_j(t)\}$, our task becomes to prove that for $\forall j$,

$$Pr\{\bar{B}_j(t_0)\} = Pr\{\bigcap_{t>t_0} B_j(t)\} = 1.$$

2. By DeMorgan's law, $Pr\{\bar{B}_j(t_0)\} = Pr\{\bigcap_{t \geq t_0} B_j(t)\} = 1 - Pr\{\bigcup_{t \geq t_0} B_j(t)^c\}$, where $c$ denotes the complement operation. We thus need to prove $Pr\{\bigcup_{t \geq t_0} B_j(t)^c\} = 0$.

3. Let $n_j(t)$ denote the number of times $\alpha_j$ has been selected up to time $t$, then

$$Pr\{\bigcup_{t \geq t_0} B_j(t)^c\} \leq \sum_{t \geq t_0} Pr\{B_j(t)^c\}$$

$$= \sum_{t \geq t_0} \left( \sum_{n=0}^{t} Pr\{q_j(t) \leq 1 - \delta' | n_j(t) = n\} \times Pr\{n_j(t) = n\} \right)$$

$$= \sum_{t \geq t_0} \left( \sum_{n=0}^{t} Pr\{Pr\{|\hat{d}_j(t) - d_j| \geq \frac{w}{2}\} \geq \delta' | n_j(t) = n\} Pr\{n_j(t) = n\} \right).$$

By applying the Hoeffding's inequality [18]: $Pr\{|\hat{d}_j(t) - d_j| \geq \frac{w}{2}\} \leq 2e^{-\frac{nw^2}{2}}$, hence,

$$Pr\{\bigcup_{t \geq t_0} B_j(t)^c\} \leq \sum_{t \geq t_0} Pr\{B_j(t)^c\}$$

$$\leq \sum_{t \geq t_0} \left( \sum_{n=0}^{t} Pr\{2e^{-\frac{nw^2}{2}} \geq \delta'\} \times Pr\{n_j(t) = n\} \right)$$

$$= \sum_{t \geq t_0} \left( \sum_{n=0}^{t} Q_e \times Q_{n_j} \right), \qquad (5)$$

where $Q_e = Pr\{2e^{-\frac{nw^2}{2}} \geq \delta'\}$, and $Q_{n_j} = Pr\{n_j(t) = n\}$.

4. It is easy to conclude that

$$Q_e = Pr\{2e^{-\frac{nw^2}{2}} \geq \delta'\} = \begin{cases} 1, & \text{when } n \leq \frac{-2\ln\frac{\delta'}{2}}{w^2}, \\ 0, & \text{when } n > \frac{-2\ln\frac{\delta'}{2}}{w^2}. \end{cases} \qquad (6)$$

Besides, the quantity $Q_{n_j}$ is the probability of $\alpha_j$ being selected for $n$ times within the given time instant $t$. As $n$ could be any integer from $[0, t]$, we have $\sum_{n=0}^{t} Q_{n_j} = 1$.

If we further assume that till the time instant $t$, $\alpha_j$ has been selected *at least* $x_0$ times, i.e. $n_j(t) \geq x_0$, then

$$Q_{n_j} \begin{cases} = 0, & \text{when } 0 \leq n < x_0, \\ \in [0, 1], & \text{when } x_0 \leq n \leq t, \end{cases} \quad \text{and} \quad \sum_{n=x_0}^{t} Q_{n_j} = 1. \qquad (7)$$

5. From Eq. (6) and (7), we see that if $x_0 > \lceil \frac{-2\ln\frac{\delta'}{2}}{w^2} \rceil$, then $\sum_{n=0}^{t} Q_e \times Q_{n_j} = 0$, whence

$$Pr\{\bigcup_{t \geq t_0} B_j(t)^c\} \leq \sum_{t \geq t_0} Pr\{B_j(t)^c\} \leq \sum_{t \geq t_0} \left( \sum_{n=0}^{t} Q_e \times Q_{n_j} \right) = 0.$$

Obviously, the above arguments apply to $\forall j \in (1, 2, ..., r)$. We thus proved that $\forall j$, $Pr\{\bar{B}_j(t_0)\} = Pr\{\bigcap_{t > t_0} B_j(t)\} = 1$, which leads to the result that $Pr\{\bar{B}(t_0)\} = 1$.

## 4.3 $\{p_m(t)_{t > t_0}\}$ Is a Submartingale under the CPA

We now prove the submartingale properties of $\{p_m(t)_{t > t_0}\}$ for the CPA.

**Theorem 4.** *Under the CPA, the quantity* $\{p_m(t)_{t > t_0}\}$ *is a submartingale.*

**Sketch of Proof:** Firstly, since $p_m(t)$ is a probability, we have $E[p_m(t)] \leq 1 < \infty$. Secondly, we proceed to explicitly calculate $E[p_m(t)]$. Using the CPA's updating rule:

$$E[p_m(t+1)|P(t)] = p_m(d_m(q[(1-\lambda)p_m + \lambda] + (1-q)[(1-\lambda)p_m]) + (1-d_m)p_m) +$$

$$\sum_{j \neq m} p_j(d_j(q[(1-\lambda)p_m + \lambda] + (1-q)[(1-\lambda)p_m]) + (1-d_j)p_m)$$

$$= p_m + \lambda(q - p_m) \sum_{j=1...r} p_j d_j,$$

where $p_m(t)$ and $q(t)$ are concisely written as $p_m$ and $q$ respectively. Then,

$$Diff_{p_m(t)} = E[p_m(t+1)|P(t)] - p_m(t) = \lambda(q(t) - p_m(t)) \sum_{j=1...r} p_j(t)d_j.$$

Invoking the definition of a submartingale, we know that if for all $t > t_0$, we have $Diff_{p_m(t)} > 0$, i.e., $q(t) - p_m(t) > 0$, then $\{p_m(t)_{t>t_0}\}$ is a submartingale. We now invoke the terminating condition for the CPA, in which we consider the learning process to have converged[1] if $p_j(t) > T = 1 - \varepsilon, (j = 1, 2, ..., r)$. Therefore, if we set the quantity $(1 - \delta)$ defined in Theorem 3 to be greater than the threshold $T$, then as per Theorem 3, there exists a time instant $t_0 < \infty$, such that for every single time instant subsequent to $t > t_0$, $q(t) > (1 - \delta) > T > p_m(t)$, which, in turn, guarantees that $\{p_m(t)_{t>t_0}\}$ is a submartingale.

## 4.4  $Pr\{p_m(\infty) = 1\} \to 1$ under the CPA

We can now finally prove the ε-optimality of the CPA.

**Theorem 5.** *The CPA is ε-optimal in all stationary random Environments. More formally, let $T = 1 - \varepsilon$ be a value arbitrarily close to 1, with $\varepsilon$ being arbitrarily small. Then, given any $\delta$ satisfying $(1 - \delta) > T$, there exists a positive integer $\lambda^* < 1$ and a time instant $t_0 < \infty$, such that for all learning parameters $\lambda < \lambda^*$ and for all $t > t_0$, $q(t) > 1 - \delta$, $Pr\{p_m(\infty) = 1\} \to 1$.*

**Sketch of Proof:** According to the submartingale convergence theory [17], $p_m(\infty) = 0$ or 1. If we denote $e_j$ as the unit vector with the $j^{th}$ element being 1, then our task is to prove the convergence probability

$$\Gamma_m(P) = Pr\{p_m(\infty) = 1|P(0) = P\} = Pr\{p(\infty) = e_m|P(0) = P\} \to 1. \qquad (8)$$

To prove Eq. (8), we shall use the theory of Regular functions, and arguments analogous to those used in [17] for the convergence proofs of Absolutely Expedient schemes.

According to theory of Regular functions, $\Gamma_m(P)$ can be bounded from below by a subregular function of $P$, denoted as $\Phi(P)$, if $\Phi(P)$ meets the boundary conditions:

$$\Phi(e_m) = 1 \text{ and } \Phi(e_j) = 0, (\text{for } j \neq m). \qquad (9)$$

Our task is thus to find such a subregular function of $P$ to investigate $\Gamma_m(P)$ indirectly. If we define a function $\Phi_m(P)$ as $\Phi_m(P) = e^{-x_m P_m}$, where $x_m$ is a positive constant, and then define an operator $U$ as:

$$U\Phi_m(P) = E[\Phi_m(P(n+1))|P(n) = P],$$

then, under the CPA,

$$U(\Phi_m(P)) - \Phi_m(P) = E[\Phi_m(P(n+1))|P(n) = P] - \Phi_m(P)$$

$$= E[e^{-x_m P_m(n+1)}|P(n) = P] - e^{-x_m P_m}$$

$$= \sum_{j=1...r} p_j d_j e^{-x_m P_m} \left( q e^{-x_m(1-P_m)\lambda} + (1-q)e^{x_m P_m \lambda} - 1 \right).$$

---

[1] In practice, $T$ is the threshold used to determine when we say that the LA has been "absorbed" into one of the absorbing barriers. This quantity is arbitrarily close to unity, say, 0.999.

Omitting the algebraic manipulation, we get the result that if

$$0 < x_m \leq \frac{2(q(1-p_m)+p_m(1-q))}{\lambda(q-2qp_m+p_m^2)}, \tag{10}$$

then

$$U(\Phi_m(P)) - \Phi_m(P) \leq 0,$$

which, according to the definition of (sub/super)regular functions, indicates that $\Phi_m(P)$ is superregular. Moreover, if we denote:

$$x_{m_0} = \frac{2(q(1-p_m)+p_m(1-q))}{\lambda(q-2qp_m+p_m^2)},$$

we have $x_{m_0} > 0$, implying that when $\lambda \to 0$, $x_{m_0} \to \infty$.

We now introduce another function

$$\phi_m(P) = \frac{1-e^{-x_m p_m}}{1-e^{-x_m}},$$

where $x_m$ is the same as defined in $\Phi_m(P)$. According to the definition of (sub/super)regular functions in [17], the $x_m$, as defined in Eq. (10), renders $\Phi_m(P)$ to be superregular, also makes the $\phi_m(P)$ be subregular.

Moreover, $\phi_m(P)$ meets the boundary conditions of Eq. (9), and therefore, according to the theory of regular functions [17], we have

$$\Gamma_m(P) \geq \phi_m(P) = \frac{1-e^{-x_m p_m}}{1-e^{-x_m}}. \tag{11}$$

As Eq. (11) holds for every $x_m$ bounded by Eq. (10), we take the greatest value $x_{m_0}$. Moreover, as $\lambda \to 0$, $x_{m_0} \to \infty$, whence $\Gamma_m(P) \to 1$. We have thus proved that $Pr\{p_m(\infty) = 1\} \to 1$, showing that the CPA is $\varepsilon$-optimal.

More detailed discussions about this proof and its implications are found in [19].

## 5 Conclusions

Estimator algorithms are acclaimed to be the fastest Learning Automata (LA), and within this family, the set of *Pursuit* algorithms have been considered to be the pioneering schemes. The $\varepsilon$-optimality of Pursuit algorithms are of great importance and has been studied for years. The proofs in almost all the existing papers have a common flaw which was discovered by the authors of [16], whom we applaud for this.

This paper aims at correcting the flaw by providing a new proof. Rather than examining the monotonicity property of the $\{p_m(t)_{(t>t_0)}\}$ sequence as done in the previous papers and in [16], our current proof studies the *submartingale* property of $\{p_m(t)_{(t>t_0)}\}$. Thereafter, by virtue of the submartingale property and the consequent weaker convergence condition, the new proof invokes the theory of Regular functions, and does not require the learning parameter to decrease gradually.

Further, as opposed to the proof found in [16], we believe that our proof can be easily extended to formally demonstrate the $\varepsilon$-optimality of other Estimator Algorithms, without the requirement of continuously changing the scheme's learning parameter.

# References

1. Oommen, B.J., Granmo, O.C., Pedersen, A.: Using stochastic AI techniques to achieve un-bounded resolution in finite player goore games and its applications. In: IEEE Symposium on Computational Intelligence and Games, Honolulu, HI (2007)
2. Beigy, H., Meybodi, M.R.: Adaptation of parameters of bp algorithm using learning automata. In: Sixth Brazilian Symposium on Neural Networks, JR, Brazil (2000)
3. Granmo, O.C., Oommen, B.J., Myrer, S.A., Olsen, M.G.: Learning automata-based solutions to the nonlinear fractional knapsack problem with applications to optimal resource allocation. IEEE Transactions on Systems, Man, and Cybernetics, Part B 37(1), 166–175 (2007)
4. Unsal, C., Kachroo, P., Bay, J.S.: Multiple stochastic learning automata for vehicle path control in an automated highway system. IEEE Transactions on Systems, Man, and Cybernetics, Part A 29, 120–128 (1999)
5. Oommen, B.J., Roberts, T.D.: Continuous learning automata solutions to the capacity assignment problem. IEEE Transactions on Computers 49, 608–620 (2000)
6. Granmo, O.C.: Solving stochastic nonlinear resource allocation problems using a hierarchy of twofold resource allocation automata. IEEE Transactions Computers 59(4), 545–560 (2010)
7. Oommen, B.J., Croix, T.D.S.: String taxonomy using learning automata. IEEE Transactions on Systems, Man, and Cybernetics 27, 354–365 (1997)
8. Oommen, B.J., Croix, T.D.S.: Graph partitioning using learning automata. IEEE Transactions on Computers 45, 195–208 (1996)
9. Dean, T., Angluin, D., Basye, K., Engelson, S., Aelbling, L., Maron, O.: Inferring finite automata with stochastic output functions and an application to map learning. Maching Learning 18, 81–108 (1995)
10. Thathachar, M.A.L., Sastry, P.S.: Estimator algorithms for learning automata. In: The Platinum Jubilee Conference on Systems and Signal Processing, Bangalore, India, pp. 29–32 (1986)
11. Oommen, B.J., Lanctot, J.K.: Discretized pursuit learning automata. IEEE Transactions on Systems, Man, and Cybernetics 20, 931–938 (1990)
12. Lanctot, J.K., Oommen, B.J.: On discretizing estimator-based learning algorithms. IEEE Trans. on Systems, Man, and Cybernetics, Part B: Cybernetics 2, 1417–1422 (1991)
13. Lanctot, J.K., Oommen, B.J.: Discretized estimator learning automata. IEEE Trans. on Systems, Man, and Cybernetics, Part B: Cybernetics 22(6), 1473–1483 (1992)
14. Rajaraman, K., Sastry, P.S.: Finite time analysis of the pursuit algorithm for learning automata. IEEE Transactions on Systems, Man, and Cybernetics, Part B: Cybernetics 26, 590–598 (1996)
15. Oommen, B.J., Agache, M.: Continuous and discretized pursuit learning schemes: various algorithms and their comparison. IEEE Transactions on Systems, Man, and Cybernetics, Part B: Cybernetics 31(3), 277–287 (2001)
16. Ryan, M., Omkar, T.: On ε-optimality of the pursuit learning algorithm. Journal of Applied Probability 49(3), 795–805 (2012)
17. Narendra, K.S., Thathachar, M.A.L.: Learning Automat: An Introduction. Prentice Hall (1989)
18. Hoeffding, W.: Probability inequalities for sums of bounded random variables. Journal of the American Statistical Association 58, 13–30 (1963)
19. Zhang, X., Granmo, O.C., Oommen, B.J., Jiao, L.: A Formal Proof of the ε-Optimality of Continuous Pursuit Algorithms Using the Theory of Regular Functions. The Unabridged Version of this Paper (Submitted for Publication. It can be made available to the Referees if needed)

# Online Exploratory Behavior Acquisition of Mobile Robot Based on Reinforcement Learning

Manabu Gouko[1], Yuichi Kobayashi[2], and Chyon Hae Kim[3]

[1] Department of Mechanical Engineering and Intelligent Systems,
Faculty of Engineering, Tohoku Gakuin University, 1-13-1, Chuo, Tagajo-shi,
Miyagi, 985-8537, Japan
gouko@tjcc.tohoku-gakuin.ac.jp
[2] Department of Mechanical Engineering, Shizuoka University, 3-5-1, Johoku,
Naka-ku, Hamamatsu, Shizuoka, 432-8561, Japan
tykobay@ipc.shizuoka.ac.jp
[3] The Honda Research Institute Japan, 8-1 Honcho, Wako-shi,
Saitama 351-0188, Japan
tenkai@jp.honda-ri.com

**Abstract.** In this study, we propose an online active perception system that autonomously acquires exploratory behaviors suitable for each embodiment of mobile robots using online learning. We especially focus on a type of exploratory behavior that extracts object features useful for robot's orientation and object operation. The proposed system is composed of a classification system and a reinforcement learning system. While a robot is interacting with objects, the classification system classifies observed data and calculates reward values according to the cluster distance of the observed data. On the other hand, the reinforcement learning system acquires effective exploratory behaviors useful for the classification according to the reward. We validated the effectiveness of the system in a mobile robot simulation. Three different shaped objects were placed beside the robot one by one. In this learning, the robot learned different behaviors corresponding to each object. The result showed that the behaviors were the exploratory behaviors that distinguish the difference of corner angles of the objects.

**Keywords:** Exploratory behavior, reinforcement learning, mobile robot.

## 1  Introduction

Over the past few decades, many studies have been carried out on general-purpose mobile robots, which provide support for everyday human activities. If the robots have the autonomous recognition capability for general surrounding objects, they can perform a wider variety of tasks, because object recognition capability is essential for a robot to orientate itself, operate objects, and establish superior knowledge. The databases for object category knowledge play an

M. Ali et al. (Eds.): IEA/AIE 2013, LNAI 7906, pp. 272–281, 2013.

important role in the object recognition of mobile robots. In these databases, a set of similar object features is corresponding to a category.

Traditionally, there are two approaches for formulating a category database according to robot motion. The first approach passively uses the robot's sensory information without robot motion. The second one uses the sensory information while the robot is moving. In this case, the robot actively moves to extract the object features. Such movements are known as exploratory behaviors. Active perception is defined as the recognition of categories on the basis of exploratory behaviors. Previous studies suggest that creatures can distinguish objects as a result of their active perception capabilities [1].

In recent years, several studies have been conducted on the realization of active perception by robots. S. Griffith et al. proposed an approach for interactive object categorization, which utilizes multiple exploratory behaviors and their resulting acoustic signatures in order to form object categories [2]. In their demonstration, an upper-torso humanoid robot drops or shakes objects in order to categorize each of them according to their resulting sounds.

In [3], a robot swing objects to make their sounds. The movement information of the objects is used to organize the category [4]. In this study, the humanoid robot pushed many household objects and recorded their movement. Next, the category was determined using the differences in the movement of each object.

However, in these studies, exploratory behaviors (e.g., shaking, dropping, and tapping objects) are designed in advance by the researchers/robot designers. To design exploratory behaviors, it is necessary to know the different kinds of features that can be extracted from target objects. Thus, the robot designers have to consider the objective features and the robot's characteristics (e.g., the shape of the body and the type of sensor) to design exploratory behaviors. On the other hand, this type of consideration is difficult for robot designers, because they are unaware of the exact situations in which the robots will be used.

In this paper, we propose an active perception system that autonomously learns exploratory behaviors. In this system, multiple objects are presented to a robot one by one, and the exploratory behaviors are automatically acquired through interaction between the objects and the robot. Effective feature extraction behaviors are obtained using reinforcement learning (RL). In this system, the category of the presented object is given for the robot while learning its behaviors. Next, the robot learns effective behaviors that discriminate the presented objects. The behaviors lead to differences in the features, which are extracted from the objects in different categories.

Several studies have addressed the learning of exploratory behaviors. S. Nolfi and D. Marocco also used genetic algorithms to acquire behaviors of a robot arm in the simulation [5]. They used a 6 degrees-of-freedom (DoF) robot arm with extremely coarse touch sensors to categorize objects with different shapes. Shibata et al. performed visual sensor motion that identifies multiple patterns [6]. They carried out recognition tasks using partially different patterns. In this study, RL is used to formulate a sensor motion, which discovers differences in these patterns.

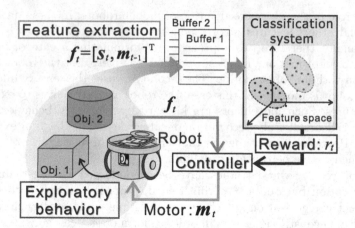

**Fig. 1.** Proposed system. This system is composed of a robot, a controller, buffers, and classification system (feature space).

However, these studies do not address online system, because their system needs to wait for classification results (success or failure information) in order to start the process for the acquisition of exploratory behaviors. On the other hand, our proposed system learns behaviors without waiting for the results using the relative data positions of categories.

In this paper, we show that it is possible to acquire an exploratory behavior by applying the proposed system to a mobile robot simulation. We analyzed the acquired motion patterns, which differed according to the objects. The remainder of this paper is organized as follows: Section 2 describes our proposed system. Section 3 provides experimental results and discussion, while Section 4 presents a summary and outlines future works.

## 2    Proposed System

### 2.1    Outline of the System

In this section, we explain the proposed active perception system. Figure 1 shows the system. The system is composed of a robot, a controller, buffers, and categories database in feature space. The controller generates the robot's exploratory behaviors. The buffers temporarily store features of the object in memory. In the feature space, a category is formed using data from the object features stored in the buffer. Several different objects are repeatedly presented one by one to the robot at constant time intervals as the robot learns its behaviors. All the objects used for the study belong to different categories. The total number of categories is given as $n$ for the robot.

## 2.2 Object Feature

It is assumed that the robot has multiple sensors, and the number of sensors is $J$. The sensor output at time $t$ is defined as $s_t = [s_t^1, s_t^2, \cdots, s_t^J]^T$, where $s_t^j$ is the output of the $j$-th sensor. Also, the motor command is defined as vector $m_t = [m_t^1, m_t^2, \cdots, m_t^K]^T$, where $m_t^k$ is the motor command for the $k$-th motor. At time $t$, the robot's controller outputs a motor command $m_t$ based on $s_t$ and $m_{t-1}$. The robot moves to a given object $i$ and extracts its feature. The object $i$ belongs to category $i$. An extracted feature at time $t$ is defined as a feature vector $f_t$ as follows.

$$f_t = [s_t, m_{t-1}]^T = [s_t^1, s_t^2, \cdots, s_t^J, m_{t-1}^1, m_{t-1}^2, \cdots, m_{t-1}^K]^T \tag{1}$$

## 2.3 Classification System

The system has buffers in which the feature vectors are memorized. Each buffer can store up to $L$ feature vectors. The buffer $i$ indicates the buffer corresponding to category $i$. The feature vectors extracted from the interaction between the robot and an object $i$ are stored in the buffer $i$. The data stored in the buffers are updated sequentially. Before learning, the randomly determined feature vectors are set to each buffer.

In the feature space, each category is formed on the basis of data stored in each buffer. In this study, each form is assumed to be a Gaussian distribution $p_i()$. $p_i(\mu_i, \Sigma_i)$ is the Gaussian distribution corresponding to the category $i$. $\mu_i$ and $\Sigma_i$ are the mean and the variance-covariance matrix of category $i$, respectively. $\mu_i$ and $\Sigma_i$ are calculated by the feature vectors memorized in buffer $i$ when the data in the buffer $i$ change.

## 2.4 Learning of an Exploratory Behavior

RL is applied to the learning of a controller [7]. In the RL framework, a robot learns a suitable state-action mapping without prior knowledge of the dynamics of its environment. The robot obtains a reward depending on its behavior. When a robot repeats a trial and error operation, an appropriate mapping for performing the task is constructed by the controller.

To learn exploratory behaviors, we apply the Actor-Critic learning method, which is a RL algorithm that is able to handle a continuous state and action spaces. This method needs a critic, which estimates a reward expectation from a state. It also needs an actor as a controller. An actor outputs a motor command on the basis of the state. In this study, a state in time $t$ is defined as vector $[s_t, m_{t-1}]^T$. Both of them are learned simultaneously. These two units are made by feed forward neural networks.

This study focuses on the acquisition of an exploratory behavior that fulfills the following two conditions:

1. Similar features should always be acquired from an object belonging to the same category.
2. The difference in the features obtained from the objects belonging to different categories should be large.

To establish such an exploratory behavior, we define rewards on the basis of the within-class variance $\sigma_W^2$ and between-class variance $\sigma_B^2$ of categories.

$$\sigma_W^2 = \frac{1}{nL} \sum_{i=1}^{n} \sum_{l=1}^{L} (d_i^l - \mu_i)^T (d_i^l - \mu_i) \tag{2}$$

$$\sigma_B^2 = \frac{1}{n} \sum_{i=1}^{n} (\mu_i - \mu_{all})^T (\mu_i - \mu_{all}) \tag{3}$$

Here, $\mu_{all}$ is an average of all the feature vectors that are stored in each buffer, $d_i^l$ is the $l$-th feature vector memorized in a buffer $i$, $\sigma_W^2$ represents the average spread of a category, and $\sigma_B^2$ represents the average spread between categories. On the basis of these variances, the reward at time $t$ is defined as follows:

$$r_t = \sigma_B^2 / \sigma_W^2 \tag{4}$$

This reward becomes large when $\sigma_B^2$ is large and $\sigma_W^2$ is small. By using this kind of reward, the acquisition of an exploratory behavior, which satisfies the conditions described above, can be expected.

## 3   Experimental Results and Discussions

### 3.1   Simulation Setup

We applied the proposed system to the mobile robot simulation, after which we verified behaviors that were acquired by the robot. The exploratory behavior learning was executed using three different shaped objects.

The exploratory behavior learning is carried out as follows. First, the robot generates actions on the given object and observes the object features as vectors. Next, the extracted features are temporarily stored in the buffer. In the feature space, the form of the category is updated sequentially with stored features. A reward that is used for RL of the controller is calculated on the basis of the forms of the categories and the relative positions of each category. Finally, the controller is updated on the basis of the reward. By repeating such learning, the behaviors, which increase the difference in features extracted from the different objects, is learned.

This simulation was performed using the software Webots 6. We used a cylindrical robot. The left part of this figure 2 shows the top view of the robot. It has eight distance sensors and two wheels; all the sensors have the same specifications. The detection range of the sensors was 40 mm. The actual range of the distance sensor data was 0-4000, where 4000 indicates that the robot was very close to an object and 0 indicates that there was no object within the sensor's detection range.

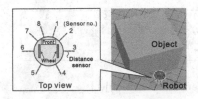 

**Fig. 2.** The left part of this figure shows the mobile robot and an object. The right part of this figure shows three objects were used in the simulation.

$s_t = [s_t^1, s_t^2, \cdots, s_t^8]^T$ is an 8-dimensional vector that contains elements corresponding to each distance sensor. $m_t = [m_L, m_R]^T$ is a 2-dimensional vector that contains elements corresponding to the motor command of the left and right wheels as $m_L$ and $m_R$, respectively. The capacity $L$ of each buffer is set to 100.

The actor and the critic are constructed by three-layer feed forward neural networks (input, hidden and output layer). In the actor network, there are 10 neurons in input layer, 20 neurons in hidden layer, 2 neurons in output layer. In the critic network, there are 10 neurons in input layer, 20 neurons in hidden layer, 1 neurons in output layer.

The right part of this figure 2 shows the objects used in this simulation. We used three different shaped objects. The square pole (object 1), triangle pole (object 2), and cylinder (object 3) are presented to the robot in series while the robot learns its behaviors. The square pole, triangle pole, and cylinder belong to categories 1, 2, and 3, respectively.

When the learning begins, the robot is placed near the object. This position is called the start position, and is unique for all objects. The triangle pole is the only object with two start positions, which are used alternately. The robot learns the exploratory behavior according to the following procedures.

1. The robot is put at the start position of object $i$.
2. On the basis of the output of sensor $s_t$, the feature vector is modified $f_t = [s_t, m_{t-1}]^T$. When the time $t$ is 0, $m_{t-1}$ is set to $\mathbf{0}$.
3. The feature vector $f_t$ is stored to the buffer $i$, and the mean $\mu_i$ and distribution $\Sigma_i$ are calculated using all data stored in the buffer $i$.
4. A reward $r_t$ is calculated by equation (4). The learning of neural networks corresponding to the critic and the actor is executed.
5. The robot moves according to the command $m_t$ which is calculated at the controller.

The procedures above were executed for every time step. The time step of this simulation $t$ is 320 ms. During the learning, the robot interacts continuously with the same object for 100 steps. This interaction between the robot and the same object is called one trial. After each trial, the object is changed. Next, the robot is placed at the start position and the next trial begins. In this simulation, the learning was performed repeatedly for 9000 trials. For each object, the robot performed the interaction in 3000 trials. If the robot does not detect the object using its sensors, the trial is stopped. A new object is then given to the robot and a new trial begins.

**Fig. 3.** Typical behaviors were observed in each object

## 3.2 Acquired Behaviors

To confirm the kinds of behaviors that were obtained, we presented the objects to the robot that had finished its learning. Figure 3 shows the typical behaviors that were observed for each object. First, the robot remained in the start position. Next, it turned toward the object and went straight to it. The robot slightly touched the object with a left-leaning posture (figure 3, S-1, T-1, C-1). Such contact behavior was observed for all three objects.

The robot continued to go straight along the side of each object. At the cylinder, the posture of the robot changed from C-2 to C-3 after coming into contact with the object. Next, the robot continued to maintain the posture of C-3. At the square pole and the triangle pole, the robot forced its frontal side onto the objects (S-1, T-1). Although the robot went in a straight direction, its body slid leftward (S-2, T-2) because the robot had touched the sides of these objects with a left-leaning posture. The robot reached the corners of these objects by sliding. The robot turned at the corner of the square pole, making contact with the front of the object (S-3). On the other hand, the robot rotated at the corner of the triangle pole (T-3).

## 3.3 Pushing Behavior

For all objects, the robot forced its front on their sides. To determine the kind of feature that is extracted by this pushing behavior, the object used in the simulation can be classified in two ways according to the side. The side of the cylinder has a curved surface, and the square pole and triangle pole have a plane side. Using forceful behavior, the robot makes contact with the front of its body on the objects, and a relatively large number of distance sensors exist there. This pushing behavior means that the robot extracts different features to distinguish between the cylinder and other shapes.

Another meaning of this behavior is found by moving the robot to the corner. It is impossible to distinguish between the square pole and the triangle pole by only pushing the sensors along the side of the object. In order to differentiate between them, the robot has to move to the corners, which show the differences in the form of each object. The robot comes in contact with the object with a left-leaning posture. The slide to the left occurs while making both wheels roll forward at uniform velocity when the robot is in such a posture. This slide makes the robot move to the corner.

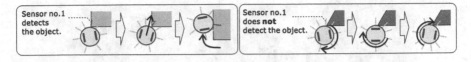

**Fig. 4.** The different behaviors were observed at the square pole and triangle pole

## 3.4 Behaviors at Corner

In order to distinguish between the square pole and triangle pole, the robot should only determine whether the angle at the corner is 90 degrees. Different behaviors were observed at the square pole and triangle pole. The robot turned at the corner of the square pole, making contact with its own front on the object (Figure 3, S-3). On the other hand, the robot rotated at the corner of 30 and 60 degrees of the triangle pole (T-3).

To determine the reason for the generation of these behaviors, we investigated the learned neural network used as the controller, and found that there were two main actions performed by the robot: forward motion and clockwise rotation. These actions are alternated by information on sensor no. 1 only.

Figure 4 shows the robot's complete behaviors observed at the corners. It is important to determine whether or not sensor no.1 has detected the object. The left part of this figure indicates the behavior of the robot at the corner of the square pole. At the corner of the 90 degree angle, sensor no. 1 continually detects the object. The robot continues to rotate with both wheels in a forward direction at the square pole. Next, the robot continues to slide along the side of the square pole and passes the corner.

The right part of this figure indicates the behavior of the robot at the corner of the triangle pole. At the corners of the angles of 60 and 30 degrees, sensor no. 1 temporarily failed to detect the object (the right part of Figure 4). At that time, the output of sensor no. 1 became small. Next, the robot rotates in a clockwise direction by rotating its right wheel in a backward direction. This rotation is repeatedly observed at the corner. The different behaviors observed at the corners were generated by this mechanism.

The robot generates two simple actions, forward motion, and a clockwise rotation. The different behaviors resulted from the interaction between the robot and the object using the shape of the robot's body, the arrangement of the distance sensors and the shape of objects. The robot learned these actions, which are difficult for the researcher to intuitively design beforehand.

## 3.5 Extracted Features by the Acquired Behaviors

In this experiment, we confirmed the kind of feature that is extracted from an object using learned exploratory behaviors. The robot which had finished learning and ten objects were used to carry out the experiment. These objects are shown in Figure 5. The four quadratic prisms (Q-1~Q-4), the three triangular prisms (T-1~T-3), and the three cylinders were used for this experiment. Q1,

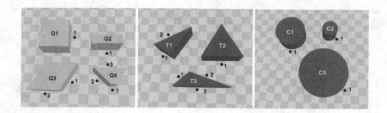

**Fig. 5.** Ten objects were used to verify the learned exploratory behaviors

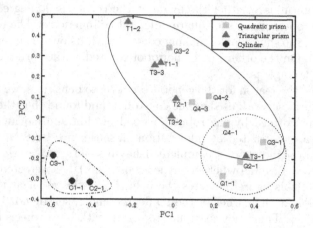

**Fig. 6.** The space is composed by the 1st (PC1) and 2nd (PC2) principal components

T1 and C1 are the same objects as those used for learning. In Figure 5, the small black dots indicate the start position of the robot. There are one or two start positions for each object.

All of these objects are given to the robot one by one. The robot was set at its start point. Next, it generates the exploratory behaviors to the object based on the learned controller, and acquires feature vectors during interaction for 100 steps. For example, the feature vectors, which are acquired by the object Q3 and the robot moving from start point no.1 are labeled "Q3-1". The mean vector $\mu$ is individually calculated for every object. After that, the principal component method was applied to these mean vectors.

Figure 6 shows the space, which is composed using 1st (PC1) and 2nd (PC2) principal components. The horizontal axis (PC1) appears to correspond to the turning of the robot at the object's corner. The vertical axis (PC2) appears to correspond to the rotation of the robot at object's corner.

The data surrounded by a one-point chain line were obtained from the interactions between the robot and the cylinders. The data surrounded by the solid line were acquired by the robot, which rotated at a corner at least one instance during the interaction. On the other hand, the data surrounded by the dotted line were acquired by the robot which turned at a corner at least one time. The data "T3-1," "Q3-1," and "Q4-1" were obtained by the robot which both turned and rotated at the object's corners. It is interesting to note that these data are located between the data surrounded by the solid line and the dotted line. The result shows that the robot learned the behaviors for distinguishing the difference in the angle of the object's corner.

# 4  Conclusions and Future Works

In this paper, we proposed an active perception system that autonomously learns exploratory behaviors. In our simulation with a mobile robot, the behaviors, which effectively extract the features of the objects, were obtained. Additionally, we confirmed the kind of feature that is extracted from an object using learned exploratory behaviors.

In future, we will analyze the form of the categories acquired by learning. Next, we plan to apply the proposed system to object recognition tasks. We will also apply the exploratory behaviors obtained by the simulation to an actual mobile robot, and will use it in an experiment to identify different objects.

**Acknowledgments.** This research was partially supported by the Ministry of Education, Science, Sports and Culture, Grant-in-Aid for Young Scientists (B), 24700196, 2012.

# References

1. Turvey, M.T.: Dynamic touch. American Psychologist 51(11), 1134–1152 (1996)
2. Griffith, S., Sinapov, J., Sukhoy, V., Stoytchev, A.: How to separate containers from non-containers? A behavior-grounded approach to acoustic object categorization. In: Proceedings of 2010 IEEE International Conference on Robotics and Automation, pp. 1852–1859 (2010)
3. Takamuku, S., Hosoda, K., Asada, M.: Object category acquisition by dynamic touch. Advanced Robotics 22, 1143–1154 (2008)
4. Nishide, S., Ogata, T., Tani, J., Komatani, K., Okuno, H.G.: Predicting object dynamics from visual images through active sensing experiences. Advanced Robotics 22, 527–546 (2008)
5. Nolfi, S., Marocco, D.: Active perception: A sensorimotor account of object categorization. In: Proceedings of the Seventh International Conference on Simulation of Adaptive Behavior on From Animals to Animats, pp. 266–271 (2002)
6. Shibata, K., Nishino, T., Okabe, Y.: Actor-q based active perception learning system. In: Proceedings of International Conference on Robotics and Automation 2001 (ICRA 2001), pp. 1000–1005 (2001)
7. Sutton, S.R., Barto, G.A.: Reinforcement Learning: An Introduction. The MIT Press (1998)

# Improved Sound Source Localization and Front-Back Disambiguation for Humanoid Robots with Two Ears

Ui-Hyun Kim[1], Kazuhiro Nakadai[2], and Hiroshi G. Okuno[1]

[1] Dept. of Intelligence Science and Technology, Graduate School of Informatics,
Kyoto University, Kyoto-shi, Japan
{euihyun,okuno}@kuis.kyoto-u.ac.jp
[2] Honda Research Institute Japan Co., Ltd., Wako-shi, Japan
nakadai@jp.honda-ri.com

**Abstract.** An improved sound source localization (SSL) method has been developed that is based on the generalized cross-correlation (GCC) method weighted by the phase transform (PHAT) for use with humanoid robots equipped with two microphones inside artificial pinnae. The conventional SSL method based on the GCC-PHAT method has two main problems when used on a humanoid robot platform: 1) diffraction of sound waves with multipath interference caused by the shape of the robot head and 2) front-back ambiguity. The diffraction problem was overcome by incorporating a new time delay factor into the GCC-PHAT method under the assumption of a spherical robot head. The ambiguity problem was overcome by utilizing the amplification effect of the pinnae for localization over the entire azimuth. Experiments conducted using a humanoid robot showed that localization errors were reduced by 9.9° on average with the improved method and that the success rate for front-back disambiguation was 32.2% better on average over the entire azimuth than with a conventional HRTF-based method.

**Keywords:** Intelligent robot audition, human-robot interaction, sound source localization, front-back disambiguation.

## 1 Introduction

Effective sound source localization (SSL) is a key to achieving more natural human–humanoid robot interaction. A humanoid robot must be able to localize sound sources to understand the acoustic scene. This enables it to face the person speaking and signal to him/her that it is ready to listen, thereby appearing to express an interest in conversing with the person. A common approach to implementing SSL is to equip the robot with many microphones [1]. However, this causes several problems, including higher maintenance costs, the use of more computational power, and degradation of the human-like appearance. Humans are binaural, which means they have two sound inputs, i.e., two ears. For a robot to appear humanoid, it should also have two sound inputs, i.e., two microphones inside two artificial pinnae, one on each side of its head like human ears. The development of a binaural robot audition method is thus important for achieving humanoid robots that are perceived to be like human beings. Moreover, such a method should not require excessive computing power.

M. Ali et al. (Eds.): IEA/AIE 2013, LNAI 7906, pp. 282–291, 2013.

Extensive studies of SSL by a number of researchers have revealed perceptual clues. They include the interaural level difference (ILD), the interaural time difference (ITD), and the spectral distortion caused by various parts of the body (the pinnae, head, shoulders, etc.). These clues are implicitly included in the head related transfer function (HRTF) [2]. The ITD, more commonly referred to as the time delay of arrival (TDOA), plays an important role in SSL; the sound signals arrive at each microphone at different times for directions other than front and back. One of the most widely used SSL methods based on the TDOA between binaural inputs is the generalized cross-correlation (GCC) method with phase transform (PHAT) weighting [3].

The use of a microphone array with many microphones has improved SSL performance on various robot platforms in actual environments. A reduction in the number of microphones generally degrades SSL performance. Since a binaural robot audition system uses only two microphones (one embedded on each side of the robot head), there are difficulties in obtaining performance as good as that with a microphone array. In the study reported here, we addressed the two main problems with SSL based on the GCC-PHAT method in binaural robot audition:

1. **Diffraction of sound waves with multipath interference caused by contours of robot head:** sound waves easily bend around the robot head, resulting in a difference in TDOA between the waves that travel around the front of the head and those that travel around the back of the head.
2. **Front-back ambiguity due to the same TDOA for the front and back:** binaural audition methods localize a sound source as coming from the front despite the actual sound source being in the rear.

The diffraction of sound waves with multipath interference degrades localization performance of binaural robot audition, especially for sound sources in the lateral direction (around ±90°). Most SSL methods are based on the assumption that a microphone array is located in a free space environment; i.e., they do not take into account the diffraction of sound waves with multipath interference in non-free space environments like robot platforms. Front-back ambiguity limits the localization range to the front horizontal space, from −90° to +90°. Current methods for solving the ambiguity problem involve using head movement [4–5] or using a specific HRTF database [6–7]. However, these methods have certain drawbacks. Using head movement does not work well for short words or phrases, such as when someone calls the robot's name, because the robot needs enough time to complete its head movement. Using an HRTF database does not work well if the system and environment change because its performance depends greatly on the system and environment.

We took different approaches to solving these two problems:

1. For the diffraction of sound waves with multipath interference, we assume that the robot head is spherical and incorporate a new time delay factor into the GCC-PHAT method to compensate for the diffraction.
2. For the front-back disambiguation, we utilize the pinna amplification effect, which creates a level difference between sound signals coming from the front and back.

These solutions were implemented and experimentally evaluated in the binaural audition system of our SIG-2 humanoid robot.

The paper is organized as follows: Section 2 summarizes SSL based on the GCC-PHAT method and describes the two main problems in binaural robot audition. Section 3 describes our solutions to the two problems. Section 4 describes our evaluation experiments and presents the results. Section 5 concludes the paper with a summary of the key ideas.

# 2       Sound Source Localization

In this section, we summarize the GCC-PHAT-based SSL method and describe the two main problems in binaural robot audition.

## 2.1       Acoustic Model

The signals received by the left and right microphones can be mathematically modeled as

$$X_l[f] = \alpha_l[f]|S[f]|\exp\left(-j2\pi\frac{f}{F}fs\tau_l\right) + N[f]$$

$$X_r[f] = \alpha_r[f]|S[f]|\exp\left(-j2\pi\frac{f}{F}fs\tau_r\right) + N[f],$$

(1)

where $X_{l/r}[f]$, $S[f]$, and $N[f]$ are the signals received by each of the two microphones (l and r), the sound source, and the uncorrelated additive noise, respectively, in frequency domain representation; $f \in \{1, ..., F\}$ denotes a frequency bin, $F$ is the width of the Fourier transform, and $fs$ is the sampling frequency; $\alpha_{l/r}$ and $\tau_{l/r}$ are the attenuation factor and time delay from the position of the sound source to each microphone.

The TDOA between the two microphones is defined by the relationship in (1) with microphone $l$ defined as a reference under the far-field assumption [8]:

$$\tau_{lr}(\theta) = \tau_r(\theta) - \tau_l(\theta) = \frac{d_{lr}}{v}\sin\left(\frac{\theta}{180}\pi\right),$$

(2)

where $\tau_{lr}$ denotes the difference between time delays $\tau_l$ and $\tau_r$, $d_{lr}$ is the distance between the two microphones, $\theta \in \{-90°, ..., +90°\}$ is the direction of the sound incidence of interest, and $v$ is the speed of sound (340.5 m/s at 15°C in air).

## 2.2       Generalized Cross-Correlation Method Weighted by Phase Transform

The direction of sound incidence in SSL with unknown parameters $\tau_l$ and $\tau_r$ is obtained using the maximum-likelihood-based GCC-PHAT method in the frequency domain [9], which is defined as

$$\hat{\theta}_{mle} = \arg\max_\theta \frac{1}{F}\sum_{f=1}^{F} G^{PHAT} X_l[f]X_r^*[f]\exp\left(j2\pi\frac{f}{F}fs\tau_{lr}(\theta)\right),$$

(3)

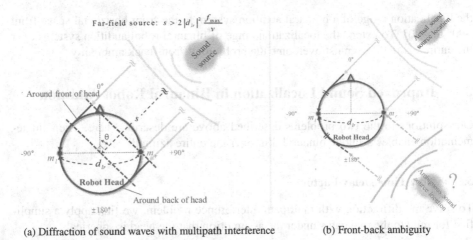

(a) Diffraction of sound waves with multipath interference     (b) Front-back ambiguity

**Fig. 1.** Two main problems of SSL in binaural robot audition

where

$$G^{PHAT} = \frac{1}{\left| X_l[f] X_r^*[f] \right|},\tag{4}$$

$G^{PHAT}$ is a normalization factor that preserves only the phase information, and $*$ represents the complex conjugate. The estimated sound incidence direction $\theta_{mle}$ can be obtained by finding the $\theta$ that maximizes the sum of the cross-power spectrum of two signal inputs with time delay factor $\tau_{lr}$ in (2) and PHAT weighting in (4).

### 2.3    Two Problems in Binaural Robot Audition

**Diffraction of Sound Waves with Multipath Interference.** The TDOAs are esti-mated under the assumption that the microphones are located in free space. However, this assumption is not applicable to TDOA estimation using two microphones in a robot head because the sound waves easily bend and spread along the contours of the robot head, which creates sound diffraction and a difference in TDOA between the waves that travel around the front of the head and those that travel around the back of the head. Figure 1-(a) illustrates the two paths created by the diffraction of the sound waves with the assumption that the robot head is spherical. It clearly shows that these two diffracted sound-wave paths and multipath interference must be considered if more accurate SSL in binaural robot audition is to be attained.

**Front-Back Ambiguity.** Normally, the direction of a sound source estimated using a microphone array corresponds to the actual direction. However, a binaural audition system has an inherent problem—a sound source appears to be at equal (mirror) an-gles in the front and rear hemi-fields due to having the same TDOA, as shown in Fig. 1-(b). For example, a sound source placed at 30° (where 0° is directly in front) is es-timated to also be at -150° in front-back ambiguity. This front-back ambiguity limits the localization range of a binaural audition system to the front horizontal space from $-90°$ to $+90°$. To extend the localization range of binaural robot audition systems over the entire azimuth, we must overcome the problem of front-back ambiguity.

## 3     Improved Sound Localization in Binaural Robot Addition

Our solutions to the two problems described above are described here. Their implementation enables accurate binaural SSL over the entire azimuth.

### 3.1     New Time Delay Factor

To solve the diffraction with multipath interference problem, we first apply a simplified formula to the two paths under the assumption that the head is spherical:

$$\tau_{front}(\theta) = \frac{d_{lr}}{2v}\left(\frac{\theta}{180}\pi + \sin\left(\frac{\theta}{180}\pi\right)\right),\tag{5}$$

$$\tau_{back}(\theta) = \frac{d_{lr}}{2v}\left(\operatorname{sgn}(\theta)\pi - \frac{\theta}{180}\pi + \sin\left(\frac{\theta}{180}\pi\right)\right),\tag{6}$$

where $\tau_{front}$ and $\tau_{back}$ are respectively the time delay for the path around the front of the head and that for the one around the back of the head for each sound incidence direction, and $sgn$ is a signum function that extracts the sign of $\theta$; i.e., if $\theta$ has a negative sign, $sgn(\theta)$ is $-1$. After formulas for the two paths are derived, the time difference between them for each sound direction is obtained using

$$\tau_{diff}(\theta) = \tau_{back}(\theta) - \tau_{front}(\theta) = \frac{d_{lr}}{2v}\left(\operatorname{sgn}(\theta)\pi - \frac{2\theta}{180}\pi\right),\tag{7}$$

where $\tau_{diff}$ is 0 when $\theta$ is $-90°$ or $+90°$. Suppose that the intensity of the multipath interference from $\tau_{back}$ for each sound direction corresponds to that of the ILD ratios between the two microphones in the robot head, where the ILD ratios represents the sine function in the ideal condition. We use $\tau_{diff}$ multiplied by the absolute sine function with attenuation factor $\beta_{multi}$ (typically set to 0.1) as the factor used to compensate for the multipath interference:

$$\tau_{inter}(\theta) = \frac{d_{lr}}{2v}\left(\operatorname{sgn}(\theta)\pi - \frac{2\theta}{180}\pi\right) \cdot \left|\beta_{multi}\sin\left(\frac{\theta}{180}\pi\right)\right|,\tag{8}$$

where $\tau_{inter}$ is the interference created by the two paths. The final time delay factor to be used instead of $\tau_{lr}(\theta)$ in (2) for binaural SSL can be derived using $\tau_{front}$ and $\tau_{inter}$:

$$\tau_{multi}(\theta) = \tau_{front}(\theta) - \tau_{inter}(\theta)$$
$$= \frac{d_{lr}}{2v}\left(\frac{\theta}{180}\pi + \sin\left(\frac{\theta}{180}\pi\right)\right) - \frac{d_{lr}}{2v}\left(\operatorname{sgn}(\theta)\pi - \frac{2\theta}{180}\pi\right) \cdot \left|\beta_{multi}\sin\left(\frac{\theta}{180}\pi\right)\right|.\tag{9}$$

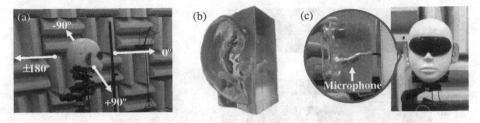

**Fig. 2.** HRTF measurement environment with the SIG-2 robot head equipped with pinnae in anechoic chamber: (a) coordinate system for SSL, (b) silicone artificial pinna (7.2 cm long × 3 cm wide × 2.3 cm high), (c) microphone location inside pinna

### 3.2    Front-Back Disambiguation by Using Amplification Effect of Pinnae

The basic function of the pinnae is to collect sound and spectrally transform it, which enable various types of audio signal processing. They collect sound through a filtering process and frequency-dependent amplification. This amplification increases the sound level by about 10 to 15 dB in the 1.5 to 7 kHz range [10].

Our approach to the front-back ambiguity problem is to utilize this amplification effect of the pinnae. To evaluate sound amplification in the frequency range over the entire azimuth, we measured HRTFs by equipping the head of the SIG-2 humanoid robot with two artificial pinnae and placing it in an anechoic chamber. Figure 2 shows the measurement environment, the shape of the pinnae, and the location of the microphone in each pinna. We used a 0.3-s time stretched pulse (TSP) signal [11] ranging from 1 to 8 kHz as a sound source and measured HRTFs over the entire azimuth in 5° steps. The measured HRTF data are illustrated in Fig. 3. The pinnae amplified the intensities of the input sound signal in the frequency band ranging from 3000 to 5300 Hz for the front directions (−90° to +90°). The left pinna amplified it from around −90° to +10° (Fig. 3-(a)), and the right one amplified it from around −10° to +90° (Fig. 3-(b)). Since this pinna effect amplifies the sound signals coming from only the front directions, the problem of front-back ambiguity can be solved by simply comparing the mean intensity of the sound signals in a specific frequency range with a threshold value, as shown in Fig.3-(d). The decision rule for front-back disambiguation using the pinna amplification effect is

$$if \quad \log \frac{1}{f_2^{FB} - f_1^{FB}} \sum_{f=f_1^{FB}}^{f_2^{FB}} |X_l[f]|^2 + |X_r[f]|^2 \; > \; \eta_{FB} \quad then \quad \text{frontal direction} \tag{10}$$

$$else \qquad\qquad\qquad\qquad\qquad\qquad\qquad\qquad\qquad \text{rear direction} \quad,$$

where $f_1^{FB} = 3000/F{\cdot}fs$ and $f_2^{FB} = 5300/F{\cdot}fs$ are the boundary frequency bins to be calculated. The input sound signals are normalized by the mean value of the intensities of all frequency bins beforehand to make the intensities consistent.

If the observed signal is determined to be behind the head by using this decision rule, the direction estimated by the SSL method is switched to the mirrored angle location in the back (e.g., +30° is switched to +150°).

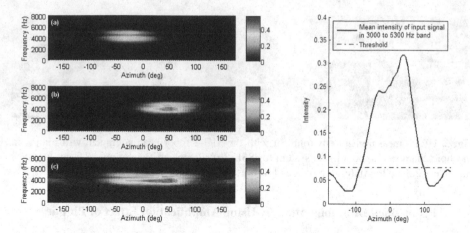

**Fig. 3.** Effects of amplification by silicone artificial pinnae: (a) amplification by left pinna, (b) amplification by right pinna, (c) amplification by left and right pinnae, (d) mean intensity of input time stretched pulse signals in 3000 to 5300 Hz frequency band.

## 4    Evaluation

We evaluated our improved SSL method under various conditions to determine whether it makes fewer localization errors than the conventional method and whether it enables more effective front-back disambiguation over the entire azimuth than the a conventional HRTF-based method. We implemented our improved method in the binaural audition system of our SIG-2 humanoid robot equipped with two Sennheiser ME 104 omnidirectional microphones. Figure 4 shows the flow of the SSL process. Since our target sound signals are human speech, we restricted localization processing to only when a human speech was detected by using the statistical model-based voice activity detection (VAD) algorithm proposed by Sohn et al. [12].

### 4.1    Experiments

The experiments were conducted in a room with a reverberation time of ~120 ms and noise from air conditioners, personal computers, background music with lyrics. The average sound pressure level (SPL) of the background noise and the average SNR of the target speech signals were about 61.2 and 23.2 dB, respectively. The SIG-2 humanoid robot was placed at the center of the room and the speakers were positioned 1.5–2.5 m from the robot. A male and then a female speaker stood at points along the azimuth from −180° to +170° in 10° steps and spoke short words or phrases (for about 1 s; e.g., calling the robot's name and expressing simple greetings) to the robot five times at each point, and the system captured their speech signals.

To accurately estimate the improvement in SSL performance with the improved method, we conducted five experiments using five different methods:

1. SSL using conventional time delay factor $\tau_{lr}$ in (2) ("conventional method").
2. SSL using derived time delay factor $\tau_{front}$ in (5) considering only front path around robot head.

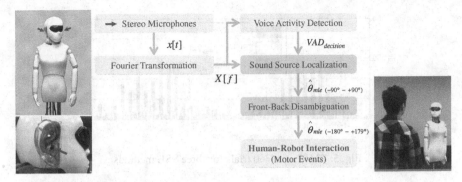

**Fig. 4.** Flow of SSL in SIG-2 humanoid robot with improved method

3. SSL using proposed time delay factor $\tau_{multi}$ in (9) considering both front and back paths around robot head.
4. Front-back disambiguation using a conventional HRTF-based method.
5. Front-back disambiguation using proposed method in (10).

The HRTF-based method used in the fourth experiment uses a decision rule in which two measures of the cross-correlation coefficient between the input signal and the HRTF data measured in the coincident front-direction or in the mirrored back-direction are compared [7]:

$$\text{if } R_{XH}^{front} >= R_{XH}^{back} \text{ then } \text{front direction } else \text{ back direction,} \tag{11}$$

where

$$R_{XH}^{front} = \frac{1}{F} \sum_{f=1}^{F} corr\left( \log\left( \frac{|X_r[f]|^2}{|X_l[f]|^2} \right), \log\left( \frac{|H_r[f,\theta_{front}]|^2}{|H_l[f,\theta_{front}]|^2} \right) \right) \tag{12}$$

$$R_{XH}^{back} = \frac{1}{F} \sum_{f=1}^{F} corr\left( \log\left( \frac{|X_r[f]|^2}{|X_l[f]|^2} \right), \log\left( \frac{|H_r[f,\theta_{back}]|^2}{|H_l[f,\theta_{back}]|^2} \right) \right),$$

$H_{l/r}$ is the HRTF data measured from each of the two microphones for the $\theta_{front/back}$ direction beforehand.

## 4.2    Experimental Results and Discussion

Figure 5 shows the root mean square error (RMSE) for the 360 trials (36 points × 5 speech signals × 2 speakers) for the three SSL methods along the azimuth from −180° to +170° in 10° steps. The SSL method using the new time delay factor had fewer localization errors than the one using time delay factor $\tau_{lr}$ derived in free space and the one using time delay factor $\tau_{front}$ derived considering only sound wave diffraction. The average RMSEs for $\tau_{lr}$, $\tau_{front}$, and $\tau_{multi}$ were respectively 11.87°, 3.40°, and 1.96°.

Figure 6 shows the rate of successful disambiguation for the 360 trials for the two disambiguation methods along the entire azimuth. The proposed method had an average success rate with 32.2% higher (92.28% vs. 69.78%) than the one using the HRTF database.

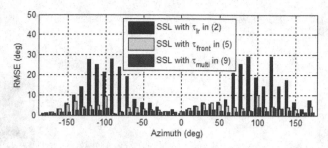

**Fig. 5.** RMSEs for 360 trials for three SSL methods

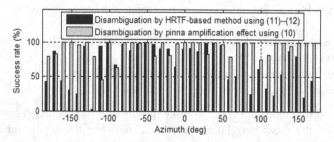

**Fig. 6.** Success rates for two disambiguation methods

As shown by the red bars in Fig. 5, considering the diffraction of sound wave with multipath interference around the robot head improves localization performance in binaural robot audition compared to the conventional SSL (blue bars) and SSL considering only the sound wave diffraction (green bars). This means that considering not only the diffraction but also the multipath interference is a key to improving localization performance in binaural robot audition.

The proposed method using the pinna amplification effect was better able to disambiguate the directions of the input sound signals coming from either the front or back, as shown in Fig. 6. The HRTF-based method had worse disambiguation performance, especially for signals coming from the back, because the measurement of the cross-correlation coefficient using the front-HRTF data was often slightly higher than or the same as that using the back-HRTF data even though the sound signals came from the back. This means that there was not much difference in the frequency properties of the front and back HRTF data for some directions.

## 5 Conclusion

We addressed the two main problems with sound source localization based on the GCC-PHAT method in binaural robot audition: 1) diffraction of sound waves with multipath interference around the robot head, which degrades SSL accuracy, and 2) front-back ambiguity, which limits the localization range to the front horizontal space. To solve the first problem, we developed a new time delay factor for use with the GCC-PHAT method that takes into account the diffraction of sound waves with multipath interference under the assumption that the robot head is spherical. To solve the second problem, we devised a front-back disambiguation method utilizing the pinna amplification effect. Experimental results demonstrated that taking the diffraction of sound waves with multipath interference into account is a key to improving localization accuracy in binaural robot audition and that utilizing the pinna amplification effect effectively solves the problem of front-back ambiguity for humanoid robots equipped with two artificial pinnae. The video demonstration is available on Youtube [13].

# References

1. Sasaki, Y., Kabasawa, M., Thompson, S., Kagami, S., Oro, K.: Spherical Microphone Array for Spatial Sound Localizationfor a Mobile Robot. In: Proc. IEEE/RSJ Inter. Conf. on Intelligent Robots and Systems (IROS), Algarve, Portugal, pp. 713–718 (October 2012)
2. Cheng, C.I., Wakefield, G.H.: Introduction to Head-Related Transfer Functions (HRTFs): Representations of HRTFs in Time, Frequency, and Space. Audio Engineering Society 49, 231–249 (2001)
3. Knapp, C.H., Carter, G.C.: The Generalized Correlation Method for Estimation of Time Delay. IEEE Trans. on Acoustics, Speech, and Signal Processing 24(4), 320–327 (1976)
4. Hill, P.A., Nelson, P.A., Kirkeby, O., Hamada, H.: Resolution of Front-Back Confusion in Virtual Acoustic Imaging Systems. Acoustical Society of America 108(6), 2901–2910 (2000)
5. Nakashima, H., Mukai, T.: 3D Sound Source Localization System Based on Learning of Binaural Hearing. In: Proc. IEEE Inter. Conf. on Systems, Man and Cybernetics (SMC), Nagoya, Japan, October 10-12, vol. 4, pp. 3534–3539 (2005)
6. Ovcharenko, A., Cho, S.J., Chonga, U.P.: Front-back confusion resolution in three-dimensional sound localization using databases built with a dummy head. Acoustical Society of America 122(1), 489–495 (2007)
7. Rodemann, T., Ince, G., Joublin, F., Goerick, C.: Using Binaural and Spectral Cues for Azimuth and Elevation Localization. In: Proc. IEEE/RSJ Inter. Conf. on Intelligent Robots and Systems (IROS), Nice, France, pp. 2185–2190 (September 2008)
8. Blauert, J.: Spatial Hearing: The Psychophysics of Human Sound Localization (Revised Edition). MIT Press, Cambridge (1997)
9. Kim, U.H., Okuno, H.G.: Improved Binaural Sound Localization and Trackingfor Unknown Time-Varying Number of Speakers. Advanced Robotics (to be published)
10. Middlebrooks, J.C.: Sound Localization by Human Listeners. Annual Review of Psychology 42, 135–159 (1991)
11. Suzuki, Y., Asano, F., Kim, H.-Y., Sone, T.: An Optimum Computer-Generated Pulse Signal Suitable for the Measurement of very Long Impulse Responses. Acoustical Society of America 97(2), 1119–1123 (1995)
12. Sohn, J., Kim, N.S., Sung, W.: A Statistical Model-Based Voice Activity Detection. IEEE Signal Processing Letters 6(1), 1–3 (1999)
13. http://youtu.be/iCE--ir-JRc

# Designing and Optimization of Omni-Directional Kick for Bipedal Robots

Syed Ali Raza and Sajjad Haider

Artificial Intelligence Lab, Faculty of Computer Science,
Institute of Business Administration, Karachi, Pakistan
{s.aliraza,sajjad.haider}@khi.iba.edu.pk

**Abstract.** The paper presents designing and optimization of key-frame based kick skills for bipedal robots. The kicks, evolved via evolutionary algorithms, allow a humanoid robot to kick in straight, sideways, backward and in angular directions. Experiments are conducted on the simulated model of Nao robot that is being used in the RoboCup Soccer 3D Simulation league. The initial sets of kicks were manually designed by human experts and were passed as seed values to the optimization process. Correctness in the kick direction and the distance covered by the ball were used as the fitness criteria. The findings of the paper not only significantly improves the capability of our RoboCup Soccer 3D team but also provides insight in the designing and optimization of key-frame based kicks that can be utilized by other teams participating in bipedal soccer.

## 1    Introduction

RoboCup Soccer competitions are held each year with the aim of promoting research and development in the fields of artificial intelligence, multi-agent systems and robotics. The main objective of these competitions is to build a team of autonomous humanoid robots that will compete to win against the human soccer champions by the year 2050. There are five categories within RoboCup Soccer and RoboCup Soccer 3D Simulation League is one of them. The Simulation 3D league provides a realistic model of Aldebaran Nao[1] robot. The environment aids in development and testing of various algorithms before deploying them on the actual robots.

In the initial years of the league, the rules and regulations were a bit relaxed. However, as the years have passed by, the rules of the Simulation 3D league have been made more challenging. For example, in the early days, the soccer field size was 12x8 meters, each side had 6 players and players had omni-vision data provided by the soccer server. The smaller ground size was to support the development of locomotion, localization and team strategy. As advancements were made in different areas of the league, rules were made more challenging in all important areas. As a result, in the 2012 World RoboCup Soccer, the field size was 30x20 meters, each team had 11 players and players had only limited vision data available. In smaller fields, the capability to dribble the ball towards opponent goal was sufficient to become a champion team.

---

[1] http://www.aldebaran-robotics.com/eng/

M. Ali et al. (Eds.): IEA/AIE 2013, LNAI 7906, pp. 292–301, 2013.
© Springer-Verlag Berlin Heidelberg 2013

However, with the increase in ground size, it has become essential to have an effective kicking and passing mechanism to have strategic advantage over other teams.

From the past matches of RoboCup Soccer 3D (and that of Standard Platform League and Humanoid League) matches, it can be easily claimed that teams with naive locomotion skills suffer badly as positioning a robot right behind the ball and keeping it aligned with kicking direction consumes precious time and efforts. Hence, designing of an omni-directional kick is an active research area within humanoid soccer.

Unlike walk motions which follow cyclic pattern and hence can be modeled via trigonometric functions (like sinusoids) [14, 15], kick routines are not cyclic in nature and thus cannot be modeled in the form of sinusoidal functions. One possibility is to generate trajectories of end effectors and to provide joint angle values computed via inverse kinematics [6]. The task, however, is not trivial as significant expertise is required in inverse kinematics formulation and generation of end effectors trajectories. Key-frame based methods provide a good alternative as they can produce reasonable kicks without going into the complexities of inverse kinematics. A key-frame refers to a pose of a robot's body at a particular time instant. It contains the description of all the joint angles that are involved in that pose. Another element associated with a key-frame is the time required for its execution. In simpler terms, it is the number of intermediary frames required to reach the key-frame.

In the RoboCup Soccer domain, a number of teams have used key-frame based technique to design kicks, dive, getup and other locomotive routines [1-4, 7-13]. In this work, we have also used key-frame based technique to enhance the ball-kicking skill of a soccer agent. We have designed an omni-directional kick engine as a set of kicks that work in different directions, such as straight, diagonal, sideways and backward. The kicks are designed for both feet and as such the engine contains 12 kicks in different directions.

The rest of the paper is organized as follows. Section 2 provides an overview of the 3D Simulation league. The proposed scheme of designing an omni-directional kick is explained in Section 3. Section 4 discusses optimization process in detail. Experimental design and results are discussed in Section 5, while Section 6 concludes the paper and provides future research directions.

## 2    Domain Description

RoboCup Soccer 3D Simulation League [1] provides a soccer simulation platform where two teams of eleven humanoid robots (soccer agents) play against each other. The simulation environment uses SimSpark, which is a generic physical multi-agent system simulator. For the simulation of rigid body dynamics, SimSpark uses an open source library ODE (Open Dynamics Engine) which provides built-in collision detection and friction model for realistic simulation.

Each agent (player) in the Simulation 3D league communicates with the soccer server on a separate process thread. In this two-way communication process, server sends game and agents' state information to the agent while the agent sends commands to control body movements as well as messages to other teammates. This communication occurs on every simulation cycle (of 20ms). An agent is provided with a restricted and noisy vision cone of 120°. Thus, it can only observe objects (flags, goal posts, teammates, opponents and ball) that are within its vision cone.

The humanoid robot model is essentially a simulated model of Aldebaran Nao robot. It has 2 joints in neck, four in each arm and six in each leg. Each joint is a hinge joint and is provided with a perceptor effector pair; a perceptor to monitor the joint's current angle and an effector to manipulate the joint to a desired angle. In addition, an agent is also equipped with a gyroscope, an accelerometer and two force resistant preceptors (one in each foot). Fig. 1(a) presents a 3D rendered model of Nao robot used in the Simulation 3D league, while Fig. 1(b) shows Nao's anatomy along with axis of rotation of each joint. The dotted circle in the fig. highlights the joints used in kicks.

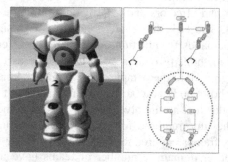

**Fig. 1.** (a) 3D rendered model of Nao robot in the Simulation 3D League environment. (b) Anatomy of Nao showing all joints and their axis of rotation.

## 3     Omni-Directional Kicks

Soccer is a highly dynamic game in which the game situation changes rapidly with changes in the position of the ball. Apart from certain facts, such as number of goals scored and the time left in the game, the position of the ball in the field derives the strategic decision-making of both teams. A player with ball possession can either dribble the ball towards opponent's goal or can dribble it away from the opponent. It can also pass the ball to a teammate or kick it towards opponent's goal. Depending upon the game situation, all of these are important actions. As is the case in human soccer, kicking skill has taken a prominent role in the Simulation 3D league specially after the introduction of bigger fields. If a robot can kick the ball in many directions without repositioning itself then it is considered a vital capability. This type of kicking skill is referred to as Omni-directional kick. In ideal conditions, a robot should be able to kick the ball in any desired direction if the ball is within its reachable space. But in reality, robots have limited degrees of freedom which do not allow them to swing kicking foot in every desired direction with the same power. Hence, there exists a trade-off between direction of the kick and power of the kick.

In this work, we have designed a set of kicks that work in different directions. For instance, for left foot, we have designed straight kick, diagonal right kick, side right kick, back kick, diagonal left kick, and side left kick. Similar kicks are obtained for right foot too. In this way, we have achieved omni-directionality. Any of the above mentioned kick can properly work for a certain range of ball positions. We have selected these positions as a rectangular region. If ball is positioned anywhere within this rectangular region the selected kick should execute properly. Fig. 2 gives a pictorial insight of these regions while their end points are provided in Table 1. These points are presented in robot's coordinate frame where the robot is at (0,0) while x and y axes are in robot's forward and left side, respectively. Given the left-right symmetry in a robot's body joints, solutions for left leg kick can be used for right leg kick except that the signs of all roll joints are inverted.

**Table 1.** Rectangular region of each kick

|  | P1 | P2 | P3 | P4 |
|---|---|---|---|---|
| Straight | (+0.18,0.08) | (+0.18,0.02) | (+0.23,0.02) | (+0.23,0.08) |
| Diagonal Right | (+0.18,-0.03) | (+0.18,-0.09) | (+0.22,-0.09) | (+0.22,-0.03) |
| Side Right | (+0.18,0.08) | (+0.18,-0.02) | (+0.23,-0.02) | (+0.23,0.08) |
| Back | (+0.18,0.08) | (+0.18,0.02) | (+0.22,0.02) | (+0.22,0.08) |
| Diagonal Left | (+0.18,0.1) | (+0.18,0.06) | (+0.22,0.06) | (+0.22,0.1) |
| Side Left | (+0.18,0.1) | (+0.18,0.05) | (+0.23,0.05) | (+0.23,0.1) |

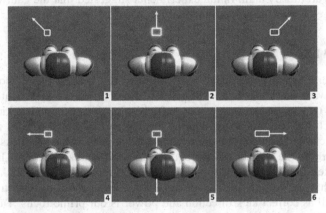

**Fig. 2.** Ranges of Different Kicks (1) DiagonalLeft (2) Straight (3) DiagonalRight (4) SideLeft (5) Back (6) SideRight

## 3.1    Kick Design

As mentioned earlier, a key-frame is essentially a posture of a robot's body at a particular time instant. If we think of the robot's action in terms of frames then each frame is a set of joint angle values and the key-frame is a particular frame whose joint angle values are defined by the designer. There can be a number of frames between any two key-frames where the joint angle values of intermediate frames are usually interpolated linearly. Additionally, a time factor is defined for every key frame that specifies the time required by the robot to reach the required body posture.

During our initial testing, it was found that some joints like head and hand joints are better to be kept stationary as their movement may cause unintentional body imbalance or losing of ball's sight. Hence, we have excluded head and hand joints from our key-frame representation.

It is typically assumed that the manual designing phase requires too much human efforts and time. We have made this process simple by analyzing each key-frame separately. For this purpose, we have divided the kick execution in two phases. In the first phase, it is ensured that the kicking leg is lifted (without disturbing the body balance) and is moved as far as possible from the ball in the opposite direction of the kick. In the second phase, the kicking leg is moved as fast as possible in the direction of the kick to hit the ball. Our mechanism of moving joints is based on P-controller. At each cycle angular speed of a joint is given by,

$$AngSpeed = \begin{cases} 0, Abs(ReqAng - CurrAng) < 1 \\ (ReqAng - CurrAng) * Gain, elsewhere \end{cases}$$

Where *RegAng* is the required angle of the joint and *CurrAng* is the current angle of the joint as reported by the joint's perceptor. Both angles are in degrees and Angular speed is in degrees per second. It should also be noticed that the torque associated with a joint is controlled by two factors $(ReqAng - CurrAng)$ and proportional gain '*Gain*'. The simulation engine sets an upper bound to angular velocity which doesn't allow it to be arbitrarily high. Initial experiments suggested that the gain value should be in the range of 0-20 for proper functioning.

## 3.2     Initial Kicks

We started our informal experiments with only two key-frames: one to move the kicking leg backward and other to move it in the direction of the kick. But we quickly realized that two key-frames were not sufficient to design complex kicks. For instance, to kick the ball sideways requires the leg to first move toward the opposite side of the kicking direction. However, before it happens, it has to lift its kicking leg without disturbing the balance which requires an additional key-frame. Thus, we added as many key-frames as needed to fulfill the requirements of each kick. It is worth mentioning that we do not consider it important for the robot to remain stable after hitting the ball as correct kick direction and power is our prime focus. Table 2 shows the number of key-frames used in each phase of a kick. The number of key-frames varies due to the complexity of motion. Fig. 3 demonstrates the body postures of the robot in five key-frames of the side kick.

**Table 2.** Number of key-frames used in two different phases of kick and number of spare key-frames for each initial kick

|  | No. of Key-frames in the first phase | No. of Key-frames in the second phase |
|---|---|---|
| Straight Kick | 3 | 3 |
| DiagonalRight Kick | 3 | 1 |
| Side Right Kick | 4 | 1 |
| Back Kick | 4 | 2 |
| Diagonal Left Kick | 3 | 1 |
| Side Left Kick | 4 | 1 |

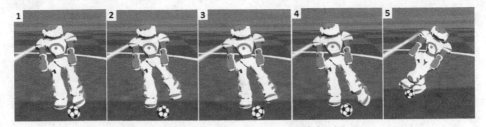

**Fig. 3.** Body posture for five key-frames in side kick

# 4    Kicks Optimization

Although, all manually designed kicks produced reasonable results, it is obvious that those were not optimal solutions. Thus, we optimized the initial kicks using evolutionary algorithms (EA). The features of the EA are described in Table 3.

**Table 3.** Features of Evolutionary Algorithm

| Feature | Technique Used |
|---|---|
| Representation | Floating Point |
| Parent Selection | Binary Tournament |
| Mutation | Non-Uniform |
| Cross-over | One-Point |
| Survival Selection | Truncation |

The mutation rate was set to 0.5 which specifies the fraction of genes mutated in a single solution. A uniform 4 unit mutation step-size was selected in the beginning for all parameters of each key-frame. However, during initial experiments we found out that uniform mutation step-size was hindering EA efforts to produce better kicks. It was realized that 4 units mutation in the first phase of the kick, in which the leg is lifted and is moved away from ball, was producing angles that were making robots instable before it enters into the kicking phase. Hence, the mutation step-size was set to 2 units for the first phase.

For each key frame there are two additional parameters in a solution. One is the gain and the other is the time duration. As mentioned earlier, gain for each joint can be selected separately. However, adding a gain parameter for each joint would have resulted in a very large number of parameters. Thus, we have used a single gain parameter for a key-frame. The other parameter is the time that defines the number of frames or the number of simulation cycles required to reach end pose of that key frame. The range of gain is 0-12, while the time parameter can have values in the range of 0-35. The ranges of joints angles are available at SimSpark's website[2]. Thus, a solution consists of concatenated frames whereas each frame is represented by joints angles, gain and the time duration to execute that frame. Fig. 4 illustrates the structure/representation of a solution.

---

[2] http://simspark.sourceforge.net/wiki/index.php/Models

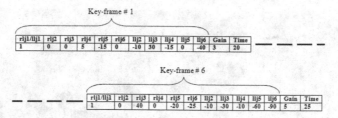

**Fig. 4.** Representation of an EA solution

Each solution corresponds to 6 key-frames. Kicks which are designed using less than six key-frames use the additional key-frames as 'spare key-frames'. That is, we fill a spare frame with the last key-frame data and the frame is made non-functional by setting time parameter to zero. The purpose of the spare key-frames is to allow EA to explore better solutions by bringing additional key-frames. By tweaking angle values of the spare key-frames and by adding time to it, there are chances of producing improved kicks.

A single fitness function has been used for all the kicks. The two important components of the fitness function are: (a) distance traveled by the ball and (b) deviation from the desired direction. Instead of handling angular deviation cost separately, we have merged it with the fitness of distance. The angular deviation is calculated in terms of distance by using arc length formula

$$S = r*\alpha$$

Where r is the distance covered by the ball and $\alpha$ is the angular deviation. Hence the overall fitness function is

$$Fitness = r - r*\alpha$$

**Table 4.** Pseudo code of evaluation procedure

```
1: Solutions • Candidate Solutions from Optimization Mod-
ule
2: Fitness • {}
3: Repeat
4:  For BallPosition from 0 to 3 do
5:    For TrailNumber from 0 to 3 do
6:      BeamAgent&Ball(AgentPos,BallPos)
7:      InitKick ()
8:      ExecuteKick()
9:      If KickExecuted
10:        BallDist • GetBallDist()
11:        BallAngle • GetBallAngle()
12:        RecordDist&Angle(BallDist,BallAngle)
13:  Fitness • Fitness ∪ {ComputeFitness()}
14: Until All Solutions are Evaluated
15: Return Fitness
```

The complete working of the kick optimization process is presented in Table 4. As mentioned earlier, each kick is tested on several points in its reachable space. We have selected the corner points of the corresponding rectangle for this purpose. A solution is run four times for each point. This redundancy ensures the consistency of a solution. The rest of the procedure is straightforward. First the robot and the ball are beamed at a pre-specified position. Then a routine is called that brings the robot into a pose that is required before the execution of a kick. The kick is executed next. When the kick is finished, its performance is recorded in terms of the distance travelled by the ball and the angle deviation. Out of four trials at each ball positions, we select the longest three distances and the lowest three angle penalties and average them. These average values are then used in the fitness function describe above.

## 5    Experimental Results

During the experiments, the evolutionary algorithm was run for 150 generations for each kick. It took almost 82 hours to complete evolution of a single kick. It must be mentioned that this time consumption is not because of EA's computation but due to the nature of the evaluation process in which a robot has to kick the ball in real time. The initial population was generated randomly by varying values of the manually designed kick. Fig. 5 shows the best-so-far curves for all kicks. The starting point in each figure represents the fitness value of the manually designed kicks. It can be observed from Fig. 5 that EA has made improvements in all manually designed kicks.

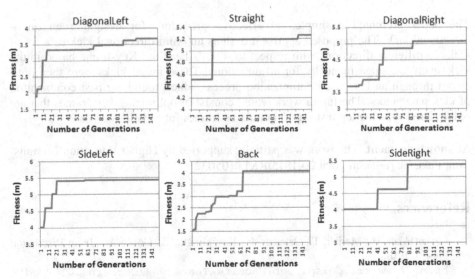

**Fig. 5.** Best-so-far Curves – fitness in meters

Although no restriction was imposed on the joint angle values, it was observed that the postures of the resultant kicks were not much different from the initial kicks. To measure optimized kicks parameters variation from the initial kicks parameters, Fig. 6

draws the absolute difference between these two set of parameters. It can be easily observed that almost all parameters were tuned and showed variation. The variations, however, are more prominent in the second phase of the kicks which, in most cases, starts after 39th parameter. Another important finding was that there was no spare key-frame left at the end of the optimization process.

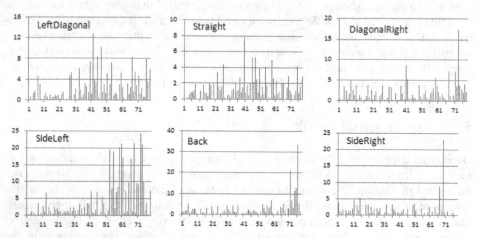

**Fig. 6.** Absolute difference between initial kick values and optimized kick values

## 6    Conclusion

The paper presented a simple yet effective procedure for designing an omni-directional kick. The proposed approach designs an omni-directional kick as a set of kicks in different directions with respect to robot's heading. Key-frame based initial kicks were designed manually for all directions. These initial kicks were then optimized through an EA-based optimization process which found the best combination of kick parameters. The future work would consist of optimizing key-frames that are only related to swing leg instead of optimizing frames for both legs.

**Acknowledgement.** The work was partially supported by Higher Education Commission, Pakistan research grant 20-1815/R&D/105032.

## References

1. Levent Akın, H., Mericli, T., Özkucur, E., Kavaklıoğlu, C., Gökce, B.: Team Discription Paper (2010),
   http://www.tzi.de/spl/pub/Website/Teams2009/Cerberus10TDP.pdf
2. Borisov, A., Ferdowsizadeh, A., Mohr, C., Mellmann, H., Martius, M., Krause, T., Hermann, T., Welter, O., Xu, Y.: Technical Report (2009),
   http://www.naoteamhumboldt.de/papers/NaoTH09Report.pdf

3. Valtazanos, A., Vafeias, E., Towell, C., Hawasly, M., Tabibian, S.B., Ivanov, P., Wilson, S., Wren, D., Austin Tobin, J., Ramamoorthy, S., Vijayakumar, S.: Team description paper (2012), http://wcms.inf.ed.ac.uk/ipab/robocup/research/ EdinfernoTDDoc2012.pdf

4. Czarnetzki, S., Hauschildt, D., Kerner, S., Urbann, O.: Team Report (2008), http://www.bredobrothers.de/lib/exe/ fetch.php?media=teamreport-08-bredobrothers.pdf

5. Xu, Y., Mellmann, H.: Adaptive Motion Control: Dynamic Kick for a Humanoid Robot. In: Dillmann, R., Beyerer, J., Hanebeck, U.D., Schultz, T. (eds.) KI 2010. LNCS, vol. 6359, pp. 392–399. Springer, Heidelberg (2010)

6. Müller, J., Laue, T., Röfer, T.: Kicking a Ball – Modeling Complex Dynamic Motions for Humanoid Robots. In: Ruiz-del-Solar, J. (ed.) RoboCup 2010. LNCS (LNAI), vol. 6556, pp. 109–120. Springer, Heidelberg (2010)

7. Urieli, D., MacAlpine, P., Kalyanakrishnan, S., Bentor, Y., Stone, P.: On Optimizing Interdependent Skills: A Case Study in Simulated 3D Humanoid Robot Soccer. In: Proc. of 10th Int. Conf. on Autonomous Agents and Multiagent Systems (AAMAS 2011), Taipei, Taiwan, pp. 769–776 (2011)

8. Rafael: Development of Behaviors for a Simulated Humanoid Robot, Master's thesis, University of Aveiro (2008)

9. Rei, L.: Optimizing Simulated Humanoid Robot Skills, Master's thesis, University of Porto (2010)

10. Hausknecht, M., Stone, P.: Learning Powerful Kicks on the Aibo ERS7: The Quest for a Striker. In: RoboCup International Symposium 2010, Singapore, pp. 254–265 (2010)

11. Zagal, J.C., Ruiz-del-Solar, J.: Learning to Kick the Ball Using Back to Reality. In: Nardi, D., Riedmiller, M., Sammut, C., Santos-Victor, J. (eds.) RoboCup 2004. LNCS (LNAI), vol. 3276, pp. 335–346. Springer, Heidelberg (2005)

12. Wama, T., Higuchi, M., Sakamoto, H., Nakatsu, R.: Realization of Tai-Chi Motion Using a Humanoid Robot. In: Rauterberg, M. (ed.) ICEC 2004. LNCS, vol. 3166, pp. 14–19. Springer, Heidelberg (2004)

13. Antonelli, M., Dalla Libera, F., Menegatti, E., Minato, T., Ishiguro, H.: Intuitive Humanoid Motion Generation Joining User-Defined Key-Frames and Automatic Learning. In: Iocchi, L., Matsubara, H., Weitzenfeld, A., Zhou, C. (eds.) RoboCup 2008. LNCS (LNAI), vol. 5399, pp. 13–24. Springer, Heidelberg (2009)

14. Shafii, N., Javadi, M.H., Kimiaghalam, B.: A Truncated Fourier Series with Genetic Algorithm for the control of Biped Locomotion. In: Proceeding of the 2009 IEEE/ASME International Conference on Advanced Intelligent Mechatronics, pp. 1781–1785 (2009)

15. Yazdi, E., Azizi, V., Haghighat, A.T.: Evolution of biped locomotion using bees algorithm based on truncated fourier series. In: Proceedings of the World Congress on Engineering and Computer Science (WCECS 2010), pp. 378–382 (2010)

# Aerial Service Vehicles for Industrial Inspection: Task Decomposition and Plan Execution*

Jonathan Cacace, Alberto Finzi, Vincenzo Lippiello,
Giuseppe Loianno, and Dario Sanzone

DIETI, Università degli Studi di Napoli Federico II, via Claudio 21, 80125, Naples, Italy
{jonathan.cacace,alberto.finzi,vincenzo.lippiello,
giuseppe.loianno,dario.sanzone}@unina.it

**Abstract.** We propose an autonomous control system for Aerial Service Vehicles capable of performing inspection tasks in buildings and industrial plants. In this paper, we present the applicative domain, the high-level control architecture along with some empirical results. The system has been assessed on real-world and simulated scenarios representing an industrial environment.

**Keywords:** Aerial Service Robotics, High level Control Architecture, Mixed Initiative Planning and Execution.

We present a high level architecture designed for an Aerial Service Vehicle (ASV) operating in close interaction with the external environment. This work is framed within the The AIRobots project [1] whose aim is to develop a new generation of unmanned service helicopters, equipped with sensors and end-effectors, and capable not only to fly, but also to achieve robotic tasks in proximity and in contact with the surface (e.g. site inspections, simple manipulations, etc.).

**Fig. 1.** Robotic Platform: ducted-fan ASV

In our scenario, the autonomous system should orchestrate a new set of operations like wall approach, docking, undocking, wall scanning etc.. These operations represent different operative modes, each associated with a different controller with specific control laws and performance the high-level control system should be aware of. Each switch from one operative mode to the other should be suitably prepared and planned to keep

---

* The research leading to these results has been supported by the AIRobots and SHERPA FP7 projects under grant agreements ICT-248669 and ICT-600958 respectively.

M. Ali et al. (Eds.): IEA/AIE 2013, LNAI 7906, pp. 302–311, 2013.

smooth control trajectories. Since the system flies close to the obstacles in cluttered and unknown environments, fast planning engines are required to generate (or to adjust) trajectories in real-time. On the other hand, the system should be able to regulate the trade off between fast planning and accurateness of the generated trajectories depending on the operative mode and the context. Moreover, since the system operates with the man in the loop, the planning/executive system should be able to manage sliding autonomy, from autonomous to teleoperated mode, depending on the humans' interventions. This applicative domain is challenging and novel and has not been investigated in depth in the UAV literature which is mainly focused on free flight tasks and simultaneous localization, mapping, and path planning problems [4,8,17]. Few high-level architectures for UAV can be found in literature [5], but none of these addresses the complexity of the operative domain proposed in this paper.

# 1 System Requirements and Architecture

The applicative scenario described so far requires a high-level control system with following features:

- The air vehicle operates in close interaction with the environment, hence reactive, adaptive, and flexible planning/replanning capabilities are needed;
- Both autonomous and human-in-the-loop control modalities should be supported to allow human interventions and teleoperation;
- High-level control strategies should be defined taking into account the low-level operative modes and constraints.

In particular, the high-level system should orchestrate the activations of a set of low-level controllers, modeled as hybrid automata [14], switching to the appropriate controller according to the operative mode and the task (see Fig. 2) feeding the selected controller with suitable data (e.g. state and references).

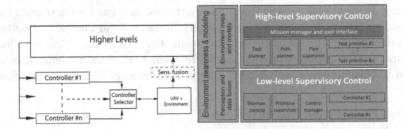

**Fig. 2.** Interaction between the high level system and the low-level controllers (left); the high level control system is composed of high-level and low-level supervisory systems

To match these requirements we proposed the layered architecture depicted in Fig. 2. Here, two layers are distinguished: the high-level supervisory system is responsible for user interaction, task planning, path planning, execution monitoring, while the low-level supervisory system manages the low-level execution of control primitives setting the controllers and providing control references. This architecture is detailed in Fig. 3.

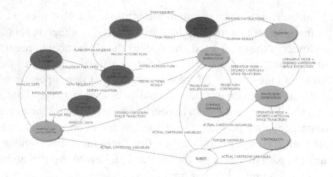

**Fig. 3.** High Level Architecture: high level, low level, and reactive level modules are respectively in blue, green, and gray

The *User* module (US) allows us to specify high-level goals (e.g. $Inspect(p)$) or lower level tasks (e.g. $TakeOff$) or to directly teleoperate. Each task/goal is delivered to the TP which expands a task into an abstract plan composed of *macro-actions*. This plan is then sent to the *Plan Supervisor* (PS) for high-level execution. Each task or macro-action can be interrupted and pre-empted by new tasks provided by the user, provoking task replanning. The PS generates, for each macro-action in the high-level plan, a set of *micro-actions* to be executed by the *Primitive Supervisor* (PR). Each *macro-action* is further decomposed into a sequence of *micro-actions* which are endowed with detailed information about the associated geometrical paths. The PR exploits the *Control Manager* (CM) to select the low-level controller responsible for the micro-action execution. Finally, the PR generates the control trajectory passing it to the *Trajectory Supervisor* (TS) to generate control references at a suitable frequency. The PR exploits concatenations of fifth-order polynomials to provide smooth trajectories between way-points [11] while ensuring the velocity and tolerance constraints. When a micro-action fails, the PS can either call the PP to generate an alternative path or call the TP to generate a different plan of macro-actions. Furthermore, it can be interrupted by the *Path Monitor* (PM) which checks for trajectory deviations and unexpected obstacles. Finally, the operator can always switch to a manual control mode, in this case the TS should monitor the trajectory provided by the *Teleman*. Once the autonomous control is restored, a replanning process is needed to recover the execution of the current task.

## 1.1 Task Planning and Executive Control

The high-level executive system coordinates task decomposition and plan monitoring. It relies on a PRS engine [9] that manages a BDI-like execution cycle [15] and hierarchical task decomposition. The high-level executive system responds to events generated by the US, PS, or TP itself by committing to handle one pending goal, selecting a method from a plan library, managing the hierarchical decomposition to extract/update the macro-actions plan. Once a plan is generated, the PS should manage the actual execution of each *macro-action* providing the action results to the TP module. During this execution process, user interventions are treated in a uniform way: at any time the user can interrupt/suspend the current task, or the execution of alternative tasks can be invoked. In this case, the executive system reacts by replanning from the current state: it selects alternative methods and generates an alternative plan. This enables mixed initiative task planning [2].

## 1.2 Path Planning and Replanning

The *Path Planner* expands each *macro-action* into a set of *micro-actions* representing a path that respects geometric and operative constraints. The path generation algorithm is based on a Rapidly-exploring Random Tree (RRT) algorithm [10] which is particularly suitable in highly unstructured and dynamic domains. In this work, the RRT algorithm generates collision-free paths composed of sequences of waypoints $(x, y, z, \theta)$, where $(x, y, z)$ is a point and $\theta$ is the yaw. More specifically, it generates a path as a sequence of $(x, y, z)$ points in a 3D search space (3D grid map), while the yaw $\theta$ is obtained as the direction pointing towards the next waypoint. The generated path should satisfy a set of additional control, safety, and temporal constraints: *Maximum angle* for pitch and yaw; *Minimal distance* from the obstacles (this parameter is also associated with the operative mode and the accuracy of the selected controller); *Maximum Time* for the path generation processes, if the algorithm cannot find a feasible path before the timeout, it should provide the best partial path. Moreover, that RRT path planner can generate several solutions to refine the path, until one of the following conditions are satisfied: *timeout*, i.e. the available time for path planning expires; *interrupt*, i.e. a replanning request or an exogenous event interrupts plan generation; *cost threshold*, i.e. as soon as the current path cost is below a suitable threshold, the generated plan is considered as satisfactory. The path planning refinement process is illustrated in the Algorithm 1 where the path generation process is iterated until the current generated path is not satisfactory. If the *timeout* occurs before the generation of the first solution, the *solveRTT* function generates the *path* that arrives closer to the target.

---

**Algorithm 1.** Refine_RRT($q_{init}, q_{goal}, threshold, timeout$)

---
  initialize(*path*,*time*);
  **while** (($time < timeout$) $\wedge$ ($preempted = false$) $\wedge$ ($pathCost \geq threshold$)) **do**
    $newPath \leftarrow$ solveRRT($q_{init}, q_{goal}, timeout$);
    **if** $C(newPath) < path$ **then**
      $path \leftarrow newPath$;
      $pathCost \leftarrow C(newPath)$;
    **end if**
  **end while**
  **return** *path*

---

The path cost is defined as follows:

$$\mathbf{c}(path) = \mathbf{c}_{lng}(path) \cdot p_{lng} + \mathbf{c}_{ang}(path) \cdot p_{ang} + \\ \mathbf{c}_{way}(path) \cdot p_{way} + \mathbf{c}_{obs}(path) \cdot p_{obs} + \mathbf{c}_{unk}(path) \cdot p_{unk} \tag{1}$$

where the $p_i$ are suitable weights and $c_i$ are defined as follows. $c_{lng}(path)$ is a cost associated with the path length; $c_{ang}(path)$ represents the cost associated with angular (yaw and pitch) variations, by minimizing this cost a straight path should be preferred to a path with angular turns; $c_{way}(path)$ counts the generated waypoints and allows us to minimize the segments in the path; $c_{obs}(path)$ is associated with obstacle proximity and penalizes paths close to obstacles; $c_{unk}(path)$ penalizes paths through -or close to- unexplored cells. Once a path is generated, the path planner defines a set of constraints $cst = (ms, md, et)$ associated with each generated segment. Roughtly, for each segment, we set the maximum speed $ms$ directly proportional to the obstacle minimal distance $mo$ along the corresponding segment; $ms$ is also associated with a proportional error $et$, therefore we set $md$ as $mo$-$et$ (if this value is not positive, the speed limit is lowered). These constraints $cst$ are also accessible to the human operator which can manually reset them. Note, that $cst$ are just rough limits used by the CM and the PR to select the right controller and to generate the trajectory associated with the path.

**Fig. 4.** (left) *Brake* to avoid collision; (center) *Escape* path to avoid the obstacle; (right) *Replan* a new path generated to reach the target

Path replanning is managed with different strategies depending on the time available for path generation. The urgency associated with the replanning activity depends on the position of the collision point $p_{obs}$ and the estimated time to collision $t_{ttc}$. This one is estimated by considering the obstacle distance $d_{obs}$ along the trajectory and the mean velocity $v_{mean}$ along the path. Given the time to collision $t_{ttc}$, we introduce two thresholds $T_b < T_e$ used to distinguish the following three cases:

- *Brake.* If $t_{ttc} \leq T_b$ then the obstacle is too close for replanning, hence the PS directly sends a *Brake* command to the PR to stop the robot in *hovering* (Fig. 4 up-left).
- *Escape.* If $T_b < t_{ttc} \leq T_e$, the PP is invoked by the PS to find an escape path that allows the robot to avoid the obstacle; the escape trajectory represents a fictituos detour that provides the planner with additional time to generate the new path on-the-fly (Fig. 4 up-right).
- *Replan.* If $t_{ttc} > T_e$ then the time is sufficient for safe replanning, hence the PS calls the PP to replan, on-the-fly, a trajectory from a suitable deviation point along the previous path (Fig. 4 down).

The PP is called in the case of *Escape* and *Replan*. In the case of *Escape*, the path planning task is simple: it is to select a close and safe target point $q_{target}$ in the free space, far enought to enable safe on-the-fly replanning, and to generate a path to reach it

(Fig. 4 right). That is, *Escape* provides a path that not only permits to avoid the obstacle, but also provides the time for replanning a new path to the goal. The interesting case is the third one, where the path planning process should find an alternative path that connects the old trajectory with a new one while the robot is flying. The replanning algorithm is illustrated in Algorithm 2. Given the target $q_{goal}$, the old path $path_{old}$, the collision point $q_{obs}$, and the $t_{ttc}$ time, the replanning process first estimates the time needed to replan $t_{rp}$ (*estimatedRepTime*); then it selects a waypoint $wp_{rp}$, along the old path $path_{old}$, from which it is possible to safely calculate the deviation $path_{new}$ from $path_{old}$ (*selectDeviationWP*); finally, upon setting a suitable threshold (*setThreshold*), the replanning process calls RRT_refine to generate the new path $path_{new}$ from the deviation waypoint $wp_{rp}$ to the target $q_{goal}$. $path_{new}$ should allow the PR to generate a new trajectory connecting the old one with a smooth deviation from $wp_{rp}$.

---

**Algorithm 2.** Replan($q_{goal}, path_{old}, q_{obs}, t_{ttc}$)

$q_c \leftarrow$ getPosition();
$t_{rp} \leftarrow$ estimatedRepTime($q_c, q_{goal}, path_{old}, q_{obs}$);
$wp_{rp} \leftarrow$ selectDeviationWP($q_c, q_{obs}, path_{old}, t_{rp}$);
$threshold \leftarrow$ setThreshold($wp_{rp}, q_{goal}, t_{rp}, t_{ttc}$);
$path_{new} \leftarrow$ Refine_RRT($wp_{rp}, q_{goal}, threshold, t_{rp}$);
**return** $path_{new}$

---

To select the deviation waypoint $wp_{rp}$ we defined the following strategy. Given the estimated time needed to replan $t_{rp}$, we estimate the robot position $q_{pr}$ at time $t_{rp}$ (assuming that it keeps following the old path $path_{old}$ during replanning), if there exists a waypoint $wp$ in $path_{old}$ that follows $p_{rp}$ and precedes $q_{obs}$ (keeping a suitable range the we assume as $maxRange$), then we select $wp$ as the deviation waypoint $wp_{rp}$, otherwise, $q_{rp}$ is on the path segment that intersects the obstacle, hence we select $wp_{rp}$ as the point $q_m$ in the middle of the segment that connects $q_{rp}$ and a point $q'_{obs}$ which is at $maxRange$ distance from $q_{obst}$. In Fig. 4 (center), we find an example of replanning from a waypoint after the collision detection (left).

### 1.3  3D Mapping

The environment for mapping and path planning is a 3D grid-map of cells which is run-time generated given the robot pose and the 3D point clouds extracted from the cameras. We deploy the well known pin-hole camera model [7]. Pose estimation of the UAV is needed to identify the 3D position of the projected camera points in the world reference frame. Our pose is either obtained by using libviso2 [6] coupled with a Kalman filter or, alternalively, by directly deploying an optitrack motion capture system. Given the pose, the associated point cloud map should be suitably processed into a 3D occupancy grid. This is obtained by discretizing the vehicle's workspace with elementary cubes of equal size. In our case, we employed a vehicle of $50 \times 50 \times 20$ cm hence, we used cubes of 10 cm. For each cube we stored: the number of inliers (3D triangulated points) fell into the cube volume, the last camera position which an inlier had been collected, and the state of the cube. The number of inliers represents the number of different points from which the same obstacle has been detected. The last camera position is required in case of hovering, to avoid that the same image feature generates dome wrong inliers, while it is possible that the same outlier is achieved from different points of view. Each cell can be associated with one of the following values: *free*, *occupied*, *obstacle*, *target*, *ignored* or *unknown*. Initially each cube is set to *free*. When a 3D point is detected to belong to a given cube, the value of the corresponding cube is set to *occupied*. When the number of points inside a cube reaches a given treshold, the state is set to *obstacle*. On the other hand, when a target is identified, the corresponding cube is set to *target*. Moreover, from each position that had generated a valid target view point, all the cubes laying along the optical rays are set to *ignored*. For wide environments, a sparse representation of the occupancy grid map is associated with a spatial/temporal vanishing criterion. This determines whether an occupancy cube is sill reliable or if it has to be discarded (depending on the distance travelled by the vehicle or on the time last after its previous update). In fact, due to the drift of the vehicle pose estimation, obstacles which have been observed a long time before or far from the current position cannot be considered reliable anymore in the current map representation, therefore they should be refreshed. With these solutions the reliability and scalability of the map representation can be suitably tuned.

## 2   Experimental Results

In this section, we present experimental results on planning, replanning, and obstacle avoidance, both in real-world scenarios and in simulated environments.

*Real-World Planning and Execution.* Our architecture has been tested in a real scenario of dimension $400 \times 400 \times 300$ $cm^3$ considering the two environments depicted in Fig. 5 (up and down). In the two testing scenarios, the task was the following: inspect a target point in pose $(380, 350, 50, 90)$ from the pose $(40, 40, 50, 0)$ with maximum and minimum speed set at $0.3\,m/s$ and $0.1\,m/s$ respectively. The obstacles are detected on the fly and this can provoke task/path replanning, escape, or brake. For each scenario, we executed each test 10 times collecting mean, max, min, and standard deviation (STD) of: time spent during planning (Tp), time spent in replanning (Tr), number of replanning episodes (Nr), length of the executed path (Lp), and total time for execution (including replanning time) (Te). For computation and simulation we used an Intel Core Duo, 1.40GHz, 3GB ram, Ubuntu 10.04. The high-level architecture was developed in ROS. As for 3D mapping, we used cameras ueye with hardware synchronized images, compressed on-board using atom 1.6 GHz, and sent to a ground station. The stereo images are streamed at around 15 Hz at the ground station. The vision algorithm can track around 120 image features correspondences on 4 images working at the streaming frequency. Each camera provides images with resolution of $752 \times 480$ and an angle of view of around $50°$.

Tab. 2 reports the results for the two scenarios (Test 1 and Test 2 in Fig. 5). For both these settings, initially, the obstacles are not visible, hence the generated plan is simple and planning time is low (Fig. 5 (left)). Once the obstacles are discovered on the fly,

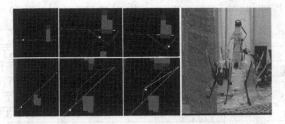

**Fig. 5.** Replanning: generated and executed path (left) real platform during plan execution (right)

**Table 1.** Planning and execution results (in seconds) in the real scenario

|  | Test 1 | | | | Test 2 | | | |
|---|---|---|---|---|---|---|---|---|
|  | Mean | STD | Max | Min | Mean | STD | Max | Min |
| Tp | 0.075 | 0.014 | 0.08 | 0.04 | 0.017 | 0.002 | 0.03 | 0.01 |
| Tr | 0.614 | 0.41 | 1.20 | 0.01 | 0.067 | 0.04 | 0.11 | 0.005 |
| Te | 60.5 | 10.12 | 75 | 42 | 49.9 | 8.18 | 60 | 40 |
| Lp | 14.4 | 1.54 | 18 | 12 | 13.18 | 1.11 | 15 | 11 |

replanning is needed to adjust on-line the trajectory. Replanning and execution time are slightly higher in the first scenario which is more complex. Instead, Tr seems negligible when compared with Te. The final trajectory length (Te) is similar in both the settings and comparable with the distance between the starting and target point, hence the final trajectory seems not affected by the continuous replanning process. In these tests, Tp and Tr are mainly due to path and trajectory planning (task planning is negligible). We never experienced brake or escape episodes. Overall, the system task/path planning performance seems compatible with the operative scenario requirements.

*Simulated Planning and Execution.* We tested our planning and execution system in simulated environments. To test continuous replanning, we considered a larger space of dimension $100 \times 100 \times 50 \ m^3$ with 4 and 9 obstacles. To decouple replanning from map bulding, we assumed a know map associated with a visibility horizon (not visible obstacles are detected on the fly causing replanning). For each test, the task was to inspect a target point in pose $(90, 90, 5, 90)$ starting from *hovering* in the pose $(5, 5, 5, 0)$ (in meters); the robot maximum and minimum velocity was set at $0.5 \ m/s$ and $0.1 \ m/s$ respectively. By changing the visibility horizon (green cells in Fig. 5) of the planner (15 or 25 m) and the complexity of the environment (4 or 9 obstacles) we obtained 4 scenarios. Tab. 2 collects means and STD of 10 tests for each entry (time and length are in *sec.* and $m$, LL, HL, etc. are for Low complexity and Low visibility, High complexity and Low visibility). Here, we can see that Tp increases with the obstacles (HL,HH) and decreases with short visibility (LL,HL). Indeed, in these cases the planning problem is simpler. However, short visibility is associated with additional replanning time which, in turn, decreases with the number of obstacles. The lower the replanning time, the lower is the execution time and the shorter the executed path. A similar effect is due to visibility: short visibility causes frequent replanning events (Nr) and longer paths (Lp)

**Table 2.** Planning and execution results(time in seconds, length in meters)

| Res/Env | LL | | LH | | HL | | HH | |
|---|---|---|---|---|---|---|---|---|
| | Mean | STD | Mean | STD | Mean | STD | Mean | STD |
| Tp | 0.21 | 0.11 | 0.39 | 0.03 | 0.25 | 0.10 | 0.31 | 0.14 |
| Tr | 0.12 | 0.03 | 0.07 | 0.01 | 0.20 | 0.04 | 0.23 | 0.03 |
| Te | 308.39 | 3.1 | 211.88 | 2.4 | 718.57 | 5.2 | 720.45 | 7.6 |
| Lp | 79.09 | 13.76 | 78.04 | 9.63 | 86.79 | 12.65 | 85.24 | 13.12 |
| Nr | 0.9 | 0.21 | 0.3 | 0.12 | 3.4 | 1.71 | 2.5 | 1.10 |

**Table 3.** Physical inspection and visual inspection

| | Physical Inspection | | | | Visual Inspection | | | |
|---|---|---|---|---|---|---|---|---|
| | Mean | STD | Max | Min | Mean | STD | Max | Min |
| Tp | 0.798 | 0.012 | 0.019 | 0.009 | 0.734 | 0.47 | 1.25 | 0.42 |
| Tm | 0.324 | 0.17 | 1.07 | 0.12 | 0.329 | 0.22 | 0.57 | 0.3 |
| Tpp | 0.473 | 0.27 | 0.71 | 0.14 | 0.405 | 0.07 | 0.49 | 0.39 |

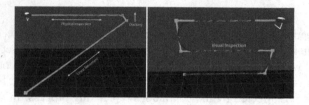

**Fig. 6.** Physical inspection (left) and visual inspection (right)

and execution times (Te). Furthermore, the variance is enhanced with short visibility that enhances the uncertainty. In these tests, the task planning time is usually negligible (Tp and Tr mainly due to path and trajectory). Also in this case, we never experienced brakes or escapes.

*Simulated Inspection.* As for operations closer to the surface, we considered two typical inspection scenarios: physical (Pi) and visual inspection (Vi). In both these cases the system has to move in a pose which faces a vertical surface hovering at a close and fixed distance (approach), in this case 50 cm. As for Pi (see Fig. 6, left), the robot executes a docking maneuver (docking) and slides (keeping the contact) along a linear trajectory (p-inspect) of 225 $cm$. In the case of Vi, an inspection trajectory (v-inspect) should be planned and executed. Here, the goal is to scan a $150 \times 100$ $cm^2$ surface with step 50 $cm$ distant 50 cm from the wall (see Fig. 6, right). In Tab. 3, we collect the results of 10 tests for each scenario considering planning time (Tp) divided in trajectory (Tm) and path planning (Tpp) time (task planning is negligible). For each test and scenario, both path and task planning times are compatible with the operative scenario requirements.

# 3  Conclusions

Aerial Service Robotics is a challenging and novel application for autonomous systems. The close and physical interaction with the environment and the frequent user interventions requires a high-level control system which integrates fast and reactive planning engines working at different levels of abstraction and sensitive to low-level operative mode constraints. We proposed a system that combines a set of AI methods (HTN planning, BDI execution, RRT path planning/replanning) showing its performance and feasibility in a industrial plant inspection case study. Future work will focus on additional real-world experiments and on the extension to multi-aerial vehicles.

# References

1. EU Collaborative Project ICT-248669, AIRobots, http://www.airobots.eu
2. Allen, J., Ferguson, G.: Human-Machine Collaborative Planning. In: NASA Workshop on Planning and Scheduling for Space (2002)
3. Bay, H., Ess, A., Tuytelaars, T., Van Gool, L.: Speeded-up robust features (SURF). Computer Vision and Image Understanding 110(3), 346–359 (2008)
4. Bloesch, M., Weiss, S., Scaramuzza, D., Siegwart, R.: Vision Based MAV Navigation in Unknown and Unstructured Environments. In: ICRA, pp. 21–28 (2010)
5. Doherty, P., Kvarnström, J., Fredrik, H.: A temporal logic-based planning and execution monitoring framework for unmanned aircraft systems. In: AAMAS, pp. 332–377 (2009)
6. Geiger, A., Ziegler, J., Stiller, C.: StereoScan: Dense 3d Reconstruction in Real-time. In: IEEE Intelligent Vehicles Symposium, pp. 963–968 (2011)
7. Hartley, R., Zisserman, A.: Multiple View Geometry in Computer Vision, 2nd edn. Cambridge Univ. Press, Cambridge (2004)
8. Hrabar, S.: Vision-Based 3D Navigation for an Autonomous Helicopter. Ph.D. Dissertation, University of S. California (2006)
9. Ingrand, F., Georgeff, M.P., Rao, A.S.: An architecture for Real-Time Reasoning and System Control. In: IEEE Expert: Intelligent Systems and Their Applications, pp. 34–44 (1992)
10. Lavalle, S.M.: Rapidly-Exploring Random Trees: A New Tool for Path Planning. Computer Science Dept., Iowa State University, Tech. Rep. (1998)
11. Macfarlane, S.E., Croft, E.A.: Jerk-bounded manipulator trajectory planning: design for realtime applications. IEEE Transactions on Robotics 19, 42–52 (2003)
12. Morales, M., Tapia, L., Pearce, R., Rodriguez, S., Amato, N.: A machine learning approach for feature sensitive motion planning. In: Int. Workshop on the Algorithmic Foundations of Robotics (2004)
13. Holmes, S., Klein, G., Murray, D.W.: A square root unscented Kalman filter for visual monoSLAM. In: ICRA, pp. 3710–3716 (2008)
14. Naldi, R., Marconi, M., Gentili, L.: Modelling and control of a flying robot interacting with the environment. Journal of IFAC 47(12), 2571–2583 (2011)
15. Rao, A.S., Georgeff, M.P.: Deliberation and its Role in the Formation of Intentions. In: UAI, pp. 300–307 (1991)
16. Robotics Operating System, ROS, http://www.ros.org
17. Stentz, A.: Optimal and efficient path planning for unknown and dynamic environments. Int. J. of Robotics and Automation 10(3), 89–100 (1995)

# A BSO-Based Algorithm
# for Multi-robot and Multi-target Search

Hannaneh Najd Ataei[1], Koorush Ziarati[1], and Mohammad Eghtesad[2]

[1] School of Electrical and Computer Engineering, Shiraz University, Shiraz, Iran
ataei@cse.shirazu.ac.ir, ziarati@shirazu.ac.ir
[2] School of Mechanical Engineering, Shiraz University, Shiraz, Iran
eghtesad@shirazu.ac.ir

**Abstract.** Swarm robots are used in robotic applications where it is difficult or impossible for a single robot to accomplish a task. In this paper, we study multi-robot, multi-target search problem in an unknown environment. Our goal is to use a group of distributed cooperative mobile robots to find position of an object which is emitting the strongest intensity of radio frequency in the environment. We propose a novel algorithm based on Bee Swarm Optimization (BSO) which is able to automatically find the object. Our experimental results, simulated on a set of random benchmarks, show that the algorithm is able to outperform the state-of-the-art techniques, in particular Particle Swarm Optimization (PSO). We show that our algorithm can be 50.6% more effective for this application in comparison to PSO.

**Keywords:** Multi-robot search, Particle Swarm Optimization, Bee Swarm Optimization.

## 1 Introduction

The problem of exploring and finding targets in an unknown environment is one of the fundamental problems in multi-agent mobile robotics. Robots can be programmed to explore the search space and to detect targets by using sensors. It is necessary to use robots instead of human for searching in some dangerous or unreachable areas. Mobile robots are used in many search applications, e.g., mine detecting [1], [2], search in damaged buildings after an earthquake [3], [4], fire fighting [5], and planetary exploration [6], [7].

One of suitable algorithms for multi-robot search is Particle Swarm Optimization (PSO). PSO [8] is an evolutionary algorithm inspired by the movement of flocking birds. It models solutions of a problem as particles of a swarm which are searching in a virtual space for good solutions. In recent years, PSO algorithm has been applied to many different problems successfully. In robotics field, Doctor et al [9], discussed on optimization of PSO parameters for multi-robot search. Pugh and colleagues [10,11], applied aspects of multi-robot algorithms to PSO and presented a simple version of PSO for multi-robot search. Couceiro et al [12] and Hereford [13] further developed a distributed particle swarm algorithm to find one target in an environment. Recently, Derr and Manic [14],

M. Ali et al. (Eds.): IEA/AIE 2013, LNAI 7906, pp. 312–321, 2013.
© Springer-Verlag Berlin Heidelberg 2013

used multiple robots to search in a noisy environment and find multiple targets when robots have only local neighborhood (each robot can communicate with a subset of robots around itself). This assumption limits their work since in some cases robots should have global neighborhood to perform their tasks; consider the application of surviving people after an earthquake. When a robot has global neighborhood, it can communicate with all of the robots in the search system and share its information. Although, PSO provides powerful method to multi-robot search, it has different problems such as premature convergence and stagnation. Especially, when the robots have global neighborhood PSO do not provide a decent performance for multi-target problems.

Robots which execute PSO algorithm with global neighborhood, could be trapped in local optima and are not capable of escaping from it. On the other hand, PSO is easy to be applied on swarm of robots due to its few adjustable parameters. So, in the case of multi-target problem, we need an algorithm that have positive features of PSO and also have the ability of exploring search space more than PSO. Bee Swarm Optimization (BSO) can be an alternative option. BSO is an optimization technique to find a solution of an optimization problem in a virtual space. The main advantage of BSO compared to PSO is the fact that BSO categorizes agents into three different types with three different patterns of searching. We expect that having three different patterns in our search system provide better results in finding the target with best fitness due to balancing exploration and exploitation. In this paper, we use swarm of realistically simulated robots and implement our BSO-based algorithm as their decentralized controller. In decentralized control method, each robot executes the algorithm by itself as well as sharing its information to achieve a global goal.

The focus of this paper is to explore an unknown environment, search for potential targets, and finally choose the target with best fitness. In this technique, fitness is the intensity of a radio frequency that each target is emitted. Consider the situation of finding the major source of fire to extinguish it. In such situations, we need to find the target with best fitness among all. Two different environments are designed and simulated to test different aspects of our BSO-based algorithm. We executed multi-robot search process using BSO and PSO algorithms in the environments with different number and initial positions of robots. Simulated robots have some information about their global positions but they do not have any map of the environment. Results show that our algorithm can find and reach to the right target 50.6% of the time more than PSO algorithm.

The rest of the paper is organized as follows. Section 2 reviews the classic BSO algorithm. Section 3 presents our BSO-based algorithm that is applied to multi-robot search. In section 4 we model the algorithm using a realistic sensor based simulation. The performance of proposed algorithm is analyzed for several numbers of robots in section 5. These results are compared to those of PSO model. Section 6 presents the conclusions.

## 2    Classic BSO Algorithm

The BSO algorithm [15] is a population based optimization technique which is inspired from foraging behavior of honey bees. A bee may be in a form of one of the following types: scout, onlooker or experienced forager. Experienced forager bees provide information about the environment and the currently discovered targets for onlooker bees and advertise them. Onlookers process the information shared by experienced foragers and select their interesting foragers. Scout bees 'fly' spontaneously and search for new targets with no knowledge about the environment and the position of discovered targets in it. The basic pseudo code for the classic BSO is presented in algorithm 1.

Initialize population with random positions and velocities;
**while** *termination condition* **do**
> *Compute fitness of each bee in swarm;*
> *Sort bees based on their fitness;*
> *Partition the swarm into the experienced forager, onlooker, and scout bees;*
> **foreach** *experienced forager bee* **do**
>> *Update the previous best position;*
>> *Select elite bee for all the experienced forager bees;*
>> *Update position of bee by (2);*
> **end**
> **foreach** *onlooker bee* **do**
>> *Select an elite bee from experienced forager bee by roulette wheel for onlooker i.;*
>> *Update position of bee by (3);*
> **end**
> **foreach** *scout bee* **do**
>> *Update position of bee by (4);*
> **end**
**end**

**Algorithm 1.** BSO Algorithm

The Euclidean distance from a bee to a target is an alternative fitness function that may be used by the BSO algorithm. The fitness function on the base of Euclidean distance is shown in (1); coordinates of the particle are $p_x$ and $p_y$, while coordinates of the target are $t_x$ and $t_y$.

$$fitness = \sqrt{(t_x - p_x)^2 + (t_y - p_y)^2} \qquad (1)$$

After sorting bees, predetermined numbers of them which have worst fitness are selected as scouts. Remaining bees are divided equally. The bees in which half with better fitness are selected as experienced foragers and the other bees are selected as onlookers. Experienced foragers and onlookers have the task of exploiting the search space precisely, like particles in PSO, while scouts help the system to have more exploration during the search process. Next positions of different bees are computed as follows:

$$p_{n+1} = p_n + c_1 * r_1 * (pbest_n - p_n) + c_2 * r_2 * (gbest_n - p_n) \tag{2}$$

$$p_{n+1} = p_n + c_2 * r_2 * (elite_n - p_n) \tag{3}$$

$$p_{n+1} = p_n + RW(\tau, p_n) \tag{4}$$

Where $p_n$ is current position of the bee, $c_1, c_2$ are acceleration constants, $r_1, r_2$ are random variables of uniform distribution in range of $[0, 1]$, *pbest* is position of the bee with best fitness value up to that moment, *gbest* is position of a bee with best fitness value in the entire swarm, *elite* is position of the interesting experienced forager bee for an onlooker which is chosen by roulette wheel, and *RW* is a random walk function that depends on the current position of the scout and the radius search $\tau$. The search process is terminated when a predetermined number of iterations or the minimum error requirement is reached.

## 3    BSO-Based Algorithm for Using in Multi Robot Search

The goal of presented multi-robot BSO-based algorithm is to find the best target in an environment. In this approach, there are number of targets at unknown locations in the search area. Each target radiates a Radio Frequency signal with different intensity that can be detected by robots. Swarm mobile robots cooperate with each other to find the target with strongest frequency intensity. Before using the proposed algorithm, we need some modification to apply aspects of multi-robot algorithms on it.

- In BSO, bees have infinite acceleration and there are no limitations on their velocity. But in real world, there are some kind of limitations on a robots velocity and acceleration due to robots capability. So, Unlike BSO, in multi-robot search algorithms it is better to calculate velocity instead of position for robots and simplify movement constraints as velocity limitations. Modified BSO equations are as follows:
  Experienced forager:

$$v_{n+1} = w * v_n + c_1 * r_1 * (pbest_n - p_n) + c_2 * r_2 * (gbest_n - p_n) \tag{5}$$

  Onlooker:
$$v_{n+1} = w * v_n + c_2 * r_2 * (gbest_n - p_n) \tag{6}$$

  Scout:
$$v_{n+1} = v_n + RW(\tau, p_n) \tag{7}$$

In these equations, $w$ is inertia coefficient which can control the exploitation during the search and $v_n$ is current bee/robot velocity. By increasing $w$, the robot can explore more places regardless of *pbest* or *gbest*. In these equations, *RW* is a pattern which is given to the scout to follow it. It could be any kind of patterns but in this research we choose a simple forward line pattern.

- In BSO algorithm, bees update their computed positions at discrete iterations. However in real world, we do not have any discrete time and robots operate in continuous time. So, instead of jumping robots we should have robots that moving continuously. Therefore, each robot move at the appropriate velocity towards its desired location until the next velocity and direction are computed. This approach requires that the swarm is synchronized.
- There are no collisions in BSO algorithm due to the assumption about their infinitely small sizes. In contrast with BSO, in multi-robot systems, stuck on surfaces for robots are probable. In this paper, we use Braitenberg obstacle avoidance to prevent robots from possible collisions. Obstacle avoidance in multi-robot systems requires amount of time for robots to adjust their headings. For simplicity, we use holonomic robots and allow them to immediately change their direction. So, this extra time is ignoring.
- In BSO algorithm, Euclidean distance is mostly used as fitness function. In multi-robot search whit no information about position of the target, this function is useless. In most of previous researches, robots approximate the distance between a target with known source power strength and themselves on the base of received signal strength. In our proposed technique, there are multiple targets with different and unknown powers. Therefore, robots cannot approximate the distance. So the fitness function in this research is the intensity of targets' signal that each robot receives. Robots prefer higher intensities, so they prefer higher fitness.

## 4   Simulation

In this section, we simulate our BSO-based multi-robot search algorithm and the PSO-inspired search with global positioning presented in [11]. We use Webots, a realistic simulator, for our robotic simulations [16]. Cylindrical robots are designed and used in this project. These robots have 8 infrared sensors around themselves to detect obstacles. They have global neighborhood and complete knowledge about their position via GPS. They are holonomic robots with two wheels; so they can change their direction easily. Radio frequency can be received by these robots and also they have wireless communication with each other for sharing their information. Common parameters for our BSO-based algorithm and the PSO-inspired algorithm which are listed in Table. 1, are set at the beginning of the simulation. We designed and simulated two environments to test the ability of the proposed algorithm.

In the first environment, robots operate in 4 m × 4 m arena with three targets with different power strength at three different corners. Different targets have different radius, depend on the power of each target. Targets' frequencies are not cover all over the search space. So, when robots cannot detect any frequency, they should search randomly.

In the second environment, we model a 4 m × 4 m arena in which the strongest target surrounded by many other less powerful targets. Our goal is finding the target which is emitting the strongest signal frequency. Robots are

**Table 1.** Special Parameters

| Parameter | BSO | PSO |
|---|---|---|
| Maximum velocity ($m/s$) | 1 | 1 |
| Inertia coefficient ($w$) | 1.2 | 1.2 |
| $c_{1,2}$ | 2 | 2 |
| Number of scouts | 20% | 0 |

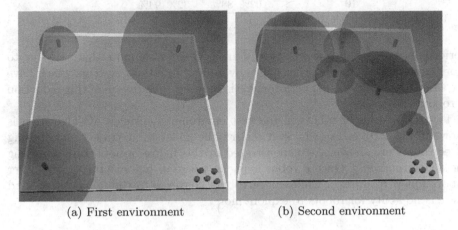

(a) First environment        (b) Second environment

**Fig. 1.** One of the random initial positions of robots in simulated environments

initially placed randomly in the search space with random headings and velocities. Figure 1 shows one of the start states for both of environments.

## 5 Results

We want to show and compare the behavior of agents using PSO and BSO algorithms during the search in both of simulated environments. Both of algorithms were executed over 10000 iterations for each test environment. The performance averaged over 50 run in the first environment with different initial positions and different number of robots is shown in Table 2. In order to avoid collisions, speed of robots will be decreased when the number of robots is increased. So, it takes more time for robots to explore in the search space to find a target. Therefore, the performance of the search system in a specified time is decreased. As shown for PSO, if we added more robots to the system in the same way then the probability of being in the range of the best target for robots initial positions would be raised. As a result, the performance is increased.

Performance of the BSO algorithm is affected by number of scout robots in the search process. In Table 2 , for 3 and 5 robots, there is one scout; for 8 and 10 robots, there are two scouts that search randomly in the environment. The

**Table 2.** Percentage of being successful in finding the best target during both of simulations with different number of robots

|  | Simulation 1 | | Simulation 2 | |
| --- | --- | --- | --- | --- |
| number of robots | PSO | BSO | PSO | BSO |
| 3 | 52% | 88% | 22% | 86% |
| 5 | 50% | 76% | 14% | 82% |
| 8 | 42% | 96% | 30% | 98% |
| 10 | 62% | 88% | 32% | 94% |

performance of the system when there are two scouts is more than one due to more exploration. On the other hand, by adding robots to systems which have the same number of scouts, performance is reduced. The reason of this reduction is robots' lower speeds.

The trajectories of robots in the first environment using PSO algorithm is shown in Fig. 2. In order to make it easier to be understood, trajectories of using three robots are shown. As demonstrated in this figure, robots start searching the space randomly. Once a robot being in range of one of targets, the other robots follow it. After a while, all of robots converge to the found target and there is no abilty for robots to explore more.

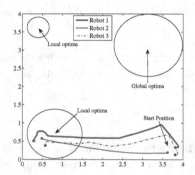

**Fig. 2.** Path taken by each robot using PSO algorithm

In contrast with PSO, BSO algorithm helps robots to have more exploration. As shown in Fig. 3, robots start searching in the environment randomly. When robot 2 is being in the range of one of targets, it becomes experienced forager and the other robots become onlooker and go through the forager. The robot which reaches to the target sooner (robot 1) is staying as onlooker while the other agent (robot 3) becomes scout and searches randomly. When a scout finds a target with less power than experienced found, it acts like an obstacle to that target; but if it finds more powerful target, scout becomes experienced and the

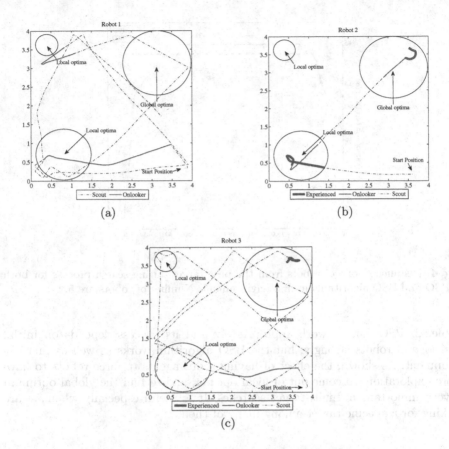

**Fig. 3.** Path taken by each robot using BSO algorithm

other robots ranked again. So, as shown, after a while robot 3 finds the global optimum and the other robots except new scout (robot 1) converge to it.

In Fig. 4, the distance of one robot and the best global target is shown for two PSO and BSO algorithms during search process in the first environment. As shown in this figure, the robot which runs the PSO algorithm is receding from the best target. Hence, it found and chose another target as its best one. In the rest of the algorithm, there is no ability for the robot to escape from local optima and the robot is trapped in it. However, for the robot which runs the BSO algorithm, in first iterations the distance to the best target is becoming large. After near 3000 iterations, this distance do not change any more which means that the robot finds a wrong target. By continuing the search process, in contrast with PSO, the robot changes its path and follows a scout robot which finds a stronger target. Therefore, the distance between the robot and the best target is being reduced until the robot reaches to the target.

In the second simulated environment, there are more local optima and therefore finding the target with strongest intensity is harder. As demonstrated in

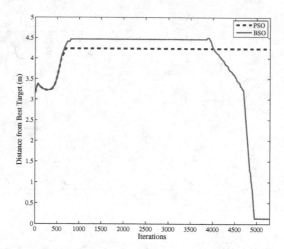

**Fig. 4.** Distance of one of robots from the best target during search process for both of PSO and BSO algorithms in first environment. Number of robots are 5.

Table. 2, PSO cannot work appropriately and its success depends on initial positions of robots strongly; but the BSO algorithm works as well as first environment. As shown, the effect of having scout agents to force robots to have more exploration, to come out of local optima, and to find the global optimum is very important in multi-target searches. It is useful especially when we are looking for a specific target among many of them.

## 6    Conclusions

In this paper, we have presented a BSO-inspired algorithm which is adopted in multi-robot search. We have simulated cooperative mobile robots and applied the proposed algorithm to explore a specific target among others. We have shown that our algorithm can be 50.6% more effective for this application in comparison to PSO.

We tested the proposed algorithm in static environments. However because of having scout robots in this algorithm, it seems that our BSO-inspired algorithm has been used successfully on dynamic environments as well, but it should be tested to determine concretely how well it is able to cope with these changes. Time of search is influenced by many issues like scout movement pattern. In this paper, we use a simple forward direction pattern for scouts. If more intelligent patterns are used, time of the search can be decreasing. However, we can run everything on a small swarm of real robots to test the algorithm in real world.

# References

1. Acar, E., Choset, H., Zhang, Y., Schervish, M.: Path planning for robotic demining: Robust sensor-based coverage of unstructured environments and probabilistic methods. The International Journal of Robotics Research 22, 441–466 (2003)
2. Gage, D.: Many-robot mcm search systems. In: Proceedings of the Autonomous Vehicles in Mine Countermeasures Symposium, vol. 9, pp. 56–64. Citeseer (1995)
3. Kantor, G., Singh, S., Peterson, R., Rus, D., Das, A., Kumar, V., Pereira, G., Spletzer, J.: Distributed search and rescue with robot and sensor teams. In: Field and Service Robotics, pp. 529–538 (2006)
4. Jennings, J., Whelan, G., Evans, W.: Cooperative search and rescue with a team of mobile robots. In: Proceedings of the 8th International Conference on Advanced Robotics, ICAR 1997, pp. 193–200. IEEE (1997)
5. Marjovi, A., Nunes, J., Marques, L., de Almeida, A.: Multi-robot exploration and fire searching. In: IEEE/RSJ International Conference on Intelligent Robots and Systems, IROS 2009, pp. 1929–1934. IEEE (2009)
6. Landis, G.: Robots and humans: Synergy in planetary exploration. In: AIP Conference Proceedings, vol. 654, p. 853 (2003)
7. Schilling, K., Jungius, C.: Mobile robots for planetary exploration. Control Engineering Practice 4, 513–524 (1996)
8. Kennedy, J., Eberhart, R.: Particle swarm optimization. In: Proceedings of the IEEE International Conference on Neural Networks, vol. 4, pp. 1942–1948. IEEE (1995)
9. Doctor, S., Venayagamoorthy, G., Gudise, V.: Optimal pso for collective robotic search applications. In: Congress on Evolutionary Computation, CEC 2004, vol. 2, pp. 1390–1395. IEEE (2004)
10. Pugh, J., Segapelli, L., Martinoli, A.: Applying aspects of multi-robot search to particle swarm optimization. In: Dorigo, M., Gambardella, L.M., Birattari, M., Martinoli, A., Poli, R., Stützle, T. (eds.) ANTS 2006. LNCS, vol. 4150, pp. 506–507. Springer, Heidelberg (2006)
11. Pugh, J., Martinoli, A.: Inspiring and modeling multi-robot search with particle swarm optimization. In: IEEE Swarm Intelligence Symposium, SIS 2007, pp. 332–339. IEEE (2007)
12. Couceiro, M., Rocha, R., Ferreira, N.: A novel multi-robot exploration approach based on particle swarm optimization algorithms. In: 2011 IEEE International Symposium on Safety, Security, and Rescue Robotics (SSRR), pp. 327–332. IEEE (2011)
13. Hereford, J.: A distributed particle swarm optimization algorithm for swarm robotic applications. In: IEEE Congress on Evolutionary Computation, CEC 2006, pp. 1678–1685. IEEE (2006)
14. Derr, K., Manic, M.: Multi-robot, multi-target particle swarm optimization search in noisy wireless environments. In: 2nd Conference on Human System Interactions, HSI 2009, pp. 81–86. IEEE (2009)
15. Akbari, R., Mohammadi, A., Ziarati, K.: A novel bee swarm optimization algorithm for numerical function optimization. Communications in Nonlinear Science and Numerical Simulation 15, 3142–3155 (2010)
16. Michel, O.: Webotstm: Professional mobile robot simulation, arXiv preprint cs/0412052 (2004)

# Economic Sentiment: Text-Based Prediction of Stock Price Movements with Machine Learning and WordNet

Arne Thorvald Gierløff Hollum, Borre P. Mosch, and Zoltán Szlávik

Department of Artificial Intelligence, Vrije Universiteit, De Boelelaan 1105, 1081 HV
Amsterdam, The Netherlands
hollum@online.no, borremosch@gmail.com, z.szlavik@vu.nl

**Abstract.** This paper explores the use of machine learning techniques in classifying financial news for the purpose of predicting stock price movements. The current body of literature on the subject is small, and the reported results are mixed. During the course of this paper we attempt to identify some causes for the divergent results, and devise experiments that account for weaknesses in existing research. A corpus of Thomson Reuter newswires was collected from Dow Jones' Factiva for seven large stocks. Each article was then linked with the associated price gap of the trading day following the article's publish date. Utilizing a sequential minimal optimization based support vector machine along with a WordNet-transformed bag-of-words representation, predictions were made in the form of long and short signals. Another variant of the system was also evaluated, wherein Latent Semantic Analysis was employed to process the input data. The signals were conditioned on a set of thresholds, meaning that trade signals were only generated when the predicted values exceeded certain threshold values. Higher thresholds were associated with higher accuracy but a lower number of trading signals. Overall the results were promising.

## 1 Introduction

Within the current body of research in the field of text mining, many studies have sought to evaluate machine learning techniques and other approaches on domains such as movie reviews [1-2], but only a small number of studies have explored economic or financial applications. Critical reviews are a popular choice since they provide convenient labeling of the data in form of review scores, allowing supervised machine learning to be performed without the chore of manually labeling the data. Interestingly, financial data offers a similar form of labeling of data through so called price gaps, i.e. the difference between the closing price of some financial instrument on a given trading day, and the opening price of that instrument on the following trading day.

Given the wide availability of economic data, it is somewhat surprising that no consensus has yet been established on the effectiveness of text mining applications in financial markets. The apparent lack of literature may have several reasons. For example, it could be that much promising research is still work in progress, or that text mining simply does not have much to offer the financial domain. However, the

M. Ali et al. (Eds.): IEA/AIE 2013, LNAI 7906, pp. 322–331, 2013.

fact that many corporations claim to have devised and made use of profitable implementations suggests that the latter is not true. Aite group [3] reports that 35% of quantitative firms use machine readable newsfeeds, compared with 2% just three years ago. Specific examples include the news giant Thomson Reuters' News Analytics service, and their main competitor Bloomberg, who recently acquired sentiment analysis based technology firm WiseWindow [4]. Taking this into account, one cannot rule out that much exciting research has been conducted in the research departments of corporations.

The main purpose of this paper is to study the application of text mining techniques on news sources with the goal of predicting stock returns. We investigate the impact of various parameters on prediction accuracy, and identify certain tradeoffs that traders who employ such methods will typically face. Choosing financial markets as a domain means that we must adapt our approach to deal with time series data. Precautions must be taken to keep intact the intrinsic order of the data. Moreover, the vocabulary of news articles changes over time, and the cyclical nature of financial markets imply that news or terms which may have carried positive connotations in one period, may have negative connotations in a another period.

To make the vocabulary more static over time, we transformed the news data by replacing words with so called cognitive synonyms using a lexical database (WordNet). In addition, we applied a popular technique known as Latent Semantic Analysis, which can identify concepts in the term-document matrix. Shrinking the vocabulary with WordNet enabled us to reduce the size of the data sets fed to the LSA algorithm, facilitating speed. In a market that is increasingly being defined by automated processes, it is not enough to have the most efficacious model; speed is also of great importance.

## 2    Previous Research

Schumaker and Chen [5] used a data set of 9211 financial news articles related to stocks on the S&P500 index during a five week period, and employed a sequential minimal optimization based support vector machine to predict the stock price twenty minutes after a news article was released in terms of direction and closeness. Historical stock data was used alongside the news articles as input. Whenever a stock demonstrated an expected movement exceeding 1%, the system would either buy or short sell the stock. Depending on the experimental setup, the system achieved between 52.4% and 57% accuracy. Gidofalvi and Elkan [6] trained a Naïve Bayes classifier to predict stock price movements based on 5500 financial news articles on 12 stocks. They compared different thresholds for the labeling of price movements, and considered different intraday price movement windows. A window of 20 minutes prior to the publication of a news, and 20 minutes after, produced the strongest correlation between the news articles and the stock price movements. However, the overall predictive power of the classifier was low.

Mittermayer and Knolmayer [7] used a novel labeling approach based on moving averages along with various classifiers to predict stock price movements. With an initial corpus of 9128 press releases from the period April 1st to December 31st 2002, the authors used heuristics to filter the most relevant items, reducing the number of articles to 989. Utilizing a threshold of 3%, the system obtained an overall classification accuracy of 82% and average roundtrip profits between 0.22-0.27% based on 10-fold cross validation.

Finally, Koppel and Shtrimberg [8] gathered 12000 articles concerning stocks on the S&P 500 index in the period 2000-2002 and trained a support vector machine to predict daily price gaps. Two labeling approaches were compared. In the first approach, the price gap was calculated from the closing price of the day preceding the date of a news article publication, and the opening price of the day following the publication (e.g. the closing price of January 14th and the opening price of January 16th for a publication on the 15th). A positive (long) threshold of 10% and a negative (short) threshold of 7.8% filtered the data down to 425 positive examples and 426 negative examples. With 10-fold cross validation they achieved a precision of 70.3%. With a simple test set approach, the accuracy decreased to 65.9%. In the second labeling approach, the price gap was calculated by taking the closing price of the day following the publication date, and the opening price of the subsequent day (e.g. the closing price of January 16th and the opening price of January 17th for a publication on the 15th). Considered a more realistic scenario for trading scenarios, the labeling approach yielded a precision of 52%. Overall, the authors found that it was easier to predict negative price movements.

To summarize, the existing literature on the subject contains divergent results. We propose some possible explanations for this divergence. First, standard cross validation might be unfit for time series prediction since it breaks the causal relation between time points, producing artificially high scores. Second, many of the aforementioned studies only considered articles that were published during a relatively short time frame, which might compromise generalizability. Third, the use of arbitrary thresholds makes it difficult to compare results. Finally, the studies employ different labeling approaches and time windows. In particular, we argue that short intraday time windows are prone to noise and therefore not ideal for the labeling of the data.

# 3     Method

In the previous section, we identified some possible reasons for the apparent conflicting results in the literature. The motivation for our methodology is therefore to account for the possible shortcomings found in existing studies. First, to maintain the intrinsic order of the data we followed a simple training and test set approach. By transforming the articles with WordNet [9], the vocabulary contained in the dataset is made less variant with respect to time, rendering the system less dependent upon being updated with recent data, which could normally be an issue with a train/test set approach.Second, to ensure that results were generalizable across time, we chose a dataset which spans over more than a decade. Third, to avoid arbitrary parameter settings we produced results for a range of thresholds. Finally, we argue that daily price gaps represent a cleaner measure of the market sentiment than intraday price movements. Given that a news event concerning some company is published after market close, the overall market's opinion regarding that event should be fully reflected by the price gap of the following day. It is more difficult to gauge the reaction to a news event taking place within market open hours since intraday price movements are small and noisy. Moreover, one has to identify the correct time window (e.g. the 1 minute return could be negative, the 5 minute return positive etc.). Our labeling approach is therefore limited to articles published outside market open hours.

Most of the studies in the literature review relied on variants of the support vector machine. Likewise, we choose a variant of the support vector machine for this study, since it enables us to more easily investigate the claims made in previous studies.

Due to technological constraints, the dimensionality of the data had to be kept within reason, so a relatively small selection of companies was chosen.

### 3.1    Data and Preprocessing

All available Reuters News wires dating from 2000-11-30 to 2012-09-08 were collected from Dow Jones' Factiva service for the following companies: Dell, Google, IBM, Microsoft, Pfizer, Verizon, and Yahoo. These companies are all large constituents of the S&P 500 index, and subject to frequent news coverage. A substantial portion of these news events relate to technological subjects, product releases and patent development, rendering the domain of the corpus relatively focused.

A total of 23997 articles were collected. All duplicates were eliminated, and the news items were structured into three parts through the Factiva service: "headline", "lead paragraph" and "text". For the text section, only sentences which referred to the main subject of the article were kept in the data set. This was done to reduce noise in the data set. All of the news articles were assigned to either of three categories and stored in a database:

1. News published before market opening times   (After midnight, before 09:30)
2. News published after market opening times (after 16:00, before midnight)
3. News published during market opening times (between 09:30 and 16:00)

For each news item in category 1 and 2, a relation was made with the associated price gap of the nearest future trading day.

Historical stock data containing the daily price movements for each of the companies was downloaded from the eSignal data provider service [10].

**Data Representation.** A bag-of-words model was used to represent the input data. The raw input data was then transformed with the WordNet lexical database, which groups nouns, verbs, adjectives and adverbs into sets of cognitive synonyms or synsets. In this study, tokens were replaced with their most common synonym. Words that were not found in the WordNet database were kept in their original form.

By using WordNet, the number of distinct tokens in the dataset could be reduced. As a consequence, the size of a given word vector or term-document matrix could be kept small while still accounting for most of the corpus. WordNet further enabled a form of stemming, since all synonyms are given in the same form and tense. The bag-of-words representation was produced with the machine learning software Weka's [11] word-vector filter, which applies term frequency–inverse document frequency (TF-IDF) to the data. The target amount of words, and the maximum number of words per term (n-grams), were set to 500 and 3 respectively, based on heuristics. In absence of the WordNet transformation, a much larger bag of words would likely be necessary to provide good coverage.

## 3.2    Algorithms

**Attribute Selection.** Latent Semantic Analysis (LSA) is a popular NLP technique that has proven successful in text mining applications [12], wherein singular value decomposition is applied on the term-document matrix. The dimensionality reduction that follows from this procedure lets us attain a set of concepts that can depend on multiple words. For example, "wine" and "beer" may be reduced to a single dimension which represents "alcoholic beverage". In a sense, both LSA and WordNet mitigate the problem of synonymy. However, we argue that the two methods can complement each other. According to [13], it can be problematic to apply LSA on large datasets due to the computational complexity of the algorithm. The WordNet data transformation thus facilitates faster computations by reducing the dimensionality of the data set on which LSA is performed.

**Regression.** The output of the LSA process was fed to a sequential minimal optimization (SMO) based support vector machine [14]. In particular, Weka's implementation of the SMO algorithm was utilized to train a support vector regression machine using a polynomial kernel, with the processed news articles as input and the price gaps as targets. Again, the algorithm was chosen due to its well-known efficacy, and its prevalence in the relevant literature.

## 3.3    Experimental Setup

The training set for all experiments consisted of 60% of the data, and the test set of the remainder (see Table 1).

**Table 1.** Description of the dataset

|              | Start date  | End date    | Number of observations |
|--------------|-------------|-------------|------------------------|
| Training set | 2000-11-30  | 2004-10-22  | 10132                  |
| Test set     | 2004-10-23  | 2012-09-08  | 6756                   |
| Total        | 2000-11-30  | 2012-09-08  | 16888                  |

To evaluate the relationship between the predicted values and the target output, linear regression was performed with the target outcome as the dependent variable, and the predicted values as the independent variable. By utilizing two separate threshold parameters, it was possible to identify the combinations that produced optimal results based on historical data. That is, for all combinations of $t_l$ and $t_s$, we regressed the predicted values against the associated price gap values and studied the resulting statistical parameters. Next, we assessed the accuracy of the system in a classification experiment. Three measures of accuracy were specified: long precision, short precision and overall precision. When the trade signal and target price gap were either simultaneously positive or negative for a given instance, the prediction was recorded as being correct, otherwise incorrect.

## 4    Results

In the following we describe the results from the experiments. First, linear regression was used to identify the impact of threshold levels on significance levels and coefficient values. In the succeeding section, we present the results for classification experiments on prediction accuracy.

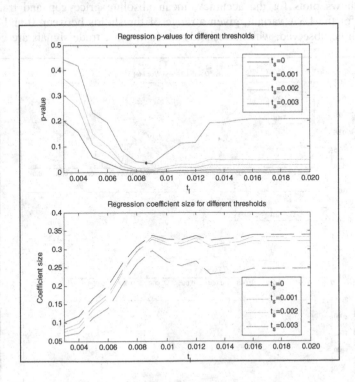

**Fig. 1.** Regression p-values and coefficient values for different threshold settings

**Table 2.** Regression: optimal p-value, coefficient and threshold values for both model variants

|          | p-value | Coefficient | $t_s$ | $t_l$ |
|----------|---------|-------------|-------|-------|
| No LSA   | 0.0019  | 0.34        | 0.009 | 0     |
| LSA      | 0.0428  | -0.8655     | 0.007 | 0.011 |

Linear regression was performed for all possible combinations of long and short thresholds between zero and the highest recorded value of prediction in the dataset. Figure 1 shows how the p-value and coefficient size vary for different values of $t_s$ and $t_l$, using the SMO model without LSA. From Table 2, the highest coefficient value (0.34) and the lowest p-value (0.0019) were obtained for a long threshold of 0 and a short threshold of 0.009. For the LSA variant, the highest absolute coefficient value (-0.8655) and lowest p-value (0.0428) were produced for $t_s = 0.007$ and $t_l = 0.011$. In an additional test wherein we predicted the magnitude of price gaps, the results remained highly significant, yielding a coefficient of 64.95% and a p-value near zero. For this test, the LSA processing step was omitted due to the detriment on performance it was shown to have in the previous test.

## 4.1    Classification

Figure 2 shows plots for the accuracy, mean absolute price gap and trade signal frequency for the LSA variant, given a range of thresholds between 0 and 0.011. A tradeoff can be observed: when thresholds are low, more trade signals are generated,

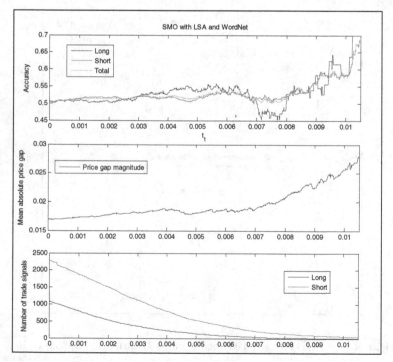

**Fig. 2.** Accuracy, mean absolute price gap and trade signal frequency for LSA variant

but at the expense of accuracy. The mean absolute price gap graph shows the following: when predicted values are high (in either direction), the price gap tends to be high. Finally, the trade signal frequency graph shows that negative price gaps are predicted more often than positive price gaps. As an example, setting the thresholds such that the amount of trade signals over the period is limited to 100 yields a total accuracy of 58.82%. 88 of those signals are short signals with an accuracy of 59.46%, while the remaining 12 purchase signals have an accuracy of 54.55%. Figure 3 shows the results from the variant without LSA. A 59% precision is achieved for long-only positions when the trade signals frequency is limited to 114. Given 85 trade signals, the long precision increases to 62%. In terms of short predictions, a 56% precision was produced across 194 trade signals.

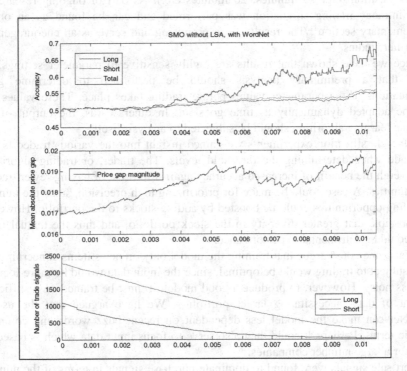

**Fig. 3.** Accuracy, mean absolute price gap and trade signal frequency for no-LSA variant

## 5    Discussion

The results of this paper suggests that financial news can contain valuable information which can be extracted with machine learning techniques and used to predict stock price movements. For a range of parameter settings, we found that the predicted values of the regression model and the target values were significantly related. This held true for both variants of the system, although the results from LSA variant were less significant. It is noteworthy that good results were obtained with a relatively small word vector based on cognitive synonyms. This suggests that news publications convey information about stock market reactions beyond buzz words and time-specific terms. Furthermore, the use of WordNet facilitated speed, which can be of great importance in financial markets.

In light of the results, the choice of labeling approach appears to have been justified. The price gap reflects a clean measure of the impact of substantial news which has been published after market open hours. While in a practical setting traders cannot profit from such price gaps, a system which has been trained according to our labeling approach could be used to predict intraday fluctuations, given that the market's reaction to news events is the same regardless of the time of publication. In a practical setting, investors must determine the optimal time window for which to hold a stock position (e.g. 10 minutes, 20 minutes etc.). As part of ongoing research, a small intraday trading simulation was performed and can be found as an online supplementary section[1]. The results were promising and serve as an encouragement for further studies.

Since we have shown that results are highly sensitive to parameter settings, we argue that a preliminary analysis should be performed to determine good combinations of the parameter values before trading takes place. These values can then be adapted dynamically as time goes on. In other words, the output of the machine learning algorithm should be subjected to analysis itself.

In the classification experiments we gained insight into the various tradeoffs that are made when determining the threshold levels. The trader, or trading algorithm, must weigh the benefit of increased accuracy against the cost of having fewer trading opportunities. A case could be made for prioritizing high precision, since the number of trading opportunities could be boosted by adding stocks to the portfolio. However, it is possible that greater diversity in the stock portfolio, and thus the textual input data, could be detrimental to performance.

In fact, one could argue that training the model only on news items concerning the stock subject to trading would be optimal, since the input data would be more focused and less noisy. However, to produce a good model, it must be trained on a sufficient amount of data, necessitating larger portfolios. We have argued that the use of WordNet can make the model less dependent on recent buzz words and company specific terms. Perhaps it could then also serve to focus input data which is based on news for a large number companies.

Short sale signals were found to dominate purchase signals in terms of the number of trading opportunities, i.e. the classifier tended to predict more substantial negative values than positive values. One possible explanation for this is that the impact of negative news is easier to predict than that of positive news, as suggested by [8]. However, the experiments also showed that the LSA variant was more precise in predicting positive price gaps, while the other variant had higher precision for negative price gaps. Since the two variants exhibited different strengths and weaknesses, a case could be made for combining them in some form of ensemble.

---

[1] https://sites.google.com/site/economicsentiment/

We conclude from the results that applications of text mining techniques in financial markets have much potential. However, this potential has not yet been fully explored by academic research, which appears to have only scratched the surface of what is possible.

# References

1. Kennedy, A., Inkpen, D.: Sentiment classification of movie reviews using contextual valence shifters. Computational Intelligence 22(2), 110–125 (2006)
2. Whitelaw, C., Garg, N., Argamon, S.: Using appraisal groups for sentiment analysis. In: Proceedings of the 14th ACM International Conference on Information and Knowledge Management, pp. 625–631 (2005)
3. Thomson Reuters Launches Social Media Sentiment Analysis Offering, http://thomsonreuters.com/content/press_room/financial/2012_03_07_social_media_sentiment_analysis_offering
4. WiseWindow Enters Strategic Relationship with Thoms Reuters to DeliverReal-Time Consumer Index Score for 300 Public Companies Covering Six Industry Groups, http://www.bloomberg.com/article/2012-02-03/aGzt5TYWYfiM.html
5. Schumaker, R.P., Chen, H.: Textual analysis of stock market prediction using breaking financial news: The AZFin text system. ACM Transactions on Information Systems 27(2) (2009)
6. Gidófalvi, G., Elkan, C.: Using news articles to predict stock price movements. Department of Computer Science and Engineering. University of California, San Diego (2001)
7. Mittermayer, M.-A., Knolmayer, G.F.: Newscats: A news categorization and trading system. In: Sixth International Conference on Data Mining, pp. 1002–1007. IEEE (2006)
8. Koppel, M., Shtrimberg, I.: Good news or bad news? let the market decide. Computing attitude and affect in text: Theory and application, pp. 297-301 (2006)
9. WordNet lexical database, http://wordnet.princeton.edu/
10. eSignal data provider service, http://www.esignal.com/
11. Weka: Data Mining Software, http://www.cs.waikato.ac.nz/ml/weka/
12. Kakkonen, T., Myller, N., Sutinen, E., Timonen, J.: Comparison of dimension reduction methods for automated essay grading. Natural Language Engineering 1, 1–16 (2005)
13. Zhang, D., Zhu, Z.: A fast approximate algorithm for large-scale Latent Semantic Indexing. In: Third International Conference on Digital Information Management, pp. 626–631. IEEE (2008)
14. Smola, A.J., Schölkopf, B.: A tutorial on support vector regression. Statistics and Computing 4(3), 199–222 (2004)

# Feature Extraction Using Single Variable Classifiers for Binary Text Classification

Hakan Altınçay

Department of Computer Engineering,
Eastern Mediterranean University, Famagusta, Northern Cyprus
{hakan.altincay}@emu.edu.tr

**Abstract.** The most popular approach for document representation is the bag-of-words where terms are considered as features. In order to compute the values of these features, the term frequencies are generally scaled by a collection frequency factor to take into account the relative importance of different terms. The term frequencies can be considered as raw data about the input document. In this study, a novel framework for feature extraction is proposed for binary text classification where feature extraction is defined as a single variable classification problem. The term frequencies are the inputs and the output of each classifier is used to define a triple of features for the corresponding term. The magnitude of the classifier output that is in the interval [0.5, 1] is an indicator for the confidence of the classifier and it is also employed in document representation together with the term frequency and the collection frequency factor.

**Keywords:** Feature extraction, single variable classifiers, document representation, term weighting, document classification.

## 1 Introduction

Grouping of the public documents on the web is needed to assist search engines for effective retrieval. Due to the exponentially increasing number of such documents, automatic text classification is generally considered as an important research direction, drawing the interest of an increasing number of researchers. This is also an important component for other tasks such as e-mail filtering, e-mail foldering, sentiment classification and biomedical information extraction.

The first stage of text classification is document representation. The most popular way of document representation is the bag-of-words (BOW) approach. A subset of discriminative terms is firstly selected using a measure such as chi-square ($\chi^2$) or information gain [1]. Then, each document is represented as a vector of the frequencies of these terms, each denoting the number of times the corresponding term occurs. The term frequencies are generally scaled by their collection frequency factors which quantify the importance of the term in the collection for determining the true category of the given document [1,2]. In general, documents are represented as high-dimensional feature vectors having

M. Ali et al. (Eds.): IEA/AIE 2013, LNAI 7906, pp. 332–340, 2013.

thousands of entries where support vector machine (SVM) with linear kernel has been show to surpass other approaches such as Bayes classifier, $k$-nearest neighbor classifier and decision trees due to its robustness to overfitting in large dimensional feature spaces [3].

The term frequency can be considered as raw data and novel features can be defined whose values will depend on the relative values of the frequencies in different classes. Classifiers built with single variables were formerly used to evaluate the prediction power of the individual features for feature selection [4]. In this study, feature extraction is defined as a single variable classification problem where the term frequencies are the inputs. By taking into account the distribution of the frequency of the term under concern in different classes, the classifier provides the most likely class together with its probability as the output. The magnitude of the classifier output that is in the interval $[0.5, 1]$ depends on the frequency of the corresponding term and it is an indicator for its support to the most likely class. Using this probability value, a weight vector is computed by employing the term frequency and the collection frequency factor as well. The weight vector represents the support to the classes under concern. A different classifier is generated for each term. Each document is then represented as a composite vector that is formed by concatenating the outputs of the individual classifiers. This approach can be considered as stacking single variable classifiers where a meta-classifier operating on the concatenated output vectors (meta-level features) is employed to combine the base classifiers [5,6].

The main strengths of this perspective can be summarized as follows: Firstly, instead of raw term frequency values, an improved representation including the confidence of the base classifier can be employed by the meta-classifier in generating the final decision. Secondly, document representation is generally considered to be a more important task compared to tuning the nonlinear kernel parameters [7] since, as it is experimentally shown, SVM with a linear kernel providing a linear decision boundary provides better performance compared to nonlinear ones [8]. The proposed representation has the potential of mapping the raw frequency based feature representation into a meta-level feature space that has a less complex decision boundary where better classification scores are achieved using the linear SVM.

The motivations for using single variable classifiers for feature extraction are presented in Section 2. The proposed document representation scheme is presented in Section 3. The datasets used, the experiments conducted and the results obtained are presented in Section 4. The last part, Section 5 summarizes the conclusions drawn from this study.

## 2   Main Motivations

One of the main problems in text categorization is document representation. The term frequencies are conventionally used for this purpose. However, they can be considered as raw inputs for classification like the pixel values in an image. The main idea is to use a single variable classifier to map the raw scores into a more discriminative space to be used as features by a meta-level classifier. With the use of a single variable classifier instead of considering a subset of features, employing other information sources related to each term such as the collection frequency factor or term occurrence probabilities in constructing the meta-level feature vectors is possible. This is important since, in text categorization, there are some terms that are much more informative than the vast majority of the others.

Extensive research has been carried out to take into account co-occurring terms which may be syntactic or statistical phrases [9,10,11]. In these studies, novel features are defined as the co-occurrences of two, three or more terms. It is generally argued that augmenting BOW with pairs of terms (bigrams) rather than replacing it leads to a slightly more powerful representation where better scores are achieved [12] but sequences of higher order are not found to provide further improvement [13]. This means that meta-feature extraction using several terms is not promising. Hence, feature extraction using the terms individually is considered in this study.

## 3   Proposed Document Representation Scheme

Consider a binary text categorization problem involving two classes namely, positive and negative. Let $S^+$ and $S^-$ denote the sets of positive and negative training documents with cardinalities $N^+$ and $N^-$ respectively. Assume that $f(t_k, d_i)$ denotes the frequency of the term $t_k$ in the particular document $d_i$. The frequencies of the selected terms are generally nonlinearly mapped using the logarithmic mapping defined as $\log(1 + f(t_k, d_i))$. Using this mapping, a large swing in the feature value of a term due to only a small change in its frequency can be avoided [1]. Our previous studies verified the advantages of this transformation [22]. Cosine normalization is then applied to the transformed term frequencies for document length normalization. In this paper, $\widehat{f}(t_k, d_i)$ is used to denote the frequencies obtained after this preprocessing. The single variable classifiers are designed to operate on the preprocessed term frequencies.

Let $\mu_{pos}(t_k)$ and $\mu_{neg}(t_k)$ denote the averages of the preprocessed frequencies of $t_k$ in the positive and negative documents, respectively which are computed as

$$\mu_{pos}(t_k) = \frac{1}{N^+} \sum_{d_i \in S^+} \widehat{f}(t_k, d_i),$$

$$\mu_{neg}(t_k) = \frac{1}{N^-} \sum_{d_i \in S^-} \widehat{f}(t_k, d_i). \tag{1}$$

Since it is obvious that the frequency of a term is document dependent, we can drop the term $d_i$ for brevity. Hence, unless necessary, the preprocessed frequency of the term $t_k$ is represented as $\widehat{f}(t_k)$ in the following context.

We studied the use of a simple classifier, namely the nearest mean classifier (NMC). The preprocessed frequency of a given term is compared with $\mu_{neg}(t_k)$

and $\mu_{pos}(t_k)$ where the absolute differences are computed. The score generated by the classifier depends on the inverses of these differences computed as

$$i_{pos} = \frac{1}{|\widehat{f}(t_k, d_i) - \mu_{pos}(t_k)|},$$

$$i_{neg} = \frac{1}{|\widehat{f}(t_k, d_i) - \mu_{neg}(t_k)|}. \tag{2}$$

Dropping $d_i$ from $\widehat{f}(t_k, d_i)$ for brevity, the score provided by the classifier is computed as

$$s_{nmc}(t_k) = \begin{cases} 0, & \widehat{f}(t_k) = 0 \\ \frac{i_{pos}}{i_{pos}+i_{neg}}, & \widehat{f}(t_k) > 0 \text{ AND } |\widehat{f}(t_k) - \mu_{pos}(t_k)| < |\widehat{f}(t_k) - \mu_{neg}(t_k)| \\ \frac{i_{neg}}{i_{pos}+i_{neg}}, & \widehat{f}(t_k) > 0 \text{ AND } |\widehat{f}(t_k) - \mu_{pos}(t_k)| \geq |\widehat{f}(t_k) - \mu_{neg}(t_k)| \end{cases} \tag{3}$$

$s_{nmc}(t_k)$ is in the interval $[0.5, 1]$. The magnitude of the score is proportional to the confidence of the classifier. The score it provides represents the support to the class having nearest mean. If the score is close to 0.5, the classifier is not very confident since $i_{pos} \approx i_{neg}$. A score close to 1.0 means that the classifier is highly confident about the label of the document. The score obtained is used to scale the term weight, $w(t_k)$ that is defined as the product of collection frequency factor, $CF(t_k)$ which takes into account the importance of the term for discriminating different classes and the preprocessed term frequency, $\widehat{f}(t_k, d_i)$ as

$$w(t_k) = CF(t_k) \times \widehat{f}(t_k, d_i). \tag{4}$$

The feature corresponding to the term under concern is defined as a triple, $\mathbf{r}$ defined as follows: If $|\widehat{f}(t_k) - \mu_{pos}(t_k)| < |\widehat{f}(t_k) - \mu_{neg}(t_k)|$, which means that the frequency of the term under concern is closer to $\mu_{pos}(t_k)$,

$$\mathbf{r} = [r_1, r_2, r_3] = \left[ s_{nmc}(t_k) \times w(t_k), \, 0, \, (1 - s_{nmc}(t_k)) \times w(t_k) \right]. \tag{5}$$

On the other hand, if $|\widehat{f}(t_k) - \mu_{pos}(t_k)| \geq |\widehat{f}(t_k) - \mu_{neg}(t_k)|$, which means that the term frequency is closer to $\mu_{neg}(t_k)$,

$$\mathbf{r} = [r_1, r_2, r_3] = \left[ 0, \, s_{nmc}(t_k) \times w(t_k), (1 - s_{nmc}(t_k)) \times w(t_k) \right]. \tag{6}$$

$r_1$ and $r_2$ represent the weights assigned to the positive and negative classes respectively whereas $r_3$ represents the weight that is assigned to the set of classes including both positive and negative classes. This is analogous to the universal set in Dempster-Shafer evidence theory [14]. It can be easily seen that the confidence of the classifier is used to distribute the term weight to the most likely class and the universal set of classes. Using this representation, the contribution to the similarity score by terms supporting different classes is reduced due to the linear kernel (inner product) used in linear SVM which is reasonable. For a better understanding, consider the case where two entries instead of three are considered, one for each class. For a given term, $t_i$ in a positive document, assume that $|\widehat{f}(t_k) - \mu_{pos}(t_k)| < |\widehat{f}(t_k) - \mu_{neg}(t_k)|$. Let

$$s_{nmc}(t_k) \times w(t_k) = a,$$
$$(1 - s_{nmc}(t_k)) \times w(t_k) = b. \tag{7}$$

$s_{nmc}(t_k) > 0.5$ implies that $a > b$. In the case of two entries, one for each class, the feature vector would be computed as $[a, b]$. Without any loss of generality, assume that the feature vector corresponding to the same term is computed as $[b, a]$ for a negative document. The similarity of these documents using two features is computed using $[a, b] \times [b, a]^T$ as $2ab$. For the proposed representation, the similarity is obtained as $[a, 0, b] \times [0, a, b]^T = b^2$ which is less than $2ab$ since $a > b$. On the other hand, consider two documents that belong to the same class (e.g. positive) where the feature vectors are computed as $[a, b]$ and $[a, 0, b]$ respectively. In this case, both representations provide the same similarity score, i.e $a^2 + b^2$. Hence, in comparing the similarity of different documents, the contribution of terms supporting different classes is reduced whereas the contribution of terms supporting the same class (positive or negative) is left unchanged. The text categorization system using the single variable classification based document representation described above will be referred as $\text{SVC}_{single}$ in the following context.

It should also be noted that, if $f(t_k, d_i) = 0$, then $\widehat{f}(t_k, d_i) = 0$. Hence, regardless of the classifier output, $\mathbf{r} = [0, 0, 0]$ when $t_k$ does not appear in the document. In other words, the terms that do not appear in the document under concern do not provide any support to any class. Let $\{t_1, t_2, t_3, \ldots, t_K\}$ denote the ordered set of selected terms ranked by $\chi^2$ to be employed in the categorization task. Assume that $\mathbf{r_i}$ denotes the triple corresponding to $t_i$. Then, the document vectors are computed as $[\mathbf{r_1}, \mathbf{r_2}, \ldots, \mathbf{r_K}]$ with this scheme.

As mentioned in Section 2, co-occurrence of two adjacent terms known as bi-grams improves the BOW representation. In this framework, this corresponds to meta-feature extraction using two terms as the classifier inputs. We are planning to work on this approach as the following research. In this study, as a preliminary experiment, we considered the combination of the feature triples obtained using different pairs of terms. More specifically, triples of the adjacent terms ranked by $\chi^2$ are added. After computing $\mathbf{r_i}$ for $i = 1, \ldots, K$, the combined triples are $\mathbf{r_i^c}$ obtained as,

$$\mathbf{r_1^c} = \mathbf{r_1} + \mathbf{r_2}$$
$$\mathbf{r_2^c} = \mathbf{r_2} + \mathbf{r_3}$$
$$\vdots \tag{8}$$
$$\mathbf{r_{K-1}^c} = \mathbf{r_{K-1}} + \mathbf{r_K}$$

The triples of features obtained are concatenated to form the document vectors as $[\mathbf{r_1^c}, \mathbf{r_2^c}, \ldots, \mathbf{r_{K-1}^c}]$. This approach combines the features generated for different term pairs, aiming to obtain a more reliable representation. This approach will be referred as $\text{SVC}_{pair}$ in the following context. It should be noted that, if $K$

terms are selected for classification, the proposed scheme generates document vectors of lengths $3K$ and $3(K-1)$ respectively for $SVC_{single}$ and $SVC_{pair}$.

Before training the meta-classifier (i.e. SVM), the document vectors obtained using the proposed approach are normalized to unit length using cosine normalization for a better similarity measurement using the linear kernel.

# 4 Experiments

## 4.1 Datasets

In this study, two widely used datasets are utilized for evaluating the proposed scheme on binary text categorization. The ModApte split of top ten classes of Reuters-21578 [15] is the first where the negative classes are defined to include documents which belong to one or more of the remaining nine categories. The same policy is applied on all datasets for forming the negative class. Due to its highly imbalanced category distribution, Reuters-21578 ModApte Top10 is a significant dataset among others. The second dataset, $CSTR^{2009}$ is composed of 625 abstracts, each belonging to a single category, from technical reports in four research areas namely "Systems", "Theory", "AI", "Robotics and Vision" published in the Department of Computer Science at the University of Rochester between 1991 and 2009[1]. The training and test sets of Reuters-21578 dataset are defined. For $CSTR^{2009}$, 4-fold cross validation is applied in the experiments and the average results are reported.

## 4.2 Preprocessing and Performance Evaluation

In the implementation of a text categorization system, the first step is the removal of stopwords. In this study, SMART stoplist is used for this purpose [16]. Then, Porter stemming algorithm is applied [17]. $SVM^{light}$ with default parameters and linear kernel is employed as the classification scheme [3,18]. For the evaluation of different approaches, precision ($P$), recall ($R$) and $F_1$ score are computed separately for each category. Let $TP$, $FP$ and $FN$ denote true positives, false positives and false negatives respectively for the category under concern [19]. Precision is defined as the percentage of documents which are correctly labeled as positive, i.e. $P = \frac{TP}{TP+FP}$. Recall provides the percentage of correctly classified positive documents, i.e. $R = \frac{TP}{TP+FN}$. Since a categorization system can be tuned to maximize either precision or recall at the expense of the other, their harmonic mean named as $F_1$ score is generally used where $F_1 = \frac{2 \times P \times R}{P+R}$ [20]. Macro-$F_1$ score is computed as the average of individual $F_1$ scores [21].

In a recent study, it is observed that the $F_1$ scores of most weighting schemes plateau after 5,000 features for SVM [1]. Because of this, top $T = 5,000$ features ranked by $\chi^2$ are considered in the experiments. However, in $CSTR^{2009}$, the total

---

[1] http://www.cs.rochester.edu/trs/

number of processed terms is less than $3,500$ in one fold. Because of this, top $T = 3,000$ terms are used for this dataset in all four folds.

In this study, the relevance frequency defined as

$$RF(t_j) = \log\left(2 + \frac{A}{max\{1, C\}}\right) \tag{9}$$

where $A$ and $C$ denote the number of positive and negative documents which contain $t_j$, respectively is considered as the collection frequency factor [1]. RF is a recently proposed scheme which is shown to surpass the conventionally used techniques such as odds ratio, information gain or inverse document frequency (idf).

## 4.3    Experimental Results

The experimental results are presented in Tables 1 and 2 respectively for Reuters-21578 and and CSTR$^{2009}$ datasets. The second columns present the $F_1$ scores achieved by the baseline system using BOW representation and $\widehat{f}(t_k) \times RF$ as the term weights. The last two columns provide the scores achieved by the proposed system.

Table 1. $F_1$ scores obtained for the top ten categories of ModApte split

| Category | $\widehat{f}(t_k) \times RF$ | SVC$_{single}$ | SVC$_{pair}$ |
|---|---|---|---|
| Earn | 98.67 | 98.31 | 98.53 |
| Acqusitions | 97.39 | 97.35 | **97.91** |
| Money-fx | 85.56 | **85.71** | **87.03** |
| Grain | 96.55 | **97.32** | 96.27 |
| Crude | 91.62 | **91.84** | 90.91 |
| Trade | 86.73 | **86.96** | 86.61 |
| Interest | 80.00 | 80.00 | **82.59** |
| Wheat | 88.11 | 87.32 | 87.59 |
| Ship | 87.27 | **88.76** | **90.17** |
| Corn | 88.89 | **92.04** | **92.04** |
| Macro $F_1$ | 90.08 | 90.56 | 90.96 |

The F-scores which are higher than the baseline are typed in boldface. It can be seen that, the proposed representation (using either SVC$_{single}$ or SVC$_{pair}$) provided better scores on eight of the ten categories of Reuters-21578 and on all categories of CSTR$^{2009}$. The experimental results clearly show that feature extraction using single variable classifiers is a fruitful approach. Based on this result, it can be argued that using a single classifier for the generation a single triple of features for each ordered pair of terms by considering their frequencies as input has the potential to provide further improvements due to employing joint information.

**Table 2.** Average $F_1$ scores obtained for $CSTR^{2009}$ using four-fold cross validation

| Category | $\hat{f}(t_k) \times RF$ | $SVC_{single}$ | $SVC_{pair}$ |
|---|---|---|---|
| Systems | 94.65 | 94.27 | **95.01** |
| Theory | 92.84 | **94.23** | **95.08** |
| AI | 63.32 | **67.31** | **69.70** |
| Robotics and Vision | 74.32 | **76.03** | 74.10 |
| Macro $F_1$ | 81.28 | 82.96 | 83.47 |

## 5   Discussions and Conclusions

In this study, the use of single variable classifiers for feature extraction in binary text categorization is proposed. The outputs of the single variable classifiers are used to form the document vectors by taking into account the term frequencies and collection frequency factors as well. It is experimentally shown that the proposed scheme provides better scores compared to the baseline system that is based on BOW representation on two different datasets.

The nearest mean classifier simply separates the one-dimensional feature space into two parts. By taking into account the term frequency of each term, the closer class is identified during testing. Other classifiers may be employed to partition the input space into larger number of decision regions for achieving better performance. For instance $k$-NN may be used for this purpose.

Combination of the features provided by consecutive terms that are ordered by $\chi^2$ simply by addition is also studied. Experimental results have shown that the average performance of the system improves further. As mentioned in Section 2, using co-occurrence of two adjacent terms known as bigrams improves the BOW representation. However, selecting the term pairs is not straightforward. In fact, a recent study reported significant improvements on four datasets by using word pairs which are not restricted to be adjacent [23]. In that study, it is mainly argued that the terms that are included in term pairs must be individually discriminative and they should co-occur in smaller number of different classes. As a matter of fact, feature extraction using term pairs should be considered as a further study where classifiers operating on their frequencies (i.e. two inputs) are employed for feature extraction. Similar to concatenating BOW based features and bigrams, the features constructed using single and double classifiers can be concatenated for improved document representation. However, investigation of different methods for selecting best term pairs is also essential for achieving further improvements.

## References

1. Lan, M., Tan, C.L., Su, J., Lu, Y.: Supervised and traditional term weighting methods for automatic text categorization. IEEE Transactions on Pattern Analysis and Machine Intelligence 31(4), 721–735 (2009)

2. Altınçay, H., Erenel, Z.: Analytical evaluation of term weighting schemes for text categorization. Pattern Recognition Letters 31, 1310–1323 (2010)
3. Joachims, T.: Text categorization with support vector machines: Learning with many relevant features. In: Proceedings of 10th European Conference of Machine Learning, pp. 137–142 (1998)
4. Guyon, I., Elisseeff, A.: An introduction to variable and feature selection. Journal of Machine Learning Research 3, 1157–1182 (2003)
5. Wolpert, D.H.: Stacked generalization. Neural Networks 5, 241–259 (1992)
6. Dzeroski, S., Zenko, B.: Is combining classifiers with stacking better than selecting the best one? Machine Learning 54, 255–273 (2004)
7. Leopold, E., Kindermann, J.: Text categorization with support vector machines. how to represent texts in input space? Machine Learning 46(1-3), 423–444 (2002)
8. Yang, Y., Liu, X.: A re-examination of text categorization methods. In: SIGIR-1999, pp. 42–49 (1999)
9. Scott, S., Matwin, S.: Feature engineering for text classification. In: Proceedings of ICML-1999, 16th International Conference on Machine Learning, pp. 379–388. Morgan Kaufmann Publishers (1999)
10. Nastase, V., Shirabad, J.S., Caropreso, M.F.: Using dependency relations for text classification. In: Proceedings of the 19th Canadian Conference on Artificial Intelligence (2006)
11. Caropreso, M.F., Matwin, S., Sebastiani, F.: A learner-independent evaluation of the usefulness of statistical phrases for automated text categorization. In: Text Databases & Document Management, pp. 78–102. IGI Publishing, Hershey (2001)
12. Bekkerman, R., Allan, J.: Using bigrams in text categorization. Technical Report IR-408, Center of Intelligent Information Retrieval, UMass Amherst (2004)
13. Fürnkranz, J.: A study using n-gram features for text categorization. Technical Report OEFAI-TR-98-30, Austrian Research Institute for Artificial Intelligence, Austria (1998)
14. Shafer, G.: A Mathematical Theory of Evidence. Princeton University Press (1976)
15. Debole, F., Sebastiani, F.: An analysis of the relative hardness of Reuters-21578 subsets. Journal of the American Society for Information Science and Technology 56(6), 584–596 (2004)
16. Buckley, C.: Implementation of the smart information retrieval system. Technical report, Cornell University, Ithaca, USA (1985)
17. Porter, M.F.: An algorithm for suffix stripping. Program 14(3), 130–137 (1980)
18. Joachims, T.: Making large-scale SVM learning practical. In: Schölkoph, B., Burges, C.J.C., Smola, A.J. (eds.) Advances in Kernel Methods - Support Vector Learning, pp. 169–184. MIT Press, Cambridge (1999)
19. Wu, C.H., Tsai, C.H.: Robust classification for spam filtering by back-propagation neural networks using behavior-based features. Applied Intelligence 31(2), 107–121 (2009)
20. Liu, X., Wu, J., Zhou, Z.: Exploratory undersampling for class-imbalance learning. IEEE Transactions on Sytems Man Cyber. Part B 39(2), 539–550 (2009)
21. Sebastiani, F.: Machine learning in automated text categorization. ACM Computing Surveys 34(1), 1–47 (2002)
22. Erenel, Z., Altınçay, H.: Nonlinear transformation of term frequencies for term weighting in text categorization. Engineering Applications of Artificial Intelligence 25, 1505–1514 (2012)
23. Figueiredo, F., Rocha, L., Couto, T., Salles, T., Gonçalves, M.A., Meira, W.: Word co-occurrence features for text classification. Information Systems 36(5), 843–858 (2011)

# An Approach to Automated Learning
# of Conceptual Graphs from Text

Fulvio Rotella[1], Stefano Ferilli[1,2], and Fabio Leuzzi[1]

[1] Dipartimento di Informatica – Università di Bari
{fulvio.rotella,stefano.ferilli,fabio.leuzzi}@uniba.it
[2] Centro Interdipartimentale per la Logica e sue Applicazioni – Università di Bari

**Abstract.** Many document collections are private and accessible only
by selected people. Especially in business realities, such collections need
to be managed, and the use of an external taxonomic or ontological resource would be very useful. Unfortunately, very often domain-specific
resources are not available, and the development of techniques that do
not rely on external resources becomes essential. Automated learning of
conceptual graphs from restricted collections needs to be robust with respect to missing or partial knowledge, that does not allow to extract a full
conceptual graph and only provides sparse fragments thereof. This work
proposes a way to deal with these problems applying relational clustering
and generalization methods. While clustering collects similar concepts,
generalization provides additional nodes that can bridge separate pieces
of the graph while expressing it at a higher level of abstraction. In this
process, considering relational information allows a broader perspective
in the similarity assessment for clustering, and ensures more flexible and
understandable descriptions of the generalized concepts. The final conceptual graph can be used for better analyzing and understanding the
collection, and for performing some kind of reasoning on it.

## 1 Introduction

Many document collections are private and accessible only by selected people.
Especially in business realities, such collections need to be managed in order to
carry out tasks as retrieval and extraction of information. Unfortunately, obtaining automatically *Full Text Understanding* is not trivial, due to the intrinsic
ambiguity of natural language and to the huge amount of common sense and
linguistic/conceptual background knowledge needed to switch from a purely syntactic representation to the underlying semantics. Nevertheless, even small portions of such knowledge may significantly improve understanding performance,
at least in limited domains. Although standard tools, techniques and representation formalisms are still missing, lexical and/or conceptual graphs[1] can provide
a useful support to many NLP tasks, allowing automatic systems to exploit

---

[1] We refer to the term 'conceptual graph' as a synonym for 'concept network', with
no reference to Sowa's formalism.

M. Ali et al. (Eds.): IEA/AIE 2013, LNAI 7906, pp. 341–350, 2013.
© Springer-Verlag Berlin Heidelberg 2013

different kinds of relationships that are implicit in the text but required to correctly understand it. Although the use of an external taxonomic or ontological resource can be very useful for these purposes, very often domain-specific resources of this kind are not available, and manually building them is very costly and error-prone. This encourages research for techniques that can work without the need for these facilities, and possibly even automatically construct them by mining large amounts of documents in natural language. Extending a previous work, this paper proposes a technique to automatically extract conceptual graphs from text and reason with them without any external knowledge. Since automated learning of conceptual graphs from small collections needs to be robust with respect to missing or partial knowledge, we first apply a clustering phase to group similar concepts, and then generalize the items in each cluster to obtain new concepts that can be used to build taxonomic relations in the graph, in some cases bridging disjoint portions of the graph. More specifically, instead of using *flat* attribute-value data we adopt a relational perspective, that allows to compare objects with more informative descriptions than those provided by their shared attributes only (that might be very few in such context).

This work is organized as follows: the next section describes related works; Section 3 outlines the proposed approach; then we present an evaluation of our solution; lastly we conclude with some considerations and future works.

## 2   Related Work

Many approaches have been attempted to build conceptual graphs, taxonomies and ontologies from text. In general there are two main strategies: the former builds them exploiting only what is contained in the texts without external knowledge, and the latter exploits some external resources to do the same. The former approach is particularly indicated in such business realities where do not exist any kind of structured and *Machine-readable* external knowledge as a domain taxonomy/ontology. In this setting [1] builds concept hierarchies using Formal Concept Analysis by grouping objects with their attributes, which are determined from text linking terms with verbs. Conversely, our approach is focused on the building of a semantic network relying on the whole graph of concepts and relationships, rather than only on shared attributes as in their taxonomical representation of concepts. Another approach to ontology discovering is proposed in [11], in which the author defines a language to build formal ontologies by deductive discovery in similar way with logic programming. In particular, the author defines both a specific language for manipulating Web pages and a logic program to discover concept lattice. Conversely, we do not limit the kind of relationships to a predefined set, new relationships are created when a never seen verbal relationship between concepts is found. The latter approach differs to the ours, since it strongly depends on external knowledge for the building of taxonomies and/or ontologies. In particular [10, 9] build ontologies by labelling taxonomic relations only; [13] builds a taxonomy considering only concepts that are present in a domain but do not appear in others; [5] uses a combination of existing linguistic resources (VerbNet [6] and WordNet [3]) to shift the representation to the semantic level.

Regarding our proposal, for the syntactic analysis of the input text we exploit the *Stanford Parser* and *Stanford Dependencies* [7, 2], two tools that can identify the most likely syntactic structure of sentences (including active/passive and positive/negative forms), and specifically 'subject' or '(direct/indirect) object' components. They also normalize the words in the input text using lemmatization instead of stemming, which allows to distinguish their grammatical role and is more comfortable to read by humans.

Since the subject of the sentences is usually written only at its first occurrence and then it is replaced by pronouns, we face the anaphora resolution task exploiting JavaRAP that is an implementation of the classic Resolution of Anaphora Procedure [14]. It resolves third person pronouns, lexical anaphors, and identifies pleonastic pronouns.

We also exploited JUNG [12] (Java Universal Network/Graph Framework), which provides a common and extendible language for the modelling, analysis, and visualization of data that can be represented as a graph or network.

Lastly, we need to assess the similarity among concepts in a given conceptual graph. This has been achieved exploiting the similarity measure between Horn clauses proposed in [4]. This measure applies a layered evaluation that, starting from simpler components, proceeds towards higher-level ones repeatedly applying a basic similarity formula, and exploiting in each level the information coming from lower levels and extending it with new features. At the basic level are terms (i.e., constants or variables in a Datalog setting), that represent objects in the world and whose similarity is based on their properties (expressed by unary predicates) and roles (expressed by their position as arguments in n-ary predicates). The next level involves atoms built on n-ary predicates, whose similarity is based on their "star" (the multiset of predicates corresponding to atoms directly linked to them in the clause body, that expresses their similarity 'in breadth') and on the average similarity of their arguments. Then, the similarity of sequences of atoms is based on the length of their compatible initial subsequence and on the average similarity of the atoms appearing in such subsequence. Finally, the similarity of clause is computed according to their least general generalization, considering how many literals and terms they have in common and on their corresponding lower-level similarities.

## 3 Proposed Approach

This proposal relies on a previous work [8], assuming that each noun in the text corresponds to an underlying *concept* (phrases can be preliminarily extracted using suitable techniques, and handled as single terms). A concept is defined by the set of the others concepts that interact with it in the world described by the corpus. The outcome is a graph, where nodes are the concepts/nouns recognized in the text, and edges represent the relationships among these nodes, expressed by verbs in the text (the direction of edges denotes the role of the associated nodes in the relationship). In particular, for each sentence we keep into account also the positive or negative valence of each verbs.

## 3.1    Conceptual Graph Construction

In [8] the input text was directly processed to build a conceptual graph, discarding the relationships (verbs) associated to pronouns because they could not be associated to specific concepts. In order to face this lack in this work, we have pre-processed the input text using JavaRAP, replacing pronouns with the corresponding nouns. Then, the set of relationships of these nouns has been extended, improving the quality of the resulting conceptual graph. The obtained text is processed using the Stanford Parser, in order to extract the syntactic structure of the sentences that make it up. In particular, we are interested only in (active or passive) sentences of the form *subject-verb-(direct/indirect)complement*, from which we extract the corresponding triples ⟨*subject, verb, complement*⟩ that will provide the concepts (the *subject*s and *complement*s) and relations (*verb*s) for the graph. Furthermore, indirect complements are treated as direct ones, by embedding the corresponding preposition into the verb: e.g., *to put, to put on* and *to put across* are considered as three different verbs, and sentence *John puts on a hat* returns the triple ⟨John,put_on,hat⟩, in which *John* and *hat* are concepts associated to attribute *put_on*, indicating that *John* can *put_on* something, while a *hat* can be *put_on*). Triples/sentences involving verb 'to be' or nouns with adjectives provide immediate hints to build the sub-class structure in the taxonomy: for instance, "The dog is a domestic animal..." yields the relationships is_a(dog, animal) and is_a(domestic_animal,animal). In this way we have two kind of edges among nodes in the graph: verbal ones, labelled with the verb linking the two concepts and encoding the assertional knowledge, and taxonomic (is_a) ones encoding the definitional knowledge. Moreover, with the aim to enrich the representation formalism previously defined, we analysed the syntactic tree to seize the sentence positive or negative form based on the absence or presence (respectively) of a *negation modifier* for the verb.

In previous works, we focused on taxonomy construction (paying attention only to the definitional portion of the network), even though we already were building a semantic network. In this work we put the emphasis on the whole network (considering also its assertional portion), even though in the non-taxonomic portion of the network, also contingent knowledge is encoded.

## 3.2    Relational Pairwise Clustering

While [8] adopted a pairwise clustering technique based on the *Hamming distance* on the feature vector of each concept, here we propose to use a relational representation for concepts, and adopt the relational similarity function presented in [4]. In order to perform the proposed clustering method, the conceptual graph has been translated in first order logic, using the relations as binary predicates and the involved concepts as arguments. More precisely, we encoded the direction of the relation in the order of the arguments (the subject in the first place

---

**Algorithm 1.** Relational pairwise clustering of all concepts in the network

---

**Input:** $O$ is the set of objects (concepts) represented as in Section 3.2; $T$ is the threshold for similarity function.
**Output:** set of clusters.

> $pairs \leftarrow empty$
> $averages \leftarrow empty$
> **for all** $O_i \mid i \in O$ **do**
>    $newCluster \leftarrow O_i$
>    $clusters.add(newCluster)$
> **end for**
> **for all** $pair(C_k, C_z) \mid C \in clusters \land k, z \in [0, clusters.size[$ **do**
>    **if** $completeLink(C_k, C_z, T)$ **then**
>       $pairs.add(C_k, C_z)$
>       $averages.add(getScoreAverage(C_k, C_z))$
>    **end if**
> **end for**
> $pair \leftarrow getBestPair(pairs, averages)$
> $merge(pair)$

$completeLink(arg1, arg2, arg3) \rightarrow$ TRUE if complete link assumption for the passed clusters holds, FALSE otherwise.
$getBestPair(arg1, arg2) \rightarrow$ returns the pair having the maximum average.

---

and the complement in the second), and the (positive or negative) valence of the action in the predicate name. Before translating the conceptual graph in a set of concepts described by the neighbors until a chosen radius, we extracted the weak components of the graph by JUNG in order to process the concepts in each separate sub-graph. A weak component is defined as a maximal sub-graph in which all its pairs of vertices are reachable from one another in the underlying undirected sub-graph. For each concept we extract the $k$-neighborhood around it, defined as the sub-graph induced by the set of concepts that are $k$ or fewer hops away from the node. A *vertex/concept-induced sub-graph* is a subset of the vertices/concepts of a graph together with any edges whose endpoints are both in this subset. In this way we have a sub-graph of concepts for which exist at least one path long at most $k$ hops between them and the root concept together with all links between each pair of concepts in this graph. This choice enriches the description language because we can portray a concept not only with the concepts along the considered paths, but with the whole graph of relations between its neighbours.

In practice, the sub-graph obtained for each concept $X$ was translated into a Horn clause of the form $concepts(X) : -rel_a(X, Y), rel_b(Z, X), rel_c(Y, T)$, where $Y, Z, T$ are the neighborhood of $X$ and the binary predicates $rel$ are verbs (positive or negative).

Pairwise clustering under the *complete link* assumption is applied to these descriptions: initially, each concept becomes a singleton cluster; then, clusters are merged while a merging condition is fulfilled (Algorithm 1). Complete link states that *the distance of the farthest items of the involved clusters must be less than a given threshold.*

### 3.3 Generalization of Cluster

In [8] the generalization task has been tackled taking advantage from an external resource. Unfortunately, for specific domains it is often unavailable. Then we need an alternative to overcome this limitation. Our proposal is to generalize each cluster using the maximum set of common descriptors of each concept. The generated concept will be the *subsumer* of the cluster. If the generated subsumer matches with an existing item, then it is promoted as subsumer, otherwise the generated concept remains without a human understandable name until a new concept will not match with it.

In order to perform the generalization phase, we apply the generalization operator proposed in [16], obtaining *least general generalizations*. For the sake of clarity, we cite its original definition:

*A least general generalization (lgg) under $\theta_{OI}$ − subsumption of two clauses is a generalization which is not more general than any other such generalization, that is, it is either more specific than or not comparable to any other such generalization. Formally, given two Datalog clauses $C_1$ and $C_2$, $C$ is a lgg under $\theta_{OI}$ − subsumption of $C_1$ and $C_2$ iff:*

1. *$C_i \leq_{OI} C, i = 1, 2$*
2. *$\forall D \ s.t. \ C_i \leq_{OI} D, i = 1, 2 : not(D <_{OI} C)$*

$$lgg_{OI}(C_1, C_2) = \{C \mid C_i \leq_{OI} C, i = 1, 2 \ and$$
$$\forall D \ s.t. \ C_i \leq_{OI} D, i = 1, 2 : not(D <_{OI} C)\}$$

The application of this operator opens to several conceptual graph refinements. In this work we focus on the insertion of new taxonomical relations. In some cases this leads to the bridging of potentially disjoint portion of the graph, but are exploitable for tasks as retrieval of documents of interest, as well as for the shifting of the representation when needed (abstraction).

### 3.4 Probabilistic Reasoning by Association

Several reasoning strategies can be exploited to extract novel information from the formalized knowledge. In particular, this framework allow to reason with the extracted knowledge, in the sense of finding a path of pairwise related concepts that establishes an indirect interaction between two concepts $c'$ and $c''$. Since real world data are typically noisy and uncertain, there is the need for strategies that soften the classical rigid logical reasoning. Hence we keep soft relationships among concepts rather than hard ones, by weighting the relationships among concepts, where each arc/relationship is associated to a weight that represents its likelihood among all *possible worlds*. Thus we deploy two reasoning strategies: the former works in breadth and aims at obtaining the minimal path between concepts together with all involved relations, the latter works in depth and exploits Problog [15] in order to allow probabilistic queries on the conceptual graph. In more details the former strategy looks for a minimal path using a *Breadth-First Search* (BFS) technique, applied to both concepts under consideration. It also provides the number of positive/negative instances, and the corresponding ratios over the total, in order to help understanding different gradations (such as permitted, prohibited, typical, rare, etc.) of actions between two objects. While this value does not affect the reasoning strategy, it allows to distinguish which reasoning path is more suitable for a given task. Conversely, the latter strategy exploits the previous values in order to prefer some paths to other ones. It exploits ProbLog for this purpose, whose descriptions are based on the formalism $p_i :: f_i$ where $f_i$ is a ground literal having probability $p_i$. In our case, $f_i$ is of the form $link(subject, verb, complement)$ and $p_i$ is the ratio between the sum of all examples for which $f_i$ holds and the sum of all possible links between $subject$ and $complement$.

## 4   Evaluation

The proposed approach has been evaluated with the aim to obtain qualitative outcomes that may indicate its strengths and weaknesses. We exploited a dataset made up of documents concerning *social networks* on socio-political and economic topic, including 695 concepts and 727 relations. The size of the dataset was deliberately kept small in order to have poor knowledge. The similarity function ranges in $]0, 4[$. We executed several experiments varying the used threshold between $[2.0, 2.3]$ with hops equal to 0.5.

The experiments concern the qualitative examination of the obtained clusters and their generalizations, with the aim to understand the behaviour of this approach using a limited collection of documents. For lack of space, we show in Table 1 only the results for the threshold 2.0. We have chosen it because summarizes the largest set of clusters. The outcomes arising from the other thresholds do not show a qualitative difference, but just a quantitative one. Analysing the Table 1, we can see that several clusters seems to be unreliable. For this reason we have inspected some outcomes in order to understand what conditions led to this results. Each case that we present is special in some way.

Let to examine the outcome 35. Applying the $lgg_{OI}$ we obtain:

$$concept(X) :- impact(Y, X), signal(Y, X), signal\_as(Y, X), do\_with(Y, X),$$
$$consider(Y, X), offer(Y, X), offer\_to(Y, X), average(Y, X),$$
$$average\_about(Y, X), experience(Y, X), flee\_in(Y, X), be(Y, X).$$

$$\theta = < \{internet/Y, visible/X\}, \{internet/Y, textual/X\} >$$

Through the substitution $\theta$ we find exactly the cluster $\{visible, textual\}$, that although seems to be unreliable, shows that in social network domain has been created on the bases of the full identity of its relations, with the same concept *internet*. This can be considered a special case because neither of the items can be promoted to subsumer, then we must leave unlabelled the subsumer.

**Table 1.** Clusters obtained processing concepts described using one level of their neighbourhood with a similarity threshold equal to 2.0

| # | Cluster |
|---|---------|
| 1 | peer, preference, picture, thing, able, close, adopter, music, life |
| 2 | content, internet |
| 3 | way, computer |
| 4 | matter, online |
| 5 | television, school, facebook |
| 6 | domain, supervision |
| 7 | screen, broadband |
| 8 | adult, transition, fourteen |
| 9 | tool, point |
| 10 | hour, time |
| 11 | paramount, today |
| 12 | percent, household |
| 13 | interest, challenge |
| 14 | role, student |

| # | Cluster |
|---|---------|
| 15 | property, executive, trouble, york, europe, fortune |
| 16 | segment, hand, america |
| 17 | relationship, capital, technology |
| 18 | question, game |
| 19 | income, female, newspaper, age |
| 20 | kid, guru, limit |
| 21 | theme, staff |
| 22 | classmate, note, connection, pass |
| 23 | magnet, comparison, relic |
| 24 | evolve, elite |
| 25 | day, view, modicum, community, station, clip, opportunity |

| # | Cluster |
|---|---------|
| 26 | environment, activity |
| 27 | average, bulk, space, creator |
| 28 | conversation, educator |
| 29 | power, payoff, classroom, distraction, unfair, media |
| 30 | value, human |
| 31 | creation, announce, july |
| 32 | myspace, destination |
| 33 | phone, visitor, letter, class, touch |
| 34 | trivial, vetere, sincere, genuine |
| 35 | visible, textual |
| 36 | shift, collection |
| 37 | desire, sphere |
| 38 | future, gain |

Now, let to consider the outcome 38. We can isolate as subsumer the concept:

$$concept(X) : - \; lead(Y,X), lead\_into(Y,X), come\_for(Y,X), come\_to(Y,X),$$
$$likely\_for(Y,X), be(Y,X), transform\_into(Y,X).$$

$$\theta = <\{youth/Y, future/X\}, \{youth/Y, gain/X\} >$$

Using the substitution $\theta$ we obtain the cluster $\{future, gain\}$, again describing their role with respect to the single concept *youth*. We want to point out that the portion of *future* left out from the $lgg_{OI}$ does not regard *youth*. The shared description that determines their similarity regards their relations with the same concept *youth* and nothing else, reinforcing their correlation. Follows the uncovered portion of *future*.

$$take\_into(plan, future), be(future, digital), see\_in(negotiation, future),$$
$$take(plan, future), be(negotiation, future), see(negotiation, future).$$

In this case we obtained a full mapping with the single item *gain*, that can label the subsumer as best representative of the cluster.

Finally, let to analyse the cluster 20.

$$concept(X) : - \; protect(Y,X), protect\_by(Y,X), become(Y,X), use(Y,X),$$
$$have(Y,X), have\_to(Y,X), have\_in(Y,X), have\_on(Y,X),$$
$$find(Y,X), go(Y,X), look(Y,X), begin(Y,X), begin\_with(Y,X),$$
$$begin\_about(Y,X), suspect\_in(Y,X), suspect\_for(Y,X).$$

$$\theta = <\{ parent/Y, kid/X\}, \{parent/Y, guru/X\}, \{parent/Y, limit/X\} >$$

The substitution $\theta$ do not fully cover any item of the original cluster. Indeed, for each concept a partial description has been left out. As for the first case, the generated concept needs to stay unlabelled. In particular, to obtain *kid* we need to add:

*teach(kid, school), launch_about(foundation, kid), teach(kid, contrast),*
*come_from(kid, contrast), launch(foundation, kid), finish_in(kid, school),*
*invite_from(school, parent), possess_to(school, parent), invite(school, parent),*
*finish_in(kid, side), come_from(kid, school), find_in(school, parent),*
*produce(school, parent), come_from(kid, side), find_from(school, parent),*
*finish_in(kid, contrast), invite_about(school, parent), come_before(school, parent),*
*release(foundation, kid), invite_to(school, parent), teach(kid, side),*
*release_from(foundation, kid).*

The same type of addition is needed for *guru*:

$$become(teenager, guru).$$

Lastly, we have *limit* through the addition of:

$$be(ability, limit), limit(ability, limit).$$

The results show that the procedure seems to be reliable in order to recognize similar concepts on the basis of their structural position in the graph. Then we believe that applying this approach to more than one level of description, can be achieved interesting results. The improvement performed in this work can be appreciated remarking the novelty in the method of description construction. Using the Hamming distance we obtained a first level *relation centric* (i.e. that describes the concept with its direct relations), meanwhile using our method we obtain a *concept centric* description (i.e. using direct and indirect relations between the first level concepts).

## 5   Conclusions

This work proposes a technique to automatically lean conceptual graphs from text, avoiding the support of external resources. We apply a relational clustering to group similar concepts, and then generalize the items in each cluster to obtain new concepts that can be used to build taxonomic relations in the graph, in some cases bridging disjoint portions of it. As planned in previous works, we used a technique of anaphora resolution to improve the quality of the semantic network. Examining the results, we have seen that the restricted collection regarding a specific domain on which the conceptual graph has been built, has affected the results. In any case, the procedure seems to be reliable in order to recognize similar concepts on the basis of their structural position in the graph. As further studies, we plan to improve the method of building of the graph, and to run further experiments representing the concepts with more than one level of neighbours.

**Acknowledgments.** This work was partially funded by Italian FAR project DM19410 MBLab "Laboratorio di Bioinformatica per la Biodiversit Molecolare" and Italian PON 2007-2013 project PON02_00563_3489339 "Puglia@Service".

# References

[1] Cimiano, P., Hotho, A., Staab, S.: Learning concept hierarchies from text corpora using formal concept analysis. J. Artif. Int. Res. 24(1), 305–339

[2] de Marneffe, M.C., MacCartney, B., Manning, C.D.: Generating typed dependency parses from phrase structure trees. In: LREC (2006)

[3] Fellbaum, C. (ed.): WordNet: An Electronic Lexical Database. MIT Press, Cambridge (1998)

[4] Ferilli, S., Basile, T.M.A., Biba, M., Di Mauro, N., Esposito, F.: A general similarity framework for horn clause logic. Fundam. Inf. 90(1-2), 43–66 (2009)

[5] Hensman, S.: Construction of conceptual graph representation of texts. In: Proc. of the Student Research Workshop at HLT-NAACL 2004, HLT-SRWS 2004, pp. 49–54. ACL, Stroudsburg (2004)

[6] Kipper, K., Dang, H.T., Palmer, M.: Class-based construction of a verb lexicon. In: Proc. of the 17th NCAI and 12th IAAI Conference, pp. 691–696. AAAI Press (2000)

[7] Klein, D., Manning, C.D.: Fast exact inference with a factored model for natural language parsing. In: Advances in Neural Information Processing Systems, vol. 15. MIT Press (2003)

[8] Leuzzi, F., Ferilli, S., Rotella, F.: Improving robustness and flexibility of concept taxonomy learning from text. In: Appice, A., Ceci, M., Loglisci, C., Manco, G., Masciari, E., Ras, Z.W. (eds.) NFMCP 2012. LNCS, vol. 7765, pp. 170–184. Springer, Heidelberg (2013)

[9] Maedche, A., Staab, S.: Mining ontologies from text. In: Dieng, R., Corby, O. (eds.) EKAW 2000. LNCS (LNAI), vol. 1937, pp. 189–202. Springer, Heidelberg (2000)

[10] Maedche, A., Staab, S.: The text-to-onto ontology learning environment. In: ICCS 2000 - Eight ICCS, Software Demonstration (2000)

[11] Ogata, N.: A formal ontology discovery from web documents. In: Zhong, N., Yao, Y., Ohsuga, S., Liu, J. (eds.) WI 2001. LNCS (LNAI), vol. 2198, pp. 514–519. Springer, Heidelberg (2001)

[12] O'Madadhain, J., Fisher, D., White, S., Boey, Y.: The JUNG (Java Universal Network/Graph) Framework. Technical report, UCI-ICS (October 2003)

[13] Cucchiarelli, A., Velardi, P., Navigli, R., Neri, F.: Evaluation of OntoLearn, a methodology for automatic population of domain ontologies. In: Ontology Learning from Text: Methods, Applications and Evaluation. IOS Press (2006)

[14] Qiu, L., Kan, M.Y., Chua, T.S.: A public reference implementation of the rap anaphora resolution algorithm. In: Proceedings of LREC 2004, pp. 291–294 (2004)

[15] De Raedt, L., Kimmig, A., Toivonen, H.: Problog: a probabilistic prolog and its application in link discovery. In: Proc. of 20th IJCAI, pp. 2468–2473. AAAI Press (2007)

[16] Semeraro, G., Esposito, F., Malerba, D., Fanizzi, N., Ferilli, S.: A logic framework for the incremental inductive synthesis of datalog theories. In: Fuchs, N.E. (ed.) LOPSTR 1997. LNCS, vol. 1463, pp. 300–321. Springer, Heidelberg (1998)

# Semi-supervised Latent Dirichlet Allocation
# for Multi-label Text Classification

Youwei Lu, Shogo Okada, and Katsumi Nitta

Interdisciplinary Graduate School of Science and Engineering,
Tokyo Institute of Technology, Tokyo, Japan
{royui,okada,nitta}@ntt.dis.titech.ac.jp

**Abstract.** This paper proposes a semi-supervised latent Dirichlet allocation (ssLDA) method, which differs from the existing supervised topic models for multi-label classification in mainly two aspects. Firstly both labeled and unlabeled learning data are used in ssLDA to train a model, which is very important for reducing the cost by manually labeling, especially when obtaining a fully labeled dataset is difficult. Secondly ssLDA provides a more flexible training scheme that allows two ways of labeling assignment while existing topic model-based methods usually focus on either of them: (1) a document-level assignment of labels to a document; (2) imposing word-level correspondences between words and labels within a document. Our experiment results indicate that ssLDA gains an advantage over other methods in implementation flexibility and can outperform others in terms of multi-label classification performance.

**Keywords:** Topic models, Semi-supervised LDA, Multi-label classification, Document modeling.

## 1 Introduction

Research on multi-label classification receives increased attention because of the need to analyze, recognize or classify the information existing on Web pages, dialogue logs, legal documents, etc. Multi-label classification methods so far are largely divided into two classes: discriminative methods such as support vector machines, and topic model-based methods. The last decade has seen increasing research on topic model-based methods, and many papers empirically show that the topic model-based methods can outperform support vector machines [1-2]. In particular, the performance of discriminative approaches tends to degrade rapidly, as the total number of labels and number of labels in a document increase [1].

However, existing topic model-based approaches usually assume that all documents in the dataset are fully labeled, and that the label-related information of a document is either a set of labels or a set of enforced direct correspondences between words and labels within a document. The first assumption is too strict in the sense that on some occasions obtaining a completely labeled dataset set is very difficult because it associates such a high cost caused by manually labeling. For example, suppose we want to build a

M. Ali et al. (Eds.): IEA/AIE 2013, LNAI 7906, pp. 351–360, 2013.
© Springer-Verlag Berlin Heidelberg 2013

multi-label classifier to automatically predict the critical arguments written in the essays collected from thousands of students taking a national examination. Despite the large amount of training data, obtaining a fully labeled dataset by manually labeling is very difficult because labeling each essay takes more time than merely reading. The second assumption makes the learning methods less flexible. According to the generative process for topic models, the independency between words is assumed and in addition each word is sampled by a certain topic (label). In fact, both document-level and word-level labeling define a correspondence between words and labels in a supervised way. Thus it is natural to develop an algorithm which can adopt these two ways of labeling at the same time.

Hence in this paper we propose a semi-supervised latent Dirichlet method, which trains a model in a semi-supervised way, namely both labeled and unlabeled data is utilized, and allows two ways of labeling a document that are mentioned above.

The contributions of this work are as follows:

1. We propose a semi-supervised latent Dirichlet allocation method which reduces the labeling cost by incorporating both the labeled and unlabeled data.
2. We introduce a general scheme to transform unsupervised topic models to supervised/semi-supervised ones by adjusting the posterior variational parameters for the target words.
3. A conditional gradient method is employed in this work, which alleviates the problem for the labeled data that the overfitting to some labels causes other labels nearly "disappearing" during the coordinate ascent within a document.
4. Two real-world datasets are used in our experiments, and the experimental results demonstrate that the proposed ssLDA algorithm outperforms the others.

## 2    Related Work

Discriminative approaches for multi-label classification, such as support vector machines attempt to reduce the task into a group of binary-classification problems, for example, by "1-vs-all" that is a most commonly used scheme. In particular, transductive support vector machines implement an idea of transductive learning , a concept that is closely related to semi-supervised learning, by including test points in the computation of the margin [5]. Additionally, k-nearest neighbor (kNN) is also found to be employed for multi-label classification in some systems [6].

On the other hand, instead of recognizing each label independently as support vector machines, supervised topic models, which are developed by adapting unsupervised PLSA or LDA to the supervised cases, seeks to predict a ranking of all possible labels within a certain document, and get a positive real value for each label. Based on the real-valued ranking, a prediction of relative labels can be produced in a way for example that the labels with a ranking value above a certain threshold are viewed as relative labels for the document. For instance, in [2] a labeled LDA model, called L-LDA is proposed. Developing a 1-1 correspondence between topics and labels, L-LDA directly learns the relation between words and each label, and is empirically proved to outperform support vector machines in terms of multi-label classification. Besides, a more recently proposed method, called dependency-LDA [1], achieves better classification results than L-LDA by introducing another topic model to capture the dependencies between labels.

It is worth mentioning that there are some existing semi-supervised topic models, but why they are so called is that they make use of label-related information during the learning process in contrast to the traditional unsupervised LDA where topics (labels) are latent variables. For instance, [4] proposes a model called Semi-LDA, which enables direct assignment of labels to words within a document, but Semi-LDA does not consider the situation where the document is only attached a set of labels like in [1-2]. In addition, a semi-supervised PLSA method is presented in [7], which is used for document clustering by utilizing the must-link or cannot-link constraints rather than for multi-label classification. Moreover, [9] presents a semi-supervised PLSA algorithm for opinion integration, which still cannot be used to solve multi-label classification problems.

Considering the existing methods discussed above, our work differs from them in mainly two aspects. First, a model is learned with ssLDA in a semi-supervised way that both labeled and unlabeled learning data can be utilized in the case that obtaining a fully labeled learning dataset is very difficult resulting in a reduction for labeling cost and an improvement of performance compared with their supervised counterparts. Second, besides the assignment of labels to each document, labeling at word-level within a document is also allowed in ssLDA, and thus ssLDA is more flexible.

# 3    Semi-supervised Latent Dirichlet Allocation

## 3.1    Notation

Basically, we follow the notation of the traditional unsupervised LDA. In particular, a corpus $C$ consists of $M$ documents as training instances which is divided into two disjoint subsets: one comprised by $M_1$ labeled documents with each of documents assigned a collection of labels denoted by $\Lambda$, and another consisting of $M_2$ unlabeled documents. A word is a unit term in a document indexed by $\{1,..., V\}$, where $V$ is the size of vocabulary. We still use the unit-basis vector form to represent each word, by which each word is a vector that has the $v$th component equal to 1 and all others zeros. A label is represented as $z$, corresponding to a latent topic in the unsupervised LDA, and the number of unique labels in the corpus is $K$.

## 3.2    Generative Model for ssLDA

As an extension of existing supervised topic models, semi-supervised latent Dirichlet allocation makes use of both labeled and unlabeled documents in the corpus, and the generative process, as shown in Figure 1 is as follows:

1. For each label:
   Draw a multinomial distribution over all words $\beta_{1:K} \sim Dir(\eta)$.
2. For each of $M_1$ labeled documents with a multi-label set $\Lambda$:
   (a) Draw a multinomial distribution over all possible labels $\theta \sim Dir(\alpha)$.
   (b) For each of the $N$ word:
       (i) Sample a label $z$ that is within the multi-label set $\Lambda$.
       (ii) Sample a word $w$ from the multinomial probability conditioned on the label $z$.
5. For each of $M_2$ unlabeled documents:
   (a) Draw a multinomial distribution over all possible labels $\theta \sim Dir(\alpha)$.
   (b) For each of the $N$ word:
       (i) Sample a label $z$.
       (ii) Sample a word $w$ from the multinomial probability conditioned on the label $z$.

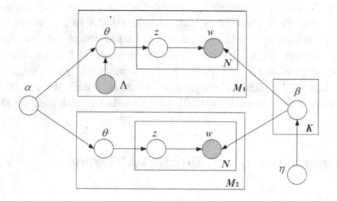

**Fig. 1.** Graphical illustration of semi-supervised LDA

It is important to emphasize that the generative process for the labeled document in ssLDA is different from that of L-LDA [2]. In L-LDA, the multinomial distribution $\theta$ for each document is generated by a Dirichlet distribution with a parameter $\alpha^{(d)}$ obtained by projecting $\alpha$ to a lower dimensional vector so that $\theta$ is only defined over the labels in $\Lambda$. It seems reasonable to do so, but if there is only one label for the document, the projection of $\alpha$ results in a single scalar which eventually violates the principle for Dirichlet distributions that the parameter should be at least two-dimensional. In order to avoid this degenerative case, ssLDA enables $\theta$ being sampled over all labels while assuming that all words in the document are drawn from labels in $\Lambda$. In fact, this assumption naturally ensures that the values of components of $\theta$ corresponding to the labels out of $\Lambda$ are so small that their influence is greatly limited, because any word in the document is not generated by any label excluded from $\Lambda$.

The structure shown in Figure 1 demonstrates that both labeled and unlabeled documents are used to learn a model in ssLDA. Although only $w$ and $\Lambda$ in labeled documents are treated as observed variables in Figure 1, ssLDA can also incorporate the assignment of labels to certain words in the document, which makes $z$ in Figure 1 representing a label also observable.

## 3.3    Parameter Estimation and Inference

In this section, we turn our focus on the issue of parameter estimation and inference for ssLDA. A number of parameter estimation algorithms have been proposed for LDA-based methods so far. For example, L-LDA employs collapsed Gibbs sampling for training, while in the traditional LDA an empirical Bayes with variational EM procedure method is presented.

For ssLDA, we adopt the same method as that of the traditional LDA, because the strategy of employing variational approach can be easily modified and can be extended to add label-related information, making ssLDA work in a semi-supervised way.

The motivation behind variational approach is to obtain a variational distribution with free variational parameters that are selected such that the Kullback-Leibler divergence between the variational distribution and the corresponding true posterior distribution is minimized. In the traditional LDA [3], the variational distribution is defined as follow:

$$q(\beta_{1:K}, \mathbf{z}_{1:M}, \boldsymbol{\theta}_{1:M} \mid \lambda, \phi, \gamma) = \prod_{i=1}^{K} Dir(\beta_i \mid \lambda_i) \prod_{d=1}^{M} q_d(\theta_d, z_d \mid \phi_d, \gamma_d), \tag{1}$$

where $q_d(\theta_d, z_d \mid \phi_d, \gamma_d)$ is the variational distribution defined for the $d$th document.

Now we discuss how to embed label-relative information to $q_d(\theta_d, z_d \mid \phi_d, \gamma_d)$ in our work. In the traditional LDA, computation for $\phi_d$ is given by:

$$\phi_{dnk} \propto \exp\left( \Psi(\gamma_{dk}) - \Psi\left(\sum_{k'} \gamma_{dk'}\right) + \sum_{v=1}^{V} w_{dn}^v \left[ \Psi(\lambda_{kv}) - \Psi\left(\sum_{v'} \lambda_{kv'}\right) \right] \right). \tag{2}$$

According to the generative process for ssLDA, if the $d$th document is labeled, words can only be sampled by the labels within $\Lambda$. Thus it is natural to restrict the variational parameters $\phi_{dnk}$ taking zero whenever the $k$th label is excluded from $\Lambda$, and to compute $\phi_{dnk}$ that corresponds to related labels conforming to Equation (2). Additionally, if it is observed that the $n$th word is explicitly generated by the $k$th label, we can fix $\phi_{dnk}$ equal to 1 and let $\phi_{dnk}$ take zero whenever $t \neq k$.

Moreover, in traditional LDA computation for $\gamma_{dk}$, which are the free parameters of variational distribution approximating $Dir(\theta \mid \alpha)$, is given as follows:

$$\gamma_{dk} = \alpha_k + \sum_{n=1}^{N} \phi_{dnk}. \tag{3}$$

We see that the value of $\gamma_{dk}$ is the sum between $\alpha_k$ and an accumulation of $\phi_{dnk}$ over the words. Since the true $\theta_d$ can be approximated by the empirical means of samples drawn from $Dir(\gamma_d)$, $\gamma_{dk}$ plays a role in indicating the value of $\theta_d$. However, as the alternation between (2) and (3) proceeds for optimization, we see that for some $k$ that corresponds to a related label, $\gamma_{dk}$ may converge to a very small number close to zero.

That means the $k$th label nearly disappears from $\Lambda$ during the inference process due to the words overfitting to other related labels. Thereby, instead of computing $\gamma_{dk}$ with Equation (3), we put constraints that $|\gamma_{dk} - \gamma_{dj}| \leq pN$, where $0 \leq p \leq 1$ for any $k$ and $j$ corresponding to a related label. Since all feasible points of $\gamma_{dk}$ under the constraints form a convex set, we can use a conditional gradient method [8] to find at least a local maximum.

In ssLDA the corpus is divided into two disjoint subsets consisting of labeled and unlabeled documents respectively, so we define a convex combination of $L_1$ and $L_2$:

$$L = \rho L_1 + (1 - \rho) L_2, \tag{4}$$

where $0 \leq \rho \leq 1$ and $L_1$ and $L_2$ corresponds to the log-likelihood of labeled and unlabeled documents respectively. The computation is similar as the unsupervised LDA, but it is complicated because of the introduction of Equation (4).

The equation for update $\lambda_{kv}$ is as follows:

$$\lambda_{kv} = \eta + \rho \sum_{i=1}^{M_1} \sum_{n=1}^{N} w_{in}^{v} \phi_{ink} + (1 - \rho) \sum_{i=1}^{M_2} \sum_{n=1}^{N} w_{in}^{v} \phi_{ink}. \tag{5}$$

The Newton's method can still be used to track $\alpha$ with the first partial derivative of $L$ with respect to $\alpha$ and Hessian matrix, which are given in Equation (6) and (7) respectively.

$$\frac{\partial L_{[\alpha]}}{\partial \alpha_k} = \Psi\left(\sum_{k'=1}^{K} \alpha_{k'}\right)[\rho M_1 + (1 - \rho)M_2] - \Psi(\alpha_k)[\rho M_1 + (1 - \rho)M_2]$$
$$+ \rho \sum_{i=1}^{M_1}\left(\Psi(\gamma_{ik}) - \Psi\left(\sum_{k'} \gamma_{ik'}\right)\right) + (1 - \rho)\sum_{i=1}^{M_1}\left(\Psi(\gamma_{ik}) - \Psi\left(\sum_{k'} \gamma_{ik'}\right)\right), \tag{6}$$

$$H = 1\left[(\rho M_1 + (1 - \rho)M_2)\Psi'\left(\sum_{k'} \alpha_k\right)\right]1^{\mathrm{T}} - diag((\rho M_1 + (1 - \rho)M_2)\Psi'(\alpha_k)). \tag{7}$$

## 4     Experiments

In the preceding section, we demonstrated the strategy of transforming the unsupervised LDA into a semi-supervised algorithm and discussed several issues related to the parameter estimation. In subsection 4.1 and 4.2, we compare the performance between ssLDA, transductive support vector machines which can be viewed as a special case of semi-supervised learning in discriminative fashion and kNN using two datasets labeled at word level and at document level respectively.

Existing evaluation metrics for multi-label prediction tasks are largely based on two perspectives: (1) document-based [10-11], in which emphasizes the prediction for each test instance; (2) label-based, where the prediction for each label is focused. In this paper, we investigate F1scores of experiment results under both perspectives. In particular, assume that there are $M$ test documents and $K$ unique labels, the computation for document-based F1 scores break the predictions within each document down into $K$ binary-classification problems, while for label-based F1 scores we first fix a label and then compute the F1 score across all $M$ documents [12]. Finally, we take the average of F1 scores across all documents and labels respectively.

## 4.1    Task Based on Word-Level Labeling

### Experimental Settings

We ran experiments on a specific corpus consisting of 116 conversation records extracted from twelve groups of mock mediation each approximately lasting forty minutes. With a vocabulary of 117 words, we count the number of the words appearing in the speech and the corpus is built under a "bag of words" scheme. Despite only 117 words selected as features to represent a speech, there are 17 predefined labels (also called issue point or factors), which are critical in dialogue analysis. Since each label usually corresponds to a certain part of the document and the direct correspondence is very easy to identify, it is natural to build such an assignment between words and labels with ssLDA. On the other hand, SVM-based algorithms in this experiment assign the labels at document level.

### Experimental Results

Figure 2 and Figure 3 present F1 scores as the number of labeled documents increases obtained by ssLDA, transductive SVM and 1NN under label-based and document-based perspective. In ssLDA, we have two parameters, $\rho$ and a threshold to decide relative labels based on the ranking of all possible labels. Here we set $\rho=0.8$, and let the threshold be 0.15. The implementation of a transductive support vector machine we use here is $SVM^{light}$ [13] with fraction of unlabeled examples to be classified into positive class as the ratio of positive and negative examples in the labeled training documents. Besides, the nearest neighbor algorithm is also compared here, and the distance metric is chosen as cosine distance between two documents under the "bag of words" presentation.

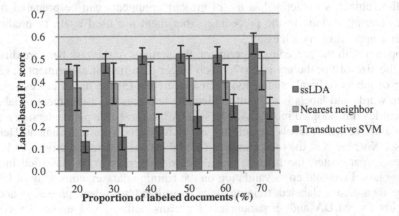

**Fig. 2.** Label-based F1 scores with respect to different proportions of labeled documents

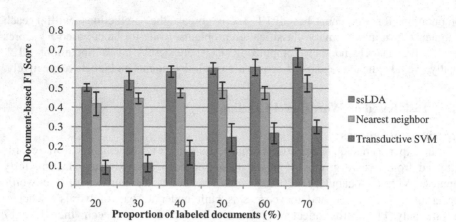

**Fig. 3.** Document-based F1 scores with respect to different proportions of labeled documents

From Figure 2 and Figure 3 we see that ssLDA achieves a better performance than transductive SVM and the nearest neighbor algorithm in label-based and document-base perspective. Furthermore, it is worth mentioning that SVM-based algorithms seem not working well in this experiment where the size of the vocabulary is fairly small while the total number of labels is relatively large. Although the nearest neighbor algorithm is the simplest one to implement, it still achieves a notably better performance than the transductive SVM.

## 4.2    Task Based on Document-Level Labeling

### Experimental Settings

We utilize another corpus in this experiment, which is comprised of 280 answers for the essay section of the Law School Admission Test.   Students are asked to state his opinions on a given debating subject, and to give reasons and analysis for his claims. Since the subject is predefined, a list of critical arguments can be prepared by the teacher. The algorithms in the preceding experiment are used again to predict the arguments appearing in each essay.

Compared with the preceding experiment, the settings of the task here are different in: (1) the size of vocabulary is 2647, much larger than that of experiment 1; (2) the number of labels (critical arguments) decreases to 10; (3) the direct correspondence between words and labels is difficult to identify, and thus the assignment of labels at document level is adopted in ssLDA. The ssLDA has three free parameters: $p$, $\rho$ and a threshold above which labels are selected as positive labels for a document. Here we fix $p$ as $1/K$ where $K$ is the total number of all labels. As for transductive SVM, there is a free $-p$ parameter, the fraction of unlabeled examples to be classified into the positive class. Five-fold cross validation on the training dataset, consisting of labeled training data and unlabeled training data is utilized to select proper $\rho$ and the thresholds for ssLDA and $-p$ parameters for transductive SVM under the average label-based F1 scores.

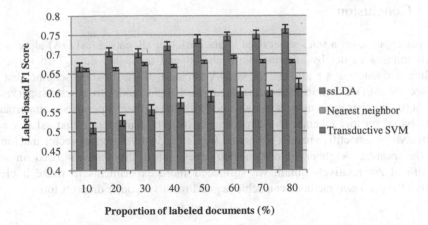

**Fig. 4.** Label-based F1 scores with respect to different proportions of labeled documents

## Experimental Results

Figure 4 shows that ssLDA gains higher mean values of label-based F1 scores than transductive SVM and nearest neighbor algorithm. On the other hands, we see in Figure 5 that under document-based perspective transductive SVM achieve higher F1 scores than ssLDA and the nearest neighbor algorithm when the proportion of labeled documents is less than or equal to 30 percent, and ssLDA surpasses the others when the proportion larger than 30 percent. It is interesting to compare the results shown in Figure 4 and Figure 5, and observe that transductive SVM has lower label-based F1 scores and yet higher document-based F1 scores than the others with relatively small proportion of labeled documents. In our experiments, it is shown that transductive SVM tends to misclassify easily those labels that have very small number of labeled training instances, resulting in lower label-based F1 scores.

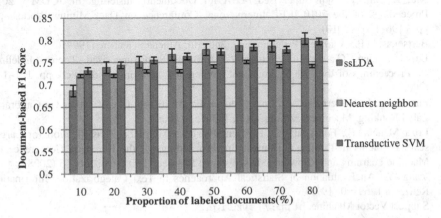

**Fig. 5.** Document-based F1 scores with respect to different proportions of labeled documents

## 5    Conclusion

This paper proposed a semi-supervised latent Dirichlet allocation (ssLDA) algorithm, which trains a model by utilizing both labeled and unlabeled data. Two ways of labeling a document are allowed in ssLDA because of the flexible scheme that is adopted by ssLDA to transform an unsupervised topic model to a semi-supervised one. Actually other topic models can also be modified to their supervised counterparts easily based on this scheme. Our experimental results demonstrate that ssLDA can achieve very competitive results compared with transductive support vector machines and the nearest neighbor algorithm. However, since the datasets used in our experiment are relatively small, we still need more experiments to make it clear whether the proposed methods can achieve good results on other datasets too.

## References

1. Rudin, T.N., Chambers, A., Smyth, P., Steyvers, M.: Statical Topic Models for Multi-label Document. Machine Learning 88, 157–208 (2012)
2. Ramage, D., Hall, D., Nallapati, R., Manning, C.D.: Labeled LDA: a Supervised Topic Model for Credit Attribution in Multi-labeled Corpora. In: Proceedings of the 2009 Conference on Empirical Methods in Natural Language Processing, Singapore, pp. 248–256 (2009)
3. Blei, D.M., Ng, A.Y., Jordan, M.I.: Latent Dirichlet Allocation. Journal of Machine Learning Research 3, 993–1022 (2003)
4. Wang, Y., Sabzmeydani, P., Mori, G.: Semi-latent Dirichlet Allocation: a Hierarchical Model for Human Action Recognition. In: Elgammal, A., Rosenhahn, B., Klette, R. (eds.) Human Motion 2007. LNCS, vol. 4814, pp. 240–254. Springer, Heidelberg (2007)
5. Chapelle, O., Scholkopf, B., Zien, A.: Semi-Supervised Learning. The MIT Press (2006)
6. Ashley, K.D., Bruninghaus, S.: Automatically Classifying Case Texts and Predicting Outcomes. Artificial Intelligence Law 17, 125–165 (2009)
7. Niu, L., Shi, Y.: Semi-supervised PLSA for Document Clustering. In: ICDMW 2010 Proceedings of the 2010 IEEE International Conference on Data Mining Workshops, pp. 1196–1203 (2010)
8. Bertsekas, D.P.: Nonlinear Programming, 2nd edn. Athena Scientific (1999)
9. Lu, Y., Zhai, C.: Opinion Integration through Semi-supervised Topic Modeling. In: Proceedings of the 17th International Conference on World Wide Web, pp. 121–130 (2008)
10. Furnkranz, J., Hullermeier, E., Loza Mencia, E.: Multilabel Classification via Calibrated Label Ranking. Machine Learning 73(2), 133–153 (2008)
11. Loza Mencia, E., Furnkranz, J.: Efficient Pairwise Multilabel Classification for Large-scale Problems in the Legal Domain. In: Proceedings of the European Conference on Machine Learning and Knowledge Discovery in Datasets (Part II), pp. 50–65 (2008)
12. Yang, Y.: An Evaluation of Statistical Approaches to Text Categorization. Information Retrieval 1, 69–90 (1999)
13. Support Vector Machine, http://svmlight.joachims.org

# Recommending Interest Groups to Social Media Users by Incorporating Heterogeneous Resources

Wei Zeng[1,2] and Li Chen[1]

[1] Department of Computer Science,
Hong Kong Baptist University, Hong Kong
[2] School of Computer Science and Engineering
University of Electronic Science and Technology, China
zwei504@gmail.com,lichen@comp.hkbu.edu.hk

**Abstract.** Due to the advance of social media technologies, it becomes easier for users to gather together to form groups online. Take the Last.fm for example (which is a popular music sharing website), users with common interests can join groups where they can share and discuss their loved songs. However, since the number of groups grows over time, users often need effective group recommendation (also called affiliation or community recommendation) in order to meet like-minded users. In this paper, based on the matrix factorization mechanism, we have investigated how to improve the accuracy of group recommendation by fusing other potentially useful information resources. Particulary, we adopt the collective factorization model to incorporate the user-item preference data, and the similarity-integrated regularization model to fuse the friendship data. The experiment on two real-world datasets (namely Last.fm and Douban) shows the outperforming impact of the chosen models relative to others on addressing the data sparsity problem and enhancing the algorithm's accuracy. Moreover, the experimental results identify that the user-item preference data can be more effective than the friendship in terms of benefiting the group recommendation.

**Keywords:** Recommending groups, matrix factorization, regularization, user-item preferences, friendship.

## 1 Introduction

In recent years, social media sites become popular among online users. Take Last.fm as a typical example, in this website, users can not only listen to music, but also be associated with different types of social relations: s/he may create a contact list including her/his friends; s/he could also join in interest groups to build membership with others whom are with some common interests in musics (thought he may not know in the offline life). Therefore, in such environment, users should be willing to receive various types of recommendation from the website so as to more effectively establish their social network. However, so far, most research focuses have been put on recommending items (such as music), but less on recommending the relationship, especially the interest groups that

M. Ali et al. (Eds.): IEA/AIE 2013, LNAI 7906, pp. 361–371, 2013.

users might be affiliated with in the social media environment. Indeed, as the available user-group data are rather sparse, purely applying the classic recommender technology (like the collaborative filtering) cannot effectively generate the group recommendation. Therefore, in this paper, we have mainly been engaged in studying how to fuse other information resources, such as the user-item preferences (i.e., users' interaction with items) and user-user friendship data, to enhance the group recommendation.

Specifically, considering that the interest group is usually created based on multiple users' comment interests in items, their ratings (or implicit interaction like clicking) on items should reveal their similarity in terms of joining the groups. On the other hand, the friendship data can be supplementary to provide the network relation between users, though they might not be stronger than user-item preferences to reflect the common interests. In our work, the respective effects of the two information resources on improving group recommendation accuracy are empirically studied. Particularly, we take into account the property of data resource (i.e., bipartite or one mode data) when choosing the proper fusion model. More notably, the work has been grounded on the Matrix Factorization (MF) mechanism owing to its well-recognized high algorithm efficiency and accuracy [7]. In the following, we will first describe related works and then in detail present our proposed fusion methods. The experiment setup and results analysis will follow. At the end, we will conclude the major findings.

## 2   Related Work

Some of related works have emphasized studying how users form group. For instance, [1] investigated the phenomenon of group formation in two large social networks and found that the tendency of an individual to join a community is influenced not only by the number of friends s/he has within the community, but also by how those friends are connected to each another. They proposed decision tree based methods to measure the movement of individuals across communities, and showed how the movements were closely aligned with change in the topic of interest of the community. As another typical work, [6] showed that a group can attract new members through revealing the friendship ties of its current members to outsiders.

As for how to recommend groups to users, in [10], two models were explored, namely the Graph Proximity Model (GPM) and the Latent Factors Model (LFM), to generate community recommendation to users by taking into account their friendship and affiliation networks. Their empirical results indicated that GPM turns out to be more effective and efficient. [3] proposed a collaborative filtering method, called Combinational Collaborative Filtering (CCF), to perform personalized community recommendation. It concretely applied a hybrid training strategy that combines Gibbs sampling and Expectation-Maximization algorithm for fusing semantic info, such as the description of communities and users. The experiment on a large Orkut dataset demonstrated that the approach can more accurately cluster relevant communities given their similar semantics.

In [2], the authors investigated two approaches to generate community recommendation: the first adopted the Association Rule Mining technique (ARM) to discover associations between sets of communities; the second was based on Latent Dirichlet Allocation (LDA) to model user-community co-occurrences with latent aspects. The experiment on Orkut data indicated that LDA consistently outperforms ARM when recommending four or more communities, while ARM is slightly better when recommending up to three communities.

However, these related works did not consider the potential value of incorporating both user-item preferences and friendship into the group recommendation process. Moreover, the model they utilized, such as the Graph Proximity Model (GPM) in [10], is inevitably with high time complexity, that is why we have chosen matrix factorization as the basis mechanism to perform the fusion.

## 3 Proposed Methodology

Given a system like Last.fm, there are two types of data available, which are: 1) *bipartite data* such as user-item interaction data (in Last.fm they are implicit binary data where 1 means users clicked the item, and 0 otherwise), since there are two types of entities involved in each relationship, and 2) *one mode data* like the user-user friendship since only one type of entity (i.e., the "user") exists. The user-group membership belongs to the first type. In order to effectively fuse the heterogenous types of auxiliary resources into the group recommendation process, we have chosen the matrix factorization technology as the basis mechanism given that it could be extended to incorporate both bipartite and one-mode data with low computation complexity.

More specifically, for the one mode data, since it describes the relation between entities which are with the same type, it can be considered as an indicator of closeness. That is, if there is a link between two entities, we can regard that the two entities are closer than the ones without the link. Because of this, most state-of-the-art works leverage regularization model to fuse the one mode data in order to minimize the gap between two entities [9,8,5].

On the other hand, for the bipartite data such as user-item preferences, we argue that it is different from the one mode data in nature since a user indicates her/his interests in the item by interacting/rating it, which is however absent in the one mode data. Therefore, such data would be more suitably addressed by the factorization model, because it can effectively factorize user-item relations into two components and obtain a user's latent factor model and an item's latent factor model simultaneously. Previously, we discussed the limitation if bipartite data were handled in the manner of regularization [11]. We also proved that the regularization is better than factorization when dealing with one mode data like friendship. In this paper, we are motivated to consolidate the findings when the recommended object is changed from item to group. That is, would one-mode data still be better handled by regularization when they are fused into the process of recommending groups, and the factorization better suits user-item preferences?

**Table 1.** Summary of notations

| Notation | Description |
|---|---|
| $m, n, l$ | the numbers of users, items and groups respectively |
| $k$ | the dimension of the factor vector |
| $X, Y, Z$ | the user-factor, item-factor and group-factor matrix respectively |
| $x_u, y_i, z_g$ | the user $u$, item $i$ and group $g$ factor vector respectively |
| $p_{ui}, p_{ug}^*, p_{uf}'$ | user $u's$ preference on item $i$, group $g$ and user $f$ respectively |
| $p(u), p^*(u), p'(u)$ | the vector that contains $u$'s the preference on all items, all groups and all friends respectively |
| $c_{ui}, c_{ug}^*, c_{uf}'$ | the confidence level indicating how much a user likes an item, a group and a friend respectively |
| $C^u, C^{*u}, C'^u$ | $C^u$ denotes the $n \times n$ diagonal matrix and $C_{ii}^u = c_{ui}$; $C^{*u}$ denotes the $l \times l$ diagonal matrix and $C_{gg}^{*u} = c_{ug}^*$; $C'^u$ denotes the $m \times m$ diagonal matrix and $C_{ff}'^u = c_{uf}'$ |
| $F(u)$ | the friend set of user $u$ |
| $\lambda_f$ | the coefficient of the regularization |
| $\alpha$ | the coefficient for the collective matrix factorization |

Table 1 first summarizes the notations used in the equations of the paper.

## 3.1 Baseline

To recommend interest groups to a user, we take the user-group matrix as the bipartite data type and use the following factorization equation as the baseline (which is without any fusions of other resources except the available user-group membership data themselves).

$$\min_{u*,g*} \sum_{u,g} c_{ug}^* (p_{ug}^* - x_u^T z_g)^2 + \lambda (\sum_u \| x_u \|^2 + \sum_g \| z_g \|^2) \tag{1}$$

where $p_{ug}^*$ equals 1 if the user $u$ joined group $g$, otherwise it is 0; $c_{ug}^*$ is the confidence level indicating how much a user prefers a group which is set as 1 if no relevant data like "visting frequency" are available.

The above cost function contains $m * l$ terms, where $m$ is the number of users and $l$ is the number of groups. To optimize it, we apply the Alternating Least Squares (ALS) [7,4], because it can help achieve massive parallelization of the algorithm by computing each $z_i$ independent of the other group factors and computing each $x_u$ independent of the other user factors. It was also demonstrated to be capable of efficiently processing the sparse binary data (such as the user-group relations in our case). Based on ALS, the analytic expressions for $x_u$ and $z_g$ that are used to minimize the above cost function are respectively:

$$x_u = (Z^T C^{*u} Z + \lambda I)^{-1} Z^T C^{*u} p^*_\cdot(u) \tag{2}$$

$$z_g = (X^T C^{*g} X + \lambda I)^{-1} X^T C^{*g} p^*(g) \tag{3}$$

To generate a top-N recommendation list for each user $u$, we assume her/his candidate group set (i.e., groups unjoined by the user) is $\phi_u$. For each group $i$ in $\phi_u$, we calculate a prediction score as follows:

$$p'_{ui} = x_u^T * z_i \tag{4}$$

where $x_u^T$ and $z_i$ are the user's latent factor model and the group's latent factor model respectively. Top-N groups with higher scores will then be included the recommendation list and returned to the target user.

## 3.2  Incorporating Friendship

To inject friendship in the above framework, we tried the factorization approach which was to factorize user-user friendship into two factor vectors. However, as mentioned before, because friendship belongs to one-mode data with only one type of entity existing, the regularization model would be more suitable [5,9]. Grounded on this model, we develop the following equation in order to minimize the gap between the taste of a user and the average taste of her/his friends:

$$\min_{u*,g*} \sum_{u,g} c_{ug}^* (p_{ug}^* - x_u^T z_g)^2 + \lambda(\sum_u \| x_u \|^2 + \sum_g \| z_g \|^2)$$
$$+ \lambda_f (\| x_u - \frac{1}{\mid F(u) \mid} \sum_{f \in F(u)} \widehat{sim}(u,f) x_f \|^2) \tag{5}$$

In this formula, $\lambda_f$ is the coefficient for the friendship regularization. $\widehat{sim}(u,f) = sim(u,f) / \sum_{v \in F(u)} sim(u,v)$ denotes the normalized similarity degree between the user $u$ and her/his friend $f$, which is used to adjust individual friends' contributions when predicting the target user's interests. It is worth mentioning that this similarity measure is a special element that we integrate into the regularization process in order to enhance its prediction power. In the experiment, we particularly compared the similarity-integrated regularization method to the one without its integration. We also tested different approaches to calculate the similarity degree, including ones based on common groups (shared by the user and her/his friend), common item preferences, and common friends. The vector space similarity (VSS) is concretely performed: $sim(u,f) = \frac{r_u r_f}{\|r_u\| \|r_f\|}$, where $r_u$ can denote the group vector, friend vector or item vector of user $u$. The experimental results show that the common-group based similarity measure performs more accurate than others (see Section 4.2).

We then adopt ALS to perform the optimization process. Due to the addition of the friendship's regularization, the analytic expression for $x_u$ is changed to:

$$x_u = (Z^T C^{*u} Z + (\lambda + \lambda_f) I)^{-1} (Z^T C^{*u} p^*(u)$$
$$+ \frac{\lambda_f}{|F(u)|} \sum_{f \in F(u)} \widehat{sim}(u,f) x_f) \tag{6}$$

The expression for the group factor $z_g$ is the same as in Equation (3).

## 3.3   Incorporating User-Item Preferences

Another auxiliary resource we considered is the user's preferences on items (that can be either explicitly stated by users via rating, or inferred from their interaction with items such as "clicking" behavior). Still, we tried both factorization and regularization approaches. We especially investigated the collective matrix factorization (CMF) technique for fusing the data, so that the user-item interaction matrix can be directly factorized into two components: the "user" latent factor and the "item" latent factor.

$$
\alpha \min_{u*,g*} \sum_{u,g} c_{ug}^*(p_{ug}^* - x_u^T z_g)^2 + \lambda(\sum_u \| x_u \|^2 + \sum_g \| z_g \|^2) +
$$
$$
(1-\alpha) \min_{u*,i*} \sum_{u,i} c_{ui}(p_{ui} - x_u^T y_i)^2 + \lambda(\sum_u \| x_u \|^2 + \sum_i \| y_i \|^2) \tag{7}
$$

where the parameter $\alpha$ is used to adjust the relative weights of user-group matrix and user-item matrix in the factorization. Similar to the definition of confidence level $c_{ug}^*$ when factorizing user-group, we introduce the $c_{ui}$ for user-item, that indicates the confidence level regarding users' preference over item. Based on ALS, the analytic expressions for $x_u$ and $y_i$ are respectively defined as:

$$
x_u = (\alpha Z^T C^{*u} Z + (1-\alpha)Y^T C^i Y + \lambda I)^{-1} *
$$
$$
(\alpha Z^T C^{*u} p^*(u) + (1-\alpha)Y^T C^u p(u)) \tag{8}
$$

$$
y_i = (X^T C^i X + \lambda I)^{-1} X^T C^i p(i) \tag{9}
$$

The expression for $z_g$ is the same as in Equation (3).

For the purpose of comparison, we also developed the regularization-based fusion method, that converts user-item matrix into user-user relationship by means of a weighted scheme:

$$
\min_{u*,g*} \sum_{u,g} c_{ug}^*(p_{ug}^* - x_u^T z_g)^2 + \lambda(\sum_u \| x_u \|^2 + \sum_g \| z_g \|^2)
$$
$$
+\lambda_f(\| x_u - \frac{1}{N(u)} \sum_{n \in N(u)} \omega_{un}^* * x_n \|^2) \tag{10}
$$

where the weight $w_{un}^* = \frac{|O_{un}|}{\sum_{i \in N(u)} |O_{ui}|}$ ($O_{un}$ is the set of common items interacted by both users $u$ and $n$, and $N(u)$ is user $u's$ neighbors who have common items with $u$).

The analytic expression for $x_u$ in respect of the above model is

$$
x_u = (Z^T C^{*u} Z + (\lambda + \lambda_f)I)^{-1}(Z^T C^{*u} p^*(u)
$$
$$
+\lambda_n \frac{1}{| N(n) |} \sum_{n \in N(n)} \omega_{un}^* x_n) \tag{11}
$$

### 3.4 Incorporating Friendship and User-Item Preferences Together

After fusing friendship and user-item preferences separately, we derive a formula to fuse them together:

$$\alpha \min_{u*,g*} \sum_{u,g} c_{ug}^* (p_{ug}^* - x_u^T z_g)^2 + \lambda(\sum_u \| x_u \|^2 + \sum_g \| z_g \|^2) +$$
$$\lambda_f(\| x_u - \frac{1}{|F(u)|} \sum_{f \in F(u)} \widehat{sim}(u,f)x_f \|^2) + \qquad (12)$$
$$(1-\alpha) \min_{u*,i*} \sum_{u,i} c_{ui}(p_{ui} - x_u^T y_i)^2 + \lambda(\sum_u \| x_u \|^2 + \sum_i \| y_i \|^2)$$

where the friendship is handled by the similarity-integrated regularization and user-item preferences are handled via the factorization. This combination was actually resulted from comparing regularization and factorization models for fusing friendship and user-item preferences respectively in the experiment (see Section 4.2). The analytic expression for $x_u$ is

$$x_u = (\alpha Z^T C^{*u} Z + (1-\alpha)Y^T C^u Y + (\lambda + \alpha\lambda_f)I)^{-1}(\alpha(Z^T C^{*u}$$
$$p^*(u) + \frac{\lambda_f}{|F(u)|} \sum_{f \in F(u)} \widehat{sim}(u,f)x_f) + (1-\alpha)Y^T C^u p(u)) \qquad (13)$$

The expression for $z_g$ is the same as in Equation (3), and for $y_i$ it is the same as in Equation (9).

## 4 Experiment

### 4.1 Dataset and Evaluation Metrics

Two real-world datasets, namely Last.fm (www.last.fm) and Douban (www.douban.com), were used to test the performance of the algorithms. The Last.fm is a worldwide popular social music site. The *membership* in the dataset refers to the user's participation in interest groups, the *friendship* was extracted from the user's friend list, and the *item* is referred to the artist (because users' preference over artists can be more stable than their preference over songs). Douban is a popular social media site in China that supports users to freely share movies, books and music. Being different from Last.fm, users of Douban can assign 5-scale ratings to items. Therefore, when assessing algorithms in both datasets, we can identify whether the performance is valid no matter whether the user-item preferences are explicit or implicit. Besides, for the sake of simplicity, we only collected users' data related to one product domain, the movie in Douban. We treat the user-item interaction matrix as 0/1, that is, the element equals to 1 if the user viewed (or rated) the item and 0 otherwise. Moreover, as Douban supports Twitter-like following mechanism, two users were treated

**Table 2.** Description of two datasets

|          | Element | Size    | Element          | Size       |
|----------|---------|---------|------------------|------------|
|          | #user   | 100,000 | #user-item pair  | 29,908,020 |
| Last.fm  | #item   | 22,443  | #friendship pair | 583,621    |
|          | #group  | 25,397  | #user-group pair | 1,132,281  |
|          | #user   | 71,034  | #user-item pair  | 12,292,429 |
| Douban   | #item   | 25,258  | #friendship pair | 273,832    |
|          | #group  | 2,973   | #user-group pair | 373,239    |

**Table 3.** Abbreviations' description

| Method | Description |
|--------|-------------|
| Group.MF | The basic matrix factorization; |
| Group.MF.F.R | Fusing the friendship by regularization; |
| Group.MF.F.F | Fusing the friendship by factorization; |
| Group.MF.I.R | Fusing the user-item preferences by regularization; |
| Group.MF.I.F | Fusing the user-item preferences by factorization; |
| Group.MF.FI | Fusing the friendship by regularization and fusing the user-item preferences by factorization; |
| Group.MF.F.FCos | Fusing the friendship by similarity-integrated regularization based on common friends; |
| Group.MF.F.GCos | Fusing the friendship by similarity-integrated regularization based on common groups; |
| Group.MF.F.ICos | Fusing the friendship by similarity-integrated regularization based on common items; |
| Group.MF.FI.GCos | Fusing the friendship by similarity-integrated regularization based on common groups and fusing the user-item preference by factorization. |

as friends only if they follow each other. The details of the two datasets are described in Table 2.

To measure the accuracy of group recommendation, we applied the *leave-one-out* evaluation scheme because user-group pairs are rather sparse so they cannot be divided into subsets to perform the cross-fold validation. Concretely, during each testing round, we randomly selected one of the user's participated groups as the target choice. The measurement goal was hence to identify whether the top-N recommendation list as generated by the tested algorithm contains this target choice or not. Correspondingly, we use the hit ratio metric $Hits@N$ to evaluate the recommendation accuracy. That is, given the total number of users $m$, $Hits@N$ is defined as $Hits@N = \sum_{u=1}^{m} hit(u)@N/m$, where $hit(u)@N$ denotes whether user $u's$ target choice was located in the recommendation list.

## 4.2  Results

We first compared regularization and factorization models for fusing friendship, and for fusing user-item preferences, respectively. We also tested the model that fuses both data resources together. In total, we assessed 10 different methods via the experiment (see Table 3).

Table 4 shows the results. It can be seen that the regularization model (Group.MF.F.R) outperforms the factorization model (Group.MF.F.F) when fusing the friendship. It further shows that the regularization model integrated with the group-based similarity measure (Group.MF.F.GCos) not only outperforms the originally non-similarity based model, but also ones integrated with other similarity measures (such as item-based Group.MF.F.ICos and friend-based Group.MF.F.FCos).

As for fusing user-item preferences, it shows that the accuracy of factorization model (Group.MF.I.F) is improved with the increase of the density level of the user-item matrix (where @train.X in Table 4 represents that $X\%$ of total user-item pairs are used). In comparison, the accuracy of regularization model (Group.MF.I.R) is lower and does not obviously change when the data density level is varied. This might be because once the user-item matrix is projected into the user-user matrix, a lot of information is lost, so the performance of Group.MF.I.R that fuses the projected matrix can not be improved even in denser user-item matrix.

The above results thus indicate that the regularization model is more suitable than the factorization for fusing one-mode data (friendship), while factorization is more suitable than regularization for fusing bipartite data (user-item preferences). In addition, the comparison between Group.MF.F.GCos (the best method regarding friendship's fusion) and Group.MF.I.F suggests that the user-item preferences act more positive than the friendship in terms of enhancing group recommendation.

Driven by the above results, we finally combined Group.MF.F.GCos and Group.MF.I.F@train.80 for fusing the two resources (friendship and user-item preferences) together, which is shorted as Group.MF.FI.GCos. From Fig. 1, it can be seen that such combination *Group.MF.FI.GCos* achieves accuracy improvement against fusing the two resources separately. Moreover, Group.MF.FI.GCos is better than an alternative combination Group.MF.FI (which is without the similarity integration).

## 5  Conclusions

In conclusion, in order to solve the user-group sparsity phenomenon that commonly occurs in social media sites, we have proposed to fuse both friendship and user-item preference data to improve the accuracy of recommending interest groups to the target user. In more detail, we explored the matrix factorization technique to incorporate both one mode and bipartite data in a collective, unified framework. We have also proved the outperforming suitability of regularization model for handling the one mode friendship data, and the factorization

**Table 4.** Algorithms' comparison results

| Method | Last.fm | | Douban | |
|---|---|---|---|---|
| | Hits@5 | Hits@10 | Hits@5 | Hits@10 |
| Group.MF (baseline) | 0.0530 | 0.0875 | 0.1995 | 0.2933 |
| *Fusing user-item preferences (via Factorization)* | | | | |
| Group.MF.I.F@train.20 | 0.0573 | 0.0899 | 0.2030 | 0.2950 |
| Group.MF.I.F@train.40 | 0.0678 | 0.1026 | 0.2102 | 0.3013 |
| Group.MF.I.F@train.60 | 0.0714 | 0.1068 | 0.2113 | 0.3079 |
| Group.MF.I.F@train.80 | **0.0722** | **0.1070** | **0.2120** | **0.3095** |
| *Fusing user-item preferences (via Regularization)* | | | | |
| Group.MF.I.R@train.20 | 0.0559 | 0.0885 | 0.2025 | 0.2932 |
| Group.MF.I.F@train.40 | 0.0559 | 0.0885 | 0.2026 | 0.2936 |
| Group.MF.I.R@train.60 | 0.0560 | 0.0886 | 0.2026 | 0.2936 |
| Group.MF.I.R@train.80 | 0.0561 | 0.0887 | 0.2027 | 0.2937 |
| *Fusing friendship* | | | | |
| Group.MF.F.R | 0.0566 | 0.0910 | 0.2072 | 0.2973 |
| Group.MF.F.F | 0.0553 | 0.0876 | 0.2038 | 0.2928 |
| Group.MF.F.FCos | 0.0549 | 0.0861 | 0.2075 | 0.2974 |
| Group.MF.F.GCos | **0.0593** | **0.0923** | **0.2093** | **0.2999** |
| Group.MF.F.ICos | 0.0569 | 0.0897 | 0.2062 | 0.2921 |

*Note:* the size of user/group latent factors (k) is 10. The other parameters were tuned with optimal values, e.g., for Group.MF.I.F@train.20 $\alpha = 0.8$ in Last.fm dataset and $\alpha = 0.9$ in Douban dataset.

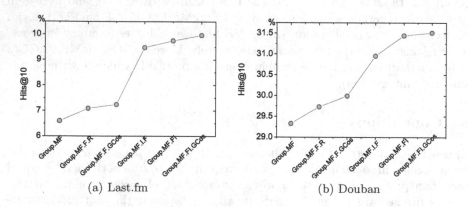

(a) Last.fm                    (b) Douban

**Fig. 1.** Comparison of different methods

model for processing the user-item bipartite data. The friendship's regularization can be further augmented by integrating the similarity measure, especially the common-group based, to distinguish different friends' contributions.

By comparing the effectiveness of the two types of auxiliary resources, we found that the user-item preferences in general perform more accurate than the friendship to benefit the group recommendation, which might be attributed to its advantage of revealing users' comment interests in items. Furthermore, combining the two auxiliary resources can further increase the recommendation accuracy. Thus, our work points out a promising trend of incorporating heterogeneous data into the group recommendation. In the future, we will continue the work by investigating other potentially useful resources.

**Acknowledgements.** This research was supported by HKBU/FRG2/11-12/049 and partially supported by ECS/HKBU211912.

# References

1. Backstrom, L., Huttenlocher, D., Kleinberg, J., Lan, X.: Group formation in large social networks: membership, growth, and evolution. In: Proc. KDD 2006, pp. 44–54. ACM, New York (2006)
2. Chen, W.-Y., Chu, J.-C., Luan, J., Bai, H., Wang, Y., Chang, E.Y.: Collaborative filtering for orkut communities: discovery of user latent behavior. In: Proc. WWW 2009, pp. 681–690. ACM (2009)
3. Chen, W.-Y., Zhang, D., Chang, E.Y.: Combinational collaborative filtering for personalized community recommendation. In: Proc. KDD 2008, pp. 115–123. ACM (2008)
4. Hu, Y., Koren, Y., Volinsky, C.: Collaborative filtering for implicit feedback datasets. In: Proc. ICDM 2008, pp. 263–272. IEEE Computer Society (2008)
5. Jamali, M., Ester, M.: A matrix factorization technique with trust propagation for recommendation in social networks. In: Proc. RecSys 2010, pp. 135–142. ACM (2010)
6. Kairam, S.R., Wang, D.J., Leskovec, J.: The life and death of online groups: predicting group growth and longevity. In: Proc. WSDM 2012, pp. 673–682. ACM, New York (2012)
7. Koren, Y., Bell, R., Volinsky, C.: Matrix factorization techniques for recommender systems. Computer 42, 30–37 (2009)
8. Ma, H., Yang, H., Lyu, M.R., King, I.: Sorec: social recommendation using probabilistic matrix factorization. In: Proc. CIKM 2008, pp. 931–940. ACM (2008)
9. Ma, H., Zhou, D., Liu, C., Lyu, M.R., King, I.: Recommender systems with social regularization. In: Proc. WSDM 2011, pp. 287–296. ACM (2011)
10. Vasuki, V., Natarajan, N., Lu, Z., Dhillon, I.S.: Affiliation recommendation using auxiliary networks. In: Proc. RecSys 2010, pp. 103–110. ACM (2010)
11. Yuan, Q., Chen, L., Zhao, S.: Factorization vs. regularization: fusing heterogeneous social relationships in top-n recommendation. In: Proc. RecSys 2011, pp. 245–252. ACM (2011)

# Recent Advances in Recommendation Systems for Software Engineering

Robert J. Walker

University of Calgary
Calgary, Canada
walker@ucalgary.ca

**Abstract.** Software engineers must contend with situations in which they are exposed to an excess of information, cannot readily express the kinds of information they need, or must make decisions where computation of the unequivocally correct answer is infeasible. Recommendation systems have the potential to assist in such cases. This paper overviews some recent developments in recommendation systems for software engineering, and points out their similarities to and differences from more typical, commercial applications of recommendation systems. The paper focuses in particular on the problem of software reuse, and speculates why the recently cancelled Google Code Search project was doomed to failure as a general purpose tool.

**Keywords:** RSSEs, overview, classification, opportunities.

## 1 Introduction

At one time, the notion of a recommendation system for software engineering (RSSE) would have been anathema: computers and software were all about correctness and precision, while recommendations bespeak fuzziness and faith. Unfortunately, many tasks that face the software engineer are inherently imprecise, staggeringly complex, and often infeasible to compute (or even formally undecidable). The software engineer must bridge the gaps between: the messy, changing needs of the real world; the interpersonal and organizational challenges involved in collaborating with other people; and the inflexibility of the computer. While correctness and precision have their place, a willingness to revert to "good enough" answers can be the difference between success and failure in the real world with its constraints on time and money.

Robillard et al. [14] have previously defined RSSEs as follows: *"An RSSE is a software application that provides information items estimated to be valuable for a software engineering task in a given context."* A key word in that definition is "estimated", because if the correct answer can be faithfully returned in all circumstances, the application is not making recommendations but computing or finding *the* answer. For example, a search for all references to a particular method within a software project is not a recommendation: there is a unique, unequivocally correct set. In contrast, finding all locations in a system that

M. Ali et al. (Eds.): IEA/AIE 2013, LNAI 7906, pp. 372–381, 2013.

should be changed when that method is changed will require recommendations, due to the infeasibility of computing the correct answer in general.

Recommendations for a software engineer play much the same role as recommendations for any user. There is a problem or task at hand. There is (often highly structured) information available about potential solutions. Some additional information about the specific task is often available—e.g., through an integrated development environment (IDE)—as is information about what other people have done in similar circumstances. The recommendation system's task is to infer the nature or details of the task, the needs and/or characteristics of the context of the task and of the engineer, to construct one or more recommendations according to some model of relevance, and to present these recommendations in a manner conducive to their timely and efficient usage.

An analysis of the differences and similarities of different recommendation systems can be facilitated by a classification scheme. Section 2 outlines three different schemes for classifying recommenders in general recommenders or RSSEs in particular: the paradigmatic classification for general recommenders provided by Jannach et al. [9], the multidimensional design classification that provided by Robillard et al. [14] for RSSEs, and a novel problem-space classification.

Section 3 examines a small set of recent RSSEs, to illustrate the breadth of applications that have been explored. Section 4 focuses more narrowly on software reuse, being a cornerstone of the author's research and a fundamental goal that led to the field of software engineering. Section 5 points to some opportunities with RSSEs that the author has seen in his own work. Section 6 summarizes and points out remaining challenges and opportunities in RSSE and potential cross-fertilization with research in recommenders from other domains.

## 2  Categorizing RSSEs

Three categorization schemes are presented here: the paradigmatic classification for general recommenders provided by Jannach et al. [9], the design-oriented classification for RSSEs developed by Robillard et al. [14], and a novel problem-space classification.

Jannach et al. [9] divide recommenders into four paradigms. Their stated focus is personalized recommendations (i.e., ones reflective of the individual's tastes, needs, habits, or characteristics) rather than generic recommendations. The archetypal example is of the online bookstore, where recommendations are made to a customer about what books may be of interest to them.

1. *Collaborative recommenders.* A typical example is, if Alice has purchased or viewed books that have been purchased or viewed by other customers, additional books purchased or viewed by those other customers are also of potential interest to Alice. Collaborative recommenders can thus be viewed as operating through analogical reasoning over different users, and extrapolation from historical data, without resorting to detailed analysis of the items being recommended.

2. *Content-based recommenders.* Again, considering the online bookstore arche-type, recommendations could be derived from the properties of the books that Alice has purchased or viewed: the author, the genre, the publication date, keywords in the title or synopsis, ... all are potential signs of what Alice cares about. Content-based recommenders then realize a similarity function (which is also a form of analogical reasoning) that depends on a specific model of the items being recommended and extrapolation from historical data.

3. *Knowledge-based recommenders.* When Alice can express properties of the books she is interested in or some of her individual properties (e.g., age, gen-der), the need for historical data about Alice might be avoidable. Knowledge-based recommenders also realize a similarity function that depends on a specific model of the items being recommended. The key difference from content-based recommenders is the avoidance of historical data.

4. *Hybrid recommenders.* Since each of the above approaches has advantages and disadvantages, a good combination of them would draw on their strengths while cancelling their weaknesses. Of course, a bad combination would do the opposite.

Robillard et al. provide a table of the design dimensions for RSSEs, reproduced in Table 1. The major dimensions focus on three issues. The *context* consists of the situation-specific information to be used by the recommender, in the form of implicit input garnered by observing the developer, explicit input provided by the developer, or some hybrid of the two. To construct its recommendations, the recommendation engine analyzes the context plus additional sources of *data*, such as the source code of a project being operated on (beyond an explicit input), a repository of change history information, or websites containing developer con-versations. While the majority of RSSEs rank their results, not all do so. The *output* involves two aspects: the mode of the recommendations may be via push, where the developer has not actually asked for help, or via pull, where an ex-plicit request was made; the presentation style may insert the recommendations within another source of information, like the developer's standard views in an IDE, or all at once in a special-purpose display. Rationales (or explanations) for the recommendations vary in nature and detail a great deal, and involve an interplay between the recommendation engine itself and the output. And finally, all manner of developer feedback can potential be accepted to refine or to change the recommendations, involving an individual's feedback affecting just their local view or everyone's.

A software engineer encounters three kinds of situations in which recommen-dations could potentially be beneficial.[1] (1) In the presence of too much infor-mation, recommendations of the most relevant cases can save time and prevent those cases from being overlooked—*the information overload situation.* (2) In the absence of specific knowledge about what information to seek or how to ex-press it, recommendations can draw on the current context (of the engineer) and

---

[1] Note that this categorization is not derived from a systematic analysis of RSSEs, so there may be other cases that do not fit here, but these three cases do occur.

**Table 1.** Design dimensions of RSSEs according to Robillard et al. [14]

| Context nature | Recommendation engine | Output form |
|---|---|---|
| *Input:* explicit, implicit, hybrid | *Data:* source, change, bug reports, mailing lists, interaction history, peers' actions | *Mode:* push, pull |
| | *Ranking:* yes, no | *Presentation:* batch, inline |
| | *Explanations:* from none to detailed | |
| *User feedback:* none, locally adjustable, individually adaptive, globally adaptive | | |

the past experiences of others—*the navigation confusion situation*. (3) When the correct answer to a problem is too complex to compute, a set of heuristically-based answers or approximations can still help the engineer to make a rational decision—*the infeasible computation situation*.

## 3 Sampling Recent RSSE Approaches

RSSEs have been realized for most activities within software engineering, from assisting with implementations, to pointing out high-risk code, to recommending people as experts, to helping plan out staged releases of features. Three specific examples are examined here.

In requirements elicitation, the problem is to discover each stakeholder's needs, preferences, and wishes in a form that can be compared and contrasted. A negotiation process then proceeds in order to resolve conflicts and ambiguities, and to reach agreement on rationales and priorities. Performing such activities in a distributed development environment is particularly challenging; in large systems, much time can be wasted involving individuals in negotiations on issues not relevant to them. The approach of Castro-Herrera et al. [2] seeks to analyze natural language descriptions of stakeholder's perspectives, in order to help clarify those perspectives and to cluster the stakeholders into specific themes for discussion. They utilize data describing the stakeholders for this purpose as well, and user feedback is used to improve the clustering process. This is a hybrid approach, involving elements of collaborative recommenders and knowledge-based recommenders. It can be seen as coping with an information overload problem (the need to read and understand the ideas of a large set of stakeholders) and a navigation confusion problem (the need to determine which stakeholders must focus on individual issues). The input context is explicit; the data source is stakeholder profiles; the output mode is pull; the presentation is batch; no explanations are provided; and the user feedback is individually adaptive, altering each user's own profile.

The YooHoo tool [6] tackles the problem of information overload for a developer whose development depends on external projects undergoing parallel development. At issue is the unpleasant surprise when one's software breaks due to an unexpected change event in an external project. While it is possible to monitor a stream of change events arriving from the external project, these will frequently be voluminous and mostly irrelevant. YooHoo attempts to classify the changes into those most likely to be impactful on the individual developer's project, those unlikely to be impactful, and those that are almost certainly irrelevant. To create its recommendations, YooHoo is configured to receive the relevant change event streams and analyzes the developer's project for external dependencies. The developer can also configure YooHoo to be more or less restrictive for particular event streams. YooHoo obtains over 99% elimination of events while maintaining a very high true positive rate. This approach deals strictly with an information overload problem. It is a knowledge-based approach. Its context is the preference settings of the developer; its other data sources are the developer's project and the external change event streams; its output is push mode (or perhaps calling it a "delayed-pull" mode would be better, since the developer has really asked for it) and batch; it does not use feedback.

Murphy-Hill et al. [12] have proposed a recommender for the unusual problem of command discovery within a developer's IDE. The issue is that modern IDEs like Eclipse suffer from a feature bloat problem: too many commands available for too much functionality. While feature-richness is important in an IDE, it can lead to inefficiencies when developers are unaware of the existence of a useful command, and they resort to more cumbersome combinations to achieve the same effect; since they are unaware of the existence of the command, there is no question of them searching for it. The approach operates atop a large repository of historical information about developers' command usage; commonality between the developer's actions and those of other developers prior to them learning about a command leads to the provision of a recommendation to use that command. The approach deals strictly with a navigation confusion problem. It is a collaborative recommendation approach. Its context is implicit in the actions of the developer; its other data source is the historical repository of command invocation sequences; its output is push mode and in-line; it does not use feedback.

## 4    RSSEs in Software Reuse

There have been many approaches over the years aimed at finding code for the developer. Traditional approaches focused on precise query languages to specify the desired semantics or other specific properties. The ultimate failure of these approaches hinges on their impracticality: industrial developers will rarely if ever be able or willing to spend the time crafting and debugging detailed, àrcane specifications. The time required to do so is liable to outweigh the benefits accrued from finding the functionality. Furthermore, such approaches are likely to succeed in delivering relevant artifacts only in the simplest of cases; in others, no perfectly matching artifacts are likely to exist. Recognizing these problems, several researchers have followed the RSSE route to make progress.

## 4.1  Test-Driven Reuse

The test-driven development paradigm [1] argues that automated test cases should be developed before the code-under-test is implemented. The supposed benefits are: (1) automated test suites should be created anyways; (2) the developer will have a better understanding up-front of what that code ought to do; and (3) since it is natural to build assumptions (sometimes false) into our code, the developer will be less prone to also build those assumptions into their tests.

The idea of *test-driven reuse* is then to leverage these test cases in order to find pre-existing artifacts that can then be reused, instead of implementing the functionality from scratch. The core issue here falls into that of the infeasible computation problem: an unequivocally correct answer would require arbitrary transformations to arbitrary code that results in compilable versions of the examples that could then be run against the automated test cases; this is clearly infeasible in the general case, so providing approximations and recommendations is a reasonable alternative. Three research groups have pursued this idea [8, 10, 13]. Many of their ideas are essentially the same: extract a set of facts from the test cases, use these to find similar artifacts in a database, and recommend the most relevant artifacts to the developer—these are knowledge-based recommendation systems. The approach of Reiss [13] is slightly different, in that he also attempts some simple transformations of the located artifacts in order to be able to run the test cases on the (transformed) artifacts.

In unpublished research, all three techniques have been found to fail badly in most realistic circumstances, as they expect the developer's tests to use *exactly* the names of types and methods that are found in the artifacts. Even Reiss's approach falls into this trap, as he uses a strictly pipelined approach to select potential examples, only then transforming and running them, and it is the first step that contains the most significant limitation. As this investigation aimed to assess the relative strengths and weaknesses of the approaches, this resoundingly negative outcome related to strong work by serious researchers is surprising![2]

All three groups apparently viewed the task as a simple information retrieval problem: extract words → search for words, where the only real question was what model of similarity to use. There is little reason in general to expect that the developer will use the same words to describe their intentions as were used in the existing artifacts (the exception being in commonly recognized domains, and even there, there will be some propensity for variation) since they would not have already settled on any specific, existing frameworks.

---

[2] In fairness, Lemos et al. [10] emphasize that their approach aims specifically at locating "auxiliary functionality", but this distinction is not particularly clear.

Interestingly, commercial approaches to code search see much of the same problem: the desire to ignore the syntax and semantics of the source code, to treat it instead as just another kind of text. To be clear, this suffices for basic searches like "find me examples of the use of IStatusLineManager" but not "find me examples where there is a stack that ignores requests to insert the same element more than once." Google Code Search never overcame such limitations and never made it out of its beta release before the project was killed, despite years of incubation. This weakness could only have contributed to its downfall.

More generally, this issue points to important lessons for RSSEs. The research failed to recognize the needs of the developer to whom the recommendations are being given, and the nature of their task. It is important to remember the original, concrete point and to evaluate that, rather than assume that an abstraction maintains the significant properties.

## 4.2 Other Approaches for Finding Code

Other kinds of input have been used to help locate relevant artifacts for reuse. The specific problems being targeted can have significant impact on the suitability of a potential approach.

*Strathcona.* In some now-not-so-recent work, Holmes et al. [7] focused on the problem of finding examples of the use of specific abstract programming interfaces (APIs) such as are provided by libraries and frameworks. The issue at hand is one of information overload: real APIs are frequently large and complex, and to utilize their functionality can sometimes involve intricate protocols that are not always well documented. As a result, the developer can easily become lost as they attempt to understand how to solve a task through the application of an API. Examples are a common means for helping to overcome such issues, but handcrafted examples are expensive to create and require someone to have recognized a specific need.

Instead, the *skeleton* expressed by the developer (a partial and uncompilable implementation), and a large repository of existing implementations, can be leveraged. Note that, unlike in the case of test-driven reuse, it *is* reasonable here to assume that the types and methods being expressed in the skeleton are identical to those in the artifacts being leveraged: the developer is aware of which API they are using. A set of structural facts (e.g., which types are referenced, which methods are called) are extracted from the skeleton. A set of heuristics-based agents each seeks examples that are similar to the extracted facts, from the perspective of their individual approach. Examples that are ranked highly by many agents are ranked highest overall, and presented first to the developer.

Strathcona is a knowledge-based recommendation system, as it requires explicit input from the developer about the properties of the items of interest and leverages those properties to find potential examples. It does utilize an element of historical data: it can only provide examples that someone else has implicitly created in the past. It is not a collaborative recommendation system in the sense of Jannach et al. since there is no comparison between the developers themselves.

Strathcona addressed both the information overload problem and the navigation confusion problem. A key property of Strathcona that was unusual at the time was its provision of rationales for its recommendations: an explanation of which properties of the skeleton were met by a given example.

*Mining feature descriptions.* The recent approach of McMillan et al. [11] aims at supporting the rapid creation of prototypes: low-quality, partially-functional software useful in the assessment of the requirements and feasibility of a full-scale development effort. They argue that rapid prototyping consists of identification of candidate features (horizontal prototyping) followed by implementation of a subset of the feature set (vertical prototyping). Since the problem is really to invest as little time and money as possible to derive a sufficient prototype for stakeholders to tinker with, reusing existing software ought to reduce the cost of the process while arriving at an acceptable artifact.

Their idea is then to leverage the cursory descriptions typically created during horizontal prototyping to provide: (1) recommendations on the relevant features detected in the description that have also been detected in a repository of software components; and (2) recommendations on components providing those features that are deemed to not be too difficult to reuse (because they contain few external dependencies). Feedback to the feature recommendations are used to update the component recommendations.

As with Strathcona, this approach makes use of an existing, developer-oriented means of describing the entities of interest, rather than demanding that the developer conform to the needs of the software. Overall, this approach is a hybrid recommendation system: the initial input involves a knowledge-based recommender, but the resulting rating is a collaborative filtering approach. Using this to refine the identified features in the repository would also lead to improved results for other developers. As with Strathcona, the approach addresses both the information overload problem (finding components in a big repository) and the navigation confusion problem (who needs to talk to whom and about what).

## 5   Some Opportunities for RSSEs

The work in the Walker's lab has not always focused on RSSEs. The desire to compute the correct answer has often led to overlooking opportunities for more completely involving the developer. Two cases are presented here.

*Jigsaw.* If one wishes to reuse a software artifact within a partially existing system, the two will need to be integrated. The Jigsaw tool [5] largely automates this process for individual methods by inferring correspondences between the reused artifact's method and the target system's method and using these to guide the transformation of the reused artifact. Naming differences, rearrangements of lines, and different nesting of structures are largely coped with. However, since computing functional equivalence is undecidable, the correct correspondence is approximated by considering structural facts augmented with some simple semantic details, and by utilizing heuristics. The result is that ambiguities can occur where the tool cannot decide on the best option; the developer is then asked for input. The lost opportunity, at this point, is that Jigsaw can decide that a correspondence exists where the developer does not agree, and yet, they have no means to provide this feedback to adjust the integration.

*DSketch.* Many software change tasks involve first understanding the dependencies between different parts of the system. The typical solution involves the creation of tool support that understands the semantics of the programming language involved, to at least indicate where an entity is declared and where it is referenced. While this is well understood theoretically, developing non-trivial tooling is difficult and expensive. This cost must be amortized in savings across multiple usages of the tooling, but for domain-specific languages, minor language variants, uncommon language combinations, and specialized bridging technology, such amortization possibilities may not exist. DSketch [3] allowed the developer to create imprecise grammars for the different parts of their software, cheaply. DSketch informs the developer about matching dependencies that it has detected as a result, and the developer can modify or add grammars until they are satisfied. In the missed opportunity, the developer could have directly pointed DSketch to false/missing matches, to have it automatically adjust the grammars.

# 6    Conclusion

Since software engineers must deal with issues of information overload, being unable to find their way, or wanting the answer to infeasible computational problems, recommendation systems play an increasingly key role in software engineering research. For the sake of providing a simple means of understanding recent research on RSSEs, three quite different classification schemes have been outlined including a novel problem-based one for RSSEs. A few recent RSSEs have been explored, with particular focus on the domain of software reuse.

The full gamut of possibilities is seen with respect to a problem-based and a paradigm-based categorization. Some combinations of dimensions in the design-based categorization were not shown in this paper; indeed some of these do not apparently exist. A common RSSE weakness is a tendency to not utilize developer feedback to improve recommendations, and especially not for other developers. In fairness, many RSSE approaches assume external iteration in cases where the developer is not satisfied, but there are potentially significant opportunities for advancement in this dimension.

Similarly, there has been little realization of the infeasible computation problem in non-software engineering applications of recommenders. There exist cases where one would like to have a more intelligent recommender, capable of dealing with a multi-objective search problem plus a fuzzy and shifting fitness function. Opportunities abound!

Going forward, one concern is the propensity of the software engineering community to remain at an "initial evaluation" stage of validation. As recently demonstrated by Cossette and Walker [4], determining the ground truth—to know what *should* be recommended—is an expensive but crucial exercise in evaluating the performance of an RSSE; otherwise, we remain at a level of myth and opinion, which are both subject to bias and manipulation. Perhaps this is less true of other domains, but scientific progress requires hard work and careful analysis of the problems at hand, before we worry too much about "improving" putative solutions that lack solid evidence.

# References

[1] Beck, K.: Test Driven Development: By Example. Addison Wesley (2002)
[2] Castro-Herrera, C., Duan, C., Cleland-Huang, J., Mobasher, B.: A recommender system for requirements elicitation in large-scale software projects. In: Proc. ACM Symp. Appl. Comput., pp. 1419–1426 (2009)
[3] Cossette, B., Walker, R.J.: DSketch: Lightweight, adaptable dependency analysis. In: Proc. ACM SIGSOFT Int. Symp. Foundations Softw. Eng., pp. 297–306 (2010)
[4] Cossette, B.E., Walker, R.J.: Seeking the ground truth: A retroactive study on the evolution and migration of software libraries. In: Proc. ACM SIGSOFT Int. Symp. Foundations Softw. Eng., pp. pp. 55/1–55/11 (2012)
[5] Cottrell, R., Walker, R.J., Denzinger, J.: Semi-automating small-scale source code reuse via structural correspondence. In: Proc. ACM SIGSOFT Int. Symp. Foundations Softw. Eng., pp. 214–225 (2008)
[6] Holmes, R., Walker, R.J.: Customized awareness: Recommending relevant external change events. In: Proc. ACM/IEEE Int. Conf. Softw. Eng., pp. 465–474 (2010)
[7] Holmes, R., Walker, R.J., Murphy, G.C.: Approximate structural context matching: An approach to recommend relevant examples. IEEE Trans. Softw. Eng. 32(12), 952–970 (2006)
[8] Hummel, O., Janjic, W., Atkinson, C.: Code Conjurer: Pulling reusable software out of thin air. IEEE Softw. 25(5), 45–52 (2008)
[9] Jannach, D., Zanker, M., Felfernig, A., Friedrich, G.: Recommender Systems: An Introduction. Cambridge University Press (2010)
[10] Lemos, O.A.L., Bajracharya, S., Ossher, J., Masiero, P.C., Lopes, C.: A test-driven approach to code search and its application to the reuse of auxiliary functionality. Inf. Softw. Technol. 53(4), 294–306 (2011)
[11] McMillan, C., Hariri, N., Poshyvanyk, D., Cleland-Huang, J., Mobasher, B.: Recommending source code for use in rapid software prototypes. In: Proc. ACM/IEEE Int. Conf. Softw. Eng., pp. 848–858 (2012)
[12] Murphy-Hill, E., Jiresal, R., Murphy, G.C.: Improving software developers' fluency by recommending development environment commands. In: Proc. ACM SIGSOFT Int. Symp. Foundations Softw. Eng., pp. 42/1–42/11 (2012)
[13] Reiss, S.P.: Semantics-based code search. In: Proc. ACM/IEEE Int. Conf. Softw. Eng., pp. 243–253 (2009)
[14] Robillard, M., Walker, R., Zimmermann, T.: Recommendation systems for software engineering. IEEE Softw. 27(4), 80–86 (2010)

# WE-DECIDE: A Decision Support Environment for Groups of Users

Martin Stettinger, Gerald Ninaus, Michael Jeran, Florian Reinfrank,
and Stefan Reiterer

Institute for Software Technology
Graz University of Technology
Inffeldgasse 16b, A-8010 Graz, Austria
firstname.lastname@ist.tugraz.at

**Abstract.** Group recommendation technologies are becoming increasingly popular for supporting group decision processes in various domains such as interactive television, music, and tourist destinations. Existing group recommendation environments are focusing on specific domains and do not include the possibility of supporting different kinds of decision scenarios. The WE-DECIDE group decision support environment advances the state of the art by supporting different decision scenarios in a domain-independent fashion. In this paper we give an overview of the WE-DECIDE environment and report the results of a first user study which focused on system usability and potentials for further applications.

**Keywords:** Group recommendation, group decision making, decision technologies.

## 1 Introduction

The quality of group decisions can be negatively influenced by various factors. For example, anchoring effects [1] are responsible for decisions which are biased by the voting of the first preference-articulating person (user). Knowledge about the preferences of other users in early phases of a decision process can lead to sub-optimal decisions [2]. Furthermore, the non-inclusion of explanations can lead to a lower level of trust [3]. A decision can also be influenced by the fact that single persons have to take a decision in place of persons who are not available for the meeting. In many cases, decision tasks are not open in the sense that it is impossible to easily integrate new decision alternatives or changed preferences within the scope of a decision process – both aspects can lead to low-quality decisions [4]. Finally, after taking a decision, the criteria for the decision remain unclear and the outcomes are not documented.

The idea of WE-DECIDE is to support a domain-independent definition of different types of decision tasks while taking into account the above mentioned risk factors. In order to achieve this goal, WE-DECIDE builds upon group recommendation algorithms [5] which are used for determining alternative solutions for the participants of a group decision process. A typical scenario for the application of WE-DECIDE technologies is the decision about which restaurant to select for a dinner or – in a scientific community – a decision regarding the selection of the destination of next year's conference.

M. Ali et al. (Eds.): IEA/AIE 2013, LNAI 7906, pp. 382–391, 2013.

The remainder of this paper is organized as follows. In the following section (WE-DECIDE *Decision Support*) we introduce the WE-DECIDE decision process on the basis of a working example. In this context we provide insights into the group recommendation approaches integrated in the WE-DECIDE environment. In Section *We-Decide Decision Scenarios* we introduce the three basic types of decision scenarios supported by WE-DECIDE. In the Section *User Study* we report the results of a first empirical study conducted with the goal of figuring out further application domains for WE-DECIDE as well as to estimate the current status of WE-DECIDE in terms of usability. We then discuss related and future work and thereafter conclude the paper.

## 2   We-Decide Decision Support

WE-DECIDE is an environment which supports the definition and solution of decision tasks in different application domains (e.g., deciding about a restaurant for a dinner, deciding about a company name or deciding about a travel destination). The system is open in the sense that it is not restricted to specific domains but rather flexible in terms of being able to support decision scenarios with different properties. In this section we will show how decision tasks can be solved on the basis of WE-DECIDE – we will do this on the basis of a working example: *deciding about a restaurant for a dinner*. This working example reflects the decision process that took place in real-world where a group of four persons selected a restaurant for a dinner to celebrate the finishing of the construction works at the house of one of the participants. Before a WE-DECIDE decision process can be started, the underlying decision task and an initial set of decision alternatives (solutions) have to be defined. Thereafter, the decision process can be triggered. Such a process consists of the phases *defining the preferences*, *aggregating individual preferences*, and *taking the decision*. In this context WE-DECIDE supports an *open decision process* [2] where users are allowed to change their mind regarding their preferences and also to include additional decision alternatives (solutions).

### 2.1   Modeling Decisions Tasks and Solutions

*Defining the Decision Task.* The first step before being able to launch a decision process is to design a decision task or – alternatively – to reuse existing decision tasks. When designing a new decision task, one has to specify the name of the decision task (in our case *"Dinner (Completion of Construction Works)"*) and further properties of the decision task (see Figure 1). The designer of the decision task (also denoted as administrator) has to decide (1) whether users (designers typically also act as users) should be allowed to *add additional information* (decision alternatives, files, links) within the scope of the decision process, (2) about the *kind of group recommendation support*, (3) about the *approach to elicit user preferences*, and (4) whether *preferences entered by users should be visible* to other users. Preference visibility for other users can have an impact on the overall decision quality since (for specific domains) it has been shown that – when confronted with the opinions of other users – the focus of the decision process is on discussing preferences and not on trying to identify an optimal solution (see, e.g., [2]). Furthermore, knowledge about the preferences of other users can significantly bias one's own preferences (see, e.g., [6]).

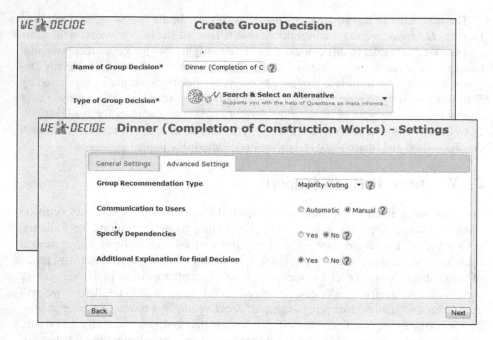

**Fig. 1.** WE-DECIDE: definition of a decision task: in this case, the title of the new decision task is set to *"Dinner (Completion of Construction Works)"* and users are supported in the process of searching and selecting a decision alternative. Furthermore, the selected group recommendation heuristic is *majority voting* and explanations have to be provided for the final decision.

In addition to these settings, the designer of the decision task is enabled to configure advanced settings such as *the way* WE-DECIDE *communicates with the participants* of the decision task (communication manually activated by the designer of the decision task vs. automated notification generated by the WE-DECIDE environment), whether the *specification of dependencies between different decision alternatives is possible* (e.g., when deciding about a new car, the option *hybrid engine* would nowadays exclude the option *gas engine*), and whether *explanations for solutions should be possible* when completing a group decision task. Explanations can play an important role in decision tasks since they are able to increase the trust of users in the outcome of a decision process [3]. This decision (include explanations or not) has to be taken by the designer of the decision task.

*Defining the Solutions.* The basis for taking a decision are solutions (decision alternatives) which are shown to the user within the scope of a decision process. Such solutions can be entered by the designer of the decision task but as well be entered or complemented by other users. Depending on the decision scenario (see Section WE-DECIDE *Decision Scenarios*), decision alternatives can additionally be annotated with meta-information (user attributes). For example, in our restaurant selection scenario each restaurant can be described in terms of it's restaurant type (see Table 1). This meta-information (user attributes) then can be used as a basis for the navigation in the set of specified decision alternatives.

**Table 1.** WE-DECIDE: evaluation of a user attribute (*restaurant type*) w.r.t. decision alternatives

| decision alternative | restaurant type |
|---|---|
| Villa Lido | Italian |
| Fallaloon | Chinese |
| Nepomuk | Austrian |
| Poseidon | Greek |

## 2.2 Taking Decisions in We-Decide

*Decision Phase 1: Collecting User Preferences.* After having completed the definition of the decision task and the corresponding (initial) set of solutions, the designer sends out an email to the participants – alternatively, WE-DECIDE can do this automatically. Each participant (user) receives an email and can then specify his/her preferences with regard to a restaurant. WE-DECIDE allows to make the preferences of other users invisible for the current user – with this feature we are able to encourage users to focus on solving the decision task and not to solely evaluate the preferences of other users [2]. In our scenario, users are specifying their preferences in terms of evaluating individual solutions on a one-five star evaluation scale. Within the scope of this preference specification (elicitation) phase users are still enabled to extend the set of decision alternatives. After each user has specified his/her preferences or a defined deadline for the completion of the individual user ratings has passed, the designer (administrator) of the decision task can forward the decision process to next phase where individual preferences are aggregated to group recommendations.

**Table 2.** WE-DECIDE: user-specific ratings with regard to the defined decision alternatives

| solution | Alex | Joe | Mary | Claire |
|---|---|---|---|---|
| Villa Lido | 5 | 3 | 5 | 4 |
| Fallaloon | 3 | 3 | 5 | 3 |
| Nepomuk | 5 | 3 | 3 | 3 |
| Poseidon | 4 | 3 | 4 | 4 |

*Decision Phase 2: Aggregating User Preferences.* The preferences of individual users can be aggregated in different ways. In our scenario, the personal restaurant evaluations of individual users are aggregated by computing the *majority vote*. This aggregation is then presented as group recommendation to each user (see Figure 2) and explained correspondingly (the explanation is related to the way the rating has been determined). In this phase of the group decision process, each user can adapt his/her own preferences and save the changes. The currently supported *aggregation functions* (group recommendation heuristics) in WE-DECIDE are the following.[1]

---

[1] For an in-depth discussion of basic types of group decision heuristics see, for example, the overview of Masthoff [5].

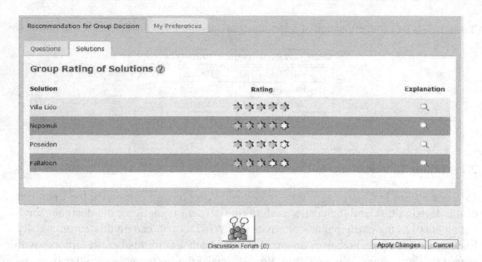

**Fig. 2.** WE-DECIDE: recommendation for a group decision (based on *majority voting*): each user can take a look at the current recommendation and adapt his/her preferences if needed.

*Majority Voting* (see Formula 1) determines the majority of votings (*d*) for a specific solution *s* where $eval(u, s)$ denotes the rating for solution *s* defined by user *u*. For example, the majority of votings for *Villa Lido* is 5 (see Table 3).

$$MAJ(s) = maxarg_{(d \in \{1..5\})}(\#(\bigcup_{u \in Users} eval(u, s) = d)) \tag{1}$$

*Least Misery* (see Formula 2) returns the lowest voting for solution *s* as group recommendation. For example, the LMIS value for the s = *Villa Lido* is 3.

$$LMIS(s) = min(\bigcup_{u \in Users} eval(u, s)) \tag{2}$$

*Most Pleasure* (see Formula 3) returns the highest voting for solution *s* as group recommendation. For example, the MPLS value for the s = *Villa Lido* is 5.

$$MPLS(s) = max(\bigcup_{u \in Users} eval(u, s)) \tag{3}$$

*Group Distance* (see Formula 4) returns the value *d* as group recommendation which causes the lowest overall change of the individual user preferences. For example, the GDIS value for s = *Villa Lido* is 5 (or, alternatively 4 – in such a case, our current approach would choose randomly among the candidate values).

$$GDIS(s) = minarg_{(d \in \{1..5\})}(\sum_{u \in Users} |eval(u, s) - d|) \tag{4}$$

**Table 3.** WE-DECIDE: supported WE-DECIDE group recommendation heuristics. MAJ = Majority Voting; LMIS = Least Misery; MPLS = Most Pleasure; GDIS = Lowest Group Distance; ENS = Ensemble Voting. This example is based on the preference information in Table 2.

| solution | MAJ | LMIS | MPLS | GDIS | ENS |
|----------|-----|------|------|------|-----|
| Villa Lido | 5 | 3 | 5 | 5 | 5 |
| Fallaloon | 3 | 3 | 5 | 3 | 3 |
| Nepomuk | 3 | 3 | 5 | 3 | 3 |
| Poseidon | 4 | 3 | 4 | 4 | 4 |

Finally, *Ensemble Voting* (see Formula 5) determines the majority of the results of the individual voting strategies $H = \{$MAJ, LMIS, MPLS, GDIS$\}$. For example, the ensemble-based majority voting for *Villa Lido* is 5.

$$ENS(s) = maxarg_{(d \in \{1..5\})}(\#(\bigcup_{h \in H} eval(h, s) = d)) \qquad (5)$$

*Decision Phase 3: Taking the Decision.* After a predefined deadline has passed or the designer of the group decision task has forwarded the decision process to the final phase (Phase 3), participants are not allowed to insert further decision alternatives and are also not allowed to adapt their personal preferences anymore. In this phase, the designer (administrator) of the group decision task is in charge of – while taking into account the group recommendation – taking the final group decision. In our scenario the administrator also has to explain the final decision textually, i.e., explanations are not generated. This definition of the group decision formally completes the group decision task – in the following, all users who participated in the group decision are notified about the final decision and also have the opportunity to take a look at the details.

## 3   We-Decide Decision Scenarios

*Scenario 1: (U)ser Attributes & (S)olutions (US).* In our working example (restaurant selection for dinner) we used the basic setting of user attributes in combination with the definition of decision alternatives. User attributes can be used for additionally describing (annotating) decision alternatives. In our case, we used the attribute *restaurant type* to additionally describe the decision alternatives.[2] In the follow-up decision process user attributes can be exploited for specifying criteria for solution search (reducing the number of solutions relevant for a specific user). For example, the restaurant *Villa Lido* is described as an Italian restaurant by instantiating the user attribute *restaurant type* (see Table 1). During solution search, this attribute is used for specifying search criteria. For example, if a user is primarily interested in selecting Italian restaurants, he/she would select the corresponding option and reduce (his/her personal) set of decision alternatives. The decision scenario *US* should be applied when there is a need for a kind of pre-selection on the basis of user attributes. Typical examples in this context are the selection of restaurants (if one is interested in supporting a decision process also on the basis of the restaurant type), car purchase (pre-selection of the car type or producer), or the selection of a tourist destination (e.g., pre-selection of the country).

---

[2] Note that a screenshot related to this description is omitted due to space limitations.

*Scenario 2: (O)nly (A)ttributes (OA).* There are scenarios where user attributes play a dominant role and a concrete set of decision alternatives is not needed. In such a case users are specifying their preferences solely on the basis of user attributes and a group decision is reflected by a complete instantiation of the defined set of user attributes. An example of such a scenario is a group-based configuration scenario where users are specifying their preferences with regard to a configuration and the system is in charge of determining a configuration (recommendation) which takes into account the preferences of the individual group members. On the basis of the answers given by the users, WE-DECIDE can generate a group recommendation. The decision scenario *OA* should be applied when the inclusion of specific solutions is not needed or the number of potential solutions is very high which makes an explicit enumeration impossible. Examples in this context are the group-based configuration of team equipment (e.g., a new soccer uniform for a football team), the group-based configuration of a family car (if a new car is purchased), the group-based configuration of the interior of a new university building, and the group-based configuration of new product lines within the scope of open innovation processes [7].

*Scenario 3: (O)nly (S)olutions (OS).* Finally, we also have to deal with situations where solutions play a dominant role and user attributes are not needed for supporting a group decision process. In such a case users are specifying their preferences directly with regard to the defined set of decision alternatives. Our working scenario of the group-based selection of a restaurant fits this scenario if the decision alternatives are clear and well-known. If users are not interested in pre-selecting decision alternatives or the number of decision alternatives is low, no attribute-based pre-selection is necessary. Application scenarios for *OS* are the selection of a birthday present for a friend, the selection of a travel destination for the next holidays (this also fits the *US* scenario), and the group-based selection of a set of games for a night-gaming session.

## 4   User Study

In order to evaluate the WE-DECIDE group decision environment, we conducted a first small-scale user study. In this study we made the software accessible to ten persons outside our university who were interested in applying a group decision support software in real-world scenarios but were not aware of the fact that the system has been developed at our university. Each of these persons defined a decision task with the support of the WE-DECIDE environment and then made accessible the decision task to a group of other persons. The average number of persons participating in a decision scenario was around 5 (the overall number of users who interacted with WE-DECIDE within the scope of our study was N=48). All the defined decision tasks were stored in the *anonymous mode*, i.e., the defined decision tasks are not allowed to be reused in future WE-DECIDE sessions. The goal of our user study was (1) to figure out how users currently evaluate the perceived usability of the WE-DECIDE environment and (2) to figure out for which domains users would apply the functionalities provided by WE-DECIDE.

First, we were interested in the overall usability of WE-DECIDE. For this purpose we elaborated a questionnaire which is based on the *System Usability Scale* (SUS) [8] (N=27 of the 48 participants of 10 different group decision tasks filled out the questionnaire). The results of this evaluation are depicted in Table 4. The overall feedback on the system was very positive and provides a good motivation to continue our work on the extension and improvement of the currently available WE-DECIDE functionalities.

Second, answers to additional questions related to the application of the WE-DECIDE environment have been collected. Table 5 summarizes the user feedback regarding potential future application domains – users could specify in which additional domains they could imagine to apply the WE-DECIDE functionalities.

**Table 4.** Results of the SUS-based usability study (N=27)

| | Strongly Disagree | Disagree | Not Sure | Agree | Strongly Agree |
|---|---|---|---|---|---|
| 1. I would use WE-DECIDE regularly | 0 | 1 | 0 | 7 | 19 |
| 2. I found it unnecessary complex | 18 | 7 | 2 | 0 | 0 |
| 3. It was easy to use | 0 | 0 | 1 | 13 | 13 |
| 4. I'd need help to use it | 15 | 9 | 3 | 0 | 0 |
| 5. The various parts of WE-DECIDE worked well together | 0 | 0 | 3 | 6 | 18 |
| 6. Too much inconsistency | 19 | 4 | 4 | 0 | 0 |
| 7. I think others would find it easy to use | 0 | 0 | 5 | 5 | 17 |
| 8. I found it very cumbersome to use | 16 | 9 | 1 | 1 | 0 |
| 9. I felt very confident using WE-DECIDE | 0 | 1 | 5 | 14 | 7 |
| 10. I needed to understand how it worked in order to get going | 12 | 13 | 2 | 0 | 0 |

**Table 5.** Preparedness to use WE-DECIDE in different application domains (N=27)

| application domain | yes | no | %would use |
|---|---|---|---|
| 1. Which cinema movie to watch with family/friends? | 23 | 4 | 85.19 |
| 2. Which holiday destination should we choose? | 27 | 0 | 100.0 |
| 3. To which restaurant should we go for dinner? | 24 | 3 | 88.89 |
| 4. Which computer should we buy? | 24 | 3 | 88.89 |
| 5. Which car should we buy? | 23 | 4 | 85.19 |
| 6. Which apartment/house should we buy? | 21 | 6 | 77.78 |
| 7. Which sports activities during the holidays? | 14 | 13 | 51.85 |
| 8. Which games for a gaming night? | 21 | 6 | 77.78 |
| 9. Which new furniture should we buy for the kitchen? | 18 | 9 | 66.67 |
| 10. Which is the best destination for the organization of the next meeting? | 25 | 2 | 92.59 |
| 11. Which present to buy for a friend? | 23 | 4 | 85.19 |

# 5 Related Work

The support of group decision processes on the basis of recommendation technologies is a new and upcoming field of research (see, e.g., Masthoff et al. [5]). The application of group recommendation technologies is still restricted to specific domains such as interactive television [9], e-tourism [10,11], software requirements engineering [6], and ambient intelligence [12].

There exist a couple of online tools which support different types of decision scenarios. *The Decider*[3] is a tool that allows the creation of issues and decision alternatives – the corresponding decision is provided to users who are articulating their preferences regarding the given decision alternatives. Rodriguez et al. [13] introduce *Smartocracy* which is a decision support tool which supports the definition of tasks (issues or questions) and corresponding solutions. Solution selection (recommendation) is based on exploiting information from an underlying social network which is used to rank alternative solutions. *Dotmocracy*[4] is a method for collecting and visualizing the preferences of a large group of users. It is related to the idea of participatory decision making – it's major outcome is a graph type visualization of the group-immanent preferences. Doodle[5] focuses on the aspect of coordinating appointments – similarly, VERN [14] is a tool that supports the identification of meeting times based on the idea of unconstrained democracy where individuals are enabled to freely propose alternative dates themselves. Compared to WE-DECIDE these tools are not able to customize their decision processes depending on the application domain and are also focused on specific tasks. Furthermore, no concepts are provided which help to improve the overall quality of group decisions, for example, in terms of integrating explanations, recommendations for groups, and consistency management for user preferences. *Hermes* [15] is a group decision support tool which focuses on argumentative support of decision makers, e.g., by detecting and helping to resolve conflicts in different user opinions. The proactive support of achieving group consensus and successfully completing decision tasks is not within the major focus of this system. Lai et al. [16] introduce an approach to the AHP (Analytic Hiercharchy Process) based selection of software (multi-media authorizing system). In their work they show how to adapt AHP for group-decision scenarios where different types of group decision heuristics (e.g., the *average* heuristic) are integrated in order to make the achievement of consensus more efficient. Morris et al. [17] introduce the *SearchTogether* environment which supports groups of users in interactive web search scenarios. For example, a couple tries to figure out interesting destinations for their next holiday trip and for this reason cooperatively searches for corresponding web content. The connection of this scenario with WE-DECIDE is that WE-DECIDE relies on solutions provided by different search processes.

## 6    Conclusions

In this paper we have presented the WE-DECIDE group decision environment which supports the flexible design and execution of different types of group decision tasks. The current version of WE-DECIDE supports three basic scenarios of group decision making which differ in the inclusion of (1) user attributes and (2) concrete decision alternatives. Compared to existing group decision support approaches, WE-DECIDE provides an end user modeling environment which supports an easy development and execution of group decision tasks. Within the scope of an empirical study we could gain first evidence of the fact that users find the current version of the system applicable and can imagine to apply WE-DECIDE functionalities in various domains.

---

[3] labs.riseup.net
[4] dotmocracy.org
[5] doodle.com

**Acknowledgements.** The work presented in this paper has been conducted in the IntelliReq research project (829626) funded by the Austrian Research Promotion Agency.

# References

1. Jacowitz, K., Kahneman, D.: Measures of Anchoring in Estimation Tasks. Personality and Social Psychology Bulletin 21(1), 1161–1166 (1995)
2. Mojzisch, A., Schulz-Hardt, S.: Knowing other's preferences degrades the quality of group decisions. Journal of Personality & Social Psychology 98(5), 794–808 (2010)
3. Felfernig, A., Gula, B., Teppan, E.: Knowledge-based Recommender Technologies for Marketing and Sales. International Journal of Pattern Recognition and Artificial Intelligence (IJPRAI) 21(2), 1–22 (2006)
4. Molin, E., Oppewal, H., Timmermans, H.: Modeling Group Preferences Using a Decompositional Preference Approach. Group Decision and Negotiation 6, 339–350 (1997)
5. Masthoff, J.: Group Recommender Systems: Combining Individual Models. In: Recommender Systems Handbook, pp. 677–702 (2011)
6. Felfernig, A., Zehentner, C., Ninaus, G., Grabner, H., Maalej, W., Pagano, D., Weninger, L., Reinfrank, F.: Group Decision Support for Requirements Negotiation. In: Ardissono, L., Kuflik, T. (eds.) UMAP 2011 Workshops. LNCS, vol. 7138, pp. 105–116. Springer, Heidelberg (2012)
7. Chesbrough, H.: Open Innovation. The New Imperative for Creating and Profiting from Technology. Havard Business School Publishing (2003)
8. Bangor, A., Kortum, P., Miller, J.: An empirical evaluation of the System Usability Scale (SUS). International Journal of Human-Computer Interaction 24(6), 574–594 (2008)
9. Masthoff, J.: Group modeling: Selecting a sequence of television items to suit a group of viewers. User Modeling and User-Adapted Interaction (UMUAI) 14(1), 37–85 (2004)
10. Jameson, A., Baldes, S., Kleinbauer, T.: Two methods for enhancing mutual awareness in a group recommender system. In: ACM Intl. Working Conference on Advanced Visual Interfaces, Gallipoli, Italy, pp. 48–54 (2004)
11. McCarthy, K., Salamo, M., Coyle, L., McGinty, L., Smyth, B., Nixon, P.: Group recommender systems: a critiquing based approach. In: 2006 International Conference on Intelligent User Interfaces (IUI 2006), pp. 282–284. ACM, Sydney (2006)
12. Perez, I., Cabrerizo, F., Herrera-Viedma, E.: A Mobile Decision Support System for Dynamic Group Decision-Making Problems. IEEE Transactions on Systems, Man, and Cybernetics 40(6), 1244–1256 (2010)
13. Rodriguez, M., Steinbock, D., Watkins, J., Gershenson, C., Bollen, J., Grey, V., de Graf, B.: Smartocracy: Social networks for collective decision making. In: HICSS 2007, p. 90. IEEE, Waikoloa (2007)
14. Yardi, S., Hill, B., Chan, S.: VERN: Facilitating Democratic group Decision Making Online. In: International ACM SIGGROUP Conference on Supporting Group Work (GROUP 2005), pp. 116–119. ACM, Sanibel (2005)
15. Karacapilidis, N., Papdias, D.: Computer supported argumentation and collaborative decision making: the HERMES system. Journal of Information Systems 26(4), 259–277 (2001)
16. Lai, V., Wong, B., Cheung, W.: Group decision making in a multiple criteria environment: A case using the AHP in software selection. Europ. Journal of OR 137, 134–144 (2002)
17. Morris, M., Horvitz, E.: Searchtogether: an interface for collaborative web search. In: 20th Annual ACM Symposium on User Interface Software and Technology (UIST 2007), pp. 3–12. ACM, Newport (2007)

# Logic-Based Incremental Process Mining
# in Smart Environments

Stefano Ferilli[1,2], Berardina De Carolis[1], and Domenico Redavid[1]

[1] Dipartimento di Informatica – Università di Bari
{stefano.ferilli,berardina.decarolis}@uniba.it
redavid@di.uniba.it
[2] Centro Interdipartimentale per la Logica e Applicazioni – Università di Bari

**Abstract.** Understanding what the user is doing in a Smart Environment is important not only for adapting the environment behavior, e.g. by providing the most appropriate combination of services for the recognized situation, but also for identifying situations that could be problematic for the user. Manually building models of the user processes is a complex, costly and error-prone engineering task. Hence, the interest in automatically learning them from examples of actual procedures. Incremental adaptation of the models, and the ability to express/learn complex conditions on the involved tasks, are also desirable. First-order logic provides a single comprehensive and powerful framework for supporting all of the above. This paper presents a First-Order Logic incremental method for inferring process models, and show its application to the user's daily routines, for predicting his needs and comparing the actual situation with the expected one. Promising results have been obtained with both controlled experiments that proved its efficiency and effectiveness, and with a domain-specific dataset.

## 1 Introduction

Pervasive and ubiquitous computing aims at integrating computation into everyday environments by hiding its complexity to the user. Intelligent interfaces are used to allow people to move around and interact more naturally with services offered by the environment. One of the main goals of a smart environment is to recognize the user's situation and to automatically adapt its behavior accordingly. This process of sensing and responding to human activity, that does not strictly follow plans but is very situation-dependent [19], can be achieved only if the environment is endowed with models of the user's routines in order to anticipate his needs and recognize problematic situations. In fact, prediction, proactivity and decision-making capabilities are important in helping users to achieve their goals in a smart environment. This assistance becomes fundamental when the smart home is inhabited by elderly people or by persons with special needs [18, 2]. To this aim there is a huge amount of work in automatically learning models of the user's routines especially in the home environment [17]. Most proposed approaches model routines of daily activities as process workflows.

M. Ali et al. (Eds.): IEA/AIE 2013, LNAI 7906, pp. 392–401, 2013.
© Springer-Verlag Berlin Heidelberg 2013

Since real-world procedures involve complex interactions of many inter-related tasks, manually producing the models is a complex, costly and error-prone activity [14], and the resulting models developed might not perfectly fit/capture the actual practices. A further problem is the need of adapting and fine-tuning existing models in dynamic environments [9]. Here we present an incremental method based on First-Order Logic for automatically learning process models, and show its application to user's daily routines.

The rest of the paper is organized as follows. Section 2 reports preliminary notions and previous approaches to process mining. Section 3 describes the proposed solution, that is discussed and evaluated in Section 4. Finally, Section 5 concludes the work.

## 2   Preliminaries

A *process* is a sequence of *events* associated to actions performed by agents [6]. A *workflow* is a (formal) specification of how a set of tasks can be composed to result in valid processes, including sequential, parallel, conditional, or iterative execution [20]. Each task may have preconditions (that should hold before it is executed) and postconditions (that should hold after execution of the task). An *activity* is the actual execution of a task. A *case* is a particular execution of actions in a specific order compliant to a given workflow, along an ordered set of *steps* (time points) [14].

Traces of cases are usually available in the form of lists of events described by 6-tuples $(T, E, W, P, A, O)$ where $T$ is a timestamp, $E$ is the type of the event (begin of process, end of process, begin of activity, end of activity), $W$ is the name of the workflow the process refers to, $P$ is a unique identifier for each process execution, $A$ is the name of the activity, and $O$ is the progressive number of occurrence of that activity in that process [1, 14]. In a hypothetical daily-routine 'morning' workflow we might have:

 - (201109280700, begin_of_process, morning, c23, start, 1)
 - (201109280700, begin_of_activity, morning, c23, wake_up, 1)
 - (201109280705, end_of_activity, morning, c23, wake_up, 1)
 - (201109280708, begin_of_activity, morning, c23, toilet, 1)
 - (201109280709, begin_of_activity, morning, c23, radio, 1)
 - (201109280720, end_of_activity, morning, c23, radio, 1)
 - (201109280722, begin_of_activity, morning, c23, book, 1)
 - (201109280735, end_of_activity, morning, c23, book, 1)
 - (201109280738, end_of_activity, morning, c23, toilet, 1)
 - (201109280740, begin_of_activity, morning, c23, wth, 1)
 - (201109280742, end_of_activity, morning, c23, wth, 1)
 - (201109280743, begin_of_activity, morning, c23, weight, 1)
 ...
 - (201109280821, end_of_process, morning, c23, stop, 1)

Simpler representations, such as pure sequences of task names referred to a single process execution, are sometimes exploited [22].

*Process discovery* [6] (or *process mining* [22]) aims at inferring workflow models from examples of cases. Desirable requirements for the learned model are [6, 1, 14, 22] completeness (it can generate all seen event sequences), irredundancy (it generates as few unseen event sequences as possible), and minimality (it is as simple and compact as possible). *Accuracy* (i.e., completeness and irredundancy [6]) is typically in contrast with minimality (a more compact model is usually more general, and thus tends to cover more cases). Additional *desiderata* are the ability to capture concurrent behavior and to deal with noise [22].

Generally speaking, a workflow can be modeled as a directed graph where nodes are associated to states or tasks/activities, and edges connecting nodes represent the potential flow of control among activities. Edges can be labeled with probabilities and/or conditions on the state of the process, which determine whether they will be traversed or not [1].

Most previous works in process mining focused on learning the graph structure. Approaches based on grammar inference [6, 7] or HMM have no inherent ability to model concurrency, a topic specifically discussed in [12]. The same Authors extend their approach to consider different occurrences of the same activity in one process [11]. All these approaches were superseded by [22], that infers causal dependencies among tasks based on several statistics about their frequency and mutual ordering. The learned models are expressed as a specialization of *Petri nets*, called *WorkFlow nets* [20] or simply WF-nets (a standard formalism to specify processes and workflows). Tests on an implementation of this technique have pointed out that it becomes less and less accurate as long as the number of parallel processes and/or nested loops increases. Some limitations of this approach were superseded by [8] using genetic algorithms, but at the cost of very long times (sometimes in the order of several hours). Much related to our representational approach is *Declarative Process Mining*. A logic-based incremental technique was proposed in [4], but it can hardly be compared to other works because, differently from the process mining mainstream research, it needs both positive and negative examples. Noise and probabilities in having some transitions are usually handled by adding a counter of occurrences to edges [1, 13]. Only a few works also envisage the possibility of mining/inducing simple boolean conditions for edges [1, 13] by learning decision tree classifiers.

In particular, in ambient intelligence contexts workflows have been used for modeling the user's activities in order to endow the home with the capability of recognizing user's needs and responding with services appropriate to the situation. Several models have been proposed in the literature for representing workflows and activities. Among the Machine Learning approaches proposed for solving this problem, naive Bayes classifiers and Hidden Markov Models (HMMs) [12] seem to be the most widely used with promising results for activity recognition [5]. Other researchers, including Maurer et al. [15], have employed decision trees to learn logical descriptions of the activities. HMM-based approaches encode the probabilistic sequence of sensor events in Markov models and dynamic Bayesian networks [21]. Cook et al. employ a boosted version of a HMM to recognize possibly-interleaved activities from a stream of sensor events.

# 3   A FOL-Based Proposal

Our proposal is based on the use of First-Order Logic (FOL) as a representation formalism, that provides a great expressiveness potential to describe both cases and their contextual information in a unified framework. In particular, we will work with the Logic Programming perspective on Machine Learning, Inductive Logic Programming (ILP) [16].

## 3.1   Describing Cases

Our descriptions are based on two predicates:

activity($S,T$) : at step $S$ task $T$ is executed;
next($S',S''$) : step $S''$ follows step $S'$.

where the vocabulary of activities is the (fixed and context-dependent) set of constants representing the allowed tasks, and each step is denoted by a unique identifier. Cases are expressed as conjunctions of ground atoms built on these predicates. Steps are timestamps associated to events. This formalism provides an explicit representation of parallel executions in the task flow (which avoids the need for inferring/guessing them by means of statistical — and hence possibly wrong — considerations) and allows to smoothly add further information and relationships concerning the steps and tasks and the context in which they appear, using domain-dependent predicates. Any trace in the previously introduced format can be automatically translated into this format.

The previous sample trace would yield the following FOL description:

$$\text{activity}(s_0,\text{wake\_up}), \text{ next}(s_0,s_1), \text{ activity}(s_1,\text{toilet}),$$
$$\text{next}(s_0,s_2), \text{ activity}(s_2,\text{radio}), \text{ next}(s_2,s_3), \text{ activity}(s_3,\text{book}),$$
$$\text{next}(s_1,s_4), \text{ next}(s_3,s_4), \text{ activity}(s_4,\text{wth}), \text{ next}(s_4,s_5),$$
$$\text{activity}(s_5,\text{weight}), \text{ next}(s_5,s_6), \text{ activity}(s_6,\text{dress}), \text{ next}(s_6,s_7),$$
$$\text{activity}(s_7,\text{tea}), \text{ next}(s_6,s_8), \text{ activity}(s_8,\text{tv}), \text{ next}(s_7,s_9),$$
$$\text{next}(s_8,s_9), \text{ activity}(s_9,\text{door})$$

## 3.2   Learning Workflow Structure and Weights

We use two predicates to describe the structure of a workflow:

- task($t,C$) : task $t$ occurs in cases $C$, where $C$ is a multiset of case identifiers (because a task may be carried out several times in the same case);
- transition($I,O,p,C$) : transition[1] $p$, that occurs in cases $C$ (again a multiset), consists in ending all tasks in $I$ and starting all tasks in $O$.

Argument $C$ represents a history of those tasks/transitions, and thus can be exploited for computing statistics on their use. A model will be expressed as a set (to be interpreted as a conjunction) of atoms built on these predicates, and will be built as reported in Algorithm 1.

---

[1] Note that the name 'transition' here has a different meaning than in Petri nets.

---

**Algorithm 1.** Refinement of a workflow model according to a new case

---

**Require:** $\mathcal{W}$: workflow model
**Require:** $c$: case having FOL description $D$
  **for all** activity$(s,t) \in c$ **do**
    **if** $\exists$ task$(t,C) \in \mathcal{W}$ **then**
      $\mathcal{W} \leftarrow (\mathcal{W} \setminus$ task$(t,C)) \cup \{$ task$(t,C \cup \{c\})$ $\}$ /* update statistics on task $t$ */
    **else**
      $\mathcal{W} \leftarrow \mathcal{W} \cup \{$ task$(t,\{c\}))$ $\}$ /* insert new task and initialize statistics */
    **end if**
    refine_precondition$(\mathcal{W}, t(s)$ :- $D|_s)$
    refine_postcondition$(\mathcal{W}, t(s)$ :- $D)$
  **end for**
  **for all** next$(s',s'') \in c$ **do**
    $I \leftarrow \{t' |$ activity$(s',t') \in c\}$
    $O \leftarrow \{t'' |$ activity$(s',t'') \in c\}$
    **if** $\exists$ transition$(I,O,p,C) \in \mathcal{W}$ **then**
      $\mathcal{W} \leftarrow (\mathcal{W} \setminus$ transition$(I,O,t,C)) \cup \{$ transition$(I,O,t,C \cup \{c\})$ $\}$
      /* update statistics on transition $p$ */
    **else**
      $p \leftarrow$ generate_fresh_transition_identifier()
      $\mathcal{W} \leftarrow \mathcal{W} \cup \{$ transition$(I,O,p,\{c\}))$ $\}$
      /* insert new transition and initialize statistics */
    **end if**
  **end for**

---

Differently from all previous approaches, this technique is *fully incremental*: it can start with an empty model and learn from one case (while others need a large set of cases to draw significant statistics), and can refine an existing model according to new cases whenever they become available (introducing alternative routes, even adding new tasks if they were never seen in previous cases, and updating the statistics). This peculiarity is a dramatic advance to the state-of-the-art, because continuous adaptation of the learned model to the actual practice can be carried out efficiently, effectively and transparently to the users.

The proposed representation also permits to easily handle complex or tricky cases in which WF-nets would require dummy or artificially duplicated task nodes, that cannot be handled by current approaches in the literature. Indeed, different transition/4 nodes can combine a given task in different ways with other tasks, or ignore a task when it is not mandatory for a specific passage.

The models learned using our technique allow to naturally handle noisy data since they implicitly express weights on the probability of transitions, that are in fact proportional to the number of cases in which the various transitions actually occurred. Indeed, each time a new training case is processed, updating the multisets of the nodes it implicitly updates the weights as a side-effect. Thus, although noisy cases are incorporated into the model, their weight will be proportional to the ratio of their occurrence with respect to the whole set of training cases. Imposing a noise tolerance $N \in [0,1]$ corresponds to just taking

$N$ as a minimum frequency threshold under which transitions are to be ignored (which is simpler and more intuitive than noise handling in other proposals).

An example of model learned from 5 examples for the 'morning' workflow is:

| | |
|---|---|
| task(stop,[1,2,3,4,5]). | transition([start]-[wake_up],p1,[1,2,3,4,5]). |
| task(door,[1,2,3,4,5]). | transition([wake_up]-[radio,toilet],p2,[1,2,3,4,5]). |
| task(tv,[1,2,4,5]). | transition([radio]-[book],p3[1,2,4,5]). |
| task(tea,[1,2,3,4,5]). | transition([book,toilet]-[wth],p4,[1,2,4,5]). |
| task(dress,[1,2,3,4,5]). | transition([wth]-[weight],p5,[1,4,5]). |
| task(weight,[1,4,5]). | transition([weight]-[dress],p6,[1,4,5]). |
| task(wth,[1,2,3,4,5]). | transition([dress]-[tea,tv],p7,[1,2,4,5]). |
| task(book,[1,2,4,5]). | transition([tea,tv]-[door],p8,[1,2,4,5]). |
| task(radio,[1,2,3,4,5]). | transition([door]-[stop],p9,[1,2,3,4,5]). |
| task(toilet,[1,2,3,4,5]). | transition([radio,toilet]-[wth],p10,[3]). |
| task(wake_up,[1,2,3,4,5]). | transition([wth]-[dress],p11,[2,3]). |
| task(start,[1,2,3,4,5]). | transition([dress]-[tea],p12,[3]). |
| | transition([tea]-[door],p13,[3]). |

It says that activity 'wake_up' was carried out in all cases, while 'weight' occurred only in cases #1, #4 and #5. Activity 'wake_up' was always followed by (possibly parallel) activities 'radio' and 'toilet', while 'tea' followed 'dress' in only one case out of five, i.e. 20% times only (hence, a noise threshold of 0.25 would ignore it). 'dress' is an optional activity, as can be noticed comparing the sequence p5-p6 to p11; indeed, these two alternatives are have complementary sets of cases, 3 for the former and 2 for the latter, which can be interpreted as a 0.6 weight for the former and a 0.4 weight for the latter.

## 3.3   Learning Conditions

Learning propositional conditions for edges in the model is a significant limitation. Our approach naturally overcomes this limitation by allowing to describe in the same FOL framework not only information about tasks and control flow, but relevant contextual observations as well. For instance, the previous 'morning' workflow case description might be extended as follows:

early($s_0$), sunny($s_0$), happy($s_0$), early($s_1$), long_duration($s_1$), early($s_2$),
news($s_2,n'$), about($n',s$), sports($s$), bad($n'$), upset($s_2$), early($s_3$),
calm($s_4$), early($s_4$), windy($s_4$), cold($s_4$), humidity_high($s_4$), early($s_5$),
short_duration($s_5$), short_interval($s_5,s_6$), on_time($s_6$), trousers($s_6$),
shirt($s_6$), pullover($s_6$), long_duration($s_6$), late($s_7$), late($s_8$),
news($s_8,n''$), about($n'',s$), updates($n'',n'$), very_interesting($n''$), late($s_9$)

meaning that the actor woke up early, when the weather was sunny and he was happy; also his activity in the toilet, during which he listened at the radio and then read a book, happened early, but took a long time. The radio gave bad news about sports, that upset the actor; however, he got calm while reading the book. After termination of the toilet activity it was still early, and the actor took a quick look at the weather, which was windy, cold and very humid. Shortly after that, he was on time while dressing trousers, shirt and pullover, which took a long time, so that in subsequent steps he was late. In particular, when listening at the TV he got more news about sports, that were a very interesting update of the previous news given by the radio.

So, while learning the workflow structure for a given case, examples for learning task pre- and post-conditions are generated as well, and provided to a learning system. For compliance with the incrementality of the proposed approach to learning the workflow structure, this learning system must be incremental as well. A suitable learner is InTheLEx [10], that is also endowed with a positive-only-learning feature [3] (useful because only examples of workflow cases actually carried out are typically available). Given a case $c$ having FOL description $D$ (including both process and contextual information), for each activity$(s,t)$ atom in $c$, two examples for learning conditions for task $t$ are created as follows:

$t(s)$ :− $D|_s$.    for the pre-condition, and
$t(s)$ :− $D$.    for the post-condition

where $D|_s$ denotes the subset of atoms in $D$ associated to steps up to $s$ only. Indeed, when applying the workflow model, pre-conditions for performing a task at a given time can only be checked against the events that took place up to that moment. Conversely, post-conditions express what must happen after a task execution, possibly depending also on specific events that took place before that task execution. In the 'morning' example, activity$(s_3,$book$)$ will generate the following example for the pre-condition of the 'book' task:

book$(s_3)$ :- activity$(s_0,$wake_up$)$, next$(s_0,s_1)$, activity$(s_1,$toilet$)$,
    next$(s_0,s_2)$, activity$(s_2,$radio$)$, next$(s_2,s_3)$, activity$(s_3,$book$)$,
    early$(s_0)$, sunny$(s_0)$, happy$(s_0)$, early$(s_1)$, long_duration$(s_1)$,
    early$(s_2)$, news$(s_2,n')$, about$(n',s)$, sports$(s)$, bad$(n')$,
    upset$(s_2)$, early$(s_3)$.

and a similar one for the post-condition, including all atoms in $D$. Together with other examples, these might lead to infer preconditions such as:

book$(Y)$ :- activity$(X,$radio$)$, news$(X,N)$, bad$(N)$, upset$(X)$,
    next$(X,Y)$, activity$(Y,$book$)$, early$(Y)$.
    (in order to read a book, the actor must be upset because of having heard bad news on the radio, and it must be early)

and postconditions such as:

book$(Y)$ :- activity$(X,$radio$)$, news$(X,N)$, about$(N,S)$,
    sports$(S)$, bad$(N)$, upset$(X)$, next$(X,Y)$, activity$(Y,$book$)$,
    early$(Y)$, next$(Y,Z)$, calm$(Z)$.
    (after reading a book when it is early, the actor is calm if he was upset due to having heard bad news about sports on the radio).

## 4   Discussion and Evaluation

The learned model can be submitted to experts for analysis purposes, for improving their understanding of the users' daily routines or for manually tailoring

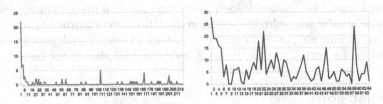

**Fig. 1.** Learning curve for Aruba (on the left) and Milan (on the right) CASAS datasets

it. It can also be used to generate possible cases of daily routines, or to trace future daily behavior of the users and checks whether it is compliant with the learned routines, raising suitable warnings otherwise. The user's response to such warnings, if any, might be exploited to fix or refine the model.

While translating a Petri net in the proposed representation is quite straightforward, the opposite translation is not trivial, because our formalism can compactly, easily and intuitively express a wider range of situations than Petri nets, in particular as regards the possibility of specifying invisible or duplicate tasks. These kinds of tasks are also problematic for the other learning algorithms.

The proposed methodology was evaluated by building 11 different artificial workflow models, each characterized by different combinations of the following complexities and potentially tricky features purposely devised to stress the learning methods: high parallelism mixed with many alternative routes, nested loops of various length. The use of artificial problems allowed to have complete control on the amount and kind of complexity to be introduced in the models, and to be able to suitably tune such complexity for analyzing specific aspects of the proposed technique. The experimental setting was as in [22]: 1000 examples were randomly generated for each experiment, and each experiment was repeated several times to ensure that the random generation did not affect the outcome. Our technique was able to learn the correct model in all cases, always converging to the correct model in a few seconds using less than 50 training cases (which would have been clearly insignificant for the statistical techniques). This behavior suggested a comparison to [22] only: indeed, older approaches were superseded by it; the approach in [4] uses negative examples that are typically unavailable in real-world process mining; the system in [8] has too long runtimes for a significant comparison. The outcome was that, even using 1000 examples, the technique in [22] was unable to learn 7 out of 11 test models, and often failed in dealing with small loops and high parallelism.

As to the specific daily routines task, two datasets taken from the CASAS repository (http://ailab.wsu.edu/casas/datasets.html) were used: Aruba (made up of 220 examples and involving 11 tasks) and Milan (including 64 examples and 15 tasks). The latter is more challenging, being smaller and with a larger number of tasks. Both were processed in less than half a second. The learning curves are shown in Figure 1: while the former shows a substantial convergence within the first 10 examples, in the latter a periodic change of behavior emerges every 20 days circa. It is not possible to foresee whether a convergence would be reached subsequently, but surely a more dynamic lifestyle of the house inhabitants is highlighted. Preconditions were learned for the Aruba dataset, returning no precondition for 3 out of 11 tasks, and two alternative preconditions for one of the remaining 8 tasks (specifically, 'enter_home').

## 5   Conclusions

While workflow management can be crucial in smart environments, manually setting up workflow models is costly and error-prone. This paper proposes a novel method for automatic workflow induction, based on First-Order Logic representations. It is more expressive than previous proposals, being able to represent cases of any complexity. Its incremental approach allows to learn from scratch and converge towards correct models using very few examples. It can also handle the context in which the activities take place, thus allowing to learn complex (and human-readable) pre- and post-conditions for the tasks, using an ILP incremental learner. It can also handle noise in a straightforward way. Both controlled hard experiments and domain-specific ones revealed that the method ensures quick convergence towards the correct model, using much less training examples than would be required by statistical techniques. Future work will consist in wrapping the learning system in a service that can be exploited by several different real-world applications for workflow learning, simulation and check. A deeper study of the results obtained in this paper is already underway, and more experiments are planned on additional Smart Environment datasets.

**Acknowledgment.** This work was partially funded by the Italian PON 2007-2013 project PON02_00563_3489339 'Puglia@Service'.

## References

[1] Agrawal, R., Gunopulos, D., Leymann, F.: Mining process models from workflow logs. In: Schek, H.-J., Saltor, F., Ramos, I., Alonso, G. (eds.) EDBT 1998. LNCS, vol. 1377, pp. 469–483. Springer, Heidelberg (1998)

[2] Bierhoff, I., van Berlo, A.: More intelligent smart houses for better care and health, global telemedicine and ehealth updates. Knowledge Resources 1, 322–325 (2008)

[3] Bombini, G., Di Mauro, N., Esposito, F., Ferilli, S.: Incremental learning from positive examples. In: Atti del 24-esimo Convegno Italiano di Logica Computazionale, CILC 2009 (2009)

[4] Cattafi, M., Lamma, E., Riguzzi, F., Storari, S.: Incremental declarative process mining. In: Szczerbicki, E., Nguyen, N.T. (eds.) Smart Information and Knowledge Management. SCI, vol. 260, pp. 103–127. Springer, Heidelberg (2010)

[5] Cook, D., Schmitter-Edgecombe, M.: Assessing the quality of activities in a smart environment. Methods of Information in Medicine (2009)

[6] Cook, J.E., Wolf, A.L.: Discovering models of software processes from event-based data. Technical Report CU-CS-819-96, Department of Computer Science, University of Colorado (1996)

[7]  Cook, J.E., Wolf, A.L.: Event-based detection of concurrency. Technical Report
     CU-CS-860-98, Department of Computer Science, University of Colorado (1998)
[8]  de Medeiros, A.K.A., Weijters, A.J.M.M., van der Aalst, W.M.P.: Genetic process
     mining: an experimental evaluation. Data Min. Knowl. Discov. 14, 245–304 (2007)
[9]  Ellis, C.A., Keddara, K., Rozenberg, G.: Dynamic change within workflow sys-
     tems. In: Proceedings of the Conference on Organizational Computing Systems,
     pp. 10–21. ACM (1995)
[10] Esposito, F., Semeraro, G., Fanizzi, N., Ferilli, S.: Multistrategy theory revision:
     Induction and abduction in inthelex. Machine Learning Journal 38(1/2), 133–156
     (2000)
[11] Herbst, J.: Inducing workflow models from workflow instances. In: Proceedings of
     the 6th European Concurrent Engineering Conference, pp. 175–182. Society for
     Computer Simulation, SCS (1999)
[12] Herbst, J.: Dealing with concurrency in workflow induction. In: Proceedings of the
     European Concurrent Engineering Conference, pp. 175–182. SCS Europe (2000)
[13] Herbst, J., Karagiannis, D.: Integrating machine learning and workflow manage-
     ment to support acquisition and adaptation of workflow models. In: Proceedings
     of the 9th International Workshop on Database and Expert Systems Applications,
     pp. 745–752. IEEE (1998)
[14] Herbst, J., Karagiannis, D.: An inductive approach to the acquisition and adap-
     tation of workflow models. In: Proceedings of the IJCAI 1999 Workshop on Intel-
     ligent Workflow and Process Management: The New Frontier for AI in Business,
     pp. 52–57 (1999)
[15] Maurer, U., Smailagic, A., Siewiorek, D., Deisher, M.: Activity recognition and
     monitoring using multiple sensors on different body positions. In: International
     Workshop on Wearable and Implantable Body Sensor Networks, pp. 4–116 (2006)
[16] Muggleton, S.: Inductive logic programming. New Generation Computing 8(4),
     295–318 (1991)
[17] Rashidi, P., Cook, D.: Keeping the resident in the loop: Adapting the smart home
     to the user. IEEE Transactions on Systems, Man, and Cybernetics, Part A: Sys-
     tems and Humans (2012)
[18] Steg, H., et al.: Ambient assisted living – european overview report. Technical
     report (September 2005)
[19] Suchman, L.A.: Plans and situated actions: The problem of human-machine com-
     munications. Cambridge University Press (1987)
[20] van der Aalst, W.M.P.: The application of petri nets to workflow management.
     The Journal of Circuits, Systems and Computers 8, 21–66 (1998)
[21] van Kasteren, T., Krose, B.: Bayesian activity recognition in residence for el-
     ders. In: 3rd IET International Conference on Intelligent Environments (IE 2007),
     pp. 209–212 (2007)
[22] Weijters, A.J.M.M., van der Aalst, W.M.P.: Rediscovering workflow models from
     event-based data. In: Hoste, V., De Pauw, G. (eds.) Proceedings of the 11th Dutch-
     Belgian Conference of Machine Learning (Benelearn 2001), pp. 93–100 (2001)

# Information Mining Processes
# Based on Intelligent Systems

Ramón García-Martínez, Paola Britos, and Dario Rodríguez

Information Systems Research Group, National University of Lanus, Argentina.
Information Mining Research Group, National University of Rio Negro at El Bolsón, Argentina
rgarcia@unla.edu.ar, pbritos@unrn.edu.ar

**Abstract.** Business Intelligence offers an interdisciplinary approach (within which is Information Systems), that taking all available information resources and using of analytical and synthesis tools with the ability to transform information into knowledge, focuses on generating knowledge that contributes to the management decision-making and generation of strategic plans in organizations. Information Mining is the sub-discipline of information systems which supports business intelligence tools to transform information into knowledge. It has defined as the search for interesting patterns and important regularities in large bodies of information. We address the need to identify information mining processes to obtain knowledge from available information. When information mining processes are defined, we may decide which data mining algorithms will support the information mining processes. In this context, this paper proposes a characterization of the information mining process related to the following business intelligence problems: discovery of rules of behavior, discovery of groups, discovery of significant attributes, discovering rules of group membership and weight of rules of behavior or rules of group memberships.

## 1 Introduction

Business Intelligence offers an interdisciplinary approach (within which are included the Information Systems), that takes all the available information resources and the usage of analytical and synthesis tools with the ability to transform information into knowledge, focuses on generating knowledge that supports the management decision-making and generation of strategic plans at organizations [1].

Information Mining is the sub-discipline of information systems which provides to the Business Intelligence [2] the tools to transform information into knowledge [3]. It has been defined as the discovery of interesting patterns and important regularities in large information bases [4]. When speaking of information mining based on intelligent systems [5], this refers especially in the application of intelligent systems-based methods to discover and enumerate the existing patters in the information. Intelligent systems-based methods [6] allow retrieving results about the analysis of information bases that the conventional methods fail to achieve [7], such as: TDIDT algorithms (Top Down Induction Decision Trees), self-organizing maps (SOM) and

M. Ali et al. (Eds.): IEA/AIE 2013, LNAI 7906, pp. 402–410, 2013.
© Springer-Verlag Berlin Heidelberg 2013

Bayesian networks. TDIDT algorithms allow the development of symbolic descriptions of the data to distinguish between different classes [8]. Self-organizing maps can be applied in the construction of information clusters. They have the advantage of being tolerant to noise and the ability to extend the generalization when needing the manipulation of new data [9]. Bayesian networks can be applied to identify discriminative attributes in large information bases and detect behavior patterns in the analysis of temporal series [10].

It has been noted the necessity of having processes [11] that allow obtaining knowledge [12] from the large information-bases available [13], its characterization [14] and involved technologies [15].

In this context, this paper proposes a characterization of the information mining process related to the following business intelligence problems: discovery of behavior rules, discovery of groups, discovery of significant attributes, discovery of group-membership rules and weighting of behavior or group-membership rules, and the identification of information-systems technologies that can be used for the characterized processes.

## 2     Proposed Techniques for Information Mining Processes

In this section, the following information-mining processes are proposed: discovery of behavior rules (Section 2.1), discovery of groups (Section 2.2), discovery of significant attributes (Section 2.3), discovery of group-membership rules (Section 2.4) and weighting of behavior or group-membership rules (Section 2.5).

### 2.1     Process of Discovery of Behavior Rules

The process for discovery of behavioral rules applies when it is necessary to identify which are the conditions to get a specific outcome in the problem domain. The following problems are examples among others that require this process: identification of the characteristics for the most visited commercial office by customers, identification of the factors that increase the sales of a specific product, definition of the characteristics or traits of customers with high degree of brand loyalty, definition of demographic and psychographic attributes that distinguish the visitors to a website.

For the discovery of behavioral rules from classes attributes in a problem domain that represents the available information base, it is proposed the usage of TDIDT induction algorithms [16] to discover the rules of behavior for each class attribute. This process and its products can be seen graphically in Figure 1.

First, all sources of information (databases, files, others) are identified, and then they are integrated together as a single source of information which will be called integrated data base. Based on the integrated data base, the class attribute is selected (attribute A in the Figure).

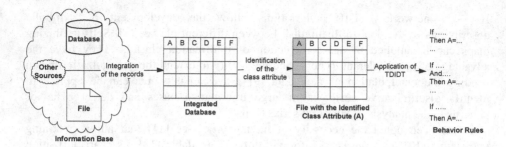

**Fig. 1.** Schema and products resulting of applying Process of Discovery of Behavior Rules using TDIDT

As a result of running the Process of Discovery of Behavior Rules, applying TDIDT to the class attribute, a set of rules which define the behavior of that class is achieved.

## 2.2   Process of Discovery of Groups

The process of discovery of groups applies when it is necessary to identify a partition on the available information base of the problem domain. The following problems are examples among others that require this process: identification of the customers segments for banks and financial institutions, identification of type of calls of customer in telecommunications companies, identification of social groups with the same characteristics, identification of students groups with homogeneous characteristics.

For the discovery of groups [17] [18] in information bases of the problem domain for which there is no available "a priori" criteria for grouping, it is proposed the usage of Kohonen's Self-Organizing Maps or SOM [19] [20] [21]. The use of this technology intends to find if there is any group that allows the generation of a representative partition for the problem domain which can be defined from available information bases. This process and its products can be seen graphically in Figure 2.

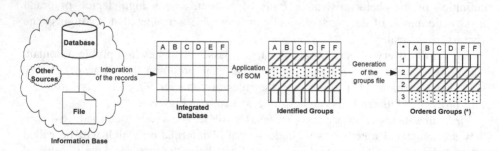

**Fig. 2.** Schema and products resulting of applying Process Discovery of Groups using SOM

First, all sources of information (databases, files, others) are identified, and then they are integrated together as a single source of information which will be called integrated data base. Based on the integrated data base, the self-organizing map (SOM) is applied. As a result of the application of Process Discovery of Groups using SOM, a partition of the set of records in different groups, that will be called identified groups, is achieved. For each identified group, the corresponding data file will be generated.

## 2.3    Process of Discovery of Significant Attributes

The process of discovery of significant attributes applies when it is necessary to identify which are the factors with the highest incidence (or occurrence frequency) for a certain outcome of the problem. The following problems are examples among others that require this process: factors with incidence on the sales, distinctive features of customers with high degree of brand loyalty, key-attributes that caracterize a product as marketable, key-features of visitors to a website.

Bayesian Networks [22] allows to see how variation in the values of attributes, impact on the variation of the value of class attribute. The use of this pocess seeks to identify whether there is any interdependence among the attributes that modelize the problem domain which is represented by the available information base. This process and its products can be seen graphically in Figure 3.

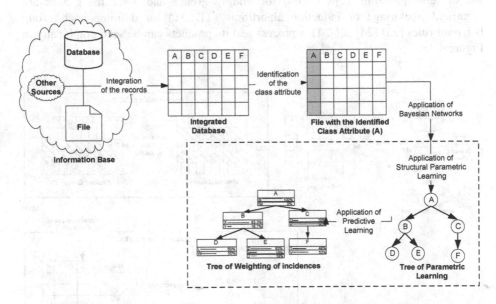

**Fig. 3.** Schema and resulting products for applying Process Discovery of Significant Attributes using Bayesian Networks

First, all sources of information (databases, files, others) are identified, and then they are integrated together as a single source of information which will be called integrated data base. Based on the integrated data base, the class attribute is selected (attribute A in the Figure 3).

As a result of the application of the Bayesian Networks structural learning to the file with the identified class attribute, the learning tree is achieved. The Bayesian Networks predictive learning is applied to this tree obtaining the tree of weighting interdependence which has the class attribute as a root and to the other attributes with frequency (incidence) related the class attribute as leaf nodes.

## 2.4     Process of Discovery of Group-Membership Rules

The process of discovery of group membership rules applies when it is necessary to identify which are the conditions of membership to each of the classes of an unknown partition "a priori", but existing in the available information bases of the problem domain.

The following problems are examples among others that require this process: types of customer's profiles and the characterization of each type, distribution and structure of data of a web site, segmentation by age  of students and the behavior of each segment, classes of telephone calls in a region and the characterization of each class.

For running the process of discovery of group-membership rules it is proposed to use of self-organizing maps (SOM) for finding groups and; once the groups are identified, the usage of induction algorithms (TDIDT) for defining each group behavior rules [23] [24] [21]. This process and its products can be seen graphically in Figure 4.

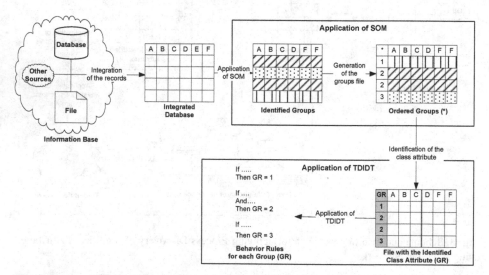

**Fig. 4.** Schema and resulting products of running Process of Discovery of Group-membership Rules using SOM and TDIDT

## 2.5    Process of Weighting of Behavior or Group-Membership Rules

First, all sources of information (databases, files, others) are identified, and then they are integrated together as a single source of information which will be called integrated data base. Based on the integrated data base, the self-organizing maps (SOM) are applied. As a result of the application of SOM, a partition of the set of records in different groups is achieved which is called identified groups. The associated files for each identified group are generated. This set of files is called "ordered groups". The "group" attribute of each ordered group is identified as the class attribute of that group, establishing it in a file with the identified class attribute (GR). Then is applied TDIDT to the class attribute of each "GR group" and the set of rules that define the behavior of each group is achieved.

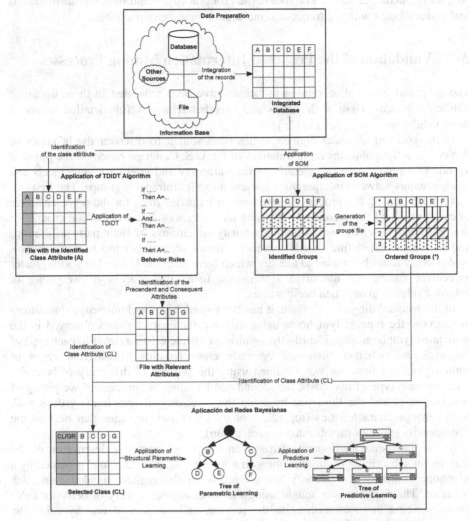

**Fig. 5.** Schema and resulting products of running process of weighting of behavior or Group-membership rules using SOM, TDIDT and Bayesian Neworkts

The procedure to be applied when there are classes/groups no identified includes the identification of all sources of information (databases, files, others), and then they are integrated together as a single source of information which will be called integrated data base. Based on the integrated data base, the self-organizing maps (SOM) are applied. As a result of the application of SOM, a partition of the set of records in different groups is achieved. These groups are called identified groups. For each identified group, the corresponding data file will be generated. This set of files is called "ordered groups". The group attribute of each "ordered group" is identified as the class attribute of that group, establishing it in a file with the identified class attribute (GR). As a result of the application of the structural learning, the learning tree is achieved. The predictive learning is applied to this tree obtaining the tree of weighting interdependence. The root is the group attribute and the other attributes as leaf nodes labeled with the frequency (incidence) on the group attribute.

# 3     Validation of the Proposed Information Mining Processes

The proposed information mining processes have been validated in three domains: political alliances, medical diagnosis and user behavior.  A full detailed report of these validations can be seen in [26].

In the political alliances domain, it has been sought to discover the behavior of democrats and republicans representatives of the U.S. Congress based on the political agenda of a regular session, identifying the intraparty and interparty agreements and disagreements between interparty groups and minority intraparty groups. The first one was obtained using the process of discovery of behavior rules for the representatives of each party, and the second one by using the process of discovery of the groups for the representatives who voted homogeneously (regardless of their party affiliation) and the rules that define that homogeneity (rules of membership to each group). Additionally it has been tried to identify which have been the law or laws with greater agreement among the identified agreements, using the process of weighting of behavior rules or group-membership rules.

In the medical diagnosis domain, it has been synthesize the knowledge that allows to diagnose the type of lymphoma using as input the characteristics observed in the associated lymphography, identify the significant characteristic related to each type of diagnosis and whether there are common characteristics to different types of pathologies. The first one was obtained using the process of discovery of behavior rules for each type of diagnosis, the second one by using the process of weighting of behavior rules and the third one by using the discovery of lymphoma groups with homogeneous characteristics (regardless of its type) and the rules that defines the homogeneity (rules of membership to each group).

In the user behavior domain, it has been sought to specify a description of the reasons for subscribing or unsubscribing to a Internet service "dial-up" provided by a telephone company and identify the reasons with the highest incidence in each behavior. The first one was sought using the process of discovery of behavior rules for subscribing or unsubscribing to the service, and the second one by using the process of weighting of behavior rules.

# 4    Conclusions

In this paper it has been proposed and described five information mining processes: discovery of behavior rules, discovery of groups, discovery of significant attributes, discovery of group-membership rules and weighting of significant atribute related to behavior or membership rules.

Each process has been associated with the following techniques: the usage of TDIDT algorithm applied to the discovery of behavior rules or group-membership rules, the usage of self-organizing maps applied to the discovery of groups, the usage of Bayesian networks applied to the weighting of interdependence between attributes, the usage of self-organizing maps and TDIDT algorithms applied to the discovery of group-membership rules and the usage of Bayesian networks applied to the weighting of significant atribute in behavior or group-membership rules.

During the documental research work it has been noted the indiscriminate use of terms "data mining" and "information mining" to refer to the same body of knowledge. However, raising this equivalence is similar to say that computer-systems are equivalent to information-systems. The first ones are related to the technology that supports the second ones and this is what makes them different.

In this context is an open problem the need of organizing the body of knowledge related to engineering of information mining, establishing that data mining is related to algorithms; and information mining is related to processes and methologies.

On the other hand, there are in the literature many papers and results about the convenience of the usage of certain data mining algorithms compared to others, but it is rarely raised the information mining process associated to these algorithms or the convenience of the usage of one algorithm compared to other for that process. In this context, is an interesting open problem the identification of the relationship between the data mining algorithm and the process of information mining.

# References

1. Thomsen, E.: BI's Promised Land. Intelligent Enterprise 6(4), 21–25 (2003)
2. Negash, S., Gray, P.: Business Intelligence. In: Burstein, F., Holsapple, C. (eds.) En Handbook on Decision Support Systems 2, pp. 175–193. Springer, Heidelberg (2008)
3. Langseth, J., Vivatrat, N.: Why Proactive Business Intelligence is a Hallmark of the Real-Time Enterprise: Outward Bound. Intelligent Enterprise 5(18), 34–41 (2003)
4. Grigori, D., Casati, F., Castellanos, M., Dayal, U., Sayal, M., Shan, M.: Business Process Intelligence. Computers in Industry 53(3), 321–343 (2004)
5. Michalski, R., Bratko, I., Kubat, M.: Machine Learning and Data Mining, Methods and Applications. John Wiley & Sons (1998)
6. Kononenko, I., Cestnik, B.: Lymphography Data Set. UCI Machine Learning Repository (1986), http://archive.ics.uci.edu/ml/datasets/Lymphography (Último acceso 29 de Abril del 2008)
7. Michalski, R.: A Theory and Methodology of Inductive Learning. Artificial Intelligence 20, 111–161 (1983)
8. Quinlan, J.: Learning Logic Definitions from Relations. Machine Learning 5, 239–266 (1990)
9. Kohonen, T.: Self-Organizing Maps. Springer (1995)
10. Heckerman, D., Chickering, M., Geiger, D.: Learning bayesian networks, the combination of knowledge and statistical data. Machine Learning 20, 197–243 (1995)

11. Chen, M., Han, J., Yu, P.: Data Mining: An Overview from a Database Perspective. IEEE Transactions on Knowledge and Data Engineering 8(6), 866–883 (1996)
12. Chung, W., Chen, H., Nunamaker, J.: A Visual Framework for Knowledge Discovery on the Web: An Empirical Study of Business Intelligence Exploration. Journal of Management Information Systems 21(4), 57–84 (2005)
13. Chau, M., Shiu, B., Chan, I., Chen, H.: Redips: Backlink Search and Analysis on the Web for Business Intelligence Analysis. Journal of the American Society for Information Science and Technology 58(3), 351–365 (2007)
14. Golfarelli, M., Rizzi, S., Cella, L.: Beyond data warehousing: what's next in business intelligence? In: Proceedings 7th ACM international Workshop on Data Warehousing and OLAP, pp. 1–6 (2004)
15. Koubarakis, M., Plexousakis, D.: A Formal Model for Business Process Modeling and Design. In: Wangler, B., Bergman, L.D. (eds.) CAiSE 2000. LNCS, vol. 1789, pp. 142–156. Springer, Heidelberg (2000)
16. Britos, P., Jiménez Rey, E., García-Martínez, E.: Work in Progress: Programming Misunderstandings Discovering Process Based On Intelligent Data Mining Tools. In: Proceedings 38th ASEE/IEEE Frontiers in Education Conference (2008) (en prensa)
17. Kaufmann, L., Rousseeuw, P.: Finding Groups in Data: An Introduction to Cluster Analysis. John Wiley & Sons (1990)
18. Grabmeier, J., Rudolph, A.: Techniques of Cluster Algorithms in Data Mining. Data Mining and Knowledge Discovery 6(4), 303–360 (2002)
19. Ferrero, G., Britos, P., García-Martínez, R.: Detection of Breast Lesions in Medical Digital Imaging Using Neural Networks. In: Debenham, J. (ed.) Professional Practice in Artificial Intelligence. IFIP, vol. 218, pp. 1–10. Springer, Boston (2006)
20. Britos, P., Cataldi, Z., Sierra, E., García-Martínez, R.: Pedagogical Protocols Selection Automatic Assistance. In: Nguyen, N.T., Borzemski, L., Grzech, A., Ali, M. (eds.) IEA/AIE 2008. LNCS (LNAI), vol. 5027, pp. 331–336. Springer, Heidelberg (2008)
21. Britos, P., Grosser, H., Rodríguez, D., García-Martínez, R.: Detecting Unusual Changes of Users Consumption. In: Bramer, M. (ed.) Artificial Intelligence in Theory and Practice II. IFIP, vol. 276, pp. 297–306. Springer, Boston (2008)
22. Britos, P., Felgaer, P., García-Martínez, R.: Bayesian Networks Optimization Based on Induction Learning Techniques. In: Bramer, M. (ed.) Artificial Intelligence in Theory and Practice II. IFIP, vol. 276, pp. 439–443. Springer, Boston (2008)
23. Britos, P., Abasolo, M., García-Martínez, R., Perales, F.: Identification of MPEG-4 Patterns in Human Faces Using Data Mining Techniques. In: Proceedings 13th International Conference in Central Europe on Computer Graphics, Visualization and Computer Vision 2005, pp. 9–10 (2005)
24. Cogliati, M., Britos, P., García-Martínez, R.: Patterns in Temporal Series of Meteorological Variables Using SOM & TDIDT. In: Bramer, M. (ed.) Artificial Intelligence in Theory and Practice. IFIP, vol. 217, pp. 305–314. Springer, Boston (2006a)
25. Britos, P., Dieste, O., García-Martínez, R.: Requirements Elicitation in Data Mining for Business Intelligence Projects. In: Avison, D., Kasper, G.M., Pernici, B., Ramos, I., Roode, D. (eds.) Advances in Information Systems Research, Education, and Practice. IFIP, vol. 274, pp. 139–150. Springer, Boston (2008b)
26. Britos, P.: Processes of Information Mining based on Intelligent Systems. PhD thesis in Computer Science. School of Computing. Universidad Nacional de La Plata (2008) (in spanish), http://postgrado.info.unlp.edu.ar/Carrera/ Doctorado/Tesis/Britos-Tesis%20

# Customer Churn Detection System: Identifying Customers Who Wish to Leave a Merchant

Cosimo Birtolo[1], Vincenzo Diessa[1], Diego De Chiara[1], and Pierluigi Ritrovato[2]

[1] Poste Italiane – Innovation and ICT Business Development
Research and Innovation - R&D Center – Piazza Matteotti 3 – 80133 Naples, Italy
{birtoloc,diessavi,dechia22}@posteitaliane.it
[2] CRMPA – Centro di Ricerca in Matematica Pura ed Applicata c/o DIEII
Department of Electronic Engineering and Computer Engineering
University of Salerno – 84084 Fisciano, Italy
ritrovato@crmpa.unisa.it

**Abstract.** Identifying customers with a higher probability to leave a merchant (churn customers) is a challenging task for sellers. In this paper, we propose a system able to detect churner behavior and to assist merchants in delivering special offers to their churn customers. Two main goals lead our work: on the one hand, the definition of a classifier in order to perform churn analysis and, on the other hand, the definition of a framework that can be enriched with social information supporting the merchant in performing marketing actions which can reduce the probability of losing those customers. Experimental results of an artificial and a real datasets show an increased value of accuracy of the classification when random forest or decision tree are considered.

**Keywords:** Churn Analysis, Social commerce, Customer Lifetime Value, Decision Tree, Data Mining.

## 1 Introduction

Within an e-Commerce platform, it is particularly useful for merchants to identify those customers who wish to transfer their habit of buying products and/or services to competing merchants. Predicting accurately customer behavior is a challenging task for merchants and companies, so that recent works tackle this problem. The idea behind the Churn Analysis is to highlight customers who are most at risk "churn"; this allows to make marketing efforts aimed at increasing customer loyalty, reducing the probability of the event of "churn".

Today, the churn analysis is applied to different sectors, such as financial markets, e-mail service providers, retailers of industrial products, telecommunications. Talking about a churn analysis in the e-commerce means avoiding that customers leave the current merchant and turn to a competitor. The purpose at the basis of this technique consists in highlighting the customers with a higher churn risk and in implementing marketing actions to increase the loyalty, reducing the probability that a churn happens.

M. Ali et al. (Eds.): IEA/AIE 2013, LNAI 7906, pp. 411–420, 2013.

In this paper we investigate the techniques in order to design a Churner Detection System (CDS) for e-Commerce on the basis of customer activities, his/her purchasing behavior and his/her habits. The goal of the proposed framework is two-fold: enriching the churn analysis technique with social information, which refers to the feeling of customers and his friends toward a merchant. The framework aims to assist the merchant in performing marketing actions able to reduce the probability of losing those customers.

The remainder of this work is organized as follows: Section 2 provides a brief overview of related literature; Section 3 describes how the problem has been modeled; Section 4 presents experimental results; and Section 5 outlines conclusions and future directions.

## 2  Related Work

The churn analysis is a technique aiming at understanding the user behavior and at predicting the moment in which the churn event will happen; the typical event of abandonment is linked to the fact that a customer stops using a specific service moving to competing one.

In the several data mining techniques applicable to the churn analysis, there are different approaches, such as the one of considering the churn like a dichotomous event by setting two different levels, either loyal or churn; in this case are applied the classification techniques such as decisional trees or the artificial neural networks. According to Dror et al. [1], churn prediction entails the definition of a set of features adopted as input in order to train a classifier or regressor for the task. These features are often related to the service and vary according the provided churn service. Indeed, the churn analysis is applied to different fields.

In the financial markets, it is interesting to identify the time at which an investor or broker decides to sell stocks, in this case the churn analysis aims to attract potential customers, increase the satisfaction of existing customers and reduce the likelihood of potential losses. In this context, Guo et al.[2] proposed RFM (Recency Frequency Monetary) - ROI (Return on Investment) model based on the use of decision trees. For internet service providers, the issue of churn is that users can stop using email offers and migrate to e-mail services by other operators. This behavior causes a decrease of the traffic on the portal and reduces advertising revenue by the provider, in this context the techniques used are mainly based on Penalized Multi-Criteria Linear Programming (PMCLP) and decision tree [3], while Dror et al. [1] performed churn prediction applying several classifiers using as experimental setup the answers of users on *Yahoo! Answers*.

A further context is the chain of retail sale of industrial products, where the customer stops buying products or services turning to a competitor. In order to avoid the abandonment of the customer, Ju et al.[4] proposed a technique based on Support Vector Machine (SVM) supported by the Principal Component Analysis. This technique is also used in a general context, with bank datasets in order to prove its efficiency [5].

In the field of telecommunications and mobile, the abandonment is based on the idea that the higher the competitiveness in the domain, the easier the loss of customers. The techniques used for the predictive analysis consist in artificial neural networks and in decisional trees through the adoption of the C5.0 algorithm [6]. Moreover, Pinheiro et al. [7] preliminary introduced the combination of predictive models, such as neural artificial networks, and pattern recognition models such as social network analysis, in order to highlight customer's behaviors when a huge amount of data is considered. They investigated the combination of scores provided by the two models in a theoretical way, without any experimental evidences. On the other hand, Richter [8] evidenced how the Social Network Analysis (SNA) is used to discover relationships and the influences of friends play a fundamental role to discover the reasons of churn and improve customer loyalty.

## 3   An Innovative Approach to Churn Analysis

Our work is aimed at developing a system able to support the merchant to identify churn customers on the basis of customer activities which derive from two main sources: orders and social network. We aim at identifying the potential churn customers and suggesting to the merchant a strategy in order to avoid the abandonment of the current e-Shop. Indeed, once the potential churn customers are identified, the system leaves to the merchant the choice of different marketing actions in order to prevent the customer abandonment.

The arising system is a result of research activities within an Italian research project conducted by Poste Italiane[1], CRMPA (Pure and Applied Mathematics Research Centre)[2], MOMA[3], ITSLab[4], and Opera 21[5]. The general aim of the project, named InViMall (Intelligent Virtual Mall), is to study and define the models, methods and technologies for e-Business based on Knowledge and on the Social Web. In details, the project focuses on the design and implementation of an electronic mall that enhances the shopping experiences of users by means of the adoption, integration and extension of existing components with some innovative features such as Personalized suggestions, Automatic generation of personalized product bundles [9], Advanced Marketing Intelligence, Faceted Navigation. Among the different advanced Marketing Intelligence features, we introduce, in this paper, the *Churn Detection System*, a system able to identify churn customers and able to support a marketing action for those customers. The proposed system is based on five modules, that will be deepened in the following paragraphs: (i) Customer Attitude, (ii) Customer Value, (iii) Customer activities (social or buying behavior), (iv) Churn Analysis Engine, and (v) Automatic Suggestions module. The first three modules perform as parallel tasks. Once the results are available, the outputs are processed by the Churn Analysis Engine. Finally, Automatic Suggestions module elaborates a strategy in order to recover the user and reduce the probability of losing those customers.

---

[1] http://www.posteitaliane.post/
[2] http://www.crmpa.it/
[3] http://www.momanet.it/
[4] http://www.itslab.it/
[5] http://www.opera21.it/

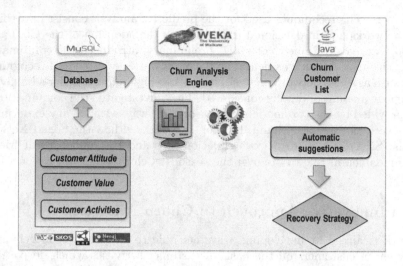

**Fig. 1.** Identification of churner customer: A system overview

## 3.1   Eliciting Customer Attitude

In the Customer attitude module, we select a set of user data that can influence a churn behavior. According to the related literature [5], personal data such as age, gender and so on can be a discriminant in the detection of a churn customer. On the other hand, the reputation of a merchant or the rating expressed by the customer can influence customer actions. Moreover customer loyalty degree can be a measurement of churn behavior. The evaluation of the customer's loyalty degree can be a problem because it is not linked to a single contact experience of the customer with the company, and it is necessary to associate the instantaneous satisfaction with a sequence of positive experiences that allow to determine a "cumulated, consecutive and uninterrupted" satisfaction. Loyalty is the result of a construction process that has fully satisfying purchasing experiences as guidelines, that is to say experiences never interrupted by negative elements that might reduce the buying loyalty level. The definition of loyalty, through the RFM analysis, requires to take into consideration three aspects: (i) *Recency*: the time elapsed since the last purchase (action) of the customer; (ii) *Frequency*: the frequency with which the customer makes purchases (the number of times that the customer has completed an action) and (iii) *Monetary*: the monetary value spent by the customer on purchases (value created by the completion of actions). Guo et al. [2] have investigated the adoption of RFM technique in the churner analysis. These techniques improve the performance of the model. In addition, according to Van den Poel et al. [10], low value can suggest a higher probability of losing those customers. So that, we choose the RFM technique as a relevant input to include as variable in the customer attitude data.

## 3.2   Estimating Customer Value

In literature Customer LifeTime Value (CLV) is a dynamic financial indicator and depends on the customer's behavior; this indicator represents an attractive metrics for the definition of marketing strategies because it provides a forecast on future costs and revenues generated by the customer. CLV represents the customer value over its entire lifecycle and can be defined like the sum of the retrospective value $V_R$ and the prospective value $V_P$ as expressed in Eq.1.

$$CLV(u) = V_R(u) + V_P(u) \tag{1}$$

where $u$ is a customer; $V_R$ is the current value of the customer (calculated using historical data); while $V_P$ is the future value that a customer will have, namely the future earnings the customer will provide to the merchant; this is based on the prediction of future purchases made by the customer through Sequential Minimal Optimization (SMO), an iterative algorithm based on support vector regression adopted for solving some optimization problems [11].

Identifying the most profitable customers, it could be possible to infer what user may be the most active customer and what user may be a churn one. Furthermore, the most profitable customers can potentially influence other customers which belong to their network of contacts or some detected churn customers.

## 3.3   Elaborating Customer Activities

An input necessary to this software module is the liveness value which allows to assess the customer's satisfaction towards a merchant.

A factor which characterizes human behavior is represented by the activities within the platform. We distinguish two kind of activities: the buying and the social one. The former is expressed by means of an index called *Liveness* which allows to assess the customer's satisfaction by measuring the frequency and the number of purchases. The latter is an alternative version of customer activities evaluation by means of social information. This is performed by means of *Intimacy*, an index we defined in order to evaluate the affection or the sentiment that each customer has towards a merchant. Intimacy of a customer towards a seller is evaluated by taking into account the ratings, the reviews, and the number "likes" expressed by the customer on some e-Shop products and, at the same time, the ratings, the reviews, and the "likes" assigned by the customer to the seller and to the provided e-Commerce service.

## 3.4   Predicting Churn Probability

The engine of the proposed system is a classifier which is able to value the churn probability for every customer. CDS can support different classifier but in our study we compared different algorithms choosing decision tree as the best classifier in the InViMall domain (i.e., e-Business).

## 3.5    Supporting Marketing Actions

Once the churn customers are detected, the framework suggests a strategy in order to perform a recovery strategy. The strategy module are out of the scope of this paper as it has been described in previous works [12,13].

# 4    Experimentation

## 4.1    Materials and Methods

In our experimentation we consider Contoso BI Demo Dataset and a transactional database of a real e-Commerce platform of Poste Italiane. Contoso BI Demo Dataset is a fictitious retail demo dataset defined by Microsoft and used for presenting Microsoft Business Intelligence products. It consists of about 19,000 customers and 2,500 products grouped in 8 categories. Poste Italiane sells different products by means of an e-Commerce platform which includes a wide set of products in different categories such as Home, Furniture, Photography, Books & magazines, Mobile phones & communication, Office equipment, containers, stamps and postal items. Therefore, the Poste Italiane dataset refers to the e-Commerce orders from January 1st, 2008 to December 31st, 2011 and consists of about 7,485 orders made by 6,820 users. In our study we consider only a subset of Poste Italiane dataset which is the result of a pre-processing aimed at eliminating users with only one purchasing order (i.e., occasional users) thus entailing a dataset of 1,104 orders made by 439 users. After the identification and analysis of the quantity and quality of available data, according to the best of our findings, we conclude that it is not possible to identify a dataset sufficiently complete and adequate to the testing needs of a CDS with social information.

In both datasets, available data lack any customer feedbacks and in particular it is not known if the customer is still active or if he does not wish to buy any other items from a specific merchant. Such information is necessary in the dataset used to train the predictive model so we consider churner customers, those customers with a period of inactivity greater than a fixed threshold which depends on the selected dataset. Therefore, we split the datasets into "Period 1" and "Period 2", the former consists of all data that the model needs, the latter defines the activity of a customer outside training period and is adopted to detect a churner customer if he/she does not buy any items in that time interval. In detail, Contoso BI demo dataset is split into two periods: one year (Period 1) and a quarter year (Period 2); while Poste Italiane dataset into two intervals: from January 1st, 2008 to December 31st, 2010 (Period 1), and from January 1st, 2011 to December 31st, 2011 (Period 2).

In order to evaluate the classification performances we measure the Correct Classification Rate (CCR) defined as the percentage of instances correctly classified, the true positive rate or *sensitivity* defined as the ratio between true positive and the expected positive, and the true negative rate or *specificity* defined as the ratio between true negative (that are the negative correctly identified) and the expected negative.

**Table 1.** Accuracy of the four different classifiers at varying the adopted dataset

| Algorithm | Contoso | | | Poste Italiane | | |
|---|---|---|---|---|---|---|
| | CCR | Sensitivity | Specificity | CCR | Sensitivity | Specificity |
| Decision Tree | 93.108 | 0.412 | 0.978 | 92.571 | 0.99 | 0.629 |
| Naive Bayes | 91.544 | 0.554 | 0.948 | 82.000 | 0.91 | 0.403 |
| Random Forest | 92.943 | 0.544 | 0.964 | 91.714 | 0.976 | 0.645 |
| Neural Network | 93.197 | 0.409 | 0.979 | 86.857 | 0.958 | 0.452 |

## 4.2  Results

According to Fig.1, we select a set of input variables which belong to the three defined categories: (i) Customer Attitude, (ii) Customer Value, and (iii) Customer Activities. In detail, the selected variables are: (i) RFM values, gender and age, (ii) $V_P$ and $V_R$ (see Eq.1), (iii) the liveness of customers which is evaluated according to the formulation provided by Fader et al.[14].

We tested prediction performance by comparing the results of four main classifiers [1], i.e. Naive Bayes, Neural Network, Decision Tree and Random Forest. We chose a typical Naive Bayes, a Multi-Layer Perceptron (MLP) with two hidden layers (we note that one hidden layer leads an inefficient classifier). Moreover, among the different decision tree algorithms, we preferred C4.5 classifier. All the classifiers were setup with standard parameters[6].

We selected a 10-fold cross-validation for comparing the different results of the four algorithms for the two different datasets (i.e., Contoso and Poste Italiane), as shown in Tab.1.

Fig.2(a) and Fig.2(b) show respectively the trend of the CCR and the sensitivity of the four classifiers at varying datasets. Looking at Fig.2, Naive Bayes provides changeable performances being the worst classifier in some cases, while Neural Network presents a low value of CCR when Poste Italiane dataset is taken into account. So that, we selected the Decision Tree and the Random Forest thus they offer better results and high performance in both cases.

Analyzing the time required to build the model, we note how it ranges between $80ms$ (Naive Bayes) and $62.84s$ (Neural Network) when Contoso dataset is kept into account, and between $20ms$ and $1.19s$ when Poste Italiane dataset is considered. Decision Tree and Random Forest require respectively $2s$ and $3.48s$ (Contoso dataset); $0.1s$ and $0.18s$ (Poste Italiane dataset). We find that Decision Tree (C4.5 algorithm) presents a good trade-off between accuracy and time performances, thus entailing a correct churn customer identification and a reasonable building time when Contoso dataset is considered.

---

[6] Parameters have been chosen by a simple qualitative analysis, according to common values adopted for them, without any in-depth quantitative analysis for their optimization.

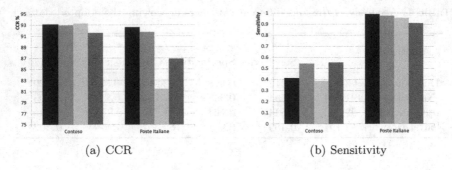

(a) CCR                                   (b) Sensitivity

**Fig. 2.** Performances of the four classifiers: Decision Tree (black bar), Random Forest (grey bar), Neural Network (light-grey bar) and Naive Bayes (dark-grey bar)

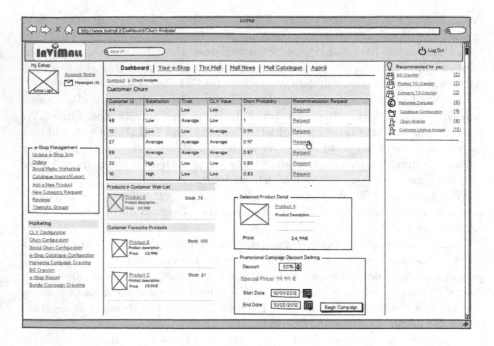

**Fig. 3.** Tool for selecting contents

## 4.3   Example of Application

In order to assist the merchant in identifying the customer churn, in InViMall project we realize a service able to elaborate churn predictions and to produce automatic suggestions thus entailing marketing actions for employing a recovery strategy. First of all, the merchant configures the service by choosing the target and the frequency of the analysis, i.e., he can specify the age of target customers and the date on which he wants to start the analysis. Therefore, the algorithm elaborate a list of potential churn customer. Finally, the merchant can navigate the results and decide the proper suggestion strategy for each detected customer. In detail, as depicted in Fig.3, a main panel shows the list of churn customers and for each of them is reported the value of probability of churn. Selecting a target customer, it is possible to discover his/her wished products. The system gives the possibility to start loyalty campaigns, suggesting to the merchant those products which the customer has added in his "wish-list" and those products he considered interesting (a product is considered interesting when there exists a high value of the rating either explicitly or implicitly defined). Consequently, the merchant can select a product from this list and start one-to-one marketing campaigns. The proposed tool gives the merchant features in the e-Commerce domain; in other words, this tool allows knowledge extraction regarding customers and suggests implicitly the products to add in some personalized marketing campaigns enhancing customer loyalty.

## 5  Conclusions and Future Work

The more e-Commerce applications increase the more the probability of losing a customer grows and data mining could play an important role in the prediction of customer behavior.

In this paper, we presented a Churner Detection System (CDS) in the e-Commerce domain. Our approach is based on the adoption of customer attitude, customer lifetime value and all the customer activities such as buying behavior in order to predict the potential customers which wish to quit the e-Shop, thus suggesting the best strategy to recovery those customers.

To sum up, the main contributions of this paper are: (i) the comparison of the main classifiers in order to perform a churn prediction in a benchmark dataset and in a real one, and (ii) the definition of a CDS framework which can capture purchasing behavior and social activities when Social Commerce data are available.

In future, we aim at extending the approach presented here in order to find the experimental evidence that the social information can provide benefits (e.g., increased accuracy of the classifier) in a real dataset of Poste Italiane Social Commerce platform as a result of the InViMall project. In detail, we will extend the proposed approach taking into account other input variables arising from social network in order to perform a more accurate measurement of social factors which can identify a churn customer.

**Acknowledgments.** This work was partially supported by MSE under the Intelligent Virtual Mall (InViMall) Project MI01-00123.

# References

1. Dror, G., Pelleg, D., Rokhlenko, O., Szpektor, I.: Churn prediction in new users of Yahoo! answers. In: WWW (Companion Volume) 2012, pp. 829–834 (2012)
2. Guo, L., Zhang, M., Sun, L., Wang, Z.: Churn analysis model of securities business based on the decision tree. In: The Sixth World Congress on Intelligent Control and Automation, WCICA 2006, vol. 2, pp. 6048–6051 (2006)
3. Li, A., Lin, Z.: Email users churn analysis based on pmclp and decision tree. In: Sixth International Conference on Fuzzy Systems and Knowledge Discovery, FSKD 2009, vol. 7, pp. 348–350 (August 2009)
4. Ju, C., Guo, F.: Research and application of customer churn analysis in chain retail industry. In: 2008 International Symposium on Electronic Commerce and Security, pp. 670–673 (August 2008)
5. Xin, Z., Yi, W., Hong-wang, C.: A mathematics model of customer churn based on pca analysis. In: International Conference on Computational Intelligence and Software Engineering, CiSE 2009, pp. 1–5 (December 2009)
6. Ghorbani, A., Taghiyareh, F.: CMF: A framework to improve the management of customer churn. In: IEEE Asia-Pacific Services Computing Conference, APSCC 2009, pp. 457–462 (December 2009)
7. Pinheiro, C., Helfert, M.: Mixing scores from artificial neural network and social network analysis to improve the customer loyalty. In: Workshop on Advanced Information Networking and Applications Workshops, 2009, pp. 954–959 (May 2009)
8. Richter, Y., Yom-Tov, E., Slonim, N.: Predicting customer churn in mobile networks through analysis of social groups. In: Proceedings of the SIAM International Conference on Data Mining, SDM 2010, April 29-May 1, pp. 732–741. SIAM, Columbus (2010)
9. Birtolo, C., De Chiara, D., Ascione, M., Armenise, R.: A generative approach to product bundling in the e-Commerce domain. In: 2011 Third World Congress on Nature and Biologically Inspired Computing (NaBIC), pp. 169–175 (October 2011)
10. Van den Poel, D., Buckinx, W.: Predicting online-purchasing behaviour. European Journal of Operational Research 166(2), 557–575 (2005)
11. Shevade, S., Keerthi, S., Bhattacharyya, C., Murthy, K.: Improvements to SMO algorithm for SVM regression. Technical report, National University of Singapore, Control Division Dept of Mechanical and Production Engineering, National University of Singapore, Technical Report CD-99-16 (1999)
12. Birtolo, C., Ronca, D., Armenise, R., Ascione, M.: Personalized suggestions by means of Collaborative Filtering: A comparison of two different model-based techniques. In: NaBIC, pp. 444–450. IEEE (2011)
13. Birtolo, C., Ronca, D., Aurilio, G.: Trust-aware clustering collaborative filtering: Identification of relevant items. In: Iliadis, L., Maglogiannis, I., Papadopoulos, H. (eds.) AIAI 2012. IFIP AICT, vol. 381, pp. 374–384. Springer, Heidelberg (2012)
14. Fader, P.S., Hardie, B.G.S., Lee, K.L.: Counting your customers" the easy way: An alternative to the Pareto/NBD model. Marketing Science 24(2), 275–284 (2005)

# Hazard Identification
# of the Offshore Three-Phase Separation Process
# Based on Multilevel Flow Modeling and HAZOP

Jing Wu[1,*], Laibin Zhang[1], Morten Lind[2], Wei Liang[1], Jinqiu Hu[1],
Sten Bay Jørgensen[3], Gürkan Sin[3], and Zia Ullah Khokhar[4]

[1] College of Mechanical and Transportation Engineering, China University of Petroleum,
Beijing, China
jinwu@elektro.dtu.dk
{zhanglb,lw,hujq}@cup.edu.cn
[2] Dept. Electrical Engineering, Technical University of Denmark, Lyngby, Denmark
mli@elektro.dtu.dk
[3] Dept. Chemical and Biochemical Engineering, Technical University of Denmark,
Lyngby, Denmark
{sbj,gsi}@kt.dtu.dk
[4] Dept. Environmental Engineering, Technical University of Denmark, Lyngby, Denmark
ziauk@env.dtu.dk

**Abstract.** HAZOP studies are widely accepted in chemical and petroleum industries as the method for conducting process hazard analysis related to design, maintenance and operation of the systems. Different tools have been developed to automate HAZOP studies. In this paper, a HAZOP reasoning method based on function-oriented modeling, Multilevel Flow Modeling (MFM), is extended with function roles. A graphical MFM editor, which is combined with the reasoning capabilities of the MFM Workbench developed by DTU is applied to automate HAZOP studies. The method is proposed to support the "brain-storming" sessions in traditional HAZOP analysis. As a case study, the extended MFM based HAZOP methodology is applied to an offshore three-phase separation process. The results show that the cause-consequence analysis in MFM can infer the cause and effect of a deviation used in HAZOP and used to fill HAZOP worksheet. This paper is the first paper discussing and demonstrate the potential of the roles concept in MFM to supplement the integrity of HAZOP analysis.

**Keywords:** Hazard identification, Multilevel Flow Modeling, HAZOP, automated HAZOP.

# 1    Introduction

In petroleum and natural gas industries, hazard identification has been advocated, and required by the government licensing authorities or clients to meet with the goals of

---

* Corresponding author.

M. Ali et al. (Eds.): IEA/AIE 2013, LNAI 7906, pp. 421–430, 2013.
© Springer-Verlag Berlin Heidelberg 2013

Health, Safety and Environment Management System (HSEMS) since 1960s. Hazard identification is included in the risk management as the first stage towards identifying and formulating major accidents scenarios needed for risk calculation. Especially, oil and gas production is becoming more and more complex with the application of monitoring and detection systems such as DSC, SCADA system, whose advantage is to strengthen the ability of automatic detection to enable adjustment of the process and to enlarge the information sources for operators. But at the same time, operators have difficulties in coping with large amounts of alarms under abnormal conditions, The insufficient support by the  control systems and data records leads to the situation that some hazards are not detected in time and propagated up to a higher level in operation hierarchy. So techniques for hazard identification recently have attracted much attention both from research communities and application companies [1].

The Definition of a Hazard in the Oxford Dictionary is " a thing that can be dangerous or cause damage ". How to systematically identify the hazards in an operating process and visualize the hazard propagation paths to help operators to make a wise decision under emergency situation is a big challenge faced by researchers. HAZOP developed in the late 1960s at Imperial Chemical Industries (ICI), among other available techniques aimed at identification of hazards is widely applied in situations where processing of hazardous materials take place, in oil and gas processing plant for instance. It is here important to ensure a high level of completeness of the set of deviations considered. Because hazards identification is the primary element of risk evaluation also the qualitative aspects of risk evaluation are dependent on an adequate description (in terms of cause and consequence models) of the hazards. The key issues for these aspects are completeness, consistency and correctness [2]. As a consequence, well-structured HAZOP methods are popular in particular in complex processes. However, the standard HAZOP methodology is lacking the consideration of the level and extent of decomposition of the plant into sub-systems. This is usually decided by the expert team leader, and to a large extent, the manual cause and consequence analysis of hazards depends on the experience and knowledge of the HAZOP team. Thus, in last two decades, improved HAZOP methods have successfully overcome some above problems, through development of different tools to assist the HAZOP studies, some of them are HAZOPExpert in continuous process [3], PHASuite in chemical processes [4], PetroHAZOP case-based reasoning in petroleum industries [5], CHECKUP tool [6] and Functional HAZOP assistant [7]. Among them, Functional HAZOP assistant proposed by Rossing et al. based on Multilevel Flow Modeling (MFM) represents functional knowledge with easier understanding of real system and provides a very efficient paradigm for facilitating HAZOP studies and for enabling reasoning to reveal potential hazards in safety critical operations. MFM is a qualitative reasoning model that could be used to assist the Hazop team by ensuring coverage and consistency and improve completeness.  MFM divides the system into subsystems according to the functions in terms of goals, relations and components and provides a set of reasoning rules which can be used to perform automatic HAZOP study and reveal the potential hazards and the casual paths of a hazard in a visual way. MFM can in this way help operators to understand the system in functional terms and give them a basis for decision making.

The purpose of this paper is to demonstrate how to apply MFM with the role concept in a case study of an offshore three-phase separation process in oil and gas industry. The remainder of this paper is organized as follows. A brief description of an offshore three-phase separation process is given in section 2. Section 3 describes after short introduction of MFM how the reference system for the present research is modeled. The hazard identification based on the HAZOP technique using the MFM editor and reasoning software is introduced in section 4. Finally, section 5 is conclusions.

## 2    Offshore Three-Phase Separation Process

The offshore three-phase separation process, used as a case study here, is commonly applied in offshore oil and gas industry. The process schematic is shown in Fig.1. The purpose of the separation system is to separate two flows of feed, mixture of crude, water and gas stream. Both feed flows have flow rate of    3600 kg/h, pressure of 56 bar and temperature of 50 °C. The components of the feed flows are water, methane, ethane, propane, butane, pentane, hexane, methanol, carbon dioxide, nitrogen, isobutene, isopentane, MEG, and four pseudo components representing higher number of higher number of hydrocarbons.

The two fluid streams are mixed before entering the three-phase separator, which is designed to separate the gas, as well as separate the oil and water. A safety valve provides protection against unwanted pressure buildup. The weir inside the separator maintains the oil level, and the level controller maintains the water level.  The oil is skimmed over the weir. The level of the oil downstream of the weir is controlled by a level controller that operates the oil export valve.  The gas flow out through a mist extractor to a pressure control valve that maintains constant vessel pressure. Then it passes to a compressor which increases the pressure of the export gas, which is driven by a variable motor speed.  At the outlet side of the compressor a heat exchanger is connected with water as cooling medium. The cooler is regulated by a temperature control loop. Also an anti-surge controller loop is installed to protect the compressor from entering a surge condition. More details about the process can be found in [8].

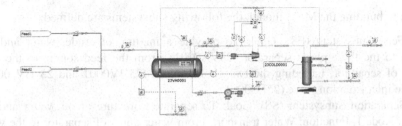

**Fig. 1.** Simplified P&ID of three-phase separation process

# 3     Multilevel Flow Modeling

## 3.1     Introduction of Multilevel Flow Modeling

Multilevel Flow Modeling (MFM) is one of the representative functional modeling methods, used to represent goals and functions of process plants involving interactions between flows of material, energy and information. Fault management of complex plants is one of the most effective and developed applications fields [9]. The concepts and symbols of MFM are shown in Fig.2.

Qualitative reasoning in an MFM model is based on representation of process knowledge on several levels of specification. The explanations generated by the model can be directly visualized in the MFM models, represented later in case study. Details concerning the automated reasoning rules about causes and consequences can be found in [10]. A newly updated MFM editor, a graphical editor, supporting the MFM method, in combination with the reasoning capabilities of the MFM Workbench developed by the Technical University of Denmark, will be used to analyze separation process equipment deviations, their root causes, and their possible consequences.

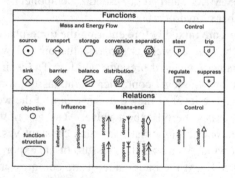

**Fig. 2.** The basic MFM symbols

## 3.2     MFM Model of Three-Phase Separation Process

Before to building the MFM model, the following subsystems are defined:

1. Feed Subsystem (FS). Goal: To provide a mixture of crude, water and gas stream to the three-phase separator (23VA0001), from the feed source to the inlet piping of separator, including the two choke valves (25HV0001 and 25 HV 0002), and one inlet isolation valve (25ES0001).

2. Separation Subsystem (SS). Goal: To separate a mixture of oil, water and gas stream. Node 1, Function: Water transport. From water outlet of separator to the water sink including the separator water level control valve (23LV0001) and related instrumentation. Node2, Function: Oil transport. From the oil outlet of separator to the oil sink including a centrifugal pump(23PA0001), motor (23EM0001) and water dump valve (23MV0003), oil level control valve(LV0002) and related instrumentation.

Node3, Function: Gas transport. From the gas outlet of the separator back to the inlet of separator including compressor 23KA0001, motor 23EM0002 and heat exchanger (23HX0001)

3. Heat Removal Subsystem (HRS). Goal: To transport the heat inside the gas to the environment.

An overall MFM model for this three-phase separation process is shown in Fig.3. In the following we will explain the MFM model by each flow structure representing functions of the subsystems.

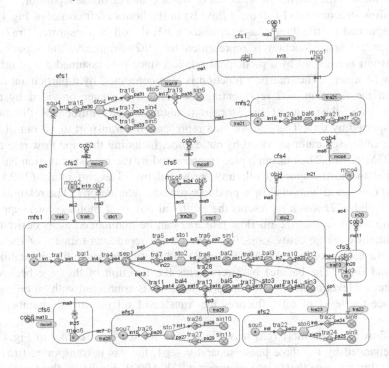

**Fig. 3.** MFM model of the three-phase separation process

### 3.2.1   Mass Flow Structures
In the flow structure mfs 1(total feed system massflow partA), the source sou1 represents a mixture flow of crude, water and gas stream. The function tra1 represents the transportation of the mixture flow realized by pipe and choke valve, connected with feed source by an influence relation since it influences the flow rate of pipe. And the balance ball represents the function of pipe between choke valve and the inlet isolation valve. The function tra4 represents the transport of mixture of crude oil, water and gas steam; finally the mixtures would flow into three phase separator, with a participant (pa4).

The flow structure mfs1 (water mass flow A) in the top left corner of Fig.4: After being separated by the three phase separator sep1, the water is transported (tra5) to the water chamber represented by the storage function sto1. The separator and the transport are related by a participant relation since it is assumed not to influence the flow of water.  The transport function is also connected by a participant relation pa6 with the function sto1 representing the storage of water provided by the water chamber since the state of the storage (amount of water) cannot influence the flow of water transported (tra5). The function tra6 represents the transfer of water out of the water chamber of the separator provided by outlet pipe, including the water level control valve (LV0001). It is connected with an influence to sto1 since the level of water influences the transfer of water out of water chamber of the separator.

The flow structure mfs1 (oil mass flow B) in the bottom left corner of Fig.4: After being separated by the three phase separator sep1, the oil is transported (tra7) to the oil chamber whose function is represented by sto2. Similarly, the separator and the transport are related by a participant relation since it is assumed not to influence the flow of water.  The transport function is also connected by a participant relation pa9 with the function sto2 representing the storage of water provided by the oil chamber since the state of the storage (amount of oil) influences the flow of oil transported by tra7. The function tra8 represents the transport of oil out of the oil chamber of the separator provided by outlet pipe, including the pipe flow rate control valve (25MV0003).Then through pipe (represented by the balance function bal2) it is connected to the transport of oil (tra9) provided by oil export pump (23PA0001), which is connected with efs2 by a producer-product relation pp1. The relation pp1 is labeled with tra22, which represents the electrical power of motor (sou6) supplied to the pump (sto6) so that the oil flow rate tra9 can be maintained at its desired value represented by the objective obj3. Tra23 and tra24 represents transfer of the energy into kinetic energy of the oil (tra23 and sin8) and friction losses in the circulation loop (tra24 and sin9). The balance bal3 represents the function of the pipe between the pump and oil level control valve tra 10(LV0002) It is connected with an influence to bal3 since the level of oil influences the transfer of oil out of oil chamber of the separator.

The flow structure mfs1 (gas mass flow C) in the top right corner of Fig.4: After being separated by the three phase separator sep1, the gas is transported (tra11) (25 ES0002) through pipe (bal4) into compressor 23KA0001. Similarly, the compressor is also driven by motor EM0002 and obtains its driving energy from the energy flow structure efs3, whose energy transformation can be referred to the flow structure efs2. The control flow structure cfs6 represents the regulation of the motor speed control the compressor. Then the gas flows through pipe (tra13) between the compressor and heat exchanger whose function here is represented by sto3. The gas production is exported out of the heat exchanger from the outlet pipe and pipe volume valve 25 ES0005 (tra14) to the next gas processing unit.

### 3.2.2    Energy Flow Structures
The energy flow structure efs1 describes the total energy flow transported from the feed subsystems to the final products. The source function (Sou 4) represents the total energy carried by mixture flow of crude, water and gas steam, which is transported (tra15) to three-phase separator (sto4). Then the energy contained in the oil and water

flows are transported (tra17) and (tra18) to the next processing units represented by the sink functions sin4 and sin5. The gas energy flow is transported to heat exchanger (sto5); the mass flow of cooling medium (from sou5 to sin 7) takes away some heat from the gas flow. A desired temperature of the gas product (sin6) is maintained by regulating of the outflow pipe valve 23TV0003 tra20 which control the outflow product of gas flow (tra 19).

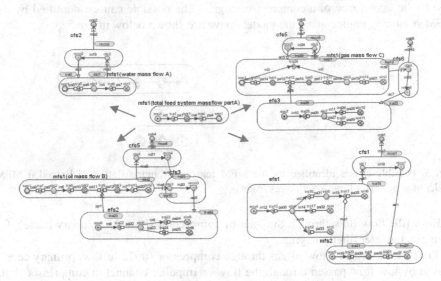

**Fig. 4.** MFM models of each subsystem

### 3.2.3    Control Flow Structures

In Fig.3, there are six control flow structures. The related control objectives and actuators are explained in Table1. Specifically, in the anti-surge control loop, by comparison between the set point of 23ASC0001set by measurement of 23PT0002, 23PT0003 and the value of flow element 23FE0001 to control the anti-surge valve 23UV0001close or open, when the measurement is higher than the set point, the valve will be closed. Also it should be noted that that the water level control valve therefore in this model has two functions, to dump outlet water (tra6) and to actuate (ac2) the water transport from separator to environment.

**Table 1.** Control Objectives

| Number | Measurement | Actuator |
|--------|-------------|----------|
| cob1 | Temperature of output gas flow | Pipe valve 23TV0003 output gas flow |
| cob2 | Water level in the separator | Dump water 23LV0001 |
| cob3 | Output flowrate of oil | Motor EM0001 speed |
| cob4 | Oil level in the separator | Dump oil 23LV0002 to next unit |
| cob5 | 23PT0002, 23PT0003 and the value of flow element 23FE0001 | Anti-surge valve 23UV0001 |
| cob6 | Output flowrate of gas | Motor EM0002 speed |

# 4    Automatic Reasoning of MFM Model with MFM Editor

## 4.1    Functions-Based Reasoning

We assume that there is a deviation of flow of gas in outlet pipe from separator i.e. the transport function is in a state of loflow (tra11, loflow) in Fig.3 This will  possibly lead to the occurrence of a compressor surge.   The possible causes identified by the MFM reasoning engine using the model above are shown below in Fig.5.

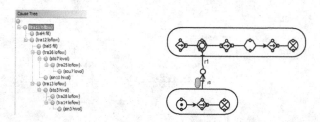

**Fig. 5.** Possible causes identified by the MFM reasoning engine (left) and Extended MFM model of water mass flow A with roles (right)

**Bal4 fill:** Low flow through the inlet of compressor (bal4 fill, primary cause), for example, the inlet filter plugging.

**Tra12 loflow:** Low flow of gas through compressor (tra12 loflow, primary cause) caused by low flow passed through the flow or impeller channel in compressor (bal5 fill, secondary cause) or low conversion of the energy into kinetic energy in compressor (tra26 loflow, secondary cause) or outlet pipe of compressor partly blocked (tra 13 loflow, secondary cause), for example, the pipe network resistance increased. The low conversion of the energy into kinetic energy in compressor caused by the low driven power stored in compressor (sto7 lovol, tertiary cause) or the overload of compressor (sin 10 hivol, tertiary cause). The low driven power stored in compressor (sto7) is caused by the low energy transported (tra25 loflow, quaternary cause), which is originally rooted in low speed of the compressor motor (sou7 lovol, quintuple cause). The low gas flow in pipe (tra13) between compressor and heat exchanger is caused by the high level in the heat exchanger tube (sto3 hivol, tertiary cause), which is originated from UV0001 valve partly closed (tra28 loflow, quaternary cause) or 25ES005 valve partly closed (tra14 loflow, quaternary cause), which is due to the gas production above the quota (sin3 hivol, quintuple cause), in another word, the overmuch deduction of gas production.

The HAZOP table of this deviation is summarized in Table 2.

## 4.2    Introducing Roles

In order to identify the relations between function and structure, the concept of roles have been introduced in MFM. The representation of roles in MFM is discussed in [11]. However, presently there is no agreement about which and how many roles are needed and no consensus has been reached on nature of the roles. However, the extension of MFM with  roles (agent, object, instrument etc.) may endow the potential for identifying hazards that are not only described with guide words in HAZOP studies such as "more", "less", "none" which is discussed above and applied in [7], but also other guide words such as" part of ", "as well as", "other than".

In the case of the three-phase separator, we will apply a patient role r1 and an agent role r2 on the separation function. The patient and the agent here mean the entity undergoing the effect of some action and the doer of the action, respectively. Specifically, in the sentence "the separator separates the mixture flow of crude oil, water and gas stream", the separator has the agent role and the mixture flow of crude oil, water and gas stream has the patient role. A patient role is adopted here instead of object role proposed in [11] because the mixture flow undergoes a change of phases and being processed by the separator, and also the separation effect is dependent on the properties of the mixture flow and agent role separator. It is suggested to distinguish between object and patient roles. When only the extrinsic properties (like location) and not the intrinsic properties of a physical are changed by the action the item is an object of the action. For example, when water is transported by a pipe, the water has an object role since it is moved from one place to another place by the agent (the pipe) without any state change. The patient role is different and very important because it allows the identification of possible hazards caused by the change of properties (composition, temperature, pressure) in mass or energy flow.

**Table 2.** HAZOP analysis of a deviation of flow of gas in outlet pipe from separator

| Function node | Deviation | Possible root causes | Possible consequence | Safety measurement |
|---|---|---|---|---|
| Transport 11 | Low gas flow | 1.The inlet filter plugging | Compressor surge occurring | 1.Inlet gas flow pressure transmitter 23PT0002 |
| | | 2.Low flow passed through the flow or impeller channel in compressor | | 2.Outlet gas flow pressure transmitter 23PT 0003 |
| | | 3.Low speed of the compressor motor | | 3.Monitoring he speed of compressor motor |
| | | 4.Overload of compressor | | 4.Improve the load capacity of the compressor |
| | | 5.UV0001 valve partly closed | | 5. Checking the anti-surge controller 23ASC0001 |
| | | 6. Overmuch deduction of gas production | | 6.Increase gas production |

# 5    Conclusions

The paper has presented a model of an offshore three-phase separation process. It is shown that the principles of Multilevel Flow Modeling for representing mass and energy flow functions and control system functions can be successfully applied for a complex system in oil and gas industries.  Also it demonstrates that the MFM functional modeling can cover shortages of standard HAZOP methodology.   The use of the automatic cause-consequence reasoning mechanism of MFM for HAZOP based on parameter deviations and hazards conditions are investigated. The role concept in MFM may be helpful for detecting hazards in lower hierarchy in HAZOP studies. This method provides a basis for key facilities in oil and gas industries for safe operation of the three-phase separation process.

**Acknowledgements.** The first author is thankful to China Scholarship Council (CSC), Ministry of Education (File No. 201206440002) for the grant of a scholarship for one year and the support from the research groups at Department of Electrical Engineering and Department of Chemical and Biochemical Engineering at DTU. This project is also supported by National Science and Technology Major Project of China (Grant No. 2011ZX05055); National Natural Science Foundation of China (Grant No. 51104168); PetroChina Innovation Foundation (Grant No. 2011D-5006-0408) and Excellent Doctoral Dissertation Supervisor Project of Beijing Grant YB20101141401.

# References

1. Dunjó, J., Fthenakis, V., Vílchez, J.A., Arnaldos, J.: Hazard and operability (HAZOP) analysis. A literature review. . Journal of hazardous materials 173(1), 19–32 (2010)
2. Rushton, A.G.: II. 3 Hazard identification techniques. Industrial Safety Series 6, 129–161 (1998)
3. Venkatasubramanian, V., Zhao, J., Viswanathan, S.: Intelligent systems for HAZOP analysis of complex process plants. Computers & Chemical Engineering 24(9-10), 2291–2302 (2000)
4. Zhao, C., Bhushan, M., Venkatasubramanian, V.: PHASuite: An automated HAZOP analysis tool for chemical processes. Part I. Knowledge engineering framework. Process Safety and Environmental Protection 83(6), 509–532 (2005)
5. Zhao, J., Cui, L., Zhao, L., Qiu, T., Chen, B.: Learning HAZOP expert system by case-based reasoning and ontology. Computers & Chemical Engineering 33(1), 371–378 (2009)
6. Palmer, C., Chung, P.W.H.: An automated system for batch hazard and operability studies. Reliability Engineering & System Safety 94(6), 1095–1106 (2009)
7. Rossing, N.L., Lind, M., Jensen, N., Jørgensen, S.B.: A functional HAZOP methodology. Computers & chemical engineering 34(2), 244–253 (2010)
8. KONGSBERG K-Spice® Tutorial, Training Manual, May 2012 © Kongsberg Oil & Gas Technologies AS
9. Lind, M.: An introduction to multilevel flow modeling. Nuclear Safety and Simulation 2(1), 22–32 (2011)
10. Lind, M.: Reasoning about causes and consequences in multilevel flow models. In: Proc. European Safety and Reliability Conference (ESREL) 2011 Annual Conference, pp. 18–22 (2011)
11. Lind, M.: Knowledge representation for integrated plant operation and maintenance. In: Seventh American Nuclear Society International Topical Meeting on Nuclear Plant Instrumentation, Control and Human-Machine Interface Technologies (2010)

# Multilevel Flow Modeling Based Decision Support System and Its Task Organization

Xinxin Zhang, Morten Lind, and Ole Ravn

Department of Electrical Engineering, Technical University of Denmark, Lyngby, Denmark
{xinz,mli,or}@elektro.dtu.dk

**Abstract.** For complex engineering systems, there is an increasing demand for safety and reliability. Decision support system (DSS) is designed to offer supervision and analysis about operational situations. A proper model representation is required for DSS to understand the process knowledge. Multilevel Flow Modeling (MFM) represents complex system in multiple levels of means-end and part-whole decompositions, which is considered suitable for plant supervision tasks. The aim of this paper is to explore the different possible functionalities by applying MFM to DSS, where both currently available techniques of MFM reasoning and less mature yet relevant MFM concepts are considered. It also offers an architecture design of task organization for MFM software tools by using the concept of agent and technology of multiagent software system.

**Keywords:** Multiagent systems, Multilevel Flow Modeling, decision support system, task organization.

## 1 Introduction

Development of decision support system (DSS) for process analysis and supervision is motivated by increasing industrial requirement of operational safety and reliability. One challenge for designing DSS for complex engineering process is to understand the process and operational knowledge and make the knowledge explicit by modeling techniques. To fully understand and represent the nowadays engineering system with sheer size and complexity, finding a plausible approach to decompose the system for modeling is considered of highly significance. Part-whole decomposition is a commonly used method to represent an engineering system for computerized tools, so that bottom-up or top-down approach such as failure mode and effect analysis (FMEA) and other diagnostic and prognostic analysis can be performed and plant knowledge can be encoding for automation system to map component states to operational situations. However, when human performing this analysis or operational tasks, common sense reasoning is often going in implicitly in the background, concepts of means and ends are involved to solve design and operational problems. This knowledge is often neglected when creating process models for automation tools. This paper proposes to use a functional modeling approach, namely Multilevel Flow Modeling (MFM) [1], as the basis for design DSS for plant supervision and analysis to complement the lack of means-end representation of the system. It also suggests a Multiagent System (MaS) as a task organization framework (on both abstract level and software level) to develop such a MFM based DSS.

M. Ali et al. (Eds.): IEA/AIE 2013, LNAI 7906, pp. 431–440, 2013.

The paper is organized as follows. In Section 2, MFM is introduced and it discusses how MFM models can be used for different computerized decision support tasks. Section 3 describes the requirement for developing an MFM based DSS. Section 4 focus on how different tasks can be organized during operation and the Multiagent MFM based DSS will be proposed.

# 2     Multilevel Flow Modeling for Decision Support

MFM represents process knowledge on multiple levels of means-end and part-whole abstractions. It is been used for modeling safety critical system in several domains such as nuclear power plant [2-3], and oil/gas gathering system[4]. These models have been explored for the purposes of causal reasoning [5] (fault diagnosis and consequence analysis), control function representation [6], operation modes representation [3, 7], alarm design [8], and safety barrier representation [9]. MFM is also used as a tool for risk assessment combining with other methodologies [10-11].

## 2.1     MFM and MFM Reasoning

MFM is a modeling method representing an industrial plant as a system which provides the means required to serve purposes in its environment. [1] MFMs incorporate goals and objectives of the system, functions and structures that describe the physical components, and relations between functions and structures. It also adopts a predefined graphical modeling language, with symbolic representation for objectives, functions and relations.

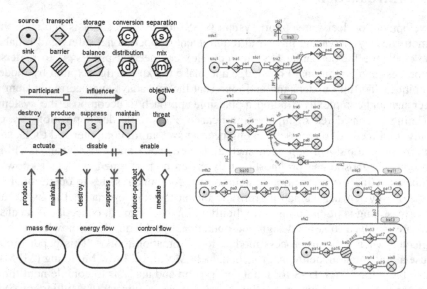

**Fig. 1.** MFM symbols and watermill model example

A list of the common symbols for MFM objectives and functions/relations with an example of complete MFM model are showed in Fig. 1. The flow structures contain partial function models to serve as means to realize functions in other flow structures or objectives. Each means-end relation across levels is labeled with a main function which directly influences the end function or objective. The whole model example in the Fig. 1 is an extended MFM model of a watermill [1], which is a commonly used example for beginners of the MFM modeling. It contains several levels of mass flow and energy flow.

As exemplified in Fig. 1, MFM constructs the model by using building blocks that correspond to functions and goals. It describes energy and mass flows in a physical system with different level of decomposition, and the representation is in an abstracted way which is independent of individual components that compose the physical system. MFM modeling is not only a way of representation, but also a convenient tool to analyze and reason about the system performance. Reasoning in MFM models is based on dependency relations between states of objectives and functions. The possible states of each MFM entity are listed in Table 1.

**Table 1.** Possible state for MFM entities

| Function | Possible States |
| --- | --- |
| Source: | normal, high volume, low volume |
| Sink: | normal, high volume, low volume |
| Transport: | normal, high flow, low flow, no flow |
| Storage: | normal, high volume, high volume |
| Barrier: | normal, leak |
| Balance: | normal, fill, leak |
| Objective: | false, true |

The dependency relations defined in MFM are independent of the particular modeling object, and only based on predefined patterns. The patterns are created by different combinations of MFM entities, states and the influence relations or means-end relations in between them. They are defined as cause-effect relations. Root cause reasoning and consequences reasoning for MFM are explained in [5]. The basic idea of reasoning propagation is that an abnormal state of one function or objective influences another function or objective's state in a MFM model along the MFM relations so that a cause or effect path can be generated by continuing inferences until the path end at a root cause or defined critical failure in operation.

## 2.2    MFM for Decision Support

A major task for DSS to function is that the system can "understand" the operational situations when an event occurs during operation, and perform analysis to support operational decisions. Diagnosis and prognosis are commonly used analysis.

MFM syntax has a relatively small set of symbols which is derived from fundamental action types, and causal reasoning based on MFM model is generic and based on function states and pattern matching. Thus a knowledge based program can be designed to perform reasoning tasks with a reasonable amount of encoding rules. A MFM model editor is developed by IFE Halden [12] and MFM models can be build and stored both as graphical data (for human users) or a structure data format (used for computerized system). A rule base for root cause and consequence reasoning based on MFM models is developed in Jess (Java Expert System Shell) by the authors' research group.

Besides the causal reasoning, recent research suggest that MFM can offer multiple representations of the same plant in different means-end and part-whole abstraction levels, or different operation modes to support the operational decision making. [7] Firstly, MFM represent operation modes based on shift of the operational objectives and goals. By studying different operation phase or different plant conditions, different MFM model can be developed and there's potential to model mode shifts by further extensions of the MFM semantics. Then a knowledge base may be developed to reason about plant mode shift based on function and objective states patterns within one mode. Secondly, within the same operation mode, models of different abstraction level might be required for different purposes. For example a more detailed and decomposed model is required for an offline diagnosis comparing with on overall online prognostic analysis when failure of objectives and goals are in focus.

Situation awareness and evaluation are of great significant for decision making process, because counter action plan are based on the understanding of the current situation and accessing the deviation from the operational goal. The reasoning of operation mode shifts will help operators to set the focus when solving problems. Customized models for different operation modes can facilitate the DSS to get more precise evaluations. Diagnosis and prognosis is more meaningful and effective when MFM describe the exact operation mode with predefined abstraction level and functional representations than performed with a general model. The analysis can also be performed upon a partial and more detailed model rather than a whole model when a preliminary analysis has already been performed.

MFM represents operation modes is one of the main research subjects within the authors' research group currently. Preliminary studies suggest that MFM can be used to reason about plant condition. Combining with the current available program package of diagnosis and prognosis analysis based on MFM models, DSS can be designed to perform customized analysis for different operational conditions.

# 3    Requirements for Software Tool

From section 2, several functionalities are identified for a MFM based DSS. In this section, the requirements for developing such a computerized system will be explored. Several tasks can be defined based on the system functionalities: 1) reason about plant situation, 2) choose the appropriate model for the current condition, 3) choose a reasoning task (or several tasks) and 4) coordinate the activities, 5) perform diagnosis and/or 6) prognosis analysis and 7) provide reasoning results and possible choices for the operator. To perform these tasks, different knowledge bases are required for the system, including plant data, model representations for each operation mode, reasoning rules about mode shift, diagnosis and prognosis rules, deductive knowledge and their interpretation and synthesis rules. This is shown in Table 2.

**Table 2.** Tasks and knowledge sets for MFM based DSS

| Tasks | Knowledge Bases |
|---|---|
| • Situation Awareness | • Plant State Knowledge |
| • Model Assembling | • Mode Change Rules |
| • Reasoning Organizing | • Model and Mode Representations |
| • Diagnosis | • Model Choosing Rules |
| • Prognosis | • Diagnosis Rules |
| • Synthesis and Decision Support | • Prognosis Rules |
| • Display | • Synthesis Rules |

The tasks listed in Table 2 are relatively independent and require different types of knowledge. However it is also requires a common understanding of the overall operation goals to perform the individual task. The problem is hard to solve in a centralized fashion with a procedural program in com due to the degree of complexity. Therefore, it is proposed to accomplish the tasks in a distributive way by several parallel subprograms. Note that here a distributive problem solving design is chosen so several tasks can be performed in parallel; however a centralized coordinator is still required for task coordination.

We propose to use the agent concept to model the task organization for MFM based DSS and use multi-agent framework to develop such a system.

# 4    Task Organization and Multiagent System

In this section, tasks that identified in Section 3 will be further examined. Based on the task organization, a multiagent framework is used for implementation of the MFM based DSS design.

## 4.1    Task Organization

A task plan for MFM based DSS is illustrated in Fig. 2. (Note that the diagram omits the interfaces between control actuators/sensors and DSS because this is not the focus of the paper. The integration between the two is still an open topic for research.)

In Fig. 2 two working routine are suggested. For no failure situation, the DSS will detect the normal operation event that happens in the system to choose suitable MFM representations and reports it to the operator through interface. This may happen during routine shift of operation modes, for example system changed from start-up phase to fully operation phase.

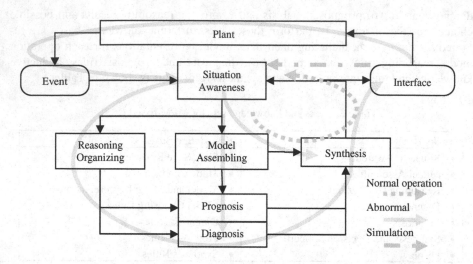

**Fig. 2.** Task planning for MFM based DSS

In a failure situation, the DSS will decide that whether the current model is still optimal representation (is there any mode changes) and what reasoning tasks that need to be performed. It will then organize the reasoning activity by using appropriate model (or models). After summarize the analysis result, the DSS may propose possible solutions to the operator through the interface so the operator may choose to intervene with the plant operation. The plant situation may shift after the operator action. Another hidden function is that the operator may test a choice of action through the DSS by simulation. In this case the reasoning activity is conducted in the same way as when abnormal states but giving operator's decision as input to the situation awareness section.

**Table 3.** Knowledge requirement for tasks

| Task | Knowledge Bases | | | |
|------|------|------|------|------|
| Situation Awareness | Model Choosing Rules | Modes | Plant States | MFM ontology |
| Model Assembling | Model Assembling Rules | Mode and Models | | |
| Reasoning Organizing | | Mode and Models | | |
| Prognosis | Prognosis Rules | Model(s) | | |
| Diagnosis | Diagnosis Rules | Model(s) | | |
| Synthesis | Synthesis Rules | Mode and Models | | |

Different knowledge bases are required for each task. The mapping from task and its required knowledge is showed in Table 3. Where MFM ontology and plant states are universal required as domain knowledge. Diagnostic and prognostic rules and single model representations are already available with software prototype, whereas the other knowledge bases require further implementations and testing.

## 4.2    Agent and Software Agent

In intelligent systems, agent, often referred to as intelligent agent, is an artifact that can sense its environment and act autonomously to meet its design goals in responding to events that happens in the environment. In [14], a conceptual agent is composed of three fundamental functional elements, which are its domain of activity, knowledge system, and preference system. These three elements describes that an agent should aware of where it acts, how to act, and when to act (how to choose a course of action under a certain circumstance).

Conceptual agents can be realized in form of software agents by developing a program that is capable to serve the role as an agent (for software agent, eventualization of an action may be done by associated hardware instead of the program itself). In the MFM based DSS, MFM ontology is the domain language shared by the agents, models and reasoning rules is coding as the agents' knowledge systems. Sub tasks are assigned to different agents so that the preference knowledge for the agents is as simple as activation rules so that the agents will know when to perform their tasks.

## 4.3    Multiagent System

MaS Engineering offers several modeling technique to design a multiagent system. The tasks and knowledge sets are already defined in section 4.1. Roles can be assigned to different agents. Roles form the foundation for software agent classes and correspond to system goals (tasks) during the design phase. [14] Four roles are defined, namely Mode Supervisor, Reasoning Organizer, Model Builder, and Reasoner, where Reasoning Organizer will perform two tasks of reasoning organizing and synthesis. The agent roles are illustrated in Fig. 3.

New reasoning tasks can be implemented as individual rule sets when the theoretical design is mature. Also note that this role model defined here can be refined into sub classes during implementations. For example, if the diagnosis tasks become too complicated to handle by one Reasoner, a group of Reasoners should be defined with another level of hierarchy. Zhang [15] introduces a multiagent planning architecture and its software prototype for a single reasoning task by using blackboard system as reasoning organizing framework. This particular framework can also be adopted in the level of reasoning organizer introduced in this paper.

The tasks and models to be used are unknown to the system before the abnormal events emerge in the system. Therefore, the tasks organization is dynamic rather than static procedures. The Reasoning Organizer has the responsibility to gather information from the Mode Supervisor and deduce the best suitable analysis to be performed and initialize other Reasoners. Meanwhile the Model Builder will determine the model(s) to be used and assign the model(s) to the Reasoners. This dynamic task organization method is very well supported by the MaS architecture, which is one of the major arguments to use MaS to implement this DSS.

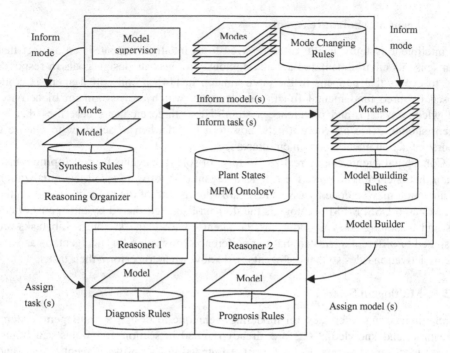

**Fig. 3.** Multiagent Role Model

## 4.4    Model Example with the MaS Framework

As mentioned in Section 2, preliminary works have been conducted for MFM representation of operation modes. A small example is used to illustrate how the proposed DSS organizes the analytic tasks. Gola [3] identified 8 control modes for pressurizer in Pressurized Water Reactor (PWR) and modeled different modes by using MFM. They are listed in Table 4.

**Table 4.** Control modes for pressurizers in PWRs

| Steady State | | | |
|---|---|---|---|
| **High Pressure Transient** | **Low Pressure Transient 1** | **High Level** | **Low Level 1** |
| **High Pressure** | **Low Pressure Transient 2** | | **Low Level 2** |
| | **Low Pressure** | | |

Fig. 4 describes the MFM models in a very abstract way. (Notice that this is NOT a MFM model but an abstract description of the models; for detailed mode explanations and MFM representations please refer to [3].) For each mode, there are different objectives and control functions to actuate different resources (spray valves, proportional heater, etc.) to control pressure and water level in the system.

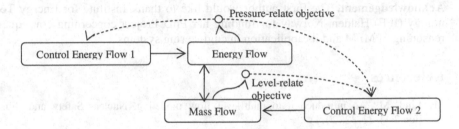

**Fig. 4.** Abstract description of pressurizer's MFM models

We consider a scenario that a high state in the energy flow is detected during normal operation. By using the DSS introduced in section 4.3, the following actions will be performed by the DSS. 1) Model Supervisor (MS) observes the high state event and the current steady state operation; it informs operator, Reasoning Organizer (RO) and Model Builder (MB). 2) MB will assemble the steady state model. RO may decide that in single function abnormal, both prognosis and diagnosis will be performed. So the RO will initiate the Reasoners (Rs) and exchange info with the MB. 3) MB informs the model for each R and the RO assign tasks to them. 4) Analysis result is gathered by RO and report to the display and MS (and signals the automatic actuator if applicable).

If the analysis predicts a possible high pressure (likely to be true under the scenario), this information and possible root causes should be presented to the operator. High pressure may be detected by the MS after a period of time is no effective control action is performed, the system may enter the high pressure transient mode by using different control resources (open the spray valves [3]). MS will update the control mode to the operator, RO and MB. The new reasoning activities will be organized according to the new situation with another mode model. And in the new case, diagnosis may not be performed again but only new predictions will be suggested. The decision making mechanism for RO can be programed according to individual system and possible mode shift sequences.

The reasoning task is very complex without a distributed solution to separate the tasks. MaS offers great flexibility for task organization so the reasoning tasks can be customized to fit a particular operation mode or operator command. Reasoning tasks can be performed in parallel or with any sequence. Most agent development environments adopt standard communication protocols so the information exchange can be realized in a formalized way. Jade (Java agent development environment) is adopted in current development with agent-based MFM software by the authors because it can be easily integrated with the existing MFM reasoning packages. MaS is also a good architecture to utilize the processing capacity of multiprocessor computer hardware.

## 5    Conclusion

This paper explores the features and potential functionalities for the development of DSSs based on MFM representations of safety critical engineering systems. It also proposes to use MaS as implementation platform so that great flexibility can be gained when designing the task organization mechanism for the MFM based DSS. The design is supported by current studies of MFM in both theoretical and software tool development.

**Acknowledgement.** The First author would like to thank Institute for Energy Technology (IFE) Halden, Norway for funding her PhD Project concerning consequence reasoning of MFM and its application in engineering systems.

# References

1. Lind, M.: An introduction to multilevel flow modeling. Nuclear Safety and Simulation 2(1), 22–32 (2011)
2. Lind, M., Yoshikawa, H., Jorgensen, S.B., Yang, M., Tamayama, K., Okusa, K.: Multilevel flow modeling of Monju Nuclear Power Plant. Nuclear Safety and Simulation 2(3), 274–284 (2011)
3. Gola, G., Lind, M., Zhang, X., Heussen, K.: Functional Representation of Process Operation with Multilevel Flow Models for Diagnostic Decision Support. In: OECD Halden Reactor Project Report. HWR-1059 (2012)
4. Wu, J., Zhang, L., Liang, W., Hu, J.: A novel failure mode analysis model for gathering system based on Multilevel Flow Modeling and HAZOP. Process Safety and Environmental Protection 91(1-2), 54–60 (2013)
5. Lind, M.: Reasoning about causes and consequences in Mulitlevel Flow Models. In: Advances in Safety, Reliability and Risk Management - Proceedings of the European Safety and Reliability Conference, ESREL 2011, pp. 2359–2367 (2011)
6. Lind, M.: Control functions in MFM: basic principles. Nuclear Safety and Simulation 2(2), 132–140 (2011)
7. Lind, M., Yoshikawa, H., Jorgensen, S.B., Yang, M., Tamayama, K., Okusa, K.: Modeling Operating Modes for the Monju Nuclear Power Plant. In: Proceedings of the 8th International Topical Meeting on Nuclear Plant Instrumentation, Control and Human Machine Interface Technologies, San Diego, CA, United States (2012)
8. Us, T., Jensen, N., Lind, M., Jorgensen, S.B.: Fundamental Principles of Alarm Design. Nuclear Safety and Simulation 2(1), 44–51 (2011)
9. Lind, M.: Modeling Safety Barriers and Defense in Depth with Mulitlevel Flow Modeling. In: Proceeding of First International Symposium on Socially and Technically Symbiotic Systems, Okayama, Japan (2012)
10. Yoshikawa, H., Yang, M., Hashim, M., Lind, M., Zhang, Z.: Design of risk monitor for nuclear reactor plants. Nuclear Safety and Simulation 3(2), 236–246 (2011)
11. Yoshikawa, H., Lind, M., Yang, M., Hashim, M., Zhang, Z.: Design Concept of Human Interface System for Risk Monitoring for Proactive Trouble Prevention. In: Proceedings of ICI 2011, Daejeon, Korea (2011)
12. Thunem, H.P., Thunem, A.P., Lind, M.: Using an Agent-Oriented Framework for Supervision, Diagnosis and Prognosis Applications in Advanced Automation Environments. In: Proceedings of ESREL 2011. European Safety and Reliability Association (2011)
13. Gadomski, A.M.: TOGA: A methodological and conceptual pattern for modeling abstract intelligent agent. In: The Proceedings of the First International Round-Table on Abstract Intelligent Agent, Rome, Italy (1993)
14. DeLoach, S.A.: The MaSE Methodology. In: Bergenti, Gleizes, Zambonelli (eds.) Methodologies and Software Engineering for Agent Systems. The Agent-Oriented Software Engineering Handbook Series: Multiagent Systems, Artificial Societies, and Simulated Organizations, vol. 11. Kluwer Academic Publishing (2004)
15. Zhang, X., Lind, M.: Agent Based Reasoning in Multilevel Flow Modeling. In: Proceedings of the first International Symposium on Socially and Technically Symbiotic System, Okayama, Japan (2012)

# Simulation-Based Fault Propagation Analysis – Application on Hydrogen Production Plant

Amir Hossein Hosseini[1], Sajid Hussain[1], and Hossam A. Gabbar[1,2]

[1] Faculty of Engineering and Applied Science, University of Ontario Institute of Technology,
2000 Simcoe St. North Oshawa, Ontario, Canada L1H7K4
[2] Faculty of Energy and Nuclear Science, University of Ontario Institute of Technology,
2000 Simcoe St. North Oshawa, Ontario, Canada L1H7K4
{Amir.Hosseini,Sajid.Hussain,Hossam.Gabbar}@UOIT.Ca

**Abstract.** Production of hydrogen from water through the Cu-Cl thermochemical cycle is a relatively new technology. The main advantages of this technology over existing ones are higher efficiency, lower costs, lower environmental impact, and reduced greenhouse gas (GHG) emissions. Considering these advantages, the usage of this technology in industries such as nuclear and oil is increasing. Due to different hazards involved in hydrogen production, design and implementation of hydrogen plants require provisions for safety, reliability, and risk assessment. However, a very less research is done from the safety point of view. This paper introduces fault semantic network (FSN) as a novel method for fault diagnosis and fault propagation analysis by using interactions among process variables. The interactions among process variables are estimated through genetic programming (GP) and neural network (NN) modeling. The effectiveness, feasibility, and robustness of the proposed method have been demonstrated on simulated data obtained from the simulation of hydrogen production process in AspenHysys software. The proposed method has successfully achieved reasonable detection and prediction of non-linear interaction patterns among process variables.

**Keywords:** Fault sematic network (FSN), Genetic Programming, Neural Network, Cu-Cl Thermochemical Cycle.

## 1 Introduction

Hydrogen is currently gaining much attention as a new energy source which can be replaced with oil and other fossil fuel in near future not only in different industries but also in transportation sector. Like other industrial processes, hydrogen production also involves different hazards that may lead to irreparable losses. The hazards, not only affect the equipment and production but also raises environmental, occupational, safety, and health related concerns. Given this, it is obvious that performing preventive actions and implementing well-established monitoring and control strategies to prevent accidents and reducing the risks associated with processes seems necessary.

Zalosh et al. made a statistical investigation in 1978 about industry related hydrogen accidents in USA, where they conclude that in about 80% of the hydrogen

M. Ali et al. (Eds.): IEA/AIE 2013, LNAI 7906, pp. 441–448, 2013.

accidents ignition occurred, 65% of them caused ignitions an explosion. Forty percent of all the hydrogen leakages were not detected prior to the accident and therefore, it was argued to install appropriate detectors in hydrogen systems [1].

There are many ways that make hydrogen different from other sources of fuel. Wider flammability limits, lower ignition energy are important factors that make the hydrogen production process more hazardous over others. Consequently, greater risk due to the increased probability of a fire and explosion would be one of the results [1]. Given this, it seems necessary to perform a detailed hazard identification method for every stage in the hydrogen production process.

This study introduces fault semantic network (FSN) as a novel method of fault diagnosis and fault propagation analysis by using interactions among process variables in Cu-Cl thermochemical Cycle. The rest of the paper is organized as follow. Section 2 describes the theory behind FSN. The hydrogen production process simulation is discussed in section 3. Section 4 presents the results obtained from process variable analysis done by GP and a dynamic neural network, the nonlinear autoregressive model with exogenous inputs (NARX). Application of the process variables interactions analysis to build the FSN and to perform the automated hazard identification is discussed in section 5 followed by the conclusion in section 6.

## 2    Fault Semantic Network (FSN)

Fault semantic network (FSN) originally described by Gabbar in [2], is a mean of representing knowledge based on the relationship between objects and concepts. The ultimate goal of this study is to develop a real-time fault propagation analysis and hazard identification method through FSN. Once process data are ready to be analyzed, the FSN module starts identifying the quantitative relationships among them and tracing deviations. The overview of research methodology and FSN module is presented in Figure 1 and the steps are explained in detail as follow [3].

1. Extracting real-time process data: History of real-time process data come from process plant. These real-time data are extracted from sensors and controllers installed on all equipment in the underlying plant to monitor the process conditions.

2. Relationship among process variables through GP & NARX: In order to uncover the relationship among process variables, GP and NARX are used as pattern recognition techniques to identify their relationship quantitatively. This step analyzes the interaction strengths among process variables.

3. Fault propagation analysis through FSN: Once the interactions among process variables recognized, it is time to trace deviations between process variables and failure modes in equipment. The FSN represents process variables and failure modes as separate nodes. To analyze propagation of faults, a database is designed that includes all process variables, all possible relations among them, different hazard scenarios, associated failure modes, and all possible causes and consequences. The FSN updates itself through real-time process data received from the plant. If any process variable is outside of the defined range (as defined in database), it will be detected. Considering the probability associated with each node, if-then rules and the defined scenarios in database, the FSN traces deviations and reason between nodes.

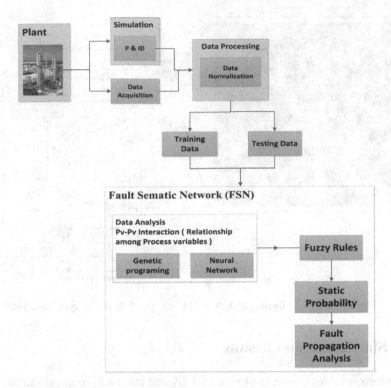

**Fig. 1.** Research Methodology & Fault Semantic Network (FSN) [3]

# 3    Process Description

Thermodynamic cycles are a novel approach in producing hydrogen. Different thermochemical hydrogen production cycles have been developed to split water thermally through chemical compounds and reactions. After considering various factors such as thermal efficiency, cost analysis and feasibility of industrialization, the Cu-Cl cycles was identified as the most promising one. The Cu–Cl cycle involves five chemical reactions as follow [4]. The conceptual layout of Cu-Cl cycle is shown in Figure 2.

$$2Cu(s) + 2HCl(g) \rightarrow 2CuCl(l) + H2(g),$$    Chlorination Step
$$2CuCl(s) \rightarrow 2CuCl(aq) \rightarrow CuCl2(aq) + Cu(s),$$    Disproportionation Step
$$CuCl2(aq) \rightarrow CuCl2(s),$$    Drying Step
$$2CuCl2(s) + H2O(g) \rightarrow CuOCuCl2(s) + 2HCl(g),$$    Hydrolysis Step
$$CuOCuCl2(s) \rightarrow 2CuCl(l) + (1/2)O2(g),$$    Decomposition Step

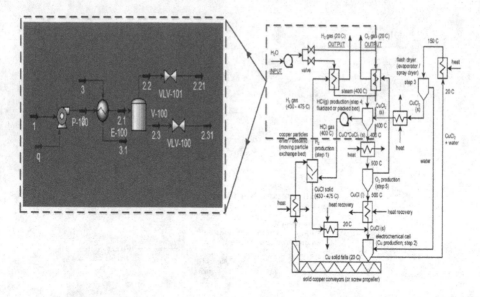

**Fig. 2.** Conceptual Layout of the Cu-Cl Cycle [4] & AspenHysys simulation

# 4     Simulations and Results

The process involves hundreds of variables divided into two groups, manipulated and measured variables. Where manipulated variables are considered as independent and measured ones as dependent. In order to see how manipulated variables affect the measured ones, they are changed either randomly or by steps. In this case, the valve opening percentage that controls flow out from the first reactor is considered as manipulated and reactor pressure considered as the measured variable, as their trends are shown in Figure 3.

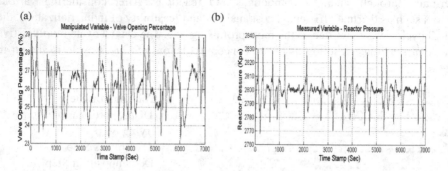

**Fig. 3.** Figure 3(a) shows the Changes in Valve opening percentage. Figure 3(b) shows Changes in reactor pressure

Figure 4 shows the actual and predicted trend of the reactor pressure as the measured variable modeled by GP and NARX.

(a)                                                    (b)

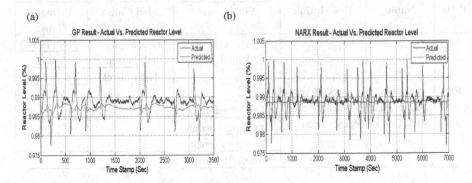

**Fig. 4.** Shows the actual and predicted trend where Figure 4(a) shows the GP result and Figure 4(b) shows NARX result

The relation between variables can be shown in mathematical formulation. The formulated relationship between Reactor pressure and valve opening percentage is presented in Table 1. Where the variable X1 is the valve opening percentage and Y1 is the reactor pressure. Subtraction, addition, sin, cosine and log are the functions that make the relationship.

**Table 1.** The nonlinear Relationship between Xi (Manipulated variable) & Yi (Measured Variable)

| $yi = Measured\ variable$ | The relation between $yi$ & $xi$ where $x1$ = Valve openning percentage as manipulated variable |
|---|---|
| $y1 = Reactor$ Pressure | $y1 = Sin[cos(log\ x1 + x1) \times log x1] + log(x1 + Sin(log x1))$ |

## 5    Fault Propagation and Hazard Identification Using FSN

As shown in the FSN module, designing a database according to the defined hazard scenarios, equipment, failure modes, and process variables is first step in building the FSN. To this, hazard scenarios should be defined for each failure mode. In order to demonstrate the application of the FSN in fault propagation analysis and identifying hazard, a hazard scenario related to the reactor consists of initial events, top event and consequences is designed in MS-Access. The designed database is shown in Table 2.

**Table 2.** Failures, Mechanism, Symptoms and Description

| F-ID | Failure Name | Description | Mechanism Final cause | Symptoms Med Cause | Primary Cause |
|---|---|---|---|---|---|
| F1 | Reactor-Leakage Through Wall Crack | Propagating crack on the reactor wall causes leakage | Vibration | Operation at High Pressure | High Flow Valve |
| | | | | High Level | High flow Valve |
| | | | Corrosion | | |
| | | | Improper Design | | |
| | | | Improper repair | | |
| | | | Structural damage | | |

LabVIEW® software is used to develop a program for the purpose of real-time fault propagation analysis and hazard identification. The FSN program queries the database to dynamically model the faults and hazards scenarios. Figure 5 represents a schematic view of the model in the LabVIEW® environment and a modeled scenario by the FSN is shown in Figure 6.

As explained in section 2, there are 2 main phases, the FSN follows to analyze faults propagation and their associated hazards. Integration of these two phases builds the prototype where phase 1 uses inter-relation pattern recognition and phase 2 uses fuzzy expert system (FES) and Bayesian Belief Network (BBN) for reasoning causes and consequences and tuning the FSN.

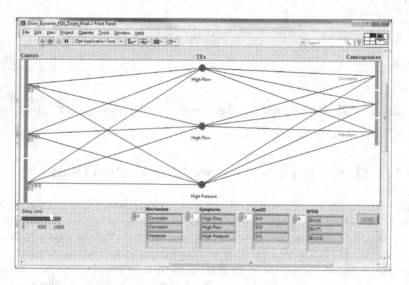

**Fig. 5.** Schematic view of the model in the LabVIEW® environment

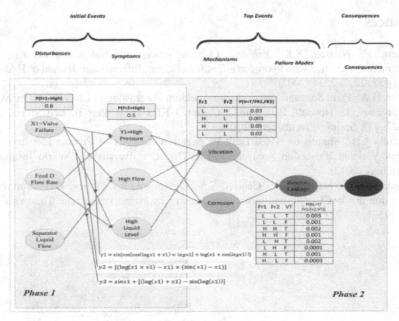

**Fig. 6.** A modeled scenario by the FSN

# 6     Conclusion

In this study, process variable interaction analysis is used to develop a simulation-based fault propagation analysis. The investigation has adopted two approaches that consisted of simulating of chemical process plant along with identifying relationships among process variables quantitatively for the purpose of finding causes and consequences and tracing faults associated with them.

In order to measure the accuracy of GP and NARX in trend estimation, mean absolute error and mean absolute percentage error is used. As the error values are presented in Table 3, accuracy of NARX in prediction of selected process variables behavior and identifying their inter-relationship pattern is better in comparison with GP.

**Table 3.** GP vs. NARX Prediction Accuracy - MAE & MAPE Values

| Process Variables | | GP Prediction | | NARX Prediction | |
|---|---|---|---|---|---|
| Manipulated | Measured | MAE | MAPE | MAE | MAPE |
| Valve Opening Percentage | Reactor pressure | 0.0057 | 0.589 | 0.0039 | 0.389 |

# References

1. Markert, F., Nielsen, S.K., Paulsen, J.L., Andersen, V.: Safety aspects of future infrastructure scenarios with hydrogen refuelling stations. International Journal of Hydrogen Energy 32, 2227–2234 (2007)
2. Gabbar, H.A.: Fault Semantic Networks for Accident Forecasting of LNG Plants. In: Setchi, R., Jordanov, I., Howlett, R.J., Jain, L.C. (eds.) KES 2010, Part II. LNCS, vol. 6277, pp. 427–437. Springer, Heidelberg (2010)
3. Hosseini, A.H.: Simulation-Based Fault Propagation Analysis of Process Industry Using Process Variable Interaction Analysis, MASc Thesis, University of Ontario Institute of Technology (2013)
4. Daggupati, V.N., Naterer, G.F., Gabriel, K.S., Gravelsins, R.J., Wang, Z.L.: Equilibrium conversion in Cu–Cl cycle multiphase processes of hydrogen Production. Thermochimica Acta 496, 117–123 (2009)

# Automatic Decomposition and Allocation of Safety Integrity Levels Using a Penalty-Based Genetic Algorithm

David Parker[1], Martin Walker[1], Luís Silva Azevedo[1], Yiannis Papadopoulos[1], and Rui Esteves Araújo[2]

[1] Department of Computer Science, University of Hull, United Kingdom
{d.j.parker,martin.walker}@hull.ac.uk,
l.p.azevedo@2012.hull.ac.uk, y.i.papadopoulos@hull.ac.uk
[2] INESC TEC, Faculdade de Engenharia, Universidade do Porto, Portugal
raraujo@fe.up.pt

**Abstract.** Automotive Safety Integrity Levels (ASILs) are used in the new automotive functional safety standard, ISO 26262, as a key part of managing safety requirements throughout a top-down design process. The ASIL decomposition concept, outlined in the standard, allows the safety requirements to be divided between multiple components of the system whilst still meeting the ASILs initially allocated to system-level hazards. Existing exhaustive automatic decomposition techniques drastically reduce the effort of performing such tasks manually. However, the combinatorial nature of the problem leaves such exhaustive techniques with a scalability issue. To overcome this problem, we have developed a new technique that uses a penalty-based genetic algorithm to efficiently explore the search space and identify optimum assignments of ASILs to the system components. The technique has been applied to a hybrid braking system to evaluate its effectiveness.

## 1 Introduction

The new ISO 26262 functional safety standard introduces new requirements for automotive engineers and designers. To comply with the standard, designs for electrical and/or electronic systems must undergo a comprehensive safety process throughout the design lifecycle. While automotive manufacturers and parts suppliers already have safety processes in place, these are not necessarily integrated or in line with ISO 26262, which aims to enforce a more holistic attitude to system safety. This means that companies must investigate new safety analysis and validation & verification (V&V) approaches and put into place new methodologies to ensure compliance. In particular, the importance of traceability has been emphasized to ensure that safety characteristics can be traced from initial safety requirements at the basic feature level right down to more detailed safety analyses at the hardware level.

One of the new features introduced by the ISO 26262 standard is that of the ASIL, or Automotive Safety Integrity Level. It is analogous to the Safety Integrity Level (SIL) defined in IEC 61508 and provides a qualitative descriptor for safety requirements of different stringencies. ASILs range from A (least strict) to D (most

M. Ali et al. (Eds.): IEA/AIE 2013, LNAI 7906, pp. 449–459, 2013.

strict). QM can also be used to represent normal Quality Management requirements on system elements with no special safety requirements.

The ASIL is the concept introduced to track the safety requirements throughout the design process. An initial hazard analysis at the most abstract level assesses possible malfunctions of the major system functionalities in various operational conditions (known as hazardous events). The standard provides guidance on how to categorize the risk of a given hazardous event on the basis of its severity, controllability, and exposure levels, and this category of criticality is represented by an ASIL. Safety goals and safe states to prevent the occurrence of the hazard are then defined, and these together with the ASIL inherited from the hazard provide the initial safety requirements for the system.

Following the system's top-down design, the ASILs of the safety goals are iteratively allocated to the functions, subsystems and finally components that by failing could corrupt these top-level safety requirements.

Since it would be prohibitively expensive (and generally impractical) for the entire system to always meet the highest safety requirements, there has to be a means of focusing the efforts of improving and ensuring safety on the most critical parts of the system. ISO 26262 provides a mechanism for this purpose in the form of ASIL decomposition, which allows the ASIL of a given system functionality to be decomposed over the various components that provide that functionality, assuming those components are arranged in an independent and parallel manner. Thus if multiple heterogeneous components must fail together in order to corrupt a safety goal, they are jointly responsible for meeting the ASIL requirement associated with that safety goal, rather than each individually having to meet that ASIL.

The nature of the decomposition is defined in ISO 26262 by an "ASIL algebra"; this algebra is formally underpinned by probability theory, and the rules for the reduction of the ASILs are defined in the standard. To formalize these rules, each ASIL is given a numeric value (e.g. A = 1, B = 2, C = 3, and D = 4; QM = 0), and simple integer addition is used to decompose the ASILs. Thus if two components were jointly responsible for meeting ASIL B, then that ASIL B could be decomposed such that each component is only required to meet ASIL A. Because a failure of both components is necessary to cause the overall hazard, and since ASIL A + ASIL A = ASIL B (i.e., 1 + 1 = 2), the high level safety requirement is being met by two lower decomposed safety requirements.

The importance of this decomposition process is that it allows system designers to meet their responsibilities to ensure the safety of the systems they create, and to comply with the standard, without incurring unnecessarily high costs. Higher ASILs inevitably mean greater expense: more safety measures need to be in place, more work has to be done, and more expensive components of a higher quality are likely to be needed. Because the type and distribution of ASILs across the system can have such a large bearing on both development and production costs, it is important to be able to efficiently allocate resources, meeting the safety requirements and yet not incurring in unreasonable expense.

The process of decomposing ASILs across the constituent elements of the system, and of determining an optimal allocation of ASILs while still meeting overall safety requirements, is not a simple one. Firstly, the information about the system necessary to allow such a decomposition to take place may be spread over multiple forms and representations, and potentially contained in different models. The increasing use of architectural description languages, or ADLs (such as AADL [1] in the aerospace domain and EAST-ADL [2] in the automotive domain), helps to overcome this problem to some extent by centralizing all of the information about the system in a single (albeit large) model and providing a common platform for analyses [3]. However, the heterogeneity of analysis tools and the different requirements from one system to the next still pose significant challenges, and the wide-ranging demands of the new ISO 26262 standard mean that many aspects of the safety process are still carried out manually, because automated tools are not available or sufficiently mature yet [4].

The second complication is a simple, inevitable consequence of the scale of the systems in question and the number of possible combinations of ASILs that could be allocated. Even a simple pair of components with an overall requirement to meet ASIL D could potentially have 5 different decomposed ASILs allocated to them, e.g. ASIL A + ASIL C, ASIL B + ASIL B and so forth. We have developed algorithms and a prototype tool, built on top of a safety analysis platform called HiP-HOPS (Hierarchically-Performed Hazard Origin & Propagation Studies) [5] that allows potential ASIL allocations to be identified and evaluated on the basis of an analysis of the system architecture and how possible failures can propagate through it. By performing this analysis, HiP-HOPS is aware of the relationships and dependencies between different elements of the system, their low-level failures, and the high-level system hazards that may be caused by combinations of those failures. It can use this information to determine which components may be responsible for causing a given hazard, and thus it can determine how the ASIL associated with that hazard may be decomposed over those components [6].

The technique is hierarchical and this enables it to be applied as part of a supply chain; thus system designers can determine an allocation of ASILs for the overall system and pass those ASIL requirements on to component suppliers, who can then employ the same techniques to ensure that their components meet the requirements passed down from the original design. This helps to manage the complexity of maintaining safety requirements across multiple design levels and multiple organizations.

However, while it is possible to exhaustively enumerate and evaluate every possible combination of ASILs that could be allocated for smaller systems, this is not practical for larger systems. The algorithms we have developed are constantly being refined and improved, but the combinatorial nature of the problem domain means that scalability is always an issue. To solve this problem, we have developed a new approach that achieves the same goal -- the optimal allocation of ASILs to system components to meet the decomposed safety requirements -- by means of automatic optimization based on genetic algorithms (GA). Meta-heuristics, such as GA, are faster than enumerative techniques as they do not need to search the entire solutions space and instead can be intelligently guided by means of heuristic evaluations. They are also known to be more efficient, for large-scale problems, than deterministic algorithms, such as Branch and Bound, that use problem-domain knowledge to direct or limit search [8]. Furthermore, in comparison with deterministic techniques, meta-heuristics are more robust as they often provide means to escape local optima; as will be explained later on, this is done in GA through chromosome mutation. It must be noted that meta-heuristics cannot usually guarantee finding optimal solutions; however, due to the high likelihood of ASIL allocation being iterated several times due to design changes and supply chain constraints, using good or near-optimal solutions provided by the efficient meta-heuristics can still offer significant improvements over manual approaches and result in a more efficient development process.

The overall goal of this work is to support designers in complying with ISO 26262 (and similar future standards) by automating the process of decomposing safety requirements over a hierarchical system architecture and evaluating the possible allocations of those requirements to determine the most promising allocations. This helps to manage the complexity of ensuring safety for modern, safety-critical systems without incurring unnecessary expense or requiring time-consuming manual analyses.

## 2    Optimization of ASILs Using HiP-HOPS

Once a system architecture has been developed, and after the system hazards have been defined and assigned ASILs on the basis of a risk analysis, those ASILs need to be decomposed across the components of that architecture according to each component's contribution to the various system hazards. This contribution is assessed via a safety analysis (e.g. fault tree analysis -- FTA), but for this to happen, the potential fault propagation of the system must be defined. This information allows us to determine the impact of individual component failure modes on the rest of the system, both singly and in combination, and thereby indicates which components may contribute to each functional failure of the system (and thus which hazards). If a component contributes to a hazard only in conjunction with other components, then it may receive a lower ASIL than a component which directly causes the hazard. Thus the design intention of components, and specifically their ability to detect, mask or propagate failures, influences ASIL decomposition across the architecture; for example, some components may be designed to fail silent in response to failure, possibly transforming a severe commission or value failure mode into a less severe omission failure mode.

This design intention can be captured in HiP-HOPS for each component as a set of failure expressions that show intended fault handling and propagation from inputs to outputs. At this early stage, we may not know what causes internal failures of each component simply because the implementation of the component is not known, but we can still know whether and how a component is supposed to mitigate, transform, and propagate input failures to outputs. HiP-HOPS can then establish the contribution of hypothesized component failures to each system hazard by using its FTA and FMEA capabilities. Making use of this information, HiP-HOPS assigns ASILs to components failures rather than components. This is to allow better refinement of requirements when a component presents more than one failure mode. Still, if a component allocation is required, the highest ASIL of the component's failure modes can be selected or some other heuristic can be used.

Although by this point we have decomposed the system ASILs across the architecture based on which failure modes contribute to the violation of which safety goals, there are still potentially many possible valid ASIL allocations for those failure modes, some more efficient than others. The next stage is therefore to use a genetic algorithm to determine the optimal allocation(s) of ASILs while respecting this prior decomposition. Genetic algorithms are meta-heuristic techniques that mimic the biological process of evolution to efficiently explore large problem spaces to find optimum solutions. The genetic algorithm described here is based on the penalty-based GA used by Coit and Smith [7].

A summary of the steps taken by the GA are as follows:

1. Randomly initialize a population of potential solutions (in this case, a solution is a set of ASILs allocated to each component failure mode).
2. Generate new candidate solutions for the next generation population by:
   a. Generating children candidates via crossover.
   b. Mutating children candidates.
3. Add next generation candidates to population.
5. Rank population by fitness.
6. Remove bottom candidates until within population limit
7. Return to step 2 unless generation limit is reached.

## 2.1    Encoding Solutions to the Problem of ASIL Allocation

The operators of GAs do not work directly on the system model, but instead use an encoding mechanism to store the variable information necessary to reconstruct and configure each candidate solution so that it can be evaluated and compared with other candidates to determine fitness.

This algorithm uses a fixed-length, real-number encoding for storing the assigned ASIL value for each failure mode in the system. The number in each 'slot' of the encoding can be between 0 and 4, relating directly to the ASILs (0=QM, 1=A, 2=B etc). For example, a system that contained 4 failure modes would have an encoding with 4 slots, and one possible candidate encoding might be (0|2|1|4), which translates to (QM|B|A|D).

When the search population is initialized, the values in the encoding are completely randomized. To configure the model using the encoding, each slot is taken in turn and its respective failure mode is assigned the ASIL indicated by the value in the slot. The model can then be evaluated according to the objective fitness evaluation algorithm given in the next section.

## 2.2    Fitness Evaluation

The goal of the optimization is to find a low cost candidate that does not violate the ASILs allocated to the safety goals of the hazards. Although other heuristics are possible, the cost heuristic used in this paper employs a logarithmic scaling:

QM = 0 cost; A = 10 cost; B = 100 cost; C = 1000 cost; D = 10000 cost.

Candidates with low ASILs will be clearly favoured; The total cost for each candidate is found by summing the ASIL cost of all decomposed assignments. For example, a candidate with encoding (0|2|1|4), which translates to (QM|B|A|D), would have the cost 10110 (0 + 100 + 10 + 10000).

This gives the base fitness for the candidate, but does not consider its feasibility. A feasible candidate is defined as one for which none of its ASIL allocations violate the decomposed safety requirements. In other words, infeasible candidates will have one or more ASIL assignments that fail to meet at least one system-level ASIL.

The feasibility checking algorithm makes use of the fault propagation information generated earlier, which is represented by the minimal cut sets (MCS) of the fault trees automatically synthesized and analysed by HiP-HOPS. Each MCS contains either a single failure mode that directly causes the hazard or a combination of two or more failure modes that act together to cause it. Each hazard has an ASIL assigned to it, and thus each MCS also has an overall ASIL that it must meet. Using the ASIL decomposition algebra: the safety requirement for the cut set is met if the sum of the ASILs assigned by the encoding to the failure modes in the cut set is greater than or equal to the ASIL of the hazard caused by that cut set. If the sum of the ASILs does not meet the total ASIL required, then an 'invalid_ASILs' counter is incremented. This is repeated for all cut sets and for all hazards.

Following this evaluation, a feasible candidate will have an 'invalid_ASILs' count of zero. Infeasible candidates will have non-zero counts that increase with the number of safety requirement violations. The 'invalid_ASILs' counter can be scaled and added as a penalty to the ASIL cost such that if two candidates have the same ASIL cost, but one violates the safety requirements, it will appear as more expensive and thus less fit. It is worth pointing out that the 'invalid_ASILs' variable serves as verification of the feasibility of the ASIL allocations in the final result. Finally, a dynamic near-feasibility threshold is used to allow a certain degree of constraint violation in the early generations of the search whilst applying harsher penalties to the later generations of the search. This is to allow for the observation that the most efficient route to a global feasible optimum can be through infeasible search space [7].

## 2.3    Selection

Candidates are selected from the population to apply the genetic operators, i.e., 'breed' the next generation. Following the "survival of the fittest" principle, the selection mechanism has a bias towards the fitter members of the population. The population members are thus first ranked according to their penalized fitness. A number is chosen from the range 1 to $\sqrt{P}$ where P is the population size. The candidate with the rank closest to the square of this random number is selected. The fitness bias can be dynamically altered by varying the power of the root. [9]

## 2.4    Genetic Operators

New candidates in the population are created through the manipulation of the encodings of existing candidates. These manipulations are specified by the actions of two genetic operators: mutation and recombination. The mutation operator randomly perturbs the population, encouraging diversity and acting to discourage the search from becoming stuck in local minima. The mutation operator acts upon each slot of the encoding according to a specified mutation probability. If a slot is selected for mutation then it is randomly changed to a number between zero and four (relating again to the different ASIL levels). The recombination operator mimics biological sexual reproduction: it produces new candidates by mixing and combining the encodings of existing candidate 'parents'. As the selection mechanism has a bias towards fitter candidates, it is more likely that the new combination of encodings will contain successful traits. Recombination promotes search convergence and acts to focus the exploration locally to known good candidates. The recombination operator in this study is uniform crossover. It works by considering two parent candidates' encodings, one value slot at a time, and generating a new child encoding by randomly selecting (with equal probability) one parent's value to populate the respective encoding slot in the child encoding.

## 3     Case Study

In the following case study, we describe a system model of a Hybrid Braking System (HBS) that we have previously applied an exhaustive search technique to [10]. A block diagram of the system is presented in figure 1. We then apply the penalty based genetic algorithm to the system in order to argue its effectiveness.

The braking system, introduced by de Castro *et al.* [11], was designed for electric vehicles whose propulsion system consists of 4 In-Wheel Motors (IWMs). It is named "hybrid" as it provides braking action through the combined effort of the IWMs and electromechanical actuators [12]. One important feature of the system is energy efficiency: while braking, the IWMs act as generators and transform the vehicle's kinetic energy into electric energy that is fed to the powertrain battery. This helps to increase the vehicle's range. The system also ensures the independent control of the braking force applied to each wheel, which improves braking performance. However, a fault in the system (for example, an incorrect torque instruction on one side) could result in an unwanted deceleration or acceleration of one wheel, leading to an unexpected yaw rate and making the vehicle uncontrollable.

The HBS is a brake-by-wire technology, as it does not include a mechanical or hydraulic connection between the braking pedal and the actuators. Instead, it integrates a redundant communication bus that enables information exchange between a central processing unit, which is fed with data from the mechanical pedal, and 4 local processing units that control the braking actuators associated with each wheel. The processing elements fall into two categories. The first type calculates braking force demands for each wheel and has a redundant configuration, as a fault would otherwise result in a catastrophic event [13]; the second type computes braking torques for each of the two actuators associated with one wheel and does not possess local redundancy. The absence of redundancy is deemed acceptable here because local processing units are expected to omit their output in the presence of a fault, and it is believed to be possible to stop the car safely with only 3 braking units operational [13]. As braking torque can be controlled independently for each wheel, the system has been modelled to consider only one braking module without harming the failure analysis (see Figure 1). As the driver presses the mechanical pedal, his actions are sensed and processed by an electronic pedal (this component includes the redundant central processing unit). The wheel braking forces are then sent through the bus system to a Wheel Node Controller (WNC) (which integrates a local processing unit).

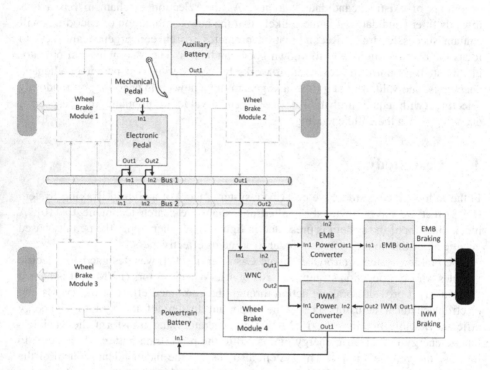

**Fig. 1.** Hybrid braking system model

Commands are generated for the power converters in order to control the 2 actuators. While braking, the auxiliary battery feeds the Electromechanical Brakes (EMBs) and the powertrain battery receives the electrical energy produced by the IWMs.

The following hazards are considered:

- No braking after command (H1)
- Wrong value braking (H2)

For this simplified case study, we have not assigned ASILs to hazards on the basis of the full ISO 26262 risk assessment procedure. Instead, they are only derived based on the hazards' severity: H1 and H2 are therefore assigned ASIL D and ASIL A respectively. H1 is caused by the omission of both braking actuators outputs, and H2 is the result of an incorrect value output of at least one of them. This is represented by the following expressions:

- H1 = Omission of EMB.out1 AND Omission of IWM.out1
- H2 = Value deviation of EMB.out1 OR Value deviation of IWM.out1

There are 24 component failure modes in the system, which results in an encoding length of 24 possible ASIL assignments. This gives a search space size of 5.96 x $10^{16}$.

HiP-HOPS performed a qualitative FTA on the hybrid braking system and determined the following MCSs:

- 19 minimal cut sets for H1: 1 single point of failure and 18 dual point failures
- 11 minimal cut sets for H2: 10 single points of failure and 1 dual point failure

The genetic algorithm was repeated 10 times, each run being 10000 generations. The population size was 500, and the mutation rate and crossover rate were 0.05 and 0.9 respectively. The results from the 10 runs can be seen below in table 1.

Table 1. Genetic algorithm runs results – solutions matching the optimum are highlighted

| ASIL costs | Failure modes and ASIL assignments | | | | | | | | | | | | | | | | | | | | | | | |
|---|---|---|---|---|---|---|---|---|---|---|---|---|---|---|---|---|---|---|---|---|---|---|---|---|
| | 1 | 2 | 3 | 4 | 5 | 6 | 7 | 8 | 9 | 10 | 11 | 12 | 13 | 14 | 15 | 16 | 17 | 18 | 19 | 20 | 21 | 22 | 23 | 24 |
| 11300 | 2 | 1 | 2 | 2 | 2 | 1 | 1 | 2 | 2 | 2 | 1 | 0 | 2 | 1 | 1 | 2 | 4 | 1 | 2 | 1 | 2 | 2 | 1 | 1 |
| 11300 | 2 | 1 | 2 | 2 | 2 | 1 | 1 | 2 | 2 | 2 | 1 | 0 | 2 | 1 | 1 | 2 | 4 | 1 | 2 | 1 | 2 | 2 | 1 | 1 |
| 14540 | 1 | 1 | 2 | 2 | 1 | 1 | 1 | 1 | 2 | 2 | 1 | 0 | 3 | 1 | 1 | 3 | 4 | 1 | 3 | 1 | 1 | 3 | 1 | 1 |
| 11390 | 2 | 1 | 2 | 2 | 2 | 1 | 1 | 2 | 2 | 2 | 1 | 0 | 2 | 1 | 2 | 2 | 4 | 1 | 2 | 1 | 2 | 2 | 1 | 1 |
| 11300 | 2 | 1 | 2 | 2 | 2 | 1 | 1 | 2 | 2 | 2 | 1 | 0 | 2 | 1 | 1 | 2 | 4 | 1 | 2 | 1 | 2 | 2 | 1 | 1 |
| 11300 | 2 | 1 | 2 | 2 | 2 | 1 | 1 | 2 | 2 | 2 | 1 | 0 | 2 | 1 | 1 | 2 | 4 | 1 | 2 | 1 | 2 | 2 | 1 | 1 |
| 11390 | 2 | 1 | 2 | 2 | 2 | 1 | 1 | 2 | 2 | 2 | 1 | 0 | 2 | 1 | 2 | 2 | 4 | 1 | 2 | 1 | 2 | 2 | 1 | 1 |
| 14540 | 1 | 1 | 2 | 2 | 1 | 1 | 1 | 1 | 2 | 2 | 1 | 0 | 3 | 1 | 1 | 3 | 4 | 1 | 3 | 1 | 1 | 3 | 1 | 1 |
| 11310 | 2 | 1 | 2 | 2 | 2 | 1 | 1 | 2 | 2 | 2 | 1 | 1 | 2 | 1 | 1 | 2 | 4 | 1 | 2 | 1 | 2 | 2 | 1 | 1 |
| 11390 | 2 | 1 | 2 | 2 | 2 | 1 | 1 | 2 | 2 | 2 | 1 | 0 | 2 | 1 | 2 | 2 | 4 | 1 | 2 | 1 | 2 | 2 | 1 | 1 |

The original exhaustive algorithm found 125 solutions ranging in ASIL cost from 11300 to 70110. In each GA run, the search converged on a single low cost solution. It is worth noting that even in the runs that did not find the optimum, the algorithm found a feasible solution at relatively low cost from the range of exhaustive solutions.

The case study provided in this paper is meant to illustrate the technique and is not large enough to demonstrate the scalability advantages of the meta-heuristic approach, as analysis times are under a second. However, prior exhaustive/deterministic techniques have failed to complete larger test models with hundreds of failure modes and many thousands of cut sets, whereas the meta-heuristic method does provide useful solutions within reasonable time frames.

# 4    Conclusion

The design of an automotive system is often distributed to various entities in the tiers of a supply chain. According to ISO 26262, safety requirements should be able to be allocated, in a top-down way, to components of an architecture using safety integrity levels (ASILs). In this paper, we extended earlier work on automating such allocation of ASILs to address scalability issues caused by the combinatorial nature of this problem. A genetic algorithm was described that seeks to find an optimum set of ASIL assignments that meet the safety requirements specified at system level. This work is currently being integrated in the context of the methods and tools that define EAST-ADL [3], an emerging automotive architecture description language that supports the lifecycle of automotive system designs from requirements to AUTOSAR implementations. Although we focused on ASILs as defined in ISO 26262, there are similarities in the definition and treatment of SILs in various other standards, and the proposed concept is applicable to other domains, e.g. aerospace.

**Acknowledgment.** This work was supported by the EU Project MAENAD (Grant 260057).

# References

1. Mian, Z., Bottaci, L., Papadopoulos, Y., Biehl, M.: System Dependability Modelling and Analysis Using AADL and HiP-HOPS. In: 14th IFAC Symposium on Information Control Problems in Manufacturing, Bucharest, Romania (2012)
2. Chen, D., Johansson, R., Lönn, H., Papadopoulos, Y., Sandberg, A., Törner, F., Törngren, M.: Modelling Support for Design of Safety-Critical Automotive Embedded Systems. In: Harrison, M.D., Sujan, M.-A. (eds.) SAFECOMP 2008. LNCS, vol. 5219, pp. 72–85. Springer, Heidelberg (2008)
3. Hillenbrand, M., Heinz, M., Adler, N., Matheis, J., Müller-Glaser, K.D.: Failure mode and effect analysis based on electric and electronic architectures of vehicles to support the safety lifecycle ISO/DIS 26262. In: Proceedings of the 21st IEEE International Symposium on Rapid System Prototyping (RSP), Fairvax, VA, USA, June 8-11, pp. 1–7 (2010) ISBN: 978-1-4244-7073-0, doi:10.1109/RSP.2010.5656351
4. Mader, R., Armengaud, E., Leitner, A., Steger, C.: Automatic and Optimal Allocation of Safety Integrity Levels. In: Proceedings of the Reliability and Maintainability Symposium (RAMS 2012), Reno, NV, USA, January 23-26, pp. 1–6 (2012) ISBN: 978-1-4577-1849-6, doi:10.1109/RAMS.2012.6175431
5. Papadopoulos, Y., Walker, M., Parker, D., Rüde, E., Hamann, R., Uhlig, A., Grätz, U., Lien, R.: Engineering Failure Analysis & Design Optimisation with HiP-HOPS. Journal of Engineering Failure Analysis 18(2), 590–608 (2011) ISSN: 1350 6307, doi:10.1016/j.engfailanal.2010.09.025
6. Papadopoulos, Y., Walker, M., Reiser, M.-O., Weber, M., Chen, D., Törngren, S.D., Abele, A., Stappert, F., Lönn, H., Berntsson, L., Johansson, R., Tagliabo, F., Torchiaro, S., Sandberg, A.: Automatic Allocation of Safety Integrity Levels. In: Proceedings of the 1st Workshop on Critical Automotive applications: Robustness and Safety (CARS 2010), Valencia, Spain, April 27, pp. 7–10. ACM, New York (2010) ISBN: 978-1-60558-915-2, doi:10.1145/1772643.1772646

7. Coit, D.W., Smith, A.E.: Reliability optimization of series-parallel systems using a genetic algorithm. IEEE Transactions on Reliability 45(2), 254–260 (1996)
8. Lin, M.-H., Tsai, J.-F., Yu, C.-S.: A Review of Deterministic Optimization Methods in Engineering and Management. Mathematical Problems in Engineering 2012, Article ID 756023, 15 pages (2012), doi:10.1155/2012/756023
9. Tate, D.M., Smith, A.E.: A genetic approach to the quadratic assignment problem. Computers and Operations Research 22, 73–83 (1994)
10. Azevedo, L.P.: Hybrid Braking System for Electrical Vehicles: Functional Safety, M.Sc. thesis, Dept. Elect. Eng., Porto Univ., Porto, Portugal (2012)
11. de Castro, R., Araújo, R.E., Freitas, D.: Hybrid ABS with Electric motor and friction Brakes. In: Presented at the IAVSD 2011 - 22nd International Symposium on Dynamics of Vehicles on Roads and Tracks, Manchester, UK (2011)
12. Savaresi, S., Tanelli, M.: Active braking control systems design for vehicles. Springer (2010) ISBN: 978-1-84996-350-3
13. Bannatyne, R.: Time triggered protocol-fault tolerant serial communications for real-time embedded systems. In: Wescon 1998, September 15-17, pp. 86–91 (1998)

# HAZOP Analysis System Compliant with Equipment Models Based on SDG

Ken Isshiki, Yoshiomi Munesawa, Atsuko Nakai, and Kazuhiko Suzuki

Graduate School of Natural and Technology, Okayama University,
3-1-1 Tsushima-naka, Kita-Ku, Okayama-shi, Okayama, 700-8530, Japan
{isshiki,fumoto}@safelab.sys.okayama-u.ac.jp
{munesawa,kazu}@sys.okayama-u.ac.jp

**Abstract.** It is important to assess the risk in chemical plants. HAZOP is widely used in the risk assessment to identify hazard. An automatic analysis system is developed to perform HAZOP effectively. In this study, semi-automatic analysis system was developed by using the Signed Directed Graph (SDG) as a deviation in the behavior of the propagation of equipment. Versatility of analysis is raised based on the propagation of deviation by adding the device in accordance with the rules. Our developed HAZOP analysis system is applied to one chemical process. And the future works for this study are explained.

**Keywords:** HAZOP analysis, Signed Directed Graph, hazard identification.

## 1 Introduction

In recent years, chemical plants increased size and complexity by improvement of the technology. So, the safety management is important to companies and society. Chemical plants have many hazards that are caused by materials, system configurations and equipment characteristics. If an accident occurs, there is a possibility of a serious impact on employees, local residents in the community. Therefore, it is more important to perform of the "risk assessment" which is assumed possible accidents and causes beforehand in industrial facilities and calculated the risk based on frequency of the accident and size of the damage when the accident occurred [1].

A variety of evaluation index is used to perform a risk assessment. Therefore, various evaluation methods are proposed in order to estimate each item as systematic and effectively. However, it is need to discuss several experts familiar with the behavior of equipment, operating conditions and the characteristics of the material. And much time, labor and cost are necessary. Furthermore, variability of evaluation results is expected by the subjectivity of the analyst. The many research institutions develop computer system to support the implementation of each evaluation method in order to solve these problems. For example, A. Meel et al. [2] developed the system used in the risk assessment to create a statistical model based on the accident database and Maryam Kalantarnia et al. [3] developed dynamic risk assessment approach based on Bayesian inference.

M. Ali et al. (Eds.): IEA/AIE 2013, LNAI 7906, pp. 460–469, 2013.
© Springer-Verlag Berlin Heidelberg 2013

## 2    Purpose and Approach

It is important to ensure the safety in industrial facilities, and hazards in the facility must be identified during the design phase of the facility. HAZOP is a technique that has been created for the purpose of identification of hazards. It has the feature that it is highly exhaustive systematic analysis compared to the other methods with the same purpose. However, the research and development of the HAZOP analysis system by the automatic processing of the computer are demanded because it requires a lot of time and labor to perform.

Various techniques are developed in each research institution in order to systematize the method the experts perform the HAZOP. For example, C. Zhao et al. [4] developed PHASuite. PHASuite models HAZOP analysis carries it out by replacing knowledge to ontology-based information and using a knowledge base and colored Petri net. K.Kawamura et al. [5] developed. HazopNavi assumes a cause and performs HAZOP analysis in conjunction with intelligent CAD system called Dynamic Flow Diagram. Dynamic Flow Diagram is CAD system to clarify the operation, behavior and the structure of the chemical process.

The System developed by various research institutions is divided into generic HAZOP analysis system and non-generic HAZOP analysis system. The generic HAZOP analysis system means that its reasoning logic can be applied to different chemical processes. For example, there are HAZID developed by S.A. McCoy et al. [6], ExpHAZOP+ developed by Shibly Rahman et al. [7] and D-higraphs HAZOP developed by Manuel Rodríguez et al. [8]. The non-generic HAZOP analysis means that its reasoning logic is chemical process specific or chemical plant specific. For example, C. Jeerawongsuntorn et al. [9] developed HAZOP application for continuous biodiesel production. Z. Švandová et al. [10] developed HAZOP using dynamic simulation. The generic HAZOP analysis system can analyze a variety of chemical process. But the generic HAZOP analysis system can't identify the inherent danger that exists in each chemical process. On the other hand, the non-generic HAZOP analysis system can identify the inherent danger that exists in the specific chemical process. But the non-generic HAZOP analysis system can analyze only one chemical process. These show that the coexistence of generic HAZOP analysis and non-generic HAZOP analysis is difficult.

The final objective of our research is the development of the HAZOP analysis system that includes both generic HAZOP analysis and non-generic HAZOP analysis. As the first, the generic HAZOP analysis system was developed in this study. The analysis model along the thought of the abnormal propagation is defined based on the logical thought procedure of an expert performing HAZOP analysis. Specifically, propagation path is built beforehand by modeling how the deviation is propagated each piece of equipment. And the deviation assumed the pipe is propagated along the propagation path. The developed system selects hazard from a database prepared beforehand by the function of equipment and the deviation. In this way, the system semi-automatically identifies the hazard of the chemical plant.

# 3     Methodology

HAZOP analysis system analyzes the abnormal propagation among each equipment in the chemical plant. This system requires to format the information about chemical plant available on the analysis system and to build a framework to use. Therefore, the equipment and process information are modeled for the analysis of HAZOP in the HAZOP analysis system. So, SDG(Signed Directed Graph) model indicates abnormal propagation within the equipment is built and stored in the database. The system analyzes the abnormal propagation by building a framework for the integration database and combining these models.

## 3.1     SDG Model

SDG technique is applied to indicate qualitatively the relationship between cause and consequence in the developed system. The basic model of SDG is shown in Fig.1.

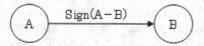

**Fig. 1.** The basic model of SDG

"A" and "B" in Fig.1 are called a node and express the vertices of the graph. This node contains a process parameter such a temperature, flow, pressure, level and composition. The node has qualitative values such as "0" "+" "-", each qualitative value expresses steady state, more steady state and less steady state. The arrow is called an arc and means association of "A" and "B". It expresses that "B" is affected by "A" in Fig.1. Sign (A-B) is expressed by two qualitative expression of "+" "-". When the Sign (A-B) is "+", the process parameter "B" has a positive correlation with respect to "A". When the Sign (A-B) is "-", the process parameter "B" has a negative correlation with respect to "A". In other words the state of node "B" is decided by a state of node "A" and the state of the arc. In the case of cause analysis, the system analysis by reversing the direction of the arc. SDG model is defined as the set of process parameters to be considered in the analysis by SDG. The developed system considers the case to be propagated to the different parameters for the deviation of the original by modeling the relationship between cause and consequence as using SDG.

## 3.2     Equipment Model

Equipment model expresses the modeled information about equipment constituting a plant. This model is consisted of SDG model to model the functions to be performed inside the equipment from input to output. And the model has a single path that is

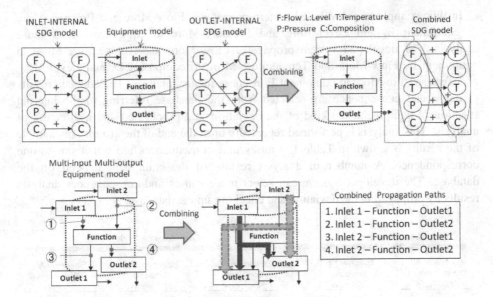

**Fig. 2.** The example of SDG model in the equipment model

connected to each of SDG model. Characterized function of equipment such as tanks and valves is expressed by equipment model. If the deviation is propagated, the change and the propagation of the deviation are defined by each of the internal function of the equipment. The change in the deviation before and after the equipment is represented by combining SDG model as shown in Fig.2. Multi-input multiple-output equipment model in Fig.2 has four SDG models according to a number of input and output. And, four propagation paths are built in this model.

### 3.3   Abnormal Propagation Analysis

#### Framework of Abnormal Propagation

The framework of abnormal propagation analysis is built as the plant model using the equipment model. The plant model is composed of a group of equipments that are passed with the same fluid. The propagation path expressed by SDG model is connected to the next equipment and then the next group. Therefore, whole plant propagation path is configured. Therefore, the developed system propagates deviation throughout the plant at the time of analysis in order to identify the hazard.

#### Cause and Consequence Analysis

Cause and consequence analysis assumes a deviation from the steady state in the pipe that is connected between the equipments. Cause analysis propagates the deviation to the upstream side along the propagation path of SDG model. Consequence analysis propagates the deviation to the downstream side deviation along the propagation path of SDG model. The flow deviation is assumed between equipment B and equipment C in Fig.3. The deviation is propagated to downstream in the path of the arrow pointing to the right, and to upstream in the path of the arrow pointing to the left.

In this example, the developed system propagates Flow-More and Pressure-More to equipment B, and Level-More and Pressure-More to equipment A at cause analysis. The developed system propagates Flow-More and Temperature-More to equipment C, Flow-More and Temperature-More to equipment D, and Level-More and Temperature-More to equipment E at consequence analysis. The analysis results of cause and consequence are selected from the database referring to the internal function of equipment, the kind of the deviation that propagated and a direction of analysis. This analysis is performed repeatedly until the end of the group. An example of the results is shown in Table 1. Causes and consequences are not a one-to-one correspondence. A number of analysis results of the equipment depend on the database. The developed system can select more causes and consequences analysis results by propagating the deviation and checking up the database.

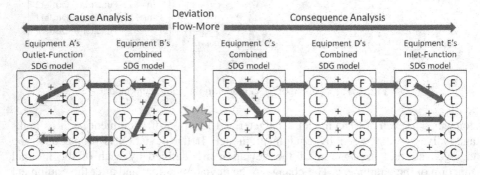

**Fig. 3.** Summary of case and consequence analysis

**Table 1.** Screen example of cause and effect analysis result

| Deviation:Flow-More | Equipment name | Equipment function | Content |
|---|---|---|---|
| Cause List | Equipment A | A's function | Cause to match the deviation-Level-More and A's function |
| | Equipment A | A's function | Cause to match the deviation-Pressure-More and A's function |
| | Equipment B | B's function | Cause to match the deviation-Flow-More and B's function |
| | Equipment B | B's function | Cause to match the deviation-Pressure-More and B's function |
| Consequence List | Equipment C | C's function | Consequence to match the deviation-Flow-More and C's function |
| | Equipment C | C's function | Consequence to match the deviation-Temperature-More and C's function |
| | Equipment D | D's function | Consequence to match the deviation-Flow-More and D's function |
| | Equipment D | D's function | Consequence to match the deviation-Temperature-More and D's function |
| | Equipment E | E's function | Consequence to match the deviation-Level-More and E's function |
| | Equipment E | E's function | Consequence to match the deviation-Temperature-More and E's function |

# 4    Case Study

The proposed method has been applied to several process plants. The developed system is applied to a part of the process to produce ethylene.

## 4.1    Analysis Object Process

Acetylene is supplied from the previous process and boosted by the compressor (see Fig.4). The range of the analysis object is as far as acetylene which has been heated in the heat exchanger enters the reactor. The range is equipment No.1-11 shown in Fig.4. There are main equipment includes a compressor(No.1), six valves(No.2,4,7-10), a tube shell heat exchanger(No.5), a reactor(No.11). In addition, a flow is divided into two with equipment No.3, and a flow is put together with one with equipment No. 6. Then, the deviation that is temperature rise is assumed in the pipe that is between the No.10 and No.6.

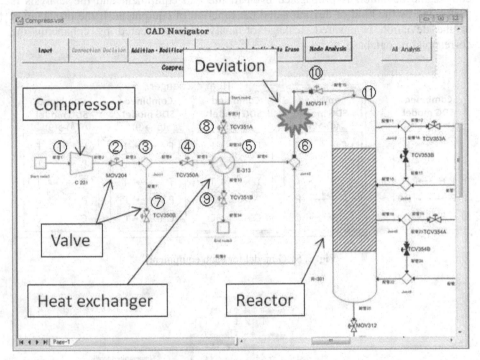

**Fig. 4.** Part of the drawing of the ethylene production process

## 4.2    Propagation Path and SDG Model of Equipment

Fig.5 shows combined SDG model for each piece of equipment to build the propagation path. All equipment except the heat exchanger is expressed by the left SDG model in Fig.5. The heat exchanger has four paths. There are four types of paths, from inlet tube side to outlet tube side (No.4 → No.5), from inlet shell side to outlet shell side (No.8→No.9), from inlet tube side to outlet shell side (No4→No.9), from inlet shell side to outlet tube side (No.8 → No.5). The heat exchanger has four SDG models because it is multi-input multi-output equipment. Two kinds of paths from inlet tube side to outlet tube side and from inlet shell side to outlet tube side are used in cause analysis. Then the two SDG models which are surrounded by broken line in

Fig.5 are used for cause analysis. The developed system builds the propagation path. Consequence analysis propagates from equipment No.10 to equipment No.11. However, cause analysis divides the propagation in two lines from equipment No.6 to equipment No.5 and from equipment No.6 to equipment No.7. Furthermore, cause analysis divides the propagation in two lines from equipment No.5 to equipment No.6 and from equipment No.5 to equipment No.8 because equipment No.5 is a heat exchanger. Cause analysis propagates to the left of the deviation as shown in Fig.6. In this case study, equipment No.1 is propagated Temperature-More, equipment No.8 is propagated Flow-More, Temperature-More and Pressure-More in cause analysis. Equipment No.11 is propagated Temperature-More in consequence analysis. Of course, the deviation is propagated even to the end equipment and the analysis is performed by the kind of deviation and the function of equipment. If the propagation of the deviation is ensured, leakage of analysis is prevented by enhancing the corresponding database.

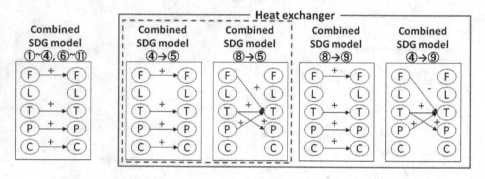

**Fig. 5.** SDG model for each equipment

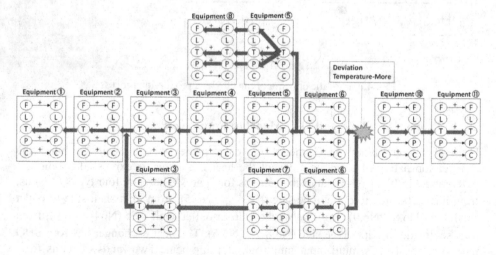

**Fig. 6.** Propagation of the deviation along the propagation path

## 4.3    Consideration

Analysis result screen by the developed system is shown in Fig.7. Cause list shows seven analysis results. Two analysis results are the cause of Temperature-More in equipment No.5. Another analysis result is the cause of Flow-More in equipment No.8. The other analysis results show that the deviation of Flow-More is propagated from upstream of equipment No.1, and the deviations of Flow-More, Temperature-More and Pressure-More are propagated from upstream of equipment No.8. Consequence list shows four analysis results. Three analysis results are the consequence of Temperature-More in equipment No.11. Another analysis result shows that the deviation of Temperature-More is propagated to downstream of equipment No.11.

In the case study, the number of cause and consequence results was less than the number of equipment that consisted of the process. It is necessary that the resulting information is stored to the database and the deviation is propagated to the equipment in order for developed system to select the result. Therefore analysis result shows cause as a result of propagation from upstream of the equipment and consequence as a result of the propagation to downstream of the equipment, the developed system shows the deviation to be analyzed is propagated to the end of the range. The developed system can contribute to identify the hazards by user enhances the database.

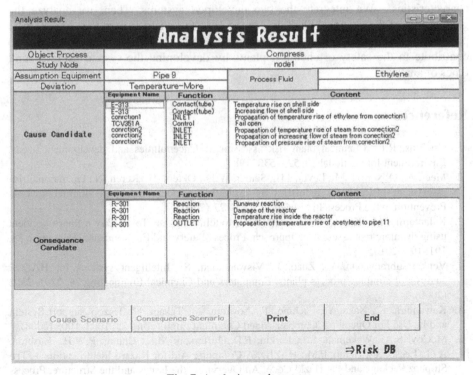

**Fig. 7.** Analysis result screen

# 5    Conclusions and Future Work

This paper proposed a method to build a propagation path between equipment inside and equipment by modeling the behavior of the equipment by SDG model in order to realize the system propagation analysis of an abnormality in the HAZOP analysis. Thereby, the system was built in order to perform semi-automated analysis by selecting the deviation to assume at the beginning of the analysis.

However, the developed system selects all cause and consequence that satisfies the condition. Analysis result of the developed system are included the results that are duplicate results and should be left out on purpose by HAZOP analysts in comparison with HAZOP which a conventional person performs.

In this situation we may overlook which analysis result is important. We think that it is necessary to sift out the results of the analysis of the system in the same thought process when people selected, and deviation dose not propagate to the end by the action of the instrumentation equipment at the time of analysis.

Our future direction is to develop the non-generic HAZOP analysis system. And we will integrate the generic HAZOP analysis system with non-generic HAZOP analysis system. We need to classify the common points and non- common points of each chemical plant for the development of non-generic HAZOP analysis system. It means that non-common points are characteristics of each chemical plant. Individual analysis logic and model are required to identify the potential hazards that exist in the characteristics.   We integrate the two developed systems. Then, we analyze the general part of the chemical plant by generic HAZOP analysis part, the characteristic part of each chemical plant by non-generic HAZOP analysis part. We want to contribute to the safety of chemical plants by developing the system using several types of analysis.

# References

1. Griffiths, R.F.: Chemical plant risk assessment: Uncertainties and development needs. Environment International 10, 523–530 (1984)
2. Meel, A., O'Neill, L.M., Levin, J.H., Seider, W.D., Oktem, U., Keren, N.: Operational risk assessment of chemical industries by exploiting accident databases. Journal of Loss Prevention in the Process Industries 20, 113–127 (2007)
3. Kalantarnia, M., Khan, F., Hawboldt, K.: Modelling of BP Texas City refinery accident using dynamic risk assessment approach. Process Safety and Environmental Protection 88, 191–199 (2010)
4. Venkatasubramanian, V., Zhao, J., Viswanathan, S.: Intelligent systems for HAZOP analysis of complex process plants. Computers and Chemical Engineering 24, 2291–2302 (2000)
5. Kawamura, K., Naka, Y., Fuchino, T., Aoyama, A., Takagi, N.: Hazop Support System and Its Use For Operation. Compter-Aided Chemical Engineering 25, 1003–1008 (2008)
6. McCoy, S.A., Wakeman, S.J., Larkin, F.D., Jefferson, M.L., Chung, P.W.H., Rushton, A.G., Lees, F.P., Heino, P.M.: HAZID, A Computer Aid for Hazard Identification: 1. The Stophaz Package and the Hazid Code: An Overview, the Issues and the Structure. Process Safety and Environmental Protection 77, 317–327 (1999)

7. Rahman, S., Khan, F., Veitch, B., Amyotte, P.: ExpHAZOP+: Knowledge-based expert system to conduct automated HAZOP analysis. Journal of Loss Prevention in the Process Industries 22, 373–380 (2009)
8. Rodríguez, M., de la Mata, J.L.: Automating HAZOP studies using D-higraphs. Computers & Chemical Engineering 45, 102–113 (2012)
9. Jeerawongsuntorn, C., Sainyamsatit, N., Srinophakun, T.: Integration of safety instrumented system with automated HAZOP analysis: An application for continuous biodiesel production. Journal of Loss Prevention in the Process Industries 24, 412–419 (2011)
10. Švandová, Z., Jelemenský, L., Markoš, J., Molnár, A.: Steady States Analysis and Dynamic Simulation as a Complement in the Hazop Study of Chemical Reactors. Process Safety and Environmental Protection 83, 463–471 (2005)

# TAIEX Forecasting Based on Fuzzy Time Series and Technical Indices Analysis of the Stock Market

Shyi-Ming Chen and Cheng-Yi Wang

Department of Computer Science and Information Engineering,
National Taiwan University of Science and Technology,
Taipei, Taiwan

**Abstract.** This paper presents a new method for forecasting the TAIEX based on fuzzy time series and technical indices analysis of the stock market. Because the proposed method uses both fuzzy time series and technical indices analysis of the stock market to analyze the historical training data in details for forecasting the TAIEX, it can get higher forecasting accuracy rate than the existing methods. The contribution of this paper is that we present a new fuzzy time series forecasting method based on the MACD index, combined with the stochastic line indices (KD indices) to forecast the TAIEX. It gets a higher average forecasting accuracy rate than the existing method for forecasting the TAIEX.

**Keywords:** Fuzzy logical relationships, fuzzy logical relationship groups, fuzzy sets, fuzzy time series, KD indices, MACD, technical indices analysis.

## 1    Introduction

The concept of fuzzy time series was proposed by Song and Chissom [6]-[8]. In recent years, some methods have been proposed for dealing with forecasting problems based on fuzzy time series [2]-[4], [9], [10]. However, in order to improve the forecasting accuracy rates of the existing methods [2]-[4], [9], [10], we need a better forecasting method to increase the forecasting accuracy rate.

In this paper, we present a new method for forecasting the Taiwan Stock Exchange Capitalization Weighted Stock Index (TAIEX) [13] based on fuzzy time series and technical indices analysis of the stock market. Because the proposed method uses both fuzzy time series and the technical indices analysis to analyze the historical training data in details for forecasting the TAIEX, it can get higher forecasting accuracy rate than the existing methods. The contribution of this paper is that we present a new fuzzy time series forecasting method based on the MACD index [1], combined with the stochastic line indices (KD indices) [5] to forecast the TAIEX. It gets a higher average forecasting accuracy rate than the method presented in [9] for forecasting the TAIEX.

The rest of this paper is organized as follows. In Section 2, we briefly review some basic concepts of fuzzy time series from [2], [6]-[8]. In Section 3, we briefly review some technical analysis indices used in the stock market. In Section 4, we present a new method to forecast the TAIEX based on fuzzy time series and technical indices analysis of the stock market. In Section 5, we make a comparison of the experimental results of the proposed method with the existing methods. The conclusions are discussed in Section 6.

M. Ali et al. (Eds.): IEA/AIE 2013, LNAI 7906, pp. 470–479, 2013.

## 2    Preliminaries

In this section, we briefly review some basic definitions of fuzzy time series from [2], [6]-[8], where the values of fuzzy time series are represented by fuzzy sets [12]. Let $U$ be the universe of discourse, where $U = \{u_1, u_2, \ldots, u_n\}$. A fuzzy set $A_i$ in the universe of discourse $U$ is defined as follows:

$$A_i = f_{A_i}(u_1)/u_1 + f_{A_i}(u_2)/u_2 + \cdots + f_{A_i}(u_n)/u_n, \tag{1}$$

where $f_{A_i}$ is the membership function of the fuzzy set $A_i$, $f_{A_i} : U \rightarrow [0, 1]$, $f_{A_i}(u_j)$ denotes the degree of membership of $u_j$ belonging to the fuzzy set $A_i$ and $1 \leq j \leq n$.

Assume that there are three triangular fuzzy sets $A_1$, $A_2$ and $A_3$, where $A_1 = (a_1, b_1, c_1)$, $A_2 = (a_2, b_2, c_2)$ and $A_3 = (a_3, b_3, c_3)$, as shown in Fig. 1. We can say that $A_2$ is located between $A_1$ and $A_3$, if $a_1 \leq a_2 \leq a_3$ and $c_1 \leq c_2 \leq c_3$.

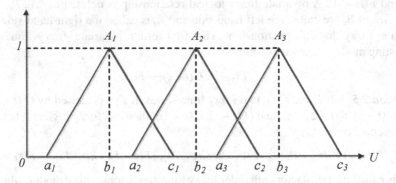

**Fig. 1.** Triangular fuzzy sets $A_1$, $A_2$ and $A_3$

***Definition 2.1:*** Let $Y(t)$ $(t = \ldots, 0, 1, 2, \ldots)$, a subset of real numbers, be the universe of discourse. Let $f_i(t)$ $(i = 1, 2, \ldots)$ be fuzzy sets defined in the universe of discourse $Y(t)$ and let $F(t)$ be a collection of $f_i(t)$ $(i = 1, 2, \ldots)$. Then, $F(t)$ is called a fuzzy time series defined on $Y(t)$ $(t = \ldots, 0, 1, 2, \ldots)$.

***Definition 2.2:*** If a fuzzy relationship $R(t - 1, t)$ exists, such that $F(t) = F(t - 1) \circ R(t - 1, t)$, where the symbol "$\circ$" denotes the max-min composition operator, then $F(t)$ is influenced only by $F(t - 1)$, denoted by the fuzzy logical relationship:

$$F(t - 1) \rightarrow F(t), \tag{2}$$

where $F(t - 1)$ and $F(t)$ are fuzzy sets. Let $F(t - 1) = A_i$ and let $F(t) = A_j$. The relationship between $F(t - 1)$ and $F(t)$ can be denoted by the fuzzy logical relationship "$A_i \rightarrow A_j$," where $A_i$ and $A_j$ are called the left-hand side (LHS) and the right-hand side (RHS) of the fuzzy logical relationship. Moreover, $F(t - 1)$ and $F(t)$ are called the "current state" and the "next state" of the fuzzy logical relationship, respectively.

We can group fuzzy logical relationships having the same LHS into a fuzzy logical relationship group (FLRG). For example, assume that the following fuzzy logical relationships exist:

$$A_i \rightarrow A_{ja},$$
$$A_i \rightarrow A_{jb},$$
$$\vdots$$
$$A_i \rightarrow A_{jm}.$$

Then, these fuzzy logical relationships can be grouped into a FLRG, shown as follows:

$$A_i \rightarrow A_{ja}, A_{jb}, \ldots, A_{jm}. \tag{3}$$

***Definition 2.3:*** Let $F(t)$ be a fuzzy time series. If $F(t)$ is caused by $F(t-1)$, $F(t-2)$, ..., and $F(t-n)$, then this fuzzy logical relationship is represented by

$$F(t-n), \ldots, F(t-2), F(t-1) \rightarrow F(t) \tag{4}$$

and it is called the $n$th order fuzzy time series forecasting model.

***Definition 2.4:*** Let $F$ and $G$ be two fuzzy time series. Assume that $F(t-1) = A_i$, $G(t-1) = B_k$ and $F(t) = A_j$. A bivariate fuzzy logical relationship is defined as "$A_i, B_k \rightarrow A_j$", where $A_i$ and $B_k$ are called the left hand side and $A_j$ is called the right-hand side of the bivariate fuzzy logical relationship. The first order bivariate fuzzy time series forecasting model is shown as follows:

$$F(t-1), G(t-1) \rightarrow F(t). \tag{5}$$

***Definition 2.5:*** Let $F$ and $G$ be two fuzzy time series. If $F(t)$ is caused by $(F(t-1), G(t-1))$, $(F(t-2), G(t-2))$, ..., and $(F(t-n), G(t-n))$, then this fuzzy logical relationship is represented by

$$(F(t-n), G(t-n)), \ldots, (F(t-2), G(t-2)), (F(t-1), G(t-1)) \rightarrow F(t) \tag{6}$$

and it is called the two-factors $n$th order fuzzy time series forecasting model, where $F(t)$ and $G(t)$ are called the main factor fuzzy time series and the secondary factor fuzzy time series, respectively ($t = \ldots, 0, 1, 2, \ldots$).

# 3     Technical Indices Analysis of the Stock Market

Trend is the moving direction of a stock's prices. As the prices of a stock continually move down, the closing price of the stock of the trading day has a tendency to crowd the lower portion of the daily range. On the other hand, as prices continually move up, the closing price of a stock of the trading day has a tendency to crowd the higher portion of the daily range. In this section, we briefly review some technical indices analysis used in stock market.

## 3.1     Moving Average Convergence/Divergence (MACD)

Moving Average Convergence/Divergence (MACD) [1] is a technical analysis indicator. It is used to spot changes in the strength, direction, momentum and duration of a trend in a stock's price. The MACD "oscillator" or "indicator" is a collection of three signals (or computed data-series), calculated from the historical price data, most often the closing price. These three signal lines are: the MACD line, the signal line (or average line) and the difference (or divergence). The term "MACD" may be used to refer to the indicator as a whole, or specifically to the MACD line itself. The first line, called the "MACD line", equals the difference between a "fast" (short period) exponential moving average (EMA) of 12 days, and a "slow" (longer period) EMA of 26 days. EMA highlight recent changes in a stock's price. The MACD line is charted over time, along with an EMA of the MACD line, termed the "signal line" or "average line". The difference (or divergence) between the MACD line and the signal line is shown as a "bar graph" or a histogram time series. A fast EMA responds more quickly than a slow EMA to recent changes in a stock's price. By comparing EMAs of different periods, the MACD line can indicate changes in the trend of a stock. By comparing differences in the change of that line to an average, an analyst can detect subtle shifts in the strength and direction of a stock's trend.

### 3.2    Stochastic Line Indices (KD Indices)

The stochastic line indices (KD indices) [5] is a momentum indicator that uses support and resistance levels. The calculation of KD indices finds the range between an asset's high and low price during a given period of time. The current securities price is then expressed as a percentage of this range with 0% indicating the bottom of the range and 100% indicating the upper limits of the range over the covered time period. The principle behind this indicator is that prices tend to close near the extremes of the recent range before turning points. The term *stochastic* refers to the location of a current price in relation to its price range over a period of time. This method attempts to predict price turning points by comparing the closing price of a security to its price range.

## 4    The Proposed TAIEX Forecasting Method Based on Fuzzy Time Series and Technical Indices Analysis of the Stock Market

In this section, we present a new method to forecast the TAIEX based on fuzzy time series and technical indices analysis of the stock market, which is a two-factors first-order fuzzy forecasting model, where the main factor is the actual TAIEX (denoted by $P_t$) and the secondary factors are the stochastic line indices (KD indices, denoted by $K_{(t)}$ and $D_{(t)}$). The proposed method is now presented as follows:

**Step 1:** The sub-steps of this step are described as follows:

**Step 1.1:** Calculate the moving average $En_{(t)}$ of the actual TAIEX $P_t$ on the trading day $t$, shown as follows:

$$En_{(t)} = En_{(t-1)} + \frac{2}{1+n}(P_t - En_{(t-1)}), \tag{7}$$

where $2/(1+n)$ is called the "smoothing coefficient", i.e., $En_{(t)}$ denotes the previous $n$-days (including the $t$th day) "exponential moving average" (EMA) of $P_t$. Usually, we let $n = 12$ to get the short period EMA $E12_{(t)}$ and we let $n = 26$ to get the long period EMA $E26_{(t)}$, defined as follows [1]:

$$E12_{(t)} = E12_{(t-1)} + \frac{2}{1+12}(P_t - E12_{(t-1)})$$

$$= \frac{11}{13}E12_{(t-1)} + \frac{2}{13}P_t, \tag{8}$$

$$E26_{(t)} = E26_{(t-1)} + \frac{2}{27}(P_t - E26_{(t-1)})$$

$$= \frac{25}{27}E26_{(t-1)} + \frac{2}{27}P_t. \tag{9}$$

**Step 1.2:** By subtracting the long period EMA $E26_{(t)}$ from the short period EMA $E12_{(t)}$, we can get the difference $MACD_{(t)}$ on the $t$th trading day to get the MACD line, where

$$MACD_{(t)} = E12_{(t)} - E26_{(t)}. \tag{10}$$

**Step 1.3:** By calculating the "exponential moving average" (EMA) of $MACD_{(t)}$ and $MACD_{(t-h+1)}$, we can get the signal line $SIG_{(t)}$ on the trading day $t$, shown as follows:

$$SIG_{(t)} = SIG_{(t-1)} + \frac{2}{1+h}(MACD_{(t)} - SIG_{(t-1)}), \tag{11}$$

where we let $h = 9$ and let the initial value of $SIG$ be the $MACD$ value of the previous day.

**Step 1.4:** Calculate the bar graph value $OSC_{(t)}$ on the trading day $t$, where

$$OSC_{(t)} = MACD_{(t)} - SIG_{(t)}. \tag{12}$$

**Step 2:** Define the universe of discourse $U$, $U = [D_{min} - D_1, D_{max} + D_2]$, where $D_{min}$ and $D_{max}$ are the minimum value and the maximum value of the historical training data of the main factor TAIEX, respectively; $D_1$ and $D_2$ are two proper positive real values to partition the universe of discourse $U$ into $n$ intervals $u_1, u_2, ...,$ and $u_n$ of equal length.

**Step 3:** Define the linguistic terms $A_1, A_2, ...,$ and $A_n$ represented by fuzzy sets of the main factor, shown as follows:

$A_1 = 1/u_1 + 0.5/u_2 + 0/u_3 + ... + 0/u_{n-2} + 0/u_{n-1} + 0/u_n,$
$A_2 = 0.5/u_1 + 1/u_2 + 0.5/u_3 + ... + 0/u_{n-2} + 0/u_{n-1} + 0/u_n,$

.
.
.

$A_{n-1} = 0/u_1 + 0/u_2 + 0/u_3 + ... + 0.5/u_{n-2} + 1/u_{n-1} + 0.5/u_n,$
$A_n = 0/u_1 + 0/u_2 + 0/u_3 + ... + 0/u_{n-2} + 0.5/u_{n-1} + 1/u_n,$

where $u_1, u_2,...,$ and $u_n$ are intervals obtained in **Step 2**.

**Step 4:** Fuzzify each historical training datum of the main factor into a fuzzy set defined in **Step 3**. If the historical training datum of the main factor of trading day $t$ belongs to $u_i$ and the maximum membership value of the fuzzy set $A_i$ occurs at interval $u_i$, where $1 \le i \le n$, then the historical training datum of the main factor of trading day $t$ is fuzzified into $A_i$.

**Step 5:** Based on the results obtained in **Step 4**, construct the first-order fuzzy logical relationships (FLRs) of the main factor (i.e., TAIEX), shown as follows:

$$A_i \rightarrow A_j,$$

where $A_i$ denotes the fuzzified linguistic term of the main factor on trading day $t-1$ and $A_j$ denotes the fuzzified linguistic term of trading day $t$. Based on the obtained FLRs, construct the fuzzy logical relationship groups (FLRGs) of the main factor (i.e., TAIEX).

**Step 6:** The sub-steps of this step are described as follows [5]:

**Step 6.1:** Calculate the Raw Stochastic Value $RSV_{(t)}$ of the trading day $t$, shown as follows:

$$RSV_{(t)} = \frac{(C_t - L_n)}{(H_n - L_n)} \times 100, \tag{13}$$

where $C_t$ denotes the closing index of the trading day $t$, $L_n$ denotes the lowest TAIEX of the previous $n$ trading days, and $H_n$ denotes the highest TAIEX of the previous $n$ trading days. In this paper, we let $n = 9$. Then, use the $3$-days exponential moving average of $RSV_{(t)}$ to calculate the value $K_{(t)}$ (called the fast $KD$ index)

$$K_{(t)} = \frac{2}{3} K_{(t-1)} + \frac{1}{3} RSV_{(t)}, \tag{14}$$

where $0 \leq K_{(t)} \leq 100$. In this paper, if there is no value of $K_{(t-1)}$, then we let $K_{(t)} = 50$.

**Step 6.2:** Use the $3$-days exponential moving average $K_{(t)}$ to calculate the value $D_{(t)}$ (called the slow $KD$ index)

$$D_{(t)} = \frac{2}{3} D_{(t-1)} + \frac{1}{3} K_{(t)}, \tag{15}$$

where $0 \leq D_{(t)} \leq 100$. In this paper, if there is no value of $D_{(t-1)}$ value, then we let $D_{(t)} = 50$. If $D_{(t)} \leq K_{(t)}$ on the trading day $t$, then the closing index of the TAIEX on the trading day $t$ moves up. Otherwise, the closing index of the TAIEX on the trading $t$ moves down.

**Step 7:** Let the universe of discourse $V = [0, 100]$. Partition the universe of discourse $V$ into $m$ equal intervals $v_1, v_2, ..., $ and $v_m$.

**Step 8:** Define the linguistic terms $B_1, B_2, ..., $ and $B_m$ represented by fuzzy sets of the secondary factors "stochastic line indices" (KD indices), shown as follows:

$$B_1 = 1/v_1 + 0.5/v_2 + 0/v_3 + ... + 0/v_{m-2} + 0/v_{m-1} + 0/v_m,$$
$$B_2 = 0.5/v_1 + 1/v_2 + 0.5/v_3 + ... + 0/v_{m-2} + 0/v_{m-1} + 0/v_m,$$

$$\vdots$$

$$B_{m-1} = 0/v_1 + 0/v_2 + 0/v_3 + ... + 0.5/v_{m-2} + 1/v_{m-1} + 0.5/v_m,$$
$$B_m = 0/v_1 + 0/v_2 + 0/v_3 + ... + 0/v_{m-2} + 0.5/v_{m-1} + 1/v_m,$$

where $v_1, v_2, ..., $ and $v_m$ are intervals obtained in **Step 7**.

**Step 9:** Fuzzify each historical training datum of the secondary factors "stochastic line indices" (KD indices) shown in Eqs. (14) and (15) obtained in **Step 6** into a fuzzy set defined in **Step 8**. If the historical training datum of the secondary factor $K_{(t)}$ of trading day $t$ shown in Eqs. (14) belongs to $v_j$ and the maximum membership value of the fuzzy set $B_j$ occurs at interval $v_j$, where $1 \leq j \leq m$, then the historical training datum of the secondary factor $K_{(t)}$ of trading day $t$ is fuzzified into $B_j$. If the historical training datum of the secondary factor $D_{(t)}$ of trading day $t$ shown in Eq. (15) belongs to $v_j$ and the maximum membership value of the fuzzy set $B_j$ occurs at interval $v_j$, where $1 \leq j \leq$ m, then the historical training datum of the secondary factor $D_{(t)}$ of trading day $t$ is fuzzified into $B_j$.

**Step 10:** /* Perform forecasting */ Assume that the testing datum of the main factor of trading day $t - 1$ is fuzzified into the fuzzy set $A_i$ and assume that we want to forecast the value of the main factor TAIEX of trading day $t$. Based on the FLRGs obtained in **Step 5** and the fuzzified "stochastic line indices" (KD indices) obtained in **Step 9**, we can forecast the TAIEX $FV_t$ of trading day $t$ according to the following cases:

**Case 1:** The right-hand side of the FLRG is $A_{j1}, A_{j2}, ..., A_{jp}$, that is, there is the following FLRG:

$$A_i \rightarrow A_{j1}, A_{j2}, ..., A_{jp}.$$

Let $B_{K(t-1)}$ be the $t-1$th day's fuzzified linguistic term represented by a fuzzy set of the secondary factor $K_{(t-1)}$, and let $B_{D(t-1)}$ be the $t-1$th day's fuzzified linguistic term represented by a fuzzy set of the secondary factor $D_{(t-1)}$. Then, for each $A_{ji}$, where $i \in \{1, 2, ..., p\}$, the forecasted TAIEX $FV_{ti}$ of trading day $t$ is calculated as follows:

  **Step 10.1:** If the fuzzy set $B_{K(t-1)}$ of the $t-1$th trading day is located between $B_{\lceil 0.2m \rceil}$ and $B_{\lceil 0.8m \rceil}$, the fuzzy set $B_{D(t-1)}$ of the $t-1$th trading day is located between $B_{\lceil 0.2m \rceil}$ and $B_{\lceil 0.8m \rceil}$, and $K_{(t-1)} \geq D_{(t-1)}$, then we let

$$FV_{ti} = m_i + |0.5 \times OSC_{(t-1)}|, \tag{16}$$

where $m_i$ denotes the middle point of the interval $u_i$, $MACD_{(t-1)} = E12_{(t-1)} - E26_{(t-1)}$, and $OSC_{(t-1)} = MACD_{(t-1)} - SIG_{(t-1)}$. Otherwise, we let

$$FV_{ti} = m_i - |0.5 \times OSC_{(t-1)}|, \tag{17}$$

where $m_i$ denotes the middle point of the interval $u_i$, $MACD_{(t-1)} = E12_{(t-1)} - E26_{(t-1)}$, and $OSC_{(t-1)} = MACD_{(t-1)} - SIG_{(t-1)}$.

  **Step 10.2:** If (the fuzzy set $B_{K(t-1)}$ of the $t-1$th trading day is located between $B_{\lceil 0.8m \rceil}$ and $B_m$ and the fuzzy set $B_{D(t-1)}$ of the $t-1$th trading day is located between $B_{\lceil 0.8m \rceil}$ and $B_m$) or (the fuzzy set $B_{K(t-1)}$ of the $t-1$th trading day is located between $B_1$ and $B_{\lceil 0.2m \rceil}$ and the fuzzy set $B_{D(t-1)}$ of the $t-1$th trading day is located between $B_1$ and $B_{\lceil 0.2m \rceil}$) and $OSC_{(t-1)} \geq 0$, then we let

$$FV_{ti} = m_i + |0.5 \times OSC_{(t-1)}|,$$

where $m_i$ denotes the middle point of the interval $u_i$, $MACD_{(t-1)} = E12_{(t-1)} - E26_{(t-1)}$, and $OSC_{(t-1)} = MACD_{(t-1)} - SIG_{(t-1)}$. Otherwise, we let

$$FV_{ti} = m_i - |0.5 \times OSC_{(t-1)}|,$$

where $m_i$ denotes the middle point of the interval $u_i$, $MACD_{(t-1)} = E12_{(t-1)} - E26_{(t-1)}$, and $OSC_{(t-1)} = MACD_{(t-1)} - SIG_{(t-1)}$.

**Step 10.3:** If $OSC_{(t-2)} \geq OSC_{(t-1)}$, then we let

$$FV_{ti} = m_i - |0.5 \times OSC_{(t-1)}|,$$

where $m_i$ denotes the middle point of the interval $u_i$, $MACD_{(t-1)} = E12_{(t-1)} - E26_{(t-1)}$, and $OSC_{(t-1)} = MACD_{(t-1)} - SIG_{(t-1)}$. Otherwise, we let

$$FV_{ti} = m_i + |0.5 \times OSC_{(t-1)}|,$$

where $m_i$ denotes the middle point of the interval $u_i$, $MACD_{(t-1)} = E12_{(t-1)} - E26_{(t-1)}$, and $OSC_{(t-1)} = MACD_{(t-1)} - SIG_{(t-1)}$.

Then, the forecasted TAIEX $FV_t$ on trading day $t$ is calculated as follows:

$$FV_t = \frac{n_{j1} \times FV_{t1} + n_{j2} \times FV_{t2} + ... n_{jp} \times FV_{tp}}{n_{j1} + n_{j2} + ... + n_{jp}}, \tag{18}$$

where $n_{j1}$, $n_{j2}$, ..., and $n_{jp}$ are the number of occurrances of $A_{j1}$, $A_{j2}$, ..., and $A_{jp}$ in the FLRG, respectively; $FV_{t1}$ denotes the forecasted TAIEX obtained from **Step 10.1**, **Step 10.2**, or **Step 10.3**; $FV_{t2}$ denotes the forecasted TAIEX obtained from **Step 10.1**, **Step 10.2**, or **Step 10.3**; ... ; $FV_{tp}$ denotes the forecasted TAIEX obtained from **Step 10.1**, **Step 10.2**, or **Step 10.3**.

**Case 2:** If the right-hand side of the FLRG is unknown (i.e., "#"), that is, there is the following FLRG:

$$A_i \rightarrow \#,$$

where $A_i$ denotes the $t-1$th day's fuzzified linguistic term represented by a fuzzy set of the main factor. Then, the forecasted TAIEX $FV_t$ on trading day $t$ is calculated as follows:

$$FV_t = m_i, \tag{19}$$

where $m_i$ denotes the middle point of the interval $u_i$ of $A_i$.

## 5    Experimental Results

In this section, we apply the proposed method to forecast the TAIEX from 1990 to 1999, where the length of each interval in the universe of discourse is equal to 100. In order to compare the experimental results of the proposed method with the ones of the methods presented in [9], we also adopt a 10 month/2 month split for training/testing, which is the same as the experimental environment shown in [9], i.e., for each year, the data from January to October are used as the training data set, and the data from November to December are used as the testing data set. We evaluate the performance of the proposed method using the root mean square error (RMSE), which is defined as follows:

$$RMSE = \sqrt{\frac{\sum_{i=1}^{n}(forecasted\ value_i - actual\ value_i)^2}{n}}, \tag{20}$$

where $n$ denotes the number of days needed to be forecasted. In Table 1 we make a comparison of the RMSEs and the average RMSE of the proposed method with the method presented in [9]. From Table 1, we can see that the average RMSEs of the proposed method is smaller than the one presented in [9].

**Table 1.** A comparison of RMSEs and the average RMSEs for different methods for forecasting the TAIEX from 1990 to 1999

| RMSE Methods / Year | | 1990 | 1991 | 1992 | 1993 | 1994 | 1995 | 1996 | 1997 | 1998 | 1999 | Average RMSEs |
|---|---|---|---|---|---|---|---|---|---|---|---|---|
| Conventional Models [9] | Average-Based Lengths | 220 | 80 | 60 | 110 | 112 | 79 | 54 | 148 | 167 | 149 | 117.9 |
| | Distribution-based Lengths | 270 | 79 | 60 | 105 | 132 | 79 | 52 | 149 | 159 | 159 | 124.4 |
| Weighted Models [9] | Average-Based Lengths | 227 | 61 | 67 | 105 | 135 | 70 | 54 | 133 | 151 | 142 | 114.5 |
| | Distribution-based Lengths | 266 | 67 | 56 | 105 | 114 | 70 | 52 | 152 | 154 | 145 | 118.1 |
| The Proposed Method | | 220.89 | 58.90 | 44.49 | 108.20 | 118.90 | 60.75 | 62.56 | 156.00 | 164.60 | 147.13 | 114.24 |

# 6    Conclusions

We have presented a new method for forecasting the TAIEX based on fuzzy time series and technical indices analysis of the stock market. Investors who rely on technical indices analysts of the stock market strive to precisely predict the price of a stock by looking at its historical prices and other trading variables. Because the proposed method uses both fuzzy time series and technical indices analysis of the stock market to analyze the historical training data in details for forecasting the TAIEX, it can get higher forecasting accuracy rate than the existing methods. The contribution of this paper is that we present a new fuzzy time series forecasting method based on the MACD index, combined with the stochastic line indices (KD indices) to forecast the TAIEX. It gets a higher average forecasting accuracy rate than the method presented in [9] for forecasting the TAIEX.

**Acknowledgements.** This work was supported in part by the National Science Council, Republic of China, under Grant NSC 100-2221-E-011-118-MY2.

# References

1. Appel, G., Hitschler, W.F.: Stock Market Trading Systems. Traders Press (1990)
2. Chen, S.M.: Forecasting Enrollments Based on Fuzzy Time Series. Fuzzy Sets and Systems 81(3), 311–319 (1996)
3. Huarng, K., Yu, T.H.K.: The Application of Neural Networks to Forecast Fuzzy Time Series. Physica A 363(2), 481–491 (2006)
4. Huarng, K.H., Yu, T.H.K., Hsu, Y.W.: A Multivariate Heuristic Model for Fuzzy Time-Series Forecasting. IEEE Transaction on Systems, Man, Cybernetics Part-B: Cybernetics 37(4), 836–846 (2007)
5. Kirkpatrick, C.D., Dahlquist, J.R.: Technical Analysis: The Complete Resource for Financial Market Technicians, 2nd edn. FT Press (2010)
6. Song, Q., Chissom, B.S.: Fuzzy Time Series and Its Model. Fuzzy Sets and Systems 54(3), 269–277 (1993)
7. Song, Q., Chissom, B.S.: Forecasting Enrollments with Fuzzy Time Series—Part I. Fuzzy Sets and Systems 54(1), 1–9 (1993)
8. Song, Q., Chissom, B.S.: Forecasting Enrollments with Fuzzy Time Series—Part II. Fuzzy Sets and Systems 62(1), 1–8 (1994)
9. Yu, H.K.: Weighted Fuzzy Time-Series Models for TAIEX Forecasting. Physica A 349(3–4), 609–624 (2005)
10. Yu, T.H.K., Huarng, K.H.: A Bivariate Fuzzy Time Series Model to Forecast the TAIEX. Expert Systems with Applications 34(4), 2945–2952 (2008)
11. Yu, T.H.K., Huarng, K.H.: Corrigendum to "A Bivariate Fuzzy Time Series Model to Forecast the TAIEX". Expert Systems with Applications 37(7), 5529 (2010)
12. Zadeh, L.A.: Fuzzy sets. Information and Control 8(3), 338–353 (1965)
13. TAIEX Web Site, http://www.twse.com.tw/en/products/indices/tsec/taiex.php

# Hierarchical Gradient Diffusion Algorithm for Wireless Sensor Networks

Hong-Chi Shih[1], Jiun-Huei Ho[2], Bin-Yih Liao[1], and Jeng-Shyang Pan[1,3]

[1] Department of Electronic Engineering, National Kaohsiung University of Applied Sciences,
Kaohsiung City, Taiwan (R.O.C.)
[2] Department of Computer Science and Information Engineering, Cheng Shiu University,
Kaohsiung, Taiwan (R.O.C.)
[3] Innovative Information Industry Research Center (IIIRC), Shenzhen Graduate School,
Harbin Institute of Technology, Shenzhen, China
hqshi@bit.kuas.edu.tw

**Abstract.** In this paper, a hierarchical gradient diffusion algorithm is proposed to solve the transmission problem and the sensor node's loading problem by adding several relay nodes and arranging the sensor node's routing path. The proposed hierarchical gradient diffusion aims to balance sensor node's transmission loading, enhance sensor node's lifetime, and reduce the data package transmission loss rate. According to the experimental results, the proposed algorithm not only reduces power consumption about 12% but also decreases data loss rate by 85.5% and increases active nodes by about 51.7%.

**Keywords:** Wireless Sensor Networks, Gradient Diffusion Algorithm, Ladder Diffusion Algorithm.

## 1 Introduction

Recent advances in micro processing, wireless and battery technology, and new smart sensors have enhanced data processing, wireless communication, and detecting capability. Each sensor node also has limited wireless computational power to process and transfer the sensing live data to the base station or data collection center. Hence, the wireless sensor network usually has a lot of sensor nodes to increase the sensor area and the transmission area.

In general, each sensor node has a low level of power, and its battery power cannot be replenished. If the energy of a sensor node is exhausted, wireless sensor network leaks will appear, and failure nodes will not relay data to the other nodes during transmission processing. Thus, other sensor nodes will be increasingly burdened with transmission processing. Given these issues, it is an important research issue about how to balance the sensor node's load and reduce the energy consumption in wireless sensor networks.

In this paper, a hierarchical gradient diffusion (HGD) algorithm is proposed to solve the transmission problem and the sensor node's loading problem in wireless sensor networks by adding several relay nodes and arranging the sensor node's

M. Ali et al. (Eds.): IEA/AIE 2013, LNAI 7906, pp. 480–489, 2013.

routing. The HGD algorithm balances sensor node's transmission loading, enhances sensor node's lifetime, and reduces the data package transmission loss rate. Moreover, the sensor node can save some backup nodes to reduce the energy for the re-looking routing by our proposed algorithm in case the sensor node's routing is broken. Finally, the HGD algorithm has less data package transmission loss rate and the hop count than the tradition algorithms in our simulate setting. Hence, in addition to balancing the sensor node's loading and reducing the energy consumption, our algorithm can send the data package to the destination node quickly and correctly.

## 2      Related Work

Firstly, we introduce the tradition directed diffusion (DD) algorithm and the ladder diffusion algorithm using ant colony optimization algorithm (LD-ACO). Our proposed algorithm is based on the tradition LD-ACO algorithm and aims to balance the loading and power consumption of sensor node.

### 2.1      Directed Diffusion

Several routing algorithms [2,6,7,10,11,12,13,14] for the wireless sensor network have been sequentially proposed in recent years. C. Intanagonwiwat et al. presented the Directed Diffusion (DD) algorithm [6] in 2003. The DD algorithm aims to reduce transmission counts of data relay for power management. Basically, the DD algorithm is a query driven transmission protocol. The collected data is transmitted only if the collected data fits the query from the destination node, hence the power consumption of the transmission is reduced. In the DD algorithm, the destination node provides its interested queries in the form of attribute-value pairs to the other sensor nodes by broadcasting the interested query packets to the whole network. Subsequently, the sensor nodes only send the collected data back to the destination node in case it fits the interested queries.

In DD algorithm, all of the sensor nodes are bound to a route when broadcasting the interested queries, even if the route is such that it will never be used. In addition, several circle routes, which are built simultaneously when broadcasting the queries, result in wasted power consumption and storage.

### 2.2      A Ladder Diffusion Algorithm Using Ant Colony Optimization for Wireless Sensor Networks

In 2011, H. C. Shih et al. [4] proposed a ladder diffusion algorithm using ant colony optimization for wireless sensor networks (LD-ACO) to solve the routing and energy consumption problem. Moreover, the LD-ACO algorithm can improve the sensor node's lifetime and data transfer effect. The LD algorithm is fast and completely creates the ladder table in each sensor node based on the entire wireless sensor network by issuing the ladder create packet that is created from the sink node. After the ladder diffusion process, the paper proposed an improved ant colony optimization algorithm to balancing the data transmission load, increasing the lifetime of sensor nodes and their transmission efficiency.

In the LD-ACO and DD algorithm, sensor nodes broadcast a package to other nodes for create their routing path according to the package transfer path and their algorithms. After the routing path created, sensor nodes send event information to sink node such as in Fig. 1 when sensor nodes detect an event.

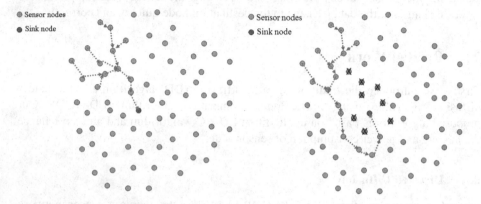

**Fig. 1.** Sensor nodes transfer event data      **Fig. 2.** Sensor nodes transfer event data, and some sensor nodes can't work in the WSN

In the wireless sensor network (WSN), reduction of energy consumption is very important for each sensor node because it can extend WSN lifetime. If some sensor nodes can't work in the WSN, the routing path will break and the detected area will have leaks. Moreover, other sensor nodes can't transfer event data to the sink node, or they need more sensor nodes to give them assistance, as shown in Fig. 2.

In this paper, sensor nodes near the sink node are called "inside node" and others are called "outside node". In Fig. 2, we can find that the outside nodes of WSN need inside nodes to give them assistance when outside nodes transfer data to the sink node. Hence, the inside nodes have huge loading, and their energy will be consumed quickly. After the inside nodes are out of energy, there is no sensor node that can transfer data to the sink node, and the WSN will be out of function.

In this paper, we proposed a hierarchical gradient diffusion (HGD) algorithm to improve traditional DD and LD-ACO algorithms which create a routing path from a single sink node and a few sensor nodes maintain entire WSN lifetime.

## 3     Hierarchical Gradient Diffusion Algorithm

This paper proposed a hierarchical gradient diffusion (HGD) algorithm based on the LD-ACO algorithm. The HGD algorithm adds some RS nodes which are relay nodes of the sink node and they can broadcast the grade creating package as the sink node. Sensor nodes can transfer data to RS nodes or the sink node to balance sensor node loading, reducing the energy consumption and enhancing WSN lifetime according to the HGD algorithm. Moreover, sensor nodes can save some backup nodes in its routing table to reduce the energy for the re-looking routing by our proposed algorithm in case the sensor node's routing is broken. The RS node s and HGD algorithm are introduced as follow.

In Fig. 3, the HGD algorithm adds some RS nodes (RS 0 ~ RS 3) in the WSN. The RS node is similar to the sink node because it doesn't have any detection ability; they can just be a data collection center for sensor nodes as well as the sink node. Moreover, the RS nodes have large transmission scale compared with sensor nodes, and they have enough energy to transfer data to real data collection center (Sink Node). Hence, events can be detected and transferred to RS nodes or the sink node by sensor node. If an RS node receives an event data, the event data will be transferred to the sink node from RS node. Hence, sensor node, RS node, and sink node become a hierarchical structure in the HGD algorithm.

In HGD algorithm, the grade creating package will be broadcasted from the sink node and RS nodes. Firstly, the sink node broadcasts grade-creating packages to create a main routing table for sensor nodes. Then, RS node broadcasts grade-creating packages again to create a backup routing table. Moreover, sensor nodes can change their main routing table and backup routing table according to the grade information received from grade-creating packages. Thus, the routing path can be cut down and the transmission loading can be reduced when the routing path from sensor node to RS node is shorter than to sink node.

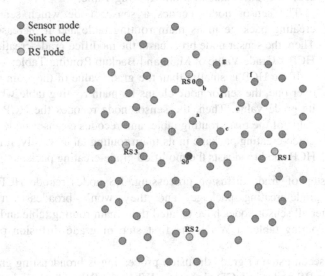

**Fig. 3.** The hierarchical gradient diffusion algorithm for wireless sensor network

Firstly, the sink node broadcasts the grade-creating package and the package format as shown in Table 1. In the grade-creating package format, the SRS mean sink node holds the value is 0, otherwise it's a grade value of RS node. The HCP means how many hop counts a sensor can transfer event data to the sink node or RS node. The DN means the destination node, and the destination is the sink node or RS node.

**Table 1.** Grade-Creating Package

| $SRS = \begin{cases} Sink\ Node & ,\quad 0 \\ Grade\ Value\ of\ RS\ node & ,otherwise \end{cases}$ | HCP | DN |
|---|---|---|

After a sensor node receives a grade-creating package, there are several different cases as follows according to its main routing and the grade-creating package.

1. The main routing table is empty:

    A sensor node recodes the HCP in its grade value of the main routing table, and recodes the sensor node that sent the grade-creating package in its main routing.

2. The main routing table isn't empty:

    a. HCP > Grade Value of Main Routing Table:

    If the HCP is bigger than the grade value of the main routing table over one, the sensor node doesn't do anything. On the other hand, if the grade value of the main routing table plus 1 is equal to the HCP, the sensor node records the sensor node that sent the grade-creating package in its set of neighbor nodes. Then, the sensor node does nothing.

    b. HCP = Grade Value of Main Routing Table:

    The sensor node recodes a sensor node which sent the grade-creating package in its main routing table and it increases the HCP. Then, the sensor node broadcasts the modified grade-creating package.

    c. HCP < Grade Value of Main and Backup Routing Table:

    If the HCP is smaller than the grade value of the main routing table over one, the sensor node cleans its main routing table which contains its grade value. Then, the sensor node recodes the HCP in its grade value of the main routing table, and recodes a sensor node that sent the grade-creating package in its main routing table. Lastly, it increases the HCP and broadcasts the modified grade-creating package.

At the first step of grade diffusion processing, RS nodes recode HCP when they receive the grade-creating package, and they won't broadcast grade-creating packages. After all sensor nodes have created their main routing table and grade value of the main routing table in WSN, the first step of grade diffusion processing is completed.

Then, the second step of grade diffusion processing is broadcasting grade-creating packages from RS nodes in HGD algorithm. When the RS nodes broadcast the grade-creating package, the package's SRS is RS node's grade value, the initial HCP is 0, and the DN is the RS node.

RS nodes broadcast grade-creating packages as sink node, but there are two different cases when sensor node receives the grade-creating package from RS nodes and the sensor nodes haven't created the backup routing table.

1.  HCP >= Grade Value of Main Routing Table:

    The sensor node records the HCP and the sensor node that sent the grade-creating package in its backup routing table because the path of the main routing table is shorter than the backup's or equal to it. Then the sensor node increases the HCP and broadcasts the modified grade-creating package.

2.  HCP < Grade Value of Main Routing Table:

    The sensor node moves the main routing table to the backup routing table and cleans the main routing. Then, the sensor node recodes the HCP and the sensor node that sent the grade-creating package in its main routing table because the RS node is closer than the sink node. After, the sensor node increases the HCP and broadcasts the modified grade-creating package.

If sensor nodes receive the grade-creating package from an RS node, it then creates the backup routing table. Then, the sensor node needs to compare the HCP with its grade value of main and backup routing table as follows.

1.  HCP > Grade Value of Main and Backup Routing Table:

    If the HCP is bigger than the grade value of the main and backup routing table over one, the sensor node doesn't do anything. On the other hand, if the grade value of the main or backup routing table plus 1 is equal to the HCP, the sensor node records the sensor node that sent the grade-creating package in its set of neighbor nodes. Then, the sensor node doesn't do anything.

2.  HCP < Grade Value of Main or Backup Routing Table:

    a.  HCP < Grade Value of Main Routing Table:

        The sensor node clean its main routing table which contains its grade value, then the sensor node records the HCP and the sensor node that sent the grade-creating package in its main routing table.

    b.  HCP < Grade Value of Backup Routing Table:

        The sensor node cleans its backup routing table which contains its grade value, then the sensor node records the HCP and the sensor node that sent the grade-creating package in its backup routing table.

        After, if the grade value of the main routing table is bigger than that of the backup's in the sensor node, the sensor node switches its main and backup routing table. This is because the main routing table is more important and we hope its grade value is smaller than the backup's. After that, the sensor node increases the HCP and broadcasts the modified grade-creating package.

3.  HCP = Grade Value of Main or Backup Routing Table:

    a.  HCP = Grade Value of Main Routing Table:

        The sensor node records a sensor node which sent the grade creating package in its main routing table, then the sensor node increase the HCP and broadcast the modified grade creating package.

    b.  HCP=Grade Value of Backup Routing Table:

        The sensor node records a sensor node which sent the grade-creating package in its backup routing table, then the sensor node increases the HCP and broadcasts the modified grade-creating package.

If the grade values of the main and backup routing tables are equal in the sensor node, the sensor node will select the main routing table in priority. This is because the main routing table is more important and we hope it has more backup sensor nodes in it.

After the hierarchical gradient diffusion process terminates, a sensor node selects a sensor node in its routing table to transfer event to the sink node or RS node when sensor node detects an event. No matter how many sensor nodes are in the main or backup routing tables, the HCD algorithm selects a sensor node according to equation 1.

$$U(k) = \frac{P_k^{-1}}{\sum_{i=1}^{k} P_i^{-1}} \quad , \quad k \in J \tag{1}$$

In equation 1, the J is a set of the sensor nodes for main or backup routing tables. The $P_i$ is the overload value of the $i^{th}$ sensor node ode and its initial value is 1 in the main and backup routing tables. The $U(k)$ is a probability of the sensor node to be selected to transfer data by $k^{th}$ sensor. When sensor node is selected to be a relay node according to equation 1, the overload value of the sensor node will be increased. Hence, equation 1 can to promote the probability of nodes when the nodes are seldom chosen.

# 4    Simulation and Analysis

In this session, we simulate the hierarchical gradient diffusion algorithm according to session 3. The experiment is designed based on a 3-dimensional space, which is defined with 100*100*100 units, and the scale of coordinate axis for each dimension is limited from 0 to 100. The radio ranges (transmission range) of the nodes were set to 15 units. In each of these simulations, sensor nodes were distributed uniformly over the space. There are three sensor nodes randomly distributed in a 10*10*10 space, and the Euclidean distance is 2 units at least between two sensor nodes. Therefore, there are 3000 sensor nodes in the 3-dimesional wireless sensor network simulator and the center node is the destination node. The data packages were exchanged between random source/destination pairs with 90,000-event data packages. In the simulation, the energies of each sensor node are set at 3600mw and they consume 1.6mw of energy when they transfer data.

In addition, we add some RS nodes at (25,25,25), (25,25,75), (75,25,75), (75,25,25), (25,75,25), (25,75,75), (75,75,75), and (75,75,25) for the HGD algorithm. When the HGD algorithm adds some RS nodes, it is hoped that this can reduce the hop count from outside nodes. In our simulation space, the maximum grade value of the main routing table is 6 and the grade value of RS nodes is 3 after the first step of grade diffusion processing. After the hierarchical gradient diffusion process terminates, some sensor nodes whose grade value of main routing table was bigger than 3, will cut down their routing path because they can transfer events to RS nodes.

Firstly, we implemented the HGD, DD, and LD-ACO algorithms, and they compared the active sensor nodes and energy consumption after 90,000 events appeared in Fig.. 4 and 5. The active nodes mean that the sensor node has enough energy to transfer data to other nodes, but some sensor nodes will be deleted from active nodes if their routing table doesn't have any sensor node that can be selected to be a relay node and they aren't in the routing tables of the other sensor nodes.

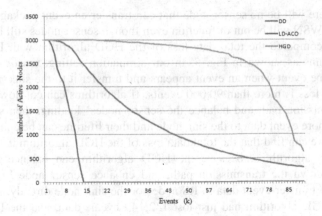

**Fig. 4.** Number of active nodes

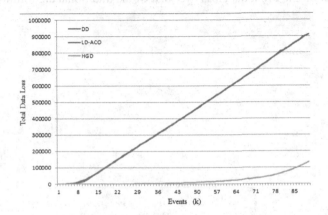

**Fig. 5.** Total data loss

After 90,000 events appeared, the HGD has 630 sensor nodes available, but the DD and LD-ACO just has 305 and 0 sensor nodes available in Fig. 4. The HCD has the most active sensor nodes compared with DD and LD-ACO because the HGD algorithm has main and backup routing tables for each sensor node. Moreover, sensor nodes can switch their main and backup routing tables, then sensor nodes having the least hop count can transfer data to the sink node or RS nodes. Besides, sensor nodes can select relay nodes in the set of neighboring nodes when the sensor node's main and backup routing tables don't have any sensor node available. Hence, the HGD algorithm can enhance at least 51.7% sensor nodes lift time over that of the traditional algorithms.

In Fig. 4, the active sensor node is zero in the LD-ACO algorithm after 14,000 events appeared because LD-ACO calculates the optimum path by ACO and sends event packages from high-grade nodes (outside nodes) to low-grade nodes (inside nodes) always. Sensor node's routing table just contains lower grade nodes than in the LD-ACO algorithm. Therefore, if the sensor nodes whose grade value is one are out of energy, there will be no sensor node that can transfer an event package to the sink node and the WSN will be out of function even though sensor nodes still have energy.

Then, we compared the total data loss of the HGD algorithm with DD and LD-ACO algorithms as shown in Fig. 5. In our simulation setting, sensor nodes might detect the same event when an event appears and transfer it to the sink node. Hence, the total data loss is more than 90,000 events. If algorithms can cut down the sensor node's transmission path and balance the sensor nodes' loading, then sensor nodes can transfer more event data to the sink node and their lifetime can be enhanced.

In Fig. 5, we can find that the total data loss of the HGD algorithm is less then DD and LD-ACO algorithms, because the HGD algorithm can balance sensor node loading, cut down the transmission path, and enhance sensor node lifetime by its algorithm and transfer event data to RS nodes or the sink node directly. After 90,000 events, the HGD algorithm had just lost 132,149 events data, and the DD and LD-ACO algorithms lost 912,462 and 913,450 respectively. Our HGD algorithm can decrease about 85.5% event data loss over that of the traditional algorithms.

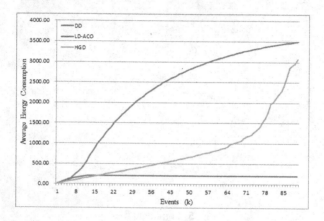

**Fig. 6.** Average energy consumption

Lastly, we compared the average energy consumption of the HGD algorithm with DD and LD-ACO algorithms as shown in Fig. 6. The average energy consumption of the DD algorithm rise suddenly after 7,000 events in Fig. 6 because the inside nodes are totally out of energy. In the DD algorithm, a sensor node can transfer event data to other sensor nodes till it's out of energy or doesn't have any neighboring nodes, and sensor node average total energy consumption is 3495.2 mw after 90,000 events. The HGD algorithm just consumes about 3077.3 mw after 90,000 events because the algorithm can balance sensor node loading and cut down the transmission path by its main and backup routing. Hence, the HGD algorithm can reduce energy consumption by out 12% compared to the DD algorithm. In Fig. 6, the LD-ACO algorithm just consumes 204.29 mw after 90,000 events, but the WSN is out of function as reiterated previously, after 14,000 events. After 14,000 events, sensor nodes don't have any available nodes in their routing table. That means sensor nodes just have energy, but they can't do anything because their routing table is empty.

# 5     Conclusion

In the real wireless sensor network, the sensor nodes use the battery to provide the power supply. Hence, its energy is limited. According to the experimental result, the HGD algorithm can increase active nodes by 51.7%, save energy by 12%, and decrease data loss rate by 85.5% compared with traditional algorithms. Hence HGD can balance sensor nodes transmission loading, enhance sensor node lifetime, and reduce data package transmission loss rate by adding some RS nodes. Besides, the HGD algorithm can send the data to destination nodes quickly and correctly.

# References

1. Carballido, J.A., Ponzoni, I., Brignole, N.B.: A graph-based genetic algorithm for sensor network design. Information Sciences 177, 5091–5102 (2007)
2. Corson, S., Macker, J.: Mobile Ad Hoc Networking (MANET): Routing Protocol Performance Issues and Evaluation Considerations. RFC 2501 (1999)
3. Hea, Z., Lee, B.S., Wang, X.S.: Aggregation in sensor networks with a user-provided quality of service goal. Information Sciences 178, 2128–2149 (2008)
4. Ho, J.-H., Shih, H.-C., Liao, B.-Y., Chu, S.-C.: A Ladder Diffusion Algorithm Using Ant Colony Optimization for Wireless Sensor Networks. Information Sciences, 204–212 (2012)
5. Hong, T.-P., Wu, C.-H.: An Improved Weighted Clustering Algorithm for Determination of Application Nodes in Heterogeneous Sensor Networks. Journal of Information Hiding and Multimedia Signal Processing 2, 173–184 (2011)
6. Intanagonwiwat, C., Govindan, R., Estrin, D., Heidemann, J., Silva, F.: Directed Diffusion for Wireless Sensor Networking. IEEE/ACM Transactions on Networking 11, 2–16 (2003)
7. Liao, W.H., Kao, Y., Fan, C.M.: Data Aggregation in Wireless Sensor Networks Using Ant Colony Algorithm. Journal of Network and Computer Applications 31, 387–401 (2008)
8. Pan, J., Hou, Y., Cai, L., Shi, Y., Shen, X.: Topology control for wireless sensor networks. In: The Ninth ACM International Conference on Mobile Computing and Networking, pp. 286–299 (2003)
9. Perkins, C.E., Royer, E.: Ad Hoc On-Demand Distance Vector Routing. In: Proceedings of IEEE WMCSA, pp. 90–100 (1999)
10. Royer, E.M., Toh, C.-K.: A Review of Current Routing Protocols for Ad-Hoc Mobile Networks. IEEE Personal Communications 6, 46–55 (1999)
11. Shih, H.-C., Chu, S.-C., Roddick, J., Ho, J.-H., Liao, B.-Y., Pan, J.-S.: A Reduce Identical Event Transmission Algorithm for Wireless Sensor Networks. In: Proceedings of the Third International Conference on Intelligent Human Computer Interaction (2011)
12. Hong, T.-P., Wu, C.-H.: An Improved Weighted Clustering Algorithm for Determination of Application Nodes in Heterogeneous Sensor Networks. Journal of Information Hiding and Multimedia Signal Processing 2(2), 173–184 (2011)
13. Liu, T.-H., Yi, S.-C., Wang, X.-W.: A Fault Management Protocol for Low-Energy and Efficient Wireless Sensor Networks. Journal of Information Hiding and Multimedia Signal Processing 4(1), 34–45 (2013)
14. Chen, C.-M., Lin, Y.-H., Chen, Y.-H., Sun, H.-M.: SASHIMI: Secure Aggregation via Successively Hierarchical Inspecting of Message Integrity on WSN. Journal of Information Hiding and Multimedia Signal Processing 4(1), 57–72 (2013)

# Constructing a Diet Recommendation System Based on Fuzzy Rules and Knapsack Method

Rung-Ching Chen, Yung-Da Lin, Chia-Ming Tsai, and Huiqin Jiang

Department of Information Management, Chaoyang University of Technology
68, Jifong East Road, Wufong Dist., Taichung City 41349, Taiwan (R.O.C.)
crching@cyut.edu.tw

**Abstract.** Many people suffer from three chronic diseases(diabetes, hypertension, cholesterol), and they often use search engine to collect related information. However, most of dietary information on the networks is not convenient for users to collect about the diet recommendations. In this paper, a diet recommendation system is suggested which can recommend a rational diet for users. We design a diet recommendation system which has the expert knowledge of three high chronic diseases. We use Protégé to establish ontology and OWL DL to construct the structure of knowledge. The system uses fuzzy logic as a guide prior to inference. According to the patient's health information, the system infers daily calories requirement, and then use JENA inference device and JENA rule format to build our knowledge of the rules. The Knapsack-like algorithm is used to recommend suitable foods for users. The system was evaluated by nutritionists to prove it is effective.

**Keywords:** Ontology languages, Dietary recommendations, Rules inference, Knapsack problem, Fuzzy set.

## 1    Introduction

Richer life brings people abundant foods and meanwhile brings a greater chance of developing certain diet-related diseases. Due to unbalanced diet, more and more people suffer from chronic diseases such as cholesterol, hyperlipidemia, and diabetes etc. So people pay more attention to healthy information including how to avoid disease, how to eat healthily and how to cure diseases. However, some diseases are difficult to cure, such as diabetes which relies on not only drug treatment but also long-term dietary control to reduce disease's injury.

Diabetes, hypertension and high cholesterol are among the top 10 causes of death[1]. In the list of causes of death, high cholesterol-related disease such as heart disease and cerebrovascular disease ranked respectively the second and third, diabetes ranked the fifth, and hypertensive diseases ranked ninth. In Taiwan, more than 60% elderly people are suffering from "three high chronic diseases", which are diabetes, hypertension and high cholesterol. In fact, these diseases can be controlled by diet. Therefore rational diet is very important. Until now there are not many researches focusing on diet recommendation but it is a meaningful work.

M. Ali et al. (Eds.): IEA/AIE 2013, LNAI 7906, pp. 490–500, 2013.

In this paper, a diet recommendation system is suggested which can recommend a rational diet for users. This system infers suitable diet on the basis of six parameters: height, weight, activity levels, kidney function, hypertension and hyperlipidemia. Activity levels are divided into four levels: confined to bed, low, medium and high activity levels. Kidney function parts are divided into: the chronic renal insufficiency and the nephritic syndrome. The hypertension parts are divided into: complex hypertension and simply hypertension. The hyperlipidemia parts are divided into: the high triglycerides and high cholesterol. The kidney functions parts are divided into: the chronic renal insufficiency and the nephritic syndrome. The system will recommend diet for users according to these six parameters.

The rest of the paper is organized as the follows. Section 2 introduces the concept of ontology, fuzzy, JENA, body information and the diet of the patient, Knapsack Problem and Recommender System. The methodology is described in Section 3. Section 4 is about the implementations and experiments of the system. The conclusions and future works are discussed in Section 5.

## 2    Literature Review

Recommender system is able to provide users decision information. The main methods of the recommendation system have several categories. The most popular filter-based are collaborative filtering and content-based filtering. Collaborative filtering is given by the human experience but content-based filtering is based on characteristic of content.

In recent years, ontology is often used in the field of computer science and artificial intelligence[1-4]. In general, the major elements of the ontology are: class, slot, instance and relationship[5]. OWL (Web Ontology Language) is the description language framework that is proposed by W3C. It is based on XML and uses the syntax of RDF. OWL can be divided into three types: OWL Full, OWL DL, OWL Lite [6][7][8]. In this paper, Protégé and OWL DL will be used to construct our system. Li et al. [9] proposed an automatic-food ontology construction method based on the information of Department of Health, Executive Yuan, R.O.C.. They used twenty-four types of nutrients provided by experts to give those weights and classification exactly, and they established a diet ontology.

The fuzzy theory was first proposed by LA Zadeh[10]. Fuzzy method solves the problems that can't be expressed by the classical logic, true or false, which is often difficult to represent concepts of real world [11]. Lee et al. [16] used Type 2 fuzzy to generate personalized diabetes diet recommended applications. They used diet ontology and Type 2 fuzzy to set applications. Six major food groups are used to make finer distinctions to achieve the recommendations of diet. And then, the system combines the ontology of personal information with the Type 2 fuzzy sets to give the information of eating more or less every day. Finally, diet plan can be given to users. Lee's method is able to recommend user' daily insufficient nutrition but it is not entirely beneficial for users.

JENA is introduced by HP Labs, which is easier for developing applications on Web semantic reasoning [17]. JENA supports a variety of different storage formats [17], such as RDF, RDFS, OWL, and the SPARQL [18]. JENA also supports rule-based inference engines [18][19]. Through JENA inference rules, we can find many previously unknown associations. The relationships of OWL and JENA are the relative relationships between ontology and knowledge logic.

Knapsack problem is a NP-Complete problem. The knapsack problem is primarily used to solve optimal selection problem. If we have a group of items, which have values and weights. The backpack container can accommodate a fixed weight, how to select these items into backpack and to get the max value. 0/1 knapsack problem: assume that the items in the group are indivisible, and to get the optimal solutions of combination [20], as shown in formula 1.

$$\text{Maximize } \sum_{j=1}^{n} p_j x_j \text{ and Maximize } \sum_{j=1}^{n} w_j x_j \leq W, \; x_j \in \{0,1\}, \tag{1}$$

where n represents the number of selected items; $p_j$ represents the profiles of the $i th$ item; $x_j$ means whether to put the jth item into the backpack, $x_j=0$ indicates to throw the i item, $x_j=1$ indicates to select the items; $w_j$ is the item i the weight; W is the weight of the largest backpack tolerate.

Greedy knapsack problem is an extension of the 0/1 knapsack problem, which assume that the items in the group can be split and get the optimal solutions of combination [22] as shown in formula 2.

$$\text{Maximize } \sum_{j=1}^{n} p_j x_j \text{ and Maximize } \sum_{j=1}^{n} w_j x_j \leq W, \; 1 \leq x_j \leq 0 \tag{2}$$

In formula 2, n represents the number of selected items, the value of $p_j$ on behalf of the profiles of i item; $x_j$ is on behalf of the item i; $x_j=0$ indicates the i item does not be selected; $x_j=1$ is the i items selected; $w_j$ is the weight of i item; W is the largest weight can be in the backpack.

Branch-and-Bound Strategy knapsack uses the upper bound of the optimal solutions of 0/1 knapsack. The lower bound is used by the optimal solutions of 0/1 knapsack and tree structure. This algorithm first selects a node to growing, that the node has bigger lower bound to get the optimal solutions [22][24]. In this paper, we use this method to solve the knapsack problem.

# 3     System Architecture and Methodology

The nutritionist calculates the patient's body mass index (BMI) based on the patient's height and weight as shown in formula 3[25].

$$\text{BMI} = \text{BW}/\text{BH}^2 \tag{3}$$

In formula 3, the unit of body weight (BW) is kilograms, and the unit of body height (BH) is meters. The BMI is to test whether BW is too heavy. The ideal body weight

(IBW) can be calculated by formula 4. Normally, the values of BMI are between 18.5 and 23.9. If the user's body weight is over 23.9, she/he is "overweight". On the other hand, if their weight is less than 18.5, they are too light. For convenience to calculate the BMI the system uses 22 as the base to calculate formula 4[26].

$$IBW = BH^2 * 22 \tag{4}$$

In order to suggest the suitable food for chronic diseases, we propose a diet recommender system, as shown in Figure 1 which has three major parts as described in the follows. This system used the Taiwanese snacks Nutrition Analysis to setup diet ontology and used fuzzy method and JENA inference for recommended services. In addition, the Knapsack-like algorithm will combine the results.

## A. Fuzzy Reasoning

In this research, the system will get the parameter of height, weight, activity levels, renal function, hypertension, high cholesterol and user's preference by user's interfaces. And then, divide the BMI into three blocks on classical logic, and use Table 1 to calculate calories of IBW's daily need. The calculation method above uses the classical logic to calculate daily needed calories of users. In the classical logic, if the weight of user alters 1 kg, the weight of user possibly changes from standard weight to overweight.

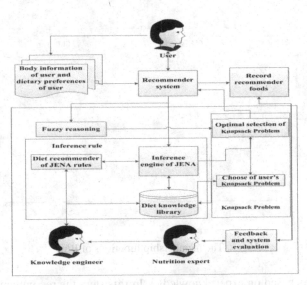

**Fig. 1.** The architecture of recommender system

For example the daily calories of user's need whose body weight is 60 kg is about 300 kilocalories, which is not reasonable and not conducive to control the patients' disease.

Therefore, we use the concept of fuzzy logic to solve the problems above. We need to fuzzy input parameter BMI .The default range of BMI is between 15 and 30. The membership function of the "BMI_low" is trapezoid, the membership function of "BMI_stand" is triangle and the membership function of "BMI_over" is trapezoid. The membership functions are shown in formula 5, formula 6 and formula 7 where 'x' is the value of an user's BMI. The corresponding membership functions were depicted in Figure 2.

$$f_{BMI-low^{(x)}} = \begin{cases} 1 & x < 17 \\ (x-17)/4 & 17 < x < 21 \\ 0 & x > 21 \end{cases} \tag{5}$$

$$f_{BMI-stand^{(x)}} = \begin{cases} 0 & x < 17 \\ (x-17)/4 & 17 < x < 21 \\ 1 & x = 21 \\ (25-x)/4 & 21 < x < 25 \\ 0 & x > 25 \end{cases} \tag{6}$$

$$f_{BMI-over^{(x)}} == \begin{cases} 0 & x < 21 \\ (x-21)/4 & 21 < x < 25 \\ 1 & x > 25 \end{cases} \tag{7}$$

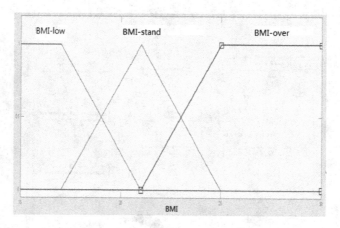

**Fig. 2.** The membership functions of BMI

Fuzzy rules is based on expert knowledge. In this step, the recommendation system requires seven parameters: height, weight, activity levels, kidney function, hypertension, hyperlipidemia and food preferences. Table I shows calorie needs of different users decided by two parameters BMI and activity level. We can see from this table BMI is divide into three types: overweight, standard and less weight. The values of overweight, standard and less weight are decided by the two thresholds: 18.5 and

23.9. The BMI value of overweight is greater than 23.9 and less weight is less than 18.5. Additionally, the standard BMI value is between 18.5 and 23.9. The activity level is divided into four levels: confined to bed, low, medium and high. Table 1 shows the needed calories per day of different people. The calories are calculated mainly depending on parameters BMI and activity level.

In this example, we can't give exact numbers to describe activity level and we just use linguistic label e.g. "low", "medium" and "high" to describe the different level. We also just give two thresholds to differentiate "over weight", "less weight" and "standard". So we consider these ambiguous parameters to use the concept of fuzzy logic which are applied here.

**Table 1.** Use activity levels and BMI to find requirement kilo calories per day (c/k)

| BMI / Activity level | Over weight | Standard | Less weight |
|---|---|---|---|
| confined to bed | 15-20 | 20-25 | 25-30 |
| low | 20-25 | 25-30 | 30-35 |
| medium | 25-30 | 30-35 | 35-40 |
| high | 30-35 | 35-40 | 40-45 |

B. Ontology and JENA Inference

The system constructed foods and patient ontology. There are three layers of food ontology: classification layer, ingredients layer and foods layer. In classification layer, the system has 18 main categories and four properties including protein(g), fat (g), carbohydrates (g) and calorie (kilocalorie). In ingredients layer, the system uses the ingredients of Taiwanese snacks to construct. The food layer uses the food entity of Taiwanese snacks Nutrition Analysis shown in Figure 3.

For patient ontology, in order to use JENA rule to map foods and users, the entity needs the following attributes: height, weight, activity levels, kidney function, hypertension, hyperlipidemia and preferences of food. Those information will be used in fuzzy inference and JEAN inference.

The JENA inference is composed of JENA rules and JENA inference. JENA inference is used to establish the most effective relationships between user's entities and foods entities. The nutritionist will give different recommendation to different users. Those knowledge rules will be obtained from nutritionist. There are the rules examples in JENA file. For example, rule: "rule1: (?a eg:p ?b) (?b eg:p ?c) -> (?a eg:p ?c)", this rule is used to set "eg: p" attributes between "a" and "c"; if "eg: p" attribute is between "a" and "b" and "eg: p" attribute is between "b" and "c", it will infer "eg:p" attribute is between "a" and "c".

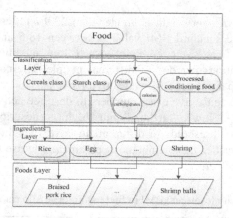

**Fig. 3.** The food ontology

The rules were designed by domain experts according to the body state and diet information. There are 13 rules in our JENA rule system totally.

### C. Knapsack Method

The knapsack algorithm is separated into two parts. The first part is to select all food of user interests. The second part uses a likely- knapsack problem to combine the fuzzy results and JENA results to recommend the better group of foods. Finally, the group of recommender results will record in our record library. The diet expert will assess and feedback to our system to let our system has more effective rules and more complete diet ontology.

The upper bound come from the 0/1 knapsack problem and the lower bound come from greedy knapsack problem in the original algorithm. In this system, we use the vector of the computing calories per day ± n% to control the growing of the tree of algorithm, where n is the range of system mistake.

Finally, the result of JENA system is used to give recommendation. If the result of food is "recommend", the weight of food is 1 otherwise the weight of food is 0. Formula 6 is a function to calculate the weight of the meals, where p is the ratio of meal's recommendation. The final result is ordered by the value of p. If p values are the same, the system orders the calories which are close to target calories. The target calories are fuzzy reasoning's result.

$$p = \frac{\sum (\text{protein} + \text{fat} + \text{carbohydrates} + \text{calories}) * jena\ weight}{tatol\ \text{protein} + tatol\ \text{fat} + tatol\ \text{carbohydrates} + tatol\ \text{calories}} \tag{6}$$

Suppose the user A's height is 150 cm, weight is 50 kg, the activity level is 7, and has not other diseases. With these data ,We can know that BMI value of the user is 22.2 ,the calories per IBW are 31.2 and the calories per day are 1560 by our fuzzy agent. The three meals proportion of 30%, 30% and 40% respectively. the proportion of protein, fat and carbohydrates are 15%, 30% and 55% respectively. The calories per gram protein have 4 calories, fat per gram protein has 9 calories and carbohydrate per gram has 4 calories. The recommendation foods of user A are listed in the Table 2. In Table 2, the field "JENA weight" is the result of JENA inference, if the value of "JENA weight" is 1, the food is recommendation to users.

**Table 2.** The example of food items list

| items | protein | fat | Carbohy-drates | calo-ries | JENA weight(J) |
|-------|---------|-----|----------------|-----------|----------------|
| A | 6.2 | 11.9 | 53.2 | 344.7 | 1 |
| B | 8.7 | 8.5 | 20.1 | 191.7 | 1 |
| C | 10.4 | 8.4 | 33.5 | 251.2 | 1 |
| D | 16.7 | 16.5 | 59.1 | 415.7 | 1 |

## 4    The Experiment and Evaluation

The system uses ASUS X52J with 2.13GHz of CPU, 3GB memory, WINDOWS 7, jdk1.6.0 and JENA 2.6.3. In our experiments there are 15 volunteers. We input the user's body and food information into our system[27].

After these experiments we give the results to nutritionist to evaluate our system. The evaluation results are shown in Table 3. Formula 8 is used to calculate the accuracy of the recommender foods. If the meal is right, the value is 1, otherwise the value is 0. For example, the user 1 is recommender 5 meals and the meals prove values are 1, 0, 0, 1 and 0 respectively. So the accuracy value is (1+0+0+1+0)/5=0.4. The formula 9 is to test the total accuracy. The system total accuracy is 72%.

$$accuracy_i = \frac{\sum meal\ prove}{the\ number of\ recommender\ meals} \tag{8}$$

$$total\ accuracy = \frac{\sum accuracy_i}{the\ number\ of\ users} \tag{9}$$

In our experiments, the user 7, 12, 13, 14 and 15 have not recommended meals because they choose the foods that are not suitable for themselves. For example, the user 7 has high cholesterol problems but his interesting foods are Noodles with goose, Pig blood soup and Thick soup noodles. Those foods have the broth; it is not suitable for the high cholesterol users. So the recommended result is proved.

In our test system, Shuijian bao is not suitable for users, who had the hyperlipidemia. So we need to change the ontology's architecture of classification layer, to recommend meals to users. After the evaluation of nutritionist, that problem of the system is that the Shuijian bao is not suitable for users who had the hyperlipidemia but the other recommended food and meals are suitable for users. In the 18 major

Table 3. The evaluation result of System recommender

| User ID | Evaluation |
|---------|------------|
| 1 | 0.4 |
| 2 | 1 |
| 3 | 1 |
| 4 | 1 |
| 5 | 0 |
| 6 | 1 |
| 7 | 1 |
| 8 | 0.4 |
| 9 | 0 |
| 10 | 1 |
| 11 | 0 |
| 12 | 1 |
| 13 | 1 |
| 14 | 1 |
| 15 | 1 |

categories Health, Executive Yuan (R.O.C), it does not have the class of fat. So we will need to check if other class is required which is not in the 18 major categories, and then add it to our ontology. If we add other class which is required but not in the 18 major categories, the system will promotes the total accuracy to 100%.

The number of selected foods had better be in a range for system to reasoning. If users select too little, the information will too little for reasoning. If users select too much, the system performances will be decreased. The suggestion is to select 6 to 10 foods.

## 5    Conclusions and Future Works

Nowadays patients hope to calculate the calories of their daily needs. Nutritionists divide the calories into the three major nutrients which is not enough, because the patients want to know what kinds of food they are suitable to eat. We propose a system to reduce the cost of health care resources of the chronic diseases. The diet knowledge ontology can reuse on the system which can also save the time of the user to find better foods recommendation. The system uses the fuzzy and JENA rules of inference and the knapsack problem algorithm to calculate the requirement calories and to recommend the food for users. The foods are JENA inference results, and the knapsack limit is fuzzy inference result. The interface is friendly which can allow nutritionist to recommend diet for users. In the experiment and evaluation, the users can get the right information of recommendation by the system. The system' total accuracy is 72%, if we add a new class fat in the 18 major categories, the system will promote the total accuracy to 100%. This research will help the patient suffering from three high chronic diseases to get diet recommendations of nutrition expert advice which is more fast than old way. The users also can select different interesting foods on the system.

In the future our work will focus on the sum of meals required for users in a day, so we need to change the number of meals in daily, and let users can give the weight of their meals. We also hope more nutritionists to evaluate our system and to compare accuracy with other diet recommender system to make the system more comprehensive.

**Acknowledgement.** This study is supported by National Science Council, Taiwan.

# References

[1] Department of Health, Executive Yuan, R.O.C (TAIWAN) - The main cause of death, http://www.doh.gov.tw/statistic/eBAS/

[2] Guo, Q.L., Zhang, M.: Semantic Information Integration and Question Answering based on Pervasive Agent Ontology. Expert Systems with Applications 36(6), 10068–10077 (2009)

[3] Jepsen, T.C.: Just What Is an Ontology, Anyway? IEEE Computer Society 11(5), 22–27 (2009)

[4] Labrou, Y., Finin, T.: Yahoo! as an Ontology: using Yahoo! Categories to Describe Documents. In: Conference on Information and Knowledge Management, pp. 180–187 (1999)

[5] Wang, M.H., Lee, C.S., Hsieh, K.L., Hsu, C.Y., Acampora, G., Chang, C.C.: Ontology-Based Multi-Agents for Intelligent Healthcare Applications. Journal of Ambient Intelligence and Humanized Computing 1(2), 111–131 (2011)

[6] Meditskos, G., Bassiliades, N.: A Rule-Based Object-Oriented OWL Reasoner. IEEE Transactions on Knowledge and Data Engineering 20(3), 397–410 (2008)

[7] Sánchez, D., Batet, M., Isern, D.: Ontology-Based Information Content Computation. Knowledge-Based Systems 24(2), 297–303 (2011)

[8] OWL Web Ontology Language, http://www.w3.org/TR/owl-features/

[9] Li, H.C., Ko, W.M.: Automated Food Ontology Construction Mechanism for Diabetes Diet Care. In: Proceedings of the Sixth International Conference on Machine Learning and Cybernetics, Hong Kong, pp. 2953–2958 (2007)

[10] Zadeh, L.A.: Fuzzy sets. Information and Control 8(3), 338–353 (1965)

[11] Hájek, P.: What is Mathematical Fuzzy Logic. Fuzzy Sets and Systems 157(5), 597–603 (2006)

[12] Lee, C.S., Wang, M.H.: A Fuzzy Expert System for Diabetes Decision Support Application. IEEE Transactions on Systems 41(1), 139–153 (2011)

[13] Ma, J., Chen, S., Xu, Y.: Fuzzy Logic from the Viewpoint of Machine Intelligence. Fuzzy Sets and Systems 157(5), 628–634 (2006)

[14] Novák, V.: Which Logic is the Real Fuzzy Logic? Fuzzy Sets and Systems 157(5), 635–641 (2006)

[15] Zadeh, L.A.: Toward Extended Fuzzy Logic a First Step. Fuzzy Sets and Systems 16(21), 3175–3181 (2009)

[16] Lee, C.S., Wang, M.H., Hagras, H.: A Type-2 Fuzzy Ontology and Its Application to Personal Diabetic-Diet Recommendation. IEEE Transactions on Fuzzy Systems 18(2), 374–385 (2010)

[17] JENA, http://JENA.sourceforge.net/

[18] JENA-inference, http://JENA.sourceforge.net/inference/index.html

[19] McBride, B.: JENA: a Semantic Web Toolkit. IEEE Computer Society 6(6), 55–59 (2002)

[20] Gorsik, J., Paquete, L., Pedrosa, F.: Greedy Algorithms for a Class of Knapsack Problems With Binary Weights. Computer & Operations Research 39(3), 498–511 (2012)

[21] Kumar, R., Singh, P.K.: Assessing Solution Quality of Biobjective 0-1 Knapsack Problem using Evolutionary and Heuristic Algorithms. Applied Soft Computing 10(3), 711–718 (2010)

[22] Ross, K.W., Tsang, H.K.: The Stochastic Knapsack Problem. IEEE Transactions on Communications 37(7), 740–747 (1989)

[23] Lee, R.C.T., Tseng, S.S., Chang, R.C., Tsai, Y.T.: Introduction to the Design and Analysis of Algorithms A Strategic Approach, pp. 157–215. McGraw Hill (2005)

[24] Chern, M.S., Jan, R.H.: Reliability Optimization Problems with Multiple Constraints. IEEE Transactions on Reliability R-35(4), 431–436 (1986)

[25] Department of Health, Executive Yuan, R.O.C. (TAIWAN) - Followed the health to travel, http://healthmap2009.doh.gov.tw/get_it.asp

[26] Clinical Nutrition Research Center, http://www2.cmu.edu.tw/~nmhls/nutritionclubibw.html

[27] Sheu, W.H.: Taiwanese snacks Nutrition Analysis in North Taiwan, Taiwanese Association of Diabetes Educators (2009)

# An Intelligent Stock-Selecting System Based on Decision Tree Combining Rough Sets Theory

Shou-Hsiung Cheng

Department of Information Management, Chienkuo Technology University,
Changhua 500, Taiwan
shcheng@ctu.edu.tw

**Abstract.** This study presents a stock selective system by using hybrid models to look for sound financial companies that are really worth making investment in stock markets. The following are three main steps in this study: First, we utilize rough sets theory to sift out the core of the financial indicators affecting the ups and downs of a stock price. Second, based on the core of financial indicators coupled with the technology of decision tree, we establish hybrid classificatory models and predictable rules that would affect the ups and downs of a stock price. Third, by sifting the sound investing targets out, we use the established rules to set out to invest and calculate the rates of investment. These evidences reveal that the average rates of reward are far larger than the mass investment rates.

**Keywords:** hybrid models, financial indicators, rough sets, decision tree.

## 1 Introduction

The problem of predicting stock returns has been an important issue for many years. Advancement in computer technology has allowed many recent studies to utilize machine learning techniques such as neural networks and decision trees to predict stock returns. Generally, there are two instruments to aid investors in predicting activities objectively and scientifically, which are technical analysis and fundamental analysis. Technical analysis considers past financial market data, represented by indicators such as Relative Strength Indicator (RSI) and field-specific charts. And it is useful in forecasting price trends and market investment decisions. In particular, technical analysis evaluates the performance of securities by analyzing statistics generated from various marketing activities such as past prices and trading volumes. Furthermore, the trends and patterns of an investment instrument's price, volume, breadth, and trading activities can be used to reflect most of the relevant market information to determine its value [1]. The fundamental analysis can be used to compare performances of a company and financial situation over a period of time by carefully analyzing the financial statements and assessing the health of a business. Using ratio analysis, trends and indications of good and bad business practices can be easily identified. Therefore, fundamental analysis is performed on both historical and

M. Ali et al. (Eds.): IEA/AIE 2013, LNAI 7906, pp. 501–508, 2013.
© Springer-Verlag Berlin Heidelberg 2013

present data in order to perform a company stock valuation and hence, predict its probable price evolution. Financial ratios including profitability, liquidity, coverage, and leverage can be calculated from the financial statements [2]. Thus, the focus of this study is not on effects but on causes that should be seeking and exploring original sources actually. Therefore, selecting a good stock is the first and the most important step for middle-term or even long-term investment planning. In order to reduce risk, in Taiwan, the public stock market observation of permit period will disclose regularly and irregularly the financial statements of all listed companies. Therefore, this study employs data of fundamental analysis by using hybrid models of classification to extract. Employing these meaningful decision rules, a useful stock selective system is proposed for middle-term - or long-term investors in this study.

In general, some related work uses a feature selection step to examine the usefulness of their chosen variables for effective stock prediction [3]. This is because not all of features are informative or can provide high discrimination power. This can be called as the curse of dimensionality problem [4]. As a result, feature selection can be used to filter out redundant and/or irrelevant features from a chosen dataset which results in more representative features for better prediction performances [5]. The idea of combining multiple feature selection methods is derived from multiple classifiers [6]. The aim of multiple classifiers is to obtain highly accurate ones. They are intended to improve the classification performance of a single classifier.

The rest of the paper is organized as follows: In Section 2 an overview of the related works is introduced, while Section 3 presents the proposed procedure and briefly discusses its architecture. Section 4 describes analytically the experimental results. Finally, Section 5 shows conclusions of this paper.

## 2     Related Works

This study proposes a new stock selective system which applies the rough set theory and decision tree algorithm to verify that whether it can be helpful in prediction the rises or falls of shares for investors. Thus, this section mainly reviews related studies of the association rules, cluster analysis and decision tree.

### 2.1     Rough Set Theory

Formally, an information system IS $= (U, A)$ , in which U is global, $U = \{x_1, x_2, \cdots, x_n\}$ , and $A = \{a_1, a_2, \cdots, a_m\}$ is for the set of attributes. Each attribute $a \in A$ defines an information function: $f_a : U \times X \rightarrow V_a$, in which $V_a$ is composed of a collection of values $a$ , known as properties of the range.

This study is based on rough set theory, carrying out the following three steps:

(1) the establishment of information system: The study analyzes a total of 993 companies as global U - 600 Taiwan's listed companies in stocks and 393 OTC companies, companies. Then, $U = \{x_1, x_2, \cdots, x_{993}\}$ , attribute sets

$A = \{a_1, a_2, \cdots, a_{15}\}$ can be further divided into condition attribute set C and decision attribute D. Use 14 properties as a condition of financial indicators, and whether the stock is rising as the decision attribute, and $V_a = \{0,1\}$ in the study.

(2) Seeking out condition attribute dependence on the decision attribute: attribute decision made under conditions of positive attributes can be defined as the region

$$pos_c(D) = \underset{X \in U/D}{\cup} \underline{C}(X)$$

(1) It means that C based on $pos_C(D)$ conducted by the division of (U / C) can be accurately classified as U / C class collection of objects. The decision attribute on condition attribute C D dependence is defined as

$$\gamma_C(D) = \frac{|pos_C(D)|}{|U|} \tag{1}$$

It means that under the condition attributes C, it can be accurately classified as decision-making in U / C that a collection of objects of the total number of objects in the ratio of the total.

(2) to search for attribute reduction and attribute set of core conditions: because the importance of each condition attribute is not necessarily the same, to test the importance of a condition of this property, we can remove it to see its influence on the C domain. If there is no impact, then this attribute is redundant and thus can be removed, and so on, to carry out attribute reduction. Importance of attributes can be defined as

$$\sigma_{(C,D)} = \frac{\gamma_C(D) - \gamma_{C-\{a\}}(D)}{\gamma_C(D)} = 1 - \frac{\gamma_{C-\{a\}}(D)}{\gamma_C(D)} \tag{2}$$

## 2.2    Decision Tree

ID3 decision tree algorithm is one of the earliest uses, whose main core is to use a recursive form to cut training data. In each time generating node, some subsets of the training input tests will be drawn out to obtain the volume of information coming as a test. After selection, it will yield the greatest amount of value of information obtained as a branch node, selecting the next branch node in accordance with its recursively moves until the training data for each part of a classification fall into one category or meet a condition of satisfaction. C4.5 is the ID3 extension of the method which improved the ID3 excessive subset that contains only a small number of data issues, with handling continuous values-based property, noise processing, and having both pruning tree ability. C4.5 decision tree in each node use information obtained on the volume to select test attribute, to select the information obtained with the highest volume (or maximum entropy compression) of the property as the current test attribute node.

Let A be an attribute with k outcomes that partition the training set S into k subsets $S_j$ (j = 1,..., k). Suppose there are m classes, denoted $C = \{c_1, \cdots, c_m\}$, and $p_i = \dfrac{n_i}{n}$ represents the proportion of instances in S belonging to class $c_i$, where $n = |S|$ and $n_i$ is the number of instances in S belonging to $c_i$. The selection measure relative to data set S is defined by:

$$Info(S) = \sum_{i=1}^{m} p_i \log_2 p_i \tag{3}$$

The information measure after considering the partition of S obtained by taking into account the k outcomes of an attribute A is given by:

$$Info(S,A) = \sum_{j=1}^{k} \frac{|S_j|}{|S|} Info(S_i) \tag{4}$$

The information gain for an attribute A relative to the training set S is defined as follows:

$$Gain(S,A) = Info(S) - Info(S,A) \tag{5}$$

The Gain(S, A) is called attribute selection criterion. It computes the difference between the entropies before and after the partition, the largest difference corresponds to the best attribute. Information gain has the drawback to favour attributes with a large number of attribute values over those with a small number. To avoid this drawback, the information gain is replaced by a ratio called gain ratio:

$$GR(S,A) = \cfrac{Gain(S,A)}{-\sum_{j=1}^{k} \frac{|S_i|}{|S|} \log_2 \frac{|S_i|}{|S|}} \tag{6}$$

Consequently, the largest gain ratio corresponds to the best attribute.

# 3     Methodology

The goal of this paper is to propose a straightforward and efficient stock selective system to reduce the complexity of investment rules.

## 3.1     Flowchart of Research Procedure

The study proposes a new procedure for a stock selective system. Figure 1 illustrates the flowchart of research procedure in this study.

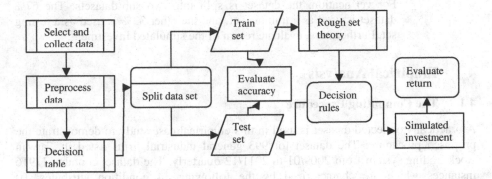

**Fig. 1.** Flowchart of research procedure

## 3.2 Flowchart of Research Procedure

This subsection further explains the proposed stock selective system and its algorithms. The proposed stock selective system can be divided into seven steps in details, and its computing process is introduced systematically as follows:

Step 1: Select and collect the data.
　　　　Firstly, this study selects target data that is collected from Taiwan stock trading system.
Step 2: Preprocess data.
　　　　It is needed to preprocess the dataset to make knowledge discovery easier. Thus, firstly delete the records that include missing values or inaccurate values, eliminate the clearly irrelative attributes that will be more easily and effectively processed for extracting decision rules to select stock. The main jobs of this step include data integration, data cleaning and data transformation.
Step 3: Build decision table.
　　　　The attribute sets of decision table can be divided into a condition attribute set and a decision attribute set. Use financial indicators as a condition attribute set, and whether the up or down of stock prices as a decision attribute set.
Step 4: Rough set theory.
　　　　To calculate the confidence of various financial ratios and stock price change by using various financial ratios and price change association rules analysis.
Step 5: Extract decision rules.
　　　　Based on condition attributes clustered in step 5 and a decision attribute (i.e. the stock prices up or down), generate decision rules by decision tree C5.0 algorithm.
Step 6: Evaluate and analyze the results and simulated investment.

For verification, the dataset is split into two sub-datasets: The 67% dataset is used as a training set, and the other 33% is used as a testing set. Furthermore, evaluate return of the simulated investment,

# 4      Empirical Analysis

## 4.1    The Computing Procedure

A practically collected dataset is used in this empirical case study to demonstrate the proposed procedure: The dataset for 993 general industrial firms listed in Taiwan stock trading system from 2009/01 to 2011/12 quarterly. The dataset contains 11916 instances which are characterized by the following 14 condition attributes: (i) operating expense ratio(A1), (ii) cash flow ratio(A2), (iii) current ratio(A3), (iv) quick ratio(A4), (v) operating costs(A5), (vi) operating profit(A6), (vii) accounts receivable turnover ratio(A7), (viii) the number of days to pay accounts(A8), (ix) return on equity (after tax) (A9), (x) net turnover(A10), (xi) earnings per share(A11), (xii) operating margin(A12), (xiii) net growth rate(A13), and (xiv) return on total assets growth rate(A14); all attributes are continuous data in this dataset. The computing process of the stock selective system can be expressed in detail as follows:

Step 1: Select and collect the data.
   This study selects the target data that is collected from Taiwan stock trading system. Due to the different definitions of industry characteristics and accounting subjects, the general industrial stocks listed companies are considered as objects of the study. The experiment dataset contains 11916 instances which are characterized by 14 condition attributes and one decision attribute.

Step 2: Preprocess data.
   Delete the 19 records (instances) that include missing values, and eliminate the 10 irrelative attributes. Accordingly, in total the data of 637 electronic firms that consist of 12 attributes and 6793 instances are included in the dataset.

Step 3: Build decision table.
   The attribute sets of decision table can be divided into a condition attribute set and a decision attribute set. Use financial indicators as a condition attribute set, and whether the stock prices ups or downs as a decision attribute set.

Step 4: Extract core attributes by rough set theory
   The use of rough set theory attributes reduction, the core financial attributes can be obtained. The core financial attributes are: (1) return on total assets, (2) the net rate of return after tax, (3) earnings per share, (4) net growth rate, (5) cash flow ratio, (6) operating profit and (7) operating margin.

Step 5: Extract decision rules.
   Based on core attributes extracted in step 4 and a decision attribute, generate decision rules by decision tree C5.0 algorithm. The decission rule set of the finacial ratios and the shares as shown in Table 2:

**Table 1.** The Definition of various Attribute

| No. | Attribute name | Notation | Definition |
|-----|----------------|----------|------------|
| 1 | Operating expense ratio | A1 | Operating expenses / net sales |
| 2 | Cash flow ratio | A2 | Net cash flow from operating activities / current liabilities |
| 3 | current ratio | A3 | current assets / current liabilities, to measure short-term solvency. Current assets (cash + marketable securities + funds + inventory + should be subject to prepayment), current liabilities within one year, short-term liabilities which must be spent on current assets to pay.    The higher the current ratio is, the better short-term liquidity is, which is often higher than 100%. |
| 4 | Quick ratio | A4 | liquid assets / current liabilities = (cash + marketable securities + should be subject to payment) divided by current liabilities = (Current assets - Inventories - Prepaid expenses) divided by current liabilities, to measure very short-term liabilities as of capacity. The higher it is, the better short-term solvency is. |
| 5 | Operating costs | A5 | because of regular business activities and sales of goods or services, enterprises should pay the costs in a period of operating time, which mainly include: cost of goods sold, labor costs. Cost of goods sold can be divided into two major categories: product cost of self-made goods and purchased products. For manufacturing sector, the former usually accounts for the most majority, while other industries the latter. By the definition of accounting costs, operating costs are cost that arises throughout the manufacturing process, also known as product costs or manufacturing costs. Manufacturing costs are composed of direct materials + direct labor + manufacturing costs (including indirect materials, indirect labor and factory operations and related product manufacturing or other indirect costs). |
| 6 | Operating profit | A6 | Operating profit / paid-up capital |
| 7 | Accounts receivable turnover ratio | A7 | net credit (or net sales) / average receivables. Measure whether the speed of the current collection of accounts receivable and credit policy is too tight or too loose. The higher receivables turnover ratio is, the better the efficiency of collection representatives is. |
| 8 | the number of days to pay accounts | A8 | Average accounts payable / operating costs * day |
| 9 | Return on equity (after tax) | A9 | Shareholders 'equity is shareholders' equity for the growth rate of the year.   The net income refers to the dividend earnings deducted the special stock, while equity refers to the total common equity. From the equity growth rate, we can see whether the company's business class objectives are consistent with shareholder objectives, based on shareholders' equity as the main consideration. Return on equity is acquired due to companies retain their earnings, and hence show a business can also promote the ability to grow their business even not to rely on external borrowing.   It is calculated as: ROE = (Net income - dividend number) / Equity |
| 10 | net turnover | A10 | net operating income / average net worth |
| 11 | Earnings per share | A11 | (Net income - Preferred stock dividends) / numbers of public ordinary shares |
| 12 | Operating margin | A12 | operating margin / revenue, often used to    compare the competitive strength and weakness of the same industrial, showing the company's products, pricing power, the ability to control manufacturing costs and market share can also be used to compare different industries industry trends change. |
| 13 | Net growth rate | A13 | the net price will fluctuate with the market increase or decrease the asset, and its upper and lower rate of increase or decrease, then is known as the net growth rate. |
| 14 | Return on total assets growth rate | A14 | which represent in a certain period of time (usually one year), companies use total assets to create profits for shareholders over the previous period the growth rate. |
| 15 | Decision attri- bute | D1 | the stock prices up or down |

**Table 2.** The Decision rule set

| No. | Decision rule set |
|-----|-------------------|
| 1 | If the return on total assets growth rate <= 0.070 and the return on total assets growth rate> -0.430 and earnings per share of $ <= 0.900 and the cash flow ratio> -2.800 operating margin <= 20.120 and the business interests <= 184,323 and operating income> -97,782 rose. |
| 2 | If the return on total assets growth rate <= 0.070 and the return on total assets growth rate> -0.430 and earnings per share of $ <= 0.900 and the cash flow ratio> -2.800 operating margin> 20.120 shares rose. |
| 3 | If the return on total assets growth rate <= 0.070 and the return on total assets growth rate> -0.430 and earnings per share of NT $ 0.900 and the cash flow ratio of> 36.240 and the return on total assets growth rate <= -0.240 then the share price rose. |
| 4 | If the growth rate of total assets> 0.070 and return on total assets growth rate <= 2.950 and Cash Flow Ratio <= 13.760 and earnings per share of RMB> -0.820 and net growth rate <= 16.710 shares rose. |
| 5 | If the growth rate of total assets> 0.070 and return on total assets growth rate <= 2.950 and the cash flow ratio <= 13.760 per share surplus element> -0.820 and net growth rate of> 16.710 and earnings per share of $ 3.750 pricerise. |
| 6 | If the return on total assets growth rate> 0.070 and return on total assets growth rate <= 2.950 and the cash flow ratio of> 13.760 and return on total assets growth rate <= 0.390 and the return on total assets growth rate <= 0.360 shares rose. |
| 6 | If the return on total assets growth rate> 0.070 and return on total assets growth rate <= 2.950 and the cash flow ratio of> 13.760 and return on total assets growth rate <= 0.390 and the return on total assets growth rate <= 0.360 shares rose. |
| 7 | If the return on total assets growth rate> 0.070 and return on total assets growth rate <= 2.950 and the cash flow ratio of> 13.760 and return on total assets growth rate <= 0.390 and the return on total assets growth rate <= 0.360 shares rose. |
| 8 | If the return on total assets growth rate> 0.070 and the return on total assets growth rate> 2.950 share price rose. |

## 4.2     Simulated Investment

The rules generating from Table 2 get down on stock selection from the listed companies rise in the rules in year 2012. There are 20 in the first quarter in line with the rise in the rules, the average quarter rate of return of 14.20 %; second quarter 2, the average quarter rate of return of 1.45%; 124 in the third quarter, the average quarter rate of return of 22.72%; 221 in the fourth quarter, the average quarter rate of return of 42.72%. The comparsion of the average quarter rate of return and the broader market quarter rate of return is shown as Table 3. These evidences reveal that the average rates of reward are far larger than the mass investment rates.

**Table 3.** The average quarter rate of return

|                | The average rates of reward | The mass investment rates |
|----------------|-----------------------------|---------------------------|
| First quarter  | 14.20 %                     | 0.32%                     |
| Second quarter | 1.45%                       | -5.25%                    |
| Third quarter  | 22.72%                      | 0.62%                     |
| Fourth quarter | 42.72%                      | 1.51%                     |

## 5     Conclusion

This paper presents a stock selective system by using hybrid models of classification. From the results of empirical analysis obtained in this study, some conclusions can be summarized as follows:

(1.) This study presents stock selection method. By the dependence of each company's financial indicators and by predicting the rises and falls of stocks, the use of rough set theory and decision tree, we can gain a simple set of classification and prediction rules.

(2.) The average rate of return derived from the empirical results shows that return on investment on stock price in the research is obvious higher than general market average.

(3.) For the welfare of majority of the investing public, the study will bring a practical and easy to understand stock selection method that can look for quality investment targets in a fast and efficient way.

## References

1. Murphy, J.J.: Technical Analysis of the Financial Markets. Institute of Finance, New York (1999)
2. Bernstein, L., Wild, J.: Analysis of Financial Statements. McGraw-Hill (2000)
3. Abraham, A., Nath, B., Mahanti, P.K.: Hybrid intelligent systems for stock market analysis. In: Alexandrov, V.N., Dongarra, J., Juliano, B.A., Renner, R.S., Tan, C.J.K. (eds.) ICCS 2001. LNCS, vol. 2074, pp. 337–345. Springer, Heidelberg (2001)
4. Huang, C.L., Tsai, C.Y.: A hybrid SOFM-SVR with a filter-based feature selection for stock market forecasting. Expert System with Applications 36(2), 1529–1539 (2009)
5. Chang, P.C., Liu, C.H.: A TSK type fuzzy rule based system for stock price prediction. Expert Systems with Application 34(1), 135–144 (2008)
6. Yu, L., Wang, S., Lai, K.K.: Mining stock market tendency using GA-based support vector machines. In: Deng, X., Ye, Y. (eds.) WINE 2005. LNCS, vol. 3828, pp. 336–345. Springer, Heidelberg (2005)

# Cloud-Based LED Light Management System and Implementation Based-on Wireless Communication

Yi-Ting Chen, Yi-Syuan Song, Mong-Fong Horng, Chin-Shiuh Shieh,
and Bin-Yih Liao

Department of Electronic Engineering
National Kaohsiung University of Applied Sciences
Kaohsiung City, Taiwan (R.O.C.)

**Abstract.** In this paper, a LED lighting management system based on wireless communication and sensor network is proposed to improve public lighting system (PLS). A web-based Human-Computer Interaction (HCI) is designed to remotely control, monitor and manage LED lights efficiently. This management system of LED lights provides energy-saving mode, periodical light-checking mode and immediate light-checking mode to reduce the energy consumption of public lighting systems and to enhance the management efficiency. This system will automatically send SMS of state notification to notify administrator when the failure exception of LED lights is occurred. The management system will reduce the maintain cost and more efficient energy-saving than traditional lightings system.

**Keywords:** Wireless Communication, Cloud Service, LED-based Public Lights, Zigbee/3G routers.

## 1    Introduction

According to the statistic from International Energy Agency (IEA) [1], the corresponding lighting-related electricity production for the year 1997 was 2016 TWh (21103 Petajoules), equal to the output of about 1000 electric power plants, and valued at about $185 billion per year. The energy consumption is quite serious. Hence, the energy consumption of street lamps has attracted a lot of attentions from academic and industry. LED lighting systems are the next generation of public lighting systems because their energy consumption is only 40% of halogen lamps. Although LED lights are with low energy consumption, there are some drawbacks in the LED public lighting application: (1) lack of management system with efficient energy-saving, (2) lack of failure-checking analysis and (3) lack of reliable system efficiency analysis.

In recent years, the progress of ICT (Information and Communication Technology) benefits the remote control of management systems. By the introduction of ICT technology to LED lighting system, the lights states will be deliver to a cloud server to improve the management of LED lighting system. Wireless communication is

M. Ali et al. (Eds.): IEA/AIE 2013, LNAI 7906, pp. 509–517, 2013.

categorized into short distance (RFID), middle distance (Zigbee, Bluetooth) and long distance (GPRS, WiFi, 3F, LTE, WiMAX). These wireless communication techniques are widely applied in entertainment, expenditure and control. And the communication content contains texe, voice and multimedia. Zigbee is low-cost, low-power, high- scalability and easy-deployment to properly apply in enviromental monitoring and sensing. And these enviromental information is gathered by router and delivered by wireless communication to analyze and manage. Cloud computing is a novel technique and attracts a lot of attention from academia and industry in recent year.

In this paper, a management system of LED lighting cloud service is implemented. This LED lighting management system combine with the wireless communication technique of Zigbee and 3G. The state of LED lighting is monitored and controled by Zigbee device. And these information of LED lighting is delivered to cloud server to analyze by 3G. These states of lighting device, routor and server connection are utilized to analyze system procedures. This method can efficinetly improve the system management and failure analysis for LED Lighting. This designed management system not only reduces the energy consumption of lighting by control function but also enchances the stability of the LED lighting system.

In the rest of this paper, we deliver the review of related work about mobile communication, sensor networks and cloud services to present the states of art in Section 2. In Section 3, the service scenario and system architecture are presented to illustrate the system design and architecture. Certainly, the implementation issues are also explored. The experimental results of the developed system are illustrated in Section 4. Finally, we conclude this work in Section 5.

## 2    Related Work

Wireless Sensor Network (WSN) increasingly developed in recent years has the advantage of low costs, high scalability, low power consumption and small size. Many studies have confirmed that wireless sensor networks are with a value of ascendency [1][2][3]. Huiping Huang et. al.[3] develop wireless sensor networks with GSM technology used in home environments Viani et. al.[4] deployed sensors to detect wild animals on the mountain roads to avoid dangerous accidents. Bin Yih Liao et. al. [5] proposed a fire detection system using wireless sensor networks to capture ambient temperature to detect flames and their positions to provide fire rescue information. In this study, the merge of mobile communication and ZigBee technology is helpful to the progress of remote management technique.

Modbus industry communication technique [5] is originally developed by Modicon Company in 1979. This communication technique is mainly applied in automatic field of industry. The Modbus communication is client-server model, and divided into master device and slave device. There is only one master device in Modbus communication network. And other devices are slave devices. If master device sends a request, the slave device will return a response. Xuyue Tu [7] proposed an operation monitor system for industrial communication technology in air compressors. The analyzed results by Modbus communication format shows in human-machine interface to provide reference for administrator. The system will notify warming to related staff when the air compressor occur failure. The Modbus communication protocol can enhance the reliability of system. Xuehua Song *et. al.* [8] used the Modbus communication technologies in information system of hybrid electric vehicle (HEV). Modbus RTU communication protocol integrates electronic control unit (ECU) of vehicle to realize the purpose of environmental monitor of vehicle by Modbus communication technique. According to above related work, the industry communication technique is highly satisfaction to apply in various applications of environmental monitor.

In this study, the cloud application service is to provide monitor software service of remote lighting system and platform service of integration different lighting platform. This system can obtain and show the operation state of lighting equipment in the distance by web interface. Kang-Min Wang [9] proposed to the handheld device platform for cloud storage media, the design of electrocardiogram activity logger through the handheld device to monitor the electrocardiogram and record carriers, in order to facilitate follow-up medical personnel can quickly learn more about portable the situation of persons, and to give appropriate to the rescue. Ting-Wey Yang [10] Peter proposed using the smart phone device, wireless Internet access and Global Positioning System (GPS) functionality, the base station positioning as the concept of cloud computing, and operational security monitoring system for the elderly and children. More and more applications and services on clouds are developed for information storage and analysis to support smart living. The application of monitor environment will be also achieved by cloud computing technique. The cloud computing technique is adopted in this study. And the server is built in management center or non-sensor area to receive the work parameters of remote lighting equipment. The administrator monitors the state of lighting equipment by network web browser to achieve the conception of cloud application service.

# 3     Cloud-Based LED Light Management System

In this section, the proposed lighting system is described in detail. And the service scenario and system operations are illustrated as depicted in Fig. 1. The system administrator can use web to monitor the state of lighting equipment in the distance. This system consists of three components: (1) LED lighting equipment and driver, (2) Zigbee and 3G router (Z3G router) and (3) cloud lighting management system. Each LED lighting is configured a driver and a Zigbee sensor node to obtain the work parameters of LED lighting, such as consumption power, voltage and current. These LED lighting make up a group. And each group configures a Z3G router to receive and transmit the Zigbee sensor information from cloud lighting management system by 3G wireless transmission. The information is stored and analyzed in the management center. The LED lighting system can achieve the efficiency of low energy consumption.

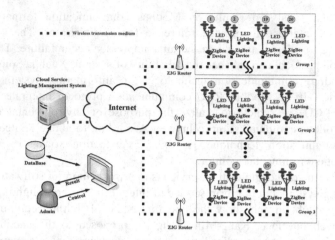

**Fig. 1.** Service scenario of LED lighting systems

## 3.1    Device Hardware System Architecture

This system is consisted of LED lighting equipment with driver, Zigbee wireless sensor network (Zigbee sensor node and Z3G router) and cloud lighting management system to detect and monitor the lighting devices in the distance. The used chip in Zigbee sensor node is CC2530 and designed by Texas Instruments (TI), the specifications of CC2530 is 256KB flash memory, 8051 microcontroller with high efficiency and low power, 8KB SRAM and so on. This chip highly suit in application area of low power. This chip can operate the sensor node by AT commands. The architecture of Zigbee sensor nodes and Zigbee/3G routers are shown in Figure 2 and 3, respectively.

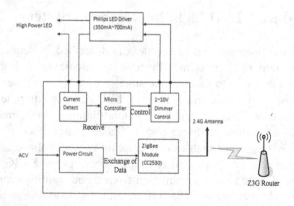

**Fig. 2.** Block diagram of ZigBee sensor nodes

The CC2530 module will deliver information to Z3G router when the microcontroller receives the work parameters from LED lighting equipment. The hardware architecture of Z3G router is shown in Figure 3. The Z3G router includes CC2530 chip module and SIM5320 chip for Zigbee and 3G communication respectively. The 3G communication module is developed by SIMCom Company for global communication. And the operation frequency band contains GSM, EGSM, DCS, PCS, WCDMA and HSDPA and is applied in global area and various applications. The microcontroller executes exchange of data and delivers to 3G wireless communication chip after CC2530 module receives the information. Then, the information is delivered through a 3G network to transmit to cloud server.

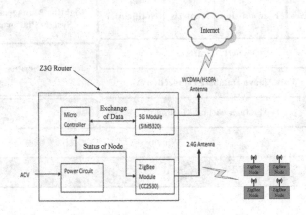

**Fig. 3.** Block diagram of a Z3G router

## 3.2    System Architecture

The software system mainly is the cloud lighting management system to as a platform of data management and LED lighting equipment. This created system is separated into background control server and lighting management system. The work parameters of LED lighting are obtained by Zigbee wireless sensor network. Then, the obtained information is delivered to cloud lighting management system by Z3G router with 3G wireless communication network. The server receives the information of work parameters and stores it in database. The state of LED lighting is analyzed by work parameters. If the LED lighting presents abnormal state, the system will send email or SMS to administrator and maintenance staff to overhaul the lighting equipment. Above status are shown in web of cloud lighting management system to provide the services of inquiry and control of LED lighting for administrator.

The architecture of the proposed cloud-based LED lights management system shown in Fig. 4. Java and Eclipse Integrated Development Environment are used to develop system. And this system comprises massage receipt, massage transition, massage analysis and massage notification. The Z3G with 3G wireless technique is used communicate between background control server and LED lighting equipment. Hence, the server has to open a communication port to continuity monitor whether the client finished connection. All information of LED lights is shown in lighting management system to provide the operational state of LED lights for system administrator. The web server of system platform is created by Apache Server. And the interface is designed by PHP programming language. The Apache web server, PHP programming language and MySQL database are used upon system level to develop system platform. This management system can show the work state of remote LED lights from current Z3G routers and record the work parameters of lighting equipment. The administrator can search the history information to obtain the statues of LED lights.

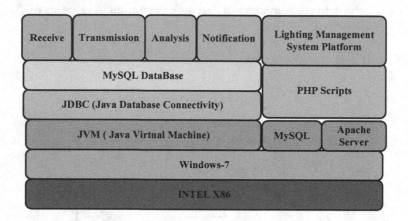

**Fig. 4.** Architecture of Cloud-based LED Light Management system

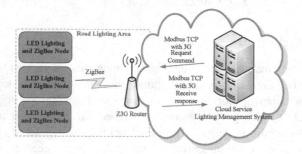

**Fig. 5.** Data transfer on Modbus TCP

The communication diagram of Z3G router is shown in Figure 6. The communication between lighting management system and lighting equipment is connected by Z3G router with wireless technique. The packets deliver the information between a Z3G router and server to achieve request command and receipt response. The industrial communication network uses Modbus TCP to connect TCP which is built by server by router. This connected Socket delivers a packet of request command. The router will return a response packet when it receives the packet of request command. And the server analyzes the packet to obtain the current state of lighting equipment.

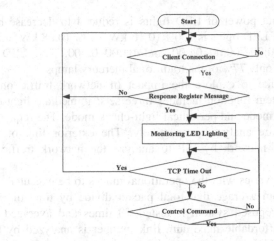

**Fig. 6.** Main system flow of cloud-based light management system

# 4    Experiments and Results

In this section, we would like to demonstrate an evaluation of energy-saving and network traffic impact. The traditional mercury lamps (0.5 kW), the LED lamps (0.16 kW) without management system and the LED lamps (0.16 kW) with proposed system are used to test the functionality and performance of the proposed system. And there is no impact of temperature factors. The comparisons of total consumption power and saving efficiency are shown the table 1. And the lighting works 12 hours in one day. The LED lights with proposed system can reduce more consumption power.

**Table 1.** Energy consumption comparisons of Various Lighting systems

| Lighting Schedule | Traditional Mercury Lamps (0.5kW) | Original LED Lamps (0.16kW) | The prsented system (0.16kW) |
|---|---|---|---|
| 18:00~20:00 | 0.5kW | 0.16kW | 0.16kW |
| 20:00~22:00 | 0.5kW | 0.16kW | 0.16kW |
| 22:00~00:00 | 0.5kW | 0.16kW | 0.16kW |
| 00:00~02:00 | 0.5kW | 0.16kW | 0.08kW |
| 02:00~04:00 | 0.5kW | 0.16kW | 0.08kW |
| 04:00~06:00 | 0.5kW | 0.16kW | 0.04kW |
| Total power consumption in a day | 6 kWh | 1.92 kWh | 1.36 kWh |
| Percentage of energy-saving | Standard | 68% | 77% |

In the night, the output power of LED lights is reduced to decrease brightness of lighting. The output of LED lights is 100% (0.16 kW), 50% (0.08 kW) and 25% (0.04 kW) during 18:00~00:00, 00:00~04:00 and 04:00~06:00. The LED lights with proposed system need only 77% of the traditional mercury lamps.

Then, we would like to explore the impact of network traffic on the system performance. The system function includes three sets: immediate light-check mode, brightness adjustment mode and periodical light-check mode. The request packet and the response packet are analyzed respectively. The exterior link of system uses 50M/5M and connect network by ISP to analyze the network traffic problem of system.

The network traffic varies with the operational modes to be executed. The resulted traffic is obtained from average that total packet divide by time difference. Each system function of proposed system is tested 10 times and averaged to calculate average traffic. The affordable maximum link number is analyzed by transmission and reception packet. The exterior connect of cloud system uses fiber-optics network with transmission and reception 50M/5M. The reception frequency of fiber optics is 50Mbps. The changed byte is about 6553600 Bytes. The transmission frequency of fiber optics is 5Mbps. The changed byte is about 655360 Bytes. The network frequency of transmission and reception is divided by average traffic to obtain maximum link number.

**Table 2.** Analysis of maximum connection number in various operational modes with an 50Mbps/5Mbps EPON link

| Operational Modes | Direction | Average flow (Bps, Bytes/Sec.) | Maximum connection number |
|---|---|---|---|
| Immediate Light-Checking Mode | Tx | 129.873 | 5046 |
| | Rx | 597.733 | 10964 |
| Dimming Mode Control | Tx | 256.42 | 2555 |
| | Rx | 13151.77 | 498 |
| Periodical Light-Checking Mode | Tx | 99.781 | 6567 |
| | Rx | 359.439 | 18232 |

The evaluation of the maximum connections is shown in Table 2. There are three modes, including Immediate Light-Checking Mode(ILM), Dimming Mode Control (DMC), Periodical Light-Checking Mode(PLCM), offered by the proposed management system. Consider with an optical link of 50Mbps, such as EPON(Ethernet passive Optical Networks), in the immediate light-checking mode, the maximum number of transmission connections is about 5046 and, the maximum number of receive connections is about 10964. The obtained maximum connection numbers of the other two modes are presented in Table 2.

From the above that the dimming mode control to receive the highest average flow, Therefore, the worst case occurs when all LED lights are operated in DMC. In this condition, there only 498 connections available for the management system and totally 9960 LED lights could be managed if each connection can monitor up to 20 lights. That will be the service capacity of the presented system.

## 5    Conclusion

A cloud-based LED light management system is presented in this paper. Through the introduction of a Zigbee/3G router and a cloud server, the developed system provides the functions of (1) reduce power consumption due to lower brightness of LED device in midnight (2) easy get the parameters of LED device in maintain (3) periodically inquire the work status of lighting to LED device and store these information in database and (4) announce administrator in LED device failure to quickly replace device. In real deployments, the system has demonstrated LED efficiency, energy consumption, network traffic analysis and reliability test. In wireless network environment, the system reliability is tested by packet transmission repeated by at least 5000 times. And successful transmission is 4967 times. Hence, the reliability of proposed system is 99.3%. This proposed system shows high reliability by tested results.

**Acknowledgment.** The authors would like to thank National Science Council, Taiwan, for the partial financial support under the grant of NSC 101-2221-E-151 -041.

## References

[1] Hong, T.P., Wu, C.H.: An Improved Weighted Clustering Algorithm for Determination of Application Nodes in Heterogeneous Sensor Networks. Journal of Information Hiding and Multimedia Signal Processing (JIH-MSP) 2(2), 173–184 (2011)

[2] Huang, H., Xiao, S., Meng, X., Xiong, Y.: A Remote Home Security System Based on Wireless Sensor Network and GSM Technology. Networks Security Wireless Communications and Trusted Computing (NSWCTC) 1, 535–538 (2010)

[3] Viani, F., Rocca, P., Lizzi, L., Rocca, M., Benedetti, G., Massa, A.: WSN-based Early Alert System for Preventing Wildlife-Vehicle Collisions in Alps Regions. Antennas and Propagation in Wireless Communications (APWC), 106–109 (2011)

[4] Liao, B.Y., Huang, W.S., Pan, J.S., Wu, H.C., Cheng, Y.M., Lee, J.K., Wang, B.S., Chen, I.L., Horng, M.F.: Research and Implementation of Service-Oriented Architecture Supporting Location-Based Services on Sensor Networks. In: Proceeding of International Conference on Software Engineering and Knowledge Engineering, SEKE 2009 (2009)

[5] Modbus-IDA, Modbus Application Protocol Specification V1.1b (December 2006)

[6] Fu, Y.: Communication Enabling Technique For Embedded Control System: Modbus-Enabled Fan Coil Unit Controllers For HVAC Systems. Master Thesis, National Taiwan university of science and technology (2004)

[7] Tu, X.: Design of Air Compressor Monitoring System Based on Modbus Protocol. In: Proceeding of International Conference on Electrical and Control Engineering (ICECE), pp. 710–713 (June 2010)

[8] Song, X.H., Lu, M., Wu, H., Wang, H., Liu, F.: The Solution of Hybrid Electric vehicle Information System by Modbus Protocol. In: Proceeding of International Conference on Electric Information and Control Engineering (ICEICE), pp. 891–894 (April 2011)

[9] Wang, K.M.: The Development of an Embedded ECG Monitoring System Based on Android and Cloud Platform. Master Thesis, ISU University, Taiwan (2011)

[10] Yang, T.W.: Using Agent and Cloud-Computing Techniques To Develop A Location Monitoring System. Master Thesis, Southern Taiwan University of Science and Technology (2010)

# A Study on Online Game Cheating
# and the Effective Defense

Albert B. Jeng[1] and Chia Ling Lee[2]

[1] Jinwen University of Science and Technology
doctor_j@just.edu.tw
[2] National Taiwan University of Information Management, Taiwan
r00725045@ntu.edu.tw

**Abstract.** Since online game becomes more and more popular and virtual assets can be transferred to real money which motivated malicious players using illegal means like "cheating" to gain the profit or superiority in the game. Therefore, we have researched existing online game cheating practices and make recommended defenses against such computer crimes. We improved existing online game cheating classification by enhancing the scope, perspective, structure and comprehensiveness of all the existing classification schemes. We also survey related defense methods against online game cheating and integrate all the public known defense strategies and methods which are applicable in each classification category and make suitable defense recommendations.

**Keywords:** Online game cheating, category, classification, defense strategy.

## 1 Introduction

Multiplayer online games become very popular nowadays and the number of players and online gaming revenue also grows rapidly. DFC Intelligence's new Online Game Market Forecasts report estimates that overall worldwide revenue for online games (for both PC and consoles) will grow from $15.7 billion in 2010 to nearly $29 billion in 2016[1].It is a good news for those game providers, but it also attracts criminals to enter these market. Many problems, security issue have arisen and that have negative impact at game providers, game players.

Recent research shows that cheating is in fact a new, major security concern for online computer games [2] but the game providers often ignore this fact. Since online games cheating has caused damages and is uprising, thus our paper will focus on this important topic.

We give a brief definition on online game cheating (adapted from Jeff Yan and Brian Randell) [3]: any behavior that people used to gain an advantage or profit over his peer players or the game providers purposely in online games is considered as cheating. To achieve their goals, the cheaters may violate the rules that the game provider set up , use items that is not allowed or make use of the weakness that has not been fixed by the game provider in the cheating process.

M. Ali et al. (Eds.): IEA/AIE 2013, LNAI 7906, pp. 518–527, 2013.
© Springer-Verlag Berlin Heidelberg 2013

## 2    An Overview on Online Games Cheating

There are all kinds of online games cheating which can achieve different goals. There are many researches who have proposed different online game cheating's classification.In [3], they classified different types of cheating into the following categories:

1. Exploiting Misplaced Trust

Because many codes and data may place in client side to reduce server's load, they exposed weakness to the cheaters.  Game providers give too much power to the clients which may be exploited by many malicious users in the games. Cheaters can use their own code or data to replace the original to gain advantage or profit. For example, using "map hack" cheaters can learn the map that will help them to make good arrangements to win the game.

2. Collusion

In many games, people can collude with their friends to take advantage. For example like cards games, friends can share information with each other that will make them easier than others to win the game.

3. Cheating Related to Virtual Assets

Since virtual assets can be sealed in real money, cheating related to virtual assets become more and more popular. It is common to see someone pay for virtual assets, but never get the items. More importantly, virtual assets could be sold that entice many people cheating in online games in order to gain virtual assets. It made online games cheating more serious.

4. Modifying Client-side System Infrastructure

Instead of modifying intrinsic game code, cheaters could modify their own machine or add additional device to help them. For example, cheaters can modify their GPU which can help them to see perspective drawing (like someone behind wall or treasury that is difficult to find out). Cheaters can also use software to gain profit. For example, in many games cheaters only need to run the software which may be provided by other cheater with sufficient instructions to help them gain lots of virtual money.

5. Exploiting a Bug or Loophole

The bug or loophole usually caused by incomplete game design or buggy game code, which were not completely debugged before its public release. Cheaters may use those bugs or loopholes to gain profit. But these kind of cheating were often discovered by game providers quickly and the problems were fixed immediately.

6. Internal Misuse

This game cheating problem is(are) attributed to the game provider's employee(s). Game provider's employees are trusted with important software, information and power. If they betrayed their trust by disclosing inside information to malicious people or abusing their power that their company gives to them, it will make a huge adverse impact on the game provider.

However the above classification is a bit unstructured, thus researchers in [4] divided online game cheating into 4 levels depending on "where the cheat targets at".

1. Game Level Cheats:

This kind of cheating is "targeting at errors on game logics or implementation," [4] A cheater discovered a weakness at the game design level (e.g., game procedure, game protocol, etc.) and use them to gain advantageous position. Cheating by game procedure abuse is an example at game level cheats. This level is focusing on the game logic part.

2. Application Level Cheats:

This kind of cheating is "either modifying the game modules or data files, interfering with the memory of the running game" [4]. Why a cheater can modify inside module or data files? It is usually due to the loophole that the game provider doesn't have good protection for those important data.

3. Protocol Level Cheats:

This kind of cheating is operated by "interfering with the packets sent and received by the game (which may be inserted, destroyed, duplicated, or modified by an attacker)" [4]. If the messages are not encrypted, then a user's personal information may be at risk of unauthorized disclosure.

4. Infrastructure Level Cheats:
   This kind of cheating is by "modifying or interfering with the software (e.g. graphic drivers) or hardware (e.g., the network infrastructure) that the game is using" [4]. Modifying client-side system infrastructure is an example of this level.

# 3     An Improved Online Game Cheating Classification

In the following, we show the pros and cons of two representative online game cheating classification methods. Then, we will propose an improved classification on online games cheating which will remediate the discrepancies of the existing classification methods.

Both classification methods have their pros and cons, so we propose a new cheating classification which is more structural and also increases the comprehensiveness compared to other existing schemes. First, we will divide the online games cheating into two major categories based on the nature of their illegal gains.

| | J Yan. et al. classification method in [3] | Van Nguyen Khanh et al. classification method in [5] |
|---|---|---|
| Pros | • provide each category a clear description.<br>• give a brief solution of each category<br>• Game providers can check their game systems following each category and keep monitoring the games | • more structure way to classify online game cheating<br>• game providers can follow these four levels' structure to design their games |
| Cons | • Unstructured<br>• Cannot included all of the online games cheating | • their definition constrain the range into a small scope<br>• Cannot included all of the online games cheating |

**Fig. 1.** The pros and cons of two online game cheating classification methods

### a. Cheating in Order to Prevail and Get Unfair Advantage in the Game

A great majority of online games cheating belongs to this type of category. The major purpose of this type cheating is to gain an advantageous position in the game by cheating. For example, modifying client-side system infrastructure is to gain the advantage in the game like having the whole map at the beginning phase. By using these kinds of cheating, cheaters will have better scores and may achieve special treasure which can be sold in real world at a good price. Therefore cheaters use any kinds of cheating methods that can help them win the advantage in this game type.

### b. Gain the Illegal Profit External to the Game

Some cheaters didn't play the games, but they just want to gain a profit outside the game itself. For example, they can steal players account by making a fishing page or steal the messages between the server and clients which may contain personal information. Then this information can be sold to other companies to acquire illegal money. A virtual asset's trade cheating is an example of this category, in which a seller receives money but never give virtual asset to the buyer. It was reported that a person earns about US$11,000 by using trade cheating [2].

In this paper, we focus on Type (a) cheating, because Type (b) is more related to e-commerce cheating or fraud domain. Thus, we go deeper by further dividing Type (a) category into two sub-categories based on the nature of the game weakness that are exploited by the cheaters.

### 1. Insecure Game Operation

This kind of online game cheating is due to the cheaters exploited the problems existing in the game provider's inside operation to commit frauds. For example, internal misuse is caused by the internal employees who disclosed inside information to malicious people or abused their power that the company gives to them. This kind of cheating is caused by the game operation's weakness which needs more procedures to monitor the insiders.

## 2.  Insecure Game Design

Incomplete game design may let cheaters have weak points to attack or to bypass the rule that the game designer sets up if the rule is not robust enough. Poor game design may lead to game disclosure.  For example, a game designer needs to have a good way to detect collusion or have a protected mechanism to alert against collusion. We can classify this category further using the previous four levels structure in Section 2 as an improvement. However, we delete the protocol level cheats and merge all its cases into the application level cheats, because it only contains a few types of online game cheating such as cheatings by interfering with packets transmission. Thus, there are only three levels used in the insecure game design category. Furthermore, we also redefine each of the three levels definition.

i.     Game Level Cheats:

This level's cheating is caused by "errors on game logics, implementation, procedures, policy, rules or the entire game design's architecture".  Previous definition only covers game logic, implementation, but it leaves out the game's rules and policy establishment. Without clear rules or a sound policy, it may result in many security problems.  Cheaters can use the gray area to gain the profit or advantage. For example, friends can win each other once a time, then both of them can acquire the price without breaking the rules. This kind of cheating can classify as collusion. These kind of illegal actions are committed because the game designer didn't set up a strict rules to take care of this kind of situation. Cheaters only find these out without telling to the providers. We can classify collusion, abusing game procedures, exploiting a bug, and timing cheating, and denying service to peer players as cases in this category.

ii.     Application Level Cheats:

Previous definition is "either modifying the game modules or data files, interfering with the memory of the running game"[4]. This definition only relates to the game data being modified or interfered. It only contains very few online games cheating types. Thus, we give a new definition to this "application level cheats" which has a little relationship with previous classification. It is defined as "any insecurity related to the program, code, game module or database sources." In other words, "any of above internal coding operation, data, communicates between the server and clients or between a class and another class, which have a relationship with the programming code can be classified to this category". It is due to the programmers didn't do a good job in programming with some leaks to be exploited by the attacker. For example, programmers develop the game often set up a hot key in order to speed up the test (e.g., press 'H' can pop up money) and this hot key's function should be removed when the game is released. However, if the programmers forget this important process, then when a cheater finds out these functions they may generate infinite money. It is a bug caused by the programmers that permits the cheaters have opportunity to cheat. We can classify the exploiting misplaced trust, compromising game server, exploiting a bug or loophole as cases to this category.

iii.    Infrastructure Level Cheats:

This category is not to change inside program or internal data; it's about adding additional device (software or hardware) or modifying an existing device to achieve the goal. It has no relationship with inside operation. This kind of online game cheating is very popular because players can easily add extra hardware or run analysis software, AI software in client-side to help them win the game. We can classify exploiting machine intelligence, modifying client-side system infrastructure, exploiting misplaced trust as cases in this category.

Here we make a table to summarize our enhanced online game cheating classification method.

| | | Cheating in order to prevail and get unfair advantage in the game | | Gain the illegal profit external to the game |
|---|---|---|---|---|
| Weakness | Insecure game operation | • Internal misuse | | • Personal information theft |
| | Insecure game design | Game level cheats | • Collusion<br>• Abusing game procedures<br>• Exploiting a bug<br>• Denying service to peer players | • Virtual treasure fraud |
| | | Application level cheats | • Exploiting misplaced trust<br>• Compromising game server<br>• Exploiting a bug or loophole | |
| | | Infrastructure level cheats | • Exploiting machine intelligence<br>• Modifying client-side system infrastructure<br>• Exploiting misplaced trust | |

**Fig. 2.** An enhanced online game cheating classification

Our enhanced classification extends the scope and the perspective of all the existing classification schemes. The new extension includes adding a new weakness such as insecure game operation and a new illegal benefit from both inside and outside the game. We still keep the structure of the previous classification methods because it lets the game providers follow these structures to inspect their games when they started the game design phase. We also include all the existing online game cheating types in our improved classification scheme. Using our improved scheme, the game providers can inspect their games from small-scale to large-scale for any security hole in developing phase to construct their architecture and game design. Furthermore, we improve the previous narrow definitions in our improved classification by redefining each level cheats which makes it more closer to the real world with broaden scope. Next chapter we will give a general defense strategy, a matter needing attention and each category's solution using right now.

# 4     An Effective Defense against Online Game Cheating

## 4.1     Existing Defense against Online Game Cheating

In this section, we give an overview of the existing defense method to deal with online game cheating today. There still aren't many researches to give an overall defense strategy against online game cheating. Many research only focus on specific domain. Thus, we survey the existing researches in this area and use our cheating classification to help solving this problem. Then, a game provider can follow our approach to inspect their game.

In the first category "gain the illegal profit independent of the game", a key point is the authentication process which must have a strict process. Research in [7] shows that "script" is full of vulnerabilities. Using script is very dangerous, malicious users can attach additional code behind the page or message, then using these additional code they can lead player to fishing page or even copy all the messages that you send and receive. They can get all the information that they want including account and password and difficult to detect those code.

The second category is "insecure game operation". Research in [8] says, company can install intrusion detection system (IDS) which can look for abnormal activities both inside and outside of the company. By using IDS, company can find malicious activities in early stage and report to the administrator to take action. Second, company should need auditing mechanism to record every action, user behaviors. Those audit logs should carefully preserved by a particular person and still need strict authentication to read those logs. Third, internal activities also need strict authentication process and authorization mechanism to prevent normal employee from access easily which means a need to have logical access controls. Forth, data need to have encryption and only an authorized person can have the right to read.

In the third category "game level cheats", it is usually caused by loopholes, bugs or procedures.  In [9], they said "it is impossible that any game is perfect without bug or loophole". It is true no matter how hard the game providers check over and over again, games still contain bugs or loopholes. The only difference is how serious the problem is. So in [9], they suggested game developer should take more time to inspect the game before its release and after that they still need to maintain the ability to fix all kinds of problems.

In the fourth category "application level cheats", it just looks like game level cheats category. Because this kind of cheating cannot avoid, game provider only can reduce the damage. First, check the entire program before releases, checking process need to set up a rule in order to avoid any detail that may generate bug. Second, keep monitoring the game's operation. For example, they can follow Q&A or message board's information to see if something abnormal or some part may have mistakes.

In the fifth category "infrastructure level cheats", many researches study this part, because more and more people using bot program to increase their experience which can let them level up or gain money. That is the cheating type "modifying client-side software". Installing bot program is easier compare to other cheating method. A player only needs to install some software then set up the environment, programming parameters before he can run the bot programs. In [6], bot detected method can classify into three ways: client-side detection, network-side detection, server-side detection. Client side detection is by installing some small program to keep monitor client's status. Challenge-response is a common method to detect the existence of a user. Network-side detection is by monitoring network traffic. Server-side detection is operating on server-side by monitoring players behaviors like log history. A game provider can use players idle time, user input behaviors, event sequence or moving path to detect bot program which are actually execute in the real game market. In [5], they used a machine learning algorithm to detect abnormal players of StarCraft which is a very popular game. It not only can detect a user using a bot program but also can detect other illegal cheating method. For example, players use map hacking by running additional software or installing extra hardware, thus when the game begins, players can straightly go to the place that filled with fog. That is an abnormal behavior, and then they can find suspicious people according to those strange behaviors. In [5], they use experts to recognize players' strategies in the early stage by observing their building. Experts will analyze those strategies and give some rules to classify abnormal behavior. Then they want to use machine learning algorithm and keep extending the rules based on those originals. They give a general method to detect all of possible irregular behavior, not just focus on specific online game cheating. Then a game provider can base on this architecture and make adjustment to fix all the game's problems property.

## 4.2    Recommend Defense in Online Game Cheating

This section, we will give some recommend defense base on previous research and our new advices. We will list the things game providers should do before and after game releases. We summarize what a game provider should do to prevent cheating and make game more secure for each category in different phase.

1. Gain the Illegal Profit External to the Game

Authentication process should use a mix of OTP, CAPTCHA, encryption technologies to increase the difficulty of stealing account or information. On the other hands, game providers can require players changing password frequently, strictly constraint password length and block potential malicious players. Also, a game provider needs to set up recovery and report mechanism in advance. After game releases, a game provider still needs to keep monitoring to detect any compromise of the player's account. Thus, a game provider needs to help victims and continue find the cheaters and make some improvement to deal with this kind of cheating.

## 2. Insecure Game Operation

Just like previous research, a game provider needs to set up an auditing mechanism and access control protocol. Only specific people can use certain data, information. Also, all of function and data should need two more people to deal with (i.e., Two-men Rule) which can prevent internal misuse. Also we can use additional software or hardware to assist manage internal operation. All of the actions should be recorded in audit trails which can help tracing if something happened. After game releases, game providers still need to keep monitoring the audit log.

## 3. Game Level Cheats

Game designers should spend more time to design the game, from policy, architecture to game's rule, all need to consider all kinds of situations and find solutions to prevent existing cheating method. That's all they can do in advance. After game releases, game provider need keep monitoring not only game's status but also information in the Internet. Because many players may share their discovery in the Internet, then game providers can use their discovery to make an improvement. This kind of information usually spread very fast, which can help game providers fix them as soon as possible.

## 4. Application Level Cheats

Game programmers should spend more time to inspect the program. They need to check program in systematic way, which can reduce possibility of ignorance somewhere. Also, all programs need to have clear structure, annotation and have a complete document which can help them maintain and check. After game releases, game provider need to monitor game's status (like Q&A or message board to see players feedback) and information in the Internet. After discovering bugs or problems, programmers can follow the recovery process that used to set up and recover quickly.

## 5. Infrastructure Level Cheats

In previous four kinds of cheating, theoretically a game provider can control in advance, because those problems are either caused by poor game logic design or incomplete game program. But in infrastructure level cheats, a game provider has less leverage to control because these kind of cheating mainly occur in client side which prevents a game provider from doing anything useful in advance. Thus, before a game is released, the game designer should consider not putting too much data into the client-side, but that will cause some adverse problem like increasing network load and server load. Each plan has its pros and cons. But storing all data at server side still may cause infrastructure level cheats despite the fact it may decrease the probability. Second, a game provider can choose a monitoring mechanism either at the network side or the server-side to detect and execute the detection plan. After game releases, game providers need to keep monitoring and detecting any abnormal behavior. They can apply previous research's detecting method and make some adjustment to fit their game. As a minimum, player monitoring is necessary and the use of those data may help the game provider collect some useful security information.

# 5    Conclusion and Future Work

Online game cheating indeed caused serious damage to the game providers and good players. To deal with this problem, we analyze existing online game cheating, defense strategies and methods in order to find a comprehensive solution to resolve this problem.We found out all existing researches still have rooms for improvement. We improved existing online game cheating classification by enhancing the scope, perspective, structure and comprehensiveness of their schemes. In addition, we reviewed the existing defense strategies against online game cheating and integrated them into a more complete overall defense strategy against online game cheating. In essence, we made a recommend defense against online game cheating in each category of our revised classification. Using our classification, the game providers have easier time to examine their games to see if there is any security hole left. We also gave a list of things the game providers should do before and after their game release to plug their security holes. In the future, we will focus on infrastructure level online game cheating defense and how to use AI technology to support such kind of defense implementation for different game types.

# References

[1] http://www.industrygamers.com/news/
    online-games-market-to-nearly-double-to-29-billion-by-2016/
[2] Yan, J., Choi, H.J.: Security Issues in Online Games. The Electronic Library 20(2) (2002); A previous version appears in Proc. of International Conference on Application and Development of Computer Games, City University of Hong Kong (November 2001)
[3] Yan, J., Randell, B.: An Investigation of cheating in online games. IEEE Security & Privacy (2009)
[4] The, L.B., Khanh, V.N.: GameGuard: A Windows-based Software Architecture for Protecting Online Games against Hackers (2010)
[5] Kim, K.-J., Cho, S.-B.: Server-side Early Detection Method for Detecting Abnormal Players of star craft. In: International Conference on Internet, ICONI (2011)
[6] Kang, A.R., Woo, J., Park, J., Kin, H.K.: Online game bot detection based on party-play log and analysis. Computers and Mathematics with Applications Journal (2012)
[7] European Network and Information Security Agency, Virtual worlds, real money security and privacy in massively-multiplayer online games and social corporate virtual worlds
[8] Grance, T., Hash, J., Peck, S., Smith, J., Korow-Diks, K.: Security Guide for Interconnecting Information Technology Systems. National Institute of Standards and Technology Special Publication 800-47
[9] Lan, X., Zhang, Y., Xu, P.: An Overview on Game Cheating and Its Counter-measures. In: International Computer Science and Computational Technology, ISCSCT 2009 (2009)

# Credit Rating Analysis with Support Vector Machines and Artificial Bee Colony Algorithm

Mu-Yen Chen[1], Chia-Chen Chen[2], and Jia-Yu Liu[1]

[1] Department of Information Management, National Taichung University of Science and Technology, Taichung 404, Taiwan, R.O.C
{mychen,s1800b112}@nutc.edu.tw
[2] Department of Management Information Systems, National Chung Hsing University, Taichung 402, Taiwan, R.O.C
emily@nchu.edu.tw

**Abstract.** Recently, credit rating analysis for financial engineering has attracted many research attentions. In the previous, statistical and artificial intelligent methods for credit rating have been widely investigated. Most of them, they focus on the hybrid models by integrating many artificial intelligent methods have proven outstanding performances. This research proposes a newly hybrid evolution algorithm to integrate artificial bee colony (ABC) with the support vector machine (SVM) to predict the corporate credit rating problems. The experiment dataset are select from 2001 to 2008 of Compustat credit rating database in America. The empirical results show the ABC-SVM model has the highest classification accuracy. Hence, this research presents the ABC-SVM model could be better suited for predicting the credit rating.

**Keywords:** support vector machine, artificial bee colony, credit rating analysis.

## 1 Introduction

The credit rating is a benchmark to measurement of the credit of a firm. Ratings also forecast the likelihood of default on financial responsibilities and the intended repayment in the case of default. Issuers seek ratings for some reasons. First, Corporate try to improve the stock pricing by rating. Second, rating will impact the profitability of firms. Third, rating may impact the corporate's trustworthiness to counterparties and debtors. Finally, the investors would invest the firms which one has high-level rating. Institutional investors, investors, regulators, and debtors use ratings as a signal of the risk assessment of security. Therefore, credit rating provide useful information can helping investors to make decision. Besides, rating also can measuring and limiting risk for regulatory institutions, such as insurance companies, commercial bank, pension funds, and money market funds. In other words, some institutions will hold securities only with investment-grade ratings.

Credit rating prediction is a classification problem, that to predict the firms will fail or not, based on financial ratios. Since the 1960s some researcher provides a statistic

M. Ali et al. (Eds.): IEA/AIE 2013, LNAI 7906, pp. 528–534, 2013.

classification method to predict corporate failure, including univariate analysis [3], discriminant analysis [1], multiple regression [12], logistic regression [11], and probit analysis [17]. However, strict traditional statistical assumptions can't overcome some problems in real world, such as some nonlinear problems [8]. Yet, since 2000s, most studied are applied Artificial Intelligence (AI) model, to credit rating predicting problems. West [16] used an artificial neural network (ANN) model to predict the credit scoring and compared with other statistical model, the result shows that, the ANN model hat highest accuracy. In addition, the logistic regression has the best result in the statistical model. Lately, the support vector machine (SVM) approach [15] has turned into popular more and more. At present, it is viewed as the state-of-the-art neural network technique for regression and classification problem solving. Mush studied had used SVM hybrid with other technique, and all the research had high accuracy and convergence capability.

The organization of this research is listed as following: Section 2 is discussed the literatures of SVM and ABC. Section 3 is proposed model in this research. Section 4 discusses the experiment from 448 firms of Compustat database. Conclusions are summarized in Section 5.

## 2     Literature Review

### 2.1     Support Vector Machine (SVM)

In recent years, new developments in statistics, including Support Vector Machines (SVM), have significantly improved the model building/learning/fitting process [6], with novel computational and machine learning methods helping to achieve the generalization of the concept of parameter estimation.

SVM is a new approach to the design of learning machines according to the principle of Structural Risk Minimization derived from computational learning theory [14]. The SVM algorithm is sufficiently simple for mathematical analysis [7], thus SVMs could potentially be used to integrate the advantages of traditional statistical methods and intelligence machine learning methods [15]. Several financial applications have recently adapted the SVM approach to solve problems related to time series classification and prediction [13]. Another paper recently used the SVM approach to select predictors for bankruptcy, with [4][5] suggesting that SVM outperformed other classifiers in terms of generalization performance. The present paper focuses on evaluating the SVM performance for credit rating prediction against that of other statistical and neural networks.

### 2.2     Artificial Bee Colony (ABC)

Artificial Bee Colony (ABC) proposed by Basturk and Karaboga in 2006, is recommended for solving numerical optimization problems. ABC is an algorithm that proved its success in handling many different kinds of domain problems. In the ABC algorithm design concept, any random solution pertaining to a specific problem conforms to a nectar source, therefore, population is made up of nectar sources [10]. The global minimizer is found and obtained by artificial bees that are assumed to look for food by flying among nectar sources that are scattered in the searching space.

The bees in ABC are divided into three types: employed bees, onlooker bees and scout bees. First, the employed bee goes to a food source and comes back to the hive to communicate information about the nectar available through a dance. Second, the onlooker bees look the dancing of employed bees, and they would select a food source according to the dancing of employed bees. Finally, the scout bees always look for food sources randomly. The onlooker and scout bees are also referred to as unemployed bees. In ABC, the location of a food source means a potential optimization problem solution, with amount of nectar in a given food source indicates the solution's quality (fitness). The amount of solutions is limited by the amount of employed bees. The process begins by generating an initial, randomly distributed population. The population then carries out multiple iterations of the three types of bees. When an employed bee finds out a new food source, it evaluates the number of nectars in the new and previous positions. If the new position has more nectar, the employed bees abandoned the old nectar. After all the bees have completed their searching jobs, they will communicate the position information of nectar with the onlookers through dance. The onlookers then choose food sources according to the number of nectars available at each. Similar to the employed bees, the onlookers update their memories to remember sources with more nectar, forgetting the older sources. The scouts then replace the abandoned sources with new ones found at random.

# 3    Research Methodology

## 3.1    Materials and Indicators

We extract our credit rating data and accounting data based on Compustat database from 2001 to 2008. In our eight-year period, there are 716 firms and 85,736 events in our all samples. After removing the event without or missing accounting data, we obtain 22,826 events to include in 448 firms. For the empirical analysis we designed, we collect the credit rating grate from AAA to CC (e.g., S&P code 2 to 23) and divide ratings into investment-grade (e.g., BBB- and higher) and non-investment-grade (e.g., BB+ and lower).

## 3.2    Principle Component Analysis and Choice of Indicators

Here we discuss the input indicators for the SVM modeling process. First, to equalize the investment-grade (BBB- and higher) and non-investment-grade (BB+ and lower) firms, we selected one investment-grade firm with one non-investment-grade firm from the same industry for each pair. Next, we initially gathered related ratios for 448 firms (224 investment-grade and 224 non-investment-grade) for empirical analysis. Third, we adopted the min-max normalization approach to demarcate these variables of the

448 firms into the range of [0, 1]. The entire data set was then divided into two unequal sub-datasets: a training dataset (80%) and a testing dataset (20%). After the normalization process, PCA was used to find out the appropriate indicators for accounting ratios and non-accounting ratios. The selection method was according to Kaiser's criteria, this criteria suggests the absolute value of the factor loadings should be greater than 0.5 and the communality should be greater than 0.8 [9].

In this research, there are one dataset to be investigated and extracted the important factors is called overall. In totally, we assembled 19 accounting ratios and 3 non-accounting ratios. The factor loadings, communality, and the explained variance information for each indicator with overall dataset are shown in Table 1. Finally, we could finally sure that the 4rd PCA is optimal, with a performance of 95.84%. Following the four times of PCA iterations, 12 indicators presented they owned the higher factor loading or communality value in the finally. Then, these indicators were selected for inclusion in the input vector of SVM model, as well as the other 10 indicators were discarded.

**Table 1.** PCA Analysis with Overall Dataset

| No. | Variables | 1st PCA Factor Loadings/ Communality | 2nd PCA Factor Loadings/ Communality | 3rd PCA Factor Loadings/ Communality | 4rd PCA Factor Loadings/ Communality |
|---|---|---|---|---|---|
| 1 | Price-Close Mnthly | 0.996/0.996 | 0.997/0.997 | 0.999/0.999 | 0.999/0.999 |
| 2 | Price-High Mnthly | 0.995/0.995 | 0.996/0.996 | 0.999/0.998 | 0.988/0.989 |
| 3 | Price-Low Mnthly | 0.995/0.996 | 0.996/0.997 | 0.998/0.998 | 0.988/0.998 |
| 4 | T-Bill-6 Month | 0.932/0.871 | 0.106/0.013 | — | — |
| 5 | Trading Volume-Monthly | 0.419/0.803 | — | — | — |
| 6 | Assets-Total Qtly | 0.987/0.938 | 0.991/0.983 | 0.995/0.990 | 0.995/0.991 |
| 7 | Common Equity-Total-Qtly | 0.930/0.913 | 0.936/0.877 | 0.947/0.898 | 0.960/0.924 |
| 8 | Current Assets-Total Qtly | 0.871/0.815 | 0.891/0.793 | — | — |
| 9 | Current Liab-Total Qtly | 0.918/0.856 | 0.934/0.872 | 0.927/0.861 | 0.928/0.863 |
| 10 | Debt - Total Qtly | 0.898/0.853 | 0.898/0.807 | 0.886 /0.785 | — |
| 11 | Invested Capital-Total Qtly | 0.973/0.955 | 0.976/0.953 | 0.981/0.962 | 0.979/0.958 |
| 12 | Liabilities-Total Qtly | 0.979/0.960 | 0.983/0.966 | 0.980/0.962 | 0.972/0.945 |
| 13 | LT Debt-Total Qtly | 0.887/0.839 | 0.882/0.780 | — | — |
| 14 | Available for Interest Qtly | 0.717/0.696 | — | — | — |
| 15 | Cash Flow Qtly | 0.821/0.797 | — | — | — |

**Table 1.** (*Continued.*)

| 16 | Earning Assets - Total Qtly | 0.942/0.916 | 0.947/0.897 | 0.954/0.910 | 0.966/0.933 |
|---|---|---|---|---|---|
| 17 | Return on Avg Assets Qtly | 0.688/0.574 | – | – | – |
| 18 | Return on Equity Qtly | 0.707/0.504 | – | – | – |
| 19 | Tot Assets/Common Eqty Qtly | 0.965/0.939 | 0.977/0.955 | 0.977/0.955 | 0.977/0.955 |
| 20 | Tot Assets/Total Equity Qtly | 0.965/0.941 | 0.990/0.981 | 0.991 /0.982 | 0.991/0.982 |
| 21 | Total Debt/Total Assets Qtly | -0.712/0.544 | – | – | – |
| 22 | Total Debt/Total Equity Qtly | 0.950/0.918 | 0.977/0.955 | 0.977/0.955 | 0.977/0.955 |
| | **Total Explained Variance** | **81.622%** | **86.377%** | **94.27%** | **95.84%** |

# 4    Empirical Analysis

## 4.1    Experimental Parameter Settings

In this research, the proposed ABC–SVM was setup the parameter values in the following. In the ABC algorithm, the maximum number of cycles (MCN) is the same as the maximum number of iterations and the colony size is the same as the population size. The ratio of onlooker bees to employed bees is 1:1, with a single scout bee. Increasing the amount of scouts stimulates found of new food sources, while increasing the number of onlookers increases the exploitation of a given food source. The searching range of parameters C and $\gamma$ were both set between 1 to 100. This research adopts the accounting and non-accounting ratios to implement a credit rating and credit changing prediction model following a four-iteration PCA. These suitable indicators would be loaded as the input nodes of SVM model.

## 4.2    Experimental Results

In this research, we collect the credit rating samples based on Compustat database from 2001 to 2008. Meanwhile, we obtain 22,826 events and include in 448 firms as our overall dataset. Then we used hybrid SVM models with optimization techniques to predict the credit ratings and its changes. These optimization techniques are including genetic algorithm (GA) and particle swarm optimization (PSO). This research used the 10-fold cross-validation procedure to evaluate the prediction performance (Chen, 2011a). In addition, in order to fair and objective evaluation for these three datasets, this research also adopted the leave-one-out cross-validation method to evaluate the prediction performance. In Table 2, it shows the prediction performance of the three hybrid models adopting both 10-fold cross-validation and leave-one-out cross-validation methods. Finally, the ABC-SVM obtained the best prediction performance in this empirically research.

Table 2. Comparison between the ABC-SVM and other models (Unit: %)

| | 10-fold Cross-Validation | | | Leave-one-out Cross-Validation | | |
|---|---|---|---|---|---|---|
| | GA-SVM | PSO-SVM | ABC-SVM | GA-SVM | PSO-SVM | ABC-SVM |
| Overall datasets | 70.34 | 74.28 | 80.42 | 69.54 | 75.62 | 81.82 |
| Ranking | 3 | 2 | 1 | 3 | 2 | 1 |

## 5 Conclusions

We applied a new bio-inspired hybrid computation method to integrate the PCA and SVM for credit rating prediction issues. This empirically research collected the datasets based on Compustat database from 2001 to 2008 as our experiment test bed. In this eight-year period, there are 22,826 events to include in 448 firms in our all samples. The empirically results presented that ABC-SVM obtained the highest accuracy and outperformed the GA-SVM and PSO-SVM. This research also finds that, the ABC-SVM significantly outperforming the other models. Finally, the ABC-SVM approach generally produces higher accuracy rate than PSO-SVM and GA-SVM models. Therefore, our proposed ABC-SVM approach is appropriate for forecasting the credit rating problems.

**Acknowledgment.** The authors thank the support of National Scientific Council (NSC) of the Republic of China (ROC) to this work under Grant No. NSC 101-2410-H-025-004-MY2 and NSC 101-2410-H-029-003.

## References

[1] Altman, E.L.: Financial ratios, discriminant analysis and the prediction of corporate bankruptcy. Journal of Finance 23, 589–609 (1968)
[2] Basturk, B., Karaboga, D.: An artificial bee colony (ABC) algorithm for numeric function optimization. In: Proceedings of the IEEE Swarm Intelligence Symposium, Indianapolis, IN, USA, May 12-14
[3] Beaver, W.H.: Financial ratios as predictors of failure. Journal of Accounting Research, 71–111 (1966)
[4] Chen, M.Y.: Using a Hybrid Evolution Approach to Forecast Financial Failures for Taiwan Listed Companies. Quantitative Finance (2011a) (forthcoming)
[5] Chen, M.Y.: Bankruptcy Prediction in Firms with Statistical and Intelligent Techniques and a comparison of Evolutionary Computation Approaches. Computers and Mathematics with Applications 62, 4514–4524 (2011b)

[6]  Galindo, J., Tamayo, P.: Credit risk assessment using statistical and machine learning: basic methodology and risk modeling applications. Computational Economics 15, 107–143 (2000)

[7]  Hearst, M.A., Dumais, S.T., Osman, E., Platt, J., Schölkopf, B.: Support vector machines. IEEE Intelligent Systems 13, 18–28 (1998)

[8]  Hua, Z., Wang, Y., Xu, X., Zhang, B., Liang, L.: Predicting corporate financial distress based on integration of support vector machine and logistic regression. Expert Systems with Applications 33, 434–440 (2007)

[9]  Kaiser, H.F.: The application of electronic computers to factor analysis. Educational and Psychological Measurement 20, 141–151 (1960)

[10]  Karaboga, D., Basturk, B.: A powerful and efficient algorithm for numerical function optimization: artificial bee colony (ABC) algorithm. Journal of Global Optimization 39, 459–471 (2007)

[11]  Martin, D.: Early warning of bank failure a logit regression approach. Journal of Banking & Finance 1, 249–276 (1977)

[12]  Meyer, P.A., Pifer, H.: Prediction of bank failures. Journal of Finance 25, 853–868 (1970)

[13]  Tay, F.E.H., Cao, L.J.: Modified support vector machines in financial time series forecasting. Neurocomputing 48, 847–861 (2002)

[14]  Vapnik, V.N.: The Nature of Statistical Learning Theory. Springer, New York (1995)

[15]  Vapnik, V.N.: Statistical Learning Theory. Wiley, New York (1998)

[16]  West, D.: Neural network credit scoring models. Computers and Operations Research 27, 1131–1152 (2000)

[17]  Zmijewski, M.E.: Methodological issues related to the estimation of financial distress prediction models. Journal of Accounting Research 22, 59–82 (1984)

# An ACO Algorithm for the 3D Bin Packing Problem in the Steel Industry

Miguel Espinheira Silveira, Susana Margarida Vieira,
and João Miguel Da Costa Sousa

Technical University of Lisbon, Instituto Superior Técnico,
Avenida Rovisco Pais N1, Lisbon, Portugal
{miguelesilveira,susana.vieira,jmsousa}@ist.utl.pt

**Abstract.** This paper proposes a new Ant Colony Optimization (ACO) algorithm for the three-dimensional (3D) bin packing problem with guillotine cut constraint, which consists of packing a set of boxes into a 3D set of bins of variable dimensions. The algorithm is applied to a real-world problem in the steel industry. The retail steel cut consists on how to cut blocks of steel in order to satisfy the clients orders. The goal is to minimize the amount of scrap metal and consequently reduce the stock of steel blocks. The proposed ACO algorithm searches for the best orders of the boxes and it is guided by a heuristic that determines the position and orientation for the boxes. It was possible to reduce the amount of scrap metal by 90% and to reduce the usage of raw material by 25%.

**Keywords:** Bin packing problem, steel cut industry, guillotine cut, Ant Colony Optimization.

## 1 Introduction

Since the formulation of cutting and packing (C&P) problems during the 50's of the 20th century, this class of problems has been receiving a growing interest by the industry, as a mean to improve profits by reducing costs [13].

However, most of the research has been focused on one or two-dimensional cutting/packing problems [1] which are a simplification of the three-dimensional cases [9]. This work focuses on cutting a set of 3D small objects from a set of 3D large objects, where both sets are strongly heterogeneous in their size. The large objects to be cut are called bins, whereas the small objects to be cut from the bins are called boxes. The objective is to minimize the scrap metal. According to the most recent cutting and packing typology [12], this is a residual bin packing problem (RBPP). However, this is a special case of the residual bin packing problem as it considers a timeline, i.e. the bins and boxes are available at a certain date.

C&P problems have been addressed by searching heuristics such as tabu-search [1,5], ACO [8] or branch-and-bound techniques [9], by simpler heuristics [4,10] or by mixed integer programming methods using column-approach [11]. However, the last ones are often applied to simpler cases, i.e. less heterogeneous set of boxes with single bin cases [11] or two-dimensional cases [7] among other

M. Ali et al. (Eds.): IEA/AIE 2013, LNAI 7906, pp. 535–544, 2013.
© Springer-Verlag Berlin Heidelberg 2013

features that reduce the number of possible cutting patterns, due to the computational burden [2]. In this way, as the addressed problem is one of the most complex type of problems found in the literature, this study focuses on heuristics and metaheuristics.

The *extreme points* heuristic (EP) is used for boxes positioning inside the bin, once it is used successfully in this type of problems [5]. In an empty bin only one extreme point exists to pack the boxes in $(0, 0, 0)$ position.

For the residual bin packing problem only two studies are known until 2004, according to Wäscher [12]. More recently, a 3D residual bin packing problem (RBPP) was studied by Bang-Jensen [2] This study was also done within the industry, proving the increasing need of more complex algorithms, i.e. three-dimensional with multiple bins, in order to solve real-world problems.

This paper proposes a heuristic algorithm (3DHA) and an ACO algorithm, where both consider the 3D packing problem of multiple boxes inside multiple bins with guillotine cut constraint. The proposed ACO algorithm searches for the best orders of the boxes and it is guided by the proposed heuristic (3DHA) that determines the position and orientation for the boxes. The orientation of a box is chosen with a new proposed method inspired on Allen et al. [1]. The algorithm is applied to a real-world problem in the steel industry.

In Section 2 the real-world problem is formulated and explained with C&P convention. Section 3 and Section 4 introduces the heuristic algorithm and the ACO algorithm, respectively. Section 5 shows the results obtained with the two algorithms and it is compared to the results obtained by an anonymous company. Finally, Section 6 presents some conclusions and future work.

## 2    Steel Cut Problem

The retail steel cut industry consists of cutting clients orders with non-standard dimension from the available steel in stock, where the optimization of the cut process can have a significant economical effect. The addressed problem consists on executing orthogonal cuts for a set of rectangular-shaped boxes from rectangular-shaped bins. Each box $j$ $(j = 1, ..., n)$ is characterized by length $l$, width $w$ and depth $d$, where $l \geq w \geq d$. The bins are characterized by length $L$, width $W$ and depth $D$, where $L \geq W \geq D$. All orientations for the boxes are allowed. The coordinates origin is located at the bottom-left-behind of the bin, and $(x_j, y_j, z_j)$ is the point where the bottom-left-behind of the box $j$ is positioned.

In order to have a feasible arrangement, the following constraints have to be met: 1) Each box is placed completely within the bin; 2) A box $i$ may not overlap another box $j$; 3) Each box is placed parallel to the side walls of the bin, which comes from the orthogonal packing.

The blocks of steel (bins) can belong to three different classes: suppliers $(P_{supp})$ that are bought from a supplier at a given date; in stock $(P_{stk})$; or scrap metal $(S)$. $P_{supp}$ and $P_{stk}$ are available to use as bins while the blocks of steel $S$ are too *small* and are sold as scrap metal. The company considers *small*, and thus scrap metal, blocks of steel that are lighter than a certain weight (10kg).

The clients orders are the boxes. Each order can have more than one identical box. The cut of steel is characterized by the following constraints: 1) All the cuts have to be edge-to-edge, also known as guillotine cut; 2) There are no cuts perpendicular to the smallest dimension of the bin $D$. In general terms the objective of the optimization is to minimize the total amount of scrap metal $S$. In practice the use of raw material $P_{supp}$ and the number of blocks in stock $P_{stk}$ are also minimized due to the methods used to sort the bins.

As the considered problem is divided into several small problems (one per day), the final stock of one day becomes the initial stock on the next day. More precisely, the residual bins of one day become the bins of the next day. The amount of scrap metal is computed at the end of each day.

## 3   3D Heuristic Algorithm

This section describes the 3D heuristic algorithm (3DHA) proposed in this paper. The algorithm starts by selecting the boxes and bins available for packing at the first processing day.

Using the available bins and boxes, the packing heuristic algorithm sorts the boxes by decreasing area $(d \times w)$. Each box in the list represents a subset of identical boxes (equal sizes). It selects the first type of box in the list and selects the *smallest bin* that produces a feasible packing with the used box, where the *smallest* bin is the one with the smallest depth, $D$.

It then chooses the best orientation of the box, and packing direction if more than one identical box is available, by maximizing the used space. The algorithm computes the new extreme points and decides how to cut the bin for each extreme point using virtual cuts as explained in Section 3.2. Thus, the *available volume* of the extreme points is also updated according to the virtual cuts. This process repeats until the selected type of box is completely packed, i.e. all boxes of that type are packed, continuing to the next type of box on the ordered list of boxes.

When the packing algorithm reaches the list end (end of a day), all the bins have to be cut (with definitive cuts). The residual bins that do not meet the minimum weight (10kg) are discarded while the others will be the bins available for the next day $(P_{stk})$ among the bins $(P_{supp})$ that will be available at the next day.

The algorithm described previously is graphically represented in Figure 1.

*Available Volume for Extreme Points:* To define the available volume for each extreme point, which is necessary to guarantee the guillotine cut constraint, it is necessary to verify what residual bin a point belongs to. Thus, the available dimensions of an extreme point will be the dimensions of the residual bin to which it belongs. In case of an empty bin (i.e. no residual bins yet) the available dimensions for the single extreme point at $(0, 0, 0)$ are the dimensions of the bin.

### 3.1   Packing Orientation and Direction

The orientation of the volume to be placed is very important as it maximizes the chances of packing that volume inside a bin. This leads to the maximization of both the number of boxes and the used volume inside a bin.

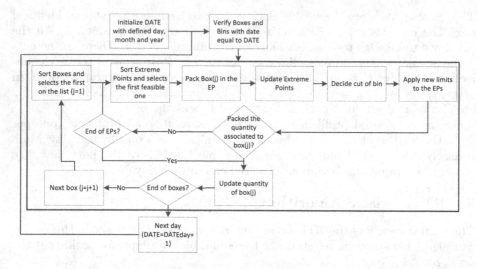

**Fig. 1.** 3D Heuristic Algorithm

This method tries every possible orientation for a box. The addition of multiple boxes is considered using a set of boxes of the same type. This way, the set of boxes is defined by multiple boxes aligned in x,y or z direction, with all the boxes having the same orientation. To evaluate how good the packing was, the fitness for each packing is computed as in Figure 2. The left example of Figure 2 packs less boxes. However, it has a perfect fit (100%) on one of its dimensions minimizing the generated scrap metal.

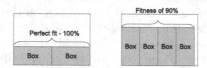

**Fig. 2.** Example of fitness

## 3.2  Guillotine Cut

For the guillotine cut constraint, it was developed the *open guillotine cut* which is concerned on the optimality of the guillotine cut as a more general approach. This method lets the algorithm to be open to different guillotine cuts for the next volume to be packed, using virtual cuts for each extreme point. To get a guillotine cut, the Equation (3.1) must be satisfied.

$$\left( \sum_{j=1}^{N} x_j \geq x + \sum_{j=1}^{N} x_j + w_j \leq x \right) \geq N \qquad (3.1)$$

Being $w_j$ the width and $x_j$ the position of box $j$, $x$ the guillotine cut position and $N$ the number of boxes. If the orientation or/and tested direction is different, $w_j$ becomes $d_j$ or $l_j$. The function is generalized and represented here for $x$ direction but is valid for every orientation and direction.

In this method, the idea is that for each candidate point it is computed how to cut the initial bin all the way to a residual bin resulted from multiple cuts. Thus, there may exist several different cutting patterns for one bin, depending on the extreme point.

It was considered that the way to cut the bin should be decided based on minimizing the reduction (in percentage) of an EP available volume. In this way, if more than one guillotine cut is available, the algorithm chooses the one that results in the lowest reduction of the EP available dimensions. This way if a cut is possible in a lower position than the EP, that cut is preferable. Also, if an available cut does not remove any box (giving no advantage and reducing the available volume), this cut is neglected by assigning the reduction as infinite as in Figure 3. Notice that it must exist at least one cut that can reduce (at least) one box. Otherwise, the guillotine cut constrain would not be met.

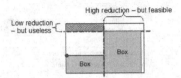

**Fig. 3.** Example of unnecessary cut reducing the available volume of an extreme point

In Figure 4 and Figure 5 it is possible to observe one example of the described above. Considering one bin (represented by the white rectangle), with a box already packed, there is (for the bi-dimensional case) two extreme points to pack the new box, represented by the dots. Each image represents two possible cuts. For the first point, Figure 4, the reduction is larger in the perpendicular dimension of the cutting direction, on the left side of the image. On the right side, on the opposite, the reduction is zero as the available limits are the limits of the bin itself.

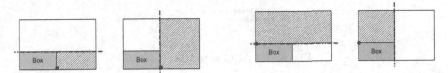

**Fig. 4.** Example of a cut for one extreme point

**Fig. 5.** Example of a cut for a different extreme point

As the cuts are only *virtual*, the reduction for the other extreme point, represented in Figure 5, is not determined by the previous cut for the first extreme point. Thus, for the second point, the best cut is the illustrated on the left side of Figure 5.

*Stock reduction:* After the packing algorithm is finished it is done the reduction of the final stock. If the final stock produced by the algorithm is higher than the stock of the company (in number of bins), the number of bins is reduced. The removed bins will be considered scrap metal as they are removed from the available bins, thus increasing the scrap metal of the algorithm. This way, it is also important that the algorithm does not produce many residual bins.

## 4    3D Ant Colony Optimization Algorithm

One of the most important subjects in cutting and packing problems is the order in which the boxes are packed [8]. ACO algorithms have proved to have a high performance for sequential ordering problems (namely the travelers man problem) [6]. This way, it is proposed an ant colony optimization algorithm combined with the packing heuristics described on the previous section. This ACO algorithm differs from the heuristic algorithm (3DHA) on the packing order of the boxes.

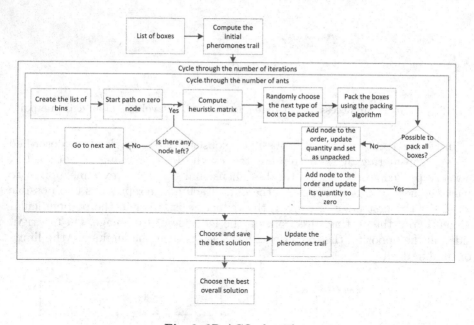

**Fig. 6.** 3D ACO algorithm

As the main goal of the (ACO) for this problem is to find the best order in which the boxes are packed, the boxes become the nodes of the ACO. The paths represent the order in which identical boxes are packed in the bin. The ACO algorithm, represented in Figure 6, is used each day of processing, searching for the best boxes order for each day.

The zero node is the starting point of each ant, representing the ants nest. In this way the path followed by the ant, is the path between the nest and the food source. The ants nest represents the initial available bins and cannot be visited again as no bins are created in this problem. If an ant cannot load all the food from one source (i.e. cannot pack all the boxes of that node) it continues to a different node other than the zero node. The ant carries all the bins for each node instead of only the last created bin as in Levine et al. [8]. This adaptation was made in contrast with the known approaches [8,3] as the objective of the present ACO approach was to solve the real-world problem of the company which has varying bins dimensions and different objectives. Here, the use of all the bins at the same time is not a problem as the packing (3DHA) heuristic selects the extreme points not by coordinates but by available volume (i.e. available depth).

*Heuristic Matrix:* The heuristic matrix introduces some heuristic information into the choice of a path. In this way, ants can know as a primitive knowledge the boxes that should be packed in the first place. As the paths are the orders representation of the boxes and are closer or not from the nest (having higher probability of being chosen) by a characteristic (volume, area or depth), these are only heuristic paths of the problem.

This virtual path with known distances between nodes can actually give a misleading path distance as the real path is defined by the occupation of the bins and by the amount of scrap metal. However, this is only a guide path for initial solutions. The ants will provide the search of a better path by the use of pheromones, quantifying how good a packing is by taking into account the bins occupation and its associated scrap metal. Based on Section 3 and on Crainic et al. [5], several characteristics may be used. The heuristic matrix parameter from packing box $j$ after box $i$ is defined as: $\eta_{ij} = d_j \times w_j$.

*Pheromone Trail:* The pheromone trail is a way of inserting the efficiency of each ant into the problem. This is done by quantifying how good a packing was when moving from node $i$ to node $j$ (packing box $j$ after box $i$) which is given by $\Delta\tau_{ij}$. The update of the pheromone trail should be done using an evaporation coefficient $\rho$. For an iteration $m+1$ it is possible to write: $\tau_{ij}^{m+1} = \tau_{ij}^m(1 - \rho) + \Delta\tau_{ij}$. Since the goal is to minimize the scrap metal, the pheromone trail becomes a measure of this parameter. So the pheromone trail given by packing box $j$ after box $i$ is:

$$\Delta\tau_{ij} = \frac{\overline{\varphi_j}^q}{(\overline{\Psi_j} \times s_j \times \sum_{j=1}^N s_j)^p} \tag{4.1}$$

Where $\overline{\varphi_j}$ is the mean of bins percentage of occupied volume the box $j$ is found; $\overline{\Psi_j}$ is the mean of bins percentage of scrap metal where box $j$ is found; $s_j$ is the mean of bins scrap metal where box $j$ is found; $N$ is the number of boxes; and $p$ and $q$ are two factors to be parametrized. It was chosen to use the mean instead of the sum as the objective is not minimizing the number of used bins. Thus, the mean is a better choice as it does not affect how many bins were used. It will only consider how much scrap metal will be generated both in percentage and in kilograms. This way, this function differentiates small and large bins. However, if a bin has no scrap metal at all, this formula will not be valid has it tends to infinity. Thus, if that is the case, it is assumed that the percentage of scrap metal is 0.1%, the scrap metal is 1 kg and the percentage of occupied volume is 100%. Thus, this will be also the value of the initial pheromone, $\tau_0 = \frac{1}{0.001^P}$.

*Cost Function:* In order to guide the algorithm towards good solutions, it is needed a way to assess the quality of the solutions, being this a cost function. For the addressed problem the number of bins is fixed and the cost function could be a percentage of the generated scrap metal by each bin. However, the objective is not minimizing the percentage of scrap metal for each bin, leading to the consideration of the total scrap metal of each bin in kg. In order to avoid solutions that do not pack all the boxes, it is included a soft constraint in the cost function. $M$ stands for the number of unpacked boxes. Thus, the cost function is defined:

$$f = \sum_{j=1}^{N} s_j + M^{10} \tag{4.2}$$

*Path Probability:* The path probability becomes a conjunction of all the factors defined before; the pheromone trail and the heuristic matrix. This way, the probability of a block of boxes $j$ being chosen after a block of boxes $i$ is:

$$p_{ij}^k = \frac{\tau_{ij}^\alpha \eta_{ij}^\beta}{\sum_{j=1}^{N} \tau_{ij}^\alpha \eta_{ij}^\beta} \tag{4.3}$$

$p_{ij}^k$- Probability of ant $k$ going from node $i$ to node $j$; In this way, it is possible to give weights to the heuristic matrix or the pheromones by tuning $\beta$ and $\alpha$. Notice that this will give relative weights between the heuristic and the pheromones but also inside them. That is, it differentiates more or less the values in the matrices as they are powered by $\alpha$ or $\beta$.

## 5   Application Example

The problem being addressed here is from an anonymous company from the steel cut industry where the objective is to reduce the scrap metal while reducing the raw material usage and blocks in stock. The used data corresponds to 3 months of operations. This section presents the characteristics of the problem in Section 5 and the computational results are shown in Section 5.

*Data:* Table 1 presents the most important characteristics of the problem. Volumetric data is in cubic millimeters. $\overline{V}$ is the average of volume. The presented data is proportional to the real data because of confidentiality issues. Looking at the relative standard deviation (RSD), it is possible to observe the heterogeneity of the boxes and bins. Both have a volume RSD above 250%.

**Table 1.** Characteristics of boxes and bins

| Types | Number | | Min. Vol. | Max. Vol. | $\overline{V}$ | $\sigma(V)$ | $\frac{\sigma}{\overline{V}}$ |
|---|---|---|---|---|---|---|---|
| Bins | 5340 | 5340 | $17 \times 10^5$ | $17 \times 10^9$ | $7 \times 10^8$ | $23 \times 10^8$ | 3.45 |
| Boxes | 2065 | 2910 | $15 \times 10^4$ | $7 \times 10^9$ | $22 \times 10^7$ | $6 \times 10^8$ | 2.87 |

*Results:* This section presents the results for both heuristic and ACO algorithms in comparison with the available results from the company. The scrap metal generated by the company is estimated to be between 5% and 10%. The necessary orders to suppliers done from the company is 195 while the final stock number is 6430. The ACO algorithm was parameterized/tuned with the following parameters: 20 ants; 40 iterations; $\rho = 0.3$; $\alpha = 2$; $\beta = 1$; $p = 1$; $q = 1$. Forty iterations were enough to stabilize the solution. In some cases, the solution was stable from the 20th iteration. A distinction is done between the results before the reduction (BR) of the final stock and the final results of the algorithms.

**Table 2.** Results for each algorithm and used method by the company

| Algorithm | Bins in stock | Scrap metal [%] | Orders to supp. | Red. of raw material [%] |
|---|---|---|---|---|
| 3DHA (BR) | 6850 | 0.09% | 150 | 23.1% |
| 3DACO (BR) | 6895 | 0.08% | 145 | 25.6% |
| Company | 6430 | 5-10% | 195 | - |
| 3DHA | 6430 | 0.48% | 150 | 23.1% |
| 3DACO | 6430 | 0.50% | 145 | 25.6% |

The results for the developed algorithms are shown in Table 2, where it is possible to observe a reduction of scrap metal and orders to the suppliers by the developed algorithms. However, the scrap metal was not reduced in the ACO approach when compared to the 3DHA after the reduction of stock. Thus, the 3DACO produced more scrap metal once it generated too many residual bins in order to reduce the scrap metal at each day using also less raw material. Nevertheless, the 3DACO reduced the number of raw material by a difference of 2% (or reduction of 3.3% bins). In fact, the difference of 0.02% of scrap metal between the two algorithms is negligible (for the company) compared to the cost of ordering five extra blocks (3.3% more bins) of steel from the supplier.

## 6   Conclusions

This paper proposes a 3DACO algorithm to solve a 3D BPP in the steel industry. This algorithm is based in a box-positioning heuristic, the extreme points proposed by Crainic et al. [5]. The 3DACO uses the 3DHA heuristic to determine the box positioning and orientation. It was shown that 3DHA is a suitable method to pack thousands of boxes in thousands of bins, so as with the metaheuristics enhancements which resulted in the 3DACO algorithm. Using the 3DACO it was possible to improve the results against the 3DHA. Both 3DHA and 3DACO algorithms provide better solutions compared with the methodologies used by the company. However, it is important to leave a final note about some practical aspects experienced day-to-day that were not formulated in this approach such as: cutting times, handling times or the difficulties when retrieving blocks of steel from large warehouses. They will be considered in future work.

**Acknowledgments.** This work is part of ToolingEdge project, partly funded by the Incentive System for Technology Research and Development in Companies, under the Competitive Factors Thematic Operational Programme, of the Portuguese National Strategic Reference Framework, and EU's European Regional Development Fund. The authors also wish to acknowledge the project partners for their contribution to the work presented in this paper. This work was also supported by Projeto Estratégico, PEst-OE/EME/LA0022/2011, from FCT (Unidade IDMEC Pólo IST, Grupo de Investigao IDMEC/LAETA/CSI) and by the grant SFRH/BPD/65215/2009 from Fundação para a Ciência e a Tecnologia (FCT), Ministério da Educação e Ciência.

# References

1. Allen, S.D., Burke, E.K., Kendall, G.: A hybrid placement strategy for the three-dimensional strip packing problem. European Journal of Operational 207, 219–227 (2011)
2. Bang-Jensen, J., Larsen, R.: Efficient algorithms for real-life instances of the variable size bin packing problem. Computers & Operations Research 39(11), 2848–2857 (2012)
3. Beirão, J.F.F.: Packing problems in industrial environments: Application to the expedition problem at indasa (November 2009)
4. Bortfeldt, M., Mack, D.: A heuristic for the three-dimensional strip packing problem. European Journal of Operational Research 183, 1267–1279 (2007)
5. Crainic, T.G., Perboli, G., Tadei, R.: Extreme point-based heuristics for three-dimensional bin packing. Informs Journal on Computing 20(3), 368–384 (2008)
6. Dorigo, M., Stützle, M.D.T.: Ant colony optimization. Mit Press, Bradford Bks (2004)
7. Gomory, P.C., Gilmore, R.E.: Multistage cutting stock problems of two and more dimensions. Operations Research 13(1), 94–120 (1965)
8. Levine, J.M., Ducatelle, F.: Ant colony optimisation and local search for bin-packing and cutting stock problems. Journal of the Operational Research Society 55(7), 705–716 (2004)
9. Martello, S., Pisinger, D., Vigo, D.: The three-dimensional bin packing problem. Operations Research 48(2), 256–267 (2000)
10. Pisinger, D.: Heuristics for the container loading problem. European Journal of Operational Research 141, 382–392 (2002)
11. Queiroz, T.A., Miyazawa, F.K., Wakabayashi, Y., Xavier, E.C.: Algorithms for 3d guillotine cutting problems: Unbounded knapsack, cutting stock and strip packing. Computers and Operations Research 39, 200–212 (2012)
12. Wäscher, G., Haußner, H., Schumann, H.: An improved typology of cutting and packing problems. European Journal of Operational Research 183(3), 1109–1130 (2007)
13. Whitwell, G.: Novel heuristic and metaheuristic approaches to cutting and packing, Phd (2004)

# Controlling Search Using an S Decreasing Constriction Factor for Solving Multi-mode Scheduling Problems

Reuy-Maw Chen[*] and Chuin-Mu Wang

Computer Science and Information Engineering, National Chin-Yi University of Technology
Taichung 411, Taiwan, R.O.C
{raymond,cmwang}@ncut.edu.tw

**Abstract.** The multi-mode resource-constrained project scheduling problem (MRCPSP) is an important issue for industry, and has been confirmed to be an NP-hard problem. The particle swarm optimization meta-heuristic is an effective and promising method and well applied to solve a variety of NP application problems. MRCPSP involves two sub-problems: the activity mode selection and the activity order sub-problems. Therefore, a discrete version PSO and constriction version PSO were applied for solving these two sub-problems respectively. Discrete PSO is utilized for determining the activity operation mode, the constriction PSO is applied for deciding the activity order. To enhance the exploration and exploitation search so as to improve search efficiency, an S decreasing constriction factor adjustment mechanism was proposed. To verify the performance of proposed scheme, instances of MRCPSP in PSPLIB were tested and comparisons with other state-of-art algorithms were also conducted. The experimental results reveal that the proposed S decreasing constriction factor adjustment scheme is efficient for solving MRCPSP type scheduling problems.

**Keywords:** Multi-mode resource-constrained project scheduling problem, Particle swarm optimization, Constriction PSO, S decreasing constriction factor.

## 1    Introduction

Resource constrained scheduling problem is often accompanied with some constraints such as limited resources and process precedence. In recent years, resource constrained problems have been widely applied in many application fields; hence have become many researchers' interested issue. Most resource constrained scheduling problem mainly considers how to effectively use limited resources, and finishes project in the shortest time. Restated, the target of the scheduling is to finish tasks and obtain minimum *makespan* with constrained resources. There is one well-known resource-constrained project scheduling problem (RCPSP). RCPSP contains two constraints: renewable resources (e.g., project human resources and equipment); activity precedence (the activity can be processed only when all predecessor activities are

---

[*] Corresponding author.

M. Ali et al. (Eds.): IEA/AIE 2013, LNAI 7906, pp. 545–555, 2013.

finished). The RCPSP is often regarded as a uni-modal model which assumes every activity has only one execution mode. Therefore, RCPSP is limited case for real-word project scheduling problem. On the other hand, multi-mode RCPSP is named MRCPSP (multi-mode resource-constrained project scheduling problem). In MRCPSP, each activity has different operation modes, and under different modes, the amount of resources required and processing time for each activity vary. Moreover, two types of resource constraints are involved, one is renewable resources and the other is non-renewable resources (e.g., capital cost and material). Thus, MRCPCP is more conform to the real-world project scheduling problem. MRCPSP has been confirmed to be an NP-hard problem [1]. Thus, many studies have suggested a variety of methods for solving MRCPSP, such as simulated annealing (SA) [2], tabu search [3], genetic algorithm (GA) [4], ant colony optimization (ACO) [5], particle swarm optimization (PSO)[6, 7, 8], etc.

PSO has some advantages over other methods: fewer parameters setting hence PSO is easier to implement, and efficient for solving problems. PSO is a meta-heuristics algorithm, a promising scheme, and has been widely applied to solve various scheduling application problems like TSP [8], flowshop [9], task scheduling in grid [6] and VRP [10]. MRCPSP can be regarded as a combinatorial problem activity operation mode and activity processing order. Restated, MRCPSP can be divided into two sub-problems: activity mode selection and activity (work) order sub-problems. This study applied two PSOs for solving these two sub-problems. Constriction version PSO is used to solve activity order sub-problem; discrete version PSO is utilized to solve activity mode selection sub-problem. The velocity update rule used in both PSOs is based on constriction version. Meanwhile, a novel constriction factor adjustment mechanism based on named S decreasing was designed to prevent falling on local solution due to too early convergence and make the algorithm obtain better search effect, i.e., starts exploration then towards exploitation. To verify the performance of the proposed scheme, instances of MRCPSP in PSPLIB were tested, and compared the results with other study results. The test results confirmed that the proposed scheme outperforms most algorithms, and MRCPSP can be solved effectively and efficiently. The remainder of this article is organized as follows. Section 2 provides a description of MRCPSP. Section 3describes the particle swarm optimization. The proposed inertia weight design and communication topology applied in PSO are presented in Section 4. Section 5, experimental results and analysis are demonstrated. Finally, the conclusions are given in Section 6.

## 2     Problem Description

In this Section, the interested MRCPSP problem is described. The multi-mode resource-constrained project problem (MRCPSP) includes both renewable and non-renewable resource and precedence constraints. Meanwhile, each activity has different operation modes associated with different processing time and resources requirement. MRCPSP aims to obtain the project schedule with the minimum *makespan*. MRCPSP is defined as follows:

- MRCPSP consists of the n activities. Activities have precedence constraint. Each activity is not preemptive, indicating that the activity should not be interrupted, and successor activities can be performed after current activity is finished.
- In MRCPSP, $M_j$ denotes the activity $j$ has several operation modes, i.e., $Mode_j=\{1,...,M_j\}$. Each operation mode has different processing time and resources demand.
- At certain time duration, the supply of renewable resources is fixed. Throughout the project, the supply of the non-renewable resources is fixed.
- In MRCPSP, the activity under different mode needs different renewable resources and non-renewable resources. Thus, a resulting schedule is said to be a feasible solution when it meets renewable resources and non-renewable resources constraints; otherwise it will be an infeasible solution.

# 3    Particle Swarm Optimization

Solving MRCPSP can be considered as solving two optimization sub-problems: activity mode selection and activity priority sub-problems. PSO has become a common meta-heuristic scheme in solving optimization problems. This study uses two PSOs to solve these two sub-problems. A discrete PSO is applied to solve activity mode selection, while the constriction PSO is utilized to solve activity priority. The constriction PSO and discrete PSO are described as follows.

## 3.1    Inertia Weight PSO

The PSO algorithm was first proposed by Eberhart and Kennedy in 1995[11]. Each particle represents a problem solution in PSO, and the PSO is initialized with a population of particles randomly spread on solution space. Through movement of particles in solution space, the optimal position (optimal solution) can be found. PSO uses two experience positions to decide particle velocity and then determine where to move. One experience is individual experience and the other is global experience. Individual experience represents particle-self obtained best position. Global experience records optimal particle solution of the swarm. In each iteration, particles use Eq. (1) to update their velocity vector, and determine new position. Restated, the velocity plays a role of movement, hence the movement to new position changes with velocity. Suppose there are N particles in a D-dimensional space representing solution consists of $D$ components, position of the i-th particle (i=1,...,N) is represented by $X_{ij}=\{X_{i1},...,X_{iD}\}$, where $X_{ij}$ denotes the component of the $j$th component of the $i$th particle. Correspondingly, the velocity of the $i$th particle is represented by $Vi=\{V_{i1},...,V_{iD}\}$. The $i$th particle's individual experience is expressed by $L_i=\{L_{i1},...,L_{iD}\}$; the global experience is expressed by $G=\{G_1,...,G_D\}$. The velocity and position update rule is displayed in Eq.(1).

$$\begin{cases} V_{ij}^{new} = w \times V_{ij}^{old} + c_1 \times r_1 \times (L_{ij} - X_{ij}) + c_2 \times r_2 \times (G_j - X_{ij}) \\ X_{ij}^{new} = X_{ij} + V_{ij}^{new} \end{cases} \tag{1}$$

Where $w$ is the inertia weight, used to control the influence of current velocity $V^{old}$ on new velocity $V^{new}$. Therefore, the use of adequate inertia weight is important for PSO algorithm. Meanwhile, $c1$ and $c2$ are learning factors. These two learning factors design is to control influence of swarm experience and self-experience on new velocity. Moreover, $r1$ and $r2$ are the random numbers distributed between [0,1].

## 3.2     Constriction PSO

In PSO, the velocity update is important and affects the solution quality. Thus, different versions of PSO were derived, which focus on velocity update. The inertia weight $w$ design was included to balance global and local search while avoiding velocity limitation by adjusting the influence of the previous particle velocities on the optimization process. At same time, another explored method of balancing global and local searches is known as constriction. Restated, the search behavior either exploration or exploitation is controlled by constriction factor. Hence, a named constriction PSO was proposed by Bratton and Kennedy [12]. The developed constriction PSO velocity update rule is listed in Eq. (2).

$$\begin{cases} V_{ij}{}^{new} = \chi \times \left( V_{ij} + c_1 \times r_1 \times (L_{ij} - X_{ij}) + c_2 \times r_2 \times (G_j - X_{ij}) \right) \\ X_{ij}{}^{new} = X_{ij} + V_{ij}{}^{new} \end{cases} \tag{2}$$

$\chi$ is constriction factor used to constrain velocity, and this velocity update rule is proposed for its stability as indicated in [12].

## 3.3     Discrete PSO

The discrete version particle swarm algorithm, proposed by Eberhart and Kennedy in 1997 [13], has been applied for solving many complicated problems including scheduling. The solution components in discrete PSO (DPSO) consist of two values either 0 or 1, $X_{ij}=0$ or $X_{ij}=1$. In this investigation, the velocity update used in discrete PSO is based on constriction velocity update rule as in Eq. (2). However, the velocity plays a different role in discrete PSO; the velocity becomes a probability of being 1 or 0 for the corresponding position component. Eberhart and Kennedy suggested that the position component $X_{ij}$ corresponding to higher $V_{ij}$ has higher probability to be equal to 1; in contrast, the position component $X_{ij}$ corresponding to lower $V_{ij}$ has higher probability to be equal to 0. Thus, discrete $X_{ij}$ determination by $V_{ij}$ is defined in Eq. (3).

$$\begin{cases} X_{ij} = 1, & if \quad rand( \quad ) < S\left( V_{ij}{}^{new} \right) \\ X_{ij} = 0, & else \end{cases} \tag{3}$$

where $S( )$ is a sigmoid function used to do probability transformation as defined in Eq. (4). The $rand( )$ is the generated random number between (0, 1), $S(V_{ij}{}^{new}) \in (0,1)$.

$$S\left( V_{ij}{}^{new} \right) = \frac{1}{1 + \exp\left( - V_{ij}{}^{new} \right)} \tag{4}$$

To prevent $S(V_{ij}{}^{new})$ from too approaching 0 or 1, $V_{max}$ can be used to constrain new velocity $V_{ij}{}^{new}$ range as displayed in Eq. (5).

$$V_{ij}{}^{new} = \chi (V_{ij} + r_1 \times (L_{ij} - X_{ij}) + r_2 \times (G_j - X_{ij}))]_{-V_{max}}^{V_{max}} \tag{5}$$

## 4    The Proposed Scheme

In this work, the activity execution mode is determined using discrete PSO, in which the velocity update is based on constriction PSO as in Eq. (2) rather than conventional PSO as in Eq. (1). The activity order is decided based on constriction PSO. This study designs a novel constriction factor adjustment mechanism which dynamically changes constriction factor during search process, allowing the algorithm to obtain better search effect from global search to local search gradually. In this Section, some other well used search control mechanisms and the suggested S decreasing constriction factor mechanism are presented.

### 4.1    Encoding Activity Mode Select and Activity Priority

In this study, solving MRCPSP is to solve activity mode selection and activity order problems. Different activity modes require different resources and execution times, thus suitable modes for activities are vital. When satisfying the resource condition constraint, the activity order has great impact on project completion time. Hence, determining the best activity order of a project is also very important. Restated, obtaining a better project schedule requires adequate mode selection and activity order. This study uses discrete POS for determining activity mode list corresponding to all activities; the activity mode list is expressed by position vector $X^M$.

In DPSO, the components of the individual experience and global experience consist of binary values. Thus, the encoding of needed mode bits ($bits_j$) for each activity $j$ ($X_j^M$) in position vector $X^M$ can be decided using the upper bound of $\log_2(M_j + \varepsilon)$; $\varepsilon$ is a small number close to 0; $M_j$ denotes the $m$ available modes for the $j$-th activity. If the components for the mode of activity $j$ in the mode list is $X_j^M = \{1\ 0\}$, this means activity $j$ is assigned to execute on mode 2.

The activity order list is presented by a position vector $X^J$ consisting of $N$ components. However, when deciding activity order, the order values are not allowed to be repeated. Thus, a random key scheme (Hartmann & Kolisch 1999[14]) is used. Each component of $X^J$ is associated with an integer key according to activity sequence. Suppose PSO position vector $X^J$ contains seven position components, that is $X^J = \{ 4.7, 7.3, 3.6, 0.16, 8.8, 0.2, 1.6 \}$; the corresponding keys order list is $\{0, 1, 2, 3, 4, 5, 6\}$. To obtain new activity order, $X^J$ are sorted by component values in ascending (or descending) order, and accordingly associated keys order also change. The sorted $X^J = \{0.16, 0.2, 1.6, 3.6, 4.7, 7.5, 8.8\}$ and obtained keys order is the activity order list that is $\{3, 5, 6, 2, 0, 1, 4\}$.

## 4.2     Search Control Schemes by Inertia Weight

The inertia weight parameter proposed by Shi and Eberhart in 1998 [15] significantly improve PSO efficiency. Restated, the balance between global and local search in conventional PSO is directly controlled via inertia weights. Starting from global search and toward to local search is convinced to provide a better search strategy. Meanwhile, an inertia weight decreased with time was confirmed to be better than a fixed inertia weight [15]. Moreover, a better search strategy is to develop in global search at first, and then in local search in middle and late period [16]. In [15], a linear decreasing inertia weight was used as shown in Eq. (6).

$$w_{iter} = \frac{iter_{max} - iter}{iter_{max}}(w_{max} - w_{min}) + w_{min} \qquad (6)$$

where, $w_{iter}$ is the inertia weight used in iteration, $iter_{max}$ is the maximum iteration number, $iter$ is the iteration number, $w_{max}$ and and $w_{min}$ are the maximum starting value of inertia weight and the minimum ending value of inertia weight. Commonly applied $w_{max}$ and $w_{min}$ are 0.9 and 0.4 respectively.

In [17], non-linear decreasing inertia weights based on decline curves (winding curves) were proposed to solve some benchmark problems; the inertia weight adjustment is based on Eq. (7). Equation (7) aims to control global search ability and local search ability and avoid too early convergence.

$$w_{iter} = \frac{1 - (iter/iter_{max})}{1 + s \times (iter/iter_{max})} \qquad (7)$$

where $s$ is a parameter controlling curvature of curves.

Nevertheless, it is difficult for local search to find an optimal solution in the end without a good search direction at the beginning of global search. Accordingly, search in solution space is not stable using nonlinear decline inertia weight design.

Thus, some studies have discussed various inertia weight adjustment ways to update velocity. A sigmoid decreasing inertia weight was proposed to increase PSO performance as in [18]. The sigmoid decreasing inertia weight design is defined in Eq. (8).

$$w_{iter} = \frac{(w_{max} - w_{min})}{1 + e^{-p \times (iter - n \times iter_{max})}} + w_{min} \qquad (8)$$

Where $p$ is a constant (usually set to 2) to vary sharpness of the function and $n$ is the constant (usually set to 0.5) to set division of the function.

## 4.3     Proposed Search Control Scheme by Constriction Factor

Both inertia weight and constriction factor are designed to balance global exploration and local exploitation abilities. This study proposes a novel velocity update rule on the basis of designed s decreasing constriction factor rather than inertia weight. In order to obtain good search control ratio of global search to local search, the designed S decreasing constriction factor mechanism is displayed in Eq. (9).

$$\chi_{iter} = 1 / \{1 + [iter_{max} - (iter_{max} - iter) / 100]^4\} \tag{9}$$

where, $\chi_{iter}$ is the constriction factor used in iteration. The designed S decreasing overcomes insufficient frequency of global search and local search as linearity decreases, lower frequency of global search in nonlinear decline and less gentle search between global and local search in sigmoid decreasing. As compared to another three decline methods, S decreasing constriction factor can highlight the search characteristic shift from global search gradually towards local search. Accordingly, the procedure of proposed enhanced particle swarm optimization is summarized as below.

| | |
|---|---|
| 1. | Initialize the particle's position and velocity |
| 2. | Iteration loop |
| 2.1 | For each particle $i$ in the swarm |
| 2.2 | Update velocity vector $V_i^{'new}$ and position vector $X_i^{'new}$ using Eq. (2) ("constriction" PSO) and Eq. (9) (S decreasing constriction factor), respectively. |
| 2.3 | Update velocity vector $V_i^{M^{new}}$ and position vector $X_i^{M^{new}}$ using Eq. (5) ("constriction" PSO) and Eq. (3) (discrete PSO), respectively. |
| 2.4 | Calculate mode vector $M$ based on $X_i^{M^{New}}$. |
| 2.5 | Calculate activity priority vector $J$ based on $X_i^{J^{New}}$ to generate activity priority list $A$ (using random key scheme) |
| 2.6 | Update $L_i$, and $G$ (gbest or lbest communication). |
| 3. | Until End condition is reached, return solution |

**Fig. 1.** The pseudo-code of proposed PSO scheme

## 5    Experimental Results

To verify the efficiency of proposed method for solving MRCPSP; the simulation instances in well-known Project Scheduling Problem Library (PSPLIB) (http://129.187.106.231/psplib/) were tested. PSPLIB includes 10 to 30 jobs (activities) and every activity is with 3 different usable execution modes. Additionally, each activity case has different project instances (ex. 536 instances for J10 case, 554 instances for J20 case and 552 instances for J30 case). The experiment using different search control schemes, constant ($\chi$ =0.7289), linear decreasing (Eq. (6)), decline curve (Eq. (7)), sigmoid decreasing (Eq. (8)) and proposed S decreasing (Eq. (9)) constriction factors, were conducted ($\chi$ =0.9~0.4 for decreasing constriction factor). Table 1 shows the experimental results of applying three different inertia weight mechanisms for solving MRCPSP for J10 case   and J20 case. The comparison is based on 5000 feasible solutions. In Tables 1, "Dev. BKS (%)" represents the average deviation from the best known solutions (BKS) as defined in Eq. (10). Meanwhile, "OPT(%)" indicates the ratio of numbers of the optimal solution found as defined in Eq. (11).

$$Dev.BKS = \sum_{instances} \left( \frac{fitness_i - best_i}{best_i} \times 100\% \right) \Big/ instances \tag{10}$$

$$OPT = (number\ of\ optimal\ solutions\ found / number\ of\ instances) \times 100\% \tag{11}$$

**Table 1.** Performance comparison on different constriction factor adjustment mechanisms with 5000 schedules on J10 and J20 ($\chi$:0.9 to 0.4)

| Instance set | J10 | | J20 | |
|---|---|---|---|---|
| | Dev.BKS(%) | OPT(%) | Dev.BKS(%) | OPT(%) |
| Constant | 0.01 | 99.63 | 0.77 | 82.85 |
| Linear decreasing | 0.02 | 99.44 | 0.71 | 82.49 |
| Decline curve | 0.02 | 99.44 | 0.73 | 82.85 |
| Sigmoid decreasing | 0.08 | 98.13 | 0.89 | 78.34 |
| S decreasing | 0.01 | 99.93 | 0.71 | 83.75 |

Moreover, Table 2 demonstrates the experiment results for solving J30 instance. For J30 instance, optimal solution may not be available; hence, make Dev.BKS impractical (or impossible). Instead, comparison criterion based on "Inc. CP (%)" is provided on the second column. The Inc. CP (%) indicates the solution is how far from the critical path (CP), and is defined in Eq. (12). The last two columns "Equal (%)" and "Worse (%)" on the Table, show the percentages of instances which result in equal and worse than the best known solution.

$$Incr.CP = \sum_{ins\,tan\,ces} \left( \left( \left( fitness_i - CP_i \right) / CP_i \right) \times 100\% \right) \Big/ instances \qquad (12)$$

**Table 2.** Performance comparison on different constriction factor adjustment mechanisms with 5000 schedules on J30 ($\chi$:0.9 to 0.4)

| | Dev. BKS (%) | Incr. CP (%) | Worse (%) | Equal (%) |
|---|---|---|---|---|
| Constant | 1.78 | 14.76 | 60.62 | 69.38 |
| Linear decreasing | 1.66 | 14.63 | 29.35 | 70.65 |
| Decline curve | 1.68 | 14.37 | 31.16 | 68.84 |
| Sigmoid decreasing | 2.04 | 14.63 | 35.15 | 64.85 |
| S decreasing | 1.66 | 14.61 | 29.89 | 70.11 |

**Table 3.** Average Dev.BKS comparison on different constriction factor adjustment mechanisms on J10, J20 and J30

| Dev.BKS (%) | Instance set | | | |
|---|---|---|---|---|
| | J10 | J20 | J30 | Avg |
| Constant | 0.01 | 0.77 | 1.78 | 0.85 |
| Linear decreasing | 0.02 | 0.71 | 1.66 | 0.80 |
| Decline curve | 0.02 | 0.73 | 1.68 | 0.81 |
| Sigmoid decreasing | 0.08 | 0.89 | 2.04 | 1.00 |
| S decreasing | 0.01 | 0.71 | 1.66 | 0.79 |

Simulation results demonstrate that the S decreasing constriction factor has better performance than constant, linear decreasing, decline curve and sigmoid decreasing constriction factors as displayed in Tables 3. Meanwhile, performance comparison of the different algorithms on the J10, J20 and J30 cases of PSPLIB was conducted, the proposed S decreasing constriction factor was used in experiment. Table 4 displays the comparison of simulation results of the J10 case and J20 case. Table 5 displays the comparison of simulation results of the J30 case. Table 6 summarizes the overall performance on J10, J20 and J30 by applied different algorithms.

**Table 4.** Comparison of algorithms with respect to 5000 schedules on J10 and J20

| Instance set | J10 | | J20 | |
|---|---|---|---|---|
| | Dev.BKS(%) | OPT(%) | Dev.BKS(%) | OPT(%) |
| This work | 0.01 | 99.63 | 0.71 | 83.75 |
| Van Peteghem et al. (2010) [4] | 0.01 | 99.63 | 0.57 | 85.74 |
| Chiang et al. (2008)[5] | 0.34 | 99.81 | 1.79 | 88.27 |
| Jozefowska et al. (2001)[19] | 1.16 | 85.60 | 6.74 | 35.70 |
| Lova et al. (2009) [21] | 0.06 | N/A | 0.87 | N/A |
| Ranjbar et al. (2009) [22] | 0.18 | N/A | 1.64 | N/A |
| Jarboui et al. (2008) [7] | 0.03 | N/A | 1.10 | N/A |

**Table 5.** Comparison of algorithms with respect to 5000 schedules on J30

| | Dev. BKS (%) | Incr. CP (%) | Worse (%) | Equal (%) |
|---|---|---|---|---|
| This work | 1.66 | 14.61 | 29.89 | 70.11 |
| Van Peteghem et al. (2010)[4] | 1.08 | 13.75 | 29.00 | 71.00 |
| Chiang et al. (2008)[5] | 2.60 | N/A | 27.36 | 72.64 |
| Jozefowska et al. (2001)[19] | 11.76 | N/A | 74.40 | 25.60 |
| CPSO[20] | N/A | N/A | 42.59 | 57.41 |

**Table 6.** Average Dev.BKS comparison on different algorithms on J10, J20 and J30

| Dev. BKS (%) | J10 | J20 | J30 | Avg |
|---|---|---|---|---|
| This work | 0.01 | 0.71 | 1.66 | 0.79 |
| Van Peteghem et al. (2010)[4] | 0.01 | 0.57 | 1.08 | 0.55 |
| Chiang et al. (2008)[5] | 0.34 | 1.79 | 2.60 | 1.58 |
| Jozefowska et al. (2001)[19] | 1.16 | 6.74 | 11.76 | 6.55 |

# 6    Conclusions

In this study, MRCPSP is studied based on discrete PSO and constriction PSO in which a proposed S decreasing based constriction factor is designed to increase both global exploration and local exploitation capabilities so as to facilitate obtaining optimal solution.

According to Tables 1 and 2, using suggested S decreasing constriction factors can obtain more than 99%, 83% and 70% optimal solution for J10, J20 and J30 cases. Meanwhile, the yielded optimal solution for J10, J20 and J30 cases are better than other schemes other than algorithms except Chiang et al. [5] as indicated in Table 4 and 5. Although, Chiang obtained more optimal solutions on J10, J20 and J30 (99.81%, 88.27% and 72.64%) than this this work did (99.63%, 83.75% and 70.11%), but the deviation on J10, J20 and J30 (0.34%, 1.79%, 2.60%) is higher than that of this study (0.01% 0.71% and 1.66%). Restated, the proposed scheme has higher stability than Chiang et al. [5]. The overall performance on J10, J20 and J30 cases of this study is ranked 2. Hence, the proposed S decreasing constriction factors applied on standard PSO is able to effectively and efficiently solve MRCPSP. It is worth noting that the sigmoid decreasing and suggested S decreasing have the similar decreasing way. However, the proposed S decreasing has much better performance than that of the sigmoid decreasing as indicated in Tables 1, 2 and 3. The performance comparison between the inertia weight and constriction factor adjustment will be conducted. To increase performance, other search control schemes will be investigated in the feature.

**Acknowledgement.** This work was partly supported by the National Science Council, Taiwan, under contract NSC 101-2221-E-167 -012 – and NSC 101-2221-E-167 -035 -.

# References

1. Hartmann, S., Kolisch, R.: Experimental evaluation of state-of-the-art heuristics for the resource-constrained project scheduling problem. European Journal of Operational Research 127, 394–407 (2000)
2. Bouleimen, K., Lecocq, H.: A new efficient simulated annealing algorithm for the resource-constrained project scheduling problem and its multiple mode version. European Journal of Operational Research 149, 268–281 (2003)
3. Nonobe, K., Ibaraki, T.: Formulation and tabu search algorithm for the resource constrained project scheduling problem. In: Ribeiro, C.C., Hansen, P. (eds.) Essays and Surveys in Metaheuristics, pp. 557–588. Kluwer Academic Publishers (2001)
4. Peteghema, V.V., Vanhoucke, M.: A genetic algorithm for the preemptive and non-preemptive multi-mode resource-constrained project scheduling problem. European Journal of Operational Research 201, 409–418 (2010)
5. Chiang, C.W., Huang, Y.Q., Wang, W.Y.: Ant colony optimization with parameter adaptation for multi-mode resource-constrained project scheduling. Journal of Intelligent and Fuzzy Systems 19, 345–358 (2008)
6. Chen, R.M., Wang, C.M.: Project Scheduling Heuristics Based Standard PSO for Task-Resource Assignment in Heterogeneous Grid. Abstract and Applied Analysis 2011, Article ID 589862, 20 pages (2011)
7. Jarboui, B., Damak, N., Siarry, P., Rebai, A.: A combinatorial particle swarm optimization for solving multi-mode resource-constrained project scheduling problems. Applied Mathematics and Computation 195, 299–308 (2008)
8. Marinakis, Y., Marinaki, M.: A Hybrid Multi-Swarm Particle Swarm Optimization algorithm for the Probabilistic Traveling Salesman Problem. Computers & Operations Research 37, 432–442 (2010)

9. Liu, B., Wang, L., Jin, Y.H.: An Effective PSO-Based Memetic Algorithm for Flow Shop Scheduling. IEEE Transactions on Systems, Man, and Cybernetics 37, 18–27 (2007)
10. Shen, H., Zhu, Y., Liu, T., Jin, L.: Particle Swarm Optimization in Solving Vehicle Routing Problem. In: Intelligent Computation Technology and Automation, pp. 287–291 (2009)
11. Kennedy, J., Eberhart, R.: Particle swarm optimization. In: IEEE International Conference on Neural Networks, vol. 4, pp. 1942–1948 (1995)
12. Bratton, D., Kennedy, J.: Defining a Standard for Particle Swarm Optimization. In: IEEE Swarm Intelligence Symposium, pp. 120–127 (2007)
13. Kennedy, J., Eberhard, R.C.: A Discrete Binary Version of the Particle Swarm Algorithm. In: IEEE Conference on Systems, Man, and Cybernetics, Piscataway, vol. 5, pp. 4104–4109 (1997)
14. Kolisch, R., Hartmann, S.: Heuristic algorithms for the resource-constrained project scheduling problem: Classification and computational analysis. In: Weglarz, J. (ed.) Project Scheduling, Recent Models, Algorithms and Applications, ch. 7, pp. 147–178. Kluwer Academic Publishers, Norwell (1999)
15. Shi, Y., Eberhart, R.C.: Parameter selection in particle swarm optimization. In: Proceedings of 7th Annual Conference on Evolution Computation, pp. 591–601 (1998)
16. Nickabadi, A., Ebadzadeh, M.M., Safabakhsh, R.: A novel particle swarm optimization algorithm with adaptive inertia weight. Applied Soft Computing 11, 3658–3670 (2011)
17. Lei, K., Qiu, Y., He, Y.: A new adaptive well-chosen inertia weight strategy toautomatically harmonize global and local search ability in particle swarm optimization. In: ISSCAA, pp. 977–980 (2006)
18. Adriansyah, A., Amin, S.H.M.: Analytical and empirical study of particle swarm optimization with a sigmoid decreasing inertia weight. In: Regional Conference on Engineering and Science, Johor (2006)
19. Józefowska, J., Mika, M., Różycki, R., Waligóra, G., Węglarz, J.: Simulated annealing formulti-mode resource constrained project scheduling. Annals of Operations Research 102(1-4), 137–155 (2001)
20. Jarboui, B., Damak, N., Siarry, P., Rebai, A.: A combinational particle swarm optimization for solving multi-mode resource-constrained project scheduling problem. Applied Mathematics and Computation 195(1), 299–308 (2008)
21. Lova, A., Tormos, P., Cervantes, M., Barber, F.: An efficient hybrid genetic algorithm for scheduling projects with resource constraints and multiple execution modes. International Journal of Production Economics 117, 302–316 (2009)
22. Ranjbar, M., De Reyck, B., Kianfar, F.: A hybrid scatter-search for the discrete time/resource trade-off problem in project scheduling. European Journal of Operational Research 193, 35–48 (2008)

# Extracting Blood Vessel in Sclera-Conjunctiva Image Based on Fuzzy C-means and Modified Cone Responses

Jzau-Sheng Lin, Yu-Yang Huang, and Yu-Yi Liao

Dept. of Computer Science and Information Engineering
National Chin-Yi University of Technology, Taichung 411, Taiwan
{Jslin,yvonne}@ncut.edu.tw, my_foint@yahoo.com.tw

**Abstract.** In this paper, we present a method of sclera-conjunctiva segmentation and blood vessel extraction to assist the doctor of traditional Chinese Medicine to diagnose patients. First, the color eye image was converted to grayscale image and clustered three classes using fuzzy c-means algorithm. Then the Sobel operator will be applied to detect the edges. Therefore, the sclera-conjunctiva region will be obtained using the morphological dilation, holes filling, and connectivity algorithm. Finally, a modified cone responses algorithm was proposed to extract the blood vessel from the sclera-conjunctiva.

**Keywords:** Sclera-conjunctiva segment, Blood vessel extract, Fuzzy C-means, Cone responses.

## 1 Introduction

The eyes reveal shows not only the true inner thought but also the health state. The doctors of traditional Chinese medicine examine the pathological changes in corresponding with bodily organs according to the changing in the shape and color of different parts of eyes. The results of diagnosis are the subjective qualitative analysis, but they always lack the objective quantitative analysis.

Wang [1] proposed a method based on the differentiation of syndromes observing eyes to assist the diagnosis disease. It is one kind of diagnosis by watching at eyes. This diagnosis method collects the characteristics of the blood vessel to discriminate the diseases in the whole body. Professor Peng [2], a well-known doctor of traditional Chinese medicine, proposed eye- acupuncture therapy to observe the changes of blood vessels on sclera-conjunctiva of eyes to diagnose diseases. He divided the sclera- conjunctiva into 8 regions equally. He observed the changes of the color and sharp of each region to diagnose the diseases by naked eyes. Professor Peng lacks objective quantitative analysis and proof, although he discovered the location and subjective qualitative analysis of blood vessels and sclera-conjunctiva.

In recent years, there are a lot of preprocessing methods for digital image such as edge detection, region growing, mathematical analysis, snakes, level-sets, and so on [6-8]. Zhu et al. [3-4] proposed the optimal threshold segmentation and morphological filtering to extract sclera-conjunctiva region. They also proposed the adaptive

M. Ali et al. (Eds.): IEA/AIE 2013, LNAI 7906, pp. 556–565, 2013.

extrapolation tracking method, which is based on Sun's model [5], to track vessels. However, their methods must manually setup the starting parameters of the tracking, and the amount of calculation is too complex. Therefore, Wu et al. [9] proposed a scan-based method to detect the edge of the blood vessels to reduce the complex computing of tracking.

In this paper, we use fuzzy c-means algorithm, Sobel operator, morphological dilation, holes filling, and connectivity algorithm to segment the sclera-conjunctive. Finally a modified cone responses method used applied to extract blood vessels. The experimental results proved that the performance of the method effectively reaches the desired results.

In the rest of this paper, Section 2 introduces the sclera-conjunctiva segmentation. Section 3 describes the blood vessels extraction. The experimental results are shown in Section 4. Section 5 makes a summary for the conclusions and future work.

## 2    Sclera-Conjunctiva Segmentation

In this paper, The eye image contains pupil, iris, sclera-conjunctive, eyelid, eyelashes and flesh around the eye. Therefore, blood vessels are hard extracted from the sclera-conjunctive. In this paper, a preprocessing with three steps including grayscale conversion, clustering, and mask was proposed to extract the sclera-conjunctiva region. The details of steps are described in following sections.

### 2.1    Grayscale Conversion

Eye images generally contain three colors: black (iris and eyelashes), white (sclera-conjunctiva) and skin (upper and lower eyelids). In order to short cut the processing time and improve the performance, we converted the true color image into grayscale, as shown in Fig. 1.

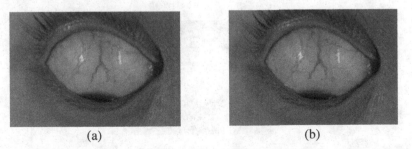

(a)                                                                    (b)

**Fig. 1.** (a) original true color image , (b) grayscale image

### 2.2    Grayscale Conversion

After converted to a grayscale image, the sclera-conjunctiva will be separated by clustering process. Fuzzy C-means (FCM) [10], a clustering algorithm, has advantages with fuzzy reasoning and easy programmable. In the fuzzy reasoning, an input vector belongs to a cluster according to the membership function. The objective function of FCM algorithm is defined in Equation (1).

$$J = \sum_{i=1}^{k} J_i = \sum_{i=1}^{k} \left( \sum_{j=1}^{n} \mu_{ij}^{m} \left\| X_j - C_i \right\|^2 \right) \tag{1}$$

where $k$ is cluster number, $n$ represents the total pixels of an image, $X_j$ is the $j$th pixel, and $C_i$ is the $i$th cluster center, $\mu_{ij}$ is the degree of membership of $x_j$ in the $i$th cluster, and $m$ is a weighting parameter. Equation (2) is used to update the degree of membership function:

$$\mu_{ij} = \frac{1}{\sum\limits_{s=1}^{k} \left( \frac{\left\| X_j - C_i \right\|}{\left\| X_j - C_s \right\|} \right)^{\frac{2}{m-1}}} \tag{2}$$

where $\mu_{ij} \in [0,1]$, $\sum_{i=1}^{c} \mu_{ij} = 1 \ \forall j$, and $0 < \sum_{j=1}^{n} \mu_{ij} < n \ \forall i$.

The initial value of $\mu_{ij}$ is set a random number between 0 and 1. The sum of $\mu_{ij}$ is 1 for the ith pixel belonging to all j clusters. The cluster center function is updated by Equation (3):

$$C_i = \frac{\sum\limits_{j=1}^{n} \mu_{ij}^{m} X_j}{\sum\limits_{j=1}^{n} \mu_{ij}^{m}} \tag{3}$$

In this paper, we used the FCM algorithm to divide the eye image into three classes such as black (iris and eyelashes), gray (skin), and white (sclera-conjunctiva). Fig. 2 shows the clustering result after applied the FCM algorithm.

**Fig. 2.** Clustering result

## 2.3   Edge Detection

Let's observe Fig. 2 closely, the blood vessels can be considered as part region of sclera-conjunctiva. The eyelids and iris are clustered in black region and they can be considered as the edge of the sclera-conjunctiva. Therefore, we use the edge detection to extract the entire sclera- conjunctiva.

Sobel operator is one of well-known mask for edge detection [11-12] and to be defined as Equation (4).

$$Edge\ (x, y) \approx \left|G_x\right| + \left|G_y\right| \tag{4}$$

Sobel operator, shown in Fig. 3, uses two 3*3 masks to calculate the edge horizontal (Gx) and vertical (Gy) directions.   The result of edge detection for Fig. 2 is shown in the Fig. 4.

| -1 | -2 | -1 |
|----|----|----|
| 0  | 0  | 0  |
| 1  | 2  | 1  |

$G_x$

| -1 | 0 | 1 |
|----|---|---|
| -2 | 0 | 2 |
| -1 | 0 | 1 |

$G_y$

**Fig. 3.** Sobel operator

**Fig. 4.** Retrieve the results of image edge

After the image edge detected, we use morphological dilation and holes fill [11-12] to obtain the overall edges that are shown as in Fig. 5.

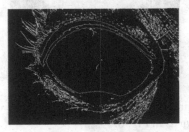

**Fig. 5.** Dilate processing image

The dilation of A by B, defined as in Equation (5), then is defined the set of all displacements z.

$$X_k = (X_{k-1} \oplus B) \cap A^c, \quad k = 1, \ 2, \ 3,\ldots \tag{5}$$

Where Xk contains the entire filled hole, after iteration step k. Ac is complement of A.

**Fig. 6.** The result of hole filling

Fig. 6 shows the result of the hole filling. It contains many white classes including the sclera-conjunctiva region. Therefore, we use the connectivity algorithm to extract the connected components for white region. Connectivity algorithm is defined as Equation 6.

$$X_K = (X_{K-1} \oplus B) \cap A, \quad k = 1, \ 2, \ 3,\ldots \tag{6}$$

where A is the input image, B is a structural element.

We then count the numbers of pixels in connected components and find the largest connected components as the sclera-conjunctiva region. The processed result was shown as in Fig. 7.

**Fig. 7.** The sclera-conjunctiva region

According to the position of sclera-conjunctiva region, we extract them from the original color image shown as in Fig. 8.

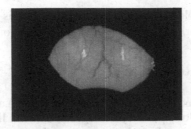

**Fig. 8.** The sclera-conjunctiva region in original color image

## 3 Blood Vessel Extraction

In this section, we will introduce blood vessel extraction. Tracking and edge detection methods [7-8] were used to extract blood vessels in some papers in recent years. The main disadvantage of the above methods is requiring a large amount of computation time. Therefore, we use the RGB color characteristics to distinguish vessel from sclera-conjunctiva for reducing the time complexity. Ruderman et al. [13] gathered cone response statistics for analyzing tree orthonormal principal axes. In RGB color space, three primary color channels have high correlation, so Gijsenij et al. [14] provided Equations (7) – (9) to decorrelate them.

$$O_1 = \frac{R - G}{\sqrt{2}} \tag{7}$$

$$O_2 = \frac{R + G - 2B}{\sqrt{6}} \tag{8}$$

$$O_3 = \frac{R + G + B}{\sqrt{3}} \tag{9}$$

where $O_1$, $O_2$, and $O_3$, are three orthonormal color channels.

The result of segmentation for the blood vessels in the sclera-conjunctiva region was not very distinct by using of Equation (7-9). The blood vessels are displayed in the sclera-conjunctiva region with red-color manner. Therefore, the equation was modified as Equation (10) in order to evidently stress red channel.

$$T = \frac{\alpha \ln R}{\beta \ln G \ \gamma \ln B} \tag{10}$$

where $\alpha$, $\beta$, $\gamma$ are constants to be set between 0.5 and 0.8.

**Fig. 9.** Extract of blood vessel results

We defined the threshold to determine the vessel. If $T$ is greater than a threshold value, the pixel will be marked as vessel shown as in Equation (11).

# 4     Experimental Results

Light source and the distance (between camera and objects) are two important factors for affecting the quality to capture eye images. First, the light source contains the intensity and the temperature. In order to remove the above two factors, we established the black curtain studio to remain stable light source. Fig. 10 (a) is the black curtain studio and a personal computer iMac OS X. Figure (b) is the equipments inside black curtain studio. There are light source, compact tripod, and facial retainer, respectively. We used the compact tripod and facial retainer to control the distance between camera and object. The standard 24-color Checker, shown as in Fig. 10 (c), was also used to correct the brightness and temperature of light source. Fig. 10 (d) shows the captured eye image. The used equipments are:

I.     Camera：Canon EOS 5D Mark II
II.    Lenses：EF 100mm MIS USM
III.   Flash：Canon MACRO TWIN LITE MT-24EX
IV.   X-rite 24-ColorChecker passport
V.    Facial retainer

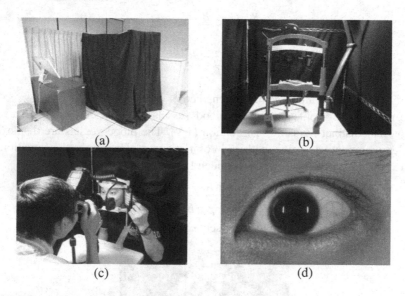

(a)                          (b)

(c)                          (d)

**Fig. 10.** (a)The shooting environment, (b) The inside of black curtain studio, (c) situation of adjusting the brightness and temperature of light source using 24-ColorChecker, (d) the result of shooting

Fig. 11 (a) and (b) show original eye images, in which the iris is on left side and right side individually. Fig. 11 (c) and (d) are the sclera-conjunctiva region of (a) and (d) by using of sclera-conjunctiva segmentation and edge detection. The blood vessels are displayed in Fig. 11 (e) and (f) by means of the blood vessel extraction algorithm.

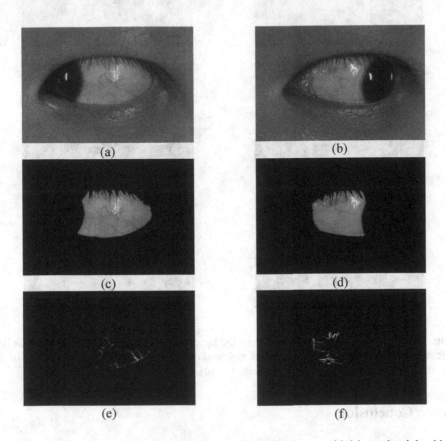

**Fig. 11.** (a) the eye image with iris on the left side, (b) the eye image with iris on the right side, (c) the sclera-conjunctiva region of (a), (d) the sclera-conjunctiva region of (b), (e) the blood vessel of (c), and (f) the blood vessel   of (d).

Fig. 12 (a) and (b) are the original eye images in which the iris are on upper side and middle region respectively. Fig. 12 (c) and (d) are sclera-conjunctiva regions after (a) and (b) through sclera-conjunctiva segmentation and edge detection algorithms. Fig. 12 (e) and (f) are blood vessels constructed by blood vessel extraction strategy from (c) and (d).

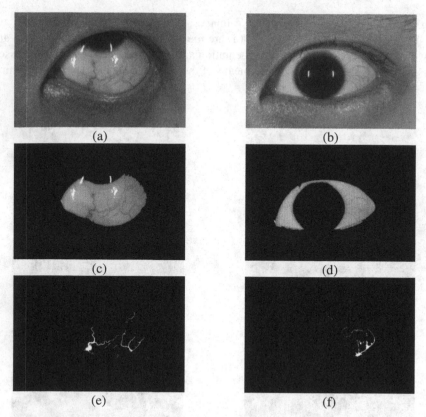

**Fig. 12.** (a) the eye image that the iris is on the upper side, (b) the eye image that the iris is in the middle side, (c) the sclera-conjunctiva region of (a), (d) the sclera-conjunctiva region of (b), (e) the blood vessel of (c), (f) the blood vessel of (d).

## 5    Conclusion

In this paper, we proposed a method to extract the sclera-conjunctiva. The experimental results showed that the method can effectively segment the sclera- conjunctiva region. The results can assist the doctors to diagnose disease in analyzing the sclera-conjunctiva.

In addition, we also proposed a vessel extraction method that is based on the concept of cone responses to decorrelate the RGB primary colors. The red-manner characteristic of blood vessel can simplify the cone-responses equation and speed up the extraction time to reduce the amount of computation.

In the future, we will continue to analysis the characteristics of the blood vessel including color, branch, width, and so on. We hope that the results can assist the traditional Chinese doctors to diagnose the different syndromes exactly.

# References

1. Wang, J.J.: The theory of Differentiation of Syndromes by Observing Eyes in Traditional Chinese Medicine. Chinese Journal of Basic Medicine in Traditional Chinese Medicine 11(5), 324–325 (2005)
2. San, P.: Peng Jing San's Eye Acupuncture by means of watching eyes to understand disease. People's Medical Publishing (2010)
3. Zhu, G.D.: Research on the Digitalization Technology of 'Differentiation of Syndromes by Observing Eyes'. Doctoral Dissertation, Graduate University of Chinese Academy of Sciences (May 2006)
4. Zhu, G.D., Shen, L., Wang, J.J.: Automatic Vessel Extraction for Sclera-Conjunctiva Images Based on Exploratory Tracking. Computer Engineering 31(17), 6–8 (2005)
5. Sun, Y.: Automated Identification of Vessel Contours in Coronary Arteriograms by An Adaptive Tracking Algorithm. IEEE Transaction on Medical Imaging 8(1), 78–88 (1989)
6. Quek, F.K.H., Kirbas, C., Charbel, F.: AIM: an attentionally-based system for the interpretation of angiography. In: Proceedings International Workshop on Medical Imaging and Augmented Reality, pp. 168–173 (2001)
7. Liu, I., Sun, Y.: Recursive tracking of vascular networks in angiograms based on the detection-deletion scheme. IEEE Transaction on Medical Imaging 12(2), 334–341 (1993)
8. Toliasand, Y.A., Panas, S.M.: A Fuzzy Vessel Tracking Algorithm for Retinal Images Based on Fuzzy Clustering. IEEE Transactions on Medical Imaging 17(2), 263–273 (1998)
9. Wu, C., Harada, K.: Study on Digitization of TCM Diagnosis Applied Extraction Method of Blood Vessel. Journal of Signal and Information Processing 2(4), 301–307 (2011)
10. Bezdek, J.C.: Pattern Recognition with Fuzzy Objective Function Algorithms. Plenum Press, NY (1981)
11. Gonzalez, R.C., Woods, R.E., Eddins, S.L.: Digital Image Processing Using MATLAB. Prentice Hall, Upper Saddle River (2003)
12. Gonzalez, R.C., Woods, R.E.: Digital Image Processing, 3rd edn. Pearson Education Taiwan and GauLih Book Co. Ltd. (2009)
13. Ruderman, D.L., Cronin, T.W., Chiao, C.C.: Statistics of cone responses to natural images: implications for visual coding. Journal of the Optical Society of America A 15(8), 2036–2045 (1998)
14. Gijsenij, A., Gevers, T.: Color Constancy Using Natural Image Statistics and Scene Semantics. IEEE Transactions on Pattern Analysis and Machine Intelligence 33(4), 687–698 (2011)

# A Spherical Coordinate Based Fragile Watermarking Scheme for 3D Models

Cheng-Chih Huang[1], Ya-Wen Yang[2], Chen-Ming Fan[3], and Jen-Tse Wang[4,*]

[1] Center of General Education, National Taichung University of Science and Technology,
Taichung, Taiwan
[2] Department of Computer Science and Engineering, National Chung Hsing University,
Taichung, Taiwan
[3] Department of Information Management, National Chin-Yi University of Technology,
Taichung, Taiwan
[4] Department of Information Management, Hsiuping University of Science and Technology,
Taichung, Taiwan
{Jimhuang,fan}@nutc.edu.tw, junkohe@cht.com.tw,
tse@mail.hust.edu.tw

**Abstract.** In this paper, a new spherical coordinate based fragile watermarking scheme is proposed. At first, the three dimensional (3D) model is translated from the Cartesian coordinate system to the spherical coordinate system. Then the quantization index modulation technique is employed to embed the watermark into the $r$ coordinate for authentication and verification. By adapting the quantization index modulation technique together with some keys in the spherical coordinate system, the distortion is controlled by the quantization step setting. Experimental results show that both the 100% embedding rate and low distortion can be achieved simultaneous in the proposed method. Moreover, the causality, convergence and vertex reordering problems can be overcome.

**Keywords:** Fragile watermarking, three dimensional (3D) models, spherical coordinate system, quantization index modulation (QIM), authentication, verification.

## 1 Introduction

Recently, the popularity of internet and the rapid development of digital content designing and processing techniques had made the distribution of the digital assets easier and faster. The copyright protection and integrity verification problems of the digital assets become more and more important. According to the extraction strategies of watermarks, watermarking techniques can be classified into blind [1-5], semi-blind [6-7] or non-blind approaches. According to the applications, watermarking techniques can be classified into two categories : robust watermarking and fragile watermarking. The robust watermarking is designed to make the embedded watermarks detectable

---

* Corresponding author.

M. Ali et al. (Eds.): IEA/AIE 2013, LNAI 7906, pp. 566–571, 2013.

against attacks. On the other hand, fragile watermarking is applied to very and locate the slightest unauthorized modification for authentication applications. A considerable progress has been made in the areas of steganogrphy [8-9] and watermarking [10-12] on 3D polygonal meshes. But, only a few fragile watermarking algorithms have been proposed to authenticate the integrity of 3D models [1- 7].

Yeo and Yeung [1] firstly proposed a fragile watermarking scheme for authenticating 3D polygonal meshes. Their scheme is both public and fragile, but there arise two problems: the causality and convergence problems. The causality problem is that the location index of a former processed vertex is changed by the perturbing of later processed neighboring vertices. The convergence problem makes the user unable to control the distortion induced by the iteratively perturbing process. Lin *et al.* [2] proposed a modified fragile watermarking scheme similar to Yeo and Yeung's method. The causality problem is conquered in this method but this scheme has the convergence problem. Chang-Min Chou and Din-Chang Tseng [3] propose a public fragile watermarking scheme based on the sensitivity of vertex geometry for 3D model authentication. [3] solved the causality problem by using the adjusting vertex method. Nevertheless, the number of embedded points is at most half of total points in this scheme.

In this paper, we propose a fragile watermarking scheme in the spherical coordinate system. First, the gravity center of the original 3D object as the origin of the spherical coordinate system is calculated. Then, the object is translated from the Cartesian coordinates system to the spherical coordinate system. The quantization index modulation technique is employed to embed and extract the watermark into the $r$ coordinate. The proposed scheme is described in Sec. 2. Experimental results and discussion are presented in Sec. 3. Finally, conclusions are provided in Sec. 4.

## 2    The Proposed Fragile Watermarking Scheme

### 2.1    The Watermark Embedding Algorithm

The watermark embedding process takes three inputs: the cover model, the watermark and the secret key to generate the stego model. The watermark embedding process embeds a watermark according to four steps as follows:

**Step 1:** Computing and recording the center of gravity. Suppose $p_C$ is the center of gravity, and it is calculated and recorded.

**Step 2:** Translating to spherical coordinate system. By using the center of gravity $p_C$ as the origin of the spherical coordinate system, the new coordinate $(r_i, \theta_i, \phi_i)$ of the point $q_i$ in the spherical coordinate system is calculated from the coordinate $(x_i, y_i, z_i)$ of the point $p_i$ in the Cartesian coordinate system.

**Step 3:** Embedding of $r$ coordinate. For security reasons, a secret key $K$ is engaged to create a random sequence of integers. Suppose a point $q_i$ is the point considered, where $i$ is an integer index and $0 < i < n-1$. Moreover, suppose a quantization step $c$ which is calculated by the equation (1), where the $r_{\max}$ is the maximum coordinate value in $r$ axis and the $r_{\min}$ is the minimum coordinate value in $r$ axis.

$$c = \left\lfloor \frac{r_{max} - r_{min}}{d} \right\rfloor \tag{1}$$

The constant value $d$ is used to control the degree of axis quantization step $c$ which decides the degree of distortion. Initially, the interval which contains point $q_i$ is decided by using the quantization step $c$ as shown in Fig. 1. Suppose a sequence of $m$ bits ($m \geq 2$, $m \in N$) should be embedded into point $q_i$. And, a sequence of $m$ bits contains $j$ bits for watermark bits and $l$ bits for verification bits, where $m$ is equal to $j + l$ ($j, l \geq 1$, $j, l \in N$). Then this interval is divided into $2^m$ equal parts before embedding the watermark. As shown in Fig. 1, if the point $q_i$ is located on the 1 subinterval, the original state $s_b$ of point $q_i$ is set to 1. And let the new quantization state $s_a$ be an $m$ bits number which is used to store two information: the watermark bits, and the verification bits. The new state $s_a$ can be calculated by equation (2), where $w_i$ is the watermark bits and $h_i$ is the verification value generated by a hash function.

$$s_a = w_i \times 2^l + h_i \tag{2}$$

$$h_i = hash(w_i, i) \tag{3}$$

Moreover, the new coordinate $r_i'$ is computed by the equation (4).

$$r_i' = (s_a - s_b) \times \frac{c}{2^m} + r_i \tag{4}$$

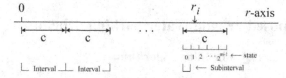

**Fig. 1.** Determination of original quantization state

**Step 4:** Translating to Cartesian coordinate system. According to the same origin $p_C$ used in Step 2, translating to Cartesian coordinate system is employed to produce the stego model. The embedded coordinate $(x_i', y_i', z_i')$ of the point $p_i'$ in the Cartesian coordinate system is calculated as the following equations :

$$x_i' = r_i' \sin \theta_i \cos \phi_i \tag{5}$$

$$y_i' = r_i' \sin \theta_i \sin \phi_i \tag{6}$$

$$z_i' = r_i' \cos \theta_i \tag{7}$$

The proposed method must record a small amount of information, including the secret key $K$; the constants $d$ and the gravity center of the cover model. This information serves as keys in the extraction stage for extracting the watermark.

## 2.2    The Watermark Extraction Algorithm

To manufacture the recovery watermark, the extraction process requires three inputs: the stego model, the secret key, and the gravity center of the original cover model. It extracts the recovery watermark according to the following two steps:

**Step 1:** Translating to spherical coordinate system. At first, by using the gravity center of the original cover model, the stego model is translated to the spherical coordinate system.

**Step 2:** Extraction. The same secret key $K$ given in the embedding stage is employed to generate the same random sequence of integers. They represent the index order for extraction. To reverse Step 3 of the embedding algorithm, the interval which contains point $q_i'$ is divided into $2^m$ equal parts. The state value $s_a'$ of a point is decided by the subinterval in which $q_i'$ is located. Moreover, the embedded hash value is extracted by the equation (8).

$$h_i' = s_a' \bmod 2^l \tag{8}$$

Then the embedded watermark bits are extracted by the equation (9).

$$w_i = \frac{s_a' - h_i}{2^l} \tag{9}$$

The verification is achieved when $h_i'$ is equal to the value $hash(w_i', i)$ by using the same hash function used in the embedding stage.

## 3    Experimental Results and Discussions

As shown in Fig. 2, a series of experiments were conducted to test the performance and validate the feasibility of the proposed method. In this paper, there is a simple root mean square (RMS) ratio to measure distortion of a 3D stego model. The RMS ratio consists of the RMS values over the diagonal length of the bounding volume for a 3D stego model. The small RMS ratios indicate insignificant positional changes during the watermark embedding.

(a)                (b)                (c)                (d)                (e)

**Fig. 2.** Three-dimensional cover models in the experiments. (a) Lion, (b) Horse, (c) Dolphin, (d) Dog, (e) Cat.

Table 1 illustrates the number of points, the watermark bits, the verification points, RMS and RMS ratio in the embedding stage of various 3D cover models. The verification operation can be performed for each point of the cover model as shown in Table 1. Moreover, it is clear that no errors are found in the recovered watermarks.

**Table 1.** Experimental results for embedding: $c = 0.0003$

| Cover Model | Number of points | Watermark bits | Verification point | RMS | RMS ratio |
|---|---|---|---|---|---|
| Cat | 41242 | 41242 | 41242 | $3.73 \times 10^{-4}$ | $4.41 \times 10^{-7}$ |
| Dog | 33144 | 33144 | 33144 | $3.72 \times 10^{-4}$ | $5.53 \times 10^{-7}$ |
| Dolphin | 32828 | 32828 | 32828 | $3.76 \times 10^{-4}$ | $1.81 \times 10^{-7}$ |
| Horse | 39217 | 39217 | 39217 | $3.75 \times 10^{-4}$ | $2.56 \times 10^{-7}$ |
| Lion | 17352 | 17352 | 17352 | $3.74 \times 10^{-4}$ | $3.49 \times 10^{-7}$ |

Fig. 3 indicates the chart of the quantization step $c$ versus the RMS ratio distortions of models with various $c$ values setting. As shown in Fig. 3, the RMS ratio is increasing when the quantization step $c$ increases. Moreover, in this paper, a vertex is embedded by a quantization index modulation embedding scheme without considering neighboring points. It makes the proposed scheme out of the problem caused by using synchronization technique. Therefore, the causality, convergence and synchronization problems can be overcome in this proposed scheme.

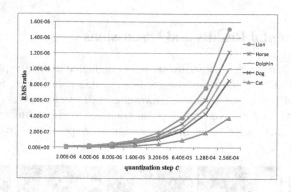

**Fig. 3.** Chart of distortions with various quantization step $c$

## 4    Conclusions

In this paper, a spherical coordinate based fragile watermarking scheme is proposed for 3D models in spatial domain. By adapting the quantization index modulation technique together with some keys in the spherical coordinate system, the proposed method is immune to vertex reordering, causality, convergence and synchronization problems. The experimental results show that every points of the cover model can be embedded for achieving verification. Moreover, the distortion of the stego model is controlled by quantization step setting.

**Acknowledgement.** This research was supported by the National Science Council R.O.C. under NSC No.101-2221-E-164-022-.

# References

1. Yeo, B.L., Yeung, M.M.: Watermarking 3-D objects for verification. IEEE Transactions on Computer Graphics and Applications 19, 36–45 (1999)
2. Lin, H.Y., Liao, H.Y., Lu, C.S., Lin, J.C.: Fragile watermarking for authenticating 3-D polygonal meshes. IEEE Transactions on Multimedia 7(6), 997–1006 (2005)
3. Chou, C.M., Tseng, D.C.: A public fragile watermarking scheme for 3D model authentication. Computer-Aided Design 38, 1154–1165 (2006)
4. Wang, J.T., Wang, P.C., Yu, S.S.: Reversible fragile watermarking scheme for three dimensional models. Optical Engineering 48, 097004 (2009)
5. Chou, C.M., Tseng, D.C.: Affine-transformation-invariant public fragile watermarking for 3D model authentication. IEEE Transactions on Computer Graphics and Application 29, 72–79 (2009)
6. Ohbuchi, R., Mukaiyama, A., Takahashi, S.: Watermarking a 3D shape model defined as a point set. In: Proc. of the International Conference on Cyberworlds, pp. 392–399 (2004)
7. Wu, H.T., Cheung, Y.M.: A reversible data hiding approach to mesh authentication. In: Proc. of the 2005 IEEE/WIC/ACM International Conference on Web Intelligence, pp. 774–777 (2005)
8. Cayre, F., Macq, B.: Data hiding on 3-D triangle meshes. IEEE Transactions on Signal Processing 51, 939–949 (2003)
9. Benedens, O., Busch, C.: Toward blind detection of robust watermarks in polygonal models. Proc. of the EUROGRAPHICS Computer Graphic Forum 19, 199–208 (2000)
10. Ohbuchi, R., Mukaiyama, A., Takahashi, S.: A frequency-domain approach to watermarking 3D shapes. Computer Graphics Forum 21, 373–382 (2002)
11. Zafeiriou, S., Tefas, A., Pitas, I.: Blind robust watermarking schemes for copyright protection of 3D mesh objects. IEEE Transactions on Visualization and Computer Graphics 11, 596–607 (2005)
12. Alface, P.R., Macq, B.: From 3D mesh data hiding to 3D shape blind and robust watermarking: A survey. In: Shi, Y.Q. (ed.) Transactions on DHMS II. LNCS, vol. 4499, pp. 91–115. Springer, Heidelberg (2007)

# Palm Image Recognition Using Image Processing Techniques

Wen-Yuan Chen[1], Yu-Ming Kuo[1], and Chin-Ho Chung[2]

[1] Department of Electronic Engineering,
National Chin-Yi University of Technology, Taichung, Taiwan, R.O.C.
[2] Department of Electronic Engineering,
Ta Hwa University of Science and Technology
cwy@ncut.edu.tw, kuoym1115@gmail.com, chc@tust.edu.tw

**Abstract.** In this research, we find the palm from the hand image firstly. And then distinguish the fingers using the triangular calculation method. In the palm detection, the color of skin, background subtraction, hand image extraction, edge detection and histogram analysis are used to achieve the goal. In fingers distinguish; we record the tips and valley of the fingers by means of calculating the histogram of the palm image firstly. Next, we find out the original point which is the center of the gravity of the palm using the area that the palm image gets rid of the fingers part. Successively, we draw the original point and center of the cut line of the palm use as the base line, means zero angle line. Meanwhile, we draw another line from tip of finger to the original point called tip line. Finally, we calculate the angle between base line and tip line use as the finger angular. Since the fingers have different angle, so the fingers are easily be distinguished.

**Keywords:** Palm image, Skin Color, Center-of-gravity, Angle Detection.

## 1 Introduction

S. C. et al. [1] proposed a contact free system for palm image recognition. Several techniques are used to achieve the goal. They use the Palm images in Red, Green and Blue spectra, a multi-spectral image acquisition method, preprocessing to dynamically locate the region of interest and image fusion using Fourier and Wavelet transform based techniques.

W. Y. Han and J. C. Lee [2] use the palm vein texture and apply texture-based feature extraction techniques to palm vein authentication. In his algorithm, a Gabor filter provides the optimized resolution in both the spatial and frequency domains. For obtaining effective pattern of palm vascular, they represent a bit string by coding the palm vein features using an innovative and robust adaptive Gabor filter method. Simultaneously, two VeinCodes are measured by normalized Hamming distance. We obtain a high accuracy and a rapid enough for real-time palm vein recognition. From

M. Ali et al. (Eds.): IEA/AIE 2013, LNAI 7906, pp. 572–580, 2013.

the experimental results, it is demonstrate their proposed approach is feasible and effective in palm vein recognition.

L. Nanni and A. Lumini [3] study the usefulness of multi-resolution analysis for the face and palm authentication problems. They adopt wavelet coefficients features for the authentication problem. Besides, several linear subspace projection techniques have been tested and compared. From the experiments results, it is carried out on several biometric datasets show that the application of Laplacian Eigen- Maps (LEM) on a little subset of wavelet sub-bands allows to obtain a low Equal Error Rate.

In this paper, the focus is on developing a scheme to distinguish the fingers of the palm images. Several researches proposed the hand gesture method can be seen in [4-8]. About histogram, skin color and angular criteria method to distinguish the exact fingers object can be finding in [9-12]. The remainder of this paper is organized as follows: in Section 2, the palm recognition algorithm. The empirical results are described in Section 3. Finally, Section 4 concludes this paper.

## 2    Recognition Algorithm

In palm recognition, several steps used to achieve the goal, it flow chart is shown in figure 1. Firstly, the background image and input image are processed by color transform. The color transform transfer the colors in RGB plans into YCbCr plans, it is superior to the skin color detection. Second, a background subtraction operation used to extract the skin color according to the background image and input image. Third, the features of the palm are computer and extracted for further fingers separation. The features include the center point of the palm image, the tip points of the fingers and the valley point of the fingers. Finally, the angles of all fingers are obtained by means of triangular method.

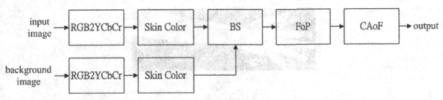

BS: Background Subtraction         FoP: Features of Plam
CAoF: Calculate the Angle of Finger

**Fig. 1.** The flow chart of the Palm recognition algorithm

### 2.1    Palm Object Extraction

Based on the concept of object extraction, a color mode is superior to the gray-scale mode in luminance variance. In the background removal stage, the background image and input image are used to extract the palm object image under the color mode. After the RGB to YCbCr color transform, we use the conditions $130 \geq Cb \geq 70$ and $175 \geq Cr \geq 130$ to extract the skin color regions in the background image and input image, respectively. Furthermore, we subtract the two images to obtain the candidate areas of the palm object.

In this paper, we focus on the concept to computer the angle of fingers. Therefore, we need a base line use as zero angle line for counting the angle of fingers. In this research, we connect the center point of the palm and the center point of cut line of the palm use as the base line. For the center point, we calculate the center-of-gravity of the palm that get ride of the finger parts, and consider it use as the center point of the palm. For the tip of fingers, we computer the histogram of the palm image, and check each finger distribution than we can easily to obtain the tip of fingers as figure 7(f) shown. Figure 3 is the histogram analysis of the hand image. The up part of the figure 3 is a binary image with hand. The down part of figure 3 is the histogram of up part. It is making easier to cut the palm part at the center valley point. Figure 4 shows palm object extraction in case stone gesture image. The up part of figure 4 a binary image and the down part is the result image, palm image, it is extracted by our method. Figure 5 shows the palm object extraction case, it is scissors gesture image. It is also obtained a correct result like the figure 3 paper case and figure 4 stone case.

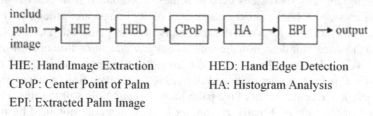

includ palm image → HIE → HED → CPoP → HA → EPI → output

HIE: Hand Image Extraction          HED: Hand Edge Detection
CPoP: Center Point of Palm          HA: Histogram Analysis
EPI: Extracted Palm Image

**Fig. 2.** Palm object extraction algorithm

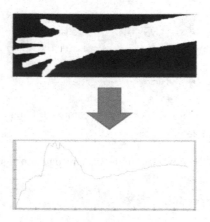

**Fig. 3.** The histogram analysis of the hand image

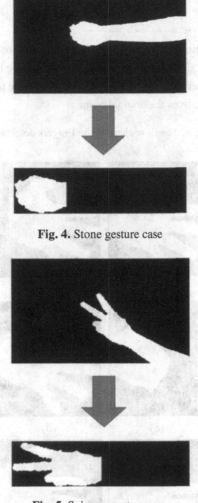

Fig. 4. Stone gesture case

Fig. 5. Scissors gesture case

## 2.2    Finger Detection

In order to obtain correct palm image recognition, the finger detection is the prime task. In finger detection, we adopt three strategies to achieve the task: 1) calculate the center-of-gravity of the palm image for use as the center of the palm for further computation of the angle of the fingers. 2) Produce a baseline as the zero-axis for angle calculation. We get the cut edge of the palm using sobel edge detection. Then, calculate and draw a line between the center-of-gravity point and center of cut line of the palm for use as a baseline; and 3) calculate the angles of the fingers using triangular method associated with the tip point of finger, center-of-gravity point and center point of the cut line of the palm. The details is describing in the figure 6.

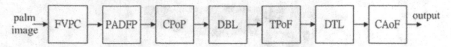

FVPC: Finger Valley Point Computing       PADFP: Plam Area Discard Finger Part
CPoP: Center Point of Palm                DBL: Draw Base Line
TPoF: Tip Point of Finger                 CAoF: Calculate the Angle of Finger
DTL: Draw the Tip Line from the original point

**Fig. 6.** The flow chart of the Fingers detection

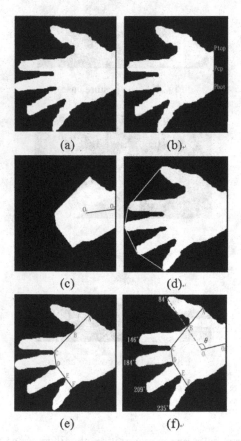

(a)                    (b)

(c)                    (d)

(e)                    (f)

**Fig. 7.** The process steps of the calculating finger angle; (a) the original binary palm image. (b) the points of the cut line of the palm. (c) the baseline connected by point O1 and O2. (d) the contour of the finger tips. (e) the valley points of the fingers. (f) the angles of the fingers calculated.

In figure 6, the finger valley point computed used to find out the valley point of the finger. The palm area discard finger parts used to obtain a center point of the palm. The draw base line operation used to create a base line for further angle computation. The tip point of the finger is needed to calculate the angle of the finger. Figure 7 shows the angle of the finger computing process. Figure 7(a) is the original binary palm image. Figure 7(b) displays the points of the cut line of the palm; there are the Ptop, Pct and Pbot. Where the Pct point, it is the kernel point used to create the base line. Figure 7(c) shows the base, it is connected by point O1 and O2, and O2 is the same as the point Pct. Figure 7(d) is the contour of the finger tips. Figure 7(e) shows the valley points of the fingers, it is used to calculate the center of the palm which is the fingers part discarded. Finally, figure 7(f) shows the how the angles of the fingers calculated.

## 3    Experimental Results

Several ( $640 \times 480$ ) test images Rock, Paper and Scissors were used in simulation to demonstrate the performance of the proposed scheme. Figure 8 shows the angle detection process of finger in scissors case. Figure 8(a) is the original binary image; it is the extracted palm image only. Figure 8(b) is the edge image corresponding to (a); it is used to get the mid point of the cut line of the palm image. Figure 8(c) shows the three point of the cut line of the palm, denoted Ptop, Pct and Pbot; where Pct also called O2 will used to draw the base line associated with center point O1. Figure 8(d) display the histogram of the (a); we use the histogram information to extract the tip points and valley points of the finger. Figure 8(e) and 8(f) show the extracted tip point and valley points of the fingers respectively. It is obviously, the green points are the tip point of the fingers in figure 8(e). Simultaneously, the valley points denote A, B, C and D are display in the figure 8(f). Form the base line; it is generated by drawing the points from O1 into O2 as figure 8(g) shown. Finally, figure 8(h) shows the obtained angles of the finger of the palm. The angular of the fingers are 143 and 181 degree respectively.

In the other hand, the simulation on right hand image case is shown in figure 9. Figure 9(a) shows the original test image, it is the color mode image which is represented in RGB planes. In order to compact the redundant describing process, we only show the finally results in figure 9(b). There are display the base line, the valley points and the angular of the fingers. The angular are 93,145, 181, 206 and 231 respectively. It is show the results are correct. Fig. 10 shows another case; the left hand image case. The same as the figure 9 figure 10(a) shows the original test image in color mode. Figure 10(b) display the obtained angle results of the fingers tip. The angular are 77. 140, 182, 207 and 232, there are also obtained the correct results. After carefully check and compare the results of the figure 9 and figure 10, we make sure our method can get correct results no matter what the hand image is right hand or left hand. From the simulation results, we find that our algorithm can exactly identify the fingers pf the palm regardless of how the hand gestures are spread.

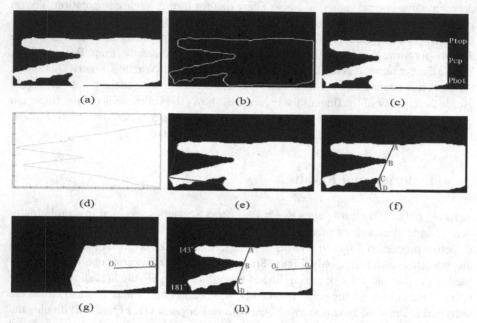

**Fig. 8.** The angle detection process of finger; (a) the original binary image, (b) the edge image of (a), (c) the three point of the cut line of the palm, (d) the histogram of the (a), (e) the tip point of the fingers, (f) the valley points of the fingers, (g) the obtained base line of the palm, (h) the obtained angles of the palm

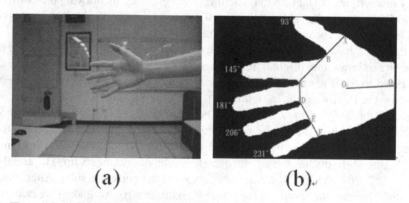

**Fig. 9.** The angles of the fingers tip on the right hand case; (a) the original image, (b) the obtained angle results of the fingers tip

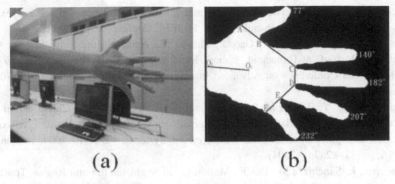

(a)                              (b)

**Fig. 10.** The angles of the fingers tip on the left hand case; (a) the original image, (b) the obtained angle results of the fingers tip

## 4    Conclusions

We consider the best method of distinguish the fingers of the palm image is the angle of tip finger because it is physical features. Besides, we computer the center-of-gravity of palm area that get rid of the fingers part used as the original point. Further, we according to the original point and tip of finger can correctly to calculate the angle of finger.

From simulation results, it is show the angle of the right hand fingers in figure 10 are 93, 145, 181, 206 and 231. Likely, it is show the angle of the left hand fingers in figure 11 are 77, 140, 182, 207 and 232. The simulation data shows the different fingers have different angles no matter what it is right hand or left hand. This also demonstrates that our method is effective and correct.

**Acknowledgments.** This work was partly supported by the National Science Council, Taiwan (R.O.C.) under contract NSC 101-2221-E-167-034-MY2.

## References

[1]  Chaudhari, S., Chandratre, S., Sale, D., Joshi, M.: Personal Identification in Contact Free Environment by Multispectral Palm Image Fusion. In: IEEE Conf., pp. 85–89 (2012)

[2]  Han, W.-Y., Lee, J.-C.: Palm vein recognition using adaptive Gabor filter. Expert Systems with Applications 39, 13225–13234 (2012)

[3]  Nanni, L., Lumini, A.: Wavelet decomposition tree selection for palm and face authentication. Pattern Recognition Letters 29, 343–353 (2008)

[4]  Hasuda, Y., Ishibashi, S., Ishikawa, J.: A robot designed to play the game "Rock, Paper, Scissors". In: IEEE 2007 Conference, pp. 2065–2070 (2007)

[5]  Yoon, H.S., Chi, S.Y.: Visual Processing of Rock, Scissors, Paper Game for Human Robot Interaction. In: SICE-ICASE International Joint Conference 2006, Bexco, Busan, Korea, October 18-21, pp. 326–329 (2006)

[6] Alexander, M., Bronstein Michael, M., Kimmel, B.R.: Rock, Paper, and Scissors: extrinsic vs. intrinsic similarity of non-rigid shapes. In: IEEE Conference, pp. 1–6 (2007)

[7] Jia, D., Huang, Q., Tian, Y., Gao, J., Zhang, W.: Hand Posture Extraction for Object Manipulation of a Humanoid Robot. In: Proceedings of the 2008 IEEE/ASME International Conference on Advanced Intelligent Mechatronics, China, pp. 1170–1175 (July 2008)

[8] Sriboonruang, Y., Kumhom, P., Chamnongthai, K.: Hand Posture Classification Using Wavelet Moment Invariant. In: IEEE International Conference on Visual Environments Human-Computer Interfaces and Measurement Systems, VECIMS 2004, Boston, MA, USA, pp. 78–82 (July 2004)

[9] Badenas, J., Sanchiz, J.M., Pla, F.: Motion-based Segmentation and Region Tracking in Image Sequence. Pattern Recognition 34, 661–670 (2001)

[10] Choi, B.D., Jung, S.W., Ko, S.J.: Motion-blur-free camera system splitting exposure time. IEEE Transactions on Consumer Electronics 54(3), 981–986 (2008)

[11] Glad, F.A.: Color temperature alignment using machine vision. IEEE Transactions on Consumer Electronics 37(3), 624–628 (1991)

[12] Romih, T., Cucej, Z., Planinsic, P.: Wavelet Based Multiscale Edge Preserving Segmentation Algorithm for Object Recognition and Object Tracking. In: International Conf. on Consumer Electronics, ICCE 2008. Digest of Technical Papers, pp. 1–2 (January 2008)

# Liver Cell Nucleuses and Vacuoles Segmentation by Using Genetic Algorithms for the Tissue Images

Ching-Te Wang[1], Ching-Lin Wang[1], Yung-Kuan Chan[2], Meng-Hsiun Tsai[2], Ying-Siou Wang[2], and Wen-Yu Cheng[3]

[1] Department of Information Management, National Chin-Yi University of Technology, Taichung, Taiwan, R.O.C.
{ctwang,clwang}@ncut.edu.tw
[2] Department of Management Information Systems, National Chung Hsing University, Taiwan, R.O.C.
{ykchan,mht}@nchu.edu.tw, larcmolly@gmail.com
[3] Molecular Biology Institute of National Chung Hsing University, Neurosurgery Department of Taichung Veterans General Hospital, R.O.C.
wycheng@vghtc.gov.tw

**Abstract.** This paper proposes image segmentation methods for cell nucleuses and vacuoles in the liver fibrosis tissue images. The novel idea is to segment the objects by extracting the image features to determine the required cell in liver fibrosis images. In the proposed segmentation phase, some image processing methods are applied to segment the objects of nucleuses and vacuoles. Run Length method makes the object regions become obviously and the noises can be suppressed. The morphological opening operation is performed to split connecting objects. For vacuole regions segmentation, the opening operation applies the mode filter to stuff up the dark holes in the objects and keep the completeness of regions. Furthermore, the proposed method uses the Genetic Algorithm to find the most appropriate parameters and weights for the region segmentation. From the experimental results, the proposed method can achieve a good performance on the segmentation of cell nucleuses and vacuoles.

**Keywords:** image segmentation, image recognition, liver fibrosis, Genetic Algorithm, image stage system.

## 1 Introduction

Nowadays, the liver cancer is the major cause of death in the world, and the liver related diseases are also threatening the people of middle-age in many developing countries [5, 8]. The liver fibrosis is the excessive accumulation of extracellular matrix proteins such as collagen [1]. It is the effects of a persistent wound healing responses to chronic liver injury from a variety of reasons including viral, drug induce, and so on. To preserve its function, the liver will generate extracellular matrix proteins [7] to repair the injury.

M. Ali et al. (Eds.): IEA/AIE 2013, LNAI 7906, pp. 581–591, 2013.

However, the overload of extracellular matrix proteins will cause the liver fibrosis. When a liver suffered severely from fibrosis, the result will be liver cirrhosis. Once the liver cirrhosis is gradually getting worse, the risk of liver cancer will increase significantly [1]. In order to study the pathogenesis of liver fibrosis, the experimental rats are made to suffer from liver fibrosis by injecting the inducing chemical. The Carbon tetrachloride ($CCl_4$) is used as the inducing chemical to inject into the rats' liver [4]. The injection is repeated to injure rats' liver and cause the liver fibrosis. In the pathogenesis of liver fibrosis, the number and area of cell nucleuses and vacuoles (fatty degeneration) have obviously appeared in each stage. In the experimental image, the cell nucleuses are main objects and the vacuoles from Stage I have become more and larger than Stage IV. However, the number and area of nucleuses and vacuoles are difficult to judge and determine the stages between two close images. These values are difficult to count and compare by the human visual interpretation. In traditional stage judgment of liver fibrosis, the medical experts have to spend a large of time to observe a huge amount of medical images and measure the nucleuses and vacuoles of liver fibrosis. To reduce the processing time, human resource, and human errors, the proposed method will develop a segmentation method to help the professionals to determine the stages. Thus, the nucleuses and vacuoles are required for segmenting and to extract their features for the experts to recognize the Stages. In this paper, we will propose a Liver Tissue Images Segmentation Method (LTISM) to identify nucleuses and vacuoles by extracting the features of the area and the number of objects. The extracting characteristics of liver tissue image can provide the experts, such as doctors, to explain images objectively and to develop a grading system for recognizing the seriousness of liver fibrosis.

## 2    Liver Tissue Images Segmentation Method

### 2.1    Nucleuses Segmentation Approach (NSA)

NSA is to segment nucleus regions in a liver tissue image. Firstly NSA has a preprocessing stage to convert a colored tissue image into a gray-level one. Then, the Contrast Enhancement Method stage sets double thresholds to stretch the contrast of image. The Run Length Enhancement stage is to highlight the nuclei and suppress noise. Furthermore, the Noise Destruction stage is performed to split connecting objects by using opening operation. Finally, The Noise Elimination removes the noise by judging the size of region. Assume that the color liver tissue image $I_{NC}$ is shown in Fig. 1, the preprocessing stage first change the image $I_{NC}$ into a gray-level image. The color image is converted into R, G, B color mode images [3]. Because the clearness of R-component image is higher than the G and B-component images, the R color component image can facilitate the segmentation of nucleuses and non-nucleuses. Let $I_R$ be the R-component image, $I_R(x, y)$ be the intensity of $(x, y)$, which is shown in Fig. 2.

### 2.1.1    Contrast Enhancement Method (CEM)

To enhance the contrast of image, the image $I_R$ is reversed to image $I'_R$, which is shown in Fig. 3. In $I'_R$, the brighter regions are nucleus specified by a arrow 4, and the darker regions are non-nucleus indicated by arrow 5. To achieve the clearness of bright and dark, NSA applies CEM to determine whether a pixel in $I'_R$ is located in the nucleus or not. The thresholds $Th_l$ and $Th_u$ are lower threshold and upper threshold of gray-level values, respectively, in the image $I'_R$. Thus, there are three situations of pixel values in the image $I'_R$. If $I'_R(x,y)$ is smaller than the threshold $Th_l$, the $I_{DC}(x,y)$ is set to 0. That is, this pixel belongs to the non-nucleus regions. Conversely, if $I'_R(x,y)$ is greater than the threshold $Th_u$, the $I_{DC}(x,y)$ is set to 255. That is, this pixel belongs to the nucleus regions. Therefore, the pixel value $I'_R(x,y)$ between the thresholds $Th_l$ and $Th_u$ are needed to strengthen the contrast by using the gamma equalization function [17]. After executed the CEM on the image $I'_R$, the enhanced image $I_{DC}$ is shown in Fig. 4.

### 2.1.2   Fracture Connection Method (FCM)

In image $I_{DC}$, the nucleus may be divided into several small regions. The proposed Fracture Connection Method (FCM) is used to connect them into one region. Suppose that the image $I_{MT}$ is a null image with the same size as the image $I_{DC}$ and it will be the output image after FCM. FCM applies the maximal filter $MF$ with size $m_l \times m_l$ to scan each pixel of $I_{DC}$ and get the maximal pixel value in the mask window. Suppose that the pixel $I_{DC}(x,y)$ is used as the center point in the filter $MF$ and $MF_{max}$ is the maximal pixel value of pixels in the mask window. This method is described as follows:

Step 1: Locate the filter window in the most left-top corner of the image, the pixel
$I_{DC}(x,y)$ is used as the center pixel in the filter, where $x = 0, y = 0$.
Step 2: Find the maximal value $MF_{max}$ of pixels in the mask window.
Step 3: Copy the maximal value $MF_{max}$ into $I_{MT}(x,y)$ at the coordinates $(x,y)$ in the image $I_{MT}$, where $0 \le x < m, \ 0 \le y < n$ .
Step 4: Shift the window one pixel from left to right and top to down, go to Step 2 until the last pixel $I_{DC}(x,y)$, i.e. $x = m-1, y = n-1$, is executed. After executed FCM, the connected image is shown in Fig. 5.

### 2.1.3   Run Length Enhancement (RLE)

Assume that the filter $RF$ is a square window with size $r_l \times r_l$ and each pixel $I_{MT}(x,y)$ is used as the center point in the filter. Then, the RLE computes the average value of pixels in the filter $RF$ and the value is denoted as $RF_{avg}$. The RLE draws eight lines, each line passes through the center point $I_{MT}(x,y)$ in the $RF$. For example, $r_l$=7, the first line $l_0$ is the same as the X-axis. The degree of the angle between $l_0$ and $l_1$ is 22.5°. That is, each line $l_e$, $e$=0,1,2,..., 7, has the degree of $22.5° \times e$ with respect to line $l_0$ (X-axis).Then, the filter is operated and scanned pixel by pixel to the image $I_{MT}$ and copy the maximal value $RFl_{max}$ or minimal value $RFl_{min}$ to the image $I_{RL}(x,y)$. The method is described as follows:

Step 1: Locate the filter window in the most left-top corner of the image $I_{MT}$, the center
pixel is $I_{MT}(x,y)$ in the filter $RF$, where $x = 0, y = 0$.
Step 2: Find the average values $RFl_{avg}$, the maximal value $RFl_{max}$ and minimal value $RFl_{min}$ in the filter window.
Step 3: Copy the filter value to the image $I_{RL}(x,y)$.
Step 4: Shift the window one pixel from left to right and top to down, go to Step 2 until the last pixel $I_{MT}(x,y)$ , i.e. $x = m-1, y = n-1$, is executed.

After performed the maximal filter for each pixel $I_{MT}(x, y)$, the RLE executes the image normalization to uniform the gray-level intensities of the image $I_{RL}$. After performed the RLE in the image $I_{RL}$, the enhanced image $I_{RLN}$ is shown in Fig. 6.

### 2.1.4    Binary Filter

To identify the objects of nucleus, NSA determines a threshold for the enhanced image $I_{RLN}$ to create a binary image $I_{BN}$. After executed the binary filter, the white pixels belong to the nucleus region while the black pixels belong to the non-nucleus region in the image $I_{BN}$. The local average filter is a cross-based filter and computes a cross average value as a threshold in the filter. This filter determines whether the pixel value is 255 (nucleus) or 0 (non-nucleus) according to the threshold. The binary stage uses the cross filter instead of the window filter to reduce the computation time. Assume that $F_{CN}$ is a cross filter with cross size $cl_N \boxplus cl_N$, where the cross symbol $\boxplus$ is operated in the cross pixels. Let the pixel $I_{RLN}(x, y)$ be the center point in the filter and $F_{CNavg}$ is the average value of pixels in $F_{CN}$. The procedure is described as follows:

Step 1: Locate the cross filter $F_{CN}$ in the left-top corner of the image $I_{RLN}$, and the
pixel $I_{RLN}(x, y)$ is used as the center pixel of the filter, where $x = 0, y = 0$.
Step 2: Evaluate the average value $F_{CNavg}$ of pixels in the cross filter $F_{CN}$.
Step 3: Copy the filter value into $I_{BN}(x, y)$ at the coordinate $(x, y)$.
Step 4: Shift the window one pixel from left to right and top to down, then go to Step 2
until the last pixel $I_{RLN}(x, y)$ is executed. The image $I_{BN}$ is illustrated in Fig. 7.

### 2.1.5    Noise Destruction

In image $I_{BN}$, the nucleuses and a few noises are white, some of noises and the nucleuses are connected. The connections between nucleuses and noise should be cut off to avert from the under-segmentation. Hence, the opening operation [15] is executed to break off the tiny links in the Noise Destruction. The opening procedure starts with erosions twice and then dilations twice. Firstly, the erosion is operated on the image $I_{BN}$ by using the 8 neighbors structure element $B$ with the size $b_l = 3$. The structure element $B$ slides pixel by pixel in the image $I_{BN}$ as the convolution operation. Assume that the expression $I_{BN} \ominus B$ is denoted as the erosion operation in the image $I_{BN}$ with structure element $B$, where the notation $\ominus$ is denoted an erosion operation. A new image $I_{NE}$ is null and will be used to store the pixel values after erosion. The erosion is described as follows:

Step 1: Locate the structure element in the left-top corner of the image $I_{BN}$, and pixel
$I_{BN}(x, y)$ is used as the center pixel in the filter, where $x = 0, y = 0$.
Step 2: Find the coinciding pixel and copy the resultant values to new image $I_{NE}$.
(i)  If the center point $I_{BN}(x, y)$ is black point let $I_{NE}(x, y) = 0$.
(ii) If the center point $I_{BN}(x, y)$ is white point, there are two situations:
A.  Let $I_{NE}(x, y) = 0$, if there is at least one of black in the neighbor of $I_{BN}(x, y)$.
B.  Let $I_{NE}(x, y) = 255$, otherwise.
Step 3: Shift the structure element one pixel from left to right and top to down, then go
to Step 2 until the last pixel $I_{BN}(x, y)$.

In order to strengthen the effect, the erosion process is performed twice and the image $I_{NE}$ is shown as Fig. 8.

Secondly, the dilation is similar to erosion operation, but the pixel values become white (pixel value is 255) instead of black (pixel value is 0). The structure element $B$ slides pixel by pixel in the image $I_{NE}$. Assume that the expression $I_{NE} \oplus B$ is denoted as the dilation operation of the image $I_{NE}$ with structure element $B$, where the notation $\oplus$ is a dilation operation. The dilation is described as follows:

Step 1: Locate the structure element in the left-top corner of the image $I_{NE}$, the center
       pixel is $I_{NE}(x, y)$ in the filter, where $x = 0, y = 0$.

Step 2: Find the coinciding pixel and copy the resultant values to new image $I_{ND}$.

   (i)   If the center point $I_{NE}(x, y)$ is black point let $I_{ND}(x, y) = 0$.

   (ii)  If the center point $I_{NE}(x, y)$ is white point, all pixels are changed to white.

Step 3: Shift the structure element one pixel from left to right and top to down, then go
       to Step 2 until the last pixel $I_{NE}(x, y)$.

Assume that the opening operation is expressed as $I_{BN} \circ B$, and the generated image is $I_{ND}$. To strengthen the effect of the image $I_{BN}$, the opening operation first executes the erosions twice and then dilations twice. The opening procedure is expressed as follows: $I_{ND} = (I_{BN} \circ B) = (((I_{BN} \ominus B) \ominus B) \oplus B) \oplus B)$.

### 2.1.6   Noise Elimination

In order to identify the connected pixel regions, the 4-connectivity filter scans the image $I_{ND}$ from left to right and then top to down. The connected component labeling operation is required two passes to execute the labeling operation. The first pass is described as followings:

Step 1: Locate the 4-connectivity filter in the left-top corner of the image $I_{ND}$, the
       center pixel is $I_{ND}(x, y)$ in the 4-connectivity filter, where $x = 0, y = 0$.

Step 2: If the pixel value $I_{ND}(x, y)$ is equal to 255, they are four situations:

   (i)  Assign a new label $p$ to the pixel $I_{ND}(x, y)$, if there are no labeled neighbors.

   (ii)  The current pixel is marked as the same label $p$, if only one of the upper or left
       pixels is 255.

   (iii)  The current pixel is marked as the same label $p$, if both of the upper and left
       pixels have the same label $p$.

   (iv)  The current pixel is marked as one of these labels, if the upper and left pixels
       have been marked and they are different labels.

Step 3: Shift the window one pixel from left to right and top to down, go to Step 2 until the last pixel $I_{ND}(x, y)$ , i.e. $x = m - 1, y = n - 1$, is executed.

After the pixels of the image are labeled, the second pass starts to operate and is described as follows.

Step 1: Scan the image, find equivalent labels.

Step 2: Integrate the equivalent labels into an equivalent class, and reassign a new label to the equivalent class.

Suppose that $R_N$ is denoted as a labeled region with pixel value 255, and $A_N$ is the amount of pixels, called an area, for the region $R_N$. The Noise Destruction sets a threshold $Th_n$ to determine whether the region $R_N$ is a nucleus or noise. If $A_N$ is smaller than $Th_n$, this region $R_N$ will be judged as a noise and the pixel values of $R_N$

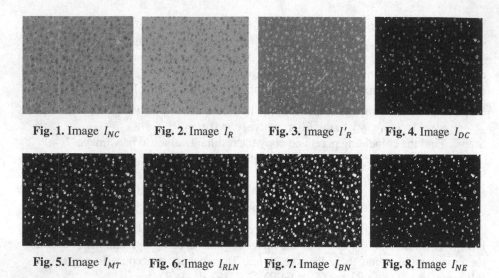

Fig. 1. Image $I_{NC}$     Fig. 2. Image $I_R$     Fig. 3. Image $I'_R$     Fig. 4. Image $I_{DC}$

Fig. 5. Image $I_{MT}$     Fig. 6. Image $I_{RLN}$     Fig. 7. Image $I_{BN}$     Fig. 8. Image $I_{NE}$

will be changed from 255 into 0. Otherwise, the region will be judged as nucleus and the pixel values will be still 255. After executed the Noise Elimination, the image $I_{ND}$ is transferred to the image $I_{NF}$.

## 2.2    Vacuoles Segmentation Approach (VSA)

The vacuoles segmentation firstly transforms the original image into a gray-level image. Assume the image $I_{VC}$ is an original color liver tissue image which is shown in Fig. 9. Because the green color component is easier to discriminate the vacuole regions, VSA would rather use G-color component than R or B-color components. Therefore, the G component image $I_G$ is selected to develop VSA and it is shown in Fig. 10. To enhance the contrast of the image $I_G$, gamma equalization [17] is performed as the CEM in NSA method. After gamma equalization, the vacuoles are more obviously as the brightness regions in the image $I_E$, which is illustrated in the Fig. 11.

### 2.2.1    Binary Filter

Suppose that $F_{CV}$ is a cross filter with the cross size $cl_V \boxplus cl_V$. To operate the filter, each pixel $I_E(x, y)$ at the coordinate $(x, y)$ is used as the center point of filter $F_{CV}$ in the image $I_E$. Let $F_{CVavg}$ be the average of pixels in the $F_{CV}$. The procedures are similar as Section 2.1.4. Obviously, the $I_E(x, y)$ will be judged as noise and $I_{BV}(x, y)$ is 0, if $I_E(x, y)$ is smaller than $F_{CVavg}$. Otherwise, the $I_E(x, y)$ will be judged as vacuoles and $I_{BV}(x, y)$ is 255. The image $I_{BV}$ is illustrated in Fig. 12.

### 2.2.2    Noise Destruction

After executed the binary filter, the white regions are still remained some noises in the image $I_{BV}$. The tiny connections between vacuoles and noises affect the quality of segmentation. Thus, the Noise Destruction of morphological operations is performed to destruct the links between noises and vacuoles. Similar to the Noise Destruction of NSA, the opening operation is also used to break off the tiny links between vacuoles and noises. The opening operation combines erosion with dilation by using the same structure element, which is a 3×3 square matrix. The operation procedure is different from the previous stage of NSA. Thus, the opening operation starts with erosion once and then dilation once. Assume that the erosion process is executed on the image $I_{BV}$ by using the 8 neighbors structure element $B$ with size $b_l = 3$. The structure element $B$ slides pixel by pixel in the image $I_{BV}$ as the convolution operation. Assume that the expression $I_{BV} \ominus B$ is denoted as the erosion operation in the image $I_{BV}$ with structure element $B$, where the notation $\ominus$ is an erosion operation. After performed the erosion, the pixel values will be stored to $I_{VE}(x, y)$. The erosion operation is the same as Section 2.1.5.

After executed the erosion operation on the image $I_{BV}$, the tiny links are broken off in the image $I_{VE}$, which is shown in Fig. 13. Next, the dilation process is similar to erosion process as above, but the pixel value are become to white (pixel value is 255) instead of the black (pixel value is 0). The dilation operation is the same as Section 2.1.5. After executed erosion and dilation sequentially, the opening operation generates the image $I_{VD}$. Assume that the opening operation is expressed as $(I_{BV} \Delta B)$ and the opening operation is shown as follows:

$$I_{VD} = (I_{BV} \Delta B) = ((I_{BV} \ominus B) \oplus B). \tag{1}$$

After performed the opening operation, the image $I_{VD}$ is shown in Fig. 14.

### 2.2.3    Spots Elimination (SE)

The VSA requires a mode filter to eliminate the bright spots and stuff up the dark holes in the image $I_{VD}$. The SE applies a mode filter to scan each pixel of the image $I_{VD}$ and get a mode value in the masked area. Assume that the mode filter $MD$ is a square size $m_d \times m_d$ and the pixel $I_{VD}(x, y)$ is the center point in the mask window. Let $n_1$ and $n_0$ be the numbers of pixel values 255 and 0, respectively, in the filter $MD$. Because the image $I_{VD}$ is a binary image, the pixel values are only two values, 255 or 0. That is, if one of them has more number than the other in the mask window, the value $I_{MD}(x, y)$ will be set to that one. The procedures of SE are described as follows:

Step 1: Locate the mode filter $MD$ in the left-top corner of the image $I_{VD}$, and the pixel $I_{VD}(x, y)$ is used as the center point of the mask window, where $x = 0, y = 0$.

Step 2: Evaluate $n_1$ and $n_0$ of pixels in the mask window.

Step 3: Copy the filter value into $I_{MD}(x, y)$ at the coordinate $(x, y)$.

Step 4: Shift the window one pixel from left to right and top to down, then go to Step 2 until the last pixel $I_{MD}(x, y)$, i.e. $x = m - 1, y = n - 1$, is executed.

To strengthen the effect of mode filter, this approach can be performed twice. After executed mode filter twice, the image $I_{VD}$ is transformed to the image $I'_{MD}$, which is shown in Fig. 15.

### 2.2.4    Noise Elimination

In the image $I'_{MD}$, VSA needs to remove the noises from the vacuoles. Similarly, Noise Elimination also uses a connected component labeling operator with 4-connectivity filter as the previous stage of NSA to obtain the area of each region. Assume that $R_V$ is denoted as a labeled region with pixel value 255, and $A_V$ is the area of region $R_V$. In order to determine the region $R_V$ is vacuole or noise, the VSA computes the mean value of all labeled regions and uses that value as a threshold $Th_v$. Firstly, this stage separates the regions into three cases according to the size of area of the labeled regions. The three cases are divides as follows:

$$\begin{cases} case\ 1, & if\ A_V > k_1 \times Th_v \\ case\ 2, & if\ k_1 \times Th_v \geq A_V > k_2 \times Th_v \ , \\ case\ 3, & if\ A_V < k_2 \times Th_v \end{cases} \tag{2}$$

where $k_1$, $k_2$, ( $k_1 > k_2$) are weight constants. After performed the three cases on the image $I'_{MD}$, the noise is removed from the vacuoles in the image $I_{VNE}$, which is shown in Fig. 16.

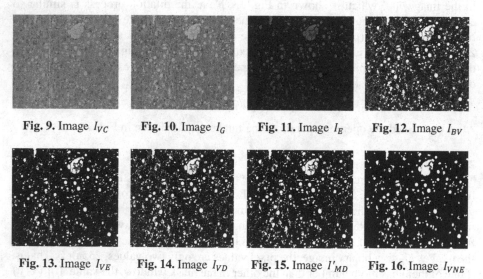

**Fig. 9.** Image $I_{VC}$    **Fig. 10.** Image $I_G$    **Fig. 11.** Image $I_E$    **Fig. 12.** Image $I_{BV}$

**Fig. 13.** Image $I_{VE}$    **Fig. 14.** Image $I_{VD}$    **Fig. 15.** Image $I'_{MD}$    **Fig. 16.** Image $I_{VNE}$

# 3    Experimental Results

The proposed method uses the Genetic-Based Parameter Detector (GBPD) to find the most appropriate parameters and weights for segmentation methods [11,12,19]. Firstly, the genetic algorithm initializes the population and selects solutions for the next population. In the experimental images, there are totally 160 rat liver tissue images, which are acquired from Laboratory Animal Center of Taichung Veterans General Hospital, Taiwan, ROC. There are four stages and 40 images in each stage. These images will be segmented in the experiments for the proposed methods.

## 3.1    Parameters of NSA

To derive the optimal performance in NSA, the parameters $Th_l$, $Th_u$, $\gamma$, $m_l$, $r_l$, $cl_N$, and $Th_n$ should be determined by GBPD. GBPD uses a chromosome of binary string, which contains seven substrings $s_1, s_2, \cdots, s_7$. These substrings are comprised of $n_1, n_2, \cdots, n_7$ binary bits, respectively. The substrings $s_1, s_2, \cdots, s_7$ are stood for the parameters $Th_l$, $Th_u$, $\gamma$, $m_l$, $r_l$, $cl_N$, and $Th_n$, respectively. For each chromosome $Ch$, the parameters $Th_l$, $Th_u$, $\gamma$, $m_l$, $r_l$, $cl_N$, and $Th_n$ are encoded as: $Th_l = 30 + (n'_1) \times 1$, $Th_u = 50 + (n'_2) \times 1$, $\gamma = 0.1 + (n'_3) \times 0.25$, $m_l = 3 + (n'_4) \times 2$, $r_l = 7 + (n'_5) \times 2$, $cl_N = 400 + (n'_6) \times 10$, $Th_n = 150 + (n'_7) \times 1$, where $n'_1, n'_2, \cdots, n'_7$ are the number of 1-bits in the substrings $s_1, s_2, \cdots, s_7$, respectively. The NSA applies GBPD to train the most appropriate values of $Th_l$, $Th_u$, $\gamma$, $m_l$, $r_l$, $cl_N$, and $Th_n$ by using the genetic algorithm. The ground truth is used as a criterion and to compare with the segmented image by NSA.

### 3.2   Parameters of VSA

Similarly, the performance of VSA is significantly affected by the parameters $\gamma_E$, $cl_V$, $m_d$, $k_1$, $k_2$, $a_1$, and $a_2$. GBPD utilizes a chromosome of binary string, which contains seven substrings $s_{11}, s_{12}, \cdots, s_{17}$. These substrings are comprised of $n_{11}, n_{12}, \cdots, n_{17}$ binary bits, respectively. The substrings $s_{11}, s_{12}, \cdots, s_{17}$ are stood for the values of $\gamma_E$, $cl_V$, $m_d$, $k_1$, $k_2$, $a_1$, and $a_2$, respectively. For each chromosome $Ch$, the parameters $\gamma_E$, $cl_V$, $m_d$, $k_1$, $k_2$, $a_1$, and $a_2$ are calculated as: $\gamma_E = 0.1 + (n'_{11}) \times 0.05$, $cl_V = 400 + (n'_{12}) \times 10$, $m_d = 3 + (n'_{13}) \times 2$, $k_1 = 0.1 + (n'_{14}) \times 0.1$, $k_2 = 0.1 + (n'_{15}) \times 0.1$, $a_1 = 0.1 + (n'_{16}) \times 0.05$, $a_2 = 0.1 + (n'_{17}) \times 0.05$, where $n'_{11}, n'_{12}, \cdots, n'_{17}$ are the number of 1-bits in the substrings $s_{11}, s_{12}, \cdots, s_{17}$, respectively. GBPD applies the VSA to train the optimal values of the parameters $\gamma_E$, $cl_V$, $m_d$, $k_1$, $k_2$, $a_1$, and $a_2$ through a genetic algorithm. Finally, GBPD uses the average of F-measure as the measure of fitness for the chromosome $Ch$ with the corresponding values of $\gamma_E$, $cl_V$, $m_d$, $k_1$, $k_2$, $a_1$, and $a_2$.

### 3.3   Accuracy of Segmentation Results

The nucleus and vacuoles regions are segmented by NSA and VSA, the optimal values of parameters are obtained by GBPD. Four images are randomly selected from all of 160 images to perform NSA and VSA, and four ground truth images are also drawn by the experts. From the experimental results, the average scores of nucleus segmentation are greater than the vacuole. The accuracy of precision, recall and F-measure are greater than 80% in the NSA. Because the cell nucleuses are ordinary and normal, the variations of cell nucleuses are seldom. On the other hand, the accuracy of precision, recall and F-measure are about 75% in the proposed VSA. Because the vacuoles are unusual and exceptional, the shapes of vacuoles and the rate of circles are variably. Averagely, the accuracy of our methods is about 80% in NSA and VSA. The accuracy is a good performance in the cell segmentation.

## 4   Conclusions

In this paper, we had proposed the image segmentation methods to determine the cell nucleuses and vacuoles. In the design procedure, the nucleus and vacuole in liver tissue image are retrieved by the proposed segmentation methods, and then the object features are used to recognize the image nucleuses and vacuoles of liver fibrosis. In the segmentation phase, a set of image processing methods are developed to segment the objects of nucleuses and vacuoles. CEM sets double thresholds and applies gamma equalization to strengthen contrast. They are simple and effective schemes for the contrast enhancement. RLE and SE are used to highlight object regions, suppress noises and reduce the influence of noises in segmentation phase. In Binary Filter stage, a cross average filter is used to reduce executive time and local region processing such that the uneven bright image can be binarized adaptively. From the experimental results, the average accuracy of the proposed methods is about 80% and shows that has a good performance in the object segmentation.

**Acknowledgements.** This work is supported partially by the National Science Council, Taiwan, ROC, under Grant No. NSC100-2511-S-167-004-MY2.

# References

[1] Bataller, R., Brenner, D.A.: Liver fibrosis. The Journal of Clinical Investigation 115(2), 209–218 (2005)

[2] Chan, Y.K., Chang, C.C.: Image matching using run-length feature. Pattern Recognition Letters 22(5), 447–455 (2001)

[3] Chaves-Gonzalez, J.M., Vega-Rodriguez, M.A., Gomez-Pulido, J.A., Sanchez-Perez, J.M.: Detecting Skin in Face Recognition Systems: A Colour Spaces Study. Digital Signal Processing 20(3), 806–823 (2010)

[4] Dai, L., Ji, H., Kong, X.W., Zhang, Y.H.: Antifibrotic effects of $ZK_{14}$, a novel nitric oxidedonating Biphenyldicarboxylate derivative, on rat HSC-T6 cells and $CCl_4$- induced hepatic fibrosis. Acta Pharmacologica Sinica 31, 27–34 (2010)

[5] Department of Health, Executive Yuan, R.O.C(TAIWAN). 2010 statistics of causes of death 2012 (2012), http://www.doh.gov.tw/ufile/doc/ 2010-statistics%20of%20cause%20of%20death.pdf

[6] Dillencourt, M., Samet, H., Tamminen, M.: A general approach to connected components labeling for arbitrary image representations. Journal of the ACM 39(2), 253–280 (1992)

[7] Friedman, S.L.: Liver fibrosis- from bench to bedside. Journal of Hepatology (38), 38–53 (2003)

[8] Griffiths, C., Rooney, C., Brock, A.: Leading causes of death in England and Wales -how should we group causes. Health Statistics Quarterly (28), 6–17 (2005)

[9] Huang, D.C., Chen, R.T., Chan, Y.K., Jiang, X.: An automatic indirect immunofluorescence based cell segmentation and counting system. National Digital Library of These and Dissertations in Taiwan (2010)

[10] MacQueen, J.: Some Methods for Classification and Analysis of Multivariate Observations. In: Fifth Berkley Symposium on Mathematical Statistics and Probability, vol. 1, pp. 281–297 (1967)

[11] Man, K.F., Tang, K.S., Kwong, S.: Genetic Algorithms: Concepts and Designs. Springer, New York (1999)

[12] Maulik, U.: Medical Image Segmentation Using Genetic Algorithms. IEEE Transactions on Information Technology in Biomedicine 13(2), 166–173 (2009)

[13] Otsu, N.: A Threshold Selection Method from Gray-Level Histogram. IEEE Transactions on System Man Cybernetics SMC-9(1), 62–66 (1979)

[14] Raghavan, V., Bollmann, P., Jung, G.S.: A critical investigation of recall and precision as measures of retrieval system performance. ACM Transactions on Information Systems 7(3), 205–229 (1989)

[15] Stevenson, R.L., Arce, G.R.: Morphological Filters: Statistics and Further Syntactic Properties. IEEE Transactions on Circuits and Systems CAS-34(11) (1987)

[16] Wikipedia, The Free Encyclopedia, F1 score,
http://en.wikipedia.org/wiki/F1_score

[17] Wikipedia, The Free Encyclopedia, Gamma correction,
http://en.wikipedia.org/wiki/Gamma_correction

[18] F precision and recall Wikipedia, The Free Encyclopedia, Sensitivity and specificity,
http://en.wikipedia.org/wiki/Sensitivity_and_specificity

[19] Yun, Y.S.: Hybrid Genetic Algorithm with Adaptive Local Search Scheme. Computers and Industrial Engineering 51(1), 128–141 (2006)

# Mouth Location Based on Face Mask and Projection

Hui-Yu Huang[*] and Yan-Ching Lin

Department of Computer Science and Information Engineering,
National Formosa University, Yunlin 632, Taiwan
hyhuang@nfu.edu.tw

**Abstract.** Facial feature detection plays an important role in biological recognition applications, such as criminal investigation, special surveillance, photographic mode, etc. In this paper, we propose an approach to achieve the mouth feature extraction and detection. This approach involves the steps of skin-color segmentation, face feature extraction, edge detection and edge projection. In order to more effectively and correctly locate the mouth position, we adopt two phases to perform our purposes, one phase is to label face mask by using our designed skin-color filter, the other phase is to detect the edge and to compute edge projection for the decided mouth region within the face mask. The present results demonstrate that our proposed system can achieve a high accuracy for month detection and can obtain coarsely the face posture estimation based on specified rule.

**Keywords:** Skin-color filter, face mask, mouth location, projection.

## 1    Introduction

As the rapid development of science and technology, it makes human life more functional and convenient. In recent years, there are many researches which focus on human biological attributes, such as face recognition or face detection, etc. As for applications, one of important schemes is eyes detection which can avoid or alarm a fatigue situation for driving a car. In addition, for mouth detection, it can provide a lip-read recognition to help the blind persons. Based on the interesting scheme, in this paper, we will focus on locate the correct mouth position based on our proposed method. Firstly, the face region in an image must be clearly labeled out. For face detection, there are many approaches which have been presented [1, 2].

Chen and Xu [1] proposed an algorithm of face detection based on the differential images and PCA in color image which modified the Eigenface technique. Hu *et al.* [2] presented an automatic face recognition solution. Authors used the AdaBoost technology to detect face region and wavelet transform and KPCA method to extract face features. These features were fed up into support vector machine for recognition. This method is superior to traditional PCA in the time of features extraction.

For face detection, one interesting topic is mouth detection such as [3-5]. Wang *et al.* [3] used Harris corner interest point on mouth detection with PCA probability model. The experiment result shows that the correct rate and the accuracy

---

[*] Corresponding author.

M. Ali et al. (Eds.): IEA/AIE 2013, LNAI 7906, pp. 592–601, 2013.
© Springer-Verlag Berlin Heidelberg 2013

of mouth position have all been enhanced. Li *et al.* [4] proposed an algorithm based on combined Adaboost algorithm and particle filter for driver's fatigue detection. Then the mouth verification was applied according to prior knowledge. Driver's state was judged by the geometrical features in a period of time. Nhan and Bao [5] proposed a mouth detection approach in color image. Firstly, authors segmented image based on skin. After skin process, they added some specific techniques to fit the mouth in color image efficiently in order to determine mouth candidates and then classify those of candidates by neural network. In this paper, the procedures of our method include face detection, skin color segmentation, Canny edge detector and projection. The skin-color segmentation is to find face mask which aims to filter out non-skin color regions and raises accuracy before processing mouth location.

The remainder of this paper is organized as follows. Section 2 presents the proposed method. Experimental results and performance evaluation are presented in Sections 3. Finally, Section 4 concludes this paper.

## 2    Proposed Approach

In the section, we will describe our approach. This approach consists of preprocess, segment the face region, edge detection, and projection. Details of procedures are described in the following subsections.

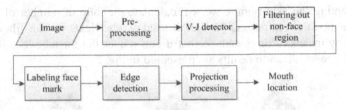

**Fig. 1.** System flowchart

### 2.1    Preprocessing

In order to decrease the computational complexity and to fast obtain the face features, we take the normalizing process for all test data. Here, we use a bi-cubic interpolation to perform the normal size. The normal size has three forms $500\times350$, $350\times500$, and $500\times500$.

### 2.2    Face Detection by Using V-J Detector

For face detection, Viola and Jones (V-J) method [6] is usually adopted to work this field that possesses a high efficiency and accuracy to detect where the face region is in an image. The V-J method consists of three phases. First, the Haar-like features is rectangular type that is obtained by integral image; Second, The Adaboost algorithm is a learning process that is a weak classification and then uses the weight value to learn and construct as a strong classification. Third, we can obtain the non-face region and face region after cascading each of strong classifiers. Details of V-J detector can study Ref. [6].

## 2.3    Filtering Out Non-faces Region

Although the face detection by used V-J detector has a higher performance, the threshold may affect the location result. Hence, we employ a ratio of skin region filtered candidates to further decide which the correct face region is, so that it can improve the face location accuracy after V-J detector. In other words, the better location of skin-color region implies a better outstanding feature for face.

Owing to the skin characteristic, we will take the different color space and threshold to decide the skin region. As previously researched, many color models, *YCbCr* or *HSV* or *HIS*, provide the advantage information in skin-color detection, in this paper, we use *YCbCr* color space to detect the skin-color area.

In *YCbCr* color space, *Y* is Luma influenced by light factor, we ignore *Y* value and adopt the chrominance components *Cb* and *Cr* to decide the skin range. Using the adaptive threshold to decide the skin region, the related researches have been published [7, 8]. According to our experiment, we redefine more appropriate skin condition range expressed as

$$\text{Skin} = \begin{cases} 1, \text{if} \begin{cases} 90 \leq Cb \leq 124, \\ 136 \leq Cr \leq 180, \end{cases} \\ 0, \text{otherwise}. \end{cases} \tag{1}$$

After segmenting the skin region, we will calculate the ratio of number of pixels belonging to skin-color region and number of total pixels. Next, we can filter out non-face region by means of threshold, so that it can reduce the mistake location of face region. The face localization results are presented in Fig. 2.

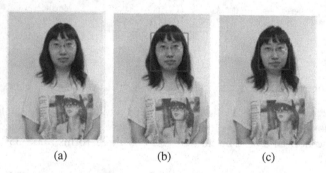

(a)                    (b)                    (c)

**Fig. 2.** Result of filtering out non-faces. (a) Original image, (b) face detection by V-J detector, (c) correct face location.

## 2.4    Labeling Face Mask

Generally, the face posture has frontal face and profile face. For mouth location, it existed high mistake on located profile face because it includes skin and non-skin color regions. In order to solve this problem, we propose a face mask to reduce mistake. The face mask is obtained by morphologic processing.

## 1)  Skin-Color Segmentation

This procedure is to transform $RGB$ into skin color before product face mask. We use the normalized $RGB$ color model $(r,g,b)$ to construct skin-color model [9-11].

Here, we present the results of the different skin color segmentation methods compared with Huang's method [10], Chen's method [11], and our proposed methods. Figure 4 shows the compared result. From Fig. 3 (d) shows that Chen's method is better than other author's method on facial contour and noise, but the median image in Fig. 3(d) has little defects which non-skin region as skin color, such as hair. Hence, based on Chen's method, in this paper, we further modify his method and redesign a more effective skin constraint to achieve the skin region for a face image, it is expressed as

$$
Skin = \begin{cases} 1, & if \begin{cases} Q_-(r) < g < Q_+(r), \\ (R-G) \ge 25, \\ w > 0.001, \end{cases} \\ 0, & otherwise, \end{cases} \tag{2}
$$

where the $Q_+$ and $Q_-$ and $w$ are depicted by [9]

$$
\begin{cases} Q_+(r) = -1.3767\,r^2 + 1.0743\,r + 0.1452, \\ Q_-(r) = -0.776\,r^2 + 0.5601r + 0.1766, \\ w = (r-0.33)^2 + (g-0.33)^2. \end{cases}
$$

From Fig. 3(d), it is evident that our proposed rule of skin color constraint is superior to Chen's method.

## 2)  Ratio Filter Design

In order to present the skin-color region in a face image, we design a novel morphological process called ratio filter. The filter aims to emphasize skin pixels and to decrease the non-skin pixel. The ratio filter is to compute number of skin pixels and non-skin pixels on $5 \times 5$ mask, and then to compare those of skin pixels.

$$
\begin{cases} P_{255} = P_{255} + 1, & if\ X = 255, \\ P_0 = P_0 + 1, & if\ X = 0, \end{cases} \tag{3}
$$

where $P_{255}$ is number of skin pixel, $P_0$ is number of non-skin pixel, and $X$ is a pixel value of the result of skin detection. The current pixel has three conditions described as follows.

$$
\begin{cases} P_{current} = 0, & if\ P_{255} < P_0, \\ P_{current} = 255, & if\ P_0 < P_{255}, \\ P_{current} = P_{original}, & if\ P_0 = P_{255}, \end{cases} \tag{4}
$$

where $P_{current}$ and $P_{original}$ denote the current and original pixel, respectively. Figure 4 presents the result after using ratio filter. The facial contour is more compact and the non-skin-color region is reduced.

**3)     Labeling**

How to label face mask is very important procedure in our scheme. It will deeply affect the accuracy of mouth location. The steps of face mask labeling consist of two phases. One phase is to find first pixel value which value is 255 on horizontal (left to right and right to left) and vertical (top to down and down to top) to obtain horizontal-based face region and vertical-based face region. The other phase is to take "AND" operation both these different face regions, thus face mask can be successfully labeled out like Fig. 5(d). The profile of this operation is shown in Fig. 5.

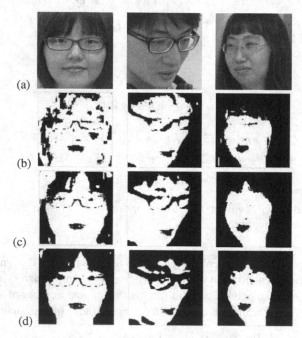

(a)

(b)

(c)

(d)

**Fig. 3.** Skin color detection results. (a) Original images, (b) Huang's method, (c) Chen's method, (d) our proposed skin-color rule.

## 2.5     Edge Projection of the Horizontal and Vertical

For mouth detection, firs, we use Canny detector to detect the edge of original image within face mask. Then, we compute the projection of horizontal and vertical edge points to locate mouth position. According to the spatial geometric relationship of facial features, in general, the mouth region is positioned between one third of face region and one quarter of face region. We define the ranges of horizontal region and vertical region described as

$$\begin{cases} H_{h1} = H \times 0.7 \\ H_{h2} = H \times 0.1, \\ W_h = W \times 0.1 \end{cases} \begin{cases} H_{v1} = H \times 0.4 \\ H_{v2} = H \times 0.1, \\ W_v = W \times 0.1 \end{cases} \qquad (5)$$

where $H_{h1}$, $H - H_{h2}$, and $W_h$ denote the first position and the last position on image height ($H$), the range for horizontal region on image width ($W$), respectively. $H_{v1}$, $H - H_{v2}$, and $W_v$ denote the first position, the last position on $H$, and the range for vertical region on $W$, respectively. Figure 6 shows this diagram of horizontal and vertical regions. The results of horizontal and vertical projection are shown in Fig. 7.

Fig. 4. (a) Non using ratio filter, (b) using ratio filter

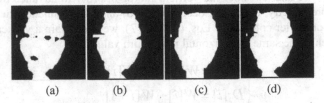

Fig. 5. The face mask. (a) The binary image by ratio filter, (b) horizontal result, (c) vertical result, and (d) face mask is to compute "And" operation both (b) and (c).

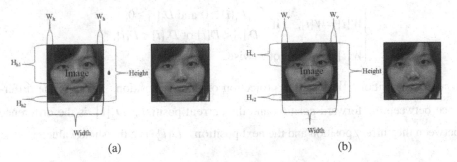

Fig. 6. The diagram of mouth region. (a) Horizontal region, (b)vertcal region.

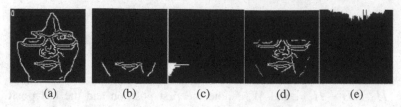

**Fig. 7.** The projections of the horizontal and vertical. (a) Canny edge, (b) the horizontal region, (c) horizontal projection (d) vertical region, (e) vertical projection.

## 2.6    Mouth Detection

Based on the previous procedures, we will compute the maximum values of horizontal and vertical projection to detect the mouth. In the following, two rules are to decide how the mouth location is located.

### 1)    Finding the Maximum Value for Horizontal

If two or more positions have the equal maximum value, we do the average position for two positions or the media position for more than two positions.

### 2)    Finding the Maximum Value for Vertical

In addition, we also consider the vertical factor in order to advance the mouth detection more exactly; hence, we design Eqs. (6) and (7) to enhance the correct vertical location. After computing Eqs. (6) and (7), we can obtain a vertical maximum value which is the same as horizontal maximum value.

$$\begin{cases} D_1[i] = W[i-1] - W[i], \\ D_2[i] = W[i] - W[i+1], \\ D[i] = W[i] \times 0.5, \end{cases} \tag{6}$$

$$\begin{cases} W[i] = W[i], & \text{if } \begin{cases} D_1[i] \geq 0 \text{ and } D_2[i] \geq 0, \\ D_1[i] < D[i] \text{ or } D_2[i] < D[i], \end{cases} \\ W[i] = 0, & \text{otherwise.} \end{cases} \tag{7}$$

where $i$ is position, $W[i]$ is the projection of current position, $D_1[i]$ is the difference between the forward position and the current position, $D_2[i]$ is the difference between the current position and the next position, $D[i]$ is a threshold value.

## 2.7    Face Posture Estimation

In addition, based on the above procedures, we can further use much advantageous information to coarsely estimate the face posture. In face posture estimation stage, we design a coarse and easy method to estimate the rate of mouth position and face edge deciding the front face or not. The steps are described as follows.

**Step 1: Face Edge Labeling**

Based on the face mask obtained the above process, we map this mark to real face image and label the red color.

**Step 2: Posture Ratio Definition and Computation**

$$\begin{cases} L_L = m_x - L, L_R = R - m_x, Length = R - L, \\ Rate_L = L_L / Length, \\ Rate_R = L_R / Length, \end{cases} \tag{8}$$

$$\begin{cases} Ratio = Rate_L / Rate_R, & \text{if } Rate_L > Rate_R, \\ Ratio = Rate_R / Rate_L, & \text{if } Rate_L < Rate_R, \\ Ratio = 1, & \text{if } Rate_L = Rate_R, \end{cases} \tag{9}$$

where $m_x$ denotes the mouth position on vertical maximum value , $L_L$ is the length between red point (left) and $m_x$ , $L_R$ is the length between red point (right) and $m_x$ , $L$ is first red point which scans from $m_x$ to left side for face edge, $R$ is first red point which scans from $m_x$ to right side for face edge.

**Step 3: Posture Estimation**

The coarsely posture estimation is defined as

$$\begin{cases} \text{Frontal face, if } Ratio \leq 2, \\ \text{Profile face, if } Ratio > 2. \end{cases} \tag{10}$$

Based on the above processes, we can coarsely estimate the face posture which is frontal or profile post in an image.

## 3    Experimental Results

Here, we use Boa database [12] and our database to verify our proposed method. The face images have 355 that contain 94 on Boa database and 261 on our database respectively. The computational time of mouth localization is about 0.75s on image size of $150 \times 150$ worked a 2.80 GHz Intel ® Core(TM) i5-2300 CPU with 4 GB RAM PC and C# language.

Based on skin-color segmentation rule which will affect the mouth location, in our experiments, the results comparing with Chen's rule and our rule about skin-color segmentation are presented the accuracy percentage of mouth detection. Table 1 presents the results. From Table 1, it is clear that our result is better than Chen's result for Bob database, it is because Chen's condition is lax, and it may obtain a large face mask which includes non-skin color region. Hence, it will cause the mistake mouth location. On the whole, the accuracy of mouth localization in our data can achieve 94%. For the facial posture estimation, the difference between the Chen's condition and our condition is 3%.

Figures 8-9 show some results of mouth location. From Fig. 8, it is clear that our result is superior to the location result of Chen's condition. Figure 10 shows some mistake cases of mouth localization.

**Table 1.** The accuracy of mouth localization and posture estimation

| Database | Samples | Mouth localization | | Face posture estimation | |
|---|---|---|---|---|---|
| | | Accuracy | | Accuracy | |
| | | Chen's | The proposed | Chen's | The proposed |
| Bob | 94 | 90% | 92% | 72% | 70% |
| Our data | 261 | 94% | 94% | 85% | 81% |
| Total | 355 | 93% | 93% | 81% | 78% |

(a)                              (b)

**Fig. 8.** Results of the mouth location. (a) Chen's condition, (b) our condition

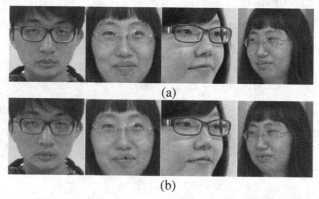

(a)

(b)

**Fig. 9.** Results of the mouth location include frontal face and profile face. (a) Original images. (b) The results.

(a)                              (b)

**Fig. 10.** Wrong localization. (a) Original image, (b) wrong result.

# 4    Conclusions

In this paper, we have presented an efficient mouth location method based on face extraction, face mask labeling, edge detector, and projection. By face mask, we can fast obtain the important facial features and detect the mouth position within mask and advance the located accuracy. The experimental results show that this approach can obtain higher accuracy than Chen's skin-color segmentation condition. For posture estimation, in spilt of the wrong case existed in current approach, it still has 78% accuracy. In the future, we will further modify the current method to improve the accuracy of face posture estimation.

**Acknowledgement.** This work was supported in part by the National Science Council of Republic of China under Grant No. NSC100-2628-E-150-003-MY2.

# References

1. Xu, Y., Chen, X.: A new algorithm of face detection based on differential images and PCA in color image. In: Proc. of the 2nd Int. Conf. Computer Science and Information, pp. 172–176 (2009)
2. Hu, T., Liu, R., Zhang, M.J.: Face recognition under complex conditions. In: Proc. of IEEE Int. Conf. on Electronics, Comm. and Control Engineering, pp. 960–963 (2010)
3. Wang, L., Ye, H., Xia, L.: Mouth detection based on interest point. In: Proc. of IEEE Conf. of Control, pp. 610–613 (2007)
4. Li, L., Chen, Y., Xin, L.: Driver fatigue detection based on mouth information. In: Proc. of the 8th Int. World Congress on Intelligent Control and Automation, pp. 6058–6062 (2010)
5. Nhan, H.N.D., Bao, P.T.: A new approach to mouth detection using neural network. In: Proc. of IEEE Int. Conf. Control, Automation and System Engineering, pp. 616–619 (2009)
6. Viola, P., Jones, M.: Rapid object detection using a boosted cascade of simple features. In: Proc. of the IEEE Computer Vision and Pattern Recognition, vol. 1, pp. 511–518 (2001)
7. Chai, D., Ngan, K.N.: Face segmentation using skin-color map in videophone applications. IEEE Trans. Circuits and System for Video Technology 9, 551–564 (1999)
8. Wu, M.W.: Automatic facial expressions analysis system, Master Thesis, National Cheng King University (2003)
9. Soriano, M., Martinkauppi, B., Huovinen, S., Laaksonen, M.: Using the skin locus to cope whit changing illumination conditions in color-based face tracking. In: Proc. IEEE Nordic Signal Proc. Symp., Kolmarden, Sweden, pp. 383–386 (2000)
10. Huang, T.S.: A smart digital surveillance system with face tracking and recognition capability. Master Thesis, Chung Yuan Christian University (2004)
11. Chen, C.T.: Multiple face recognition based on skin-color regional segmentation and principal component analysis. Master Thesis, National Taiwan Ocean University (2006)
12. Boa database,
    http://www.datatang.com/datares/go.aspx?dataid=604374

# Environmental Background Sounds Classification Based on Properties of Feature Contours

Tomasz Maka

West Pomeranian University of Technology, Szczecin,
Faculty of Computer Science and Information Technology,
Zolnierska 49, 71-210 Szczecin, Poland
tmaka@wi.zut.edu.pl

**Abstract.** In this paper, an approach to environmental sound recognition (ESR) by using properties of feature trajectories is presented. To determine the discriminative attributes of background sounds, several audio classes have been analysed. Selected groups of sounds reflect the acoustical environments that may occur in real sound acquisition situations. We proposed the feature extraction scheme, where obtained trajectories at parameterization stage are further processed in order to improve classification accuracy. A discriminatory analysis of popular audio features for ESR task has been performed. Obtained results show that proposed technique gives promising classification results and can be applied in systems where properly identified audio scene can improve other audio processing tasks.

**Keywords:** environmental sounds recognition, ESR, feature contours, audio classification.

## 1 Introduction

Environmental sounds are the fundamental part of living acoustic environment. Many applications of such sounds like automatic surveillance systems, audio scene recognition, multimedia retrieval and indexing and robotic systems stimulate rapid development of classification techniques dedicated to ESR in recent years [4,6]. The type of such sounds is strongly dependent on the number and characteristic of sound sources. Environmental sounds often have low stability and contain event sounds, therefore non-stationary analysis is performed for recognition tasks. The important part of ESR system is feature extraction and selection stage. Features set should capture most of specific properties of acoustic environment between audio classes. The number of classes can be defined in many ways – in case of environmental sounds a lot of various classes can be defined [3,1]. However, in real application of ESR, the number of classes vary between few to several due to subtle differences in many environmental sounds. The interval of sound sources activity and frequency range are the main properties determining the components of environmental sound. Therefore, to extract these properties a parameterization process involving audio features in both time

M. Ali et al. (Eds.): IEA/AIE 2013, LNAI 7906, pp. 602–609, 2013.

and frequency domain is necessary [6]. The robustness of ESR depends on many factors describing audio scene. The number of sources and their movement dynamics can introduce sound distortions as well as often increase the number of frequency components. In real audio acquisition process the weather conditions like rain or wind may deteriorate the overall quality of recordings. Thus, to compensate such situations, a multi-microphone solutions are often applied in recording chain. In the result, an additional post-processing stage involving separation and mixing procedures have to performed to obtain audio stream with moderate level of distortions which do not obscure the main properties of audio scene [14]. Many typical audio processing tasks like speech recognition, speaker identification, verification and diarization, audio indexing and retrieval, etc. are very sensitive to the environmental noise. In particular, the situation where audio query is based on recording performed in outdoor conditions needs to apply an enhancement techniques to the input signal.

Presented approach exploits the modified audio feature contour properties to identify environmental sounds. Proposed technique can be applied in many audio and speech processing tasks to identify the type of background noise, tune the parameters and improve the final performance.

## 2    Proposed Approach

The feature contours carry the properties of analysed signal calculated in different domains. Therefore, a set of feature contours may describe the properties of input signal in more details. In our approach presented in Figure 1, statistical properties of processed feature contours have been used to create descriptors.

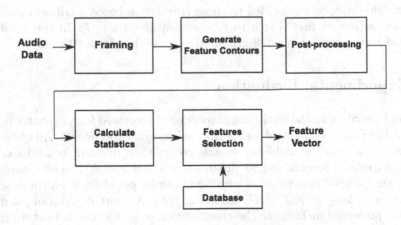

**Fig. 1.** Audio data parameterization process

At the first stage signal is divided into overlapping frames (in our case the frame length was 30 ms and overlapping equal to 50%) in order to calculate features for each frame, leading to obtain a feature contour. Then single or selected set of operations is performed on the contour. Finally, several statistics are calculated for trajectory to define resulting features vector. Such approach needs to tune some parameters like frame size and overlapping, set of audio features, post-processing operations, and statistics selection. As was mentioned earlier, features should cover properties of analysed signal in time and frequency domains. The post-processing operations set includes smoothing techniques (exponential, moving average, Gaussian, median and Savitzky-Golay filters) [11], thresholding (center clipper, soft, hard) and delta features (single and double) [13]. From each modified feature contour, a set of 16 statistics described in Table 1 is calculated.

Table 1. Statistical descriptors for feature contours

| Statistic | Description | Statistic | Description |
|-----------|-------------|-----------|-------------|
| MIN | minimum | MRG | mid-range |
| IQR | interquartile range | MOD | mode |
| KRT | kurtosis | Q1 | first quartile |
| MAD | median absolute deviation | Q3 | third quartile |
| MAX | maximum | RGE | range |
| MCC | mean-crossing count | SKW | skewness |
| MEA | arithmetic mean | SDE | standard deviation |
| MED | median | VAR | variance |

In the final step, a feature selection is performed using several recordings for each of classes. After experimenting with a few feature selection methods on audio descriptors, we decided to chose correlation-based attributes selection approach with best first search to explore feature space [7]. In the result we obtained features which have discriminative properties for ESR.

## 3    Experimental Evaluation

In order to verify that the statistical properties of processed feature contours may be useful in ESR, we prepared and performed several experiments. All tests have been performed on the ambience sounds recorded in different conditions. The database contains 5 hours and 20 minutes of audio material. The environmental sounds are grouped into 8 classes with 80 examples per class. Each single sound is 30 seconds long with 22050Hz sampling rate. A short description of audio classes is presented in Table 2. The classification stage was carried out using the Support Vector Machine (SVM) classifier with RBF kernel [2].

The first experiment was aimed at checking whether applying post-processing operations on feature contour will improve the classification accuracy. For this purpose, two feature sets (16·12 statistical descriptors of 12 MFCCs [13] contours, and 16 descriptors of signal envelope using discrete Hilbert transform [8])

**Table 2.** Characteristics of evaluation database

| Class | Description |
|---|---|
| 1: Birds | Various bird calls in the park or forest, sometimes with insect sounds. |
| 2: Passing cars | Passing cars after each other in the street or highway. Approaching cars and trucks moving at different speeds. |
| 3: Barking dogs | Single or several barking dogs in indoor and outdoor scenes. |
| 4: Footsteps | Footsteps on various surfaces (tiles, hardwood floor, pavement, sand, gravel) with different speeds and footwear (slippers, high heels, casual and gym shoes). |
| 5: Rain | Sounds of falling rain with variable intensity on various surfaces. |
| 6: Restaurant | People eating and chatting in restaurant. Sounds of babble noise, glasses, cutlery, chair movements from time to time. |
| 7: Shopping centre | Shopping scenes with some distant speeches. Many sounds of moving trolleys, cash register checkouts, sometimes moderate level of background music. |
| 8: Wind | Different types of winds whistling and rumbling. Some scenes with fierce wind blowing through a window or door. |

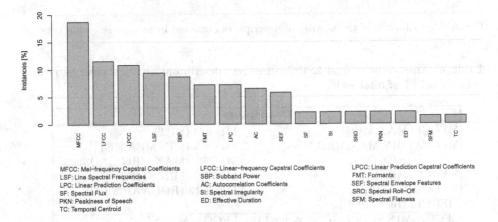

MFCC: Mel–frequency Cepstral Coefficients    LFCC: Linear–frequency Cepstral Coefficients    LPCC: Linear Prediction Cepstral Coefficients
LSF: Line Spectral Frequencies    SBP: Subband Power    FMT: Formants
LPC: Linear Prediction Coefficients    AC: Autocorrelation Coefficients    SEF: Spectral Envelope Features
SF: Spectral Flux    SI: Spectral Irregularity    SRO: Spectral Roll–Off
PKN: Peakiness of Speech    ED: Effective Duration    SFM: Spectral Flatness
TC: Temporal Centroid

**Fig. 2.** An occurrence of basic audio features used for calculation of descriptors in the final set

have been calculated and exploited to select the operations. The classification gain for four example post-processing operations (single and double delta, median and Savitzky-Goolay smoothers) is depicted in Table 3.

In the result of the analysis, only the smoothing filters (running median and Savitzky-Golay) slightly improve the accuracy in both cases (maximal improvement was 7%). Because the smoothing operation removes mainly a high-frequency components, the signal is less noisy. After extensive experimentation, we decided to utilize the running median in subsequent tests as such filter does not need a tuning phase.

**Table 3.** An example of classification gain for post-processed feature contours with different techniques

| | $\Delta$ | $\Delta\Delta$ | Median smooth | Savitzky-Golay smooth |
|---|---|---|---|---|
| MFCC contours | -12 % | -20 % | -4 ÷ +7 % | +3 ÷ +5 % |
| Hilbert envelope | -6 % | -11 % | +1 ÷ +3 % | +2 ÷ +3 % |

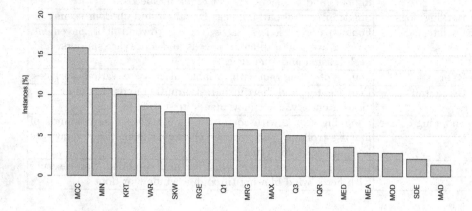

**Fig. 3.** Contribution of the statistical descriptors obtained in feature selection process

**Table 4.** An excerpt of final ESR features set (descriptors with instances exceeding more than 10% of total set)

| Descriptors | |
|---|---|
| MFCC1 (MAX,MRG) | MFCC2 (KRT) |
| MFCC3 (MIN,MAX,MRG,MEA, | MFCC5 (MIN,MEA,KRT) |
|     KRT,Q1,Q3,MCC) | MFCC7 (MAX,MRG,Q1,Q3) |
| MFCC6 (MEA) | MFCC9 (MIN,KRT) |
| MFCC8 (RGE) | MFCC12 (RGE,VAR,MCC) |
| MFCC10 (RGE,MCC) | |
| LFCC1 (MIN,MAX,RGE,SKW,KRT) | LFCC2 (MIN,KRT) |
| LFCC3 (RGE,MCC) | LFCC4 (MIN,MCC) |
| LFCC5 (MAX,Q3) | LFCC7 (VAR) |
| LFCC9 (KRT) | |
| LPCC1 (SKW) | LPCC3 (MRG,MCC) |
| LPCC4 (MIN,RGE) | LPCC5 (MCC) |
| LPCC6 (KRT, MCC) | LPCC9 (RGE,VAR,IQR,MCC) |
| LPCC10 (MIN, KRT) | |

In the next experiment, we have determined the feature vector for ESR task (Figure 1). Due to the fact, that features should describe most relevant and unique attributes of signal, a set of many audio features described in [10] and [12] have been chosen. Also, we included nine spectral envelope derived features

(SEF) to the set, as these features exhibit discriminatory power [9]. Taking the whole set of 145 contours and calculating the statistical descriptors, $145 \cdot 16 = 2320$ features in total were obtained. As a result of the attributes selection process a set containing 136 descriptors was further used for ESR.

The contribution of audio features and statistical descriptors in final set is depicted in Figure 2 and Figure 3 respectively. It is worth to notice that most occurrences in feature vector have MFCC (Mel-Frequency Cepstral Coefficients), LFCC (Linear Frequency Cepstral Coefficients) and LPCC (Linear Prediction Cepstral Coefficients) descriptors [5] in connection with MCC (mean crossing count), KRT (kurtosis), VAR (variance) and MIN (minimum) statistics as is shown in Table 4.

Finally, having obtained dedicated features set, we performed supervised classification where we have done a 50/50 split for training and test samples in each class. The first trial has been performed for descriptors without post-processing

**Table 5.** Confusion matrix for 8-class classification without (a) and with (b) post-processing stage of feature contours

| | birds | passing cars | barking dogs | footsteps | rain | restaurant | shopping centre | wind |
|---|---|---|---|---|---|---|---|---|
| birds | 90% | | | | | 10% | | |
| passing cars | | 80% | | | | | | 20% |
| barking dogs | | | 100% | | | | | |
| footsteps | | | | 100% | | | | |
| rain | | | | | 100% | | | |
| restaurant | | | | | | 80% | 20% | |
| shopping centre | | 10% | | | | 20% | 70% | |
| wind | | 10% | | | 20% | | 10% | 60% |

(a)

| | birds | passing cars | barking dogs | footsteps | rain | restaurant | shopping centre | wind |
|---|---|---|---|---|---|---|---|---|
| birds | 100% | | | | | | | |
| passing cars | | 80% | | | | | | 20% |
| barking dogs | | | 100% | | | | | |
| footsteps | | | | 100% | | | | |
| rain | | | | | 100% | | | |
| restaurant | | | | | | 80% | 20% | |
| shopping centre | | | | | | 20% | 80% | |
| wind | | 10% | | | 20% | | | 70% |

(b)

phase and the achieved classification accuracy was equal to 85%. In the second trial, feature trajectories were smoothed by exploiting 3-point running median filter before calculation of statistical descriptors. Such post-processing results in a improved accuracy equal to 88.75%. The detailed classification results are shown in confusion matrix (Table 5a and 5b).

## 4    Conclusions

In this paper an analysis of properties of audio features and importance of post-processing stage for ESR task has been performed. A combination of audio features with their statistical description for audio segments leads to classification accuracy is equal to 85%. The obtained result is partially connected with the specificity of audio classes – there were significant differences between samples belonging to the other classes. Also, even though the audio events have a important impact on classification accuracy of ESR, such events have low incidence and intensity in our database. As show experiments with post-processing stage, applying smoothing operations to feature contour before calculating final feature vector can improve overall classification efficacy – for evaluation database improvement was about 3%. Although such gain may seems not particularly important, the joint combination of several post-processing operators using some heuristics should be further investigated. Performed experiments show that proposed technique gives promising results and can be used in practice.

**Acknowledgements.** The research work presented in this work was supported by Polish National Science Centre (grant no. N N516 492240).

## References

1. Al-Zhrani, S., AlQahtani, M.: Audio Environment Recognition using Zero Crossing Features and MPEG-7 Descriptors. Journal of Computer Science 6(11), 1283–1287 (2010)
2. Chang, C., Lin, C.: LIBSVM: a library for support vector machines. ACM Transactions on Intelligent Systems and Technology 2, 27:1–27:27 (2011), Software available at http://www.csie.ntu.edu.tw/~cjlin/libsvm
3. Chu, S., Narayanan, S., Jay Kuo, C.-C.: Content analysis for acoustic environment classification in mobile robots. In: Proceedings of the AAAI Fall Symposium, Aurally Informed Performance: Integrating Machine Listening and Auditory Presentation in Robotic Systems, Arlington, Va, USA (2006)
4. Feki, I., Ammar, A., Alimi, A.: Audio stream analysis for environmental sound classification. In: Proceedings of the International Conference on Multimedia Computing and Systems (ICMCS) (2011)
5. Ganchev, T.: Contemporary Methods for Speech Parameterization. Springer, New York (2011)
6. Ghoraani, B., Krishnan, S.: Time-Frequency Matrix Feature Extraction and Classification of Environmental Audio Signals. IEEE Transactions on Audio, Speech and Language Processing 19(7), 2197–2209 (2011)

7. Hall, M.: Correlation-based Feature Subset Selection for Machine Learning. Hamilton, New Zealand (1998)
8. Han, B., Hwang, E.: Environmental sound classification based on feature collaboration. In: Proceedings of the 2009 IEEE International Conference on Multimedia and Expo (ICME), New York (2009)
9. Maka, T.: Features of Average Spectral Envelope for Audio Regions Determination. In: International Conference on Signals and Electronic Systems, ICSES 2012, Wroclaw, Poland, September 19-21 (2012)
10. Mitrovic, D., Zeppelzauer, M., Breiteneder, C.: Features for Content-Based Audio Retrieval. In: Advances in Computers Improving the Web, vol. 78, pp. 71–150 (2010)
11. Press, W., Teukolsky, S., Vetterling, W., Flannery, B.: Numerical Recipes: The Art of Scientific Computing. Cambridge University Press (2007)
12. Peeters, G.: A large set of audio features for sound description (similarity and classification) in the CUIDADO project, CUIDADO I.S.T. Project Report (2004)
13. Rabiner, L., Schafer, W.: Theory and Applications of Digital Speech Processing. Prentice-Hall (2010)
14. Rodemann, T., Joublin, F., Goerick, C.: Filtering environmental sounds using basic audio cues in robot audition. In: Proceedings of International Conference on Advanced Robotics (ICAR), Munich, Germany. IEEE-RAS (2009)

# On the Prediction of Floor Identification Credibility in RSS-Based Positioning Techniques

Maciej Grzenda

Warsaw University of Technology,
Faculty of Mathematics and Information Science,
00-662 Warszawa, ul. Koszykowa 75, Poland
M.Grzenda@mini.pw.edu.pl
and
Orange Labs Poland
02-691 Warszawa, ul. Obrzeżna 7, Poland
Maciej.Grzenda@orange.com

**Abstract.** The future of Location Based Services largely depends on the accuracy of positioning techniques. In the case of indoor positioning, frequently fingerprinting-based solutions are developed. A well known k Nearest Neighbours method is frequently used in this case. However, when the detection of a floor a mobile terminal is located at is an objective, only limited accuracy can be observed when the number of available signals is limited.

The primary objective of this work is to analyse whether the credibility of floor estimates can be a priori assessed. A method assigning weights to individual GSM fingerprints and estimating their reliability in terms of floor estimation is proposed. The method is validated with an extensive radio map. It has been shown that both low and high accuracy floor estimates are correctly identified. Moreover, the objective criterion is proposed to assess individual weight functions from a proposed family of functions.

## 1 Introduction

Among many promising directions for future development of mobile systems, Location Based Services (LBS) [1,2] play major role. These services require the location of a service user to be reliably estimated. This is mandatory in order to provide add-on functionalities such as finding the nearest shopping centre or a restaurant in this shopping centre.

In the case of mobile applications and mobile terminals, in many cases Global Navigation Satellite System (GNSS), and in particular Global Positioning System (GPS) seems to be the natural choice when positioning is needed. However, not all mobile phones include GPS receivers. Moreover, in indoor scenarios, GPS-based positioning usually is not available, hence other techniques are needed. Among other techniques, proposed for indoor scenarios, fingerprinting-based methods [2,7,8] are developed. In the latter case, first, received signal strengths (RSS) of

M. Ali et al. (Eds.): IEA/AIE 2013, LNAI 7906, pp. 610–619, 2013.

different categories of signals observed in a certain known location are collected. This process is repeated in a group of locations, to provide a *radio map* [2,9,11]. Next, in the on-line phase, the position of a terminal, such as mobile phone, is determined based on the similarity of the vector of signal strengths observed in it to a radio map. The radio map plays the role of a database of signal strength vectors linked to known indoor locations. Thus, the *database correlation* [8] is another term used to refer to a group of such techniques.

As far as on-line phase is concerned, several decisions have to be made to define the positioning method. These include, but are not limited to, the selection of metric function used to compare the RSS vectors, the selection of base method such a Nearest Neighbour (NN), k Nearest Neighbours (kNN) or its weighted variants i.e. Weighted kNN (WkNN). Moreover, in the case of the latter two techniques the value of $k$ has to be selected. Alternatively other machine learning techniques can be used to use radio map and RSS vector to estimate terminal location.

In the case of terminal-centric fingerprinting-based indoor positioning, two dominating approaches are observed. Most frequently, signal strength of WiFi Access Points (AP) is used as an input for positioning methods. Alternatively, GSM RSS i.e. the signal strengths observed in mobile stations (MS) (usu. mobile phones) are used. Among others, V. Otsason et al. analyse and compare these approaches [11] and show the benefits and limitations of GSM-based positioning with fingerprinting techniques. Among key benefits, the fact that GSM signals are easily available, also when power supply problems in a building are experienced, is emphasised. Moreover GSM signal strength was reported to be more stable than the strength of IEEE 802.11 AP. These reasons make the investigation of GSM-based positioning an interesting alternative to WiFi-based positioning.

GSM RSS-based fingerprinting in multifloor indoor conditions has been addressed to limited extent so far. Among exceptions the work of B.D.S.Lakmali and D. Dias [8] can be mentioned. However, in this case, only two floors and a limited testing area was reported. More precisely, only 80 fingerprinting locations were used. Moreover, most of the fingerprints were collected on a grid of 3.6 m x 2.4m. Only two floors were considered also by V. Otsason et al. [11]. In this case, reported floor identification accuracy was 81.2% when the signal strength information for the 6-strongest cells was used by kNN algorithm. Finally, Bento et al. [3] treated the positioning problem as a classification problem i.e. assumed that a monitored object appears in one of predefined locations and made measurements on up to two floors depending on scenario considered. However, the accuracy of floor identification was not reported.

The remainder of this work will concentrate on GSM-based indoor positioning with fingerprinting approach. The problem of floor identification will be analysed. From a machine learning perspective this can be treated as a difficult classification task due to numerous similar or even identical input vectors mapped to different classes. For this reason, it has been shown [11] that a floor of a building a MS is located at, can be identified with some limited accuracy only. This largely hinders practical application of the technique. However, the question arises whether the credibility of floor estimates produced for individual GSM RSS vectors can be reliably estimated. If so, such credibility estimates could provide a vital input for trajectory tracking techniques. In such techniques, credibility of floor estimates for individual time steps could be used to decide whether a user actually moved to another floor or not. This could accompany other techniques such as the measurement confidence selected depending on the variation of the RSS [4]. Significant accuracy improvement could be attained when tracking objects moving in a building once some points of the trajectory could be identified with high certainty.

The remainder of this paper is organised as follows. Sect. 2 presents the GSM RSS data set used in the simulations. This is followed by a discussion of GSM RSS vectors in view of floor identification task and the evaluation of distance functions to be used in the experiments. Sect. 4 discusses the proposed methods and the way they are used to deal with the task of assessing the credibility of floor estimates calculated with kNN method. Next, simulation results are discussed in Sect. 5, which is followed by the conclusions outlined in Sect. 6.

## 2  The Experimental Data

The data used for the simulations described in this work is based on extensive data set collected in a university building of the Warsaw University of Technology. The GSM RSS data set has been collected on all 6 floors of a building including ground floor and contains GSM Received Signal Strength collected at mobile terminals in a number of reference points. More precisely, the signal strengths from Base Transceiver Stations (BTS) have been collected. Every vector contains up to 7 values i.e. the signal strengths from the serving cell and up to 6 other cells. Other RSS values may be missing due to GSM standard restricting the dimension of RSS vector. They could be treated with a single imputation method or a combination of different imputation methods, such as proposed in [12].

The GSM RSS data set contains data for two categories of points. A grid of training points was planned in the building with variable resolution of 1.5 or 3 meters. A similar grid of testing points accompanies the training grid. It is shifted by half of the size of the training grid. Hence, the tests of positioning techniques can be made based on the signal strengths observed in a separate set of locations. There were 1199 training and 1092 testing locations altogether. The exact number of testing records for every floor is provided in Table 1. Furthermore, an assumption was made that the orientation of the terminal in the testing phase, when unknown position is being estimated, is not known to the algorithms. This is not to assume that a mobile phone is a smart phone providing digital compass and other accompanying functionalities such as accelerometers that could be used to estimate orientation changes.

**Table 1.** Distribution of testing GSM RSS records

| Floor | No. of records | Room categories |
|---|---|---|
| 0 | 6600 | Ground floor, entrances to the building, offices |
| 1 | 6320 | Lecture rooms |
| 2 | 15520 | Laboratories, lecture rooms incl. large lecture room |
| 3 | 4680 | Laboratories, lecture rooms |
| 4 | 6200 | Offices |
| 5 | 4480 | Offices |

# 3  Terminal Positioning with GSM RSS Data

## 3.1  Key Factors

In accordance with fingerprinting paradigm, the unknown location of a MS is determined based on the similarity of the RSS vector observed in it to a database of RSS vectors registered in a reference database constructed for training locations in the off-line phase. Unfortunately, due to signal reflections and multipath propagation, the same signal strength can be observed in a number of separate locations. It is worth noting here that while WiFi Access Points (AP) are located mostly in a building, which reduces the potential for identical RSS from an AP on different floors, most or all GSM BTSs are located outside of a building. As a consequence, the GSM signal propagation on neighbouring floors under some conditions could be similar.

## 3.2  Selection of Metric Functions

As far as fingerprinting-based positioning is concerned, a dominating approach is to use k Nearest Neighbours (kNN) technique and its variants to deal with the problem. It can be used as a reference technique, as most of the studies discuss results attained with this technique. While the selection of kNN can be made, the $k$ value has to be determined. Moreover, the question remains whether Euclidean distance or other metric functions should be used. In their recent study [10], Machaj, J. and Brida, P., evaluate a number of distance functions. These include Euclidean and Manhattan distance functions, their generalisation being Minkowski distance, and other similarity measures such as correlation distance. The evaluation is based on the simulated RSS data for IEEE 802.11 network. The distance functions are compared in view of positioning accuracy (expressed with root mean square error) and computational overhead. Based on extensive simulations, Manhattan and Sorensen distances are suggested [10]. Another alternative is the use of Euclidean distances, which provides almost identical accuracy at a higher computational cost [10]. However, the selection of appropriate distance function seems to be partly an open issue. In 2007 IEEE ICDM Data Mining Contest [5] aiming at finding the terminal location based on WiFi RSS data, the best positioning accuracy was attained with Minkowski norm with $p = 0.5$. The authors of the best solution suggest that this is due to the fact that this norm puts more importance on presence or absence of signals than on the amount of change [6]. It is not clear how these divergent conclusions can be transferred to GSM RSS positioning, as the number of GSM signals is usually much lower than the number of IEEE 802.11 signals and the GSM signals are detectable in larger area than their WiFi counterparts.

Hence, the selection of optimal distance function for GSM RSS fingerprinting has to be made. The results of this evaluation are depicted in Fig. 1. It can be observed that the accuracy of floor estimates for kNN combined with Manhattan, Euclidean and Minkowski (with $p=1.5$) distances are virtually identical. Moreover, $k > 15$ yields the best accuracy. Hence, the remaining tests were performed with Euclidean distances. In addition, a decision was made to extend the analysed maximum $k$ value to $K = 50$. Finally, it is worth adding here that the results attained for Minkowski norm with $p = 0.5$ are significantly worse, which shows that the outcomes of WiFi-oriented research not necessarily match the needs of GSM-based fingerprinting.

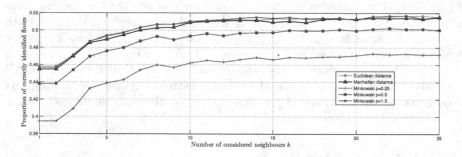

**Fig. 1.** Proportion of correctly identified floors for different distance functions and $k$ values

## 4    Credibility of Estimated Locations

The best accuracy of overall floor selection attained in evaluation performed in Sect. 3.2 is 51.7%, which is far from optimal. At the same time the base level for classification task is 35.4%, which follows directly from Table 1. Thus, the results of floor identification are significantly higher than the base level. Moreover it has been shown [11] that GSM signals when combined with IEEE 802.11 signals may provide improved position estimates. Hence, the question arises whether the problem of limited floor identification accuracy with GSM RSS data can be addressed by estimating the credibility of individual estimates. More precisely, the question is whether a floor estimate for GSM RSS vector collected in unknown location can be assigned a *credibility weight (CW)* showing the level of confidence in this value. This would mean that the CW factor could be used to decide whether the floor estimate is likely to be correct or probably not due to the fact that similar RSS vectors exist on more than one floor in the radio map.

More formally, a CW function $w : \mathbb{R}^n \longrightarrow [0, \ldots, 1]$ is needed. Let $S$ denote the set of RSS vectors i.e. $S = \{\mathbf{s}_1, \ldots, \mathbf{s}_n\}$. For any $0 \leq a_1 < a_2 < b_1 < b_2 \leq 1$, let $A = \{\mathbf{s} \in S : a_1 \leq w(\mathbf{s}) \leq a_2\}$ and $B = \{\mathbf{s} \in S : b_1 \leq w(\mathbf{s}) \leq b_2\}$, the following condition should be fullfiled by a CW function $w()$:

$$\frac{\sum_{\mathbf{s} \in A} \theta(c(\mathbf{s}))}{card(A)} < \frac{\sum_{\mathbf{s} \in B} \theta(c(\mathbf{s}))}{card(B)} \tag{1}$$

where $\theta(c(\mathbf{s})) = 1$ when classifier $c()$ produces correct floor for RSS vector $\mathbf{s}$. Otherwise $\theta(c(\mathbf{s})) = 0$. Hence, the proportion of valid classifications for heigher weight values $0 < b_1 < b_2 \leq 1$ should be higher. Moreover, ideally when $b_2 = 1$, $lim_{b_1 \longrightarrow 1} \frac{\sum_{\mathbf{s} \in B} \theta(c(\mathbf{s}))}{card(B)} = 1$. Due to limited number of elements in $S$, $\exists 0 \leq a_1 < a_2 \leq 1 \wedge a_1 \neq a_2 : card(\{\mathbf{s} \in S : a_1 \leq w(\mathbf{s}) \leq a_2\}) = 0$. Hence, condition 1 has to be replaced with

$$\frac{\sum_{\mathbf{s} \in A} \theta(c(\mathbf{s}))}{card(A) + 1} \leq \frac{\sum_{\mathbf{s} \in B} \theta(c(\mathbf{s}))}{card(B) + 1} \tag{2}$$

Let $S_L$ denote a training radio map, $S_T$ - the testing RSS vectors. Moreover, let $L_{S_L}^k(\mathbf{s}, d) \subset S_L$ denote the set of $k$ closest neighbours of $\mathbf{s}$ in radio map $S_L$, in terms of distance function $d : \mathbb{R}^n \times \mathbb{R}^n \longrightarrow \mathbb{R}_+ \cup 0$. Let us propose the following families of CW functions:

- Floor-Neighbour-Based CW (FNBCW) function $w_k(\mathbf{s})$

$$w_k(\mathbf{s}) = \frac{card(\{l \in L_{S_L}^k(\mathbf{s}, d) : z(l) = mode_{l \in L_{S_L}^k(\mathbf{s}, d)} z(l)\})}{k} \tag{3}$$

i.e. based on the mode of floors $z(l)$ associated with $k$ closest neighbours $l \in L_{S_L}^k(\mathbf{s}, d)$ of $\mathbf{s}$. In this case, the credibility of floor estimation will be based on the proportion of the most frequent floor among $k$ neighbours. Intuitively, the higher the number of neighbours sharing the same floor, the higher the expectation that this floor is correct also for RSS vector $\mathbf{s}$.

- Weight-Neighbour-Based CW (WNBCW) function. For this function to be defined, every training vector $\mathbf{s}_i \in S_L$ will be assigned its weight $w_L(\mathbf{s}_i, k)$ first. This will be done, as defined in Alg. 1. Next the WNBCW function $\tilde{w}_{k,j}(\mathbf{s})$ can be proposed. It is defined by Alg. 2. The main proposal in this case it to estimate the credibility of floor identification of a testing vector by averaging the weights of the training vectors closest to it. The weights of training vectors $w_L(\mathbf{s}_i, k)$ are calculated based on the correctness of their floor estimation while using kNN as a classifier. In particular, $w_L(\mathbf{s}_i, k) = 1$ means that for all $j = 1, \ldots, k$ the $k$NN classifier returned the same correct floor $z(\mathbf{s}_i)$. In other words, this training pattern $\mathbf{s}_i \in S_L$ is placed among many neighbours located on the correct floor. On the other hand, when $w_L(\mathbf{s}_i, k) \longrightarrow 0$, the training pattern is located among neighbours from other floors than the correct one. Hence, the floor estimates made based on such patterns are likely to be incorrect.

**Input:** $s_i \in S_L$ - training RSS vector contained in radio map, $k$ - the number of considered neighbours, $L_X^k(s_i, d) \subset X \subset S_L$ - $k$ closest neighbours of $s_i$ in $X$ in terms of distance function $d()$

**Data:** $(x_i, y_i)$ - horizontal location corresponding to training vectors $s_i$, $z_i$ - floor index corresponding to $s_i$, $z_i \in \{0, 1, \ldots, F\}$, where $F$ - total number of floors covered with radio map, $c^k(X, s)$ - a $k$NN classifier calculating the mode of floors of $k$ closest neighbours of $s$ in $X$

**Result:** $w_L(s_i, k)$ - the weight of training vector $s_i$

**begin**

$\qquad X = \{s_j \in S_L : x_i \neq x_j \wedge y_i \neq y_j\};$

$\qquad X = L_X^k(s_i, d) ;$

$\qquad w_L(s_i, k) = \frac{\sum_{s \in X} \theta(c^k(X, s))}{card(X)};$

**end**

**Algorithm 1.** The calculation of weights for training vectors

**Input:** $s \in S_T$ - testing RSS vector, $k$ - the number of considered neighbours, $j$ - the category of considered weights of these neighbours

**Result:** $\tilde{w}_{k,j}(s)$ - the credibility weight assigned to testing vector $s$

**begin**

$\qquad X = L_{S_L}^k(s, d) ;$

$\qquad \tilde{w}_{k,j}(s) = \frac{\sum_{s_i \in X} w_L(s_i, j)}{k} ;$

**end**

**Algorithm 2.** The calculation of Weight-Neighbour-Based CW function

## 5    Simulation Results

To estimate the merits of different CW functions, several experiments were made. First of all, maximum $K = 50$ was selected for the experiments. Hence, before the actual tests were made, Alg. 1 was used to calculate $w_L(s_i, k)$ for $k = 1, \ldots, K, s_i \in S_L$. Next for all testing RSS vectors $s_i \in S_T$, the values of FNBCW functions $w_k(s_i), k = 1, \ldots, K$ and WNBCW functions $\tilde{w}_{k,j}(s_i), k, j = 1, \ldots, K$ were calculated. In order to allow further processing and visualisation of function values, the number of testing patterns assigned weights from certain ranges and assigned correct floor was developed. In the remainder of this study, 11 ranges of weight values being $[0, 0.1), [0.1, 0.2), \ldots, [1, 1]$ are used to evaluate and visualise different weight functions. A sample analysis for $\tilde{w}_{20,20}()$ has been shown in Fig. 2. It can be observed that the proportion of correctly determined floors is the lowest for $\tilde{w}_{20,20}() \in [0, 0.1)$. It is equal to 0.2371 i.e. the testing RSS vectors a priori assigned so low credibility, actually were correctly matched with their floors in only 23.71% of cases. On the other hand, for high values of weights $\tilde{w}_{20,20}() \in [0.9, 1]$ more than 80% of testing vectors were actually correctly classified i.e. assigned the correct floor.

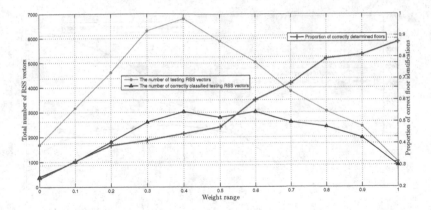

**Fig. 2.** Total number of patterns assigned weights in different ranges and total number of correct floor identifications for $\tilde{w}_{20,20}()$

Similarly, the results for FNBCW function are shown in Fig. 3. It can be observed that this method produces also satisfactory results. An important factor to consider is the number of input vectors assigned high weights and actually correctly matched with their patterns. In this case it is even higher than in the case of $\tilde{w}_{20,20}()$.

Next, the question of the best weight function can be raised. While the precise answer depends on the way individual weight values are used to determine the trajectory of a moving object, some proposals can be formulated. Let $A_{\Omega}(a_1, a_2, k, j) = \{\mathbf{s} \in \Omega : a_1 \leq \tilde{w}_{k,j}(\mathbf{s}) < a_2\}$. In formula 4 an objective function $\beta_{\Omega}(k, j)$ is proposed. The first part of the formula is aiming to maximise the proportion of RSS vectors correctly identified as unreliable i.e. resulting in wrong floor estimates. The other part reaches its maximum for possibly high proportion of vectors resulting in correct floor estimates and assigned high weight values. These two parts correspond to the concept of *odds* used in logistic regression. However, by formulating a group of weights functions, we can select the most appropriate function in terms of its impact on trajectory estimation algorithm. In particular, the selection of optimal weight function can be made in wrapper manner. The investigation of $\beta_{\Omega}(k, j)$ values for $\Omega = S_{\mathrm{T}}$ is shown in Fig. 4. These values include the results of the evaluation of both categories of functions proposed above. The results of $w_k(\mathbf{s})$ evaluation by formula 4 were shown for $j = 51$. It follows from the evaluation shown here that the best function is in this case $\tilde{w}_{40,7}(\mathbf{s})$ i.e. a function taking into account $k = 40$ neighbours and their weights of $j = 7$ category. This means this function produces the best credibility estimates in terms of formula 4. What is even more important, the dominating factor here is the number of neighbours taken into account, not the category of their weights.

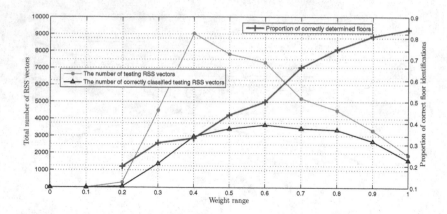

**Fig. 3.** Total number of patterns assigned weights in different ranges and total number of correct floor identifications for $w_{20}()$

$$\beta_\Omega(k,j) = \sum_{i=1}^{2} e^{1+\frac{k-1}{10}} \frac{card(A_\Omega(\frac{i-1}{10}, \frac{i}{10}, k, j)) - \sum_{\mathbf{s} \in A_\Omega(\frac{i-1}{10}, \frac{i}{10}, k, j)} \theta(c(\mathbf{s}))}{1 + \sum_{\mathbf{s} \in A_\Omega(\frac{i-1}{10}, \frac{i}{10}, k, j)} \theta(c(\mathbf{s}))}$$
$$+ \sum_{i=8}^{11} e^{1+\frac{k-1}{10}} \frac{\sum_{\mathbf{s} \in A_\Omega(\frac{i-1}{10}, \frac{i}{10}, k, j)} \theta(c(\mathbf{s}))}{1 + card(A_\Omega(\frac{i-1}{10}, \frac{i}{10}, k, j)) - \sum_{\mathbf{s} \in A_\Omega(\frac{i-1}{10}, \frac{i}{10}, k, j)} \theta(c(\mathbf{s}))} \tag{4}$$

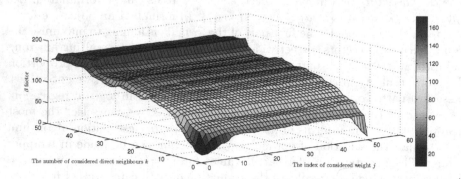

**Fig. 4.** The evaluation of credibility weight functions

## 6    Summary

Fingerprinting approach to GSM-based floor identification in a multifloor building is a demanding task. Due to limitations imposed by the GSM standards, only up to 7 signal strengths can be used for this tasks. This increases the risk of matching the RSS vector with similar vectors, but observed on a different floor. The credibility weights proposed in this study make it possible to a priori assess whether a testing RSS vector will be matched with its correct floor or not. Hence, the input for trajectory tracking techniques is provided.

What is important, the proposed techniques use only training data contained in radio map and signal strengths of testing data and can be used with any fingerprinting approach. It is worth noting here that, while kNN is frequently not the best classifier, the technique proposed here can be applied to estimate the credibility of testing RSS vectors that will be supplied to other models, such as random forests or neural networks. This issue is also planned to be analysed.

# References

1. Baranski, P., Polanczyk, M., Strumillo, P.: Fusion of Data from Internal Sensors, Raster Maps and GPS for Estimation of Pedestrian Geographic Location in Urban Terrain. Metrology and Measurement Systems XVIII(1), 145–188 (2011)
2. Benikovsky, J., Brida, P., Machaj, J.: Localization in real GSM network with fingerprinting utilization. In: Chatzimisios, P., Verikoukis, C., Santamaría, I., Laddomada, M., Hoffmann, O. (eds.) MOBILIGHT 2010. LNICST, vol. 45, pp. 699–709. Springer, Heidelberg (2010)
3. Bento, C., Soares, T., Veloso, M., Baptista, B.: A Study on the Suitability of GSM Signatures for Indoor Location. In: Schiele, B., Dey, A.K., Gellersen, H., de Ruyter, B., Tscheligi, M., Wichert, R., Aarts, E., Buchmann, A.P. (eds.) AmI 2007. LNCS, vol. 4794, pp. 108–123. Springer, Heidelberg (2007)
4. Evennou, F., Marx, F.: Advanced Integration of WiFi and Inertial Navigation Systems for Indoor Mobile Positioning. EURASIP Journal of Applied Signal Processing, 1–11 (2006)
5. http://www.cse.ust.hk/~qyang/ICDMDMC07/
6. Kashima, H., et al.: A Semi-supervised Approach to Indoor Location Estimation (2007),
   http://www.geocities.jp/kashi_pong/publication/ICDMDMC2007slide.pdf
7. Krishnakumar, A.S., Krishnan, P.: The Theory and Practice of Signal Strength-Based Location Estimation. In: International Conference on Collaborative Computing: Networking, Applications and Worksharing (2005)
8. Lakmali, B.D.S., Dias, D.: Database Correlation for GSM Location in Outdoor & Indoor Environments. In: 4th International Conference on Information and Automation for Sustainability, ICIAFS 2008, pp. 42–47 (2008)
9. Lee, M.K., Han, D.S.: Dimensionality reduction of radio map with nonlinear autoencoder. Electronic Letters 48(11), 655–657 (2012)
10. Machaj, J., Brida, P.: Performance Comparison of Similarity Measurements for Database Correlation Localization Method. In: Nguyen, N.T., Kim, C.-G., Janiak, A. (eds.) ACIIDS 2011, Part II. LNCS (LNAI), vol. 6592, pp. 452–461. Springer, Heidelberg (2011)
11. Otsason, V., Varshavsky, A., LaMarca, A., de Lara, E.: Accurate GSM Indoor Localization. In: Beigl, M., Intille, S.S., Rekimoto, J., Tokuda, H. (eds.) UbiComp 2005. LNCS, vol. 3660, pp. 141–158. Springer, Heidelberg (2005)
12. Zawistowski, P., Grzenda, M.: Handling Incomplete Data Using Evolution of Imputation Methods. In: Kolehmainen, M., Toivanen, P., Beliczynski, B. (eds.) ICANNGA 2009. LNCS, vol. 5495, pp. 22–31. Springer, Heidelberg (2009)

# A Study of Vessel Origin Detection on Retinal Images

Chun-Yuan Yu[1,2], Chen-Chung Liu[3], Jiunn-Lin Wu[1], Shyr-Shen Yu[1,*],
and Jing-Yu Huang[1]

[1] Department of Computer Science and Engineering, National Chung-Hsing University,
Taichung, Taiwan
[2] Department of Digital Living Innovation, Nan Kai University of Technology, TsaoTun, Nantou
County, Taiwan
[3] Department of Electronic Engineering, National Chin-Yi University of Technology, Taiping,
Taichung, Taiwan
t112@nkut.edu.tw, ccl@ncut.edu.tw, {jlwu,pyu}@nchu.edu.tw,
qzmpfish@msn.com

**Abstract.** Parabolic model is commonly used for fovea and macular detection. But, the center of an optic disc is mostly taken as the vertex of the parabola-like vasculature. Since vessels generate out from the vessel origin, taking vessel origin as the vertex can provide better fovea localization than taking optic disc center. Recently, the vessel origin is also used to detect vessels within an optic disc. However, there is no published research for finding the exact vessel origin position. This paper proposed a novel method to locate the position of vessel origin. First, a retinal image is processed to get the vascular structure. Then, four features based on the characteristic of vessel origin are selected, and Bayesian classifier is applied to locate the vessel origin. The proposed method is evaluated on the publicly available database, DRIVE. The experimental results show that the average Euclidean distance between the vessel origin and the one marked by experts are 13.3 pts, which are much better than other methods. This can further provide a more accurate vessel and fovea detection.

**Keywords:** Feature selection, vessel origin, Bayesian classifier, optic disc, retinal image.

## 1    Introduction

Diabetic retinopathy is a chronic disease which will harm the vision of the sufferers, even make them blind. Early detection of diabetic retinopathy by computer assistant examination system is an efficient disease detection method with low cost. While developing a computer assistant system for detecting retinal disease, to locate the anatomic structures detected according to the region of fovea is an important task.

Parabolic model is commonly used for fovea localization. Usually, the center of an optic disc is taken as the vertex of the parabola-like vasculature. Foracchia, et al. [1]

---

* Corresponding author.

M. Ali et al. (Eds.): IEA/AIE 2013, LNAI 7906, pp. 620–625, 2013.
© Springer-Verlag Berlin Heidelberg 2013

presented a fovea locating method based on the geometrical directional pattern of the retinal vascular system. Tobin, et al. [8] employed this model to identify the angle between the horizontal raster and the line separating the superior and inferior retinal regions of the retina. Niemeijer, et al. [6] combined the local vessel geometry and image intensity information. Li, et al. [4] extracted 30 landmark points to describe the main courses of the blood vessels, which are used to fit a parabolic structure. Fleming et al. [12] described an elliptic form of the major retinal vessels. The fovea is constrained to a circular region (with 1.6 disc diameter), whose center is between the optic disc and the ellipse center and is located 2.4 disc diameter from the optic disc.

According to the anatomic aspect, main retinal vessels grow out from the vessel origin. Using the retinal vessel origin as the parabolic vertex can provide a more precise fovea localization than using the center of optic disc. In order to correctly detect vessel origins from retinal images, a novel method for vessel origin localization is proposed in this paper. A diagram of the proposed method is shown in Fig. 1. Section 2 presents the image preprocessing and the procedures of vessel segmentation. Section 3 depictures the feature selection and vessel origin localization. Section 4 is the experimental results. Conclusions are given in Section 5.

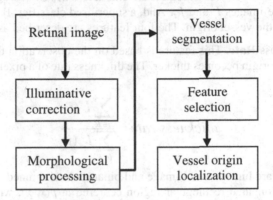

**Fig. 1.** A diagram of the proposed method

## 2     Vessel Segmentation

The illumination of a retinal image may not be even. For adjusting the uneven background intensity, an illumination correction operator [3] is used. The illumination correction operator works on the green band image, because the green band exhibits the best contrast in the retinal image. The illumination corrected image is denoted as $I_{ic}$.

Next, the morphological closing operator [2] is used to eliminate vessels and leave the optic disc and background on the illumination corrected image. The structuring element is a disc, with a diameter 11 pixels (the largest vessel width). Then the illumination corrected image is subtracted by the vessel-removed image, and the difference shows a crude vascular structure image. And, the Otsu algorithm [2] is applied to this crude vascular structure image to obtain a binary vessel image denoted as $I_v$. Fig. 2 illustrates the examples of vessel segmentation.

Furthermore, the common morphological thinning operator [2] is employed on $I_v$ to get a binary vessel thinned image, $I_{vt}$.

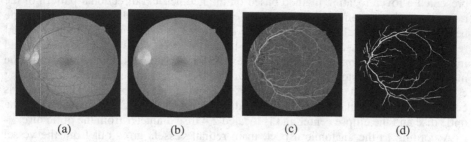

     (a)               (b)               (c)               (d)

**Fig. 2.** Examples of vessel segmentation, (a) illumination corrected image, (b) vessel-removed image, (c) difference image between (b) and (a), (d) Otsu binary image of (c)

## 3    Feature Selection and Vessel Origin Localization

In this section, four features based on the characteristics of vessel origin are selected according to the images $I_v$ and $I_{vt}$. And, a supervised classifier, Bayesian classifier, is adopted to find the vessel origin. The four features are described as follows:

**Vessel Thickness Rate.** This feature is based on the observation that the vessel being closer to vessel origin becomes thicker. The thickness rate of a pixel $p(x, y)$ is defined as follows,

$$thickness\ rate = \frac{\sum\limits_{i \in R} I_v(i)}{\sum\limits_{i \in R} I_{vt}(i)} \tag{1}$$

where $I_v$ and $I_{vt}$ are binary vessel image and binary vessel thinned image, respectively. $R$ denotes an elongated rectangular region centered at $p(x, y)$, which is set as $d \times 2d$. Here, $d$ represents the diameter of the optic disc.

**Vessel Density.** The vessel density of a pixel $p(x, y)$ is defined as the number of vessel pixels within a rectangle mask centered at $p(x, y)$ on the image $I_v$. The value of vessel density near the vessel origin is larger than that in others.

**Vessel Radiation.** The radiation of a pixel $p(x, y)$ is defined as follow,

$$radiation = \sum\limits_{i \in R} \left| \vec{t_i} \cdot \frac{\vec{a_i}}{a_i} \right| \tag{2}$$

where $R$ is a $r \times r$ region centered at $p(x, y)$ on the image $I_v$. $r$ represents the radius of optic disc. $\vec{t_i}$ denotes the vessel flux through the $i_{th}$ element within $R$, which is tangent to the vessel segment. $\vec{a_i}$ is the vector from the $i_{th}$ element to the pixel $p(x, y)$, and $a_i = |\vec{a_i}|$. The dot product of $\vec{t_i}$ and $\vec{a_i} / a_i$ means the component of $\vec{t_i}$ in $\vec{a_i}$ direction.

**Vessel Average Intensity.** The definition of vessel average intensity of a pixel $p(x, y)$ is the average gray level in a 5×5 window size. This feature is obtained from the illumination corrected image $I_{ic}$.

These four features selected are illustrated in Fig. 3 exhibited with visual feature maps. According to these features mentioned above, each pixel is associated with a 4-component feature vector **x**. Then, the Bayesian classifier is used to classify a pixel as a vessel origin (VO) candidate or a non-vessel origin (NVO) point.

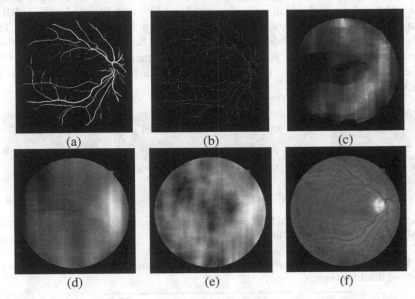

**Fig. 3.** Feature selection, (a) binary vessel image, (b) binary vessel thinned image of (a), (c) thickness rate, (d) vessel density, (e) vessel radiation, (f) average vessel intensity

The Bayes formula can be expressed as the following [11]:

$$P(C_i \mid \mathbf{x}) = \frac{P(\mathbf{x} \mid C_i) P(C_i)}{P(\mathbf{x})} \quad \text{for } i = 1, 2 \tag{3}$$

where $P(C_i|\mathbf{x})$ and $P(C_i)$ are the *a posteriori* probability and prior probability, respectively. $P(\mathbf{x} \mid C_i)$ is the likelihood of $C_i$ with respect to **x**.

The multivariate Gaussian function is used to model the likelihood. According to Bayes formula, the likelihood ratio of posterior probabilities at a pixel $p(x, y)$ can be expressed as follows [8, 11],

$$LR(x, y) = \frac{p(\mathbf{x} \mid C_{VO}) \cdot p(C_{VO})}{p(\mathbf{x} \mid C_{NVO}) \cdot p(C_{NVO})} \tag{4}$$

where $C_{VO}$ and $C_{NVO}$ are the state of vessel origin and of non-vessel origin, respectively. Finally, the vessel origin $(h, k)$ is defined as the point $(x,y)$, which has the maximal average confidence value by an $r \times r$ average filter. Here, $r$ is the radius of optic disc.

## 4    Experimental Results

This method for vessel origin detection is evaluated on the public database, DRIVE [9], which includes 40 images. 20 images are randomly selected for training, and the surplus 20 images are used for testing.

As mentioned earlier, applying the vessel origin for parabolic model is more suitable than the optic disc center. Fig. 4 shows the position of vessel origin obtained by our proposed method and of optic disc obtained by other literatures [5, 7, 10]. As one can see, the measured point marked with "∎" is almost on the vessel origin. Table 1 is the average value of Euclidean distance (in pixels) between the optic disc by each method and that by an expert. According to the experiment results, the location of the result obtained by the proposed method is closest to that given by an expert, with 13.3 pixels in DRIVE database.

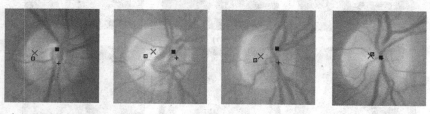

**Fig. 4.** "+" Mahfouz et al. method [5], "⊡" Walter et al. method [10], "×" Sinthanayothin et al. method [7], and "∎" proposed method

**Table 1.** The average of Euclidean distance (in pixels) between the optic disc measured by each method and that by an expert

| Method | DRIVE |
|---|---|
| Proposed method | 13.3 |
| Mahfouz et al.[5] | 18.7 |
| Walter et al.[10] | 23.1 |
| Sinthanayothin et al.[7] | 18.6 |

## 5    Conclusions

In this paper, a method for vessel origin localization on retinal images is presented. To find the vessel origin, four features (vessel thickness rate, density, radiation, average intensity) based on the characteristic of vessel origin are employed. According to these features, each pixel is classified as a vessel origin candidate or a non-vessel origin point using Bayesian classifier. Then, the retinal vessel origin is obtained with a maximal average confidence value.

The experimental results show that the average Euclidean distance between the vessel origin and the one marked by experts are 13.3 pixels in DRIVE database, which are much better than other methods. This leads to a better result for further fovea detection.

**Acknowledgement.** This research was supported by the National Science Council R.O.C. under the Grants NSC 101-2221-E-005-089 and NSC 101-2221-E-167-038.

# References

1. Foracchia, M., Grisan, E., Ruggeri, A.: Detection of Optic Disc in Retinal Images by Means of a Geometrical Model of Vessel Structure. IEEE Transactions on Med. Imaging 23, 1189–1195 (2004)
2. Gonzalez, R.C., Woods, R.E.: Digital Image Processing, 3rd edn. Pearson, New Jersey (2010)
3. Hsiao, H.K., Liu, C.C., Yu, C.Y., Kuo, S.W., Yu, S.S.: A Novel Optic Disc Detection Scheme on Retinal Images. Expert Systems with Applications 39, 10600–10606 (2012)
4. Li, H., Chutatape, O.: Automated Feature Extraction in Color Retinal Images by a Model Based Approach. IEEE Transactions on Biomedical Engineering 51, 246–254 (2004)
5. Mahfouz, A.E., Fahmy, A.S.: Fast Localization of the Optic Disc Using Projection of Image Features. IEEE Transactions on Image Processing 19, 3285–3289 (2010)
6. Niemeijer, M., Abràmoff, M.D., Van Ginneken, B.: Automated Localization of the Optic Disc and the Fovea. In: IEEE EMBS Conf. on 30th Annual International, pp. 3538–3541 (2008)
7. Sinthanayothin, C., Boyce, J., Cook, H., Williamson, T.: Automated Localization of the Optic Disc, Fovea, and Retinal Blood Vessels from Digital Colour Fundus Images. Br. J. Ophthalmology 83, 902–910 (1999)
8. Tobin, K.W., Chaum, E., Govindasamy, V.P.: Detection of Anatomic Structures in Human Retinal Imagery. IEEE Transactions on Med. Imaging 26, 1729–1739 (2007)
9. University Medical Center Utrecht, Image Sciences Institute, http://www.isi.uu.nl/Research/Databases/DRIVE
10. Walter, T., Klein, J.C., Massin, P., Erginay, A.: A Contribution of Image Processing to the Diagnosis of Diabetic Retinopathy Detection of Exudates in Color Fundus Images of the Human Retina. IEEE Transactions on Med. Imaging 21, 1236–1243 (2002)
11. Duda, R.O., Hart, P.E., Stork, D.G.: Pattern Classification, 2nd edn. Wiley, New York (2001)
12. Fleming, A.D., Goatman, K.A., Philip, S., Olson, J.A., Sharp, P.F.: Automatic Detection of Retinal Anatomy to Assist Diabetic Retinopathy Screening. Phys. in Med. & Biology 52, 331–345 (2007)

# The Tires Worn Monitoring Prototype System Using Image Clustering Technology

Shih-Yen Huang[1], Yi-Chung Chen[1], Kuen-Suan Chen[2], and Hui- Min Shih[2]

[1] Department of Computer Science and Information Engineering
National Chin-Yi University of Technology
[2] Department of Industrial Engineering and Management,
National Chin-Yi University of Technology Taichung 411, Taiwan, R.O.C
{syhuang,kschen}@ncut.edu.tw, style214209@gmail.com

**Abstract.** In order to improve traffic safety, researchers studied the driver's physical or mental monitoring, vehicle structure, airbags, brake systems and tires and so on for continuous improvement. The tires need to carry the vehicle loading, to enhance grip, to improve the drainage ability and to reduce the friction noise. Accordingly, tire wear will affect the aforementioned features. This study had designed an experiment platform which can detect the main tread depth, applying image clustering technique, under conditions of low tire speed. In addition, the proposed image clustering algorithm FCM_sobel, could measure the depth of the main tread, at $\alpha = 0.5$ (the influence weighting of the neighboring pixels) and rotating cycle equal 2.5 seconds/rotation. The implemental results show that the precision rates were 93.41%, 96.86 % for the depths of the main tread I and II respectively. Consequently, detected the depth of the main tread I, the precision rate improved 3% compared with FCM_S1.

**Keywords:** Tire wear, image clustering technique, FCM, FCM_sobel.

## 1    Introduction

Transport vehicles are human beings engaged in all kinds of activities, such as trucks to transport materials, public transport buses and even personal use passenger cars. Most of the vehicles rotating the wheel to drive vehicles, but also the use of the brake stop the wheels to stop the vehicles. Therefore, the wheels need to carry the weight of the vehicle, and is exposed to contact with the ground surface of the friction. And shall take into account: The grip to prevent slipping, noise reduction and flexibility to improve driving comfort. Generally, the tire-covered wheel provides the above functions. However, the friction between the tire and the ground caused the tire worn and endangering traffic safety. According to the statistics, traffic accidents accounted for 12.6% caused by tire over worn or wheels came off in Taiwan, in 2008-2010.

According to the above description, to detect tire wear state, is important for traffic safety. The tire surface needs to take into account the pressure of carrying vehicles, and friction between the need to withstand the ground, and the need to have drainage,

M. Ali et al. (Eds.): IEA/AIE 2013, LNAI 7906, pp. 626–634, 2013.

low noise. Usually the tire tread are designed, as shown in Figure 1, the tread having a specific block and texture to provide the above functions. The deepest treads, in Figure 1, named the main treads in this paper. There several convex, on the main treads ditch, are the wear limit marks. While the tire worn reaching these marks, the tire unserviceable. In other words, the depths of these main treads provide the status of the tire worn. Accordingly, this study had implemented a platform to analysis image of the tire tread. In addition, the proposed FCM_sobel algorithm to segment and enhance edge, tire treads from tire images. Consequently, this paper implemented a system to detect the main tread depth to monitoring tire tread wear status.

The remainder of this paper is organized as follows. Section 2, tread imaging experiment platform is introduced. Section 3, we proposed a FCM_sobel algorithm. Section 4, show the experimental results. Section 5, give the conclusion.

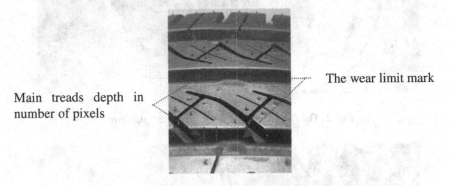

Main treads depth in number of pixels

The wear limit mark

**Fig. 1.** Tire static image

## 2     Tread Imaging Experiment Platform

To repeat the experimental tires image operations, to shoot the actual size and rotation wheel tread, this study had implement a tread imaging experiment platform as shown in Figure 2,3. This platform tire speed is adjusted manually and has a 7-segment display show the speed of the tire. The lowest rotation cycle is 240 seconds / turn. There are equipped with two cameras and supplemental light source. The distance between camera and tread is 7.5 cm, FOV (Field Of View) is 6.5 × 4.3 cm, and the focal length is 20.

The wheel rotation period was about 4 to 5 seconds/rotation, of small passenger cars, during street parking, reversing warehousing. The wheel rotation period is 5 to 6 seconds / rotation, nearing completion parking. In this platform, equipped the Microsoft LifeCam Studio webcam , specifications are as follows : 1920 × 1080 pixels resolution , 0.1 m to ≥ 10 m AF or manual focus , 30 frames per second. The stationary and rotating image of tire are shown in Figure 1 and 4 respectively.

**Fig. 2.** The front view of the experimental platform

**Fig. 3.** The side view of the experimental platform

# 3     FCM_Sobel Algorithm

Fuzzy C-means clustering [1] is a traditional image clustering algorithm, but suscept-ible to noise impact is its main drawback. To relieve the noise impact, the researchers incorporating the spatial neighborhood information to alleviate the biased intensity due to the structured light source [2], or to weaken the noise sensitive [3]. M. N. Ahmed et al. [4] proposed FCM_S1, the degree of influence of the neighboring pixels is involved while clustering pixel. S. C. Chen et al. [5] proposed a FCM_S1 can elim-inate the Gaussian noise while clustering pixel. However, FCM_S1 need to search the weighting of the influence of the neighboring pixels, range from 0.2 to 8 (increment 0.2). To simplified the weighting search, W. Cai et al. [6] proposed algorithm for the weight calculation of the degree of influence of the neighboring pixels, significantly reduced to 0.5 to 6 (increment 0.5). Since the edge is significant for the tire tread im-age, this paper proposed FCM_sobel algorithm which involved the convolution opera-tion of sobel filter to enhance edge. The below describe the details.

## 3.1     Fuzzy C-Means (FCM)

FCM clustering method is a based on C-means algorithm derived through the concept of fuzzy, hope to further enhance the effect of clustering. Its objective function, such as equation (1) , wherein $c$ is a grouping number, $N$ for clustering pixels , $x_j$ is the $j$-th pixel value, $v_i$ is the $i$-th group of the pixel center, $u_{ij}$ the fuzzy membership of the $x_j$ pixels attributable to the group $v_i$ vesting degrees, $m$ is the weighting exponent of $u_{ij}$ ,

$J_m$ is the objective function. By via iterative calculations, $J_m$ converges to the mini-mum, to complete the pixel divided into $c$ clusters.

$$J_m = \sum_{i=1}^{c} \left( \sum_{j=1}^{N} u_{ij}^m \left\| x_j - v_i \right\|^2 \right) \tag{1}$$

## 3.2     FCM with Spatial Constraints (FCM_S1)

Chen et al. [5] proposed FCM_S1 that integrated FCM through the mean filter into the equation (2), for clustering neighboring pixels of the pixel.

$$J_m = \sum_{i=1}^{c} \sum_{k=1}^{N} u_{ik}^m \left\| x_k - v_i \right\|^2 + \alpha \sum_{i=1}^{c} \sum_{k=1}^{N} u_{ik}^m \left\| \bar{x}_k - v_i \right\|^2 \tag{2}$$

Equation (2) $x_k$ the value for the $k$-th pixel, $u_{ik}$ is the fuzzy membership degree of the $k$-th pixel $x_k$ to the $i$-th group $v_i$, $\alpha$ is the weight value of the degree of influence of the neighboring pixel. The $\alpha$ term in equation (2) is noted energy term of adjacent pixels, referred EAP. Let $N_A$ as the pixel set for the neighboring pixels around the clustering pixels $x_k$ , $\bar{x}_k$ is the average value of all the pixels except the pixel $x_k$ in the set $N_A$ . The EAP is the attribute of the distance between the neighboring pixels average values $\bar{x}_k$ and the $i$-th group center $v_i$ , $J_m$ is the objective function. This EAP is possible to reduce the impact of noise, but not conducive to the strengthening of the edge. Since the tread edge trace is important feature, this paper proposed EAP_s substituted for EAP, to combine sobel operator to strengthening edge and named this algorithm as FCM_sobel.

# 4    FCM with Sobel Operator (FCM_sobel)

In order to strengthen the edge pixel, convoluting the pixels $x_k$ within $N_A$ via sobel operator, and then get the convolution value $\tilde{x}_k$ . In other words, $\bar{x}_k$ in equation (2) (FCM_S1) is substituted by $\tilde{x}_k$ , is the proposed FCM_sobel , as shown in Equation (3) below.

$$J_m = \sum_{i=1}^{c} \sum_{k=1}^{N} u_{ij}^m \|x_k - v_i\|^2 + \alpha \sum_{i=1}^{c} \sum_{k=1}^{N} u_{ij}^m \|\tilde{x}_k - v_i\|^2 \qquad (3)$$

## 4.1    The Tread Extraction Process

In order to select the wheel rotation period while taking the tread images of the wheel, the authors measured the wheel rotation period of the small passenger cars during street parking, reversing warehousing. According our experiments the wheel rotation period were about 4 to 5 seconds /cycle. While nearing completion parking, wheel rotation cycle growth to about 5 to 6 seconds / cycle. Accordingly, set the rotating cycle at 2.5 seconds to captured tread image and shown in Figure 4.

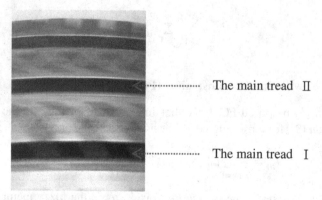

.................... The main tread  Ⅱ

.................... The main tread   I

**Fig. 4.** The tread image captured at 2.5 seconds / cycle

In the experiment, the first pre-processing was convert the color tread images to grayscale, and then applied the proposed FCM_sobel clustering tire tread as shown in Figure 5, 6.

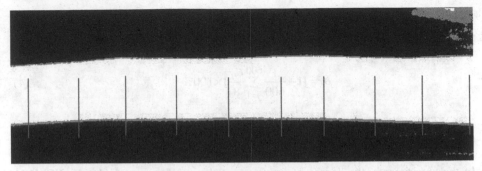

**Fig. 5.** FCM_sobel clustering the main tread   I

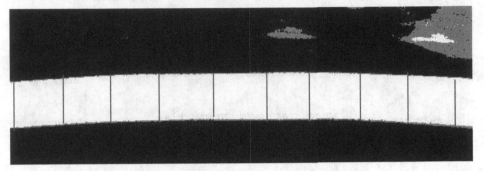

**Fig. 6.** FCM_sobel clustering the main tread   II

## 4.2    The Tread Depth Measurements

In order to compare the difference between the actual tread depth and the measured values those were evaluated by the proposed technique. The authors continuously took 60 images while the tire rotated cycle at 2.5 seconds and the resolution is 800 × 600 pixels of each image. And then took 10 measurement points from each image (on the red line in Figure 5, 6). In other word, there are 600 measurement points for each major tread. In the experiment, we used a new tire which had 8mm tread depth. Further, since the experimental environment and camera are fixed, the two main tread depths are the number of pixels 75 and 60 respectively that could be measure from the static image as shown in figure 1. In this paper, the distance between each pixel in the t-th main tread is defined as $\beta_t$ ( $mm/pixel$ ). Therefore, the distance between each pixel is calculated and noted as $\beta_1 = 0.107$ , $\beta_2 = 0.133$ , for the first and the second main tread, respectively.

In order to analyze performance of FCM_sobel clustering technique, the main tread in the tire images are clustered first. Then using the above $\beta_1$, $\beta_2$ calculate the depth of each main tread depth $d_{1j}$, $d_{2j}$, where $j \in [1,600]$. Therefore, the error of each measurement point is $\varepsilon_{tj} = |d_{tj} - 8|$. Therefore, the precision rate $\delta_t$ for the $t$ main tread is defined as equation 4.

$$\delta_t = [1 - \frac{1}{600} \sum_{j=1}^{600} \frac{\varepsilon_{tj}}{8}] \times 100\% \qquad (4)$$

## 5     The Experimental Results

In the experiments, the degree of influence of the neighboring pixels, the FCM_S1 and FCM_sobel weighting $\alpha$ is set to 0.5. To evaluate the performance of the proposed algorithm, the precision rate is also compared with FCM, FCM_S1 and FCM_sobel and shown in Table 1.

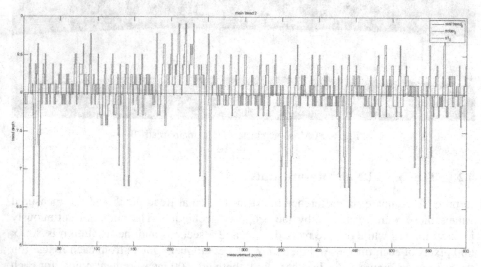

**Fig. 7.** The depth $d_{1j}$, $j \in [1, 600]$ of the main tread I for all measure points, where $\alpha = 0.5$

Figure 7—8 shown the experimental results of the depth $d_{tj}$ (unit is $mm$) for the main tread I and II. The line in the middle on each figure is the real depth of the main tread. Each chart has six regional measurement depth are less than 7mm, because the bottom convex which is the wear limit mark, on main tread.

The main tread measurement precision rates were got by calculating equation (4) for each of the main tread. Table 1 show the error rate which from the experiments. Obviously, the proposed FCM_sobel classification technology at $\alpha = 0.5$, the precision rate is increased than FCM_S1 about 3%.

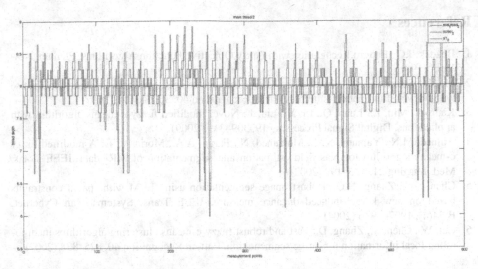

**Fig. 8.** The depth $d_{2j}$, $j \in [1, 600]$ of the main tread II for all measure points, where $\alpha = 0.5$

**Table 1.** The precision rate of the proposed FCM_sobel and another algorithms

|  | The first main tread | The second main tread |
|---|---|---|
| **FCM** | 92.63 | 97.07 |
| **FCM_S1_0.2** | 91.67 | 96.94 |
| **FCM_sobel_0.2** | 91.87 | 97.07 |
| **FCM_S1_0.5** | 90.5 | 96.9 |
| **FCM_sobel_0.5** | 93.41 | 96.86 |

## 6    Conclusions

In this study, an experiment platform was implemented which can operate in low tire speed conditions and apply image clustering technology to cluster the main treads. Then the depths of the main treads were detected automatically. Under conditions in the wheel rotating at 2.5 seconds / cycle and the weighting of influence of the neighboring pixel values is selected as $\alpha = 0.5$. Appling the proposed FCM_sobel, the depth of the main treads were detected with precision rates were 93.41%, 96.86 % for the main tread I and II respectively. Wherein, the precision rate of the main tread I improved 3% than FCM_S1. Has been preliminarily confirmed that the experiment platform and the proposed FCM_sobel can auto detect the depth of the main tread.

**Acknowledgement.** This paper was partly supported by the National Science Council, Taiwan (R. O. C.), under contract NSC 100-2221-E-167-015-MY2.

# References

1. Bezdek, J.C.: Pattern Recognition with Fuzzy Objective Function Algorithms. Plenum, New York (1981)
2. Ma, L., Staunton, R.C.: A modified fuzzy C-means image segmentation algorithm for use with uneven illumination patterns. Pattern Recognition 40, 3005–3011 (2007)
3. Kang, J., Min, L., Luan, Q., Li, X., Liu, J.: Novel modified fuzzy c-means algorithm with applications. Digital Signal Processing 19, 309–319 (2009)
4. Ahmed, M.N., Yamany, S.M., Mohamed, N., Farag, A.A., Moriarty, T.: A modified fuzzy c-means algorithm for bias field estimation and segmentation of MRI data. IEEE Trans. Med. Imaging 21, 193–199 (2002)
5. Chen, S.C., Zhang, D.Q.: Robust image segmentation using FCM with spatial constraints based on new kernel-induced distance measure. IEEE Trans. Systems Man Cybernet. B 34(4), 1907–1916 (2004)
6. Cai, W., Chen, S., Zhang, D.: Fast and robust fuzzy c-means clustering algorithms incorporating local information for image segmentation. Pattern Recognition 40, 825–838 (2007)

# Quality Assessment Model for Critical Manufacturing Process of Crystal Oscillator

Kuen-Suan Chen[1], Ching-Hsin Wang[2,*], Yun-Tsan Lin[3],
Chun-Min Yu[1], and Hui-Min Shih[1]

[1] Department of Industrial Engineering and Management,
National Chin-Yi University of Technology
[2] Institute of Project Management, National Chin-Yi University of Technology
[3] Department of Leisure Industry Management, National Chin-Yi University of Technology
No. 35, Lane 215, Section 1, Chung Shan Road, Taiping, Taichung, 411 Taiwan, ROC
{kschen,yuntsan}@ncut.edu.tw, thomas_6701@yahoo.com.tw

**Abstract.** Quartz crystal is an electronic component made of "quartz" element. In the beginning, it was used in timepieces as the basis of reference for time keeping. Because the crystal has excellent features of stability in temperature change and low wearing, crystal component-based piezoelectric oscillators like crystal, crystal oscillator, crystal filter and optical device have become indispensible passive components for communication in the telecom industry. Furthermore, Of these processes, the wiring process is one of the crucial processes throughout the packaging, where if the quality of the wire for packaging the product is poor, it is very likely to cause poor contact between signal connectors and gold wire on IC or broken gold wires in the course of product transfer, sealing and baking as well as in the course of the product being bonded to the substrate, resulting in the whole IC-packaged product unable to function normally. This article, thus, will develop a model of assessing and testing process capabilities with process capability index, $C_{pl}$, specifically for the larger-the-better quality characteristic of wire process. The article will also provide the assessment procedure, whereby the industry can effectively evaluate whether the process capabilities of their products meet the benchmarks they are supposed to.

**Keywords:** crystal oscillator (CXO), wire welding process, process capability index.

## 1 Introduction

Consumer electronic products have become an indispensable part of living for people with the development of information technologies. In this century of well developed communication technology, the trend of electronic product development is toward compactness and light weight. Portable consumer electronic products, such as the cell phone, Bluetooth, GPS, W-LAN, digital camera, wireless phone, and notebook

---

* Corresponding author.

M. Ali et al. (Eds.): IEA/AIE 2013, LNAI 7906, pp. 635–643, 2013.
© Springer-Verlag Berlin Heidelberg 2013

computer are increasing in demand; as a result, the frequency control components are being paid more attention—the crystal oscillator (CXO), for instance, is widely used in communications related industries because it has excellent characteristics, such as temperature stability and a low loss.

Quartz crystals are electronic passive components made of quartz that offer frequency stability during high frequency oscillation. Presently, quartz crystals are widely used in all kinds of electronic products and systems, including those in military, communications, and consumer electronic categories. Crystal oscillator is composed of the quartz crystal and the IC that controls oscillation circuit (as Fig.1 shows). As CXOs are applied in communication products of higher precision, they require very high precision frequencies. In normal room temperature, the frequency precision can be as low as 100 ppm; but if the output frequency of quartz crystal deviates or is unstable, its influence on the quality and functions of the expensive communication product would be severe. Consequently, each step of the manufacturing processes of a CXO is crucial.

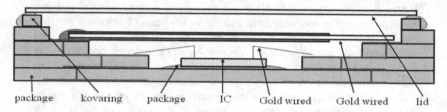

package    kovaring    package    IC    Gold wired    Gold wired    lid

**Fig. 1.** Basic structure of CXO packaging

The process of communication passive components can be divided in two stages in tandem. The process of the former includes the cutting, grinding and cleaning of crystal bars into blanks, and that of the latter begins by blank cleaning, evaporation deposition, precision IC laying, followed by using the wire-bonding process which welds gold wire on the IC and the ceramic package, IC laying, frequency adjustment, packaging to final testing and packing and shipping. Of these processes, the wiring process is one of the crucial processes throughout the packaging, where if the quality of the wire for packaging the product is poor, it is very likely to cause poor contact between signal connectors and gold wire on IC or broken gold wires in the course of product transfer, sealing and baking as well as in the course of the product being bonded to the substrate, resulting in the whole IC-packaged product unable to function normally.

Furthermore, the chief CXO components are the quartz crystal, controller IC, and ceramic package. The CXO manufacturing processes can be divided into front end, where the frequency is adjusted and back end processes, which provide function testing. Of these processes, the wiring process is one of the crucial processes throughout the packaging. he wire process involves the insertion of gold wires into the welding needle with spared tails, which are burned with extremely high voltage to become spherical, and, after their relative positions are calculated by machine visual recognition system, the wire welding. The purpose of wire quality verification is to ensure its quality meets the requirements of the specifications, in which wire pull test and ball shear test are two important tests. The former measures to find whether the strength of the wire meets the lower specification limit, LSL, and the latter assesses whether the bonding between the welding point and the pad interface is sufficiently strong. Poor wire strength or a insufficient bonding strength between welding point and pad interface affects the wire quality, which in turn affects the reliability of the frequency controlling component as a product.

As seen in Fig.1 of CXO, the two wires on the right having a diameter of 1.2 MIL should have wire pull strength reaching at least 5g and ball shear strength reaching at least 40g. Also, the four wires on the left having a diameter of 1.0 MIL should have wire pull strength reaching at least 4g and ball shear strength reaching at least 30g. As such, the key quality characteristics of these two welding points all are larger-the-better specifications, as listed in Table 1.

Table 1. Wire pull test, ball shear test and criteria of requirements

| Wire diameter | Benchmark | Index |
|---|---|---|
| 1.2 MIL | $LSL_{11} = 5$ | $C_{pl11}$ |
| | $LSL_{12} = 40$ | $C_{pl12}$ |
| 1.0 MIL | $LSL_{61} = 4$ | $C_{pl61}$ |
| | $LSL_{62} = 30$ | $C_{pl62}$ |

As stated earlier, Kane (1986) proposed a process capability index for larger-the-better quality characteristics, $C_{pl}$, which has a one-to-one correlation to the process yield, whereby to reflect both process yield and process loss. This article thus will use it to develop an appropriate assessing model for evaluating and improving wire quality, also provide an evaluating procedure, whereby the industry can effectively assess whether the product PCI meets the criteria as it is supposed to.

## 2    Quality Assessment Model for Processing Crystal Oscillator

As the wire can be divided by its diameter, 1.0MIL and 1.2MIL, in the processes of pull strength and the interface bonding strength, which both are larger-the-better quality characteristics, we can evaluate the wire quality using the larger-the-better PCI Kane (1986) proposed. That PCI is defined as follows:

$$C_{plj} = \frac{\mu_j - LSL_j}{3\sigma_j}, j = 1, 2, 3, 4. \tag{1}$$

Where, $LSL_j$ is the lower specification limit, $\mu_j$ the process mean, and $\sigma_j$ the standard deviation for process. $j$ stands for the process of 1.2- or 1.0-MIL wire diameter. Obviously $C_{plj}$ is an index of unilateral specification, whereas, in general terms, the yield rate of a larger-the-better quality characteristic of unilateral specification is:

$$P_j = \Phi(3C_{plj}) \tag{2}$$

Where is the cumulative distribution function of the random variable, Z. Obviously the index, $C_{plj}$, and yield rate, $p_j$, have a one-to-one mathematical relationship. According to the concept of Chen et al. (2003) and Chen et al. (2011), it is allowed to assume each process is independent of another in terms of the data on sampling tests. Hence, assuming the yield rates of different quality characteristics are independent of each other, the yield rate of each welding point, p, is:

$$p = \prod_{j=1}^{4} P_j = \prod_{j=1}^{4} \Phi(3C_{plj}), \quad j = 1,\ldots,4. \tag{3}$$

From the above analysis of yield rate and PCI, we conclude, by summarizing the two quality characteristics of each model, a product PCI, , for the welding points that reflects the product yield rates as follows

$$C_T = \frac{1}{3}\Phi^{-1}\left[\prod_{j=1}^{4} \Phi(3C_{plj})\right], \quad j = 1,\ldots,4 \tag{4}$$

From Equations (4) above, the process yield rate of the welding wire (P) and the PCI, $C_T$ ,can be rewritten as follows (both sides of Eq.(4) being multiplied by 3 and $\Phi$):

$$p = \Phi(3 C_T) \tag{5}$$

Obviously, as PCI ($C_T$) for wire process is larger, so is the process yield rate (P). For example, when for wire welding process, its PCI ($C_T$) = 1, it is guaranteed that the product yield rate, P = 99.87%. According to Chen et al. (2006) and Wang (2011), Six Sigma is the application of quality methods and also statistical tool. It can be used to analyze data and solve problems. In addition, to ensure that products and services meet customer expectations ,improve product quality, increasing the financial benefits, many businesses, such as the U.S. General Electric Company, Motorola, SONY, IBM and other international big business in order to make enterprises more competitive, have to apply this techniques to reduce costs and ensure the quality of the product or service.

According to studies by Pearn and Chen (1997) and Chen et al. (2006), the Motorola 6-sigma quality level means a quality level of 6-sigma is reached only when the standard deviation of a process is $\sigma = d/6$ and the allowable process variation is $1.5\sigma$ .Hence, the process yield rate of wire process can be expressed by k-Sigma follows:

$$C_{pl}(k) = \frac{\mu - LSL}{3\sigma} = \frac{d - (T - \mu)}{3\sigma} = \frac{d - 1.5(d/k)}{3(d/k)} = \frac{k - 1.5}{3} \tag{6}$$

Where T is the midpoint of the specification interval, d represents the specification limits. When  , we can calculate the value corresponding to k-Sigma, as listed in Table 2.

**Table 2.** Comparison Table of index and k-Sigma

| $C_{pl}(k) = u$ | k-Sigma | Yield% | $C_{pl}(k) = u$ | k-Sigma | Yield% |
|---|---|---|---|---|---|
| 1.50 | 6.00 | 0.999997 | 1.20 | 5.10 | 0.999841 |
| 1.45 | 5.85 | 0.999993 | 1.15 | 4.95 | 0.999720 |
| 1.40 | 5.70 | 0.999987 | 1.10 | 4.80 | 0.999517 |
| 1.35 | 5.55 | 0.999974 | 1.05 | 4.65 | 0.999184 |
| 1.30 | 5.40 | 0.999952 | 1.00 | 4.50 | 0.998650 |
| 1.25 | 5.25 | 0.999912 | 0.95 | 4.35 | 0.997814 |

According to Table 2, we can draw a diagram by set the X-axis as and Y-axis as k-Sigma as Fig.2. Obviously, when the value of going higher, and the quality value, (k-Sigma), will be higher. It stands that the good yield rate of processing is high, too.

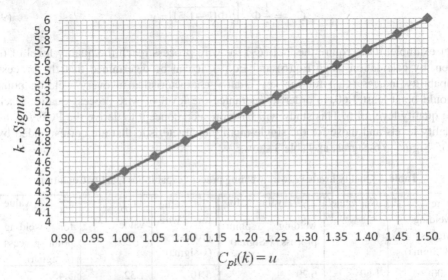

**Fig. 2.** Diagram of index value, and quality value (k-Sigma)

# 3    The Relationship between Six Sigma and Manufacturing Process Ability Index of Welding Wire

The wire process can be divided by its diameter (1.0MIL and1.2MIL)in the processes of pull strength and the interface bonding strength. According to Huang et al. (2002), Chen et al.(2006) and Wang et al. (2011), as the welding point process involves several key quality characteristics, the PCI for each key quality characteristic should be above that for the welding point process in order for the PCIs for the welding point process to reach the level required by the customer. When the requirement for PCI for welding point process is $C_T = u$.

$$C_T = \frac{1}{3}\Phi^{-1}\left[\prod_{j=1}^{4}\Phi(3C_{plj})\right] = u \qquad (7)$$

From which if the PCIs for every welding point process are equal and are v, then the value of w can be deduced as follows:

$$C_T = \frac{1}{3}\Phi^{-1}\left[\prod_{j=1}^{4}\Phi(3v)\right] = u$$

$$\Rightarrow v = \frac{1}{3}\Phi^{-1}\left[\sqrt[4]{\Phi(3u)}\right] \qquad (8)$$

Besides, the corresponded values for k-Sigma, values for welding point process capability index could be calculate according to Eq.(6) as follows:

$$C_{plj} = \frac{1}{3}\Phi^{-1}\left[\sqrt[4]{\Phi(k-1.5)}\right] = v \qquad (9)$$

For example, if to require the PCI ($C_T$) for wire process reach the quality level of 6, then, by calculation using Eq.(6) and Eq.(9), it can be determined that the process capability indices ($C_{plj}$) for the key quality characteristics in every welding point should be at least 1.60, in order to guarantee the level of wire process quality reach the quality level of 6. This study will base on such concept to derive the PCIs for the quality levels and those for the single-quality characteristic that are corresponded by 4.35-Sigma to 6-Sigma, as Table 3 shows.

**Table 3.** Quality levels vs. wire welding process capabilities Quality Condition

| wire welding process (k-Sigma) | $C_{plj}$ value | $C_{T_j}$ value for welding point process (k-Sigma) | $C_T$ value for wire process (k-sigma) | $C_{plj}$ value | $C_{T_j}$ value for welding point process (k-sigma) |
|---|---|---|---|---|---|
| 6.00 | 1.60 | 6.29 | 5.10 | 1.32 | 5.45 |
| 5.85 | 1.55 | 6.14 | 4.95 | 1.27 | 5.31 |
| 5.70 | 1.50 | 6.00 | 4.80 | 1.22 | 5.17 |
| 5.55 | 1.45 | 5.86 | 4.65 | 1.18 | 5.03 |
| 5.40 | 1.41 | 5.72 | 4.50 | 1.13 | 4.90 |
| 5.25 | 1.36 | 5.58 | 4.35 | 1.09 | 4.77 |

# 4     Estimation and Testing of Process Capability Index, $C_{pl}$

According to Cheng (1994-1995), because the parameters of a process are unknown, it needs to use samplings to derive the estimates for the indices. Also, as errors are present in sampling, it is not objective to judge whether the process capability meets customer requirement with only the estimates for the indices. As such, we can assess by the statistic testing, which is one of the objective methods of evaluating process capabilities, whether the process capability of each model of product meets customer requirement. From the above statement, in practice we can require, based on Eq.(6) and Eq.(9), that the PCI ($C_{plj}$)for each key quality characteristic be r to ensure the PCI for the whole wire process is c.

Therefore, when desiring to test whether the PCI for the whole wire process is greater than or equal to c, we should begin by testing whether the PCI for each quality characteristic is greater than or equal to r, in which the hypothesis for the testing can be expressed as follows

$$H_0: C_{plj} \geq r$$

$$H_a: C_{plj} < r$$

We can evaluate whether the PCIs reach the levels which they are supposed to by using unbiased estimator, $\hat{C}_{plij}$, for the index $C_{plj}$ as the test statistic, which can be expressed as follows

$$\hat{C}_{plj} = \left( \frac{\bar{X}_j - LSL_j}{3S_j} \right), \quad j = 1, 2, 3, 4 \tag{10}$$

Where $\bar{X}_j = (n)\text{-}1(\sum_{j=1}^{n} X_j)$ and $S_j = ((n-1)\text{-}1\sum_{l=1}^{n}(X_j - \bar{X}_j)^2)^{1/2}$ are the sample mean and standard deviation of the random sample, $X_{j1}, \ldots, X_{jn}$, respectively, for estimating $\mu_j$ and $\sigma_j$, $j = 1, 2, 3, 4$. In effect, on the supposition of normal condition, $\hat{C}_{plj}$ is the minimum variance unbiased estimator (UMVUE) for $C_{plj}$. Because the distribution of $(3\sqrt{n})\hat{c}_{plj}$ is a non-central t-distribution with degree of freedom at n – 1, its non-central parameter, which is $\delta = 3\sqrt{n} C_{plj}$, can be written as $t'_{(n-1;\delta)}$. If the test statistic value obtained by calculation on the observed random sample is $\hat{C}_{plj} = w_j$, then we can calculate for the p-value of every key quality characteristic process as follows:

$$pv_j = P\{\hat{c}_{plj} \leq w_j \mid C\ C_{plj} = r\}$$
$$= P\{(3\sqrt{n})\hat{C}_{plj} \leq (3\sqrt{n})w_j \mid C_{plj} = r\}$$
$$= P\{t_{n-1}(\delta = 3\sqrt{n}\,r) \leq (3\sqrt{n})\,w_j\} \text{。} \tag{11}$$

We can determine, based on sampling data, the value of test statistic, $\hat{C}_{plj} = w_j$, followed by selecting level of significance (producer risk), $\alpha$, and determine the p-value for the quality characteristic of each individual model of product based on sample size n, the calculated value of test statistic, $w_j$, and the required PCI for each quality characteristic, r, to evaluate whether the process capability meets the requirement of the specification. This article will set forth the complete assessment procedures based on above theory as follows:

Step 1: Determine the quality level of the wire process and look up, with quality levels in Table 3, the corresponded values for wire process capability index ($C_T$) and single-quality characteristic process capability index ($C_{pij}$).

Step 2:   Decide the level of significance $\alpha = 0.05$.

Step 3: Take sample number and calculate for sample mean and sample standard deviation.

Step 4:   Calculate $\hat{C}_{pij} = w_j$ and $p_{vj}$ values based on $\bar{x}_j$ and $S_j$.

Step 5:  Determine whether the process capability of key quality characteristic in the wire.

## 5    Conclusions

Process yield and process expected losses are two basic tools for evaluating product quality and performance. Process capability indices (PCIs) combine the advantages of these two assessment methods. A greater PCI indicates a higher process yield and lower process loss. Hence, the article provide the assessment procedure, whereby the industry can effectively evaluate whether the process capabilities of their products meet the benchmarks they are supposed to. In this article, the process of CXO was first analyzed to find out the key processes affecting final products. The development of process evaluation model with larger-the-better process capability index, Cpl, , followed to define the analyzing model for the processes of the wire, the welding points and the key quality characteristics. Also, the method of statistic testing was used to evaluate the process capabilities, enabling the wire process to meet the expected quality levels to ensure the whole product meets customer demands and further to elevate the competitive edge of related industries globally.

**Acknowledgments.** The authors would like to thank the National Science Council of the Republic of China for financially supporting this research.

# References

1. Buck, D.L.: Digital design for a self-temperature compensating oscillator. IEEE International Frequency Control and PDA Exhibition, 615–643 (2002)
2. Chen, K.S., Pearn, W.L., Lin, P.C.: Capability measures for processes with multiple characteristics. Quality & Reliability Engineering International 19, 101–110 (2003)
3. Chen, K.S., Wang, C.H., Chen, H.T.: A MAIC approach to TFT-LCD panel quality improvement. Microelectronics Reliability 46, 1189–1198 (2006)
4. Deno, S., Hehnlen, C., Landis, D.: A low cost microcontroller compensated crystal oscillator. In: IEEE Frequency Control Symposium, pp. 954–960 (1997)
5. Cheng, S.W.: Practical implementation of testing process capability indices. Quality Engineering 7, 239–259 (1995)

6. Kane, V.E.: Process capability indices. Journal of Quality Technology 18, 41–52 (1986)
7. Huang, M.L., Chen, K.S., Hung, Y.H.: Integrated Process Capability Analysis with an Application in Backlight. Microelectronics Reliability 42, 2009–2014 (2002)
8. Wang, C.C., Chen, K.S., Wang, C.H., Chang, P.H.: Application of 6-sigma Design System to Developing an Improvement Model for Multi-process Multi-characteristic Product Quality. Proceedings of the Institution of Mechanical Engineers, Part B, Journal of Engineering Manufacture 225, 1205–1216 (2011)
9. Pearn, W.L., Chen, K.S.: Multi-process performance analysis: A case study. Quality Engineering 10, 1–8 (1997)

# Local Skew Estimation in Moving Business Cards

Chun-Ming Tsai[*]

Department of Computer Science, Taipei Municipal University of Education,
No. 1, Ai-Kuo W. Road, Taipei 100, Taiwan
cmtsai2009@gmail.com

**Abstract.** Current methods to help visually impaired persons read brief text like menus, business cards, and book covers are problematic because they assume both the user and the captured scene are stationary and they do not tell the visually impaired user if the target is captured by the camera. Further, these methods cannot estimate whether the text is locally skewed. This paper presents an intelligent system to estimate movement, thumbs, motion blur, text, and local skew in moving business card targets. Experimental results show that the proposed method can reduce time complexity, obtain high text detection rates, and achieve high local skew estimation rates.

**Keywords:** Local skew estimation, text detection, moving business cards, visually impaired, post-processing.

## 1 Introduction

Computer applications that provide support to visually impaired persons have become an important topic [1]. In Taiwan, there are now some 56,000 visually impaired persons and the number is increasing annually. For visually impaired persons, inability to read business cards (BCs) has a huge impact on interpersonal relations. Several devices have been designed to help them "read" business cards using an alternative sense such as sound or touch, but these developments are still at an early stage. This paper proposes an intelligent system to estimate whether the text in moving business cards is locally skewed and thereby help visually impaired persons to read the text in moving business cards.

Previously proposed reading systems. A two-mode text-reading system was proposed by Ezaki et al. [1]. In the first mode, a camera on the user's shoulder tries to acquire a scene image automatically and find small characters. Then, it zooms to take higher resolution images necessary for character recognition and reads them aloud via a voice synthesizer. In the second mode, their system "reads" a menu or book cover. The user guesses approximately where the menu or the book is, and uses the camera as a hand scanner. However, the first mode assumes both the user and the captured scene are standing still, but in many real applications, like reading a BC. The BC

---

[*] This work is supported by the National Science Council, R.O.C., under Grants NSC 100-2221-E-133-004- and NSC 101-2221-E-133-004-.

M. Ali et al. (Eds.): IEA/AIE 2013, LNAI 7906, pp. 644–653, 2013.

usually is moving, which grabbed by thumb and forefinger. In the second mode, the visually impaired person must both guess where the menu or the book is and know that it is captured by the camera. Furthermore, their system cannot detect whether the text is skewed and then deskew.

A prototype device was described by Zandifar et al. [2] for scene text acquisition and processing, using a camera-based OCR method in a head-mounted smart video camera system. This system can detect and recognize text in the environment and then convert text to speech. However, it detects the textual information in every frame-- which is time-consuming and repetitive. Similarly, their system cannot detect skew and then deskew.

A wearable, head-mounted camera system was developed by Tanaka and Goto [3] that can automatically find and track text on signboards in a hall way using a revised DCT feature to extract text regions and then group them into image chains by a text tracking method based on particle filtering. However, when the camera makes quick movements, text detection and tracking are not robust. And in many situations, the user needs to hear the text messages while close to a signboard not after passing it. Moreover, this system cannot deal with blurred frames and it track the text regions in every frame, which is time-consuming.

An improved wearable camera system [3] was also presented by Goto and Tanaka [4]. It used automatic text image selection adequate for character recognition. This proposed method outperforms their previous method [3], but duplicate or non-text images are seen; the system still tracks the text regions in every frame; and it cannot deal with moving business cards or cannot detect text skew and then deskew.

Local Skew. Conventions describe local skew as a situation in which particular document blocks have different slant angles from those of other blocks that are caused intentionally by the document layout designer. Antonacopoulos [5] proposed a local skew estimation method based on a description of the background space (white tiles) in printed regions. The method proceeds from the set of coordinates of all white tiles to identify suitable sub-sets of points to which straight lines can be fitted. The orientation of each of the fitted lines is then examined and the mean of the most frequently occurring angles is taken as the orientation of the region.

Messelodi and Modena [6] combined projection profiles with a clustering procedure based on simple heuristics to overcome the problems of limited angle range. The projection profile of a set of components X with respect to an angle $\theta$ is defined as the histogram of the projection of each point in X performed over an axis with a slope of $\theta+\pi/2$. Although this method is robust for document images with internal skew and small interlining spacing, it was only tested for small images on book covers that included a few lines of text.

Pal et al [7] proposed a multi-skew detection of Indian script documents. The detection accuracy is high. However, when the above-mentioned local skew estimation was applied for moving scanned documents, such as hand-held business cards, their method did not work satisfactorily.

In sum, time-consuming detection, skew text estimation, and recognition of duplicate text strings that appear in consecutive video frames are problems for all proposed systems. Moreover, the visually impaired user would not want to hear a synthesized voice repeating the same text. Furthermore, the captured video frames are blurred by motions of the camera or the BC or by non-motion blur even when the camera and BCs are stable.

## 2    Detecting the Moving Business Card

This paper proposes a skew text estimation system, equipped with a "glasses" video camera, for helping the visually impaired to "read" the text in the moving business cards. In the proposed system, the camera scans the BC as it is grasped in the thumb and forefinger, capturing many video frames of the BC. For this, the BC must be in the frontal-flat view of the glasses camera. To detect whether the BC has moved into the frontal-flat view, a modified two-frame differencing method is used.

### 2.1    Modified Two-Frame Differencing (M2FD)

Take a 220-frame video, simulating a visually impaired person trying to "read" a BC, captured by a glasses video camera. The two-frame differencing (2FD) method [8] is used to obtain the differencing images, and the variance for each differencing image is computed. Several of the variances are near zero, which means there is no card in the frontal-flat view of the camera. The conventional 2FD algorithm will process all these frame images with no card, which is time-consuming. Thus, to speed performance, it is better to remove such near-zero variances by using a modified 2FD (M2FD) algorithm [9]. Figure 1 shows an example: The previous frame (#79), the current frame (#80), and the detection result of the moving business card are shown in Fig. 1(a), 1(b), and 1(c), respectively.

### 2.2    Detecting Skin Region

The BC is assumed to be grasped by the right thumb and forefinger. So a modified skin detection algorithm [9], based on the Wong et al. [10], is used to detect the skin region finding that the distributions of skin color on the $Cb$-$Cr$ plane under different illumination are not the same. As in Wong's method, here the skin-color distribution ($Cb$-$Cr$) is divided into six groups of different luminance, and the distribution of $Cb$ and $Cr$ are recomputed. Note that, the $Cb$-$Cr$ values for achromatic colors (black, gray, and white) are between 127 and 128. In Wong's method, these achromatic colors will be detected as skin colors. To avoid this, here the normal *red* color, $Rn$, is used [9]. More detail can be found in [9].

### 2.3    Identifying Thumb Skin

This modified skin detection detects many skin regions in the video, but the detected skin region may not all be thumb skin. To identify whether the detected skin region is in fact the thumb, size and aspect ratios are used as thumb identifying criteria. More detail can be found in [9]. In the Fig. 1 example, we successively apply to the current frame (#80) by using the modified skin detection method, connected component

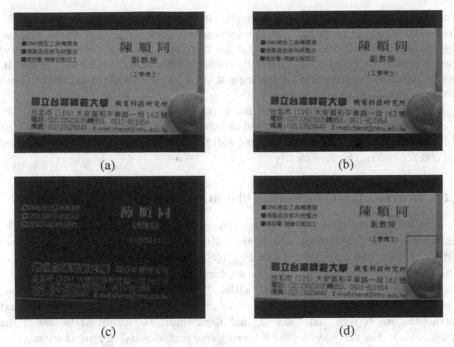

**Fig. 1.** Example of moving business card detection. (a) 79th frame (b) 80th frame (c) Result obtained by applying M2FD method (d) Result by applying skin detection and thumb skin identification.

labeling, a geometry-based BB merging operation, and finally thumb skin identification methods, yielding the final result for skin region shown in Fig. 1(d).

## 3    Detecting Text in Moving Business Card

When the visually impaired person is "seeing" the BC, many of the captured video frames are blurred by camera or BC motions or by non-motion blur even when both are stable. Only in the non-motion blur BCs is text clear and easy to detect. To save execution time the motion blur BC must be detected and removed using motion blur BC detection approach described below.

### 3.1    Detecting Motion Blur BCs

An input frame image with thumb skin is motion blur if the following detection criterion is satisfied:

$$UM < T_{UM},$$ (1)

where $UM$ and $T_{UM}$ represent the uniformity measure and the threshold value for uniformity measure, respectively. More detail can be found in [9]. UM is based upon the region property. If the differencing image for two successive frames is motion blur, its uniformity measure value is smaller. In this circumstance, the text in the BC cannot be detected properly, and the subsequent steps to process these detected regions are very difficult. Conversely, if the differencing image for two successive frames is non-motion blur, its uniformity measure value is larger, the text in the BC can be detected properly and the subsequent steps to process these detected regions are not difficult.

Figure 1 shows an example where BC is non-motion blur with $UM = 0.792$. The #79 frame, #80 frame, differencing, and thumb images are shown in Figs. 1(a), 1(b), 1(c), and 1(d), respectively. The #79 and #80 frame images are non-motion blur and the text regions can be seen clearly.

## 3.2     Detecting Text in Non-motion Blur BCs

If, after BCs detected as motion blur are removed, a BC shows non-motion blur, the text in it will be detected as follows: the Otsu thresholding method [12] is used to threshold the differencing image (see Fig. 1(c)) into the binary image (Fig. 2(a)), to which connected component labeling is applied, and many text candidates (Fig. 2(b)) are identified by the bounding-boxes (BBs). These BBs may be strokes, Chinese words, chars, alphabets, symbols, skin, or noises. However, some of the extracted text candidates are broken and touched, and these must be identified to detect the complete text. For this purpose, the following five text-identifying criteria are used.

The feature of the first text-identifying criterion is thumb region ($S_{BB}$), which is satisfied if:

$$TC1_{BB} = TC0_{BB} - \{TC0_{BB} \cap S_{BB}\} \tag{2}$$

where $TC0_{BB}$ is the original BB set of the text candidates (Fig. 2(b)), $S_{BB}$ is the BB of the thumb (Fig. 1(d)), and $\cap$ is the intersection operation. $TC0_{BB} \cap S_{BB}$ represents the intersection set of the original BB set of the text candidates and the BB of the thumb. $TC1_{BB}$, from the original BB set of such candidates, is the first BB set of the text candidates. This criterion is used to remove the small BBs in the thumb region. As an example, Figure 2(b) shows the connected component image, while Figure 2(c) is the result obtained by using this criterion, in which many BBs in the thumb region are removed.

The feature of the second text-identifying criterion, which is used to remove thin horizontal and thin vertical noises, is aspect ratio ($AR_{BB}$). It is satisfied if:

$$TC2_{BB} = T_{AR3} \leq \{AR_{BB} \quad in \quad TC1_{BB}\} \leq T_{AR4} \tag{3}$$

where $AR_{BB}$ represents the BB aspect ratio in the input set of the text candidates (Fig. 2(c)), and $T_{AR3}$ and $T_{AR4}$ are threshold values, derived from an unsupervised learning method that accepts training BC video frames as input and are set as 0.096 and 11.0, respectively.

The feature of the third text-identifying criterion is geometry type, including (1) intersection at the edge; (2) intersection at the corner; and (3) embedding. For this criterion, a geometry-based BB merging operation [9] is applied to merge the touching BBs to form the complete text.

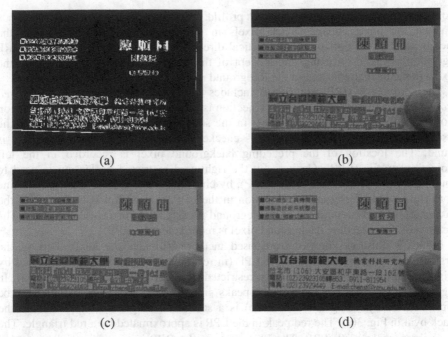

(a)

(b)

(c)

(d)

**Fig. 2.** Example of text detection in moving business card. (a) Binary image (b) Connected component image. (c) Result of removing small BBs with thumb region. (d) Result of removing small BBs with aspect ratio, merging touched BBs with geometry-based BB merging operation, merging broken BBs with BB-based closing operation, and detecting text after removing small noise BBs.

The feature of the fourth text-identifying criterion is also geometry type. A BB-based closing operation [11] is used to merge the broken BBs to form the complete text or text region.

The feature of the fifth text-identifying criterion is area ($A_{BB}$). It is satisfied if:

$$A_{BB} \geq T_A \tag{4}$$

where $A_{BB}$ represents the area of the BB and $T_A$ is derived from an unsupervised learning method that accepts training BC video frames as input and is set as 16.

The result of applying the second text-identifying criterion (aspect ratio), the third (geometry-based BB merging operation), the fourth (BB-based closing operation), and the fifth (area filter) in Fig. 2(c) is shown in Fig. 2(d). In Fig. 2(d), one thin horizontal BB and one thin vertical BB are removed, some touched BBs have been merged, some broken BBs have been merged, some small BBs have been removed. Furthermore, one text region containing three text lines, three characters for a Chinese name, two text lines, and a text region containing four text lines have been extracted.

## 4    Local Skew Estimation

The two text regions that are extracted in Fig. 2(d) have a little local skew. Herein, the *horizontal restricted projection* (HRP) algorithm is proposed to estimate the angle of the local skew. To obtain the projection profile, conventional projection (CP) methods scanned the binary image (black pixels in the foreground, white pixels in the background) in the horizontal and vertical directions alternately. The proposed HRP algorithm is based on using the concept of the CP method, but scanning only the pixels in the background and in the foreground region boundary.

The proposed HRP algorithm includes *leftward* and *rightward restricted projection*. The leftward restricted projection is proposed to obtain the left projection profile (LPP), by checking the pixels in the scanning line from left to right. If the pixel is background, the next pixel is checked, until the first foreground pixel is found. The location of the preceding background pixel is recorded in the left projection profile (LPP). Conversely, the rightward restricted projection is used to obtain the right projection profile (RPP), by checking the pixels in the scanning line from right to the corresponding location in the LPP. If the pixel is background, the next pixel is checked, until the first foreground pixel is found in a similar manner. The location of the preceding background pixel is recorded in the RPP.

Figure 3 illustrates how the proposed method estimates the local skew of the largest text region in Fig. 2(d). The LPP (in red) and RPP (in blue) are obtained by using the leftward and the rightward restricted projection, as shown in Fig. 3(a). In Fig. 3(a), there are two significant peaks in the LPP and RPP, which can be approximated by triangles. Figure 3(b) is a schematic diagram to approximate the black oval in Fig. 3(a). The red peak in the LPP is approximated by a red triangle. The local skew angle is $\theta_0\_LPP$. The blue peak in the RPP is approximated by a blue triangle. The local skew angle is $\theta_1\_RPP$.

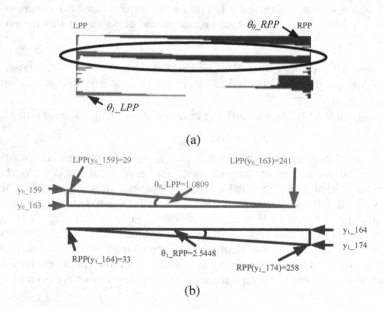

(a)

(b)

**Fig. 3.** Example of estimating the local skew angle. (a) LPP and RPP. (b) Schematic diagram for estimating the local skew angle from LPP and RPP.

To estimate the local skew angle, $\theta$, the following equation is used.

$$\theta = average(\min(\theta\_LPP), \min(\theta\_RPP)), \tag{5}$$

where $\min(\theta\_LPP)$ and $\min(\theta\_RPP)$ are the minimum local skew angles, estimated from $\theta\_LPP$ and $\theta\_RPP$, respectively. $\theta\_LPP$ and $\theta\_RPP$ are local skew angle sets which are estimated from $LPP$ and $RPP$, respectively and can be represented by the following equations:

$$\theta_i\_LPP = \tan^{-1}\left(\frac{\Delta y_i}{\Delta x_i\_LPP}\right), \tag{6}$$

$$\theta_i\_RPP = \tan^{-1}\left(\frac{\Delta y_i}{\Delta x_i\_RPP}\right), \tag{7}$$

where $\Delta y_i$ is the $i$th difference between two coordinates. For example, $y_0\_159$ and $y_0\_163$, $y_1\_164$ and $y_1\_174$ in Fig. 3(a) are used to obtain $\Delta y_0$ and $\Delta y_1$, respectively. $\Delta x_i\_LPP$ and $\Delta x_i\_RPP$ are the $i$th difference between two values in $LPP$ and $RPP$. For example, $LPP(y_0\_159)$ and $LPP(y_0\_163)$ in $LPP$ and $RPP(y_1\_164)$ and $RPP(y_1\_174)$ in $RPP$ are used to obtain $\Delta x_0\_LPP$ and $\Delta x_1\_RPP$, respectively. In Fig. 3(b), $\theta_0\_LPP$ and $\theta_1\_RPP$ are equal to 1.0809 and 2.5448, respectively. Similarly, the other two angles in LPP and RPP are computed, $\theta_1\_LPP$ and $\theta_0\_RPP$ are equal to 2.1409 and 2.5385, respectively. Thus, the local skew angle, $\theta$, for the largest text region in Fig. 2(d) is 1.8097.

Figure 4 shows an example of results for local skew estimation and deskew. The text region in the left-top of Fig. 2(d) has local skew angle 1.0609, as shown in Fig. 4(a), and the deskew result shown in Fig. 4(b). The text region in the bottom of Fig. 2(d) has local skew angle 1.8097 (Fig. 4(c)) and the deskew result shown in Fig. 4(d).

## 5    Experimental Results

The local skew estimation method proposed was implemented as a Microsoft Visual C# 2010 Windows-based application on an Intel(R) Core(TM) i7-3667U CPU @ 2.00GHz Notebook, carried out on a video clip with 700 frame images. This video clip, captured by a "glasses" camera with 320 x 240 pixel resolution, simulated a

(a)                                                        (b)

(c)                                                        (d)

**Fig. 4.** Example results of local skew estimation and deskew. (a) Text region with local skew angle 1.0609. (b) Deskew result of (a). (c) Text region with local skew angle 1.0809. (d) Deskew result of (c).

visually impaired person "reading" three BCs. The clip was divided into a training and a testing set of 220 and 480 frame images, respectively.

Table 1 shows that the proposed method reduced overall time complexity. After M2FD and thumb detection methods were applied, 75 (training) and 147 (testing) frames were reserved. When the non-motion blur detection method was applied, only 33 and 109 frames were detected in the two sets. Thus, the total reduced rates for training and testing were 83.18% and 77.08%, respectively.

**Table 1.** Performance of the proposed method to reduce overall time complexity

| Video clip | Original frames | M2FD | Thumb | Non-Motion blur | Total reduced rate |
|---|---|---|---|---|---|
| Training set | 220 | 112 | 75 | 33 | 85.0% |
| Testing set | 480 | 312 | 147 | 109 | 77.29% |

Table 2 shows the performance of the proposed text detection method. The number of original texts in the training and testing sets are 4726 and 16725, respectively. The proposed method detected 4526 and 15835 in the respective sets, i.e., detected rates for the proposed method in two sets of 95.76% and 94.67%, respectively. The reason for the non-detected texts is that they do not appear in the frontal-flat view of the glasses camera. However, if the texts in moving BCs appear in the frontal-flat view of the camera, the proposed text detection method can detect them.

**Table 2.** Performance of the proposed text detection method

| Video clip | Original frames | Original texts | Detected texts | Detected rate |
|---|---|---|---|---|
| Training set | 33 | 4726 | 4526 | 95.76% |
| Testing set | 109 | 16725 | 15835 | 94.67% |

Table 3 shows the performance of the proposed local skew estimation methods. There are 30 and 105 original frames with local skew in the training and testing sets, respectively. The proposed method made 27 and 93 local skew estimation, respectively, i.e., detected rates for the proposed method in two sets were 90.00% and 88.57%, respectively. The reason for the non-estimated frames is that they include the circle logo. In the future, the estimated rate will be enhanced.

**Table 3.** Performance of the proposed local skew estimation method

| Video clip | Original frames | skew frames | Estimated skew | Estimated rate |
|---|---|---|---|---|
| Training set | 33 | 30 | 27 | 90.00% |
| Testing set | 109 | 105 | 93 | 88.57% |

Table 4 shows the execution time performance of the proposed method. The sizes of video one and video two are 320 x 240 and 360 x 240, respectively. Both have 700 frame numbers. The total execution times for the two videos are 8.05 and 9.05 seconds, respectively, and the FPSs are 86.95 and 77.35, respectively.

**Table 4.** Execution times (second) performance of the proposed method

| Video (700 frames) | M2FD | Thumb | Motion blur | Text | Skew | whole | FPS |
|---|---|---|---|---|---|---|---|
| Video 1(320 x240) | 1.12 | 5.03 | 0.13 | 1.40 | 0.37 | 8.05 | 86.95 |
| Video 2 (360 x 240) | 1.29 | 5.77 | 0.14 | 1.47 | 0.38 | 9.05 | 77.35 |

# 6    Conclusions

A method is proposed to estimate the text region with local skew in moving BCs. If a BC is moving into the frontal-flat view of the glasses camera, the BC is grasped by right-hand thumb, and the BC is not motion blurred, the proposed method is triggered to detect the text. Next, the local skew angle is estimated for the text regions. Experiments show that the proposed method is effective and efficient, reducing time complexity and avoiding the visually impaired person hearing a synthesized voice repeating the same text. In the future, the deskew text regions will be recognized and translated into voice to form the full system to "read" business cards.

# References

1. Ezaki, N., Bulacu, M., Schomaker, L.: Text detection from natural scene images towards a system for visually impaired persons. In: Proceeding of ICPR 2004, Tampa, FL, vol. 2, pp. 683–686 (2004)
2. Zandifar, A., Duraiswami, R., Chahine, A., Davis, L.: A video based interface to information for the visually impaired. In: Proceeding of IEEE 4th International Conference on Multimodal Interfaces, pp. 325–330 (2002)
3. Tanaka, M., Goto, H.: Text-tracking wearable camera system for visually-impaired people. In: Proceeding of ICPR 2008, Tampa, FL, pp. 1–4 (2008)
4. Goto, H., Tanaka, M.: Text-tracking wearable camera system for the blind. In: Proceeding of ICDAR 2009, Barcelona, Spain, pp. 141–145 (2009)
5. Antonacopoulos, A.: Local skew angle estimation from background space in text regions. In: Proceedings of ICDAR 1997, pp. 684–688 (1997)
6. Messelodi, S., Modena, C.M.: Automatic identification and skew estimation of text lines in real scene images. Pattern Recognition 32, 791–810 (1999)
7. Pal, U., Mitra, M., Chaudhuri, B.B.: Multi-skew detection of Indian script documents. In: Proceedings of ICDAR 2001, pp. 292–296 (2001)
8. Kim, C., Hwang, J.: Fast and automatic video object segmentation and tracking for content-based applications. IEEE Trans. on CSVT 12(2), 122–129 (2002)
9. Tsai, C.M.: Non-motion blur detection for helping blind persons to "see" business cards. In: ICMLC 2012, pp. 1901–1906 (2012)
10. Wong, K.W., Lam, K.M., Siu, W.C.: A robust scheme for live detection of human face in color images. Signal Processing: Image Communication 18(2), 103–114 (2003)
11. Tsai, C.M.: Intelligent post-processing via bounding-box-based morphological operations for moving objects detection. In: Jiang, H., Ding, W., Ali, M., Wu, X. (eds.) IEA/AIE 2012. LNCS (LNAI), vol. 7345, pp. 647–657. Springer, Heidelberg (2012)
12. Otsu, N.: A threshold selection method from gray-level histogram. IEEE Trans. SMC 9, 62–66 (1979)

# Imbalanced Learning Ensembles for Defect Detection in X-Ray Images*

José Francisco Díez-Pastor, César García-Osorio, Víctor Barbero-García,
and Alan Blanco-Álamo

University of Burgos
{jfdpastor,cgosorio}@ubu.es, {vbg0011,aba0034}@alu.ubu.es

**Abstract.** This paper describes the process of detection of defects in
metallic pieces through the analysis of X-ray images. The images used
in this work are highly variable (several different pieces, different views,
variability introduced by the inspection process such as positioning the
piece). Because of this variability, the sliding window technique has been
used, an approach based on data mining. Experiments have been carried out with various window sizes, several feature selection algorithms
and different classification algorithms, with a special focus on learning
unbalanced data sets. The results show that Bagging achieved significantly better results than decision trees by themselves or combined with
SMOTE or Undersampling.

**Keywords:** Non Destructive testing, ensemble learning, X-ray, Bagging,
Undersampling, SMOTE.

## 1 Introduction

The inspection of defects is a very important task to ensure the quality of industrial processes. Quality control has become an essential prerequisite for companies to remain competitive. The Non-Destructive Testings [1](NDT) are among
the most commonly used tests, because they do not destroy or alter the material
on which they are applied. The Radiography is one of the oldest NDT methods
but it remains the most widely used. The inspection performed by human operators have the advantage that can be adapted to many different situations, many
of which have not been previously seen [2]. Unfortunately, human inspection
is not as consistent as an automated process since it is a process that requires
high concentration, besides it is affected by the occurrence of fatigue, different
levels of skill, experience or ways of working of each operator. In an extreme
case, the inspection of a casting by an operator may determine that a pore or
bubble is large enough to cause the breakage of the casting and another operator
may determine that the casting is correct. There is a need of objective automatic inspection systems. It is considered that there are two types of problems,
the detection of defects (defect or non-defect) and the classification of defects
(porosity, lack of penetration, etc.)[3], in this case it is the first.

---

* This work was supported by Project Magno MAGNO2008-1028-CENIT. We also
thanks to Grupo Antolin for providing us the X-ray images used in the experiments.

M. Ali et al. (Eds.): IEA/AIE 2013, LNAI 7906, pp. 654–663, 2013.
© Springer-Verlag Berlin Heidelberg 2013

## 2    Problem Description and Methodology

This paper describes the process of detecting defects in magnesium alloys castings. Magnesium alloys are approximately 60% lighter than aluminium casting alloys and 80% lighter than steel, in spite of this advantage its adoption is slowed due to the internal porosity in the magnesium casting components [4]. In this paper, we analize the problem of defect detection in the context of high variability images: pieces of different types, multiple views for each piece, variability between views of the same piece due to the manual process of positioning the piece in the X-ray inspection system and the mechanical imprecision of the positioning system. In this regard the images used in this work are very different from the images used in previous works, as seen in Figure 1. Along recent years, a large number of methods for automatic detection of defects in X-ray images have been developed, but many of them do not work properly when the variability in the images is too high. The defect detection methods can be classified into: a) methods based on the subtraction of a reference image (this reference image can be obtained by applying specific image processing operations to the image where defects are being sought [5], obtaining the reference image can automatically be obtained from a set of images [4]), b) methods based on digital image processing: automatic thresholding [6,7], Mathematical morphology [8], or watershed [9,10]. The first approach was tested, obtaining a reference image from the median of a set of images of the same view, to mitigate the problem that the images are not taken from exactly the same prespective due to manual placement process and positioning system imprecision a stitching process was attempted, using the algorithm SURF [11], however the quality expected in the alignment was not achieved. The methods in the second approach were also considered, but these methods are highly dependent on the type of images and could not find one that worked well for processing all the pieces and views used in this work.

In many computer vision problems, the solution has evolved from an approach based on image processing tailored to the specific characteristics of the problem at hand, to a more general approach called "Appearance-based method" in which learning methods are applied to a data set, transforming the detection problem into a binary classification problem. This approach allows to work with images of poor quality and can deal with more generic problems. For example, a system can detect vehicle license plates using image processing and heuristics techniques as [12] or detect faces using a skin color model [13] or it can detect license plates [14] or faces [15] transforming the problem into a classification problem. This Appearance-based approach is receiving considerable attention and has been applied to the detection of defects in radiography images in [16] and [17].

### 2.1    Sliding Window

The sliding window technique consists of extracting features from a subimage, the window, of the image that is being procesed. This window is sistematically moved along the image. The features extracted together with a label, the presence or not of defects inside the window, are used to train a classifier. In the predictions

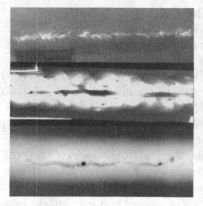 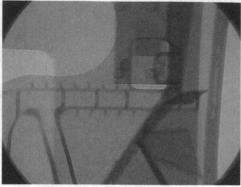

(a) Images used in previous works          (b) Images used in this work
[10,18,17]
**Fig. 1.** Differences between images

stage the classifier is used to detect windows with defect inside them. Figure 2 describes how the technique works, there is an image $I$, which is scanned with a window of size $N \times N$, starting from an initial position it moves horizontally at intervals of *step_h* and vertically at intervals of *step_v*. This window is passed to one or more feature extractors. The set of one or more feature extractors returns a vector of features for each of the windows. This technique can be applied on the raw grayscale image or on the result of applying some processing to this image. The window at coordinates $x, y$ can be represented as the concatenation of the feature vector extracted from the original grayscale image and the feature vector extracted from images obtained from processing the original image. As in [17], in this work characteristics have been extracted both on the original image in grayscale and on the saliency Map [19] (an image transformation based on a biologically inspired attention system) of the grayscale image.

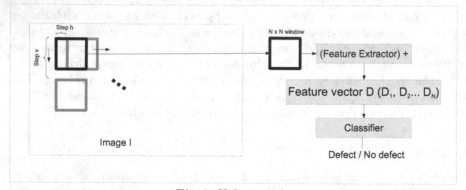

**Fig. 2.** Sliding window

**Table 1.** Features used in the experiments

| Type | Names | Number |
|---|---|---|
| Standard | 1. mean, 2. standard desviation, 3. first derivative, 4. second derivative | 4 |
| Haralick | 1. Angular Second Moment 2. Contrast, Correlation, 4. Sum of squares, 5. Inverse Difference Moment, 6. Sum Average, 7. Sum Entropy, 8. Sum Variance, 9. Entropy, 10. Difference Variance, 11. Difference Entropy, 12-13. Information Measures of Correlation, | 13 features x 5 different pixel distances x 2 (mean and range of each vectors) =130 |
| LBP | LBP(1) ... LBP(59) | 59 |
| | | 193 features per channel |
| | | 2 channel (normal and saliency) x 193 = 387 features |

## 2.2  Features

In artificial vision problems is unusual to train classifiers directly with the intensity values of the pixels. Typically, different types of features are extracted from these intensity values. These characteristics are often dependent on the images and type of problem. In this work, a subset of the characteristics used in [17] has been selected (see Table 1).

Standard features includes average and standard deviation of the intensities of the region and the average of the first and second derivative of the region. Haralick features [20], represent textural information, these features are computed from the co-occurrence matrix that represents second order texture information, these features are 14, but Maximal Correlation Coefficient was excluded due to its elevated computation time and because the system is desired to operate in near real-time. The local Binary Patterns texture descriptors are extracted from an histogram elaborated from the relationship between each pixel intensity value with its eight neighbors.

## 2.3  Atribute Selection

Due to the large number of attributes, an attribute selection process is necessary to reduce the classifiers training time and for improving its performance. We tested three methods:

1. Correlation Feature Selection (CFS)[21] in conjunction with best first search. Evaluates a subset of attributes by considering the individual predictive ability of each feature along with the degree of redundancy between them. Subsets of features that are highly correlated with the class while having low intercorrelation are preferred.
2. FCBF [22] (Fast Correlation-Based Filter Solution) is a feature selection method based on correlation measure, relevance and redundancy analysis especially oriented to sets of high dimensionality.
3. SVM attribute evaluator [23] evaluates attributes by using an SVM classifier.

**Table 2.** Features selected characteristics of each type

| Dataset | FCBF | | | | | | |
|---|---|---|---|---|---|---|---|
| | Total | Standard | Haralick | LBP | Standard(S) | Haralick(S) | LBP(S) |
| Size 8 | 4 | 1 | 1 | 1 | 1 | 0 | 0 |
| Size 16 | 7 | 0 | 1 | 5 | 0 | 1 | 0 |
| Size 24 | 4 | 0 | 0 | 1 | 0 | 1 | 2 |
| Size 32 | 4 | 0 | 0 | 1 | 0 | 1 | 2 |
| Size 40 | 5 | 1 | 0 | 2 | 0 | 1 | 1 |

| Dataset | SVM Att Eval | | | | | | |
|---|---|---|---|---|---|---|---|
| | Total | Standard | Haralick | LBP | Standard(S) | Haralick(S) | LBP(S) |
| Size 8 | 30 | 0 | 8 | 0 | 3 | 18 | 1 |
| Size 16 | 30 | 0 | 11 | 0 | 2 | 15 | 2 |
| Size 24 | 30 | 0 | 8 | 3 | 2 | 13 | 4 |
| Size 32 | 30 | 0 | 11 | 2 | 2 | 13 | 2 |
| Size 40 | 30 | 0 | 7 | 3 | 1 | 17 | 2 |

| Dataset | CFS-Best First | | | | | | |
|---|---|---|---|---|---|---|---|
| | Total | Standard | Haralick | LBP | Standard(S) | Haralick(S) | LBP(S) |
| Size 8 | 52 | 1 | 17 | 0 | 0 | 20 | 14 |
| Size 16 | 39 | 0 | 15 | 0 | 1 | 13 | 10 |
| Size 24 | 31 | 2 | 11 | 0 | 1 | 11 | 6 |
| Size 32 | 33 | 1 | 13 | 1 | 0 | 11 | 7 |
| Size 40 | 31 | 0 | 11 | 6 | 1 | 9 | 4 |

The first two methods return a subset of attributes. The third elaborates a ranking, so it is necessary to specify $N$, the number of selected attributes; $N = 30$ was used in the experiments. To perform the atribute selection, a dataset for each window size was elaborated, this dataset was obtained from 10 different images. In each image 500 samples were obtained through a random window. Half the windows including defects, the other half without any defect inside them. Table 2 show the total number of selected attributes and the number of selected attributes for each type of features and attribute selection algorithm. The suffix (S) indicates that these characteristics have been calculated on the saliency Map. It can be seen that FCBF is a very aggressive attribute selector that selects very small subsets. Attributes selected in a greater number according to the CFS and SVM Att EVal algorithms are those of Haralick. It is generally observed that the characteristics calculated on the saliency map are selected slightly more times.

## 2.4   Ensemble Learning for Inbalanced Datasets

The class-imbalance problem occurs when there are many more instances of some classes than others [24]. X-ray images are unbalanced datasets because the proportion of regions with defects is much smaller than regions without them. This proportion must also exist in the training set, because it is not a good practice to train a classifier with a dataset with a very different proportion of classes that the proportion that will be found in the exploitation phase.

**Table 3.** Instances of the datasets

| Dataset | Non-defect Windows | Defect Windows | Imbalance Ratio |
|---------|-------------------:|---------------:|:---------------:|
| Size 8  | 205702 | 10618 | 19,3730 |
| Size 16 | 201566 | 14754 | 13.6618 |
| Size 24 | 197600 | 18720 | 10.5556 |
| Size 32 | 193647 | 22673 | 8.5409 |
| Size 40 | 189774 | 26546 | 7.1489 |

In unbalanced data sets, precision should not be used, since this measure provides the same value to the hits and misses, regardless of the distribution of classes. Commonly used measure of performance for imbalanced data is the Area Under the ROC (Receiver Operation Characteristic) curve [25]. There are several strategies for dealing with unbalanced sets, the most used is to preprocess the data set to reduce its imbalance, either by removing random instances of the majority class (Random undersampling) or by adding artificial instances of the majority class the technique is more representative of the latter is SMOTE [26]. In Machine Learning, the ensembles of classifiers are known to increase the performance of single classifiers by combining several of them. The ensemble of classifiers can be combined with the previous preprocessing techniques, to handle the problem of the imbalance better than any individual classifier.

## 2.5   Experimental Setup

A total of 15 different data sets were elaborated using 5 different window sizes and 3 attribute selection algorithms for each of them. These data sets were obtained from 10 different images, using the sliding window procedure, with $step\_h$ y $step\_v = 4$. Table 3 shows the number of instances of each class and the imbalance ratio for each window size. And as previously mentioned, the tables 2 show the number of attributes depending on the attributes selection algorithm used.

Weka [27] was used for the experiments. J48, the Weka's re-implementation of C4.5 [28], was chosen as the base classifier in all ensembles. As recommended for imbalanced data [29], it was used without pruning and collapsing but with Laplace smoothing at the leaves. C4.5 with this options is called C4.4 [30]. We tested various ensembles methods such as Bagging [31], Bagging + undersampling (eliminating as many instances of the majority class as necessary to achieve an imbalance ratio IB = 1 and IB = 2), Bagging + SMOTE (generating an artificial number of instances equal to 100% and 200% of the size of the minority class). The size of the ensembles was 20. We also tested the performance of J48 by itself and in combination with undersampling and SMOTE, with the same settings as those used in the ensembles.

The Area Under the ROC results were obtained with a $5 \times 2$-fold cross validation [32]. The data set is halved in two folds. One fold is used for training and the other for testing, and then the roles of the folds are reversed. This process is repeated 5 times. The results are the averages of these 10 experiments. Cross validation was stratified.

## 2.6   Results

The results of the area under the ROC for J48 and Bagging of J48, with the features selected by different feature selection methods and different window sizes are shown in Figure 3 It can be seen how the classifiers built with features extracted by CFS + Best First and SVM Att Eval perform significantly better than those built with features extracted by FCBS, which is understandable since FCBS is a very strict attribute selector which selects a very small number of attributes.

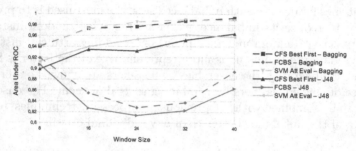

**Fig. 3.** Area Under ROC vs. Window Size for Bagging and J48 classifiers

It is also noted that for window size 8, Bagging of J48 built with FCBS features get surprisingly good results despite using only 4 attributes, beating J48 built with CFS and SVM Att Eval features.

In general, it is observed that, as the window size increases the area under the ROC also increases, which can be explained in part because as seen in Table 3, as window size increases the imbalance between classes is reduced. Given that by increasing window size, the chances of a window covering a defect is greater. However increasing the window size leads to a less accurate defect detection.

Figure 4 show the difference (improvement or deterioration) between the area under ROC obtained by J48 with respect to that obtained by Undersampling + J48 (A and B), SMOTE (C and D), Bagging (E), Bagging + Undersampling (F and G) and Bagging + SMOTE (H and I). Bagging of J48 is significantly better than J48 in each of the data sets. Contrary to what would be expected, neither SMOTE nor undersampling, J48 improve performance for virtually any of the combinations of window size and the set of attributes used. Combinations of Bagging + Undersampling and Bagging + SMOTE outperforms Bagging in some datasets, Bagging + smote 100% is in general the classifier which obtains better results, although the differences are not significant compared to Bagging.

The results of the predictions of the classifiers are combined as shown in Figure 5. When a pixel is covered by a window which the classifier predicts as defect,

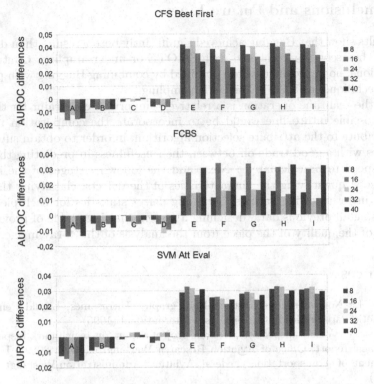

**Fig. 4.** Area under ROC Difference of between J48 and the other classifiers

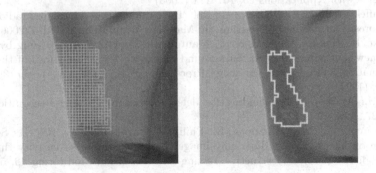

**Fig. 5.** Results of the detection process

this window receives a vote. Following this, a threshold is set to determine the minimum number of votes required to consider that a pixel actually belongs to a defect. This threshold depends on the size of the window, the larger the window, the higher the threshold, and in the current implementation is not calculated automatically, instead, the user can adjust it, making the predicted region widens or fit to the true defect.

# 3  Conclusions and Future Lines

The results show that Bagging achieved significantly better results than decision trees by themselves or combined with SMOTE or undersampling. Contrary to expectations, no improvements are obtained by combining Bagging with preprocessing techniques as SMOTE or Undersampling.

Since the real-time operation is a relatively important constraint for the system, a possible future line would be to incorporate the computation time of each attribute to the attribute selection algorithm, in order to obtain subsets of attributes with a good trade-off between their usefulness in predicting the class and its speed to be calculated. Next, add the following stages of the inspection process: quantifying the characteristics of the defects, classifying the type of defect in terms of its features, obtaining defect statistics for each piece and finally using all previous data, designing a system that is capable of providing a measure of the quality of the piece from the analysis of this x-ray image.

# References

1. Cartz, L.: Nondestructive Testing: Radiography, Ultrasonics, Liquid Penetrant, Magnetic Particle, Eddy Current. Asm International (1995)
2. Spencer, F.: Visual inspection research project report on benchmark inspections. Technical report, Office of Aviation Research Washington, D.C. 20591: U.S. Department of Transportation, Federal Aviation Administration,Washington, DC (1996)
3. Liao, T.: Classification of welding flaw types with fuzzy expert systems. Expert Systems with Applications 25, 101–111 (2003)
4. Rebuffel, V., Sood, S., Blakeley, B.: Defect detection method in digital radiography for porosity in magnesium casting. In: Materials Evaluation, ECNDT 2006 (2006)
5. Hanke, R., Hassler, U., Heil, K.: Fast automatic x-ray image processing by means of a new multistage filter for background modelling. In: Proceedings of the IEEE International Conference on Image Processing, ICIP 1994, vol. 1, pp. 392–396. IEEE (1994)
6. Ng, H.: Automatic thresholding for defect detection. Pattern Recognition Letters 27, 1644–1649 (2006)
7. Saravanan, T., Bagavathiappan, S., Philip, J., Jayakumar, T., Raj, B.: Segmentation of defects from radiography images by the histogram concavity threshold method. Insight-Non-Destructive Testing and Condition Monitoring 49, 578–584 (2007)
8. Anand, R., Kumar, P., et al.: Flaw detection in radiographic weld images using morphological approach. NDT & E International 39, 29–33 (2006)
9. Wang, M., Chai, L.: Application of an improved watershed algorithm in welding image segmentation. Transactions China Welding Institution 28, 13 (2007)
10. Anand, R., Kumar, P., et al.: Flaw detection in radiographic weldment images using morphological watershed segmentation technique. NDT & E International 42, 2–8 (2009)
11. Bay, H., Tuytelaars, T., Van Gool, L.: Surf: Speeded up robust features. In: Leonardis, A., Bischof, H., Pinz, A. (eds.) ECCV 2006, Part I. LNCS, vol. 3951, pp. 404–417. Springer, Heidelberg (2006)

12. García-Osorio, C., Díez-Pastor, J.F., Rodríguez, J.J., Maudes, J.: License plate number recognition - new heuristics and a comparative study of classifiers. In: Proceedings of the Fifth International Conference on Informatics in Control, Automation and Robotics, Robotics and Automation, ICINCO 2008, vol. 1, pp. 268–273 (2008)
13. Jones, M., Rehg, J.: Statistical color models with application to skin detection. In: IEEE Computer Society Conference on Computer Vision and Pattern Recognition, vol. 1, IEEE (1999)
14. Dlagnekov, L.: License plate detection using adaboost. Computer Science and Engineering Department, San Diego (2004)
15. Viola, P., Jones, M.: Rapid object detection using a boosted cascade of simple features. In: Proceedings of the 2001 IEEE Computer Society Conference on Computer Vision and Pattern Recognition, CVPR 2001, vol. 1, pp. 511–518. IEEE (2001)
16. Wang, Y., Sun, Y., Lv, P., Wang, H.: Detection of line weld defects based on multiple thresholds and support vector machine. NDT & E International 41, 517–524 (2008)
17. Mery, D.: Automated detection of welding discontinuities without segmentation. Materials Evaluation, 657–663 (2011)
18. Belaifa, S., Tridi, M., Nacereddine, N.: Weld defect classification using em algorithm for gaussian mixture model. In: SETIT Tunisia 2005 (2005)
19. Montabone, S., Soto, A.: Human detection using a mobile platform and novel features derived from a visual saliency mechanism. Image and Vision Computing 28, 391–402 (2010)
20. Haralick, R., Shanmugam, K., Dinstein, I.: Textural features for image classification. IEEE Transactions on Systems, Man and Cybernetics, 610–621 (1973)
21. Hall, M.: Correlation-based feature selection for machine learning. PhD thesis, The University of Waikato (1999)
22. Yu, L., Liu, H.: Feature selection for high-dimensional data: A fast correlation-based filter solution. In: Machine Learning-International Workshop Then Conference, vol. 20, pp. 856–863 (2003)
23. Guyon, I., Weston, J., Barnhill, S., Vapnik, V.: Gene selection for cancer classification using support vector machines. Machine Learning 46, 389–422 (2002)
24. Chawla, N., Japkowicz, N., Kotcz, A.: Editorial: special issue on learning from imbalanced data sets. ACM SIGKDD Explorations Newsletter 6, 1–6 (2004)
25. Fawcett, T.: An introduction to roc analysis. Pattern Recognition Letters 27, 861–874 (2006)
26. Chawla, N.V., Bowyer, K.W., Hall, L.O., Kegelmeyer, W.P.: SMOTE: Synthetic Minority Over-sampling Technique. Journal of Artificial Intelligence Research 16, 321–357 (2002)
27. Hall, M., Frank, E., Holmes, G., Pfahringer, B., Reutemann, P., Witten, I.H.: The weka data mining software: an update. SIGKDD Explor. Newsl. 11, 10–18 (2009)
28. Quinlan, J.: C4.5: Programs for Machine Learning. Morgan Kauffman (1993)
29. Cieslak, D.A., Hoens, T.R., Chawla, N.V., Kegelmeyer, W.P.: Hellinger distance decision trees are robust and skew-insensitive. Data Min. Knowl. Discov. 24, 136–158 (2012)
30. Provost, F., Domingos, P.: Tree induction for probability-based ranking. Machine Learning 52, 199–215 (2003)
31. Breiman, L.: Bagging Predictors. Machine Learning 24, 123–140 (1996)
32. Dietterich, T.: Approximate statistical tests for comparing supervised classification learning algorithms. Neural Computation 10, 1895–1923 (1998)

# Improvements in Modelling of Complex Manufacturing Processes Using Classification Techniques

Pedro Santos, Jesús Maudes, Andrés Bustillo, and Juan José Rodríguez

Department of Civil Engineering, University of Burgos, Spain
C/ Francisco de Vitoria s/n, 09006, Burgos, Spain

**Abstract.** The improvement of certain manufacturing processes often involves the challenge of how to optimize complex and multivariable processes under industrial conditions. Moreover, many of these processes can be treated as regression or classification problems. Although their outputs are in the form of continuous variables, industrial requirements define their discretization in compliance with ISO 4288:1996 Standard. Laser polishing of steel components is an interesting example of such a problem, especially its application to finishing operations in the die and mould industry. The aim of this work is the identification of the most accurate classifier-based method for surface roughness prediction of laser polished components in compliance with the aforementioned industrial standard. Several data mining methods are tested for this task: ensembles of decision trees, classification via regression, and fine-tuned SVMs. These methods are also tested by using variants that take into account the ordinal nature of the class that has to be predicted. Finally, all these methods and variants are applied over different transformations of the dataset. The results of these methods show no significant differences in accuracy, meaning that a simple decision tree can be used for prediction purposes.

**Keywords:** ensembles, ordinal classification, discretization, process optimization, laser polishing.

## 1 Introduction

The use of high-power lasers in the manufacturing industry has been expanding for over 20 years. The main laser applications for the manufacture of vehicle components in the 1990s involved cutting 2D or 3D metallic sheets. More recently, these applications have not only been used for rapid manufacturing of car components, but also for new tools such as moulds and dies [28].

Over the last ten years, various rapid manufacturing techniques have been developed: Laser Engineered Net Shaping, Selective Laser Sintering (SLS) or Selective Laser Melting. All of these techniques can manufacture fully functional metal parts from raw materials in the form of powder. They achieve significant reductions in manufacturing-time and build parts with very complex geometries that are beyond the scope of traditional manufacturing techniques. Despite these advantages, rapid manufacturing techniques present a fundamental disadvantage for their industrial application: the poor surface quality of the manufactured component [22].

M. Ali et al. (Eds.): IEA/AIE 2013, LNAI 7906, pp. 664–673, 2013.

The laser polishing process melts the surface of a component using a laser beam as an energy source. If the process parameters are correctly selected, the laser beam will only melt the peaks of the surface and the melted material will run into the valleys, achieving a smoother topography than the initial one. Experimental results have demonstrated that laser polishing can obtain high-quality roughness surfaces on parts manufactured with SLS that present a very high initial roughness. However, the roughness reduction ratio depends on many variables. Under laboratory conditions, the laser polishing process depends mainly on three factors: the energy density of the laser beam, the surface material, and its initial roughness. Although measurement of the last two factors can be reasonably accurate, the first one is often unknown under real industrial conditions. Furthermore, other process parameters could vary under industrial conditions, such as assistance gas, the beam incidence angle, etc. The influence, under real industrial conditions, of all these parameters has yet to be established, which makes it difficult to optimize the process parameters in the laser polishing process. In consequence, industrial demand for models that take all of these influences into account is high.

Different approaches have been proposed to build a suitable model for this industrial application. Dubey has divided these approaches into three categories: empirical models, analytical models, and artificial intelligence models [8]. This paper falls into the third category. The most common artificial intelligence techniques applied to laser milling and laser polishing include Artificial Neural Networks [2,6], connectionist techniques [3] and particle swarm [6]. To improve the results of a single artificial intelligence model, a combination of two or more models can be built that improves overall model accuracy. These combinations are called ensembles. The prediction capability of an ensemble is built by merging the predictions of the combined models. Ensembles have demonstrated their superiority over single models in many applications such as roughness prediction in milling operations [4], burr detection in the drilling process [10], monitoring of lubricating oil quality [5] and wind turbines fault diagnosis [27]. The artificial intelligence models in this study, have been built using ensembles.

There is another open question related to the artificial intelligence model to be used. Rather than roughness prediction as a continuous variable, the final industrial application will require its accurate prediction in terms of a discretized scale of levels specified in an International Standard, ISO 4288:1996 [16]. Therefore, two methodologies may be followed: i) predict the continuous roughness of the polished component and then discretize the predicted value for each process condition using ISO 4288:1996, or ii) discretize the roughness and train the artificial intelligence model with this new dataset. This open issue has been also studied for other industrial tasks such as roughness prediction in face milling [7], drilling [14] or burr size in drilling [10].

This paper is organized as follows. Section 2 introduces the experimental procedure for data collection and finishes with the dataset description. Section 3 describes the possibilities of ensemble modelling, considering the specific nature of the laser polishing process. Section 4 presents the results of the ensemble modelling and other classification techniques that consider roughness as an ordinal class. Finally, the conclusions and future lines of work are summarized in Section 5.

## 2   Experimental Procedure and Dataset Description

The following experimental set-up had the objective of testing a broad range of process parameters used in combination to provide the classifier model with sufficient information on all the relationships between process parameters in the polishing process. A more detailed description of these experiments has previously been published [2].

A $CO_2$ laser (model Rofin Sinar DC 025) was used to perform all tests. This laser achieves up to 2.5 kW power in continuous mode in an almost Gaussian energy distribution with a 0.4 mm diameter spot. Two different materials were tested for laser polishing: a commercial alloy LaserForm™ST-100 and Orvar Supreme steel. LaserForm ST-100 is typically used to build up commercial components by SLS. Orvar Supreme is typically used in mould manufacturing. To consider a typical surface, the test blanks were previously milled by a 12mm diameter ball end mill with a variable radial step, in order to obtain a surface quality of a typical milling-operation. The surface topography, in terms of roughness $R_a$ parameter, can be calculated using an equation developed by Quintana [25].

Different energy densities were tested, in order to adjust the conditions to each material. Under industrial conditions [8,6], the energy density depends on three main factors: feed rate of the laser head, power of the laser, and focal offset of the beam. Therefore the experimentation was developed on the basis of a factorial DoE of three factors at three levels, and includes parameter combinations to get results under and over the optimal value presented in the bibliography [21]. The tests consisted of a series of single polishing paths of 20mm in length. Table 1 resumes the parameter values during testing. Three different conditions regarding assistance gas were tested: no assistance gas, air and argon. Altogether, 178 different conditions were tested, the results of which generated a dataset with 178 instances. Finally, the roughness $R_a$ parameter of the laser polished blanks was measured following industrial standard procedure ISO 4287:1997 [17]. Once the experimental measurements of roughness had been made, the dataset for the artificial intelligence model was generated. The variables included in the dataset and their variation ranges, see Table 2, are: material, focal offset distance, spot diameter, laser power, feed rate, energy density, assistance gas, initial roughness and final roughness. Roughness was discretized according to industrial standard ISO 4288:1996 [16]. Five different levels of roughness were identified in the dataset: from level 3 to level 7 of standard ISO 4288:1996. But only 30 of the 178 instances referred to roughness classes 3 and 7, therefore these instances were combined with classes 4 and 6, respectively. This decision was taken on the assumption that the dataset would not be suitable to predict process conditions in these outer limits and should concentrate on correctly

**Table 1.** Process parameter selective for laser polishing tests

|  | LaserForm™ ST-100 | | | Orvar Supreme | | |
|---|---|---|---|---|---|---|
|  | Level1 | Level 2 | Level 3 | Level 1 | Level 2 | Level 3 |
| Power [W] | 600 | 800 | 1,000 | 1,200 | 1,400 | 1,600 |
| Feed Rate [mm/min] | 1,200 | 1,500 | 1,800 | 1,100 | 1,300 | 1,500 |
| Focal Offset [mm] | 20 | 27 | 34 | 20 | 27.5 | 35 |

**Table 2.** Variables, units and ranges used during the experiments

| Variable (Units) | Input / Output | Range |
|---|---|---|
| Materia | Input variable | 1 Orvar, 2 LaserForm |
| Focal offset distance (mm) | Input variable | 20 - 35 |
| Spot diameter (mm) | Input variable | 1.1-1.9 |
| Laser Power (W) | Input variable | 600 -1,600 |
| Feed rate (mm/min) | Input variable | 1,100 -1,800 |
| Energy density (J/mm²) | Input variable | 10.5 - 81.5 |
| Assistance gas | Input variable | 0 none, 1 Argon, 2 Air |
| Initial roughness (μm) | Input variable | 1.0 - 6.8 |
| Final roughness (μm) | Output variable | 0.2 - 6.2 (continuous) or 4-5-6 (discretized) |

dividing the middle roughness levels 4-6. Within this dataset the three different levels of roughness, were balanced: 49 (Class 3-4), 75 (Class 5) and 54 instances (Class 6-7).

## 3  Data Mining Techniques Applied in the Study

As detailed in the previous section, the aim is to predict final roughness. Final roughness was categorized using three ordinal classes which label roughness to smoothness on a descending scale (a roughness level of 4 is smoother than 5, and so on).

Ordinal classification methods assume that an order exists among the classes. Classification of laser polishing is an ordinal class problem, where classes represent a grade of final roughness. There are methods for ordinal classification [11] [19] that exploit the information derived from such orders to arrive at greater accuracy.

The standard approach to ordinal classification probably consists of predicting a continuous variable (i.e., numeric final roughness in this problem), which is then discretized into a categorical set of prediction classes [19]. This standard approach is ideal for our problem, as (i) in [4], experimental tests confirmed that Bagging [1] of regression trees gave better results than other state-of-the-art regressors, and (ii) a discretization method [16] exists to map this numeric variable onto the classes that need to be predicted. Therefore this technique is tested in the study and denoted as *Classification-Via-Regression*.

In [11], the ordinal multiclass problem with $k$ classes is transformed into $k-1$ binary problems. The $ith$ problem tries to predict whether the class is bigger than the $ith$ ordinal class value. Therefore $k-1$ binary classifiers have to be trained, and their probability estimations are used to compute the final prediction.

The main advantage of this approach is that it can combine any kind of binary classifiers to tackle the $k-1$ individual binary sub-problems. For example, in [11] C4.5 decision trees are used, a technique that is labeled OCC (i.e., *Ordinal Class Classifier*) in this work.

As stated in [11,12], an analogous strategy to OCC can be applied to input attributes. Each input attribute $a_i$ can be discretized using a supervised discretization method that maps it into $m$ nominal categories. Each category covers a range for continuous values of $a_i$, so there are $m - 1$ splitting points delimiting the $m$ ranges. Hence, the resulting $m$ categories can be grouped into $m - 1$ binary features that signify that $a_i$ is either larger or smaller than one of these splitting points. The discretization method used in [12] was the Fayyad-Irani supervised discretization [9].

The main advantage of constructing these new boolean features is that learning algorithms can be sensitive to the order in the categories obtained by the discretization. Moreover, some algorithms such as decision trees can use the split points given by the discretization method and its global view from the dataset, mapped into a local region formed by the instances belonging to a branch. These combinations of a global and local view from the dataset are useful to reduce classifier overfitting.

As in [12], the original dataset is denoted as (RAW) in this study. The discretized version of the dataset using the Fayyad-Irani method over all numeric attributes is denoted as (DISC), and finally, a discretized dataset where discrete attributes have been transformed into a set of binary attributes representing their ordinal nature is denoted as (ORD). There are another two dataset versions used in the study, according to whether original numeric features are maintained alongside the discretized features. DISC_KEEP and ORD_KEEP are respectively the DISC and ORD versions that retain the original numeric features.

Some of the more representative classification methods were tested. In a first group, two singleton methods are considered:

1. Decision Trees: C4.5 Quinlan decision tree was used [24]. As previously pointed out, decision trees can take advantage of binary features computed on ORD and ORD_KEEP dataset versions. Another two remarkable issues are that decision trees can process directly nominal features, such as those computed in the DISC and DISC_KEEP versions and can also work on multiclass problems.
2. Support Vector Machine (SVM) [29]. Radial Function Basis Function Kernel was used for this classifier. SVM is one of the most popular and successful state-of-the-art classifiers [30]. It reports very competitive results provided that its parameters ( i.e., the slack variable and gamma) have been properly tuned. SVM is a binary classifier (i.e., it only can predict on two classes problems). Multiclass problems for SVM are transformed into $k$ binary problems, where $k$ is the number of classes. Each binary problem predicts if an instance belongs to one of the classes vs. the rest (i.e., one vs. all approach). Another interesting method for SVM to tackle with the three classes in the polishing-laser dataset is OCC. OCC also divides the multiclass problem into a few binary problems and takes its ordinal nature into account.

The other group of classifier methods to consider are the ensembles [20]. Ensembles gather predictions from a set of so-called base classifiers. The final predictions are made by simple majority voting, weighted voting, or any other combination schema. Most popular ensembles combine base classifiers trained using the same algorithm. Hence, a fundamental element of ensembles is their strategy that ensures the resulting base classifiers are different. If all base classifiers agree, there would be no difference between using only one base classifier or an ensemble of base classifiers. This property of ensembles is known as the *diversity* of base classifiers [20]. Diversity may be reached by introducing some kind of perturbation in the version of the dataset that is used to train each base classifier. Hence, base classifiers that are highly sensitive to changes in the training dataset (i.e., *unstable* base classifiers) are appropriate for ensembles. Decision trees are an example of unstable classifiers whereas SVM is an example of a stable one.

In this study we have used the following ensembles, taking C4.5 trees as base classifiers:

1. Bagging [1]: the diversity of this ensemble depends on training each base classifier by sampling with replacement from the original dataset. Bagging reduces the component of the error debt to the casual distribution of the instances forming the training dataset (i.e., variance error component [18]).
2. Boosting [13]: in Boosting, each base classifier is trained by focusing on training instances misclassified by the previous base classifier. A weight is given to each instance using the prediction of the previous base classifier. This weight increases when the instance has been misclassified. Final prediction is made by using a weighted voting schema where the larger training error on each base classifier results in a lower weight. There are a lot of Boosting variants. In this paper, AdaBoost M1 [13] is used because it is the most popular.
3. Rotation Forest [26] trains each base classifier by grouping their attributes into subsets (e.g., subsets of three attributes are usually taken). Then PCA (*Principal Component Analysis*) is computed for each group using a subsample from the training set. The whole dataset is transformed according these projections and can then be used to train a base classifier.

## 4  Results and Discussion

WEKA [15] that was used for the experimental validation provides implementations for all the methods and data transformations in use. Eight methods were tested: C4.5 decision tree, SVM with Radial Basis Function kernel, Bagging, AdaBoost M1, Rotation Forest, and Classification via Regression applying results in [4].

The default parameters of these methods were used, except that:

1. The size of the ensembles was set at 100. This setting was chosen because in [4] Bagging for regression used 100 regression trees and classification via that regression will be tested, so the rest of ensembles take that size to ensure comparable results.
2. The base classifiers for all the ensembles was a C4.5 decision tree, without pruning.
3. SVM was always tuned for each training data partition in the validation. The slack variable and gamma parameters were tuned.

These methods were tested using the following dataset versions, RAW, DISC, DISC_KEEP, ORD and ORD_KEEP described in the previous section.

All the methods, except for classification via regression, were tested using these dataset versions. The supervised discretization was computed for each training fold.

Finally, because the dataset has an ordinal class, OCC was tested. So, all the configurations were validated with and without OCC.

10×10 cross-validation was used in the experimental validation as in [4]. Table 3 shows the accuracy results for all the methods, except for the classification via regression the accuracy of which was 79.26%.

**Table 3.** Accuracies reached in the datasets tested

| | No OCC RAW | OCC RAW | No OCC DISC | OCC DISC | No OCC DISC_K | OCC DISC_K | No OCC ORD | OCC ORD | No OCC ORD_K | OCC ORD_K | Avg |
|---|---|---|---|---|---|---|---|---|---|---|---|
| C4.5 | 82.73 | 82.73 | 82.78 | 82.73 | 83.23 | 82.18 | 82.62 | 82.73 | 82.4 | 82.68 | 82.68 |
| SVM | 78.27 | 77.14 | 82.61 | 82.18 | 81.52 | 81.17 | 82.23 | 81.95 | 81.26 | 81.24 | 80.96 |
| Bagging | 82.51 | 82.29 | 82.67 | 82.57 | 82.67 | 82.12 | 82.56 | 82.73 | 82.23 | 82.06 | 82.44 |
| M1 | 81.33 | 80.92 | 82.45 | 81.17 | 80.81 | 80.75 | 82.16 | 80.88 | 81.04 | 80.69 | 81.27 |
| Multiboost | 81.89 | 82.5 | 80.92 | 80.6 | 82.45 | 82.21 | 80.53 | 80.94 | 82.5 | 82.32 | 81.69 |
| RotForest | 82.17 | 83.01 | 82.56 | 82.17 | 82.9 | **83.75** | 81.89 | 82.23 | 82.29 | 82.28 | 82.53 |
| Average | 81.7 | 81.43 | 82.44 | 81.9 | 82.56 | 82.03 | 82.17 | 81.91 | 82.06 | 81.88 | |

The best accuracy (83.75%), in bold in Table 3, was reached by Rotation Forest using OCC from DISC_KEEP version, and the worse result (77.14%), by SVM using OCC on RAW version. Classification via regression performance (79.26%) is therefore closest to the worse method.

The resampled *t-test* [23], applied between the best and the worse configurations, showed no significant difference (significance 5%). There were therefore no significant differences between the methods. Moreover, a single C4.5 without OCC on the DISC_KEEP version reached 82.73%. The decision tree may be the best classifier option, because it is the fastest and simplest model and because its decision nodes can be interpreted on sight to extract valuable knowledge.

OCC does not increase accuracy, as was expected. This could be because the problem has only three classes. The three classes generate only two binary problems, which are also imbalanced problems.

Average accuracies on SVM, and AdaBoostM1 are worse than average accuracies of the rest of the methods. This might suggest that data is somewhat noisy as these methods are more sensitive to learn the noise. This idea is reinforced by the low average accuracy on data versions without any discretization (i.e., RAW).

Regarding data versions, DISC_KEEP appeared to be the version that yielded the best results. The two top configurations are computed over this version, and the average accuracy of DISC_KEEP was the best. Discretization is used to avoid overfitting and to increase classifiers generalization. Keeping the original attributes gives a broader range of choices in decision trees branching. On the other hand, average accuracy in ORD versions does not increases accuracy as expected.

## 5    Conclusions and Future Lines of Work

This research has presented the results of an investigation to identify the most appropriate methodology to solve a real-life industrial problem concerning the laser polishing of metallic components. This industrial task, like many others, can be treated as a regression or a classification problem, because the industrial requirement is based on an ISO Standard that discretizes the output, even though it is a continuous variable. Several ensembles were investigated to achieve the best practical solution to this interesting problem. The two proposed methodologies were: building a regression model and discretizing its output using the industrial Standard ISO 4288:1996 and discretizing the output variable of the dataset using this Standard and then building a classification model afterwards.

A real dataset generated under real industrial conditions was used to validate this approach. The dataset included 178 instances with 8 input variables and 1 output, the final roughness of the polished blank. The correct evaluation of three different levels of roughness with a balanced number of instances within them were researched. This research shows that there are no significant differences between the machine learning techniques tested. For this reason the simplest method (i.e., ensembles of C4.5 decision trees) is probably the best solution for the problem, as it is faster than the other methods and is the only method that provides insight on the classification criteria. Ordinal classification was treated using OCC but the results did not improve, maybe because of the reduced number of classes. Techniques that preprocess data including some discretization variants were also tested. It appears that discretization slightly helped to increase accuracy. The best results in the experimental validation were reached by merging the discretized attributes and the original features.

Future work will consider other ensemble methods, using ensembles from other methods instead of decision trees and studying the use of non-homogeneous ensemble models. These ensembles are built by combining different methods, (e.g., SVM and Decision Trees) and could improve final model accuracy by specializing each base learner in an area of the problem space. With regard to preprocessing, the use of supervised projections should be tested, as Rotation Forests, a method that uses an internally unsupervised projection, gave some of the best results. A methodology improving accuracy results could also be applied to other industrial problems that predict surface roughness in compliance with Industrial Standard ISO 4288:1996.

**Acknowledgments.** This investigation has been partially supported by the Projects CENIT-2008-1028, TIN2011-24046, IPT-2011-1265-020000 and DPI2009-06124-E/DPI of the Spanish Ministry of Economy and Competitiveness. This work has been made possible thanks to the support received from University of the Basque Country, which provided the laser polishing data and performed all the experimental tests. The authors would especially like to thank Dr. Aitzol Lamikiz and Dr. Eneko Ukar for their kind-spirited and useful advice.

# References

1. Breiman, L.: Heuristics of instability and stabilization in model selection. The Annals of Statistics 24(6), 2350–2383 (1996)
2. Bustillo, A., Díez-Pastor, J., Quintana, G., García-Osorio, C.: Avoiding neural network fine tuning by using ensemble learning: application to ball-end milling operations. The International Journal of Advanced Manufacturing Technology 57(5), 521–532 (2011)
3. Bustillo, A., Sedano, J., Villar, J.R., Curiel, L., Corchado, E.: AI for modelling the laser milling of copper components. In: Fyfe, C., Kim, D., Lee, S.-Y., Yin, H. (eds.) IDEAL 2008. LNCS, vol. 5326, pp. 498–507. Springer, Heidelberg (2008)
4. Bustillo, A., Ukar, E., Rodriguez, J., Lamikiz, A.: Modelling of process parameters in laser polishing of steel components using ensembles of regression trees. International Journal of Computer Integrated Manufacturing 24(8), 735–747 (2011)

5.  Bustillo, A., Villar, A., Gorritxategi, E., Ferreiro, S., Rodríguez, J.J.: Using ensembles of regression trees to monitor lubricating oil quality. In: Mehrotra, K.G., Mohan, C.K., Oh, J.C., Varshney, P.K., Ali, M. (eds.) IEA/AIE 2011, Part I. LNCS, vol. 6703, pp. 199–206. Springer, Heidelberg (2011)

6.  Ciurana, J., Arias, G., Ozel, T.: Neural network modeling and particle swarm optimization (PSO) of process parameters in pulsed laser micromachining of hardened AISI H13 steel. Materials and Manufacturing Processes 24(3), 358–368 (2009)

7.  Díez-Pastor, J., Bustillo, A., Quintana, G., García-Osorio, C.: Boosting projections to improve surface roughness prediction in high-torque milling operations. In: Soft Computing-A Fusion of Foundations, Methodologies and Applications, pp. 1–11 (2012)

8.  Dubey, A., Yadava, V.: Laser beam machining – A review. International Journal of Machine Tools and Manufacture 48(6), 609–628 (2008)

9.  Fayyad, U., Irani, K.: Multi-interval discretization of continuous-valued attributes for classification learning, pp. 1022–1027 (1993)

10.  Ferreiro, S., Sierra, B., Irigoien, I., Gorritxategi, E.: Data mining for quality control: Burr detection in the drilling process. Computers & Industrial Engineering 60(4), 801–810 (2011)

11.  Frank, E., Hall, M.: A simple approach to ordinal classification. In: Flach, P.A., De Raedt, L. (eds.) ECML 2001. LNCS (LNAI), vol. 2167, pp. 145–156. Springer, Heidelberg (2001)

12.  Frank, E., Witten, I.: Making better use of global discretization (1999)

13.  Freund, Y., Schapire, R., et al.: Experiments with a new boosting algorithm. In: Machine Learning-International Workshop, pp. 148–156. Morgan Kaufmann Publishers, Inc. (1996)

14.  Grzenda, M., Bustillo, A., Zawistowski, P.: A soft computing system using intelligent imputation strategies for roughness prediction in deep drilling. Journal of Intelligent Manufacturing, 1–11 (2012)

15.  Hall, M., Frank, E., Holmes, G., Pfahringer, B., Reutemann, P., Witten, I.: The WEKA data mining software: an update. ACM SIGKDD Explorations Newsletter 11(1), 10–18 (2009)

16.  International Organization for Standardization: ISO-4288. Geometrical Product Specifications (GPS): Rules and procedures for the assessment of surface texture (1996)

17.  International Organization for Standardization: ISO-4287. Geometrical Product Specifications (GPS) — Surface texture: Profile method — Terms, definitions and surface texture parameters (1997)

18.  Kohavi, R., Wolpert, D., et al.: Bias plus variance decomposition for zero-one loss functions. In: Machine Learning-International Workshop, pp. 275–283. Morgan Kaufmann Publishers Inc., San Francisco (1996)

19.  Kramer, S., Widmer, G., Pfahringer, B., Groeve, M.: Prediction of ordinal classes using regression trees. Fundamenta Informaticae 47(1-2), 1–13 (2001)

20.  Kuncheva, L.: Combining pattern classifiers: methods and algorithms. Wiley-Interscience (2004), http://books.google.es/books?id=9TJ6igZtqWAC

21.  Lamikiz, A., Sanchez, J., Lopez de Lacalle, L., Arana, J.: Laser polishing of parts built up by selective laser sintering. International Journal of Machine Tools and Manufacture 47(12), 2040–2050 (2007)

22.  Lü, L., Fuh, J., Wong, Y.: Laser-induced materials and processes for rapid prototyping. Springer (2001)

23.  Nadeau, C., Bengio, Y.: Inference for the generalization error. Machine Learning 52(3), 239–281 (2003)

24.  Quinlan, J.R.: C4.5: Programs for Machine Learning. Morgan Kaufmann (1993)

25.  Quintana, G., De Ciurana, J., Ribatallada, J.: Surface roughness generation and material removal rate in ball end milling operations. Materials and Manufacturing Processes 25(6), 386–398 (2010)

26. Rodríguez, J., Kuncheva, L., Alonso, C.: Rotation forest: A new classifier ensemble method. IEEE Transactions on Pattern Analysis and Machine Intelligence 28(10), 1619–1630 (2006)
27. Santos, P., Villa, L.F., Reñones, A., Bustillo, A., Maudes, J.: Wind turbines fault diagnosis using ensemble classifiers. In: Perner, P. (ed.) ICDM 2012. LNCS, vol. 7377, pp. 67–76. Springer, Heidelberg (2012), doi:10.1007/978-3-642-31488-9_6
28. Tuck, C., Hague, R., Ruffo, M., Ransley, M., Adams, P.: Rapid manufacturing facilitated customization. International Journal of Computer Integrated Manufacturing 21(3), 245–258 (2008)
29. Vapnik, V.: The nature of statistical learning theory. Springer (1999)
30. Wu, X., Kumar, V.: The top ten algorithms in data mining, vol. 9. Chapman & Hall/CRC (2009)

# A Study on the Use of Machine Learning Methods for Incidence Prediction in High-Speed Train Tracks

Christoph Bergmeir[1], Gregorio Sáinz[2,3],
Carlos Martínez Bertrand[4], and José Manuel Benítez[1]

[1] Department of Computer Science and Artificial Intelligence,
CITIC-UGR, University of Granada, Spain
[2] CARTIF Centro Tecnológico, Parque Tecnológico de Boecillo 205,
47151 Valladolid, Spain
[3] University of Valladolid, Systems Engineering and Control Department,
School of Industrial Engineering, 47011 Valladolid, Spain
[4] VÍAS Y CONSTRUCCIONES, S.A. (Grupo ACS)
{c.bergmeir,j.m.benitez}@decsai.ugr.es
http://dicits.ugr.es

**Abstract.** In this paper a study of the application of methods based on Computational Intelligence (CI) procedures to a forecasting problem in railway maintenance is presented. Railway maintenance is an important and long-standing problem that is critical for safe, comfortable and economic transportation. With the advent of high-speed lines, the problem has even more importance nowadays. We have developed a study, applying forecasting procedures from Statistics and CI, to examine the feasibility of predicting one-month-ahead faults on two high-speed lines in Spain. The data are faults recorded by a measurement train which traverses the lines monthly. The results indicate that CI methods are competitive in this forecasting task against the Statistical regression methods, with $\epsilon$-support vector regression outperforming the other employed methods. So, application of CI methods is feasible in this forecasting task and it is useful in the planning process of track maintenance.

**Keywords:** time series, machine learning, railway maintenance, prediction, track maintenance.

## 1 Introduction

Since the early beginnings of railway transport, track maintenance is a central problem in the railway sector. The rails are continuously exposed over years to harsh weather and temperature conditions. In addition, the trains traveling on them cause periodical strong pressures and forces. This leads to fatigue and faults, which could finally result in failure and line breaks. Omitted maintenance results in a low quality of service, uncomfortability of the users, delays, and, in general, a loss of competitiveness for companies involved in the railway system.

M. Ali et al. (Eds.): IEA/AIE 2013, LNAI 7906, pp. 674–683, 2013.

Throughout the decades, rails have undergone constant enhancement, from the early days when cast iron or even wood was used as rails, to steel with specialized production techniques, specialized welding techniques, etc. Nonetheless, though the rail has undergone constant enhancement, also the requirements have grown constantly, with modern high-speed tracks where trains pass at over 300 km/h, and with tracks where freight trains pass with axle loads of up to 30 tons and more. The interaction of the wheel with the rail is complex, and the faults that may occur are diverse [2]. Another important issue is that there is usually no redundancy of tracks, i.e., if a track needs to be repaired, no trains can pass during this time (compare this to the trains, where simply another train can be used if one breaks down).

So, railway maintenance is a central task of every railway infrastructure company, and a main economic factor, causing high costs. Nowadays, with opened railway markets in most European countries, it becomes even more important to maintain tracks economically, so that the tracks can be used safely, with high comfort, and at high velocities, during a high percentages of the time.

In order to guarantee proper functioning of the track network, periodic rail inspections are undertaken. A specialized measurement train travels on the track and takes continuous measurements of parameters of the track, using measurement techniques such as ultrasound [9], radiography, Eddy current testing, visual inspection, and acceleration detection devices [2]. The faults that are detected may be of different severity, which especially means that the faults are closely connected to the train speed on the track. An irregularity in the rail may be no problem at slow speeds, but may cause problems at higher speeds. So, on high-speed tracks, rail maintenance is especially important, as irregularities in the rail yield rapidly to speed limitations, causing delays and hindering proper functioning of the railway traffic.

In this work, our aim is to use data from a measurement train to predict future faults or the railway infrastructure. With proper predictions, maintenance can be performed before the problem in the rail actually arises. So, maintenance can be managed in a better way, with the final goal to have an optimized maintenance procedure. Here, it is essential to have reliable forecasts of the expected faults available.

Some work can be found in the literature where statistical methods are applied in related problems, in order to improve the maintenance procedure. Zhao et al. [13] present a "probabilistic model for predicting rail breaks and controlling risk of derailment," and Dick et al. [5] propose a "multivariate statistical model for predicting occurrence and location of broken rails." Furthermore, Podofillini et al. [10] use Computational Intelligence (CI) methods, namely a multi-objective genetic algorithm, to optimize a maintenance process with ultrasonic inspection trains. Besides general optimization tasks, also for general prediction, CI methods have become increasingly popular throughout the last years [4].

Concretely, the problem to be solved consists of predicting the number of incidents in two Spanish high-speed lines. For this, our aim is not only to predict the number of global incidents in each pathway, but we also perform local predictions for different regions of the track. These regions will be called *sections* in the following. The sections are obtained by partitioning the track at different levels of granularity. Lengths of 10, 20, and 30 km are considered. The purpose of the sections is not only to perform global, but also local predictions, so that each local maintenance station has forecasts available for their own planning. We perform a study applying several well-known CI methods to this prediction problem, thus assessing the feasibility of the use of such methods in this particular forecasting problem.

The rest of the paper is organized as follows: Section 2 presents the maintenance framework and the forecasting problem in detail. Section 3 describes the applied models, and the way they are used. Section 4 presents the empirical study and the results, and Section 5 concludes the paper.

## 2    Maintenance Framework and Forecasting Problem

The data we use are obtained from two Spanish high-speed lines, which due to confidentiality reasons cannot be identified. We call them A and B in the following. As there are tracks for both directions, this gives us four tracks in total. The tracks are periodically controlled with a measurement train for incidents of certain types. The train undertakes measurements of the following types:

- Bogie acceleration
- Lateral body acceleration
- Vertical body acceleration
- Vertical axle box 1 acceleration
- Vertical axle box 2 acceleration

If one of these variables reaches a value outside their normal working range, it is considered an event to be recorded, together with the kilometer, speed, and current date, where it was detected. The available data are measurements acquired along 12.5 years (138 months) for the B, and 4 years (50 months) for the A track. The time interval between two measurements varies from about three weeks to one month.

Based on the available data, for this work we developed a set of time series where each value $x(t)$ is the number of incidents in a given month $t$. For our experiments, we use the following concepts to construct time series for the prediction purposes:

**Line.** We distinguish two large sets of time series, corresponding to the entire lines A and B.

**Track.** Within each line, two large time series representing the entire tracks are used.

**Section.** The tracks are divided into sections of different granularity, with lengths of 10, 20, and 30 km, to make predictions about each section independently. The A line is about 200 km in length, B is more than double.

**Type of Incidence.** The five types of incidents are considered collectively and separately, so that all time series for the whole tracks and the sections exist in six different versions.

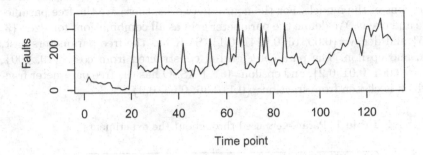

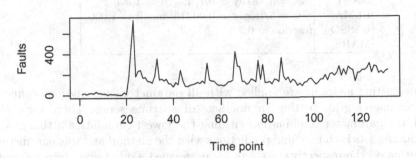

**Fig. 1.** Monthly recorded faults (all types of faults) by the measurement train for the two B tracks, during the period of 138 months

In total, there are 1476 time series in consideration. Fig. 1 shows the series of all fault types for the whole tracks of B.

## 3   Forecasting Models

This section describes the different CI methods employed, and the ways they are used in our study. All experiments are performed using the R programming language [11]. We use an implementation of a multi-layer perceptron with back-propagation learning, available in the package RSNNS [1]. This method will be called MLP in the following. Furthermore, we use a multi-layer perceptron trained with the BFGS algorithm, present in the nnet package [12] (we call this method NNET in the following), and an implementation of Elman recurrent networks [7] (ELMAN), also available in RSNNS. Apart from the different neural network types, we also use $\epsilon$-support vector regression (SVR), present in the package e1071, which wraps an SVR standard implementation [3]. For comparison purposes, we use two statistical autoregression models, namely lasso regression as proposed by Efron et al. [6] (LASSO), and multi-adaptive regression splines (MARS), as proposed by Friedman [8].

All of the methods but MARS have free parameters to choose. Therefore, we define parameter grids for each of the methods, containing typical parameter combinations of the models. For the neural networks, there are the free parameters size and decay. We define the parameter grid as all combinations for size={3, 5, 9, 15} and decay={0.00316, 0.0147, 0.1}. SVR has the free parameters cost, gamma, and epsilon. The parameter grid is constructed from cost={10, 100}, gamma={0.001, 0.01, 0.2}, and epsilon=0.1. LASSO has the free parameter fraction, which is chosen from fraction={0.1, 0.36, 0.63, 0.9}.

**Table 1.** Parameters used throughout the experiments

| Algorithm | Parameters |
|-----------|------------|
| SVR | cost = 100, gamma = 0.001, epsilon = 0.1 |
| MLP | size = 15, decay = 0.00316, maxit = 1,000 |
| NNET | size = 3, decay = 0.1, maxit = 1,000 |
| ELMAN | size = 5, decay = 0.00316, maxit = 1,000 |
| LASSO | fraction = 0.9 |
| MARS | – |

All forecasting methods are applied with all parameter combinations defined in the parameter grids to the training sets of all time series. Then, for each method, the parameter combination yielding the lowest cumulative fitting error on all training sets is chosen, and predictions with the method and this parameter combination on the respective test sets are performed. Only taking into account the fitting error for model selection might cause over-learning, but as the amount of data is limited, withholding another amount of data as a validation set for parameter estimation would lead to even less data being available, so we prefer to control for over-training by limiting the number of iterations in the learning algorithms, and to choose one parameter combination that is used for all time series. The parameters that were chosen in this way and finally used are shown in Tab. 1.

## 4   Empirical Study

This section presents the experiments carried out for the prediction of events on the A and B lines. The forecasting methods presented in Section 3 have been applied to the time series described in Section 2.

### 4.1   Experimental Setup

The series are monthly fault recordings, and forecasting is performed one-step-ahead, i.e., the forecasting target is the next month. So, maintenance planning is supported in this way by insights for next months requirements. Given that the series are monthly, reasonable embedding dimensions are yearly (we use the past

11 values for forecasting) and quarterly (we use the past 3 values for forecasting). But taking into account the short length of the series, especially for the A case where the series only cover 4 years, forecasting with yearly embedding leads not to satisfactory results, so that in the following we use quarterly embedding. The series are partitioned into training and test set, where we use always the last 12 values (i.e., one year) as a test set. In this way, though the time period of available data is much shorter for the A track, we keep the test set at a constant size.

Errors on the test sets are computed using the root mean squared forecaster error (RMSFE), defined as follows.

$$\text{RMSFE} = \sqrt{\frac{1}{n} \sum_{t=1}^{n} (y_{t+h} - \hat{y}_{t+h})^2},$$

where $y_{t+h}$ is the target value, and $\hat{y}_{t+h}$ is the forecast. The variable $t$ runs over all time points in the test set, and $n$ is the overall amount of values in the test set. The horizon $h$ is one in our case.

Other error measures commonly applied in forecasting such as relative or percentage measures (MAPE, sMAPE), which normalize by values from the series cannot be applied in our case, as a typical value in the data is "zero", when no incidents are reported. Instead, the series are normalized to zero mean and unitary variance. In this way, it becomes possible to compare and use the RMSFEs across series in the analysis of the results.

During the experiments, we do not take into account the time series that do not provide enough information for applying the forecasting methods, e.g., series of a single type of incident and which only take into account a small section of 10 km may be just a constant zero, as this type of incident simply never occurred in that section during the considered time period. The best possible forecast for such a series is a constant zero, and no complex forecasting methods need to be applied.

## 4.2   Results

This section presents the results obtained by calculating RMSFEs on the test sets, for each method with the parameters adjusted in the way discussed in Section 3. Separate studies are performed for the series encompassing complete tracks on the one hand, and for series of track sections, on the other hand.

Tab. 2 and Fig. 2 show the results. Fig. 2 shows box plots for the RMSFE obtained with each of the six models used, for the whole A and B tracks, respectively. The boxes show the first and third quartiles, and the middle line is the median. The whiskers extend to the maximal values, but are no longer than 1.5 times the size of the box. Values that are further off are shown as outliers.

From Tab. 2, we can see that ELMAN and SVR greatly outperform the other methods, and especially the statistical models, both in terms of ranking as in terms of averaged RMSFE. Especially for the data of line A, SVR performs well. Fig. 2 illustrates and confirms these findings.

**Table 2.** RMSFE results for the forecasting of whole track faults. Where b: Bogie acceleration, ab1: axle box 1 acceleration, ab2: axle box 2 acceleration, l: lateral body acceleration, v: vertical body acceleration, all: all types of faults. Mean is the averaged RMSFE across all series, and rank is the averaged rank.

|  |  | MARS | LASSO | NNET | ELMAN | MLP | SVR |
|---|---|---|---|---|---|---|---|
| Line A, Track 1 | b | 0.55 | 0.62 | 1.18 | 0.54 | 0.94 | **0.47** |
|  | ab1 | **1.07** | 1.44 | 1.90 | 1.22 | 1.43 | 1.19 |
|  | ab2 | 1.43 | 1.38 | 2.03 | **1.11** | 1.32 | 1.17 |
|  | l | 0.64 | 0.56 | 0.54 | **0.50** | 0.64 | 0.51 |
|  | v | 0.43 | 0.50 | 0.29 | 0.46 | 0.41 | **0.26** |
|  | all | 1.76 | 1.22 | 1.02 | **0.99** | 1.39 | 1.01 |
| Line A, Track 2 | b | **0.54** | 0.75 | 0.63 | 0.55 | 0.59 | 0.54 |
|  | ab1 | 1.69 | 2.74 | 1.44 | 0.88 | 1.67 | **0.78** |
|  | ab2 | 0.79 | 1.24 | 0.77 | 0.48 | 0.91 | **0.45** |
|  | l | 3.07 | 1.48 | 3.24 | 1.27 | 2.36 | **1.07** |
|  | v | 0.27 | **0.20** | 0.27 | 0.28 | 0.31 | 0.23 |
|  | all | 1.06 | 1.27 | 0.74 | **0.63** | 1.01 | 0.78 |
| Line B, Track 1 | b | 0.98 | 1.03 | 1.00 | **0.97** | 0.99 | 1.11 |
|  | ab1 | 0.91 | 0.73 | 0.88 | 0.75 | 0.99 | **0.68** |
|  | ab2 | 0.82 | 0.80 | **0.63** | 0.63 | 0.86 | 0.71 |
|  | l | **0.26** | 0.27 | 0.28 | 0.29 | 0.30 | 0.26 |
|  | v | **0.74** | 0.77 | 1.05 | 0.84 | 0.80 | 0.83 |
|  | all | 1.29 | **0.66** | 0.84 | 0.99 | 0.87 | 0.66 |
| Line B, Track 2 | b | 0.58 | **0.40** | 0.48 | 0.43 | 0.48 | 0.41 |
|  | ab1 | 0.69 | 0.62 | 0.53 | **0.51** | 0.54 | 0.53 |
|  | ab2 | 0.55 | 0.53 | **0.41** | 0.55 | 0.54 | 0.52 |
|  | l | 0.28 | 0.29 | 0.23 | 0.24 | 0.28 | **0.16** |
|  | v | **0.42** | 0.45 | 0.44 | 0.49 | 0.59 | 0.59 |
|  | all | 0.53 | 0.52 | 0.53 | 0.57 | 0.56 | **0.50** |
|  | mean | 1.01 | 0.97 | 0.99 | 0.76 | 1.01 | **0.70** |
|  | rank | 4.12 | 3.81 | 3.54 | 2.88 | 4.54 | **2.12** |

We see from Fig. 1, that the amount of faults for line B ranges from 0 to over 300. The averaged normalized RMSFE of SVR in Tab. 2 is 0.7, which leads to a denormalized error of 19.7 faults.

Tab. 3 shows the results for the sections of 30, 20, and 10 km, and for the A and B tracks, respectively. We see from Tab. 3 that the ELMAN method does not perform so good any more, and instead LASSO performs better. The good performance of the SVR is still present, being the best method in terms of averaged RMSFE in 3 out of 6 cases, and in terms of averaged rank in 4 out of 6 cases. So, for this forecasting task, the SVR seems to be the most adequate method, outperforming in many cases the statistical methods and the other CI methods.

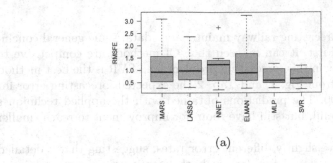

(a)

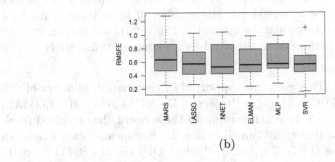

(b)

**Fig. 2.** Box plots of the results of the experiments. (a): Line A, faults of complete track (b): Line B, complete track.

**Table 3.** Results for the sections of length 30, 20, and 10 km. Averaged mean and averaged rank for all time series are shown.

|  |  | MARS | LASSO | NNET | ELMAN | MLP | SVR |
|---|---|---|---|---|---|---|---|
| **Mean** |  |  |  |  |  |  |  |
| Line A, Sections | 30 km | 1.06 | 1.07 | 1.14 | 1.39 | 0.82 | **0.80** |
|  | 20 km | 0.92 | 0.93 | 1.01 | 1.14 | 0.76 | **0.73** |
|  | 10 km | 0.86 | 0.89 | 0.98 | 1.04 | 0.73 | **0.69** |
| Line B, Sections | 30 km | 0.90 | **0.87** | 0.98 | 0.92 | 1.02 | 0.87 |
|  | 20 km | 0.93 | **0.90** | 0.99 | 0.95 | 1.02 | 0.90 |
|  | 10 km | 0.93 | **0.92** | 1.00 | 0.98 | 1.05 | 0.93 |
| **Ranks** |  |  |  |  |  |  |  |
| Line A, Sections | 30 km | 3.67 | 4.01 | 4.03 | 4.44 | 2.58 | **2.26** |
|  | 20 km | 3.77 | 3.89 | 4.14 | 4.12 | 2.78 | **2.30** |
|  | 10 km | 3.56 | 4.00 | 4.21 | 4.05 | 2.90 | **2.28** |
| Line B, Sections | 30 km | 3.18 | **2.82** | 4.16 | 3.43 | 4.33 | 3.07 |
|  | 20 km | 3.24 | **3.04** | 4.00 | 3.41 | 4.24 | 3.07 |
|  | 10 km | 3.11 | 3.20 | 3.92 | 3.56 | 4.23 | **2.98** |

# 5   Conclusions

After this study on forecasting railway maintenance data, some general conclusions can be drawn. First, it can be seen that CI methods are competitive to the statistical methods in this forecasting problem, and SVR is the best method for this task. We see from Fig. 1, and Tab. 2 the expected forecasting error for line B is 19.7 over 300. The predictions obtained with the applied techniques are acceptable and useful, but still have room for improvement to reach smaller errors.

Also, both lines have slightly different error rates, suggesting that a detailed study of the characteristics of the data in both lines is necessary to minimize the error. Data preprocessing techniques could be further explored to enhance the results. Another issue to consider is the number of lagged values used for prediction. Expert knowledge may be critical in this effort, to complement the massive data analysis performed in this study with computational tools.

**Acknowledgements.** This work was supported by the Spanish Ministry of Science and Innovation (MICINN) under Projects TIN-2009-14575 and GEOMAF, IPT-2011-1656-370000. The research leading to these results has received funding from the European Research Council under the European Union's Seventh Framework Programme (FP7/2007-2013), Project OPTIRAIL, Ref. FP7-SST-2012-RTD-1, ERC grant agreement no. [314031]. C. Bergmeir holds a scholarship from the Spanish Ministry of Education (MEC) of the "Programa de Formación del Profesorado Universitario (FPU)".

# References

1. Bergmeir, C., Benítez, J.M.: Neural Networks in R Using the Stuttgart Neural Network Simulator: RSNNS. Journal of Statistical Software 46(7), 1–26 (2012)
2. Cannon, D.F., Edel, K.-O., Grassie, S.L., Sawley, K.: Rail defects: An overview. Fatigue and Fracture of Engineering Materials and Structures 26(10), 865–886 (2003)
3. Chang, C.-C., Lin, C.-J.: LIBSVM: a library for support vector machines (2001), Software available at http://www.csie.ntu.edu.tw/~cjlin/libsvm
4. Crone, S.F., Hibon, M., Nikolopoulos, K.: Advances in forecasting with neural networks? empirical evidence from the nn3 competition on time series prediction. International Journal of Forecasting 27(3), 635–660 (2011)
5. Dick, C.T., Barkan, C.P.L., Chapman, E.R., Stehly, M.P.: Multivariate statistical model for predicting occurrence and location of broken rails (2003)
6. Efron, B., Hastie, T., Johnstone, I., Tibshirani, R., Ishwaran, H., Knight, K., Loubes, J.-M., Massart, P., Madigan, D., Ridgeway, G., Rosset, S., Zhu, J.I., Stine, R.A., Turlach, B.A., Weisberg, S., Hastie, T., Johnstone, I., Tibshirani, R.: Least angle regression. Annals of Statistics 32(2), 407–499 (2004)
7. Elman, J.L.: Finding structure in time. Cognitive Science 14(2), 179–211 (1990)
8. Friedman, J.H.: Multivariate adaptive regression splines. The Annals of Statistics 19, 1–67 (1991)

9. Highs, D., Coal, P.: The potential of ultrasonic surface waves for rail inspection. In: AIP Conference Proceedings, vol. 760, pp. 227–234 (2005)
10. Podofillini, L., Zio, E., Vatn, J.: Risk-informed optimisation of railway tracks inspection and maintenance procedures. Reliability Engineering and System Safety 91(1), 20–35 (2006)
11. R Development Core Team. R: A Language and Environment for Statistical Computing. R Foundation for Statistical Computing, Vienna, Austria (2009)
12. Venables, W.N., Ripley, B.D.: Modern Applied Statistics with S, 4th edn. Springer, New York (2002) ISBN 0-387-95457-0
13. Zhao, J., Chan, A.H.C., Burrow, M.P.N.: Probabilistic model for predicting rail breaks and controlling risk of derailment (2007)

# Estimating the Maximum Shear Modulus with Neural Networks

Manuel Cruz[1,*], Jorge M. Santos[1], and Nuno Cruz[2]

[1] DMA - School of Engineering, Polytechnic of Porto and LEMA, Portugal
[2] Mota-Engil and Universidade de Aveiro, Portugal
{mbc,jms}@isep.ipp.pt, nbdfcruz@gmail.com

**Abstract.** Small strain shear modulus is one of the most important geotechnical parameters to characterize soil stiffness. In-situ stiffness of soils and rocks is much higher than was previously thought as finite element analysis have shown. Also, the stress-strain behaviour of those materials is non-linear in most cases with small strain levels. The commun approach for getting the small strain shear modulus is usually based on measure of seismic wave velocities. Nevertheless, for design purposes is very useful to derive that modulus from correlations with in-situ tests output parameters. In this view, the use of Neural Networks seems very appropriate as the complexity of the system keeps the problem very unfriendly to treat following traditional data analysis methodologies. In this work, the use of Neural Networks is proposed to estimate small strain shear modulus for sedimentary soils from the basic or intermediate parameters derived from Marchetti Dilatometer Test.

## 1 Introduction

Maximum shear modulus, $G_0$, is nowadays a key geotechnical parameter in soil stiffness evaluation. The standard way to measure it is to evaluate compression and shear wave velocities and thus obtain results supported by theoretical interpretations. Despite the advantages appointed by the scientific community (e.g. [1, 2] ), this approach has a drawback that is mainly appointed by the industrial counterpart: the use of seismic measures implies a specific and more expensive test than the ones in old-fashioned way. As a result, many authors have dedicated their efforts to correlate other in-situ test parameters with $G_0$. Among others, the works from Peck, Lunne, Marchetti or Cruz do it for the Standard Penetration Test (SPT) [3], Piezocone Test (CPTu) [4] or Marchetti Dilatometer Test (DMT) [5–7].

In this context, the DMT seems a very appropriate equipment to accomplish that task with success. That may be explained as follows:

1. DMT measure a load range related with a specific displacement ($E_D$)
2. $E_D$ may be used to deduce highly accurate stress-strain relationship, supported by the Theory of Elasticity

---

* The author was partially supported by both DMA and LEMA.

M. Ali et al. (Eds.): IEA/AIE 2013, LNAI 7906, pp. 684–693, 2013.

3. The type of soil can be numerically represented by DMT Material Index, $I_D$
4. The in situ density, overconsolidation ratio (OCR) and cementation influences can be represented by lateral stress index, $K_D$

which allows for high quality calibration of the stress-strain relationship [7].

In this paper, an estimation of $G_0$ derived from the DMT basic and intermediate parameters using neural networks is presented.

## 2 $G_0$ Prediction by DMT

Marchetti dilatometer test, commonly designated by DMT, has been increasingly used and it is one of the most versatile tools for soil characterization. The test was developed by Silvano Marchetti [5] and can be seen as a combination of both Piezocone and Pressuremeter tests with some details that really makes it a very interesting test available for modern geotechnical characterization [7]. The main reasons for its usefulness on deriving geotechnical parameters are related to the simplicity and the speed of execution generating quasi-continuous data profiles with high accuracy and reproducibility.

In its essence, dilatometer is a stainless steel flat blade with a flexible steel membrane in one of its faces. The blade is connected to a control unit on the ground surface by a pneumatic-electrical cable that goes inside the position rods, ensuring electric continuity and the transmission of the gas pressure required to expand the membrane. The equipment is pushed (most preferable) or driven into the ground, by means of a CPTu rig or similar, and the expansion test is performed every 20cm. The (basic) pressures required for lift-off the diaphragm ($P_0$), to deflect 1.1mm the centre of the membrane ($P_1$) and at which the diaphragm returns to its initial position ($P_2$ or closing pressure) are recorded. Due to the balance of zero pressure measurement method (null method), DMT readings are highly accurate even in extremely soft soils, and at the same time the blade is robust enough to penetrate soft rock or gravel. The test is found especially suitable for sands, silts and clays.

Four intermediate parameters, Material Index ($I_D$), Dilatometer Modulus ($E_D$), Horizontal Stress Index ($K_D$) and Pore Pressure Index ($U_D$), are deduced from the basic pressures $P_0$, $P_1$ and $P_2$, having some recognizable physical meaning and some engineering usefulness [5], as it will be discussed below. The deduction of current geotechnical soil parameters is obtained from these intermediate parameters covering a wide range of possibilities. In the context of the present work, besides the basic pressures, only $E_D$, $I_D$ and $K_D$ have a physical meaning on the determination of $G_0$, so they will be succinctly described as follows [7]:

1. Material Index, $I_D$: Marchetti [5] defined Material Index, $I_D$, as the difference between $P_1$ and $P_0$ basic measured pressures normalized in terms of the effective lift-off pressure. In a simple form, it could be said that $I_D$ is a "fine-content-influence meter" [7], providing the interesting possibility of defining dominant behaviours in mixed soils.

2. Horizontal Stress Index, $K_D$: The horizontal stress index [5] was defined to be comparable to the at rest earth pressure coefficient, $K_0$, and thus its determination is obtained by the effective lift-off pressure ($P_0$) normalized by the in-situ effective vertical stress. $K_D$ is a very versatile parameter since it provides the basis to assess several soil parameters such as those related with state of stress, stress history and strength, and shows dependency on several factors namely cementation and ageing, relative density, stress cycles and natural overconsolidation resulting from superficial removal, among others.

3. Dilatometer Modulus, $E_D$: Stiffness behaviour of soils is generally represented by soil moduli, and thus the base for in-situ data reduction. Generally speaking, soil moduli depend on stress history, stress and strain levels drainage conditions and stress paths. The more commonly used moduli are constrained modulus ($M$), drained and undrained compressive Young modulus ($E_0$ and $E_u$) and small-strain shear modulus ($G_0$), this latter being assumed as purely elastic and associated to dynamic low energy loading.

Maximum shear modulus, $G_0$, is indicated by several investigators [2, 7, 10] as the fundamental parameter of the ground. It can be accurately deduced through shear wave velocities,

$$G_0 = \rho v_s^2 \tag{1}$$

where $\rho$ stands for density and $v_s$ for shear wave velocity.

However, the use of a specific seismic test imply an extra cost, since it can only supply this geotechnical parameter, leaving strength and insitu state of stress information dependent on other tests. Therefore, several attempts to model the maximum shear modulus as a function of DMT intermediate parameters for sedimentary soils have been made in the last decade. Hryciw [11] proposed a methodology for all types of sedimentary soils, developed from indirect method of Hardin & Blandford [12]. This methodology ignores dilatometer modulus, $E_D$, commonly recognized as a highly accurate stress-strain evaluation, and also lateral stress index, $K_D$, and material index, $I_D$, which are the main reasons for the accuracy in stiffness evaluation offered by DMT tests [6]. Being so, the most common approaches [13–15] with reasonable results concentrated in correlating directly $G_0$ with $E_D$ or $M_{DMT}$ (constrained modulus), which have revealed linear correlations with slopes controlled by the type of soil. In 2006, Cruz [6] proposed a generalization of this approach, trying to model the ratio $R_G \equiv \frac{G_0}{E_D}$ as a function of $I_D$. In 2008, Marchetti [16] using the commonly accepted fact that maximum shear modulus is influenced by initial density and considering that this is well represented by $K_D$, studied the evolution of both $R_G$ and $G_0/M_{DMT}$ with $K_D$ and found different but parallel trends as function of type of soil (that is $I_D$), recommending the second ratio to be used in deriving $G_0$ from DMT, as consequence of a lower scatter. In 2010, using the Theory of Elasticity, Cruz [7] approximate $G_0$ as a non-linear function of $I_D$, $E_D$ and $K_D$, from where a promising median of relative errors close to 0.21 with a mean(standard deviation) around 0.29(0.28) were obtained. It is worth mention that comparing with the previous approach - $R_G$ - this approximation, using the same data, lowered the mean and median of relative errors in more than 0.05 maintaining the standard deviation (Table 2).

In this work, to infer about the results quality it will be used some of the same indicators used by Hryciw, Cruz and others that are: the median, the arithmetic mean and standard deviation of the relative errors

$$\delta^i_{\widetilde{G_0}} = \frac{|\widetilde{G_0}(i) - G_0(i)|}{|G_0(i)|}; i = 1, 2, ..., N \tag{2}$$

where $\widetilde{G_0}(i)$ stands for the predicted value and $G_0(i)$ for the measured value given by seismic wave velocities (which is assumed to be correct). A final remark to point out that since in this work the no-intercept regression is sometimes used, the $R^2$ values will not be presented as they can been meaningful in this case [17]. It is also worth to remark that in the context of DMT and from the engineering point of view, median is the parameter of choice for assessing the model quality [7] since the final value for maximum shear modulus relies on all set of results obtained in each geotechnical unit or layer.

## 3    Data Sets, Experiments and Results

### 3.1    The WDS and PsS Data Sets

In the forthcoming experiments there was used one subset of the WDS data set named PsS data set. The WDS data set was used in the development of the non-linear $G_0$ approximation done by Cruz in [7], resulting from 860 DMT measurements performed in Portugal by Cruz and world wide by Marchetti et al. [16] (data kindly granted by Marchetti for the work presented in [7]), which included data obtained in all kinds of sedimentary soils, namely clays, silty clays, clayey silts, silts, sandy silts, silty sands and sands. Afterwards was used again as base for the work by Cruz et al [8] where the DMT intermediate parameters are used to estimate $G_0$. Since the Marchetti data does not include the record of the basic parameters, $P_0$, $P_1$ and $u_0$, only the Portuguese subset (denoted by PsS) will be used when trying to predict $G_0$ from those parameters.

In order to have some comparisons between the present work and the one made in [8], the WDS main statistical measures with respect to $I_D$, $E_D$, $K_D$ and $G_0$ parameters are given in Table 1 (in parenthesis the same measures for PsS). Figures 1 and 2, where data from WDS and PsS, respectively, is represented using MatLab function *plotmatrix*, aims a clear view of variables dispersion. This is important in a Geotechnical point of view as it may show the (very) different types of soil who serve has base to this work. It should be noted that in Figure 2 the additional parameters $P_0$, $P_1$ and $u_0$ are presented.

### 3.2    $G_0$ Prediction by DMT Parameters: A NN Approach

In addition to the work reviewed in Section 2, in 2011, Cruz *et al* [8] went a little further reported the fitting of $G_0$ through the DMT intermediate parameters $E_D$, $I_D$ and $K_D$ based on the use of different types of Least Square Non-Linear Regression and Neural Networks (NN). Using the WDS dataset, an attempt to

**Table 1.** Sample WDS (PsS) statistical measures rounded to 4 significant digits

| Values | $I_D$ | $E_D$ | $K_D$ | $G_0$ |
|--------|-------|-------|-------|-------|
| min | 0.05070 (0.05070) | 0.3644 (0.3644) | 0.9576 (0.9576) | 6.430 (12.71) |
| max | 8.814 (8.814) | 94.26 (85.00) | 24.61 (24.61) | 529.2 (110.6) |
| median | 0.5700 (0.2192) | 13.44 (4.372) | 3.575 (3.136) | 77.91 (34.51) |
| mean | 0.9134 (1.063) | 18.83 (9.963) | 4.916 (3.808) | 92.52 (38.81) |
| std | 1.074 (1.946) | 18.83 (13.08) | 3.608 (2.791) | 69.61 (19.37) |

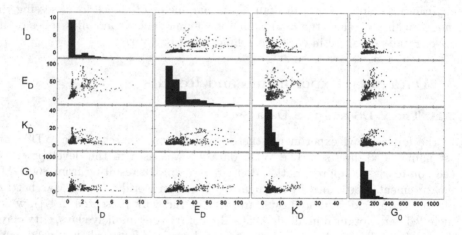

**Fig. 1.** Sample WDS: values for $I_D, E_D, K_D$ and $G_0$

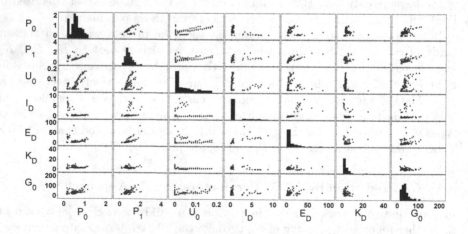

**Fig. 2.** Sample PsS: values for $P_0, P_1, u_0, I_D, E_D, K_D$ and $G_0$

improve the quality of these results was carried out by using Support Vector Regression (SVR). Support Vector Machines [20] are based on the statistical learning theory from Vapnik and are specially suited for classification. However, there are also algorithms based in the same approach for regression problems known as Support Vector Regression. The performed experiments with SVRs were carried out using LIBSVM [21] for Matlab. Two different kinds of SVR algorithms: $\epsilon$-SVR, from Vapnik [22] and $\nu$-SVR from Schölkopf [23] were applied, which differ in the fact that $\nu$-SVR uses an extra parameter $\nu \in (0, 1]$ to control the number of support vectors. For these experiments a search for the best results was made in the $C$, $\epsilon$ ($\nu$) space and so different values for the parameter $C$ (cost) and for parameters $\epsilon$ and $\nu$ were used.

The best results obtained with both $\epsilon$-SVR and $\nu$-SVR with the radial basis function kernel reveal slightly better results when compared with those obtained with the fitting neural network and better than those obtained with the other MLP's and the traditional regression algorithms.

In order to have an easier reading of the present paper, a summary of the results achieved in [8] is presented in Tables 2 and 3.

**Table 2.** Sample WDS: Relative Error Results (Median/Mean(std)) obtained with $\widetilde{G_0} = f(I_D, E_D, K_D)$ [8]

| | Type | Hidden neurons | Median/Mean(std) |
|---|---|---|---|
| Non-Linear Regression | $G_0 = \alpha E_D (I_D)^{\beta}$ | - | 0.28/0.34(0.29) |
| | $G_0 = E_D + E_D e^{(\alpha + \beta I_D + \gamma \log(K_D))}$ | - | 0.21/0.29(0.28) |
| | Quasinewton | 50 | 0.20/0.38(0.72) |
| | Conj.Grad. | 100 | 0.19/0.30(0.38) |
| Neural | SCG | 40 | 0.20/0.28(0.33) |
| Networks | MLP-Bayesian | 20 | 0.20/0.29(0.30) |
| | RBF | 200 | 0.20/0.31(0.39) |
| | Fitting | 60 | 0.17/0.27(0.29) |

**Table 3.** Sample WDS: Relative Error Results (Median/Mean(std)) obtained with Support Vector Regression $\widetilde{G_0} = f(I_D, E_D, K_D)$

| Type | Cost/$\epsilon(\nu)$ | Median/Mean(std) |
|---|---|---|
| $\epsilon$-SVR | 200/0.1 | 0.16/0.27(0.43) |
| $\nu$-SVR | 200/0.8 | 0.16/0.27(0.41) |

Despite all the work reviewed in Section 2 it hasn't been already tried to model $G_0$ as a straightforward function of the DMT basic parameters $P_0$, $P_1$ and $P_2$. In addition, the promising results showed in Table 3 led the authors to go further and to try that approach. However, there are some difficulties in the interpretation of $P_2$ values, since it can represent very distinctive situations in different type of soils, as explained below:

- In sands the parameter can be roughly compared to the pore pressure resulting from the hydrostatic level, in equilibrium. In fact the pressure on the membrane is that of the water in the pores.
- In clays $P_2$ parameter represents a mixed of both water and soil pressures, and thus it should only be used qualitatively, as sustained by Marchetti [16].
- Furthermore, in soils with intermediate behaviours (silts, sandy clays or clayey sands) the problem is even worse than with clays creating some important problem for a reasonable interpretation [7].

As a consequence of these, it was considered more appropriate to work with equilibrium pore-pressures ($u_0$), calculated from the position of water level externally obtained, instead of $P_2$. Thus, in the next experiments the objective is to model $G_0$ as function of $P_0$, $P_1$ and $u_0$ parameters, avoiding the need for special interpretations, which turns to be much more efficient to include in mathematical operations. With the characterization of the PsS data set presented on Table 1 and Figure 2 it can be seen that this subset is comparable to the WDS in terms of variables distribution and limits in exception of the $G_0$ parameter where the available data is restricted to the range 12-110, where in WDS it goes 6-530. This is relevant, as the conclusions about this experiments must take this into account.

The straight application of the expressions calculated in [7] for the regression applied to this subset returned the relative error parameters shown in Table 4, and the recalculation of the regression constants and subsequent relative error evaluation lead to the results shown in Table 5. Comparing the variability of these results with the ones showed in Table 2 highlights the advantage of using cross validation in experiments.

Concerning the $G_0$ prediction using the ($P_0$,$P_1$,$u_0$) parameters, the schema was similar to the one described in the previous subsection for the intermediate parameters. A traditional regression approach was first used and then several Neural Network experiments were made. Two sets of input parameters were used: one using $P_0$, $P_1$ and $u_0$ and other neglecting $u_0$.

Regarding traditional regression, the least squares method returned some interesting results that can be seen on Table 6. Those results are the best when considering all the possible combinations of the transformations exponential, square root, logarithmic and square to the dependent and independent variables.

It should be noted that in Table 6, $\delta \approx 0.448$, which combined with the range of values for $u_0$ (roughly say, [0,0.2] ) results on a multiplicative effect in the prediction $\widetilde{G_0}$ - that is $e^{\delta u_0} \times f(P_0, P_1)$ - of approximately [1,1.1]. Thus, it was expectable that the introduction of the $u_0$ parameter didn't bring too much improvement to our previous result as it happened.

For all the experiments using NN's or SVR's the 10 fold cross validation method with 20 repetitions was used, since this is the most common and widely accepted methodology to guarantee a good neural network generalization [19].

**Table 4.** Subset PsS: Relative Error Results (Median/Mean(std)) obtained with non-linear regression $\widetilde{G_0} = f(I_D, E_D, K_D)$ using the $(\alpha, \beta, \gamma)$ calculated in Table 2

|  | Type | Median/Mean(std) |
|---|---|---|
| Non-Linear | $G_0 = \alpha\, E_D\, (I_D)^\beta$ | 0.32/0.55(0.63) |
| Regression | $G_0 = E_D + E_D\, e^{(\alpha + \beta I_D + \gamma \log(K_D))}$ | 0.34/0.50(0.49) |

**Table 5.** Subset PsS: Relative Error Results ( /Mean(std)) obtained with non-linear regression $\widetilde{G_0} = f(I_D, E_D, K_D)$ revaluating the $(\alpha, \beta, \gamma)$ parameters

|  | Type | Median/Mean(std) |
|---|---|---|
| Non-Linear | $G_0 = \alpha\, E_D\, (I_D)^\beta$ | 0.26/0.34(0.33) |
| Regression | $G_0 = E_D + E_D\, e^{(\alpha + \beta I_D + \gamma \log(K_D))}$ | 0.14/0.18(0.16) |

**Table 6.** Subset PsS: Relative Error Results (Median/Mean(std)) obtained with non-linear regression $\widetilde{G_0} = f(P_0, P_1)$ and $\widetilde{G_0} = f(P_0, P_1, u_0)$

|  | Type | Median/Mean(std) |
|---|---|---|
| Non-Linear | $G_0 = \alpha\, e^{\beta P_0 + \gamma P_1}$ | 0.22/0.28(0.23) |
| Regression | $G_0 = \alpha\, e^{\beta P_0 + \gamma \sqrt{P_1} + \delta u_0}$ | 0.22/0.28(0.23) |

**Table 7.** Sample PsS: Relative Error Results (Median/Mean(std)) obtained with SVR's

| Input | Type | Cost/$\epsilon(\nu)$ | Median/Mean(std) |
|---|---|---|---|
| $(P_0, P_1)$ | $\epsilon$-SVR | 40/0.0001 | 0.24/0.29(0.22) |
| $(P_0, P_1)$ | $\nu$-SVR | 20/0.9 | 0.25/0.31(0.25) |
| $(P_0, P_1, u_0)$ | $\epsilon$-SVR | 40/0.0001 | 0.24/0.29(0.22) |
| $(P_0, P_1, u_0)$ | $\nu$-SVR | 20/0.9 | 0.25/0.31(0.25) |

For each NN a huge set of experiments was performed, varying the involved parameters such as the number of neurons in the MLP hidden layer, the number of epochs or the minimum error for stopping criteria. The results here presented are therefore the best ones for each regression algorithm and represent the mean of the $10 \times 20$ performed tests for each best configuration. It is also important to stress the fact that, when compared to traditional approaches where all the data is used to build the model, this methodology tends to produce higher standard deviations since in each experiment only a fraction of the available data is used to evaluate the model. Several exploratory experiments were performed with different kinds of MLPs and SVRs. Results from these preliminary experiments show that the best ones were also obtained with SVRs with the radial basis function kernel and for that reason we focus on more detailed experiments using this combination. Results from the SVRs with radial basis function kernel are presented in Table 7, where the $u_0$ parameter also seem to be negligible in terms of $G_0$ prediction.

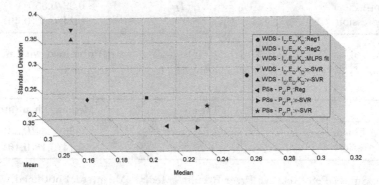

**Fig. 3.** Best Results: values for median, mean and std of $\frac{|G_0 - \widetilde{G_0}|}{G_0}$

## 4    Conclusions

Figure 3 summarizes the results presented in the previous subsections and represents the quality parameters of some of the best results on the estimation of $G_0$ via DMT's basic and intermediate parameters.

This emphasizes the good results of applying Neural Networks to predict maximum shear modulus by DMT. Based on performed experiments it is possible to outline the following considerations:

- Neural Networks and/or SVR's improve the state-of-the-art in terms of $G_0$ prediction. The results show that, in general, NNs and/or SVR's lead us to much smaller medians, equivalent means and higher standard deviations in respect to relative errors, when compared to traditional approaches.
- Regarding the problem characteristics the SVR approach gives, on the prediction with DMT intermediate parameters, the best results considering the median as the main quality measure as discussed earlier.
- When compared with the intermediate parameters, the results show that the basic input parameters $(P_0, P_1)$ does not improve the fitness of $G_0$.
- In addition to the previous sentence, the inclusion of $u_0$ as third input parameter does not seem to improve the fitness. Future work should consider other auxiliar data, mainly measured depth, depth of water level, and/or $P_2$.
- The available unbalanced data, regarding $G_0$ distribution, suggests that more tests should be made using $G_0$ values of higher magnitude ($>110$).

# References

1. Clayton, C., Heymann, G.: Stiffness of geomaterials at very small strains. Géotechnique 51(3), 245–255 (2001)
2. Fahey, M.: Soil stiffness values for foundation settlement analysis. In: Proc. 2nd Int. Conf. on Pre-failure Deformation Characteristics of Geomaterials, vol. 2, pp. 1325–1332. Balkema, Lisse (2001)
3. Peck, R.B., Hanson, W.E., Thornburn, T.H.: Foundation Engineering, 2nd edn. John Wiley & Sons (1974)
4. Lunne, T., Robertson, P., Powell, J.: Cone penetration testing in geotechnical practice. Spon E & F N (1997)
5. Marchetti, S.: In-situ tests by flat dilatometer. Journal of the Geotechn. Engineering Division 106(GT3), 299–321 (1980)
6. Cruz, N., Devincenzi, M., Viana da Fonseca, A.: DMT experience in iberian transported soils. In: Proc. 2nd International Flat Dilatometer Conference, pp. 198–204 (2006)
7. Cruz, N.: Modelling geomechanics of residual soils by DMT tests. PhD thesis, Universidade do Porto (2010)
8. Cruz, M., Santos, J.M., Cruz, N.: Maximum Shear Modulus Prediction by Marchetti Dilatometer Test Using Neural Networks. In: Iliadis, L., Jayne, C. (eds.) EANN/AIAI 2011, Part I. IFIP AICT, vol. 363, pp. 335–344. Springer, Heidelberg (2011)
9. Marchetti, S.: The flat dilatometer: Design applications. In: Third Geotechnical Engineering. Conf., Cairo University (1997)
10. Mayne, P.W.: Interrelationships of DMT and CPT in soft clays. In: Proc. 2nd International Flat Dilatometer Conference, pp. 231–236 (2006)
11. Hryciw, R.D.: Small-strain-shear modulus of soil by dilatometer. Journal of Geotechnical Eng. ASCE 116(11), 1700–1716 (1990)
12. Hardin, B.O., Blandford, G.E.: Elasticity of particulate materials. J. Geot. Eng. Div. 115(GT6), 788–805 (1989)
13. Jamiolkowski, B.M., Ladd, C.C., Jermaine, J.T., Lancelota, R.: New developments in field and laboratory testing of soils. In: XI ISCMFE, vol. 1, pp. 57–153 (1985)
14. Sully, J.P., Campanella, R.G.: Correlation of maximum shear modulus with DMT test results in sand. In: Proc. XII ICSMFE, pp. 339–343 (1989)
15. Tanaka, H., Tanaka, M.: Characterization of sandy soils using CPT and DMT. Soils and Foundations 38(3), 55–65 (1998)
16. Marchetti, S., Monaco, P., Totani, G., Marchetti, D.: Situ Tests by Seismic Dilatometer (SDMT). In: Crapps, D.K. (ed.) From Research to Practice in Geotechnical Engineering, vol. 180, pp. 292–311. ASCE Geotech. Spec. Publ. (2008)
17. Huang, Y., Draper, N.R.: Transformations, regression geometry and R2. Computational Statistics & Data Analysis 42(4), 647–664 (2003)
18. Netlab, http://www1.aston.ac.uk/eas/research/groups/ncrg/resources/netlab/
19. Bishop, C.: Neural Networks for Pattern Recognition. Oxford University Press (1996)
20. Cortes, C., Vapnik, V.: Support-vector Networks. Journal of Machine Learning 20(3), 273–297 (1995)
21. Chang, C.-C., Lin, C.-J.: LIBSVM: a library for support vector machines. ACM Transactions on Intelligent Systems and Technology 2, 27:1–27:27 (2011)
22. Vapnik, V.: Statistical Learning Theory. Wiley, New York (1998)
23. Schölkopf, B., Smola, A., Williamson, R.C., Bartlett, P.L.: New support vector algorithms. Neural Computation 12, 1207–1245 (2000)

# Author Index